TRANSACTIONS
OF THE
INTERNATIONAL
ASTRONOMICAL UNION

VOLUME XVI B – PROCEEDINGS

LEO GOLDBERG

PRESIDENT OF THE INTERNATIONAL ASTRONOMICAL UNION
1973–1976

INTERNATIONAL ASTRONOMICAL UNION
UNION ASTRONOMIQUE INTERNATIONALE

TRANSACTIONS
OF THE
INTERNATIONAL ASTRONOMICAL UNION
VOLUME XVI B

PROCEEDINGS OF THE SIXTEENTH
GENERAL ASSEMBLY
GRENOBLE 1976

Edited by

EDITH A. MÜLLER
General Secretary of the Union

ARNOST JAPPEL
Executive Secretary

D. REIDEL PUBLISHING COMPANY
DORDRECHT-HOLLAND/BOSTON-U.S.A.

1977

ISBN-13: 978-94-010-1259-1   e-ISBN-13: 978-94-010-1257-7
DOI: 10.1007/978-94-010-1257-7

Published on behalf of
the International Astronomical Union
by
D. Reidel Publishing Company, P.O. Box 17, Dordrecht, Holland

Sold and distributed in the U.S.A., Canada, and Mexico
by D. Reidel Publishing Company, Inc.
Lincoln Building, 160 Old Derby Street, Hingham,
Mass. 02043, U.S.A.

PREFACE

The General Assemblies of the International Astronomical Union are landmarks in the life of the world-wide astronomical community, as they review, at triennial intervals, the progress made in this scientific field, promulgate the most spectacular astronomical achievements, formulate scientific programmes for the years to come and, last but not least, deal with the administration and finances of the IAU.

The Reports on Astronomy 1976, published as Transactions XVIA (in 3 volumes) before the XVIth General Assembly, are a synopsis of the work done in astronomy from 1973 to 1975.

The volume "Highlights of Astronomy , as presented at the XVIth General Assembly of the IAU in Grenoble, 1976" includes some selected scientific topics, and will appear in the first half of 1977. Apart from the Invited Discourses and the Proceedings of the seven Joint Discussions, the Highlights volume No.4 contains the proceedings of two Joint Commissions Meetings.

The present volume, subtitled "Proceedings of the XVIth General Assembly, Grenoble, 1976" contains the full record of the Inaugural Ceremony, the report of the two sessions of the XVIth General Assembly, the resolutions it adopted, the slate of the Executive Committee, the Presidents, Vice-Presidents and general members of the Union's 40 Commissions, and the alphabetical list of IAU Members as of December 1976. The volume also contains the reports of Commission meetings held during the assembly and the report of the Working Group for Planetary System Nomenclature. In compliance with the request of the organizers and editors of three Joint Commissions Meetings, (1. "Topics in Interstellar Matter," 2. "CNO Isotopes," 3. "Supernovae,") their detailed proceedings are being published in separate volumes by the D. Reidel Publishing Company; and the discussion on Solar Flares ("How can solar flares be understood?") held by Commission 10 will be published in an issue of Solar Physics. The report of the Working Group for Planetary System Nomenclature (WGPSN) gives details on the topographic nomenclature of the Moon and Planets, as formulated at the various meetings of the WGPSN since its creation in 1973 in Sydney, and as endorsed by the XVIth General Assembly. It also reproduces a table of names of Apollo Landing Sites, as approved by the XVth General Assembly, but not published in Transactions, volume XVB.

It should be pointed out that the Report of the Executive Committee on the Union's activity and finances for 1973-1975 has been published in Information Bulletin No.36 which appeared in June 1976. It is reproduced herein on p. 565, as an appendix to Part 1.

The size of the volume corresponds to the intensified activity of Commissions and Working Groups in Grenoble and also reflects the increased number of IAU Members, their alphabetical list covering more than 150 pages.

The present book is the first Transactions volume produced by the method of offset printing from camera-ready manuscripts. It is the result of a combined effort of the outgoing and incoming Commission Presidents, Commission Secretaries, Chairmen of Working Groups, and the IAU Secretariat. On behalf of the Union I wish to express my sincere thanks to everyone who made a contribution to this volume. My thanks are due in particular to my secretaries Mme R. Bertschi and Mme R. Läubli for their painstaking and careful editorial work.

Edith A. Müller
General Secretary

C O N T E N T S

Part 3

Reports of Meetings of Commissions

Nos.   Page No.

Part 4

Appendix

P A R T   1

REPORT OF THE EXECUTIVE COMMITTEE

1973-1975

The report of the Executive Committee to the Sixteenth
General Assembly of the International Astronomical Union for
the term of office 1973-1976 has been published in Information
Bulletin No. 36. In order to maintain continuity in the IAU Transactions,
the report also appears at the end of this volume, p. 565.

The narrative report of the General Secretary for the
period from 1 January 1976 until the first day of the XVIth General
Assembly is included in the report of the General Assembly, p. 22.

PART 2

INAUGURAL CEREMONY

REPORT OF THE GENERAL ASSEMBLY

CEREMONIE INAUGURALE

Mardi 24 août, à 9^h 30^m

La Cérémonie Inaugurale s'est tenue à la Patinoire de Grenoble, devant und assemblée distinguée de représentants du gouvernement de la République Française, du Conseil Général de l'Isère, de la ville de Grenoble, et de l'Université Scientifique et Médicale de cette ville.

Madame Alice Saunier-Séité, Secrétaire d'Etat aux Universités, Monsieur Louis Mermaz, Président du Conseil Général de l'Isère, Monsieur Hubert Dubedout, Député de l'Isère, Maire de Grenoble, et Monsieur Gabriel Cau, Président de l'Université Scientifique et Médicale de Grenoble, ont honoré la cérémonie de leur présence.

Le fauteuil présidentiel a été occupé par Monsieur Jean Kovalevsky, Président de l'Association UAI-France 76.

Après un interlude musical, le Président ouvre la séance et donne la parole à Monsieur Dubedout, Maire de Grenoble, qui souhaite la bienvenue à l'Assemblée Générale de l'UAI au nom de la ville de Grenoble.

Le Président, après avoir remercié Monsieur le Maire Dubedout, prie Monsieur L. Mermaz de prendre la parole.

ALLOCUTION DE MONSIEUR L. MERMAZ, PRESIDENT DU CONSEIL GENERAL DE L'ISERE, DEPUTE:

Madame le Ministre, Monsieur le Préfet, Monsieur le Maire, Messieurs les Présidents, Mesdames, Messieurs,

Le Conseil Général, assemblée représentant les habitants du département de l'Isère, avait décidé de subventionner l'organisation de votre 16ème assemblée générale. Il a aussi demandé à son président de venir vous saluer ce matin.

En participant ainsi directement à vos travaux, le Conseil Général veut souligner d'abord combien il apprécie l'honneur qui est fait au département d'avoir à accueillir les astronomes du monde entier; il veut également montrer notre fierté d'abriter le remarquable potentiel universitaire grenoblois; il veut, enfin, signifier l'intérêt que veut porter une instance régionale à la recherche scientifique.

C'est en effet sur le territoire du département de l'Isère que vous allez être amenés à travailler et aussi à vous détendre durant plus de dix jours: les structures d'acceuil de la ville de Grenoble, auxquelles la municipalité a attaché un intérêt méritoire, le domaine universitaire de Saint-Martin-d'Hères et Gières, avec ses installations scientifiques et d'accueil, constitueront le cadre de vos travaux et de votre vie quotidienne. Les organisateurs de votre assemblée générale ont également prévu une série de manifestations culturelles, de promenades et d'excursions qui permettent aux participants et accompagnants de mieux connaître le département. Nous vous souhaitons donc ici une très cordiale bienvenue dans notre département de l'Isère.

La deuxième raison qui a incité le Conseil Général à marquer un intérêt particulier à votre 16ème assemblée générale est son souci de manifester son attachement au potentiel universitaire grenoblois. Sans parler des palmarès établis récemment par divers hebdomadaires et revues françaises et qui placent très souvent Grenoble au premier plan des universités de ce pays, la notoriété des maîtres de nos universités grenobloises a acquis une dimension internationale que le Conseil Général se devait aussi de saluer: abriter un tel potentiel pédagogique et scientifique donne en effet au département et à ses instances des devoirs particuliers en la matière.

La troisième raison de la part que prend le Conseil Général à cette manifestation est l'intérêt qu'il entend manifester à la recherche scientifique quel que soit, en effet, le rôle irremplaçable d'instances nationales de coordination et d'impulsion de la recherche, comme la D.G.R.S.T. et le C.N.R.S. Il importe que soit accru le rôle des universités dans le devenir de chaque région. L'autonomie a souvent per-

mis aux nouvelles universités de consacrer une part grandissante de leur activité
de recherche scientifique au mieux être économique, social et culturel des habitants
de leur région. On se doit d'insister sur la nécessité de donner en France la priori-
té à la recherche scientifique.

Mais, au-delà de ces considérations qui ont animé notre Conseil Général de l'Isère,
vous me permettrez de vous dire, Mesdames et Messieurs, combien l'Assemblée générale de l'union astronomique internationale et cette science elle-même, dont vous
êtes ici les meilleurs représentants, nous paraissent porteurs d'avenir. Depuis les
sages de la Chaldée, l'homme cherche à percer le mystère de l'univers. Si l'astronomie a pu ainsi naître et présenter aujourd'hui le fantastique développement que
nous lui connaissons c'est parce qu'elle a su se nourrir de deux complémentarités:
la complémentarité des hommes d'abord, les Egyptiens, les Grecs, les Chinois, les
Indiens, les Arabes, les Européens, se passèrent tour à tour le flambeau de l'astronomie pour aboutir aujourd'hui à cette magnifique communauté scientifique qui couvre
le monde entier. La complémentarité des sciences aussi, puisque l'astronomie n'a
pu se développer qu'en faisant appel à de multiples disciplines et en donnant nais-
sance à son tour à de nouvelles sciences. En effet, si l'astronomie utilise toutes
les sciences et fait appel à toutes les techniques, elle contribue également à
toutes les sciences et à toutes les techniques.

L'astronomie moderne n'a pas pu se développer sans la contribution décisive de
la mécanique, de la physique, de la physique nucléaire, de la chimie, de la géolo-
gie et bien d'autres. Le développement considérable de l'astronomie a rendu possible
des changements qualitatifs décisifs dans les domaines de la géographie, la météo-
rologie, la radio-astronomie, la photographie et la conquête de l'espace qui n'au-
rait jamais pu naître et se développer sans l'astronomie.

C'est cette double complémentarité, dont l'astronomie a toujours su donner le
meilleur exemple, qui nous paraît la plus réconfortante. Et c'est finalement à
travers vous, ce symbole de progrès et d'espoir que je tenais surtout à saluer
ce matin.

Le Président remercie Monsieur le Député Mermaz et passe la parole à Monsieur
G. Cau.

ALLOCUTION DE MONSIEUR G. CAU, PRESIDENT DE L'UNIVERSITE SCIENTIFIQUE ET
MEDICALE DE GRENOBLE:

Madame le Ministre, Monsieur le Préfet, Monsieur le Président du Conseil Général,
Monsieur le Maire, Monsieur le Président de l'U.A.I., Mesdames, Messieurs,

Après la cité, par la voix de son Maire, et après le département par celle du
Président du Conseil Général, l'Université Scientifique et Médicale de Grenoble
se doit de souhaiter la bienvenue à la XVIe Assemblée Générale de l'Union Astronomique Internationale.

Je mesure tout l'honneur qui échoit à notre Université de servir de siège à un
congrès qui rassemble plus de 40 délégations, honneur d'autant plus grand que l'Astronomie ne figure pas encore dans les disciplines enseignées à Grenoble, honneur d'autant plus périlleux que désormais il nous faut nous en montrer dignes en favorisant
l'implantation de recherches astronomiques dans l'environnement favorable d'une
Physique auréolée de la présence de prix Nobel tels que Louis NEEL et MOSSBAUER,
de mathématiciens de notoriété internationale.

Au nom des Universités de GRENOBLE, je remercie très vivement les responsables
de l'Union Astronomique Internationale d'avoir arrêté leur choix sur notre ville,
sur les installations du campus de St. Martin d'Hères. Je tiens à témoigner ma gratitude à mon prédécesseur, Monsieur le Professeur SOUTIF, qui a su déployer énergie
et diplomatie pour orienter et pour obtenir la décision de tenir pour la seconde
fois en FRANCE l'Assemblée Générale de l'U.A.I.

Ce choix suscite une grande espérance pour notre Université: l'aboutissement du
projet franco-allemand d'un grand interféromètre millimètrique installé sur le pla-

teau de Bure. Le bouillonnement d'idées qui pendant deux semaines fait de GRENOBLE la capitale mondiale de l'Astronomie ne peut que laisser un foyer dynamique de recherche astronomique. Il appartient à nos autorités de tutelle, à vous, Madame le Ministre, qui présidez à la destinée de l'Université Française, de répondre à cette espérance et d'adjoindre une nouvelle marque de rayonnement scientifique à GRENOBLE grâce à la réalisation de ce projet.

Les travaux de cette XVIe Assemblée Générale abordent les multiples aspects de l'astronomie qui a vu s'étendre son champ d'étude au-delà de l'imaginable. Grâce aux progrès techniques prodigieux, les limites de l'inconnu reculent en entraînant en un premier temps la stupéfaction, l'admiration, puis avec l'accélération de la communication, un intérêt moins soutenu pour aboutir à l'indifférence. Vos découvertes interfèrent en la vie de chacun tantôt de façon spectaculaire, tantôt de façon insidieuse: chaque planète apporte sa moisson. Que de réflexions suscitées à propos de la présence ou de l'absence de vie sur Mars à la fois sur le plan scientifique et sur le plan philosophique! N'est-ce pas le moment face à cette révolution des connaissances en astronomie d'évoquer cette pensée de TEILHARD de CHARDIN: "L'univers ne tend aucunement, comme nous pourrions le craindre, à écraser, mais au contraire à exalter par son énormité nos valeurs individuelles" en y ajoutant une prise de conscience des responsabilités.

A l'instar de l'Antiquité qui nous légua la règle selon laquelle rien de ce qui touche l'homme ne saurait laisser indifférent, notre époque par l'astronomie nous oblige à n'être indifférent ni au monde, ni à l'univers. L'Université a pour mission d'aider à la diffusion des connaissances sur les systèmes planétaires, stellaires ou galactiques autant que sur l'organisation de la matière ou de l'homme, dans le souci d'élever le niveau scientifique au profit de l'épanouissement de l'individu, au profit du progrès social et non dans le but de former des hommes sur les bases de besoins apparents du moment et de critères d'utilitarisme discutable.

Puissiez-vous trouver dans le cadre du campus universitaire les conditions favorables au déroulement agréable des nombreux colloques programmés! Puissiez-vous garder le meilleur souvenir de ce séjour à GRENOBLE et vous remémorer tel contact, telle discussion, telle approche scientifique en évoquant le site de nos montagnes! Puissiez-vous à l'issue de vos travaux avoir le désir de revenir pour constater que vous avez suscité un nouveau centre d'Astronomie!

Je vous redis à nouveau, au nom de l'Université Scientifique et Médicale de GRENOBLE, bienvenue à l'Union Astronomique Internationale, et, voeux les plus ardents pour le plein succès de cette XVIe Assemblée Générale.

Le Président, après avoir remercié Monsieur Cau, s'adresse à l'Assemblée lui-même comme suit:

ALLOCUTION DE MONSIEUR J. KOVALEVSKY, PRESIDENT DE L'ASSOCIATION UAI-FRANCE 1976:

Madame le Ministre, Monsieur le Préfet, Monsieur le Recteur, Monsieur le Député-Maire, Messieurs les Présidents, Mes Chers Collègues,

Au nom de l'Académie des Sciences et au nom du Comité National Français d'Astronomie, j'ai le grand honneur, et aussi le grand plaisir de vous accueillir et de vous souhaiter la bienvenue à la XVIe Assemblée Générale de l'Union Astronomique Internationale.

L'Académie des Sciences, qui a délégué au Comité National Français d'Astronomie ses responsabilités pour l'organisation de ce Congrès, avait été extrêmement heureuse que votre Union ait accepté l'invitation qu'elle vous a adressée il y a 3 ans de vous réunir en France.

Notre pays a une longue tradition astronomique, depuis la création de l'Observatoire de Paris il y a près de 310 ans. Les découvertes astronomiques de Cassini, Laplace, Le Verrier, Jansen et de bien d'autres encore sont bien connues de vous tous. Mais c'est surtout l'importance des apports actuels faits en France dans le domaine de l'Astronomie, que nous aimerions vous présenter au cours de votre séjour en France et vous permettre ainsi d'en juger les résultats. Tous les congres-

sistes trouveront, parmi les documents qui leur sont distribués, un panorama de
l'Astronomie Française, rédigé par l'Institut National d'Astronomie et de Géophysique. Un certain nombre d'entre vous visiteront, ou ont déjà visité quelques-uns
de nos observatoires. Nous vous invitons cordialement à profiter de votre séjour
en France pour le faire.

La dernière fois que l'UAI s'est réunie en France, fut en 1935, à Paris, pour
votre 5e Assemblée Générale. Quelques-uns d'entre vous s'en souviennent, mais vous
êtes très peu nombreux dans ce cas. Ceci parce que, au cours de ces quarante dernières années qui ont vu tant d'évènements graves se dérouler, nombreux sont vos
collègues de 1935 qui nous ont quitté. Mais c'est aussi parce qu'ils n'étaient
que 300, tandis que nous serons plus de 2000, ces jours-ci à Grenoble pour échanger
les résultats de nos derniers travaux et pour organiser de nouveaux programmes de
recherche en coopération internationale.

Si ce bond dans le nombre de congressistes témoigne de l'essor de l'astronomie
au cours des dernières décennies, ce fait a aussi pour conséquence que l'organisation de cette assemblée générale est devenue très complexe. Nous avons eu, pour
cela, le concours des astronomes, des autorités nationales et locales, politiques,
administratives et universitaires, ainsi que d'un certain nombre d'entreprises
privées.

Les astronomes qui vous accueillent, travaillent au sein d'une association appe-
lée UAI-France 76 qui est une émanation du Comité National Français d'Astronomie.
De nombreux collègues ont consacré beaucoup de leur temps au sein de cette organisation et je tiens à les en remercier ici tout particulièrement. Ils ont fait ce
travail en collaboration avec l'organisation locale "Grenoble Accueil" qui n'a
pas ménagé ses efforts sur le plan de l'organisation sur place.

Le gouvernement français a reconnu l'importance de cette Assemblée Générale.
Je suis heureux de remercier ici Madame le Secrétaire d'Etat aux Universités
pour l'important effort financier consenti par son Ministère pour l'organisation
de ce Congrès. Je remercie aussi le Ministère des Affaires Etrangères pour les
bourses de séjour et de voyage qu'il a accordées à de nombreux collègues étrangers.

L'enthousiasme, exprimé par les diverses autorités locales à l'idée de réunir à
Grenoble cette assemblée générale a certainement été pour beaucoup dans le choix
de cette ville pour vous rassembler. L'aide qu'elles ont toutes apportée a été
considérable. C'est avec gratitude que je remercie ici Monsieur le Président du
Conseil Général de l'Isère et Monsieur le Député-Maire de Grenoble pour les importantes contributions du Département de l'Isère et de la ville de Grenoble à l'organisation de ce Congrès.

Je remercie aussi MM. Les Présidents de l'Université Scientifique et Médicale
de Grenoble, de l'Institut National Polytechnique de Grenoble et de l'Université
des Sciences Sociales de Grenoble, ainsi que M. le Directeur de la Bibliothèque
Inter-Universitaire de Grenoble pour avoir mis à notre disposition les magnifiques
locaux du Domaine Universitaire. Je remercie aussi M. le Directeur du Centre Ré-
gional des Oeuvres Universitaires et Scolaires pour la part que ses services ont
pris pour résoudre les problèmes matériels de la vie quotidienne des astronomes.
Nous remercions enfin les établissements et organismes publics et privés qui ont
participé au financement de ce congrès.

J'espère que tous ces efforts n'auront pas été vains et que vous vous plairez à
Grenoble. Je vous renouvelle donc, au nom de l'Academie des Sciences, du Comité
National Français d'Astronomie et de l'Association UAI-France, nos souhaits de
bienvenue et nos voeux les meilleurs pour le succès de vos travaux scientifiques.

Ensuite, la parole est donnée à Monsieur L. Goldberg, Président de l'Union
Astronomique Internationale.

ALLOCUTION DU PROFESSEUR L. GOLDBERG, PRESIDENT DE L'UAI

Madame le Ministre, Monsieur le Préfet, Monsieur le Recteur, Monsieur le Président
de l'Université, Monsieur le Président du Conseil Général, Monsieur le Maire, Professeur Kovalevsky, Mesdames, Messieurs,

L'Union Astronomique Internationale est très honorée d'avoir été invitée à tenir
sa seizième Assemblée Générale en France, ce pays qui symbolise le progrès et la
coopération internationale en astronomie. Nous éprouvons une profonde gratitude
pour la chaleur et la générosité de votre bienvenue et nous nous réjouissons gran-
dement de l'agréable semaine que nous allons passer ici.

Il y a 41 ans que l'U A I s'est réunie en France pour la dernière fois. Durant
cette période, l'astronomie a subi de nombreuses et remarquables modifications qui,
pour la plupart, n'auraient guère pu être prédites. En 1935, notre vénéré collègue
Bernard Lyot nous décrivit les observations de la couronne solaire qu'il avait
effectuées à l'Observatoire du Pic-du-Midi avec le coronographe qu'il avait récemment inventé. Lyot avait fait à plusieurs reprises l'ascension du Pic du Midi en
skis, en portant le télescope attaché sur son dos. A présent, de nouvelles géné-
rations d'instruments puissants et sophistiqués, couvrant le domaine complet du
spectre électromagnétique, peuvent être installés à peu près n'importe où dans le
système solaire, sur orbite terrestre, dans le milieu interplanétaire et, même,
sur les surfaces de la lune, de Vénus et de Mars. Ces progrès étroitement liés
avec les nouveaux développements en physique atomique et nucléaire et en astrophysique théorique, combinés aux progrès des ordinateurs ont suscité une véritable révolution dans nos connaissances et créé pour les astronomes, maintes sources de
perplexité.

Pourtant, tout comme nous sommes émerveillés par les nouveaux développements ré-
cents, les astronomes de 1935 considéraient avec admiration, la richesse de *leurs*
propres expériences. Le Professeur Ernest Esclangon, Directeur de l'Observatoire
de Paris, qui devait être élu Président de l'U A I , faisait la déclaration suivante:
"Mais le présent siècle devait, dans le domaine de l'Astronomie, nous réserver des
révélations aussi extraordinaires qu'inattendues, de nature à troubler de surprises
et d'émerveillement notre esprit confondu. A ce titre, on peut dire qu'une science
nouvelle est née, comme si, par une sorte de dédoublement, notre vieille astronomie
avait enfanté une astronomie nouvelle, débordant de force, de jeunesse, de vie,
forçant, dès ses premiers pas, l'admiration la plus enthousiaste et la plus unanime".*

Si le contenu scientifique de l'Astronomie peut changer, en revanche le sens de
l'enthousiasme avec lequel les astronomes poursuivent leur travail ne diminue pas,
tout comme l'esprit d'amitié et de générosité unissant les astronomes de pays divers
maintient la coopération dans les réunions internationales.

Tous les astronomes sont les bienvenus à notre union internationale, même si,
pour des raisons indépendantes de notre volonté, certains collègues ne peuvent par-
ticiper à des réunions en dehors de leur pays.

Monsieur le Maire,
Nos collègues français n'auraient pu choisir un site plus agréable que Grenoble,
pour tenir notre seizième Assemblée Générale. C'est une ville belle et, même, fascinante, moderne, à l'avant du progrès et, cependant, pleine de souvenirs d'une

---

* Yet the present century was going to give rise in the field of Astronomy, to
  revelations which were as extraordinary as they were unexpected, and which filled
  our mind with surprise and wonder. A new science was indeed borne, as if, through
  some kind of doubling, our old astronomy had given birth to a new astronomy,
  overflowing with strength, youth and vitality, creating the most enthusiastic
  and unanimous admiration".

histoire   ancienne et glorieuse. C'est un homage à l'ingéniosité et au courage de
votre peuple, qu'une telle ville ait pu se développer dans un site aussi sauvage,
au centre de massifs montagneux. A première vue, il peut sembler surprenant de trouver un des centres culturels importants de France, dans un tel environnement,
jusqu'à ce qu'on réalise la coopération unique qui existe entre l'industrie et les
activités scientifiques des universités et laboratoires de la ville.

Monsieur le Préfet,
   Nous, astronomes, nous ressentons des liens particulièrement étroits avec Grenoble et le Département de l'Isère, car un des premiers préfets fut ce grand ma-
thématicien du début du 19e siècle, Jean Baptiste Joseph Fourier qui fit de Grenoble sa résidence officielle pendant son mandat de 1802 à 1815. C'est ici qu'il
développa ses importantes recherches sur la conduction de la chaleur et qu'il dé-
couvrit la fameuse série mathématique portant maintenant son nom et constituant à
présent un des outils fondamentaux pour l'analyse des observations astronomiques.

Monsieur le Recteur,
   Nous vous sommes profondément reconnaissants d'avoir mis à notre disposition
les magnifiques installations de l'Université scientifique et médicale, une des
quatre grandes universités constituant l'Académie de Grenoble. Votre Université
possède une réputation mondiale et est particulièrement renommée parmi les astronomes pour son éminent Département de Physique et pour son Ecole d'Eté de Physique
théorique et d'Astrophysique des Houches.

Madame le Ministre,
   L'aide généreuse que le gouvernement français a bien voulu octroyer à de nombreux
astronomes, à titre d'intervention dans leurs frais de voyage, est très appréciée
par l'Union.
   Le peuple français et ses dirigeants ont toujours manifesté une attitude éclairée
vis-à-vis de l'astronomie, depuis que Louis quatorze décida la construction de
l'Observatoire de Paris et nomma Giovanni Domenico Cassini en qualité de premier
directeur. Telle était la puissance et l'influence du Grand Roi Soleil, que pendant
les soixante-douze années de son règne, de 1643 à 1715, très peu de taches solaires
se manifestèrent. L'inauguration de l'Observatoire de Paris établit une tradition
d'éminentes contributions et de collaborations internationales qui n'eut aucune
interruption jusqu'à nos jours. Les grands mathématiciens et astronomes français
des 17e, 18e et 19e siècles ont contribué à bâtir les fondations de l'astronomie
classique; leurs travaux constituent une base solide pour la formation de tous les
astronomes professionnels. Avec une telle tradition, il était normal que les recherches en mécanique céleste et en astronomie de position, jointes à l'astronomie
solaire devaient continuer à recevoir une grande attention au 20e siècle. Par
exemple, Dr. Benjamin Baillaud fut l'organisateur et la personnalité la plus mar-
quante dans le grand projet de la Carte du Ciel. Elu premier président de l'U A I ,
Baillaud fut un ardent apôtre de la coopération internationale. Le professeur
André Danjon qui servit plus tard également en tant que président de l'U A I in-
venta un instrument de transit bâti sur un nouveau principe, grâce auquel des positions d'étoiles purent être déterminées avec une précision remarquable.
   Depuis la fin de la deuxième guerre mondiale, l'astronomie française a subi une
renaissance majeure, tout en transférant l'accent de l'astronomie classique, vers
l'astrophysique, la radioastronomie et les recherches spatiales, tout en maintenant
sa puissante tradition en physique solaire.
   Il est impressionnant de constater le très rapide développement de l'astronomie
française dans maints observatoires. J'espère que de nombreux astronomes profite-
ront de leur présence en France pour visiter les observatoires à Paris, Meudon,
Pic-du-Midi, Nançay, Haute Provence, Nice, Lyon et Bordeaux.
   Lors de la cérémonie inaugurale de la cinquième Assemblée Générale à Paris, en
1935, le Président Frank Schlesinger exprimait ses vues sur l'absence presque to-

tale de construction de grand télescope en Europe occidentale, résultant de la
1ère Grande Guerre.
Schlesinger disait:
"I believe that there is not a similar period in which so few large telescopes have
been constructed in these centers of civilization".*

Heureusement, la situation de nos jours peut être décrite en changeant simple-
ment les mots "si peu" ("so few") par "tant" ("so many") et nulle part la construction de télescope n'a été plus active qu'en France; en ce moment, en tout cas, où
bientôt, les Français auront en opération le télescope de 3.6 m en collaboration
avec le Canada et Hawaii, un télescope de 2 m, un télescope de Schmidt à Nice, un
télescope de 1 m pour l'Observatoire de Lyon (au Gornergrat en Suisse), un instrument semblable au Chiran (pour l'O.H.P.), un interféromètre dans la région millimétrique.

Nos collègues français ont toujours fait preuve d'un grand talent dans la construction d'instrumentations nouvelles. Je me contenterai de citer les techniques
connues comme les lentilles de Fresnel et de Fabry, le foyer cassegrain, le dépôt
chimique d'argenture, le test de Foucault, les étalons Fabry-Perot, l'optique
Chrétien-Ritchey, le spectro-héliographe, le coronographe, l'astrolabe impersonnel,
les cameras électroniques, les spectromètres à transformer de Fourier, l'inter-
férométrie spiculaire, etc....

Dr. Kovalevsky,
Je voudrais vous remercier encore, ainsi que les astronomes français pour la
façon magistrale avec laquelle vous avez organisé la plus grande Assemblée générale dans l'histoire de l'U A I .Puis-je aussi, au nom de l'U.A.I., exprimer les
plus chaleureuses félicitations pour les progrès magnifiques que l'astronomie fait
dans votre pays, ainsi que nos voeux les plus sincères pour l'avenir de l'astronomie française.

Monsieur Kovalevsky remercie Monsieur le Président Goldberg et donne la parole
à Madame Alice Saunier-Séité.

ALLOCUTION DE MADAME ALICE SAUNIER-SEITE, SECRETAIRE D'ETAT AUX UNIVERSITES

Monsieur le Député-Maire, Monsieur le Préfet, Monsieur le Président du Conseil
Général, Messieurs les Présidents, Mesdames, Mesdemoiselles, Messieurs,

L'importance de votre congrès illustre la vitalité et l'universalité de votre
union. Elle est à la mesure de l'intérêt accordée à l'astronomie dans les grands
pays scientifiques.

Pendant une semaine vous allez débattre des données d'une science fascinante
qui a, plus qu'aucune autre et depuis les origines les plus lointaines, préparé
et accompagné le développement de l'humanisme.

Depuis toujours, l'homme se préoccupe de son milieu, de son environnement et
essaye de comprendre et d'expliquer. Les différents aspects de la voute céleste
par ses grandes régularités et ses subtiles différences ont, probablement très tôt,
stimulé ses observations et ses analyses.

Tour à tour le ciel a été pour l'homme le symbole de la perfection, la source
de la vie et de son inspiration poétique. Il y a vu l'écriture de son destin. Il
y a placé ses espoirs, l'origine de ses misères, le siège de ses divinités enfin le
support de ses méditations les plus sûres et les plus hardies sur lesquelles est
venu se fonder votre astronomie.

---
* Traduction: "Je crois qu'il n'y a pas eu de période semblable au cours de laquelle
             si peu de grands télescopes ont été contruits dans les centres civi-
             lisés d'Europe occidentale".

Pour vous, aujourd'hui, le ciel est surtout le premier et le plus complet des laboratoires où s'élaborent les théories les plus avancées. Les astronomes ont beaucoup de chance de posséder un laboratoire aussi étonnant que l'espace, où se trouvent rassemblées les conditions extrêmes de températures, de densité, de gravitation, de rayonnement, de dimensions.

Grâce à lui, l'astronomie ne cesse d'apporter aux autres sciences des éléments nouveaux et extraordinaires et par delà les découvertes du phénomène de la vie et de la conception de l'univers, elle permet de relier l'homme à celui-ci.

C'est vrai que l'on ne peut rester indifférent à l'idée qu'une étoile mène une existence compliquée depuis la contraction qui l'a vu naître jusqu'à l'effondrement qui annonce sa mort.

Et c'est un des plus beau triomphe de la physique que d'avoir réussi à fonder la connaissance de cette évolution sur la seule étude d'un rayonnement émis par un point dans l'espace il y a quelques millions d'années.

En révèlant l'unité de la matière, l'astronomie confère à la science et à ses efforts des valeurs universelles, elle est un instrument de culture incomparable.

Plusieurs personnalités et savants eminents vous ont présenté avant moi l'astronomie, ses acquisitions, ses problèmes, votre Union Internationale si active et qui est un élément moteur indispensable à l'épanouissement de votre science. Quant à moi je tiens à rappeler que cette science fascinante, carrefour de nombreuses disciplines est une de celle qui nécessite des instruments puissants et où la coopération est la plus nécessaire car ces moyens coûteux ne peuvent être le fait d'un seul pays, même quand celui-ci soutient un effort important dans le domaine scientifique.

La coopération internationale est une des constantes de la politique astronomique de la France. Notre pays prend part activement à plusieurs grandes opérations. C'est ainsi que nous participons pour un tiers à l'Observatoire Européen Austral pour lequel un télescope de 3,60 m a été construit et son installation au Chili, à peu près terminée. Avec le Canada, et les Etats-Unis, le C.N.R.S. construit à Hawaii un télescope de 3,60 m en voie d'achèvement. Un projet franco-allemand de radioastronomie millimétrique est étudié par le C.N.R.S. et la Max Plank Gesellschaft; ce projet extrêmement prometteur du point de vue scientifique doit réussir. Enfin les moyens spatiaux pour l'étude des rayonnement X ou γ sont utilisés grâce à notre participation à l'effort spatial européen au sein de l'Agence Spatiale Européenne qui a plusieurs programmes de satellites astronomiques déjà réalisés ou en projet et dans le cadre de notre coopération avec l'Union Soviétique.

Cette énumération illustre assez bien les possibilités de l'astronomie française et sa place dans le domaine international à côté de ce qui est réalisé au sein de nos propres établissements.

La France abrite aussi à l'Observatoire de Paris le bureau International de l'Heure. C'est le plus important des services internationaux placé sous le patronage de l'Union Astronomique Internationale. Il m'est agréable aujourd'hui de confirmer que la France met à votre disposition des locaux pour installer à l'Observatoire de Paris votre secrétariat permanent.

Le Congrès International de l'Union Astronomique qui se tient en France, pour la première fois depuis 1936 a pour cadre Grenoble et son Université. Je n'insisterai pas sur le dynamisme et les atouts de cette université prestigieuse qui vous acceuille et qui fait le renom de cette ville. Monsieur le Député-Maire nous l'a rappelé il y a quelques instants. Je n'insisterai pas non plus sur les illustres prédécesseurs qui dans cet établissement firent tant pour la science et plus particulièrement la vôtre, qui regroupe aujourd'hui plus de 3000 participants représentant plus de 30 nations. Aucune autre ville en France après Paris ne méritait d'assurer une telle charge ni une telle responsabilité.

Au cours de ces journées, vous allez réfléchir, beaucoup discuter, et essayer d'expliquer ce qu'est votre science, ce que sont vos travaux, ce qui vous guide, pourquoi et comment. Cet effort d'explication est un devoir essentiel du savant. L'homme est inquiet devant le progrès scientifique, il ne comprend pas, il n'en voit bien souvent que les effets négatifs. Les effets directs et indirects de l'explosion scientifique sur la vie des sociétés sont considérables. Il est du devoir du savant d'y réfléchir et d'aider l'homme à comprendre, à ne pas rejeter cette connaissance, à ne pas accroître son trouble et son angoisse.

Ces idées qui semblent si actuelles sont depuis longtemps débattues, puisqu'au VIème siècle avant J.C., Théopompe de Lamsaque disait: "Tout théoricien pur est un praticien social qui s'ignore".

Plus que jamais le savant doit sortir de sa tour d'ivoire. La science conditionne le devenir de la société, elle ne peut lui rester étrangère.

Après les paroles de Madame le Ministre Alice Saunier-Séité, et l'exécution d'un interlude musical très apprécié, le Président Kovalevsky lève la séance.

REPORT OF THE XVIth GENERAL ASSEMBLY OF THE INTERNATIONAL ASTRONOMICAL UNION
(RESOLUTIONS INCLUDED)

AGENDA

First Session

1. Formal opening of the General Assembly, by the President
2. Appointment of official interpreters
3. Report of the Executive Committee:
   (a) Discussion of the printed report, as published in Information Bulletin No. 36
   (b) Report of decisions taken at meeting in Grenoble, in particular of admission of new Members
4. Report by the General Secretary
5. Report by the President of the proposals for membership in the Executive Committee
6. Announcement of:
   (a) The names of representatives of Adhering Countries, empowered to vote on their behalf
   (b) The names of representatives to serve on the Nominating Committee
   (c) The names of Acting Presidents of Commissions
7. Appointment of the Finance Committee
8. Appointment of the Resolutions Committee
9. Consideration of proposal for resolution submitted by the French National Committee of Astronomy as follows:
   "L'Assemblée Générale de l'Union Astronomique Internationale, considérant l'érosion monétaire, estime qu'une augmentation de 20% des cotisations est nécessaire pour éviter une dégradation des services qui sont offerts par l'UAI à la Communauté Astronomique Internationale". The proposal will be submitted to the Finance Committee.
   Other Adhering Countries did not present motions for resolutions.
10. Consideration of proposals for resolutions submitted by Commissions

(a) *Commission No. 5*

Résolution 1

L'Assemblée Générale de l'UAI
Compte tenu du travail excellent et complet, accompli par le Groupe de Travail désigné pour procéder à la révision de la Classe Astronomique de la Classification Décimale Universelle (UDC), en coopération avec la Fédération Internationale de Documentation (FID)

recommande l'adoption par tous les astronomes des propositions de ce Groupe de Travail, concernant UDC 52, telles qu'elles sont décrites dans le document PC 75-7 de la FID en date du 31 mars 1975,

recommande la publication de ce document et sa diffusion de façon appropriée par l'UAI, et

recommande à tous les éditeurs de journaux primaires ou secondaires et à toutes les bibliothèques astronomiques d'adopter ce texte, de le divulguer et de l'appliquer aussi vite et aussi largement que possible, ainsi que les révisions de manuel de rédaction de l'UAI, établies par la Commission 5 et son Groupe de Travail sur la Politique d'Edition.

(b) *Commission No. 8*

Résolution 1
on long-term planning of meridian astronomy
1. *Noting* the great improvements of accuracy and efficiency of night-time ob-

servations by means of photographic and photoelectric techniques obtained during
the recent decade,
  recognizing that even greater improvements must be expected in the coming de-
cate, both for night and daytime absolute observations by photoelectric techniques,
  recognizing the importance of continued first class visual absolute observations for a period of overlap, it is
  recommended that meridian departments review their long-term plans and make
these known to the commission.
  2. *Noting* the expected, revolutionary progress of astrometry through space
techniques, it is
  recommended that space astrometry should be developed and performed as soon as
possible, and it is
  recommended that this should not affect planning of groud based techniques
before the accuracy, reliability and long-term planning of space astrometry have
been certified.

Resolution 2
on values of precession (also for Commission 4)

Noting the great improvements of accuracy and efficiency of meridian observations
obtained during the recent decade and expected during the next,
  noting the expected progress through space astrometry and radio astrometry,
  noting the relatively large estimated errors (10%) of the proposed changes of
the precessional values, and reminding that the error of such unique quantities
often prove to be greater than estimated,
  noting that a change of the values will cause numerous trivial, but annoying
mistakes in computing during many coming decades,
  noting that the improved values given by Fricke can be easily introduced in proper motions when it is of importance, it is
  anticipated that a change now would be regretted and probably changed again in
only 20 years, it is
  recommended to make no change now, and it is
  recommended to print in all catalogues explicitly the formulae and constants
which were used to transfer between different equinoxes.

  (c) *Commission No. 10*
Resolution 1

  Reflecting upon the continuing contributions from research at ground-based observatories, plasma laboratories and theoretical institutes,
  recognizing the large gaps in solar observations that inevitably will occur
between individual space missions, and being
  deeply concerned about the many recent abandonments of ground-based programs
due to lack of funding.
  recommends that the various national funding agencies take care to support a
balanced program of solar research with appropriate recognition of the vital role
played by ground-based observatories, theoretical institutes and plasma laboratories.

Resolution 2
on Flare reporting

IAU Commission No. 10. recommends
1. That the solar community, during the next IAU Assembly, states the present
   scientific utility of the flare patrol with regard to solar research and observations, and endorses a program of flare patrol devoted to the *next solar cycle*.

2. That on a regional basis, the solar astronomers reach an agreement on the solar
   institutes which, in fact, will carry out this "quasi-permanent" program.

3. That in addition to this permanent program there will be plans for a few months each year for some kind of temporary "subflare campaign" in which more of the observatories would participate, and for which there would be an increase in the benefit of those who need a very complete survey of small subflares.

11. Consideration of proposal for resolution submitted by the Executive Committee.
On the proposal of the President of IAU Commission No. 5 the Executive Committee resolved to submit to the General Assembly the following rewording of the second sentence of article 12 (a) of the By-laws for approval:
"This (Special Nominating Committee) consists of the President and past President of the Union, a member proposed by the retiring Executive Committee, and four members elected by the Nominating Committee from among twelve Members proposed by Presidents of Commissions. Other than the President and immediate past President, present and former members of the Executive Committee shall not serve on the Special Nominating Committee".
To become valid, this motion requires the approval by an absolute majority of the votes of Adhering Countries.
Proposals of a financial character have been incorporated in the draft budget to be considered by the Finance Committee. Other proposals for resolutions have been presented in draft form and will first be discussed by the Commissions. No proposals for resolutions have arrived from Inter-Union Commissions.

Second Session

12. Report of the Finance Committee
    (a) Accounts for 1973-1975
    (b) The Residual budget for 1976
    (c) The budget for the ensuing period, as proposed by the Executive Committee:
        Draft Budget for 1977-1979
        Unit of Contribution 1237.50 gold francs
        (1.462.92 Swiss francs at 1 April 1976)

Receipts in Swiss francs

1. Contributions from Adhering Countries	943.585,34	
2. Revenue from publications	61.260,03	
3. Interest on accounts	33.924,00	
4. UNESCO subvention through ICSU	92.520,00	
Total receipts	1.131.289,37	
Total payments	1.131.289,37	
Balance	0	

Projects under Payments 4.2.

Commission	Swiss francs
4. Ephemerides	2.487,76
6. Telegram Bureau	5.597,46
16. Planetary documentation, Meudon	5.286,49
17. Lunar documentation, Meudon	5.286,49
20. Minor Planets	6.219,40
carried forward:	24.877,60

Payments in Swiss francs

1.	Administrative Office	436.900,00
2.	Subvention to ICSU	23.589,63
3.	Commission expenses	10.280,00
4.	Projects of Commissions	
4.1.	Exchange of Astronomers	60.000,00
4.2.	Other projects	34.206,70
5.	General Assembly	136.210,00
6.	Publication	38.550,00
7.	Publications for developing countries	50.000,00
8.	Executive Committee	51.400,00
9.	Officers' meetings	20.560,00
10.	Symposia, Colloquia	117.540,74
11.	Inter-Union Commissions	51.400,00
12.	Executive Committee projects	12.850,00
13.	Representation	27.404,30
14.	Bank charges	3.558,00
15.	Astronomers' schools	26.000,00
16.	Regional meetings	30.840,00
	Total payments	1.131.289,37

brought forward		24.877,60
27. Variable Stars Catalogue		9.329,10
	Total	34.206,70

13. The unit of subscription for 1977-1979. The Executive Committee will propose to the Finance Committee that the unit of subscription be increased by 10% from 1125 gold francs to 1237,50 gold francs.

14. Appointment of the Special Nominating Committee

15. Proposals for resolutions presented by the Executive Committee

16. Proposals for resolutions presented by Commissions, subject to the recommendation by the Resolutions Committee

17. (a) New Adhering Country
    (b) Announcement of new individual Members of the Union

18. Commissions:
    (a) Structural changes in Commissions, if any
    (b) The election of Presidents and Vice-Presidents
    (c) The membership of Organizing Committees

19. Place and date of the XVIIth General Assembly
    Consideration of the Canadian invitation to hold the XVIIth General Assembly at Montreal, Canada, in 1979

20. Election of President, three Vice-Presidents, General Secretary, and Assistant General Secretary

21. Addresses by the retiring and newly elected Officers

22. Closing Ceremonies

At the first session of the General Assembly, to be held on Tuesday, 24 August, items 1 through 11 will be considered. The remaining items, and any that may have been adjourned from the first session, will be considered at the second session, to be held on Thursday, 2 September.

First Session
Held in the "Patinoire" in Grenoble, France, on Tuesday, 24 August 1976 at 10.30. Professor L. Goldberg, President, in the chair.

1. Formal Opening. The President welcomed the Members of the IAU, Invited Participants, Registered Guests, representatives of Adhering Countries and National Committees of Astronomy, representatives of Scientific Unions and other organizations, and formally opened the first session of the sixteenth General Assembly of the Union.

He wished a particularly hearty welcome to the official representatives of ICSU and sister Unions as follows:

ICSU (International Council of Scientific Unions)	: Prof. C. de Jager
BIH (Bureau International de l'Heure)	: Dr. B. Guinot
BIPM (Bureau International des Poids et Mesures)	: Dr. J. Terrien
CODATA (Committee on Data for Science and Technology)	: Prof. P. Melchior
COSPAR (Committee on Space Research)	: Prof. C. de Jager
FAGS (Federation of Astronomical and Geophysical Services)	: Ing. Gen. G. R. Laclavère

ICSU-AB (ICSU Abstracting Board)   : Prof. J.-C. Pecker
IPMS (International Polar Motion Service)   : Dr. S. Yumi
IUAA (International Union of Amateur Astronomers)   : Mr. K. E. Chilton, Dr. V.
                                                      Barocas, Dr. C. Baldinelli
IUGG (International Union of Geodesy and Geophysics) : Prof. P. Melchior
IUHPS (International Union of the History and
      Philosophy of Science)   : Prof. R. Taton
SCOPE (Scientific Committee on Problems of the Envi-
      ronment)   : Prof. J.-C. Pecker
URSI (International Union of Radio Science)   : Prof. W. N. Christiansen
ICTS (ICSU Committee on the Teaching of Science)   : Dr. D. McNally
SCOSTEP (Special Committee on Solar-Terrestrial
      Physics)   : Dr. Z. Svestka (Dr. E. R.
                                                      Dyer)

   The President continued by suggesting that messages of good wishes should be
sent to former Presidents and General Secretaries of the IAU prevented from attending the present meeting. This suggestion was approved by acclamation.
   The President then took the word and addressed the General Assembly as follows:
   Before proceeding to the business of the General Assembly, there are two matters
on which I should like to comment. First, I am especially happy to see so many of
our young colleagues here - astronomers who are attending a General Assembly for
the first time, and it is to them that I wish to address a few words on what a General Assembly is and what role it plays in the life of the Union.
   As you know, the aims of the I A U are (1) to facilitate relations between astronomers of different countries where it is useful or necessary to organize international co-operation and (2) to promote the study and development of astronomy in
all its branches. The work of the Union is centered in 40 Commissions, of which a
very few are administrative in character, and the others cover all of the sub-dis-
ciplines of astronomy, a small number by technique and the majority by subject
matter. During the intervals between General Assemblies, the Commissions, which
are vested with a high degree of autonomy, will normally organize about six specialized symposia per year and an equal number of colloquia. Moreover, regional
meetings, covering a fairly broad range of subject, are now being scheduled in
different parts of the world at the rate of one or two in each of the years in
which no General Assembly is held.
   The General Assembly, however, sympolizes the unity and shared common goals
of astronomy as well as the often unpredictable interdependence of superficially
unrelated branches of astronomy, e.g. the solar corona and quasars, black holes
and binary stars, radio astronomy and astrometry, solar oscillations and stellar
evolution, etc. The Plenary Sessions, of which this is the first of two, are entirely administrative in nature, terribly important for the smooth operation of
the Union but also, I must confess, somewhat ceremonial and dull. Some of you may
find it interesting to learn how your affairs are conducted by the officers of
the I A U and the official delegates of the countries but all of you are, I am
sure, looking forward to the next seven or eight days. The menu will be a rich
one. First of all, each of the 40 commissions will be meeting singly and jointly
for an average of about 9 hours each. A certain number of commission meetings
will deal with routine administrative matters, such as elections, reports of working groups, resolutions, international co-operation, etc., but the majority will
be scientific in content. The scientific meetings have been organized well in advance, in most cases with a well defined program of invited talks, and feature general review papers that will help the young astronomers particularly to learn
about the latest advances in a given field of research.
   In addition to the commission meetings, seven Joint Discussions have been organized, most of them lasting an entire day, and all of them covering broad areas of
research that interest two or more commissions. Most of the programs of the scientific meetings will be found in the final printed program and therefore you will
have an opportunity to plan each day in advance. On any one day, the choices among

a joint discussion and a dozen commission meetings will require the application of a stern will, but no more so than you require nowadays when selecting topics for your own research. Finally, the Invited Discourses, which are intended to display the latest ideas and results in various branches of astronomy, will climax the week-long presentation of astronomical progress.

You will also be tempted by many entertainments of a non-scientific nature during the days ahead - receptions, excursions, fine restaurants, matchless scenery. Do not slight these opportunities altogether, for the importance of personal contacts during General Assemblies is just as great today as 41 years ago when the I A U last met in France, and the attendance was less than 300. At that time, the Chairman of the French National Committee, M. A. de la Baume Pluvinel, reminded his audience that "these meetings are the occasion, for the members of the great astronomical family, to know one another better and to form durable ties of friendship with each other. Exchanges of views take place between astronomers who were acquainted with each other only by their work and these particular conversations are sometimes as profitable as the work done in the Commissions". I would like to add my personal observation that the ever widening contacts between scientists at international meetings, especially since 1950, have created an enormous amount of understanding and good will among all countries concerned, East and West and North and South. The exchanges of scientific information and ideas among astronomers have been immeasurably freer and generous because we have come to like and respect each other as identifiable human beings and not only as names on scientific papers. And so I urge the young astronomers to take full advantage of the opportunity to make friends while they are here. The rewards will be immediate and in many cases they will last for a lifetime.

I should like now to speak about the problem of Chinese membership in the IAU, which is a matter of the utmost concern to all of us. At the close of the last General Assembly in Sydney, I expressed the view that the Executive Committee should have no task more urgent than that of restoring membership in the IAU to the People's Democratic Republic of China. Unfortunately, I cannot report that we have been successful in this endeavour, although we, together with ICSU, have made earnest efforts to dissolve the deadlock in which we find ourselves. Up until now, the Executive Committee has found no way acceptable to the Chinese of reinstating their membership in the IAU without expelling Taiwan, an action that would violate the statutes of the Union. Considering that the membership of the IAU has tripled since China's withdrawal in 1960, it is obvious that most of those present here will not be familiar with the issues relating to the Chinese problem and the attention being given by the Executive Committee to its solution, and therefore a few words of explanation are in order.

May I point out at the outset that the IAU has never regarded Taiwan as in any sense substituting for the People's Democratic Republic of China. As you know, China adhered to the IAU in 1935 but ceased temporarily to participate in the work of the Union following the revolution and the founding of the People's Republic in 1949. In 1955, Professor Otto Struve, then President of the IAU, wrote Professor Chang, President of the Chinese Astronomical Society, inviting China to resume its membership in the IAU. As a result, the People's Republic of China sent delegations to the general assemblies in 1955 and 1958, in Dublin and in Moscow, respectively. In 1959, when Taiwan applied and was accepted for membership in the IAU, the Executive Committee passed a resolution specifying that it would be representing astronomy *only on the island of Taiwan*. Nevertheless, China withdrew in 1960 and thus its seat in the IAU is not being filled by anyone at the present time - to the great sorrow of all astronomers, it is empty.

Why did China leave the I A U ? Because, if I may quote from the joint statement issued at the conclusion of President Nixon's visit to Peking in 1972, the Chinese government "firmly opposes any activities which aim at the creation of 'one China, one Taiwan', 'one China, two Governments', 'two Chinas', and independent Taiwan' or advocate that 'the status of Taiwan remains to be determined'". They insist that Taiwan is a province of China and therefore that the Academia Sinica of Peking must be recognized as the sole adhering organization representing all of China. Furthermore, they would require the I A U to deny admission to representatives from Taiwan to all I A U sponsored conferences. For the I A U to accept these conditions would not only violate both its principles and statutes, but would establish a precedent that could lead in the future to the politically motivated expulsion of other members of the Union. It has been our view that the movement of scientists into and out of Taiwan is at present under the control of the government in Taipei and that this *de facto* situation must be recognized by separate representation for Taiwan in the I A U We do not advocate or favor any particular political relationship between Taiwan and China, but neither do we demand that Taiwan incorporate itself into China as a condition for membership by its scientists in the international scientific community. That is not properly the business of the I A U

Some of our colleagues are growing restive at the continued absence of Chinese astronomers from I A U meetings and would like quick action to speed their return. For example, I have heard it said that: "if the U. N. can dismiss Taiwan, why can't the I A U do the same?" The answer is simple: the I A U believes that all astronomers have the right to be represented in the organization, regardless of political considerations. The U. N. chooses its membership on political grounds; the I A U does not. I need only remind you that in 1955 the People's Democratic Republic of China was welcomed into the I A U fifteen years before its acceptance by the U. N.

I would like to reiterate that, at present, the readmission of China to the I A U would require a change in the statutes of the Union. The Executive Committee cannot simply decree that the Academia Sinica of Peking should replace the Academia Sinica of Taipei as the rightful adhering organization of China because the latter represents only Taiwan and the statutes make no provision for cancelling memberships. If the statutes were to be changed to allow cancellations of membership, the criteria would have to be clear and unambiguous. Two proposals have been made: (1) to restrict membership to countries that belong to the U. N. or U N E S C O or (2) to require a certain minimum level of astronomical activity as a condition for membership. In my judgment, both of these proposals are in conflict with the aims of the Union and are therefore unacceptable, the first for reasons to which I have already alluded and the second because it would force more than one country out of the Union and would stifle the progress of astronomy in developing countries.

Despite these difficulties, I believe that the present impasse is not insoluble and that we must continue to strive for an arrangement that does not require China to modify its political position towards Taiwan and yet at the same time makes it possible for astronomers on Taiwan to participate in the work of the I A U

Irrespective of whether China rejoins the I A U in the immediate future, we astronomers should try to accelerate the frequency of contacts with our Chinese colleagues and to develop co-operative projects on an individual or country-to-country basis. There are now many signs that the Chinese have resumed the expansion and development of their activity in astronomy and that co-operation and exchanges with astronomers of other countries would be welcomed by them. As an example, I should mention that at the request of the Chinese, the Kitt Peak Observatory has recently sent to the Peking Observatory copies of drawings of its two largest telescopes and has invited the Chinese astronomers to engage in joint discussions on telescope design. I would also urge astronomers, as the General Secretary urged all National Committees one year ago, to use every opportunity to invite Chinese astronomers to conferences in their own countries and to let them know how eager we are to communicate and work with them.

After these words, the President asked those present to stand while the General Secretary would read the names of Members who had died since the XVth General Assembly, or whose deaths had not been known at the time of that meeting. The General Secretary then read the following list:
C. G. Abbot, J. S. Astapovich, P. F. Bok, P. E. Bourgeois, J. A. Carroll, R. P. Cesko, G. A. Chebotarev, N. A. Chudovicheva, G. M. Clemence, P. Collinder, E. U. Condon, H. Daene, R. E. Danielson, L. D. de Feiter, L. Detre, A. J. Dos Santos,

V. C. A. Ferraro, G. Gjellestad, P. M. Gorshkov, A. Gougenheim, B. J. Harris, F.
Henn, S. Herrick,H. G. Hertz, W. R. Hindmarsh, J. Hopmann, T. E. Houck, H. M.
Jeffers, N. S. Kalikhevich, H. Kienle, K. O. Kiepenheuer, G. F. G. Knipe, F.
Koebcke, G. P. Kuiper, E. M. Lindsay, A. W. Lines, T. A. Littlefield, M. R. Madwar,
V. Maître, W. J. Miller, R. L. Minkowski, D. Mugglestone, A. T. Nesmyanovich, E. V.
Novopashennyj, C. P. Olivier, J. Paton, S. B. Pikel'ner, E. K. Rabe, R. O. Redman,
B. Rosen, S. V. Rublev, E. W. Salpeter, A. Schmitt, F. P. Scott, B. M. Shchigolev,
W. M. Smart, N. N. Sytinskaya, G. van Biesbroeck, W. H. van den Bos, M. J. Verbaan-
dert, A. N. Vyssotsky, S. Wierzbinski, R. Wildt, J. Witkovski, K. Wurm, N. A. Yakov-
kin, F. Zagar, W. Zonn, F. Zwicky.

2. *Appointment of official interpreters and tellers*. The General Assembly appointed by acclamation B. E. J. Pagel as official interpreter from French to English,
L. N. Houziaux and J. P. Swings as official interpreters from English to French,
and Claude Froeshle and A. F. Hayli as official tellers.

3. *Report of the Executive Committee*. The President passed the word to the General Secretary who invited the national representatives to discuss the report of
the Executive Committee as printed in Information Bulletin No. 36. There was no
discussion and the report was approved unanimously.
On the proposal of the General Secretary, the Assembly then approved unanimously to put on the agenda the application of Iraq for admission to the IAU as Adhering Country.

4. *Report of the General Secretary*. Invited by the President, the General Secretary read his addendum to the report of the Executive Committee, as follows:
"The present report is to provide supplementary information on the activities
of the Union for the period from 1 January 1976 until now, and on its finances until 30 June 1976.
The President, General Secretary, and the Assistant General Secretary, together with the Staff of the Secretariat, met in Athens, Greece, on 16 February
and, without the Staff, in Tucson, Arizona, U.S.A., on 1 June 1976. Dealt with
were the most pressing problems of the IAU, such as its finances, the Reports on
Astronomy 1976, the proposals for new Presidents and Vice-Presidents of IAU Commissions and the representation of the Union in other international bodies. Particular attention was paid to the forthcoming ordinary General Assembly: the invited
speakers were reviewed, the programmes of the Joint Discussions were re-examined,
and the Commissions meetings, their subjects included, were co-ordinated to avoid
overlaps as much as possible. It was agreed to move the IAU office to Lausanne,
Switzerland, well before the General Assembly.
The Executive Committee had to continue its work by correspondence. It approved
the triennial report to the General Assembly, including the budget to be presented to the Finance Committee, and the Agenda for the XVIth General Assembly.
In view of the savings resulting from the generous allocation by the French Government of free air-tickets to some members of the Executive Committee and Com -
missions Presidents, the Executive Committee resolved to increase the number of
young astronomers'grants, each of $ 100, to one hundred.
The General Secretary maintained close contact with the French Local Organizing
Committee, especially as regards the distribution of the French and IAU grants to
young astronomers, the allocation of meeting rooms, and the printing of the Final
Programme. He found the preparations for the General Assembly to proceed quite
satisfactorily, thanks to the untiring efforts of the Local Organizing Committee.
Presidents of IAU Commissions formulated the programmes of the meetings of Commissions to be held in Grenoble, and forwarded to the General Secretary their requirements for meeting rooms and times. They subsequently actively participated
in finalizing the listing of Commissions meetings as printed in the Final Programme. A few last minute changes will be made public on the announcements board.
Financial considerations made it necessary to publish the Reports on Astronomy
1976 in three volumes, each volume grouping the reports of Commissions with simi-

lar or complementary activities. It is hoped that this will make the reports more accessible than if they were published in one volume. Most of the reports of Commissions sent to the General Secretary for inclusion in the volumes are of a remarkably high standard and show that the decision to continue the publication of the Reports on Astronomy was correct. I wish to thank the Presidents of Commissions for their outstanding work, and to express my gratitude also to the Vice-Presidents, and all Members who contributed to the success of the Reports on Astronomy 1976.

Two IAU Symposia and 7 Colloquia are being held this year in conjunction with the General Assembly, and I wish to thank all those who spared no effort in making these meetings successful, and especially the Assistant General Secretary for her untiring work, as the meetings were her responsibility. I wish to thank also particularly the organizers of the two Regional Meetings under the auspices of the I A U , held in Trieste and in Tbilisi, in 1974 and 1975 respectively. These meetings, following the Athens meeting of 1972, have now established a new tradition of IAU activity, which is to be applied in the near future not only in Europe, but in other parts of the world as well, such as South America and New Zealand.

My short summary would not be complete if I did not mention the finances of the IAU. Really, there is not much to say, and this is good - for they are good. I hope the Finance Committee will be able to confirm my statement. Moreover, I am happy to report that only 2 Adhering Countries, all in category one, failed to pay their contributions for 1975. (Four other countries paid the greater part of their contributions for 1975, and the fact that they are somewhat behind is outside their control, since they spared no effort to pay their dues completely). Very encouraging are the payments for 1976: 26 out of 47 Adhering Countries have already paid their contributions, and have thus positively responded to the appeal of the Executive Committee to pay promptly in the year of the General Assembly.

On the other hand, as explained in the main report of the Executive Committee, the continuing inflation makes it necessary to increase the unit of contribution for the years to come by 10%. However, the proposal of the Executive Committee will be changed slightly so that the increase that will be asked will be 10.16% *in Swiss Francs*. If it had been in gold francs, as it was stated originally, it would be about double. The reasons are due to the subtleties of the money market, which will be explained in detail to the Finance Committee. Furthermore, as our budget is in Swiss Francs, the Executive Committee proposes at the same time to abandon the gold franc as basis for the unit of contribution and adopt the Swiss Franc, which is also based on gold, being equal to 0.2175926 grams of pure gold.

Finally, I am happy that the application of Iraq to become a Member Country of the IAU has been placed on the agenda. The Executive Committee considers this application as fulfilling the requirements of the Union and recommends the admission of Iraq to the Union.

Now that my term of office is drawing to an end, I wish to thank heartily all those who contributed to the successful work of the Union during that period. Thanks to them the international collaboration in astronomy, on a global scale, has been so fruitful. The concerted efforts of our colleagues not only led to many exciting discoveries, but also gave us the unique feeling of security which one has within a big family that works towards a noble common goal."

5. *Report of the President on the proposals for membership in the Executive Committee*. The President called the attention of the Assembly to article 12 (a) of the By-laws which provides that proposals for the elections to the President of the Union, six Vice-Presidents, the General Secretary and the Assistant General Secretary are submitted to the General Assembly by the Special Nominating Committee. The Special Nominating Committee in office, consisting of the present President, the former President B. Strömgren, a member nominated by the former Executive Committee, H. Elsässer, and four members selected by the Nominating Committee: A. Blaauw, J.-C. Pecker, L. Perek and A. B. Severny, had made the following proposals as to the composition of the next Executive Committee:

As President for the term   1976-1979:   Prof. A. Blaauw, The Netherlands

As Vice-Presidents for the term 1976-1982:   Dr. D. S. Heeschen, U.S.A.
                                                   Prof. E. K. Kharadze, U.S.S.R.
                                                   Prof. S. van den Bergh, Canada

As General Secretary for the term 1976-1979:   Prof. E. A. Müller, Switzerland

As Assistant General Secretary for the term   Prof. P. A. Wayman, Ireland
                             1976-1979:

The full Executive Committee will thus consist
of the above six together with:
Continuing Vice-Presidents: Mr. J. G. Bolton, Australia
                                                   Prof. Ch. Fehrenbach, France
                                                   Prof. W. Iwanowska, Poland

In an advisory capacity: Prof. L. Goldberg, former
                                                      President, U.S.A.
                                                   Prof. G. Contopoulos, former
                                                      General Secretary, Greece

According to the By-laws the formal election will take place at the final session
of the General Assembly.

    6. *Announcements*. Called on by the President, the General Secretary announced
(a) the names of the representatives of Adhering Countries to vote at the General
    Assembly and
(b) the names of representatives of Adhering Countries on the Nominating Committee:

Country	*(a) Official Representative*	*(b) Representative on Nominating Committee*
Arab Republic of Egypt	A. S. Asaad	A. S. Asaad
Argentina	J. A. Lopez - absent	J. A. Lopez - absent
Australia	R. G. Giovanelli	W. N. Christiansen
Austria	H. Haupt	H. Haupt
Belgium	P. Swings	M. S. Arend
Brazil	J. A. de Freitas Pacheco	S. Ferraz-Mello
Bulgaria	N. Nicolov	B. Kovachev
Canada	D. A. MacRae	D. F. Gray
Chile	C. Anguita	F. Noël
Colombia	J. Arias de Greiff	J. Arias de Greiff
Cuba	not represented	not represented
Czechoslovakia	V. Guth	V. Bumba
Denmark	A. Reiz	O. Møller
Finland	J. Tuominen	J. Tuominen
France	J. F. Denisse (J.-C.Pecker)	M. J. Boulon
G.D.R.	G. Ruben	G. Ruben
Germany, F.R.	W. Priester	G. Traving
Greece	J. Xanthakis (M.Mousoulas)	S. Svolopoulos
Hungary	L. Dezsö	B. Seidl
India	V. Radhakrishnan	G. Swarup
Iran	H. Zomorrodian	H. Zomorrodian
Ireland	P. K. Carroll	P. K. Carroll
Israel	Y. Ne'eman	G. Shaviv
Italy	L. Gratton	L. Gratton
Japan	Z. Suemoto	W. Unno

Korea DPR	not represented	not represented
Korea Republic	not represented	not represented
Mexico	P. Pişmiş	E. Mendoza
Netherlands	H. van der Laan	E.P.J. van den Heuvel
New Zealand	P. J. Edwards	P. J. Edwards
Norway	Ø. Elgarøy	Ø. Elgarøy
Poland	W. Iwanowska	A. Opolski
Portugal	not represented	not represented
Roumania	not represented	not represented
South Africa	M. W. Feast	E. E. Baart
Spain	D. A. Orte	L. Quijano
Sweden	T. Elvius	B. Höglund
Switzerland	U. W. Steinlin	B. Hauck
Taiwan	Chun-Shan Shen	Chun-Shan Shen
Turkey	N. Gökdogan	Z. Tüfekçioglu
United Kingdom	M. J. Seaton	R. J. Tayler
Uruguay	not represented	not represented
U.S.A.	B. F. Burke	F. J. Kerr
U.S.S.R.	A. B. Severny	A. G. Massevich
Vatican City State	P. J. Treanor	M. F. McCarthy
	(M. F. McCarthy)	
Venezuela	J. Stock	J. Stock
Yugoslavia	G. Teleki	M. Vukicevic

Note: In some cases National Committees designated substitute delegates. Their names are listed in parentheses.

(c) The General Secretary announced that the Executive Committee had asked H. van der Laan to act for Yu. N. Parijskij as President of Commission No. 40 for the duration of the General Assembly.

7. *Appointment of the Finance Committee.* In accord with Article 18(a) of the By-laws the General Assembly appointed the following Finance Committee consisting of one representative of each Adhering Country:

Country	*Representative in Finance Committee*
Arab Republic of Egypt	S. Yousef
Argentina	J. A. Lopez - absent
Australia	W. H. Robertson
Austria	H. Haupt
Belgium	F. Bertiau
Brazil	P. Kaufmann
Bulgaria	M. Popova
Canada	J. L. Locke
Chile	C. Anguita
Columbia	J. Arias de Greiff
Cuba	not represented
Czechoslovakia	M. Kopecký
Denmark	K. Gyldenkerne
Finland	K. Mattila
France	N. Bel
G.D.R.	G. Ruben
Germany, F.R.	W. Fricke
Greece	L. N. Mavridis
Hungary	B. Balázs
India	S. D. Sinvhal
Iran	H. Zomorrodian

Ireland	P. A. Wayman
Israel	G. Shaviv
Italy	G. Godoli
Japan	Z. Suemoto
Korea DPR.	not represented
Korea Republic	not represented
Mexico	M. Peimbert
Netherlands	J. van Nieuwkoop
New Zealand	P. J. Edwards
Norway	Ø. Hauge
Poland	S. Piotrowski
Portugal	not represented
Roumania	not represented
South Africa	M. W. Feast
Spain	D. J. Cardús
Sweden	B. Westerlund
Switzerland	M. C. Huber
Taiwan	Ting, Yeou-Tswen
Turkey	K. Özemre
United Kingdom	D.W.N. Stibbs
Uruguay	not represented
U.S.A.	L. W. Fredrick
U.S.S.R.	G. S. Khromov
Vatican City State	P. J. Treanor
Venezuela	J. Stock
Yugoslavia	G. Teleki

Note: Substitutions designated by a few National Committees during the General Assembly have been omitted in the above list.

8. *Appointment of the Resolutions Committee.* At the request of the President, the General Secretary informed the assembly that the Resolutions Committee is to advise the Executive Committee as regards the proposals for resolutions submitted to the General Assembly for consideration. The General Assembly appointed M. K. V. Bappu and L. Dezsö to serve on the Resolutions Committee, with the General Secretary and the Assistant General Secretary attending in an advisory capacity.

9. through 11. *Considerations of proposals for resolutions submitted by the French National Committee of Astronomy, by Commissions, and by the Executive Committee.* The General Secretary said that the proposals for resolutions as given under points 9., 10. and 11. of the agenda as well as proposals for resolutions submitted during the General Assembly would first be dealt with by the Resolutions Committee and then voted on, with due regard to the recommendations of that Committee, by the General Assembly.

The General Secretary continued by announcing that all other items of the agenda would be deferred until the next session of the General Assembly.

The President then formally adjourned the meeting at 12.00.

FINAL SESSION

Held in the Weil Amphitheatre of the Scientific and Medical University, Grenoble, on Thursday, 2 Septembre 1976, at 10.00.

Professor L. Goldberg, President, in the chair

Before passing to the agenda, the General Secretary, called upon by the President, read to the Assembly a message of good wishes received from the People's Republic of Korea.

12. and 13. *Report of the Finance Committee and consideration of the unit of contributions.* The Finance Committee, as appointed at the first session, had to examine the accounts of the Union for the years 1973-1975, the budgets of the IAU for 1976 and for the period 1977-1979, and to consider the unit of contribution payable to the IAU by Adhering Countries from 1 January 1977 on. The President called upon Professor G. Teleki, member of the Finance-Subcommittee, to present to the General Assembly the report prepared by the Chairman of the Finance Committee Professor D. W. N. Stibbs. Dr. Teleki read as follows:

REPORT OF THE FINANCE COMMITTEE TO THE GENERAL ASSEMBLY

1. At the First Session of the XVIth General Assembly on 24th August 1976, the Finance Committee, consisting of the nominated representatives of Member Countries, was duly appointed to perform its duties in accordance with Article 18(a) of the By-laws. The Finance Committee met on that date when the President welcomed the members and outlined the function of the Committee. The President explained that for each ordinary General Assembly the Executive Committee prepares an estimate of the budget for the period to the next ordinary General Assembly which it submits to the Finance Committee, together with the accounts for the preceding period. The Finance Committee examines the accounts and the estimate of the budget on which it prepares reports and submits them to the General Assembly for approval.

2. The Finance Committee appointed D. W. N. Stibbs as its Chairman, and it appointed a Sub-Committee to make a detailed examination of the accounts and the estimate of the budget. The Sub-Committee consisted of the Chairman ex officio, Miss M. Bel (France), L. W. Fredrick (U.S.A.), G. S. Khromov (U.S.S.R.), G. Teleki (Yugoslavia), with G. Contopoulos, Miss E. A. Müller and A. Jappel members ex officio as General Secretary, Assistant General Secretary and Executive Secretary of the Union respectively. The Sub-Committee was convened by the Chairman on the 26th and 27th August, and it presented a report to the Finance Committee at its second meeting on 28th August.

3. The Chairman reported to the Finance Committee that the accounts for the period 1973-75, summarized in the narrative report of the Executive Committee to the General Assembly, had been subjected to detailed examination by the Sub-Committee with verifications supplied by the Secretariat during the examination. It was noted that the accounts had been kept at a very high standard and that all the relevant documents were available. The Sub-Committee had concluded that the accounts were in order, both on the basis of its own inspection and on the summary statement of the Auditor to the Union, M. Roger Bacle of Paris.
The Finance Committee, on the basis of the inspection of the accounts by the Sub-Committee, recommends to the General Assembly that the accounts for 1973-75 be approved.

4. The budgetary estimates for 1976, summarized in the Report of the Executive Committee, were examined by the Sub-Committee which noted that the projected expenditure was within the financial provision of the Budget accepted by the XVth General Assembly, and that it was consistent with the policy of the Union embodied in the Budget.
The Finance Committee recommends to the General Assembly that the budgetary estimates for 1976 be accepted.

5. The Sub-Committee reported to the Finance Committee that it had examined in detail (a) the comprehensive budgetary proposals for the period 1977-79 submitted by the Executive Committee, and (b) the proposal from the French National Committee that the Unit of Contribution be increased by 20 per cent.
The Sub-Committee recommended to the Finance Committee that, as proposed by the

Executive Committee, the Unit of Contribution be increased by 10.16 per cent, and that it should be expressed in the reference currency of the Union, the Swiss franc (one Swiss franc being equal in value to 0.2175926 g of fine gold). This recommendation was accepted by the Finance Committee by an absolute majority.

The Finance Committee recommends to the General Assembly that the Unit of Contribution be increased by 10.16 per cent and that it be expressed in the reference currency of the Union, namely the Swiss franc.

6. The Finance Committee considered the Budget for 1977-79 based on the new Unit of Contribution in Swiss francs and by an absolute majority accepted the revised budget put forward by the Sub-Committee.

The Finance Committee recommends for adoption by the General Assembly the following budget:

BUDGET FOR 1977-79

Unit of Contribution 1465.00 Swiss Francs
(One Swiss Franc equal to 0.2175926 g of fine gold)

Receipts in Swiss Francs		*Payments in Swiss Francs*	
1. Contributions from Adhering Countries	944.925	1. Administrative Office	449.900
2. Revenue from publications	60.000	2. Subscription to ICSU	23.625
3. Interest on accounts	51.925	3. Commission expenses	10.280
4. UNESCO subvention through ICSU	92.520	4. Projects of Commissions 4.1. Exchange of Astronomers	67.405
		4.2. Other projects	34.205
Total	1.149.370	5. General Assembly	136.210
Payments	1.143.585	6. Publications	38.550
		7. Publications for developing countries	50.000
Excess of receipts over payments	5.785	8. Executive Committee meetings	51.400
		9. Officers' meetings	20.560
		10. Symposia and Colloquia	117.540
		11. Inter-Union Commissions	41.400
List of projects under Payments: Item 4.2		12. Executive Committee projects	10.000
Commissions and projects	Swiss Francs	13. Representation	20.000
		14. Bank charges	3.660
4. Ephemerides	2.485	15. Young Astronomers' schools	38.850
6. IAU Telegram Bureau	5.595	16. Regional meetings	30.000
16. Planetary Documentation, Meudon	5.290		
17. Lunar Documentation, Meudon	5.285	Total	1.143.585
20. Minor Planet Center	6.220		
27. Variable Star Catalogue	9.330		
Total	34.205		

7. The Sub-Committee, noting that more funds of the Union were located in Current Accounts than was necessary for its recurrent expenses, recommended that 60.000 dollars be transferred from those accounts into Savings Accounts where the funds by interest would augment the revenue of the Union by at least 18.000 Swiss francs by the end of the budgetary period 1977-79.

The Finance Committee accepted the recommendation unanimously, and so recommends to the General Assembly.

8. The Sub-Committee reported to the Finance Committee that, on the basis of a detailed examination of the running costs of the Administrative Office of the Union, it was unreasonable at the present time to expect the host institution to contribute more than 27.000 Swiss francs to the total costs during the budgetary period, and that 37.000 Swiss francs of the reserves of the Union should be available to meet the remainder of expenses not specifically included in the budgetary provision if the need arose.

The Finance Committee accepted the recommendation unanimously, and so recommends to the General Assembly.

9. The Sub-Committee, recognizing the need for flexibility in indenting charges against payment items in the Budget, recommended to the Finance Committee that it should be formally stated that the Executive Committee would be expected to make adjustments in the total charges against items should the need arise, without indenting charges against reserves except where specified, provided such adjustments were consistent with the policy embodied in the Budget and were within the overall financial provision.

The Finance Committee, noting that the recommendation would be particularly relevant to Items 4.1. and 15 in the Budget if the Commission proposals were not satisfactory to the Executive Committee, unanimously accepted the recommendation on the adjustment of payments, and so recommends to the General Assembly.

10. The Sub-Committee considered the arguments for and against the augmentation of the receipts of the Union by the introduction of an individual membership fee, and it concluded that, for the time being, this possible source of funds should not be considered.

The Finance Committee concurred with the view of the Sub-Committee, and so recommends to the General Assembly.

11. The Sub-Committee considered the question of the investment of some of the funds of the Union, and recommended that speculative or growth investment should not be entertained at the present time. However, it considered that some of the funds of the Union might profitably be placed in higher interest bearing accounts than are currently operated by the Union.

The Finance Committee unanimously accepted these recommendations, and so recommends to the General Assembly.

12. In conclusion, the Finance Committee warmly supported the wish of the Sub-Committee to place on record its gratitude to the Secretariat of the Union for their courtesy and helpfulness in making available their meticulously kept documents on the finances of the Union and for their expert professional advice which greatly assisted the Committee in its work.

D. W. N. Stibbs
Chairman, Finance Committee

The President thanked Professor Teleki and said that since the unit of contribution and the comprehensive budget for 1977-1979, as proposed by the Finance Com-

mittee, differed from those mentioned in the Agenda, it would be necessary first
to put these two amendments on the Agenda by the approval of at least two thirds
of the votes of Adhering Countries represented at the General Assembly, and then
to vote on the report of the Finance Committee itself. The General Assembly placed
the two amendments on the agenda by a majority of more than two thirds of votes.

The President then proceeded to take the vote on the individual recommendations
of the Report of the Finance Committee. They were all approved by an absolute majority of votes, as called for by Statute 11(b). The President therefore declared
the Report of the Finance Committee approved and thanked Professor Stibbs, the
members of the Finance Committee and the General Secretary for their work.

14. *Appointment of the Special Nominating Committee.* The President referred
to Article 12 of the By-laws which provides that proposals for elections to the
President of the Union, six Vice-Presidents, the General Secretary, and the Assistant
General Secretary are submitted to the General Assembly by the Special Nominating
Committee which consists of the President and the past President of the Union, a
member proposed by the retiring Executive Committee from Members not belonging to
the Executive Committee, both immediately past and present, and four members selected by the Nominating Committee from among twelve Members proposed at the meeting of Presidents of Commissions.

The President then moved that the General Assembly appoint the following Mem-
bers to serve on the Special Nominating Committee:
A. Dollfus (France)
B. A. Lindblad (Sweden)
A. G. Massevich (U.S.S.R.)
A. Reiz (Denmark)
J. P. Wild (Australia)
This list was unanimously approved.

15. and 16. *Resolutions.* The General Secretary reported that the proposals for
resolutions submitted to the General Assembly by the Executive Committee, by National Committees of Astronomy and by Commissions had been passed on to the Resolutions Committee for consideration. Thus, before voting on the resolutions, Professor M. K. V. Bappu would present to the General Assembly the report of the Resolutions Committee.

The President then called upon Professor Bappu, Chairman of the Resolutions Committee, to take the floor.

REPORT OF THE RESOLUTIONS COMMITTEE TO THE GENERAL ASSEMBLY

The proposals for Resolutions examined by the Committee include those received
from the Executive Committee, the French National Committee, Commissions of the
Union and the Working Group for Planetary System Nomenclature. We have now examined all proposals carefully, and in some cases discussed these with the appropriate Commission Presidents. Some of these Resolutions have been presented to the
General Assembly at its First Session. We recommend that the proposal of the Executive Committee listed under 11 of the Agenda, and which refers to the Special
Nominating Committee, be adopted. The proposal is to reword the second sentence
of article 12(a) of the By-laws as follows:
This consists of the President and past President of the Union, a member proposed
by the retiring Executive Committee, and four members elected by the Nominating
Committee from among twelve Members proposed by Presidents of Commissions. Other
than the President and immediate past President, present and former members of
the Executive Committee shall not serve on the Special Nominating Committee.

The proposal for a resolution made by the French National Committee of Astronomy
is financial in scope and as indicated at the First Session, has been submitted to
the Finance Committee. The General Assembly has already accepted the report of the

Finance Committee, which has considered the French proposal. A proposal for a resolution on values of precession, proposed by Commission 8, has been withdrawn by
the Commission. All other proposals for resolutions that have appeared on the
Agenda at our first meeting have been taken as resolutions of Commissions and will
be covered by Resolution 10 that this Committee will propose later.

I shall now read the text of each resolution that the Resolutions Committee
wishes to place before you, for adoption by the General Assembly. This list covers proposals adopted by Commissions during the General Assembly and which, in
our opinion, are of sufficient interest and importance for consideration by the
General Assembly.

RESOLUTION NO. 1

Proposée par les Commissions Nos. 4, 8 et 31 de l'UAI

L'Union Astronomique Internationale soutient les recommandations contenues
dans le Rapport Commun des Groupes de Travail de la Commission 4 sur:

> le Système des Constantes Astronomiques de l'UAI (1976),
> le nouveau standard pour l'époque et pour l'équinoxe,
> le système de référence fondamental,
> les procédés de calcul des positions apparentes et de
>   réduction des observations,
> les échelles de temps pour les théories dynamiques et les
>   éphémérides,
> et les autres données nécessaires à la préparation des éphémérides;

et recommande de les utiliser pour la préparation du catalogue fondamental K5 et,
à partir de l'année 1984, pour celle des éphémérides nationales et internationales,
ainsi que pour tout autre travail astronomique impliquant ces éléments.

RESOLUTION NO. 1

Proposed by IAU Commissions 4, 8 and 31

The International Astronomical Union endorses the recommendations given in the
Joint Report of the Working Groups of Commission 4 on:

> the IAU (1976) System of Astronomical Constants,
> the new standard epoch and equinox,
> the fundamental reference frame,
> the procedures for the computation of apparent places
>   and the reduction of observations,
> time scales for dynamical theories and ephemerides,
> and other quantities for use in the preparation of ephemerides;

and recommends that they shall be used in the preparation of the fundamental catalogue FK5 and of the national and international ephemerides for the years 1984 onwards, and in all other relevant astronomical work.

RESOLUTION NO. 2

Proposée par la Commission No. 5 de l'UAI

L'Assemblée Générale de l'Union Astronomique Internationale s'inquiète de la forte augmentation du prix des publications qui, liée à l'expansion continue de la littérature scientifique, rend difficile aux jeunes astronomes et aux astronomes des pays en voie de développement l'accès à la connaissance astronomique, et *reconnaissant* que le Comité Exécutif a déjà pris, avec succès, des mesures en vue de réduire le coût des publications,
demande
(*i*) au Comité Exécutif de continuer à veiller par tous les moyens possibles à ce que les publications de l'UAI soient disponibles à des prix fortement réduits,
(*ii*) aux éditeurs d'ouvrages astronomiques de faire tous efforts pour produire des publications à un prix moins élevé et pour utiliser de nouveaux moyens de diffusion des informations.

RESOLUTION NO. 2

Proposed by IAU Commission No. 5

The General Assembly of the International Astronomical Union *notes* with concern the large increases in prices of publications, which, together with the continuous expansion in scientific literature, makes accessibility of astronomical knowledge difficult to young astronomers and astronomers in the developing countries, and *recognizing* that the Executive Committee has already taken action, with success, to reduce the cost of publications,
requests
(*i*) the Executive Committee to take whatever further action may be possible to ensure that IAU publications are made available at greatly reduced prices
(*ii*) publishers of astronomical literature to make every effort towards the production of less expensive publications and to employ new ways of disseminating information.

RESOLUTION NO. 3

Proposée par la Commission No. 5 de l'UAI

L'Assemblée Générale de l'Union Astronomique Internationale considère que Astronomy and Astrophysics Abstracts (AAA) correspond parfaitement aux besoins spécifiques exigés par un service analytique en Astronomie et en Astrophysique et recommande vigoureusement que AAA continue à être préparé, sous les auspices de l'UAI, par l'Astronomisches Rechen-Institut, Heidelberg, République Fédérale l'Allemagne.

RESOLUTION NO. 3

Proposed by IAU Commission No. 5

The General Assembly of the International Astronomical Union considers that Astronomy and Astrophysics Abstracts (AAA) fulfils excellently the specialized needs for an abstracting service in Astronomy and Astrophysics and recommends strongly that AAA continue to be produced by the Astronomisches Rechen-Institut, Heidelberg, Germany Federal Republic, under the auspices of the IAU.

RESOLUTION NO. 4

Proposée par les Commissions Nos. 12, 14 et 29 de l'UAI

L'Union Astronomique Internationale tient en haute estime les activités du Bureau National des Standards (NBS) des Etats-Unis, consacrées à la compilation et à l'évaluation critique des données atomiques et moléculaires, et considère qu'elles sont essentielles aux progrès de l'Astronomie.

RESOLUTION NO. 4

Proposed by IAU Commissions 12, 14 and 29

The International Astronomical Union highly values the activities of the United States National Bureau of Standards in the compilation and critical evaluation of atomic and molecular data, and considers these activities essential for the advancement of astronomy.

RESOLUTION NO. 5

Proposée par les Commissions Nos. 16 et 17 de l'UAI

L'Assemblée Générale de l'Union Astronomique Internationale prenant note de l'intérêt que plusieurs de ses Commissions portent au développement d'un Programme International du Système Solaire, appuie la proposition faite par le COSPAR de créer un comité directeur, comprenant des représentants de l'UAI, et chargé de développer ce programme.

RESOLUTION NO. 5

Proposed by IAU Commissions 16 and 17

The General Assembly of the International Astronomical Union noting the interest of several of its Commissions in the development of an International Solar System Programme, supports the COSPAR proposal for the establishment of a steering committee, including IAU representatives, to develop this programme.

RESOLUTION NO. 6

Proposée par les Commissions Nos. 19 et 31 de l'UAI

L'Union Astronomique Internationale
reconnaissant
que le Service International du Mouvement du Pôle et le Bureau International de l'Heure ont des activités complémentaires et que tous deux contribuent d'une façon essentielle à la détermination et à la compréhension du mouvement du pôle, et
reconnaissant
que les nouvelles techniques radio et laser apporteront une contribution importante à l'étude du mouvement du pôle, mais qu'il est encore trop tôt pour définir le profil d'un nouveau service fondé sur l'utilisation de ces techniques, et
notant
avec satisfaction que la détermination du mouvement du pôle issue des différentes stations du Service International du Mouvement du Pôle a atteint la précision exigée pour la solution de problèmes qui se posent depuis longtemps,
recommande
que le Service International du Mouvement du Pôle continue à fonctionner sous sa forme actuelle et que le Conseil Scientifique du Service International du Mouvement du Pôle et le Comité de Direction du Bureau International de l'Heure continuent à rechercher en commun les possibilités d'utilisation des techniques modernes sur une base permanente, et
insiste
auprès des organismes nationaux et internationaux concernés pour qu'ils maintiennent leur aide au Bureau Central du Service International du Mouvement du Pôle et à chacun des observatoires qui coopèrent avec ce service.

RESOLUTION NO. 6

Proposed by IAU Commissions 19 and 31

The International Astronomical Union
recognizing
that the activities of the International Polar Motion Service and of the Bureau International de l'Heure are complementary, and that they both make essential contributions towards the determination and understanding of the motion of the pole, and
recognizing

that the new laser and radio techniques will make an important contribution
to the study of polar motion but that it is at present too early to determine the
form of a new service based on these techniques, and
noting
with satisfaction that the International Polar Motion Service multi-station derivation of polar motion has attained the precision needed to resolve long-standing
problems,
recommends
that the International Polar Motion Service continue to operate in its present
form, and that the Scientific Council of the International Polar Motion Service
and the Directing Board of the Bureau International de l'Heure jointly keep under
continuous review the possibility of the utilization of modern techniques on a
permanent basis, and
urges
that the international and national agencies concerned continue their support
of the Central Bureau of the International Polar Motion Service and of each cooperating observatory.

RESOLUTION NO. 7

Proposée par les Commissions Nos. 19 et 31 de l'UAI

L'Union Astronomique Internationale ayant passé en revue les diverses fonctions
du Bureau International de l'Heure, BIH, telles qu'elles sont définies dans les
Transactions de l'UAI, Vol. XIIIA, 1967, et prenant en considération l'accroisse-
ment de ces activités consécutif aux responsabilités supplémentaires confiées au
BIH,
recommande
maintenant que soit adopté pour le BIH le texte de référence suivant:
Les fonctions du BIH sont
(a) d'établir l'échelle du Temps Atomique International TAI, en accord avec les
    décisions de la 14ème Conférence Générale des Poids et Mesures et de concert
    avec le Bureau International des Poids et Mesures;
(b) d'établir à partir de toutes les données pertinentes, et de publier, les
    valeurs courantes du Temps Universel et de la vitesse angulaire de la rota-
    tion de la Terre et, également, les coordonnées operationnelles du pôle
    utilisées à cet effet;
(c) de rendre effectif le système du Temps Universel Coordonné UTC en diffusant
    tous les renseignements nécessaires à la coordination des émissions des
    signaux horaires et à la synchronisation des pendules sur l'échelle UTC;
(d) de diffuser les informations importantes pour les utilisateurs scientifiques
    du temps, et de fournir sur demande les données disponibles concernant le
    temps;
(e) d'effectuer les recherches scientifiques nécessaires aux progrès du service.

RESOLUTION NO. 7

Proposed by IAU Commissions 19 and 31

The International Astronomical Union having reviewed the functions of the Bureau International de l'Heure, BIH, which were defined in the Transactions of the IAU, Vol. XIIIA, 1967, taking account of subsequent developments which have resulted in the BIH being entrusted with additional responsibilities, it now
recommends
that the following terms of reference of the BIH be adopted:
The functions of the BIH shall be
(a) to establish the scale of the International Atomic Time TAI, in accordance with the decisions of the 14th Conférence Générale des Poids et Mesures and in conjuction with the Bureau International des Poids et Mesures;
(b) to establish, from all relevant data, and to publish the current values of the Universal Time and of the angular velocity of the Earth's rotation and, in addition, the operational coordinates of the pole used for this purpose;
(c) to implement the system of the Coordinated Universal Time UTC by the distribution of all necessary information for the coordination of time signal emissions and the synchronization of clocks on the UTC scale;
(d) to distribute information important for scientific users of time, and to supply on request the available data on the subject of time;
(e) to perform scientific research as necessary for the improvement of the service.

RESOLUTION NO. 8

Proposée par la Commission No. 40 de l'UAI

L'Union Astronomique Internationale
reconnaissant
(a) que les résultats scientifiques de l'exploration de l'univers obtenus grâce à la radioastronomie sont importants pour l'humanité;
(b) que le spectre radio est de plus en plus utilisé, principalement par des émetteurs localisés dans l'espace ou la haute atmosphère;
recommande
1. aux ingénieurs responsables de la conception des futurs services d'émission à partir du sol, de la haute atmosphère ou de l'espace, de veiller à ce que les effets de l'interférence dans les bandes allouées, les bandes adjacentes et les bandes harmoniques restent en-dessous des limites nuisibles d'interférence telles qu'elles sont spécifiées dans le Rapport CCIR 224-3;
2. à toute la communauté astronomique de continuer à insister fermement auprès des utilisateurs effectifs du spectre radio pour qu'ils respectent ces limites.

RESOLUTION NO. 8

Proposed by IAU Commission No. 40

The International Astronomical Union
recognizing
(a) the value to mankind of the scientific results achieved by radio astronomy through the exploration of the universe;
(b) the increasing use of the radio spectrum, especially by space and air-borne transmitters;
recommends
1. that designers of future ground, airborne and space-based transmitting services, ensure that the effects of in-band, adjacent-band and harmonic interference are below the harmful interference limits as specified in CCIR Report 224-3;
2. that efforts continue by the entire astronomical community to stress that the active users of the radio spectrum should adhere to these limits.

RESOLUTION NO. 9

Proposée par la Commission No. 50 de l'UAI

L'Union Astronomique Internationale s'inquiète vivement de l'augmentation des niveaux d'interférence avec les observations astronomiques, qui résulte de l'illumination artificielle du ciel nocturne, des émissions radio, de la pollution atmospérique et du survol par les avions des sites d'observatoires.

En conséquence, l'UAI demande instamment aux autorités civiles responsables d'entreprendre une action urgente afin de préserver de telles interférences les observatoires existants ou en projet. Dans ce but, l'UAI se charge de fournir, par l'intermédiaire de la Commission 50, tous renseignements concernant les limites acceptables d'interférence et les moyens possibles de contrôle.

RESOLUTION NO. 9

Proposed by IAU Commission No. 50

The International Astronomical Union notes with alarm the increasing levels of interference with astronomical observation resulting from artificial illumination of the night sky, radio emission, atmospheric pollution and the operation of aircraft above Observatory sites.

The IAU therefore urgently requests that the responsible civil authorities take action to preserve existing and planned observatories from such interference. To this end, the IAU undertakes to provide through Commission 50 information on acceptable levels of interference and possible means of control.

A large number of Resolutions have been formulated by several different Commissions of the IAU. It would be impractical to give to each one of these the attention of the General Assembly. These Resolutions have been formulated in each Commission with care and after much deliberation. The Union has complete confidence in the functioning of its Commissions and I present to you one additional Resolution, proposed by the Resolutions Committee, that will endorse this view.

RESOLUTION NO. 10

Proposée par le Comité des Résolutions

L'Assemblée Générale de l'Union Astronomique Internationale cautionne les Résolutions adoptées individuellement par chacune de ses Commissions et recommande aux astronomes de mettre ces Résolutions en application.

RESOLUTION NO. 10

Proposed by the Resolutions Committee

The General Assembly of the International Astronomical Union endorses the Resolutions adopted by its individual Commissions and recommends that astronomers give effect to these Resolutions.

Dr. M. K. V. Bappu
Chairman, Resolutions Committee

All proposals of the Resolutions Committee as presented by Dr. Bappu were carried by an absolute majority of votes.

17(a). *Admission of Iraq to the IAU*. The General Secretary referred to his narrative report and reaffirmed that the Executive Committee had carefully examined the application of Iraq for admission to the Union and found that the degree of astronomical development in this country satisfied the standards required for IAU membership. He therefore moved that the General Assembly admit Iraq as a new Adhering Country to the Union. This motion was carried unanimously.
The President invited the official representative of Iraq Dr. May A. Kaftan to be seated among the representatives of Adhering Countries. Y. Ne'eman the official representative of Israel, welcomed on behalf of his country Iraq as a Member Country of the IAU and emphasized the importance of scientific co-operation of the countries in the Near East.
17(b). *New Members of the Union*. The General Secretary reported that the Executive Committee had, on the proposal of adhering bodies and on the advice of the Nominating Committee, admitted 724 new Members to the Union, and deleted 31 Members from its membership list. The names of the new Members had been displayed in a prominent place and the General Secretary did not propose to read them. He informed the Assembly that these names would be incorporated in the alphabetical list of IAU Members to appear in print.

18. *Commissions*. The General Secretary presented, on behalf of the Executive Committee, the following list of Presidents and Vice-Presidents of Commissions for election by the General Assembly:

No.	Commission Name	President	Vice-President
4	Ephemerides	V. K. Abalakin	A. M. Sinzi
5	Documentation	J.-C. Pecker	W. D. Heintz
6	Astronomical Telegrams	E. Roemer	J. Hers
7	Celestial Mechanics	V. Szebehely	Y. Kozai
8	Positional Astronomy	R. H. Tucker	E. Høg
9	Astronomical Instruments	J. Ring	E. H. Richardson
10	Solar Activity	G. Newkirk, Jr.	V. Bumba
12	Solar Atmosphere	M. K. V. Bappu	Y. Uchida
14	Spectroscopic Data	E. Trefftz	J. G. Phillips
15	Physics of Comets, Minor Planets, and Meteorites	N. B. Richer	B. D. Donn
16	Planets and Satellites	T. C. Owen	B. A. Smith, V. G. Tejfel'
17	Moon	E. Anders	K. P. Florensky
19	Rotation of the Earth	R. O. Vicente	P. E. G. Pâquet
20	Motions of Minor Planets, Comets, and Satellites	B. G. Marsden	G. Sitarski
21	Light of the Night Sky	R. Dumont	H. Tanabe
22	Meteors, Interplanetary Dust	I. Halliday	W. G. Elford
24	Photographic Astrometry	C. A. Murray	H. K. Eichhorn v. W.
25	Stellar Photometry, Polarimetry	M. F. McCarthy	J. A. Graham
26	Double Stars	P. Muller	O. G. Franz
27	Variable Stars	J. Smak	J. D. Fernie
28	Galaxies	B. E. Markarian	B. E. Westerlund
29	Stellar Spectra	M. Hack	W. K. Bonsack
30	Radial Velocities	A. H. Batten	M. Duflot
31	Time	A. Orte	S. Iijima
33	Galactic Structure and Dynamics	F. J. Kerr	G. G. Kuzmin
34	Interstellar Matter and Planetary Nebulae	G. B. Field	V. Radhakrishnan
35	Stellar Constitution	B. Paczynski	R. J. Tayler
36	Stellar Atmospheres	D. Mihalas	G. Traving
37	Star Clusters	S. van den Bergh	G. Lyngå
38	Exchange of Astronomers	D. A. MacRae	J. Delhaye
40	Radio Astronomy	H. van der Laan	G. Swarup
41	History of Astronomy	J. Dobrzycki	M. A. Hoskin
42	Close Binary Stars	G. Larsson-Leander	B. Warner
44	Observations Outside the Atmosphere	R. M. Bonnet	R. J. van Duinen
45	Spectral Classification	B. Hauck	A. Slettebak
46	Teaching of Astronomy	E. V. Kononovich	D. Wentzel, L. Houziaux
47	Cosmology	I. D. Novikov	G. O. Abell
48	High-Energy Astrophysics	I. S. Shklovski	F. Pacini
49	Interplanetary Plasma	A. Hewish	H. J. Fahr
50	Protection of Observatory Sites	R. Cayrel	F. G. Smith

This list was approved by the General Assembly.

The General Secretary continued by showing the panels of the Organizing Committees of IAU Commissions, as approved by the Executive Committee, and announced that the composition of Commissions would be published in the IAU Transactions volume XVIB.

19. *Place and date of the XVIIth General Assembly*. The President informed the Assembly that the Union had been invited by the Canadian National Committee of Astronomy to hold the XVIIth General Assembly in Montreal, Canada, from 14 to 24 August 1979, and asked Professor D. A. MacRae to present the invitation. Professor MacRae spoke, first in English and then in French, as follows:

"Mr. President, Ladies and Gentlemen:
    As Chairman of the Canadian National Committee I extend a cordial invitation to the International Astronomical Union, to hold its seventeenth General Assembly in the city of Montreal, Canada. This invitation is issued by our adhering organization, the National Research Council of Canada, on behalf of Canadian astronomers. Canada will welcome all members of the Union and invited participants who come as individuals to attend the General Assembly.
    Canada is a young country. Astronomy in recent years has blossomed in a vigorous proliferation of research centres across our land. Our astronomers have the enthusiasm and ambition of youth. They look forward eagerly to playing host at the first IAU General Assembly to be held in our country.
    The city of Montreal is proud to have been chosen as the proposed site. It is one of our eastern cities and its latitude is almost identical with that of Grenoble. Montreal is easy to reach by air, by land or by water. It is our largest city. There are ample accommodations and adequate facilities for attending to IAU business, and also for diversions. The city authorities, the University of Montreal, and the Province of Quebec will do their utmost to welcome the IAU and make the seventeenth General Assembly a success.
    Canada is a large country. We hope you will also plan visits to our astronomical observatories and research institutions. As many of you know, we have a number of optical, solar and radio observatories, astrophysical laboratories, and centres for theoretical studies. Several are within easy reach of Montreal. Others you can visit when you travel westward via the St. Lawrence River and Lake Ontario to Toronto, Niagara Falls and beyond. The Great Lakes are North America's fresh-water equivalent of the Mediterranean Sea. Several of our major research centres are to be found dotted across the Prairies, located among our Canadian Rocky Mountains, and perched on our Pacific Ocean coast.

Monsieur le Président, Mes chers Collègues,
    Je voudrais déclarer de nouveau que le Conseil national de recherches du Canada vous invite cordialement à venir à notre pays en été 1979. Cette invitation est appuyée par l'Université de Montréal, par la ville de Montréal, par la belle province de Québec, et par tous les astronomes du Canada. Nous espérons vous avoir au Canada à l'occasion de la dix-septième Assemblée Générale de l'Union Astronomique Internationale."

The motion of the President that the Canadian invitation be accepted was carried unanimously. The President thanked Professor MacRea for his kind words.

    20. *Election of the new Executive Committee*. The Chairman formally moved that Professor Adriaan Blaauw be elected the new President of the Union for the term 1976-1979. This was approved by acclamation.
    The Chairman then called the attention of the Assembly to the names put forward at the first session, and formally proposed that Dr. D. S. Heeschen, Professor E. K. Kharadze and Professor S. van den Bergh be elected Vice-Presidents until 1982. This was unanimously approved. Next, the Chairman formally proposed that Professor Edith A. Müller be elected General Secretary in place of Professor G. Contopoulos, and Professor P. A. Wayman Assistant General Secretary in place of Professor Edith A. Müller. These proposals were approved by acclamation.
    The Chairman then invited Professor Blaauw, Dr. Heeschen, Professor Kharadze, Professor van den Bergh and Professor Wayman to join the Executive Committee on the

platform, and invited the former IAU President P. Swings to do so too.

21. *Adresses by retiring and newly elected Officers.* The retiring President, Professor Leo Goldberg, spoke as follows:

"We are drawing to the close of the administrative part of this XVIth General Assembly.

Before calling on Professor Blaauw, I should like to express my sincere thanks and appreciation to my good friends and colleagues on the Executive Committee whose hard work and devotion to the I A U have made my task as President an easy one during the past three years. There have been many outstanding Executive Committees in the history of the I A U , but this one has achieved a very special destinction. In a period of otherwise severe inflation, in prices, in budgets, in numbers of astronomers, this Committee succeeded in reducing the customary length of its meetings by nearly a factor of two. I have felt proud to be associated with such an efficient and harmonious body. I particularly want to thank the outgoing vice-Presidents Bart Bok, Per Olof Lindblad, Bernard Lovell and Evald Mustel for their many years of valuable service and co-operation.

The I A U also owes thanks to the outgoing advisors of the Executive Committee, Bengt Strömgren and Kees de Jager, from whose wisdom and experience we have benefited richly.

Next, I would like the members of the I A U to know how much they owe to the General Secretary, the Assistant General Secretary and their staff for the loving care with wich they handle the administration of the Union's affairs in accordance with the decisions of the General Assembly and the Executive Committee. During the last three years, I have watched with wonder and admiration as George Contopoulos, with the loyal and hardworking assistance of the Executive Secretary Arnost Jappel and the administrative assistant Jarka Dankova, have handled literally thousands of pieces of correspondence, administered the funds of the Union, prepared and issued the publications and worked out the intricate details of the program of scientific and administrative meetings during the General Assembly. What is most remarkable about your performance, George, is that you have been able to continue your outstanding research in galactic structure, as those of us who have heard you lecture recently can testify. Finally, I want George Contopoulos to know how much I appreciated the tact and diplomacy with which he educated me to the responsibilities and especially to the prerogatives of our respective offices.

Edith Müller, as Assistant General Secretary, has had the responsibility for overseeing the planning of colloquia, symposia and regional meetings and in so doing has left no doubt that she is prepared to uphold the long standing I A U tradition of distinguished general secretaries. May I ask the General Assembly to show its appreciation of the Secretariat in the usual way?"

The final remarks of the retiring President were followed by great applause.

Then the new President, Professor Adriaan Blaauw, took the word and said:

"Dear Colleagues, Ladies and Gentlemen,

Now that the General Assembly has elected me President of the Union, I wish to convey to you my feelings of profound gratitude. Gratitude first of all, for the confidence you thereby have expressed in the guidance which, from now on, I may contribute in the handling of the affairs of the Union - and gratitude also because it is a very great honour indeed to be elected to this office. As we all know, among those who have held it have been some of the most distinguished astronomers of this century. It was therefore not without embarrassment that I received the request for being a candidate for this post. I accepted it, in the awareness that in the course of my career I have had the privilege of becoming well acquainted with astronomers and their institutes in many parts of the world, and have been in a position to gain valuable experience in the management of a large international organization, during the years of my directorate of ESO. I trust that these experiences may help me meeting the expectations you expressed by electing me.

Reading reports of past Union Assemblies - as one tends to do in such circumstances - I was struck by the words spoken by that remarkable astronomer and former
president of the Union, Otto Struve, when he offered among his reasons for accepting the post the following:
   "... I have become a confirmed internationalist, and believer in the neccessity
   for international co-operation. I feel most at home in an organization such as
   the IAU."
I wish to take up this task in the same spirit.

The affairs of the IAU are manyfold, but a good deal of them remain remote from
the membership in their daily work in research or teaching. For most of the members,
it is only occasionally, and especially at our Assemblies, that they become more
closely acquainted with the role of the IAU. Yet, there are important matters that
do require the regular attention of the Executive Committee, and many smaller ones
that require the constant care of the Union's secretariat. This situation is a natural one, like that on a big summer cruiser where the steering is the business of
a small crew only, while the passengers have a good time - each one in his way.
   We have now again arrived at the moment of a change of this crew - at least partially - and I want to, first of all, thank you, Leo Goldberg, for your leadership
during these past three years. Your outstanding record of both scientific and administrative achievements made you particularly fitted for this task. We admire
your work in many fields of astrophysics and we equally admire the talents you
showed in your administrative career, from your directorate at Harvard to your
present position as director of Kitt Peak Observatory - with many simultaneous
activites including leadership in areas of space research. It was especially in
this "Kitt Peak phase" that I have the pleasure of learning to know you more in-
timately, for which I am very grateful. The Union thanks you for all you did for
it, not forgetting your serving 6 years as a vice-president, and we consider ourselves very fortunate that for the years to come we may continue to benefit from
your wise counsel as a member of the Executive Committee.
   To you, George Contopoulos, who has now served the IAU for six years with so
much devotion and enthusiasm, may I tell you how much it means to us that you,
with your profound knowledge of the Union, will remain with us in the Executive
Committee.
   To the younger ones among us - and we are glad that there are so many - who have
attended only the last few assemblies, it may seem that there is a fairly fixed
pattern in the Union's activities: like these assemblies with the regular commission
meetings, the planning of symposia and colloquia, the regional meetings, decision
on membership and travel grants, etc. But viewed over a longer time span we see
how some of the things we now take for granted, at one time were major items for
consideration or surprising developments. It is interesting to note that when Otto
Struve spoke the words I just quoted, he also noted the following three great
events in the progress of the Union: the admission of Germany to the Union's membership, the impact of the scientific contributions from the Soviet Union, and the
rise of the IAU symposia. Other striking developments could be recalled for more
recent years.
   So, we may ask ourselves which will be the major developments of tomorrow and
try to be prepared for them. Among the highlights of this Assembly was Dr. Sagan's
fascinating discourse on the exploration of the planets; we note how in these days
the notion of possible extraterrestrial life enters our domain as a matter of course,
and we cannot but guess at the enormous impact this may have on our future research
and teaching. An item of deep concern has already been raised earlier at these
meetings: the absence of astronomers from the Chinese continent. Let us hope this
sad incompleteness will soon be remedied. But then, I believe it is equally important that we try to ensure that the participation from nations that did since long
adhere to the IAU and contributed so much to it, will not diminish; that the Union
preserve the truly world-wide international support it already has, and which makes it such a precious tool for our science. - To mention another topic: that of
the preservation of the very best sites for ground based observations, now being

discussed rather inconspicuously in Commission 50, we may ask: may it not become
a matter of serious concern for the Union in the foreseeable future?

As regards the nature of our assemblies, I believe some soul searching is in
order. The lower than expected attendance of this Assembly - notwithstanding the
inviting and beautiful surroundings where it was held - may perhaps be due to more
than the conflicting academic duties and shortage of travel funds; could it perhaps be that the interest shifts more to the symposia and colloquia than to the
scientific sessions of our committee meetings?

With a few such questions in mind, and a good number of smaller ones at hand,
I am looking forward to a profitable collaboration with the new Executive Committee.

Si, maintenant, j'ajoute encore quelques mots en français, ce n'est pas tellement parce qu'il est aussi une des langues de notre Union, mais bien plutôt pour
exprimer ma joie du fait, que je commence mes nouvelles fonctions dans ce pays où,
pendant mon mandat à l'ESO, de nombreux et éminents collègues me sont devenus des
amis chers, sans parler des amitiés profondes qui remontent à beaucoup plus loin.
Et c'est dans ce même esprit que je vais retrouver à Genève notre sympatique et
dévouée Secrétaire Générale. Permettez-moi, chère Edith, de vous assurer que le
Comité Exécutif et moi-même nous nous réjouissons de cette future période de collaboration fructueuse pour le bénéfice de l'Union."

22. *Closing Ceremonies.* The Chairman called upon Professor M. Migeotte to propose a comprehensive vote of thanks. Prof. M. Migeotte spoke as follows:
"Le Comité Exécutif m'a invité à m'adresser tout spécialement à nos collègues
français, responsables de l'organisation de cette Assemblée Générale, pour les
remercier au nom de tous les autres astronomes et astrophysiciens qui y ont participé.

Après avoir bien réfléchi, j'ai compris pourquoi j'avais été choisi pour cette
importante mission. Certains d'entre-vous savent que je professe à l'Université de
Liège, en Belgique. Beaucoup d'entre-vous sans doute ont lu des romans de Georges
Simenon, qui est liégeois et Docteur Honoris Causa de notre Université. Sans doute
peu d'entre-vous savent que c'est en lisant une partie de l'oeuvre de cet auteur
que notre président sortant Léo Goldberg a perfectionné son français, ce qui lui a
permis d'impressionner tous ses auditeurs lors de son discours d'ouverture prononcé
à l'occasion de cette Assemblée Générale. On peut donc en conclure que l'Université
de Liège, grâce à un de ses éminents Docteur Honoris Causa, a joué un rôle important dès le début de notre Assemblée.

Dans son allocution, Léo Goldberg a signalé que Louis XIV était tellement puis-
sant que, pendant son règne, il était parvenu à imposer au Soleil, un minimum de
taches! Au cours de ces dix derniers jours, nous avons pu nous rendre compte de la
puissance extraordinaire de certains collègues français. Notons, par exemple, que
dans le ciel habituellement serein de la Haute Provence, Charles Fehrenbach est
parvenu à accueillir par un formidable coup de tonnerre les congressistes qui sou-
haitaient visiter l'O.H.P. De son côté, Jean Rösch a réussi à faire neiger pendant
la visite de l'Observatoire du Pic-du-Midi. Ceci pour vous dire que cette Assemblée
Générale a vraiment été préparée sur une très grande échelle.

Grâce à une excellente organisation, les présidents de toutes les commissions
ont trouvé immédiatement des locaux adéquats. Lors des grandes conférences publiques,
nous avons pu admirer la maîtrise de M. Wlérick pour canaliser de grandes foules.

L'exposition astronomique à l'Université de Grenoble a certainement intéressé
beaucoup d'entre nous, tandis que des expositions, en ville, ont permis de sensibili-
ser le grand public à certains aspects de l'astronomie.

Les excursions dans les environs de Grenoble nous ont donné la possibilité d'admirer la nature et les richesses de cette région, tandis que les visites dans différents grands observatoires ont permis de juger du développement de l'astronomie
française.

Nous avons aussi été très sensibles aux manifestations artistiques. Enfin, hier

soir, après l'excellent et copieux dîner, la piste de danse est devenue rapidement
un foyer intense d'émissions de radiations infrarouges d'où sortiront sans doute
beaucoup d'étoiles naissantes.

Bref, pour nous, Grenoble n'est plus seulement la "Capitale de la noix" et le
siège d'une importante université mais c'est une ville où nous avons passé des
heures heureuses, enrichissantes et stimulantes.

Nous n'oublierons pas que nous les devons à tous nos collègues français qui
ont consacré une grande partie de leur temps à la préparation de cette Assemblée.
Permettez-moi de citer en particulier M. KOVALEVSKY,. Président du Comité National
Français d'Astronomie et d'U A I   France 1976, M. WLERICK, Secrétaire Général,
Melle AVIGNON, Melle DEBARBAT, MM. PECKER, BOISCHOT, MORANDO, CHAPRONT et LEQUEUX
et toutes leurs collaboratrices, notamment Melle DROUIN, Mmes MANNING et GOUTAS.

Au nom de tous les autres participants à la XVIème Assemblée Générale de l'U A I
et en nom personnel, j'ai l'honneur et le plaisir de les remercier chaleureusement
et de leur témoigner notre profonde reconnaissance."

On behalf of the ladies, Mrs. Charlotte Goldberg expressed her thanks as
follows:

"My dear friends,

I take great pleasure to speak for the many guests of the IAU who have been
entertained and taken care of so graciously by the French Ladies Committee of the
General Assembly.

We know that it took a combined effort of so many of you working with your tire-
less Chairman, Mrs. Delhaye, to plan these days with very interesting trips to
introduce us to this most beautiful part of your country and opportunities to see
and visit with each other again.

This has been a memorable visit for us and we say a most sincere thank you to
all of you from all of us."

The Chairman then called upon the retiring General Secretary, Professor George
Contopoulos, who spoke as follows:

"Perhaps I am the most happy person of this General Assembly at this moment,
as my task is over.

In the past 6 years I tried to serve the Union as best as I could. It was an
exciting, although sometimes difficult, work. Every day many letters with beautiful stamps would come from all the corners of the world, from the far east to the
far west. In the peak of the work before the General Assembly, we reached 40 letters a day. And every letter required some action. Arnost Jappel and myself tried
hard to reply immediately and satisfy, as much as possible, every wish of our colleagues all over the world. But our work was not always easy. There was no respite,
whether we had a revolution, a threat of war, or the students threatening to occupy
our building in the University. The mail would come, more or less regularly, and if
you dared to take one day's vacation it would pile up to dangerous heights. Then
there was the difficulty in travelling. Several times during the junta period I
went to the airport and I was not sure whether I would be allowed to pass through.

However, the work of the Union continued normally. But for that I must say that
I am very grateful to many people, too many to be mentioned here separately. Thus
I will mention only a few of them.

I had a very close and fruitful collaboration first of all with the Executive
Committee, with the president of the IAU Commissions, the Organizers of IAU Meetings, especially the French Organizers of the General Assembly, with the National
Committees and the IAU representatives in various international bodies. They all
provided useful information and advice and, more important, they did most of the
work. It is the greatest pride of the IAU that most of its services are provided
free by a large number of devoted people.

Then our Secretariat had a substantial support from the Greek Ministry of Culture and Science and the staff of the Department of Astronomy of the University of
Thessaloniki and Athens.

More important was the contact and help from the Officers of the IAU, the President and the Assistant General Secretary. It was a great comfort to me that I could rely on Leo Goldberg in every difficulty I had to face. And I was sure that he would take over himself the most difficult cases. Sometimes, as I was going to bed at night the telephone would ring. And there was Leo Goldberg who would say, very cheerfully, "Good morning George". Of course it was then noon time in Arizona.

As regards Edith Müller, she has taken up, with complete success, one of the most important tasks of the Union: Scientific Meetings and Publications. It is characteristic of Edith Müller that she managed to write more letters than myself as Assistant General Secretary. I am therefore quite sure that I leave the work of the IAU in very competent hands.

Finally I should thank Arnost Jappel and Jarka Dankova. They have been so devoted to the work of the Union that they have become *the* living tradition of the Union. They stayed always in their post, even during the most difficult period of the IAU Secretariat.

I remember one particular Sunday. Arnost and Jarka took their car for an excursion. They noticed that the town was empty and they said to each other "How nice it is to drive early on Sunday". Then they saw some tanks. They thought that a parade was going on and paid no attention. Further on they saw a soldier and they asked him how to reach the highway. (By the way Arnost has managed to learn some Greek during his stay in Greece). But the answer of the soldier was "forbidden". "Why forbidden?" asked Arnost. And the answer was "Papadopoulos (the Greek dictator) is down. Another Papadopoulos has taken over". In fact the tanks were not on parade, they were simply overthrowing the previous government.

I could tell many stories like that. But I would like to finish with a small personal story. As I was absorbed by the work of the IAU together with my other duties, research, administration and travel, my children complained that they did not see enough of their father. And during one of my travels, my little daughter improvised a new kind of prayer. "Let daddy come back to us, and not be lost in some foreign airport". Well, thank God, I was not lost anywhere and I am happy to be here with you to-day. But I am also happy to return to my family and to my research.

As regards the IAU, it is now in the most competent hands of the new Executive Committee. Thus I have no doubt for its progress.

Many thanks to all of you."

The Chairman then invited the newly elected General Secretary, Professor Edith A. Müller, to say a few words. Professor Müller spoke as follows:

First of all I wish to thank the General Assembly for the confidence given to me. It is a great honour for me to serve the Union as its General Secretary. The IAU is a large and, in fact, rapidly growing family and, thus there will be an increasing amount of work and problems that the General Secretary will have to face. But I am not worried because I know that I can count on the co-operation of all of you, on the expert advice of my predecessors and good friends George Contopoulos, Kees de Jager, Lubos Perek, Jean-Claude Pecker and Donald Sadler, and on the fine collaboration of the executive secretary Arnost Jappel and the administrative secretary Jarka Dankova. The Union now starts a new three-years term during which, I am sure, many new scientific achievements and discoveries will be made about which you will be reporting at the XVIIth General Assembly. I wish you all a lot of success in your work and life and I look forward to seeing all of you and many more in Montreal three years from now.

The Chairman closed the meeting at twelve o'clock wishing all a good journey home.

PART 3

REPORTS OF MEETINGS OF COMMISSIONS

COMPTES RENDUS DES SÉANCES DES COMMISSIONS

COMMISSION 4: EPHEMERIDES (EPHEMERIDES)

Report of Meeting 25 August 1976

PRESIDENT: R. L. Duncombe   SECRETARY: A. T. Sinclair

The Chairman asked for a moment of silence in memory of four
members of the Commission who had died since the last IAU meeting:
G. A. Chebotarev, G. M. Clemence, S. Herrick, and H. Hertz.

The Chairman announced the names recommended by the committee
for the new officers of the Commission for the next three years:
President: V. K. Abalakin
Vice President: A. M. Sinzi
Organizing Committee: T. Lederle, J. H. Lieske, B. Morando,
                        P. K. Seidelmann, G. A. Wilkins, and
                        R. L. Duncombe
The meeting agreed that these names should be put to the Executive
Committee.

The Chairman regretted that the report of the Commission had not
been made available before the meeting, due to printing delays. He
assured the meeting that all information communicated to him had been
included in the report.

G. A. Wilkins gave a summary of the activities of the IAU working
group on Numerical Data during the last three years. The main activity had been the organization of IAU Colloquium No. 35 held at
Strasbourg the previous week. He felt that there was a need in
Astronomy for a guide on the presentation of data, similar to a
bulletin published by CODATA.

The Chairman announced a list of names of people who had been
proposed as new members of the Commission: B. D. Yallop, G. M. R.
Winkler, P. M. Janiczek, K. J. Johnston, and Y. T. Ting. The meeting
agreed that their names should be put to the Executive Committee.

The Chairman drew the attention of the meeting to the following
recommendation on the Physical Ephemeris of Mars.
"Considering that recent new determinations of the rotational
elements of Mars indicate the need for a revision of the elements
currently adopted in the physical ephemeris of Mars, and that a new
approach to the definition of the origin of areographic longitudes
appears useful (G. de Vaucouleurs, M. E. Davies and F. M. Sturms,
Jr., J. Geophs. Res. 78, 4395, 1973), Commission 4 and 16 recommend
(1) that the tie between the new and current physical ephemeris of
     Mars be firmly established by appropriate comparisons between
     ground-based and Mariner coordinate systems, and
(2) that new elements and a new definition of the origin of the
     areographic longitudes consistent with the results of (1) above
     and the definitions adopted previously (IAU Trans XVB, 1973, 107)
     be incorporated in the physical ephemeris of Mars as soon as
     deemed practicable in the judgement of the cognizant Directors
     of the National Ephemerides Offices."

P. K. Seidelmann said that a similar resolution had been passed previously, and work was still underway to carry out that resolution.

    The Chairman said that the present resolution was to re-emphasize
the need for this work.

    The resolution was approved by the meeting.

    The Chairman then introduced the following recommendation con-
cerning cartographic coordinates and rotational elements of the
planets and satellites.
"Noting that
    (a)   confusion exists regarding the present rotational elements
           of some of the planets
    (b)   extensive amounts of new data from radar observations and
           by direct imaging from spacecraft have made cartography
           of the surfaces of the Moon, Mercury, Venus and Mars a
           reality
    (c)   there will be an extension of these techniques to the map-
           ping of larger satellites of Jupiter and Saturn in the near
           future
assert that
           to avoid a proliferation of inconsistent cartographic and
           rotational systems, there is a need to define the rotational
           elements of the planets and satellites on a systematic
           basis and to relate the new cartographic coordinates rig-
           orously to the rotational elements
and therefore recommend that
           Commission 4 (Ephemerides) and Commission 16 (Physical
           Study of Planets and Satellites) establish a Joint Working
           Group to study the cartographic coordinates and rotational
           elements of the planets and satellites and to report
           recommendations thereon at the next general assembly of
           the IAU."

    G. A. Wilkins questioned whether observers would in practice
be prepared to change the systems they use at present. He quoted
the fate of a recommendation made in 1884 that longitudes on the
Earth should be measured positive towards the East.
The recommendation was approved by the meeting, with the understanding
that the incoming and outgoing presidents of Commission 4 and 16 would
arrange the membership of the Working Group.

    P. K. Seidelmann spoke about plans to modify the layout and con-
tents of the Astronomical Ephemeris and the American Ephemeris from
1981 onwards. Other publications would also be involved. The next
publication of "Planetary Coordinates" would cover the years 1980-1984,
and would be called "Planetary and Lunar Coordinates". It would give
low precision planetary and lunar data, to be used for planning purposes, and also by those countries who wished to produce their own
navigational almanacs. There would no longer be a distribution of the
"advanced data" previously used for these purposes. However it was
hoped that countries producing their own almanacs would make use of
reproducible data provided by the U.S.A. and U.K. Nautical Almanac
Offices. At present many countries use the "advanced data" to obtain
information about astronomical phenomena for publication in diaries,
calendars, etc. In future the publication "Astronomical Phenomena"
will be expanded, and it is hoped that it can meet all these needs.
The "Astronomical Ephemeris" and the "American Ephemeris" will be
replaced by a single publication. The principal changes in the contents will be:
    The hourly lunar ephemeris will be replaced by the coefficients
of Chebyshev or economized polynomial series, which would represent

the Moon's position for one day;
    Differences would be eliminated throughout the volume;
    Satellite data would be expanded to cover all natural satellites;
    More stellar data would be included;
    An ephemeris of the barycentre of the Solar System would be
included for use in pulsar observation reductions;
    The volume would be re-organized in a more logical manner.
These new arrangements would be introduced in 1981, but the basis of
the ephemerides would still be the same as at present. However from
1984 onwards it was hoped that the ephemerides would be on a new,
improved basis, and the star positions would be based on FK5.
    A description of the basis of the new ephemerides would be
published so that other people could duplicate the computation of the
ephemerides if they wished. The ephemerides would also be available
on magnetic tape, to facilitate comparisions.

    T. Lederle asked what sort of positions would be given in the
star lists, and if Day Numbers would still be given.

    P. K. Seidelmann replied that probably the mean position for the
beginning of the year would be given, and Besselian Day Numbers, but
not Independent Day Numbers.

    G. A. Wilkins said that some thought had been given to the idea
of giving star positions for the middle of the year in question. The
Day Numbers would also be referred to the middle of the year, and so
the volume would contain all the information necessary to compute
apparent places for any time during the year. The present situation
is that for the second half of the year the Day Numbers are referred
to the beginning of the next year, and so next year's volume is needed
to obtain the mean places of the stars. A. M. Sinzi pointed out that
the middle of the year is used in the Japanese Ephemeris.
    T. Van Flandern said that he had understood that a very precise
means of computing apparent places was to be included in the new
volume.

    G. A. Wilkins replied that for each day the coefficients would
be given of the matrix needed to apply precession and nutation from
a fundamental fixed equator and equinox to the true equator and
equinox of date. Aberration could be computed from the tabulations
of the velocity of the Earth relative to the barycentre of the Solar
System.

    P. K. Seidelmann said that it was also planned to modify the form
of the list of observatories given in the Ephemeris. It would be
restricted to professional observatories, and an improved index would
be given. At intervals a more detailed list would be given, including
such information as the types and positions of the telescopes at the
observatories.

    V. K. Abalakin suggested that the geodetic datum to which an
observatory's position was referred should be specified.

    P. K. Seidelmann said that the form of the eclipse maps would be
altered somewhat, to facilitate their production by a computer and
graph plotter.

    A. M. Sinzi commented that at Tokyo they already had a computer
program to produce eclipse maps similar to those in the Ephemeris.

B. Morando gave a report on the work of the International Information Bureau on Astronomical Ephemerides, which had been created at
the 1970 IAU General Assembly. Details of each set of data were
included on a card, and these cards were sent to people who requested
them. At present 158 names were on the mailing list, and 125 cards
had been issued. The production of about a further 50 cards was in
hand. The Bureau received a small grant from the IAU, but the bulk of
its expenses was met by the Bureau des Longitudes. He announced that
the present postal address of the Bureau was

                              B.I.I.E.A.
                              77 Avenue Denfert-Rochereau
                              75014-Paris
                              FRANCE

G. A. Wilkins questioned whether the issue of all of the cards
to all of the users was the most efficient procedure. He suggested
that the information could be given in the regular bulletins of the
Stellar Data Centre, or the issue of cards could be restricted to
sending only the relevant cards to people who requested information
data of a particular type.

J. D. Mulholland said that he found the cards convenient to use,
and he liked to have the complete set as it gave him information
about data that he would not have expected to be available, and so
would not have thought to ask about.

J. Kovalevsky reminded the meeting that it was necessary to
renew the membership of the Bureau's advisory committee. He also
proposed the following resolution.
    "Commission 4 recognizing the usefulness of the International
    Information Bureau on Astronomical Ephemerides (BIIEA) considers
    it important to continue its operation during the coming three
    years and requests the IAU to increase its subvention to this
    Bureau to $400 per year."
The proposal was seconded by R. L. Duncombe, and approved by the
meeting. The Officers of Commission 4 then requested the following
persons to serve on the Scientific Advisory Board of the International
Information Bureau on Astronomical Ephemerides: T. Lederle, G. A.
Wilkins, V. K. Abalakin, P. K. Seidelmann, and B. Morando, ex officio.
A COSPAR representative will be appointed by that organization.

REPORT OF JOINT MEETING OF COMMISSIONS 4, 8, 31 ON THE NEW SYSTEM OF ASTRONOMICAL CONSTANTS

Report of Meeting 26 August 1976

PRESIDENT: R. L. Duncombe   SECRETARIES: A. T. Sinclair,
                                                R. H. Tucker

The Joint Meeting of Commissions 4, 8, and 31 was called to
consider and, if possible, to resolve the acceptance of the Report
prepared by the Commission 4 Working Groups on Precession, Planetary
Ephemerides, Units and Time Scales. The Chairman proposed to accept
only questions of fact after each presentation on the agenda, and to
delay general discussion until the whole report had been presented.

G. A. Wilkins described briefly the form of the new system of
astronomical constants and related quantities; it follows the system
adopted in 1964 with a few exceptions. It gives explicitly the

relationships between the astronomical units of length, mass,
and time, and units (metre, kilogram, and second) of the International
System (SI). The choice of defining constants (arbitrary values),
primary constants (best values from observations) and derived constants
(calculated from the defining and primary constants) has been changed
in a few cases. The astronomical unit of time (the day) is now based
on the atomic second, rather than on the tropical year. Correspondingly, the new time-scale for use in apparent geocentric ephemerides
is based on this unit and is offset from the scale of International
Atomic Time (TAI) so that it will be continuous with the scale of
ephemeris time for all practical purposes; it is defined precisely
and unambiguously and it will be suitable for use with relativistic
theories. The velocity of light (c) has a special significance since
it is understood that its value will not be changed if the definition
of the metre is changed; it would not be appropriate for astronomers
to specify c as a defining constant since this would effectively redefine the metre. The system of constants is completed by adding
constants for: the size, gravity field and shape of the Earth; the
principal coefficients for precession and nutation; and the masses
of the Moon and planets. Other coefficients and constants required
for the reduction of celestial positions and the computation of
ephemerides are given separately. A. H. Cook asked if the geodetic
quantities were based on a mass for the Earth which included the
atmosphere, and was told they were. He suggested that the point
should be made more explicitly since the mass recommended by the IAG
did not do so. R. d'E. Atkinson pointed out the need to correct the
constant of aberration, as aberration was now to be computed using
the Earth's velocity relative to the barycentre of the Solar System
rather than the Sun. Wilkins agreed to consider this.

     G. M. R. Winkler dealt with the problems and details of the new
time scale proposed in the Report. It was emphasized that parts a
and b provide the rate and epoch of a unique time scale for apparent
geocentric ephemerides while part c provides for a family of dynamical
time scales for equations of motion referred to the barycentre of the
solar system and dependent upon the relativistic theory being used.
The new dynamical time scale for apparent geocentric ephemerides is
continuous with the present ephemeris time scale. The terminology
and notation for the dynamical time scales will require further
consideration. G. Becker raised the question of a name for the new
time scale, and suggested "Astronomical Reference Time". G. M. R.
Winkler said that no name had yet been decided, and meanwhile he
thought it best to use the name "Dynamical Time Scale."

     W. Fricke in a report on the activities of the Working Group on
Precession presented the main arguments in favor of the adoption of
a new value of the general precession. These arguments will be
published in Astronomy and Astrophysics. In another paper the methods
of determining lunisolar precession will be summarized and the observational material which was used in the determinations will be
given. Fricke explained the reasons for the introduction of the
new standard epoch and equinox 2000 and the use of the Julian century
instead of the tropical century in the fundamental formulae for
precession. Furthermore, he reported on the specification of the new
fundamental reference system, the FK5, which will be compiled at
Heidelberg.

     T. Lederle reported on the proposed changes in the procedures for
the computation of apparent places. The following modifications will
be made simultaneously with the introduction of the new system of
constants and the FK5: (a) the aberration terms which depend on the

eccentricity of the Earth's orbit shall be shifted from the mean places
to the reduction from mean to apparent places; (b) the nutation
terms shall be slightly modified in such a way that the apparent
places are more directly comparable with observations; (c) the
reductions to apparent places shall be computed rigorously and
directly from the mean places for the standard equinox.

In the course of his presentation he announced a revised
wording for Recommendation 4(b) submitted by Atkinson and accepted
by the Chairman on behalf of the Working Groups. T. Van Flandern
proposed an addition to the notes, suggesting that there should
be a possibility of making further amendments to the recommendations
concerning nutation as a result of a Symposium on the subject to
be held in Kiev in 1977.

J. H. Lieske discussed the expressions for the precession
quantities. He stated that in comparing astronomical observations with calculated places of celestial objects, reductions have
to be made to refer either the observed or the calculated positions
to the same reference coordinate system. Such reductions include
such well-known effects as aberration, parallax, precession and
nutation. Although rather lengthy numerical expressions are
usually given for the parameters describing precession, the expressions, in fact, only depend upon a rather limited set of basic
parameters or fundamental constants and are thus amenable to revision.
He examined the structure of the expressions usually employed in
calculating the effects of precession and outlined the method by
which the expressions are revised to account for changes in the
fundamental astronomical constants. It was shown that the basic
set of parameters, upon which the lengthy polynomials for calculating
the mean obliquity of date and the elements of the precession matrix
depend, consists of the mean obliquity and the speed of general
precession in longitude at a fixed epoch E_c together with the system
of planetary masses. Expressions for the precession quantities
enabling one to precess to and from an arbitrary epoch were developed
as a function of the fundamental astronomical constants. The expressions with numerical values of the coefficients are given relative to epoch J2000.0. They must be used with the introduction of
the new constants into the ephemerides and in constructing the new
fundamental reference system, the FK5. The developments presented
are applicable for revising relevant precession quantities whenever
the system of astronomical constants is changed.

P. K. Seidelmann discussed the numerical values of the astronomical constants proposed for adoption in the Report. The values,
ranges of uncertainties, and sources of the specific values of
Recommendation 1 on the IAU (1976) System of Astronomical Constants
and of Recommendation 6 on other quantities for use in the preparation of the Ephemerides were presented. It was pointed out that the
following quantities have not been specified: the value for the
correction to the equator and equinox of the FK4; the secular acceleration of the Moon, i.e., the contribution to the mean longitude of
the Moon due to tidal friction; and the rotational elements of the
planets, i.e., the periods of rotation and the location of the poles
of rotation.

The Chairman opened the Report to general discussion remarking
that when the time came to vote for acceptance of the Report this
would give formal approval to the Recommendations only, leaving the
Notes for further revision by the signers of the Report in the light

of comments received. S. Vasilevskis proposed an amendment to the
wording of Recommendation 3(b), so that it should read "a correction
to the motion of the equinox of the FK4 shall be derived from relevant modern observations." This was in view of a paper presented by
K. C. Blackwell on equinox motion at a meeting of Commission 8.
W. Fricke welcomed the change of wording, as it gave him more
freedom in the compilation of FK5. P. J. Melchior remarked that
the constant of nutation proposed in the Report was not consistent
with the proposed constants and any model of the rigid body nutation
of the Earth. W. Fricke thought that a new value for the constant
could be considered at the Kiev Symposium, where the effect of non-
rigidity would be discussed. C. A. Murray asked if the previous
adherance to a rigid Earth model for nutation would be abandoned at
the Kiev Symposium. P. J. Melchior commented that the present nutation constant did not conform to a rigid body model, and if there was
sufficient agreement the Symposium may well recommend the inclusion
of further non-rigid-body terms. D. Eckhardt remarked that the proposed set of constants for the gravity field of the Moon was over-
determined. One of the constants could be derived from the others.
J. G. Williams said that the recommended values were self consistent
under the restriction imposed by Eckhardt's own theory. Y. Kozai
questioned the value of 0.0167 for the J_2 coefficient of Saturn; he
would prefer 0.016. It was thought that this could have been a typing
error, and the Chairman assured the meeting that the final version of
the Recommendations would be thoroughly checked. J. D. Mulholland
remarked that the clear intention of the IAU in 1973 was to adopt the
speed of light as a defining constant, which would be unchangeable.
J. Terrien replied that the General Conference of Weights and Measures
was expected to adopt this procedure eventually, but that they saw no
urgency in the matter. Meanwhile he thought that the procedure adopted
in the Report was the correct thing to do. P. Herget asked who would
have the power to amend the Recommendations to this Report following
the Kiev Symposium. The Chairman replied that this would be under
the control of Commission 4. S. Aoki pointed out that the values
given in the Report for the constants defining the size and gravity
field of the Earth would soon be superseded. I. Mueller remarked that
this was an unfortunate situation which could not be avoided. He
remarked further that the revised wording of Recommendation 4(b)
appeared to be in conflict with Woolard's use of the terms "axis of
figure" and "axis of rotation." R. d'E. Atkinson said that he did
not think there was any ambiguity, because the recommendation referred
to the forced periodic terms in the motions of these axes. B. Guinot
said that a convenient name was needed for the new axis, and suggested
"mean diurnal axis". I. Mueller suggested that the name "Eulerian
Pole of Rotation" was suitable. R. d'E Atkinson said that Woolard
did not use this name explicitly in his work on nutation, but used
the term "Eulerian place of the pole of rotation" in the caption to
one of his diagrams. The Chairman suggested that the chairmen of
the Working Groups should try to find a suitable name, and if they
did it could be included in the Notes.

    B. K. Bok, representing Commission 33, was invited to address the
meeting. He expressed the great interest of astronomers of all
disciplines in the matters being discussed and said that he had been
well satisfied with the care and thoroughness with which the Report
had been prepared. He hoped to see the FK5 published now, on a well
defined system, as a good fundamental reference system was very
important. He said Commission 33 hoped for a firm decision on this
Report from this meeting.

The Chairman then asked the meeting to vote for the acceptance of the Resolution contained in the Report, on the understanding that Recommendations 3(b) and 4(b) would be amended as discussed during the meeting, and all known errors would be corrected. The Resolution was proposed and seconded, and carried by general consent, with no objections.

JOINT REPORT OF THE WORKING GROUPS OF IAU COMMISSION 4 ON PRECESSION, PLANETARY EPHEMERIDES, UNITS AND TIME-SCALES

I. INTRODUCTION

The report contains recommendations for: the revision of the IAU System of Astronomical Constants that was adopted in 1964; the introduction of a new standard epoch; the improvement of the specification of the fundamental reference system and related computational procedures; time scales for dynamical theories and ephemerides; and the values of other data on the planets and satellites. These changes are intended to provide an agreed basis for a new fundamental catalogue (FK5) and for a new set of high-precision lunar and planetary ephemerides. The new catalogue and ephemerides are required in order to meet the greater demands on positional and dynamical astronomy being made by new techniques of observation, such as radar and laser ranging and long baseline radio interferometry, as well as by improvements in the classical techniques of astrometry. These studies should lead to advances in our understanding of the dynamics of the Earth and of the solar system, the theory of relativity, and both galactic and extragalactic astronomy. It is hoped that these recommendations will be adopted by all who are distributing ephemerides or observational data for general use. The recommendations in this report have been developed from the recommendations of IAU Colloquium No. 9, which was held at Heidelberg in August 1970 (Celestial Mechanics 4, 128-280, 1971). At the IAU General Assembly that followed immediately afterwards, Commission 4 set up three working groups, which included members of other Commissions having an interest in these matters (Trans. IAU XIVB, 81-83, 1971). The preliminary reports of the working groups (Trans. IAU XVA, 9-10, 1973) were discussed at the next General Assembly at Sydney in August 1973 at a Joint Discussion (Highlights of Astronomy 3, 209-232, 1974), and by Commission 4, which decided that the groups should continue their work (Trans. IAU XVB, 70, 1974). Members of the groups and some other astronomers met in Washington in October 1974 at a small discussion meeting arranged by the President of Commission 4. The recommendations of this meeting were considered further by the working groups and others, and provided the basis for this joint report. The chairmen of the three groups met at Herstmonceux at the beginning of October 1975 to prepare the draft of the report, which was circulated widely for comment. The authors met in Washington in June 1976 and prepared the revised report for submission to a Joint Meeting of Commissions 4, 8 and 31 to be held during the IAU Assembly at Grenoble in August 1976.

The new system of astronomical constants is intended to replace the system adopted at Hamburg in 1964. At that time it was recognized that the precessional constants and the planetary masses differed significantly from the most likely values, but it was considered that it would be desirable to continue to use the current values until new values could be introduced with greater confidence in a new fundamental catalogue and in ephemerides based on improved

theories and new comparisons with observation. Whenever appropriate
the values of the constants have been expressed in the units of the
International System (SI).

The opportunity has been taken to introduce improved values of
the constants for the Earth for dynamical purposes; the values given
are the currently representative estimates of geodetic parameters as
recommended by the International Association of Geodesy in 1975. No
attempt has been made to introduce a full set of parameters to define
a standard model for the Earth. Until such a model and a new theory
of nutation can be adopted, consideration will be given to the possibility of applying correction terms determined from observations.
A change in the value of the general precession in longitude has become necessary because of the introduction of new observational
methods, such as the direct measurement of distances in the planetary
system and long baseline radio interferometry. The error of about
$1\overset{''}{.}1$ per century in the present value of the general precession and a
spurious motion of the equinox of about the same size manifest themselves as unacceptable discordances between the observed and theoretical motions of celestial objects and introduce spurius contributions
into the proper motions of stars.

The recommendations for the use of the Julian century, rather
than the tropical century, which is both variable and theory dependent,
and for the adoption of the new standard epoch and equinox of 2000
are intended to simplify precessional computations and to unify the
standard epochs of the fundamental catalogue and planetary theories.

A dynamical time scale for apparent geocentric ephemerides is
specified in such a way as to provide continuity with the current
values and practice in the use of Ephemeris Time. It is based on
the SI second as the unit of time and is distinct from the practical
realization of International Atomic Time (TAI) which is only available
from 1958 onward. It facilitates the comparison of observations with
theories of all types.

The values given for the masses of the minor planets and satellites, and the other data on the Sun, Moon and planets are considered
to be the current best estimates of these quantities and to provide a
suitable basis for the construction of ephemerides except in those
cases where the requirements for extreme accuracy are more important
than the advantages of standardization. It is intended that these
values will be used in the international ephemerides except where new
information indicates that the value given here is significantly in
error.

We regard it as essential that decisions on the matters discussed
in this report be made at the IAU General Assembly in 1976 in order
that the new constants and procedures can be used in the preparation
of the new fundamental catalogue FK5 and new high-precision lunar and
planetary ephemerides. We consider that the annual publications
based on these new data should be issued for the year 1984 and we see
no need for the publication of differential corrections for the preceding years. The individual recommendations are presented in section
III of this report and notes on them are given in section IV. These
notes are not intended to form part of the recommendations, but are
given by way of explanation or justification.

II. RESOLUTION

     We request that the following draft resolution be submitted to
the Joint Meeting of IAU Commissions 4, 8 and 31 with the view of its
being adopted at the Sixteenth General Assembly of the IAU.
"The IAU endorses the recommendations given in the Joint Report of the
Working Groups of Commission 4 on:
                    the IAU (1976) System of Astronomical Constants,
                    the new standard epoch and equinox,
                    the fundamental reference frame,
                    the procedures for the computation of apparent
                       places and the reduction of observations,
                    time scales for dynamical theories and ephemerides,
                    and other quantities for use in the preparation of
                       ephemerides;
and recommends that they shall be used in the preparation of the
fundamental catalogue FK5 and of the national and international
ephemerides for the years 1984 onwards, and in all other relevant
astronomical work.

III. RECOMMENDATIONS TO IAU GENERAL ASSEMBLY, 1976

RECOMMENDATION 1: IAU (1976) SYSTEM OF ASTRONOMICAL CONSTANTS

     It is recommended that the following list of constants shall be
adopted as the "IAU (1976) System of Astronomical Constants".

Units

     The units metre (m), kilogram (kg) and second(s) are the units
of length, mass and time in the International System of Units (SI).
     The astronomical unit of time is a time interval of one day
($\underline{D}$) of 86400 seconds. An interval of 36525 days is one Julian century.
     The astronomical unit of mass is the mass of the Sun ($\underline{S}$).
     The astronomical unit of length is that length ($\underline{A}$) for which the
Gaussian gravitational constant ($\underline{k}$) takes the value 0.017 202 098 95
when the units of measurement are the astronomical units of length,
mass and time. The dimensions of $\underline{k}^2$ are those of the constant of
gravitation ($\underline{G}$), i.e., $L^3 M^{-1} T^{-2}$. The term "unit distance" is also
used for the length $\underline{A}$.

Defining constants

1. Gaussian gravitational constant $\underline{k}$ = 0.017 202 098 95

Primary constants

2. Speed of light $\underline{c}$ = 299 792 458 m s^{-1}
3. Light-time for unit distance τ_A = 499.004 782 s
4. Equatorial radius for Earth $\underline{a}_e$ = 6 378 140 m
5. Dynamical form-factor for Earth $\underline{J}_2$ = 0.001 082 63
6. Geocentric gravitational constant $\underline{GE}$ = 3.986 005 X 10^{14} m^3 s^{-2}
7. Constant of gravitation $\underline{G}$ = 6.672 X 10^{-11} m^3 kg^{-1} s^{-2}
8. Ratio of mass of Moon to that of
       Earth μ = 0.012 300 02
9. General precession in longitude,
       per Julian century, at standard
          epoch 2000 $\underline{p}$ = 5 029''096 6

10. Obliquity of the ecliptic, at stand-
     ard epoch 2000 ε = 23°26' 21".448
11. Constant of nutation, at standard
     epoch 2000 $\underline{N}$ = 9".2109

Derived constants

12. Unit Distance $\dfrac{c\,\tau_A}{\pi_{\odot}}$= $\underline{A}$ = 1.495 978 70 X 10^{11}m
13. Solar parallax   arcsin $(\underline{a}_e/\underline{A})$ = $\pi_{\odot}$= 8".794 148
14. Constant of aberration, for
     standard epoch 2000 κ = 20".495 52
15. Flattening factor for the
     Earth $\underline{f}$ = 0.003 352 81 = 1/298.257
16. Heliocentric gravitational
     constant $\underline{A}^3\underline{k}^2/\underline{D}^2$ = $\underline{GS}$ = 1.327 124 38 X $10^{20}\mathrm{m}^3\mathrm{s}^{-2}$
17. Ratio of mass of Sun to
     that of Earth $(\underline{GS})/(\underline{GE})$= $\underline{S}/\underline{E}$= 332 946.0
18. Ratio of mass of Sun to
     that of Earth+Moon $(\underline{S}/\underline{E})/(1+\mu)$ = 328 900.5
19. Mass of the Sun $(\underline{GS})/\underline{G}$ = $\underline{S}$ = 1.989 1 X 10^{30} kg

System of planetary masses

20. Ratios of mass of Sun to those of the planets and their satellites

Mercury	6 023 600	Jupiter	1 047.355
Venus	408 523.5	Saturn	3 498.5
Earth+Moon	328 900.5	Uranus	22 869
Mars	3 098 710	Neptune	19 314
		Pluto	3 000 000

RECOMMENDATION 2: THE NEW STANDARD EPOCH AND EQUINOX
It is recommended that:

     (a) the new standard epoch (designated J2000.0) shall be 2000
January 1^{d}5, which is JD 2 451 545.0, and the new standard equinox
shall correspond to this instant;
     (b) the unit of time for use in the fundamental formulae for
precession shall be the Julian century of 36525 days; and
     (c) the epochs for the beginning of year shall differ from the
standard epoch by multiples of the Julian year of 365.25 days.

RECOMMENDATION 3: THE FUNDAMENTAL REFERENCE FRAME
     It is recommended that:
     (a) the fundamental reference frame defined by the positions and
centennial variations in the FK5 shall correspond as closely as possible
to the dynamical reference frame;
     (b) a correction to the zero point of right ascensions of the
FK4 (equinox correction) and a correction to the motion of the equinox
of the FK4 shall be derived from relevant modern observations;
     (c) the expression for Greenwich mean sidereal time at 0^hUT shall
be amended by the same equinox correction and motion as adopted for the
FK5 in order to avoid a discontinuity in UT.

RECOMMENDATION 4: THE PROCEDURES FOR THE COMPUTATION OF APPARENT
PLACES AND THE REDUCTION OF OBSERVATIONS
     It is recommended that:
     (a) stellar aberration shall be computed from the total
velocity of the Earth referred to the barycentre of the Solar
System, and mean places shall not contain E-terms;
     (b) the tabular nutation shall include the forced periodic

terms listed by Woolard for the axis of figure in place of those
given for the instantaneous axis of rotation, and the two calibrations performed by him shall be revised accordingly, taking
account of the change in the adopted precession;
    (c) reductions to apparent place should be computed rigor-
ously and directly, without the intermediary of the mean place
for the beginning of year, whenever high-precision is required.

RECOMMENDATION 5: TIME-SCALES FOR DYNAMICAL THEORIES AND
EPHEMERIDES
    It is recommended that:
    (a) at the instant 1977 January $01^d00^h00^m00^s$TAI, the value
of the new time-scale for apparent geocentric ephemerides be 1977
January $1.^d000$ 372 5 exactly;
    (b) the unit of this time-scale be a day of 86400 SI seconds
at mean sea level;
    (c) the time-scales for equations of motion referred to the
barycentre of the solar system be such that there be only periodic
variations between these time-scales and that for the apparent geocentric ephemerides; and
    (d) no time-step be introduced in International Atomic Time.

RECOMMENDATION 6: OTHER QUANTITIES FOR USE IN THE PREPARATION
OF EPHEMERIDES
    It is recommended that the values given in the following list
should normally be used in the preparation of new ephemerides.
1. Masses of minor planets

Minor Planet	Mass in solar mass
(1) Ceres	5.9×10^{-10}
(2) Pallas	1.1×10^{-10}
(4) Vesta	1.2×10^{-10}

2. Masses of satellites

Planet	Satellite	Satellite/ Planet mass
Jupiter	Io	4.70×10^{-5}
	Europa	2.56×10^{-5}
	Ganymede	7.84×10^{-5}
	Callisto	5.6×10^{-5}
Saturn	Titan	2.41×10^{-4}
Neptune	Triton	2×10^{-3}

3. Equatorial radii in km

Mercury	2 439	Jupiter	71 398	Moon	1 738
Venus	6 052	Saturn	60 000	Sun	696 000
Earth	6 378.140	Uranus	25 400		
Mars	3 397.2	Neptune	24 300		
		Pluto	2 500		

4. Gravity fields of the planets

	J_2	J_3	J_4
Earth	+0.001 082 63	-0.254×10^{-5}	-0.161×10^{-5}
Mars	+0.001 964	+0.000 036	
Jupiter	+0.014 75		-0.000 58
Saturn	+0.016 45		-0.001 0
Uranus	+0.012		
Neptune	+0.004		

	C_{22}	S_{22}	S_{31}
Mars	-0.000 055	+0.000 031	+0.000 026

5. Gravity field of the Moon

$\gamma = (B-A)/C = 0.000\ 227\ 8$ $C/MR^2 = 0.392$
$\beta = (C-A)/B = 0.000\ 631\ 3$ $I = 5\ 552''7 = 1° 32' 32''7$
$C_{20} = -0.000\ 202\ 7$ $C_{30} = -0.000\ 006$ $C_{32} = +0.000\ 004\ 8$
$C_{22} = +0.000\ 022\ 3$ $C_{31} = +0.000\ 029$ $S_{32} = +0.000\ 001\ 7$

$$S_{31} = +0.000\ 004 \qquad\qquad \begin{aligned} C_{33} &= +0.000\ 001\ 8 \\ S_{33} &= -0.000\ 001 \end{aligned}$$

IV. NOTES ON RECOMMENDATIONS

NOTES ON RECOMMENDATION 1

Units

The constants of this revised system are generally expressed in terms of the SI units in order to ensure compatibility with the usage in related sciences. In astronomy it is, however, necessary to use the astronomical system of units of length, mass and time. The astronomical unit of time (day) is redefined in terms of the SI second, which was itself defined so as to be equal to the ephemeris second to within the error of the determination. Specifically it is the SI second at mean sea level.

1. The Gaussian gravitational constant serves to define the astronomical unit of length when the corresponding astronomical units of time and mass are already defined. The value for k is that adopted by the IAU in 1938. The value (rounded) of $k/86400$ is 1.990 983 675 X 10^{-7}.

2. The value for the speed of light is that recommended by the fifteenth General Conference on Weights and Measures in 1975. It is understood that this value will be unchanged even if the metre is redefined in terms of a different wave-length from that now used.

3. The value of the light-time for unit distance (1 astronomical unit of length) is based on radar measurements of planetary distances. It is numerically equal to the number of light-seconds in 1 astronomical unit of length. Its reciprocal (0.002 003 988 81) is equal to the speed of light in astronomical units of length per second. The speed of light in astronomical units per day is 173.144 633.

4. The term "equatorial radius for Earth" refers to the equatorial radius of an ellipsoid of revolution that approximates to the geoid. (See also note 15.) The values given for constants 4, 5, 6 and 15 are those recommended by the International Association of Geodesy at Grenoble in 1975 as currently representative estimates of fundamental geodetic parameters.

5. The term "dynamical form-factor for Earth" refers to the coefficient of the second zonal harmonic in the expression for the Earth's gravitational potential as defined in Trans. IAU XIIB (1964), 117-8, 1966. (See also notes 4 and 15.)

6. The geocentric gravitational constant is appropriate for use for geocentric orbits when the units are the metre and the second; E denotes the mass of the Earth including its atmosphere. (See also note 4.)

7. The value for the constant of gravitation is that given in the CODATA system of physical constants of 1973 (CODATA Bull. No. 11).

8. The value for the mass ratio is based on recent data from lunar and planetary spacecraft. (Reference as Note 3.) The mass of the

Earth includes the atmosphere. The reciprocal of 0.012 300 02 is
81.3007.

9. The value of the general precession in longitude has been
derived on the basis of recent determinations of the correction
to Newcomb's value of the lunisolar precession (Ref: Fricke, W.,
1967 Astron. J. 72, 1368; 1971 Astron. & Astrophys. 13, 298) and on
the basis of a new value of the planetary precession derived from the
new planetary masses. For convenience of those making differential
corrections the exact value Δp_1 = +1".10 has been adopted in computing
the new value of p for 2000. The four decimal places of p are
required in order to secure consistency with computations based on
the current value for 1900 and the correction Δp_1. A corresponding
set of numerical formulae for precessional reductions will be made
available. (Ref: Lieske, J., in press.)

10. The value of the obliquity of the ecliptic results from applying
secular terms computed with the new values of the planetary masses
to the current value for 1900. (Ref: Lieske, J., in press.)

11. The value of the constant of nutation results from applying the
secular term given by Woolard (Astron. Papers American Ephemeris
15, 153, 1961) to the current value for 1900. The value will have
to be changed as soon as it is possible to adopt a new theory of
nutation based on a non rigid model of the Earth. In consideration
of the symposium on nutation to be held in Kiev in 1977, it was
agreed at Grenoble that the adopted recommendations concerning nutation may be amended by Commission 4.

12.-19. The values of the derived constants have been computed from
the values of the defining and primary constants. All the values are
consistent with those determined more directly from observations.

12. The number of metres in one astronomical unit of length is now
treated as a derived constant.

13. The rounded value 8".794 for the solar parallax may be used except where extra figures are required to ensure numerical consistency.

14. The constant of aberration is the ratio of the mean speed of the
Earth to the speed of light, and is conventionally expressed in
seconds of arc. It is calculated in radians from the expression
$F\ k\tau_A/86400$ where F is the ratio of the mean speed of the Earth to the
speed of a hypothetical planet of negligible mass moving around the
Sun in a circular orbit of unit radius. The value of F for epoch
2000 is 1.000 1414, and is given by
$$F\ k = n\ a\ (1-e^2)^{-\frac{1}{2}}$$
where n is the sidereal mean motion of the Sun in radians per day, a
is the perturbed mean distance of the Sun in astronomical units of
length, and e is the mean eccentricity of the Earth's orbit.

The rounded value 20".496 may be used for κ except where the extra figures are required to achieve numerical consistency.

15. The flattening factor for the Earth is derived from the adopted
values of the primary geodetic parameters using the condition that
the corresponding ellipsoid of revolution shall be an equipotential
surface. (See Note 4.) (Ref: Geodetic Reference System 1967, IAG
Spec. Pub. No. 3, 1971)

16. The heliocentric gravitational constant is appropriate for use for heliocentric orbits when the units are the metre and the second.

17.-20. The values given for the reciprocal masses of the planets include the contributions from atmospheres and satellites. For Mercury, Venus and Mars values close to the best spacecraft determinations are adopted. For the Earth the mass is that derived from the adopted values of A, GE and μ. For Jupiter, Uranus and Neptune the modern determinations do not indicate the necessity to change the Newcomb values. The value for Saturn is the unweighted mean of the most reliable determinations. The value for Pluto is based on analyses of the motion of Neptune.

(Ref: Howard H. T., et al, Science 185, 12 July 1974; Anderson, J., Trans AGU 55, May 1975; Duncombe, R. L., et al, Highlights of Astronomy 3, 1973.)

The values given for the reciprocal masses are to be treated as exact, except that for Earth+Moon the gravitational constant should be calculated from GE(1+μ), that is from the exact values of the primary constants, if numerical consistency is required.

The mass of the Sun in kilograms is given to indicate the relationship between the astronomical and SI units of mass; it is known only to the low precision with which the constant of gravitation is known in SI units. The corresponding values of the masses of the planets are:

Mercury	$3.302\ 2 \times 10^{23}$ kg		Jupiter	$1.899\ 2 \times 10^{27}$ kg
Venus	$4.869\ 0 \times 10^{24}$ kg		Saturn	$5.685\ 6 \times 10^{26}$ kg
Earth	$5.974\ 2 \times 10^{24}$ kg		Uranus	$8.697\ 8 \times 10^{25}$ kg
Moon	$7.348\ 3 \times 10^{22}$ kg		Neptune	$1.029\ 9 \times 10^{26}$ kg
Mars	$6.419\ 1 \times 10^{23}$ kg		Pluto	7×10^{23} kg

Ranges of uncertainty for the constants

The true values of the primary constants are believed to lie between the following limits:

c:	299792456.8 - 299792459.2	G:	$(6.668 - 6.676) \times 10^{-11}$
τ_A:	499.004 776 - 499.004 788	μ:	0.012 300 06 - 0.012 299 97
a_e:	6 378 135 - 6 378 145	p:	5028.95 - 5029.25
J_2:	0.001 082 62 - 0.001 082 64	ε:	23°26'21".35 to ... 21".55
GE:	$(3.986002 - 3.986 008) \times 10^{14}$	N:	9".200 - 9".211

Correspondingly, the limits for the derived constants are:

A : $(1.495\ 978\ 68 - 1.495\ 978\ 72) \times 10^{11}$
$\pi_{\odot}$: 8".794 141 - 8".794 155
κ : 20".495 518 - 20".495 520
f : 0.003 352 79 - 0.003 352 83
GS : $(1.327\ 124\ 33 - 1.327\ 124\ 43) \times 10^{20}$
S/E: 332 945.7 - 332 946.3
S : $(1.9879 - 1.9903) \times 10^{30}$

The limits for the reciprocal masses of the planets are believed to be:

Mercury:	6 020 000 - 6 027 000	Jupiter:	1 047.330 - 1 047.380
Venus :	408 521 - 408 526	Saturn :	3 497 - 3 500

Earth +				
Moon:	328 900.0 - 328 901 .0	Uranus :	22 650 - 23 100
Mars :	3 098 600 - 3 098 760	Neptune:	19 300 - 19 450
		Pluto :	2 000 000 - 15 000 0C

NOTES ON RECOMMENDATION 2

1. The new standard epoch is one Julian century after 1900 January $0\overset{d}{.}5$, which corresponds to the fundamental epoch of Newcomb's planetary theories. The new standard epoch is expressed in terms of dynamical time instead of Universal Time. Specifically for precise planetary and lunar theories, it is expressed in terms of the time scale of the equations of motion with respect to the barycentre of the Solar System.

2. In the new system a Julian epoch is given by
$$J2000.0 + (JD - 2\ 451\ 545.0)/365.25,$$
where JD symbolizes the Julian date. If the Besselian epoch is still required, it is given by
$$B1900.0 + (JD - 2\ 415\ 020.313\ 52)/365.242\ 198\ 781.$$
The Besselian year is here fixed at the length of the tropical year $(365\overset{d}{.}242\ 198\ 781)$ at B1900.0 (JD 2 415 020.313 52).

The prefixes J and B are used to distinguish Julian and Besselian epochs; they may be omitted only where the context, or precision, makes them superfluous.

NOTES ON RECOMMENDATION 3

1. Failure to distinguish between the catalogue equinox of the FK4 (that is its zero point of right ascension on the equator) and the dynamical equinox (the crossing point of the ecliptic on the equator) has been the cause of much difficulty. The FK4 equinox was based on determinations of the dynamical equinox before 1930.

Recent determinations of the equinox corrections have to be taken into account in the determination of the system of the FK5 such that the equinox error will be removed as far as currently possible together with the removal of an erroneous motion of the equinox of the FK4. The corrections to equinox motion and precession must be applied together to avoid introducing an additional fictitious rotation into the stellar proper motions in right ascension.

2. The current expression for Greenwich mean sidereal time is given in the Explanatory Supplement to the Astronomical Ephemeris on page 75.

NOTES ON RECOMMENDATION 4

1. The elliptic component in the Earth's velocity has traditionally been omitted in the computation of the day numbers, and the so-called E-terms of aberration have remained imbedded in the mean places of celestial objects. This practice has caused much confusion and it is convenient to use the occasion of other changes to remove E-terms from mean places and to include them in the reduction from mean to apparent place so that the apparent places will not be changed. The mean places in the FK5 will not contain E-terms and so tables will be given in the FK5 for reducing mean places with E-terms included (for example, mean places in the N30 catalogue) to mean places without E-terms.

2. Nutation is taken into account in the current procedure for computing true places by a reduction from the mean celestial pole of
date to a celestial pole which approximates to the direction of the
instantaneous axis of rotation of the Earth. It has, however, been
demonstrated that observations give the place of a pole whose position
with respect to the mean pole of date can be obtained by the procedure
described in Recommendation 4 (b). This pole may continue to be
called the true celestial pole of date. The prescribed procedure
may be achieved by removing the seven small forced periodic-terms in
Woolard's equations 55 (Astron. Papers American Ephemeris 15, 133,
1953), by substituting the corresponding terms in equations 54 (page
132), and by scaling.

     The equation of the equinoxes (nutation in right ascension on the
equator) causes periodic variations in the location of the true
equinox of date and hence a variation in apparent sidereal time.
In certain applications it may be convenient to remove the effects
of these periodic variations by subtracting the equation of the
equinoxes, but the origin of apparent right ascension shall continue
to be the true equinox of date.

3. It is intended that formulae and tabulations will be made available for use in rigorous, direct reductions without the intermediate
formation of the mean place for the beginning of the year. For users
who do not require the highest precision, Besselian day numbers will
still be provided.

NOTES ON RECOMMENDATION 5

1. The time-like arguments of dynamical theories and ephemerides are
referred to as dynamical time-scales. While it is possible, and
desirable to base the unit of a dynamical time-scale on the SI second
(which is used in the draft IAU (1976) system of astronomical constants), it is necessary to recognize that in relativistic theories
there will be periodic variations between the unit of time for an
apparent geocentric ephemeris and the unit of the corresponding
time-scale of the equations of motion, which may, for example, be referred to the centre of mass of the Solar System. (In the terminology
of the theory of general relativity such time-scales may be considered
to be proper time and coordinate time, respectively.) The time-scales
for an apparent geocentric ephemeris and for the equations of motion
will be related by a transformation that depends on the system being
modelled and on the theory being used. The arbitrary constants in the
transformation can be chosen so that the time-scales have only periodic
variations with respect to each other. Thus, it is sufficient to
specify the basis of a unique time-scale to be used for new, precise,
apparent geocentric ephemerides.

     The dynamical time scale for apparent geocentric ephemerides of
Recommendation 5(a) and (b) is a unique time-scale independent of
theories, while the dynamical time-scales referred to the barycentre
of the Solar System are a family of time-scales resulting from the
transformations of various theories and metrics of relativistic
theories.

2. This recommendation specifies a particular dynamical time-scale
for apparent geocentric ephemerides that is effectively equal to
TAI + $32^{\text{s}}184$. (There are formal differences arising from random
and, possibly, systematic errors in the length of the TAI second

and the method of forming TAI, but the accumulated effect of such
errors is likely to be insignificant for astronomical purposes over
long periods of time.) The scale is specified with respect to TAI
in order to take advantage of the direct availability of UTC (which
is based on the SI second and is simply related to TAI), and to provide continuity with the current values and practice in the use of
Ephemeris Time. Continuity is achieved since the chosen offset betweer
the new scale and TAI is the current estimate of the difference
between ET and TAI, and since the SI second was defined so as to
make it equal to the ephemeris second within the error of measurement. It will be possible to use most available ephemerides as
if the arguments were on the new scale. Before 1955, when atomic
time is not available, the determinations of ET can be considered
to refer to the new scale. The offset has been expressed in the
recommendation as an exact decimal fraction of a day since the
arguments of theories and ephemerides are normally expressed in days.

3. In view of the desirability of maintaining the continuity of TAI
and of avoiding the confusion that could arise if it were to be
redefined retrospectively, no step in TAI is proposed. Although
the recommendation is in terms of TAI, in practice astronomers will
use UTC and convert directly to the dynamical time-scales.

4. The terminology and notation for dynamical time-scales require
further consideration in due course.

5. Recognizing that the TAI second differed from the SI second
between 1969 and the present by $(10\pm2)\times10^{-13}$, a step will be introduced in the scale interval of TAI. Therefore, the epoch of the
dynamical time-scale for apparent geocentric ephemerides was adjusted
to 1977 from 1958 at a subsequent meeting of Commissions 4 and 31.

NOTES ON RECOMMENDATION 6

1. There are not enough independent determinations of the masses
of Ceres, Pallas and Vesta to derive ranges of uncertainty, but the
internal standard errors are $\pm(0.3, 0.2, 0.1) \times 10^{-10}$, respectively.
(Ref: Schubart, J., Astron. & Astrophys. 30, 289, 1974 and 39, 147,
1975; Hertz, H. G., Science 160, 19 April 1968.)

2. Masses of satellites of Jupiter are derived from Pioneer 10.
Mass of Titan is derived from the motion of Iapetus. Mass of Triton
is estimated from the motion of Neptune. (Ref: Anderson, J. D.,
et al, J. Geophys. Res. 79, 3661, 1974; Duncombe, R. L., et al,
Fundamentals of Cosmic Physics 1, 119, 1973.)

3. For Mercury, Venus, Earth, Mars and Moon the values refer to the
planet's crust. A value for Venus including the height of the cloud
layer is 6110 km. The radius of the Moon implicit in Watts' profile
of the lunar limb is 1 738.065 km. For Jupiter the value is based
on determinations from Pioneer 10 and 11. For Saturn, Uranus, and
Neptune the values are means of the best optical measures by double
image micrometer and heliometer. The value for Pluto is a crude estimate. (Ref: Howard, H. T. et al, Science 185, 12 July 1974; Bulletin
Geodesique 118, 365, Dec. 1975; Anderson, J. D., Review of Geophysics
and Space Physics 13, July 1975; Null, G. W., Anderson, J. D., Wong,
S. K., Science 188, 476, 1975; Anderson, J. D., EOS of Amer. Geophys.
Union 55, May 1974; Kaula, W., et al, Geochimica et Cosmochimica
Acta 3, 3049, 1974; Dollfus, A., Surfaces and Interiors of Planets

and Satellites, Academic Press, New York, 1970.)

4. For notation see Eckhardt, D. H., The Moon 6, 127, 1973.
Earth - coefficients are given for the only three terms that have
a significant effect on the orbital motion of the Moon. They should
not be considered as defining the dynamical model of the Earth. They
are consistent with the first three terms of the presently available
models of the Earth's potential, but end figures are subject to change.
Mars - derived from Mars-orbiter data. The coefficients given are the
ones having a significant effect on the orbital motion of satellites.
Jupiter - derived from Pioneer 10 and 11 results,
Saturn and Neptune - derived from motions of their nearby satellites.
Uranus - based on optical measures of the flattening and is in rea-
sonable agreement with the dynamical determination. (Ref: Bulletin
Geodesique 118, 365, Dec. 1975; Anderson, J. D., EOS of Amer. Geophys.
Union 55, May 1974; Null, G. W., Anderson, J. D., Wong, S. K.,
Science 188, 476, 1975.)

5. The values are best estimates based on lunar laser ranging data
and spacecraft data. (Ref: Liu and Laing, Science 173, 1017, 1971;
Sjogren, W., J. Geophys. Res. 76, 7021, 1971; Gapcynski, et al, Geophys.
Res. Lett. 2, 353, 1975; Williams, J., EOS Trans. AGU 56, 236, 1975.)

6. Several additional quantities must be specified before ephemerides
can be completed, but definitive values cannot be selected at the
present time. Such quantities are the secular acceleration of the
Moon, the corrections to the equator and equinox of the FK4, and the
rotational elements of the planets.

V. ACKNOWLEDGEMENTS

    We wish to acknowledge the assistance given to us by all who have
contributed to the preparation of this report through correspondence
and discussions. In particular, V. K. Abalakin, J. Anderson,
S. Aoki, R. d'E. Atkinson, G. Becker, H. Enslin, B. Guinot, P. Herget,
P. Janiczek, H. Kinoshita, W. Klepczynski, J. Kovalevsky, R. Laubscher,
T. Lederle, J. Lieske, J. Lorell, B. Morando, Y. Kozai, I. Mueller,
J. D. Mulholland, C. A. Murray, C. Oesterwinter, D. H. Sadler,
J. Schubart, I. I. Shapiro, A. T. Sinclair, A. M. Sinzi, T. C.
Van Flandern, R. O. Vicente, J. Williams and G. Winkler have given
generously of their time and ideas.

                                          R. L. Duncombe
                                          W. Fricke
                                          P. K. Seidelmann
                                          G. A. Wilkins

COMMISSION 5 : DOCUMENTATION (DOCUMENTATION)

Report of Meetings, Friday 27 and Saturday 28 August 1976

PRESIDENT : J.-C. Pecker   SECRETARY : L. Remy-Battiau

1. PRESIDENT'S REPORT
    The report is adopted on the motion of W. Fricke.

2. MEMBERSHIP
    The loss of two deceased members of the Commission : P. Bourgeois and F. Henn
was noted with regret.
    Besides the two new members co-opted since the Sydney meeting and whose mem-
bership has already been approved by the Executive Committee : R.S. Dixon and L.
Remy-Battiau, the following proposals were accepted, subject to approval by the
Executive Committee : W. Bidelman, J.O. Fleckenstein, R.F. Griffin, S.A. Mitton,
F. Ochsenbein, L. Schmadel, C.E. Worley as new members. G. Bérardini, A. Berthelot,
P. Dale, J. Duncan, G. Feuillebois, G. Grassi Conti, M. Guidoni, D.A. Kemp, N.A.
Lavrova, J. Mead, P. Morholt, R.B. Rodman, R.A. Seal, V. van Brunt, K. Zadla were
elected or re-elected as consulting members, for the next three years.

3. OFFICERS OF THE COMMISSION, 1976-1979
    According to the rules of the Commission, J.-C. Pecker and W.D. Heintz are re-
conducted for another three years period respectively as President and Vice-President. The Commission adopted the following composition of the Organizing Committee:
D.A. Kemp, J. Kleczek, P. Lantos, J.R. Shakeshaft, T.S. Sherbina-Samojlova, G.A.
Wilkins.

4. SECONDARY PUBLICATIONS
    The account by W. Fricke that the Heidelberg Institute may in the future be
discharged of the responsability of producing Astronomy and Astrophysics Abstracts,
led to the formulation of a resolution which the Commission voted unanimously, including the consulting members, and which was so important that it was voted separately and specifically by the General Assembly (Resolution No. 3).

5. ICSU AB
    W.D. Heintz reported on the last General Assembly of ICSU AB, held in Washing-
ton D.C., which he has attended as representative of the Commission. This concerned large fields of information and the major point of the meeting concerned the
planning of interdisciplinary projects in cataloguing. No mention was made of UDC 52.
    J.-C. Pecker reports that ICSU AB has put forward a classification for Physics
in which Astronomy, Geophysics,... are considered as parts of Physics and not separately as in UDC. However, comments from members of the Commission to improve
ICSU AB classification and to have it differ as less as possible from UDC 52 have
been taken into consideration. Since the two classifications are essentially different in nature, it remains that we have now two different classifications. The
Commission believes that UDC 52 corresponds better to the needs of astronomers.
    Despite this difficulty the Commission believes that the IAU should keep its
membership within ICSU AB in order to be able to express its views on astronomical
and technical problems.
    The Commission regrets that the relations between the different International
Unions are not closer. In particular, it is inadvisable that many different organizations should each develop their own independent systems for classifying and
indexing knowledge, whether this knowledge is in the form of publications or data
banks.

6. WORKING GROUP ON NUMERICAL DATA

Report for 1973-76

The Chairman of the Working Group on Numerical Data, G.A. Wilkins, gave a brief summary of the report of the Working Group on Numerical Data for the period 1973-75 (see Transactions of the IAU, vol. XVIA, part I, pp. 193-194). In addition, he had amended the "Survey of Astronomical Data Activities" that had originally been prepared for the Sydney Meeting ; the IAU Executive Committee had declined on grounds of cost to print it in an IAU Information Bulletin, but he was grateful to C. Jaschek for publishing it in the Information Bulletin of the Stellar Data Center at Strasbourg. He had attended the 1976 General Assembly of CODATA as the delegate of the IAU and had been appointed Chairman of the Committee on the Geosciences. CODATA has set up a Task Group on the "Methodology of handling space and time-dependent data" and he hoped that astronomers would both contribute to and benefit from its investigations and recommendations.

IAU Colloquium No. 35

G.A. Wilkins also gave a short account of the IAU Colloquium No. 35 on "The compilation, critical evaluation, and distribution of stellar data" which had been held at Strasbourg on 19-21 August, 1976. He was grateful to C. Jaschek and other members of the staff at Strasbourg for all the work that they had done to organize the Colloquium. There had been five sessions each with an invited speaker and contributed papers. The principal topics discussed and the invited speakers were : standards for the presentation of data (M.S. Davis), acquisition and processing of techniques (G. Westerhout), the critical evaluation of data (Miss A. Underhill), the distribution of data (B. Hauck), and exerting facilities (C. Jaschek) ; the last session had included a general discussion on current problems and future development, and as a result several proposals would be put forward for consideration by Commission 5.

Organisation

G.A. Wilkins reported that he had had discussions about the future organisation of the activities of the Working Group and as a consequence he suggested : (a) that it be renamed the "Working Group on Astronomical Data" ; (b) that it have a small organising committee ; (c) that specific projects should be carried out by subgroups as required ; and (d) that information about the data activities of the Group and of the Commissions should be published in the Information Bulletin of the Stellar Data Center at Strasbourg. These suggestions were accepted by the Commission, which then went on to consider the resolutions and projects that had been discussed at the Strasbourg Colloquium.

Bibliographic catalogues

The following resolution on the transcription to machine-readable form of certain handwritten bibliographic catalogues was adopted unanimously :
    Commission 5 recognizes the great value of the handwritten bibliographic catalogues on the spectra of stars, compiled by W.P. Bidelman of the Warner and Swasey Observatory, and on eclipsing binaries maintained by F.B. Wood at the University of Florida, Gainesville, and recommends that these catalogues be transcribed into machine-readable form as soon as possible so that the information in them can be made available to astronomers throughout the world.

Use of Computers

It was agreed that the Chairman should set up a small sub-group on computer

technology and standards, to study new developments of interest to astronomers
and to recommend standards to facilitate the exchange of information in machine-
readable form.

Guide on Presentation

The Commission confirmed the general view at the Strasbourg Colloquium that
there is a need for a "Guide on the Presentation of Astronomical Data in the Primary Literature", on the lines of CODATA Bulletin No. 9 for experimental data ;
such a Guide could be incorporated in the IAU Style Book. G.A. Wilkins agreed to
continue the preparation of such a Guide in consultation with the members of the
Working Group on Editorial Policy and others ; it would be referred to Presidents
of Commissions for comment.

Designation of astronomical objects

There was a general discussion on the problems of the designation of astronomical objects ; at present, one object may have several designations, or none,
and one designation may refer to several different objects. In order to alleviate
the first problem, W.P. Bidelman suggested that there should be an order of preference for existing catalogues, and that the positions of objects not yet included in catalogues should be used for identification purposes. H.K. Eichhorn-von
Wurmb suggested that the positions should be expressed in galactic coordinates. It
was pointed out that it would be difficult to get agreement to an order of preference, since the appropriate catalogue number would depend on the field of interest,
and that the precision required for the position would depend on the type of object and the wavelength of observation. There was general agreement to the suggestion by C. Jaschek that ambiguity would be avoided if Editors would ensure that
all abbreviations to catalogues were clearly identified in each paper. It was also
agreed that the Chairman should set up a sub-group to study these problems, consult the Commissions concerned and prepare recommendations for consideration at
the next General Assembly.

Draft Organisation of the Working Group on Astronomical Data at September 1976

Chairman : G.A. Wilkins.
Vice-Chairman : B. Hauck.
Organising Committee : R.S. Dixon, C. Jaschek, Z. Zadla, (ex-officio) W.P.
Bidelman, M.S. Davis, J.-C. Pecker.
Subgroup 1 : Computer technology and standards. Chairman : M.S. Davis.
Members : F. Ochsenheim and two others.
Subgroup 2 : Designation of astronomical objects. Chairman : W.P. Bidelman.
Members : R.S. Dixon, A. Moffat, J.R. Shakeshaft, F. Spite.
Subgroup 3 : Presentation of astronomical data. Chairman : G.A. Wilkins.
Members : R.H. Garstang, C. Jaschek, J.R. Shakeshaft.

7. WORKING GROUP ON EDITORIAL POLICY

After an exchange of correspondence, it has been agreed that the new Chairman
of the Working Group will be J.R. Shakeshaft, from the day of this meeting, in
replacement of A. Maxwell who has asked to resign.
J.R. Shakeshaft reports as follows :
"The efforts of the Working Group during the past three years have been devoted
to the preparation of a revised version of the IAU Style Manual which, it is hoped,
will be adopted by many astronomical periodicals as their Style Manual. A draft
has been circulated to the Working Group and the comments received, together with
those forthcoming from members of Commission 5 at the General Assembly, will be
incorporated in a further draft to be circulated more widely. If possible, a version upon which there is a broad agreement among members of Commission 5, and

editors of astronomical journals will be produced before the next General Assembly.
     It should be emphasised that it is NOT the aim of the Working Group to elimi-
nate the individuality of the various journals as regards the essential matter of
content, but rather to promote a common policy with respect to the "mechanical"
matters of references, abbreviations, symbols and units, in order to save the time
of authors and editors, and to reduce ambiguities and uncertainties for the readers.
     Many of the changes from the earlier versions of the IAU Style Manual follow
from adoption of recommendations by the International Standards Organization.
Potentially the most contentious is the recommendation of the use of S.I. Units,
which are not the units being taught to science students in most schools and
universities, following adoption of the International System by the General Conference on Weights and Measures."
     Then the problem of refereeing was discussed : as a result of the instructions
given by the editors towards some severity, controversial papers are often rejected,
although their stimulating role should not be neglected. It has been suggested that
editors remind the referees of the limitations of their role.
     A short discussion on the cost of IAU publications led the Commission to adopt
a resolution which was voted by the General Assembly specifically because of its
wide importance (Resolution No. 2).

8. WORKING GROUP ON UDC 52

     Chairman : D.A. Kemp
     The Universal Decimal Classification, Class 52 (Astronomy) has been revised
in view of three different aims, that it could be used for arranging books on library shelves, as entries for catalogues and bibliographies and in conjunction with
computational retrieval systems. In its revised form UDC 52 is a so-called "facettes"
classification based primarily on astronomical bodies. It is described in Document
PC 75-7, 1975 issued by the "Fédération Internationale de Documentation".
     Moreover a preliminary edition of the "Handbook on the use of UDC in Astronomy"
realized by D.A. Kemp, G.A. Wilkins and V. Bacau, will be available by the end of
1976 from the Royal Greenwich Observatory. A copy of this document will be sent to
all members of Commission 5. It contains primarily instructions and suggestions on
how the classification may be used. The Commission carried unanimously the following
resolution :
     The Commission 5,
     in view of the excellent and comprehensive work achieved by the Working Group
     appointed to undertake the revision of the Astronomical Class of Universal
     Decimal Classification (UDC), in cooperation with the International Federation
     of Documentation (FID),
     recommends the formal approval by the General Assembly of the IAU of its pro-
     posals concerning UDC 52, as described in the preliminary document PC 75-7 of
     FID, dated March 31st, 1975, and in other forthcoming documents,
     recommends the publication of this document, and its appropriate diffusion by
     the IAU, and
     recommends all publishers of primary and secondary journals, and all astrono-
     mical libraries, to adopt the UDC 52 as revised, to make it known and to put
     it in operation as much and as soon as feasible, as well as revisions of the
     IAU style book to be elaborated by Commission 5 and its Working Group on Edi-
     torial Policy.
     The future tasks of the Working Group will be the following. First, to prepare
an index of UDC 52 based on alphabetical order of key words. A list of concordance
between UDC 52 and the thesaurus of Astronomy and Astrophysics Abstracts should
also be settled in view of future computering work.
     A short colloquium on UDC 52, open to astronomers and librarians should be or-
ganized, may be in Edinburgh. It could be sponsored by the IAU.
     Another task for the Working Group should be the implementation and diffusion
of UDC 52. A letter should be sent broadly telling about the Handbook on UDC 52
and how to purchase it.

Class 51 (Mathematics) has been revised and a "Permuted index of terms" has just been published by the University of Grenoble Science Library. Class 53 (Physics) has not been revised ; it would be advisable for the Working Group to participate in the revision of Class 53 and to advice the best classification to use in astronomical libraries in the meanwhile.

9. BRIEF REPORTS

P.G. Kulikovskij has announced the publication of the first volume of the "Bibliography of books and papers published on the History of Astronomy" edited jointly by IAU Commission 41 and the Moscow Academy of Sciences. Advices and comments are hoped from Commission 5.

G. Feuillebois reports on the publication at the end of 1976 by Mansholt Publishing Co. of : "Catalogue des ouvrages d'astronomie des 15ème et 16ème siècles conservés dans les bibliothèques d'observatoires européens" by G. Grassi-Conti. Mrs Grassi Conti is now preparing the continuation of this work for publications of the 17th century.

M.D. Heintz reports that no progress has been made in the completion of the "New Directory of Astronomical Institutes in the World". The need for such a work is now greatly diminished since most Astronomical Institutes addresses may already been found in the "Bibliography of Non-Commercial Publications of Observatories and Astronomical Societies" published by the Utrecht Observatory. On the other hand C. Jaschek is ready to send to anybody who requests it a list, realized by the Strasbourg Stellar Data Center, of all Institutions where there is at least one IAU Member.

G. Feuillebois and D.A. Kemp will write a note to appear in the Strasbourg Information Bulletin on Information retrieval systems usable in Europe.

Joint Meeting of Commissions 5 and 41, 31 August 1976

Chairman : J.-C. Pecker

RARE BOOKS IN OBSERVATORY LIBRARIES
The following papers have been presented :
The Census of 16th Century Books in Observatory Libraries (G. Grassi Conti)
La Bibliothèque de l'Observatoire de Paris (G. Feuillebois)
The Uppsala Observatory Collection (N. Olander)
The Crawford Collection in Edinburgh (M. Smyth)
The San Fernando Observatorio Collection in Cadiz (Almorza)
The Census of Copernicus "De Revolutionibus" (O. Gingerich)
A. Mikhailov, absent from Grenoble, could not deliver his paper on The Pulkovo Observatory Collection.

COMMISSION 6: ASTRONOMICAL TELEGRAMS (TÉLÉGRAMMES ASTRONOMIQUES)

Report of Meeting, 30 August 1976

PRESIDENT: P. Simon.   SECRETARY: B. G. Marsden.

After welcoming those in attendance, the President announced the proposals of
E. Roemer for incoming President and J. Hers for incoming Vice-President, while F.
Biraud, C. U. Cesco, E. Everhart, J. Grindlay, K. A. Pounds and L. Rosino were pro-
posed as new members of the Commission. These proposals were unanimously accepted.

The President had previously submitted to the Finance Committee his recommendation that the Central Telegram Bureau continue to receive a subvention from the
IAU. (The Finance Committee accepted this recommendation, the amount of the subvention approved for the triennium 1977-79 being S Fr. 5595.-)

The Secretary then brought up the continuing problem of the proliferation of
material received for publication on the IAU Circulars. The resolution adopted by
the Commission in 1973 had helped to curb the amount of this material actually
published, but the problem was obviously a more wide-ranging one. Of course, from
the point of view of the Central Bureau, the problem could be resolved rather
easily by an even stricter application of the 1973 resolution. On the other hand,
the Bureau recognized that there is a need for the moderately rapid dissemination
of important new observations of phenomena that are not necessarily transient:
telescope time has to be scheduled several months in advance, and since it also
takes several months to get a paper published in even the most cooperative journals, a whole observing season can pass before readers can make supplementary observations. Several points were made during the ensuing discussion. It appears
that some of the conventional astronomical journals tend to discriminate against
papers that are entirely observational in nature and expect each unusual observation to be accompanied by some kind of interpretation. Commissions 40 and 48 were
also concerned with the problem, and while it was agreed that Commission 27's
Information Bulletin on Variable Stars provided a suitable outlet for many optical
astronomers, radio and x-ray observations were not being taken care of in this
way; and although Commissions 40 and 48 could consider instituting their own publications, such a move would be regressive when so much of modern astrophysical
research involves the coordination of optical, radio and x-ray data. What seemed
to be needed was a publication, separate from but perhaps somehow associated with
the IAU Circulars, where short observational papers would receive a bare minimum
of refereeing and editing, where they would be photo-offset from the original
typescript and airmailed to subscribers in batches within three or four weeks of
receipt by the editor. An ad hoc committee, consisting of Biraud, Grindlay,
Marsden (Chairman), Martynov and Rosino, was appointed to examine the situation
further and to make some appropriate recommendations.

F. Biraud enquired whether it would be possible to produce cometary ephemerides in which aberration had been allowed for completely. The Secretary replied
that this was really a matter for Commission 20 to consider, but that he was prepared to provide individual ephemerides, corrected also for parallax, on a limited
basis. The Secretary also agreed to implement P. Wehinger's suggestion that it
would be useful to include the Central Bureau's telex (and telephone) number on
each IAU Circular. Finally, the incoming President proposed a vote of thanks to
the outgoing President and the Secretary for their services to the Commission.

COMMISSION 7 : CELESTIAL MECHANICS
(MECANIQUE CELESTE)

Report of Meetings, 25 and 28 August 1976

PRESIDENT: P. J. Message   SECRETARY: A. H. Jupp

25 August 1976

BUSINESS SECTION

1. Membership

It was agreed that the following persons should become members of Commission 7:
V.K. Abalakin, K. Aksnes, G. Balmino, D.G. Bettis, O. Calame, J.R. Dormand,
R. Dvorak, H.K. Eichhorn, A. Fiala, C. Froeschlé, N.P. Grouschinsky, D.C. Heggie,
J. Henrard, V.K. Kholshevnikov, D.G. King-Hele, P. Lala, J.P. Lazovic, V. Matas,
T.B. Omarov, W.J. Robinson, H. Scholl, P.K. Seidelmann, A.T. Sinclair, J.W. Siry,
J. Yoshida.
It was agreed that the following persons should be invited to become
consulting members, for a period of three years: R.R. Allan, V.R. Bond, J. Roels,
C.J. Brookes, C. Marchal, J. Moser.

2. Election of Officers

President V. Szebehely, Vice President Y. Kozai, Organising Committee
E.P. Aksenov, V.A. Brumberg, M.S. Davis, A. Deprit, G.N. Duboshin, G.E.O. Giacaglia,
E.A. Grebenikov, M. Henon, G. Hori, P. Kustaanheimo, P.J. Message, B. Popovic,
J. Schubart, P.K. Seidelmann.

3. Working Group on Program and Data Banks

The Chairman, Dr. M.S. Davis, stated that the Working Group had not done much
since its inception, but the time now seems propitious to go forward with its
work for a number of reasons. The Commission was reminded of the purposes of the
Working Group: 1. To collect and disseminate information on program and data
relating to celestial mechanics. 2. To recommend standards for programs and data.
3. To serve as a clearing house for questions related to 1. and 2. above.
The Working Group in its earlier report (see the IAU Proceedings of the
XIVth General Assembly at Brighton, 1970, p.93) defined three kinds of data:
1. Observational data. 2. Numerical data derived from theories. 3. Analytical
data.
It was pointed out that observational data is now the domain of the Centre de
Données Stellaires (CDS) in Strasbourg, while the data described by 2. and 3. are
still of interest to members of the celestial mechanical community. In view of
this, Dr. Davis remarked that the name of the Working Group adequately describes
its functions and recommended no change of title.
Dr. Davis further remarked that he had recently spoken with Dr. Jaschek,
Director of CDS, who said that his Centre was willing to store all programs and
data fields submitted to it by Commission 7, as well as distributing directories,
programs and data files. He indicated that there are now 500 subscribers so that
information and files are assured of wide distribution in the astronomical
community.
The Working Group will issue a questionnaire to complete an existing,

rudimentary directory of programs and data. The completed directory will be available through CDS and, later, in a general directory including all the sciences involved in CODATA (Committee on Data in Science and Technology, sponsored by the International Council of Scientific Unions, of which the IAU is a member) which has already published one directory and is now completing a more up-to-date one.

For the next term of the Working Group, the following slate of members was recommended: V. Brumberg, J. Chapront, M. Davis (Chairman), A. Deprit, G. Janin, W. Jefferys, H. Kinoshita. Dr. Davis also recommended that the chairman of the Working Group have the authority to co-opt other members who might be useful to the efforts of the group. These recommendations were accepted without dissent.

4. The Three-Body Problem

It was agreed to set up an ad hoc committee to look at ways of presenting numerical results in the general gravitational problem of three bodies. (It is hoped that eventually a consensus can be reached in regard to a recommended style of presentation). Dr. Hadjidemetriou agreed to act as chairman of this committee and it was decided to ask Drs. Henon, Broucke and Grebenikov to serve as members. It was understood that the committee will present a report in due course.

REVIEW PAPERS

1. The General Gravitational Problem of Three Bodies (V. Szebehely)

Previous major reviews of the subject were prepared by Marcolongo, covering the period 1686 to 1919, by Whittaker, for 1868-1898 and by Leimanis whose partial review of some aspects of the problem emphasizing Soviet contributions is the most recent available.

The literature of the problem during the last three or so years has been enriched by major contributors whose partial list includes Aarseth, Aqekian, Aksenov, Anosova, Arenstorf, Batrakov, Benest, Bettis, Bhatnagar, Boggs, Bozis, Broucke, Brumberg, Christides, Contopoulos, Delie, Deprit, Dunham, Duboshin, Easton, Efimov, Elmabsout, Erdi, Feagin, Ferraz-Mello, Froeschle, Garfinkel, Giacaglia, Goudas, Grebenikov, Guillaume, Hadjidemetriou, Hagihara, Hamid, Harrington, Heggie, Hénon, Henrard, Herget, Hori, Janin, Jefferys, Kozai, Kunitsin, Marchal, Martinez, Markellos, McGehee, Merman, Message, Michalodimitrakis, Monaghan, Moser, Musen, Nacozy, Nahon, Pollard, Robinson, Roels, Saari, Scholl, Schubart, Sconzo, Shenal, Sharma, Siegel, Simo, Smale, Solovaya, Sperling, Standish, Stiefel, Subba-Rao, Szebehely, Tung, Vagner, Valtonen, Vidyakin, Waldvogel, Williams, Yoshida, Zare, and Zikides.

These papers make use of the essential classical results of Birkhoff, Chazy, Euler, Hill, Jacobi, Lagrange, Poincaré, Stromgren, Sundman, Whittaker and Wintner.

The three major areas of progress during the last three years may be found in, (i) partitioning the phase space, (ii) establishing new families of periodic orbits and, (iii) the studies related to triple collision and regularization. In the following these three areas are discussed in detail.

(i) Partitioning of the phase space. The principal problem in dynamics may be considered the proper and detailed partitioning of the phase space regarding the behaviour of the dynamical system. In other words, a qualitative (if not quantitative) description of the general behaviour of the system is desired for any given set of initial conditions. Several attempts to offer such partitionings based on establishing bifurcation sets, where the topological classification of the manifold of motion changes, appeared in the mathematical literature since 1970, but projections into the configuration space have been offered only recently. Existence-proofs and descriptions of regions in the configuration space where motion cannot occur are available today for the general problem of three bodies, quite similarly to the so-called Hill surfaces or surfaces of zero velocity that are well known for the restricted problem of three bodies. This breakthrough in the qualitative classification of possible motions of the general problem was

achieved by three different approaches, almost simultaneously. The parameters
controlling the surfaces separating forbidden and possible regions of motion are
the angular momentum (c) and the total energy (h) of the system of three bodies.
The dimensionless combination $(c^2 h)/G^2 m^5)$ plays the role of the Jacobian constant
and the bifurcation sets are established by evaluating the above parameter at the
libration points – again quite similarly to the restricted problem. Here G is the
constant of gravity and m the average mass of the system. An important result is
that for all values of h and of c the zero-velocity surfaces are open and,
therefore, escapes are always possible. This fact supports Birkhoff's original
conjecture according to which the set of points of the phase space corresponding
to non-escaping orbits has zero measure. Another important result is that the
stability of classical triple stellar systems and of planetary systems may be now
studied with the new approach of establishing forbidden regions of motions. For
instance, previous investigation of the stability of the Sun-Jupiter-Saturn system
by means of the restricted problem showed that Saturn may not enter the regions of
motions of the Sun and Jupiter, but the validity of this result is questionable
since the model of the restricted problem is not truly applicable. On the other
hand, when the same problem is investigated using the model of the general
problem, the result becomes dynamically meaningful. In fact it can be established
that Saturn may escape the system but the Jupiter-Sun binary system is permanent.
The way Birkhoff's conjecture was modified by Szebehely does not allow the escape
of Saturn since according to the modified conjecture a triple close approach is
necessary for escape. Such a triple close approach is not allowed by the zero-
velocity surfaces, therefore, the stability of this simplified model of the solar
system seems to be assured, in a way quite similar to Hill's result concerning the
stability of the moon's orbit. The stability of classical triple stellar systems
becomes a study of the possibility of exchanges. As long as the outer member of
the triplet does not interrupt the motion of the inner binary we speak of
stability – remembering that the escape of the outer member is always possible
since the surface is open. On the other hand, if the surfaces do not allow
"interplay" or "mixing" or "exchanges" then no close triple approaches can occur
and the outer member will not escape. The ratio of the pericenter distance of the
outer star $q_2 = a_2(1-e_2)$ to the semi-major axis of the inner binary a_1 is the
parameter controlling the stability of the system. With equal masses and for
circular orbits, numerical integration gives the condition for stability that
$q_2/a_1 \geq 3.5$ while from the study of forbidden regions we have $q_2/a_1 \geq 3.2$; indeed
a remarkable agreement considering the significant difference between the two
approaches. (Unfortunately the agreement breaks down for infrequently occurring
counter-rotational systems, and the zero-velocity surfaces give a much higher
q_2/a_1 value than the numerical integration).

        (ii) Periodic orbits. During the past three years, an impressive number of
families of periodic orbits were discovered. The development in this field seems
to be so rapid that it is difficult to account for all the results. It is hard to
resist to note that in 1953 an (erroneous) non-existence proof for other than
Lagrangian periodic orbits appeared in the literature. This was followed by the
discovery of single periodic orbits with collisions in a fixed system of
coordinates. The next step in the development was the establishment of pseudo
families of periodic orbits with variable masses, also in a fixed system. With
all three masses having fixed values, no families can be established in a fixed
system, therefore, the discovery of true families of periodic orbits made use of
rotating systems – quite similarly to the restricted problem of three bodies. The
existence of such families has been shown by analytic continuation from the
restricted problem and the actual computation of these families is presently in
progress. The difficulty of organizing the results of numerical integrations is
considerable as may be shown in the following way. We may start from one member
of one family of the restricted problem for a given value of the mass parameter μ.
(There exist at least 20 well established families of two-dimensional symmetric
periodic orbits for a given value of μ and the asymmetric, multiple periodic and
three-dimensional families have only been sampled up to date). This periodic orbit

of the restricted problem may be continued (unless its period is $2k\pi$) by increasing
the mass of the third body from $m_3 = 0$ to $m_3 > 0$ and a group of periodic orbits of
the general problem may be obtained with varying values of m_3. We may now take one
member of this group with, say $m_3 = a$, and using this periodic orbit of the general
problem we may establish a family of periodic orbits, all members having the same
masses, that is m_1, m_2 and $m_3 = a$. In this way a true family of the general
problem is generated. This process now may be repeated for all members of the
group mentioned above. But every group comes from one periodic orbit of the
restricted problem whose orbit is a member of a certain given family. The families
change with μ and the number of all possible families is not established. Therefore, the number of periodic orbits of the general problem that may be established
in this way is at least ∞^6. A few conclusions in the form of conjectures rather
than in the form of well established theorems may be stated: (a) All families are
composed of relative (to the rotating frame) one parameter periodic orbits.
(b) These orbits may be perturbed resulting in quasi-periodic orbits corresponding
to the class of orbits known as interplays. (c) Some of the families terminate
in classical triple systems with the third star at infinity. (d) The stable
behavior of some of the established periodic orbits indicates that other than
classical triple systems should exist. The stability of some of these orbits
excludes the possibility of escapes, contradicting Birkhoff's conjecture. But since
no stable families with triple close approaches have been established yet,
Szebehely's conjecture seems to be valid. It must be remarked that at present
there are only a few scattered results concerning the set of ∞^6 periodic orbits.
Not even all classes of Stromgren's orbits of the restricted problem have been
continued yet, not much is known about the effect of μ, only two-dimensional
symmetric orbits have been studied in detail, termination principles have not been
established and very few families are known that were not generated by continuation.

(iii) <u>Triple collision and regularization</u>. When the total energy of the system
of three bodies is negative, escape is proceeded by a triple close approach and it
is followed by the formation of a binary. The escape velocity has been shown to be
arbitrarily large and the associated binary to be arbitrarily close, (i.e., with
an arbitrary small semi-major axis) if the triple approach is arbitrarily close.
The fact that triple collisions in general cannot be regularized, points out the
difficulty encountered in the numerical integration of triple close approaches. The
numerical and analytical problems of triple close approaches on one hand and their
importance in producing escapes and in forming binaries on the other hand, may be
considered another case when nature successfully hides its true face. Significant
analytical results regarding the one-dimensional problem and semi-analytical
techniques for the one-dimensional case have been accomplished recently, so that
escape velocities and directions may now be predicted. Binary collisions have been
regularized for about 75 years and some of these methods are presently being applied
to the regularization of the three-dimensional general problem of three bodies. Local
regularizations (in the sense of Birkhoff) are available with regularization of the
two smallest distances. The methods may require the relabeling of the particles as
the motion proceeds. Global regularizations (i.e., elimination of the singularities
for all possible binary collisions) also became available recently.

2. The Motion of Artificial Satellites (Y. Kozai)

In the first few years after the Sputnik various theories on motion of artificial
satellites were published by several authors, most of them treating the main problem
and/or effects due to the air drag on the orbits. In most of the papers for the
main problem, solutions with the accuracy of ten to the minus six during about one
thousand revolutions were derived by using the disturbing functions containing
short-periodic terms of the first order and secular and long-periodic terms up to
the second order. Luni-solar gravitational, solar radiation pressure and other
zonal, as well as non-zonal gravitational perturbations, were derived with the same
accuracy in several papers during the same period.

Since then the tracking accuracy for satellites has been increased by

introducing radio doppler and laser ranging techniques, and it has been claimed that
the ranging accuracy by laser can be as high as a few centimeters, corresponding
to ten to the minus nine for close satellites, and the position accuracy by doppler
tracking is a few decimeters. Therefore, in order to extract full information from
tracking data, theories on artificial satellite motions with ten to the minus nine
accuracy are needed.

Kozai (1962) published a second-order theory of the main problem, in which
perturbations referred to Keplerian ellipses in Delaunay variables were derived by
use of Hamiltonians including short-periodic terms up to the second order and
secular and long-periodic terms up to the third order. Since von Zeipel
transformations were applied to eliminate two angular variables one by one from the
Hamiltonians the old sets of the angular variables and the transformed sets of the
action variables appear in the solutions, and, therefore, the solutions were
transformed so that the old variables are expressed as functions of the transformed
variables. That is, the osculating elements are expressed as functions of slowly
varying elements including secular and long-periodic terms (primed variables), and
the long-periodic terms are expressed as functions of mean elements changing
secularly (doubly primed variables). In this way, expressions for the short-periodic
terms became much simpler, although those for the long-periodic terms could not be
simplified.

The expressions can be written in closed form with respect to the inclination
and the eccentricity. Later P. Sconzo checked Kozai's expressions by an IBM computer
with FORMAC language. M. Gaposchkin and his colleagues reformulated them by their
computer algebraic program called SPASM.

K. Aksnes (1970) published another second-order theory by use of Hill variables.
He adopted as the intermediary orbit a rotating ellipse which is the exact solution
of the equation of motion with the force function $(GM/r)\left[1-J_2P_2/(pr)\right]$. Therefore,
his disturbing function is $(GMJ_2P_2/r^2)\left[(1/r)-(1/p)\right]$. He applied Lie-Hori transformations to the Hamiltonian to eliminate periodic terms. Since the Lie-Hori
transformation is not an implicit one, it is not necessary to make any inversion to
derive the final solution. Further, since he used Hill variables, he could derive
the perturbations directly in coordinate and velocity and could thus avoid the
singularity at zero eccentricity. It should be mentioned that his expressions are
very compact compared with those by Kozai. His solutions were compared with those
by numerical integrations, and it was found that the discrepancy is within 30m for
the first-order theory and 30cm for the second-order theory after 9,000 minutes of
time.

A. Deprit and A. Rom (1970) published their third-order theory of the main
problem. They developed their theory with the computer algebraic program MAO by
applying Lie transformations to the Hamiltonian. Since they could not find a way
to express the solution in a closed form of the eccentricity, the solutions had to
be developed into power series of the eccentricity up to the sixteenth power. They
used modified Delaunay variables, F=ℓ+g,h, C=ecosg, S=esing, L and H to avoid the
singularity at zero eccentricity. They also expanded the solutions into power
series of $(1-5H''^2/L''^2)$ which appears as a divisor for most of the other theories.
Their solutions were compared with those by numerical integrations for satellites
ANNA 1B and Relay II. For ANNA 1B the deviations are within 20cm along track, and
5cm across track after 200 days; for Relay II they are within 2.4m along track, and
10cm across track after 350 days. Kinoshita and Kutuzov checked their solutions with
their computer programs.

H. Kinoshita (1976) developed his third-order theory with the computer algebraic
program SPASM by including J_3 and J_4 terms in the geopotential. His expressions are
expressed in the modified Delaunay variables, and there d'Alembert characteristics
hold as he applied Hori's transformations to the Hamiltonian. His solutions were also
compared with those by numerical integrations for three satellites, Geos 3, Starlette
and Lageos. The deviations in the short interval of time, corresponding to two
revolutions, are extremely small; namely, 0.2mm for Starlette. However, the discrepancy becomes 1cm after 40 days in track component. Therefore, it can be concluded
that the third-order theory is enough for short-periodic perturbation computations

even if the tracking accuracy is as high as 1cm. Without the third-order terms the
deviations are as large as 10 to 20 cm.

For deriving long-periodic perturbations some difficulties for small eccentricity
may arise by most of the methods, as X. Berger (1975) pointed out. Berger solved the
Lagrange equations by a computer algebraic program, and found that as he proceeded
to higher-order perturbations higher powers of the eccentricity appeared as divisors
in his expressions of long-periodic perturbations. Therefore, the convergency of
the expressions is very slow for the small eccentricity case. The same difficulty
arises if von Zeipel's transformation is applied to the transformed Hamiltonian to
eliminate g'. However, if the equation with the transformed Hamiltonian is solved
directly, under the assumption that the eccentricity is so small that the cubic
power can be neglected in the Hamiltonian, then the exact solution can be derived as,

$$e \cos g' = (1-A) \, e_0 \cos g'',$$
$$e \sin g' = (1+A) \, e_0 \sin g'' + B,$$

where e_0 is an integration constant, A is a constant with J_2 as a factor and B is
another constant with J_3 as a factor. We can use these expressions for an extremely
small eccentricity case. If e and g-g" are expressed as functions of g or g" by
solving the expressions, the convergency is found to be very slow when e_0 and B are
of the same order of magnitude. This difficulty has been avoided in Kinoshita's
solution, as he applied Hori's transformation and d'Alembert characteristics hold in
any order.

As the tracking accuracy has been increased, very small perturbations which had
been neglected in earlier days should be taken into account. Namely, effects due
to the precession and nutation motions of the earth's equatorial plane, the earth
tides, ocean tides and so on. These effects were also formulated by several authors.
Generally speaking, the perturbations due to the solid earth are one tenth of the
corresponding direct luni-solar perturbations for close satellites, and ocean tidal
effects are roughly one tenth of the solid tidal effects.

The accuracy to compute the luni-solar perturbations has been increased by
several authors, and the solar radiation pressure perturbations have been treated
more carefully than before by including the earth albedo perturbations. However,
it is still very difficult to compute the radiation pressure perturbations with
sufficient accuracy. Therefore, now satellites with very small values of area-to-
mass ratio, such as Starlette and Lageos, and a surface force free satellite, Triad,
are in orbit to make easier accurate orbit computations.

However, the most difficult problem still has not been solved. Nobody has ever
tried to solve the equations of motion analytically, including all the forces
together.

References
Aksnes, K.: 1970, Astron. J. 75, 1066.
Berger, X.: 1975, in G.E.O. Giacaglia (ed.), Satellite Dynamics, Springer Verlag,
            Berlin, Heidelberg, New York, p.111.
Deprit, A. and Rom, A.: 1970, Celes. Mech. 2, 166.
Kinoshita, H.: 1976, Celes. Mech. (in press).
Kozai, Y.: 1962, Astron. J. 67, 446.

3. Impact of lunar laser ranging on Celestial Mechanics (J.D. Mulholland)

28 August 1976

Members were asked to sign letters of best wishes which Dr. Message was
sending to former presidents of Commission 7 Professor Hagihara and Professor
Duboshin, who were unable to be at the General Assembly.

REVIEW PAPERS

1. Planetary Theory Developments (P.K. Seidelmann)

   The developments in planetary theories during the period 1973 to 1976 are
reviewed. The emphasis in numerical integrations has been upon the methods of
obtaining the specified accuracy through automatic adjustments of the step-size
and determinations of the order required. Thus, for current numerical integrations
the problem and the accuracy desired are specified, where as previously, the
problem, step-size and the order had to be specified.
   A number of approaches are being pursued to develop computer methods for
generating general theories for the planets. At the Bureau des Longitudes,
general planetary theories in terms of the osculating elements are being developed
in both algebraic and numerical expressions and by both ordered and iterative
approaches. At the Institute of Theoretical Astronomy, a mathematical construction
of the series for eight major planets in accordance with the series of Krasinsky
is completed. Brumberg continues to investigate iterative methods. At the Naval
Observatory, a combination of iterative and ordered approaches is being used to
generate numerical general theories, particularly for the inner planets.
   The emphasis in theoretical developments concerning the motion of minor
planets has centered on motions near commensurabilities, stability of planetary
systems, and the distributions of the minor planets near the Kirkwood Gaps.

2. Evolution of Orbits in the Solar System due to Tidal Friction and other
   Long-Term Effects (S.F. Dermott)

   Direct calculation of the long-term evolution of orbits in the solar system
due to either point-mass gravitational interactions or dissipative effects is not,
at present, possible and we are forced to consider indirect approaches to the
problem. A useful approach is to analyse the gross dynamical structure and to seek
out those features which are characteristic of a dynamically evolved system. A
simple, but highly successful, example of this approach involves the rotational
periods of the various bodies in the solar system. These bodies can be divided
into two groups: those for which the rotational period is about 8 hours and all
others. The deviants from the isochronism of spin rule must be suspected of
dynamical evolution and in all cases, except that of Mars which is anomalous in
this respect, there are various other features which support the hypothesis that
the spins have been braked by tidal interactions.
   The features of the dynamical structure of the solar system which one would
expect to yield information on its evolutionary history are: the distribution of
mean motions; the distribution of mass, particularly in the satellite systems; and
various properties of the 7 or 8 resonances (or orbit-orbit couplings as they are
sometimes called) which exist in the solar system - such as the probability of
capture into resonance, the conditions for stability, the present amplitudes of
libration and the present values of the inclinations and eccentricities.
   The distribution of mean motions is of particular importance. In 1954,
A.E. Roy and M.W. Ovenden showed that the number of occurrences of commensurabilities
between pairs of mean motions in the solar system as a whole is greater than would
be expected to occur by chance if the mean motions were randomly distributed.
Their analysis was modified by P. Goldreich and by myself, but the result still
stands; thus we must suppose either that for some reason the satellites and planets
were formed preferentially at the commensurabilities, perhaps, but not necessarily,
in a state of stable resonance, or that since the time of formation orbital
evolution has occurred.
   Recent observations of the terrestial planets have revealed that Mercury does
not keep one face permanently turned towards the sun but has a rotational period
which is exactly 2/3 of its orbital period. The spin of Venus was shown to be
retrograde and almost exactly an integral multiple of its synodic period. Spin-orbit
coupling, as the phenomenon is now called, is another example of resonance and the
dynamics has many features in common with that used to describe the orbit - orbit
resonances observed in the satellite systems. Roy and Ovenden's paper and obser-

vations awakened interest in the dynamical evolution of the solar system and much
recent research has been concerned with the origin and evolution of the observed
resonances.

In analysing the preference for near-commensurability amongst pairs of mean
motions, I consider it important to distinguish between firt-order resonance and
higher-order resonance. It is also desirable to treat separately those ratios
involving satellites and those ratios involving planets. The near-commensurate
ratios are not the only ratios that can yield information on the dynamical
evolution of the various subsystems in the solar system.

Let n'/n $(n>n')$ be a ratio of mean motions and let the definition of c, a
measure of the preference for near-commensurability, be given by $c = 2(a-b)$ with

$$a = \{n/n-p'/(p'+q')\}/\{p/(p+q)-p'/(p'+q')\}, \quad b = 0 \text{ if } a \leqslant \tfrac{1}{2}, \ 1 \text{ if } a > \tfrac{1}{2}$$

where $p/(p+q)$ and $p'/(p'+q')$ are the two fractions which bound n'/n from above and
below and $\max(q, q')$ is the maximum order of resonance that is being considered.
Then $-1 \leqslant c \leqslant +1$ and the preference for near-commensurability increases as c tends
to zero. If the distribution of n'/n is random then the distribution of c is
rectangular.

T. Gold and P. Goldreich suggested that the preference for exact resonance
in the satellite systems of Jupiter and Saturn is the result of orbital evolution
due to tidal friction. This is a good example of a process which will certainly
result in a characteristic distribution of ratios of mean motions. Tidal forces
are highly dependent on the distance of the satellite from the planet and thus we
must expect n'/n to increase with time (n is the mean motion of the inner
satellite). After appreciable orbital evolution we should observe not simply a
preference for $c = 0$ but also an absence of small positive values of c. This is
just the distribution of c that is observed - and only in the satellite systems.

In the planetary system there is only a single example of exact resonance
(that involving Neptune and Pluto whose mean motions are almost exactly in the
ratio 2 to 3), but there is a definite preference for near-commensurability amongst
the pairs of mean motions of the outer planets, that is, Jupiter, Saturn, Uranus,
Neptune and Pluto. This preference obviously cannot be accounted for by the tidal
hypothesis and we must suppose either that it is the result of orbital evolution or
that the outer planets were formed in near-commensurate orbits. The latter hypothesis
could be extended to include the satellites if we add that only in the latter systems
has orbital evolution since formation driven the bodies into stable resonances.

Two-body resonance is not the only form of orbit-orbit resonance possible.
Resonances of the type $An_1-(A+B)n_2+Bn_3 = 0$ also exist. I have pointed out that
this relation implies that ratios of relative mean motions are commensurate, and I
have proved that in the solar system there is a strong preference for near-commensurability amongst ratios of this kind. Again, this observation cannot be
accounted for by the tidal hypothesis.

The tidal hypothesis can be resolved into a number of problems all of which
demand a solution. We must show that tidal forces are capable of changing <u>ratios</u>
of mean motions. Changes of at least 7% are needed to account for the observed
distribution of c. We must show that capture into resonance is not only possible
but also probable and that the resonances are stable under the action of tidal
forces. If we allow, as we must, that our position in time is not special, then
the ages of the resonances, as deduced from the rates of change of the amplitudes
of libration (or some other parameter characteristic of the age of the resonance),
must not be too small compared with the age of the solar system. Finally, we must
identify an adequate source of energy dissipation.

The rate of change of the orbital radius a of a satellite of mass m is given by

$$\frac{da}{dt} = \frac{f(q)}{Q} \times \frac{m}{a} \, 11/2$$

where Q is the tidal dissipation function and $f(q)$ a function of parameters of the

planet only. As a first step, it is reasonable to take Q as proportional to the
pth power of the amplitudes of the tidal force, which itself is proportional to
m/a^3. The integration of the differential equation then gives the appropriate
result

$$\log a_o = \left(\frac{1-p}{13/2-3p}\right) \log m + \left(\frac{constant}{13/2-3p}\right)$$

in which a_o is the present orbital radius.

    For the Saturnian satellites, Mimas, Enceladus, Tethys, Dione, Rhea and Titan,
a plot of log a_o against log m is a good straight line of slope corresponding to
$p \sim -2$. This feature would appear to be strong evidence in favour of the tidal
hypothesis, but there are serious problems associated with the implied amplitude
dependence of Q. Using the stability condition of the Mimas-Tethys and Enceladus-
Dione resonances, which is $(n'<\dot{n}>)<(n<\dot{n}>)$, and a more general form of Q which
incorporates a factor equal to the sth power of the frequency, it can be deduced
that

$$-1/6 < p \leq 0,$$

$$-2/3 < s \leq 0.$$

This suggests that Q is amplitude and frequency independent. In which case neither
the observed mass distribution nor the formation of the Titan-Hyperion resonance
can be accounted for.

    It is tempting to speculate that p and s are not zero and that a correct
treatment of the non-linear tidal interactions would take care of all the problems.
I do not agree with that view. I consider that there is evidence in favour of the
tidal hypothesis but that Q is probably amplitude and frequency independent. I
list the following points:

1. the observed distribution of ratios of mean motions in the satellite systems
   suggests that these ratios (of n'/n) have increased with time;
2. the values of Q, as deduced from the present orbits of the inner satellites,
   are so very high ($\sim 10^5$) that it would be surprising, particularly if parts of
   the planets are solid, if tidal evolution had not occurred;
3. the large amplitude of libration or equivalently, the low age ($\sim 4.10^8$y) of the
   Mimas-Tethys resonance and the low age ($\sim 4.10^8$y) of the Enceladus-Dione
   resonance are best accounted for by the tidal hypothesis with Q amplitude
   independent;
4. all three resonances in Saturn's satellite system have been shown to be stable
   under the action of tidal forces provided that Q is amplitude and frequency
   independent.

    It needs to be pointed out that tidal forces are not always efficient at
changing ratios of mean motions. As the orbital radii of a pair of satellites,
which are not in resonance, separately increase with time, the ratio of their mean
motions tends to a constant value. Further, da/dt itself decreases rapidly as a
increases and thus, for orbital evolution due to tidal friction, we should expect
the commensurabilities to be encountered in the early stages of evolution and for
the resonances to be nearly as old as the solar system. The only exceptions should
be those cases in which the ratio a'/a is consistent with the aforementioned
constant value relation, as the commensurability will then be approached comparatively slowly. Once in resonance, the rate of change of the amplitude of libration,
will also be correspondingly slow. The Mimas-Tethys and Enceladus-Dione resonances
are young. The amplitude of libration of the Mimas-Tethys resonance is still large
and the eccentricity of Enceladus (the feature which characterizes the age of the
Enceladus-Dione resonance) is still small. Both of these pairs almost satisfy the
previously mentioned constant value relation, with p = 0. I consider this to be
strong observational evidence in favour of the tidal hypothesis with Q amplitude
independent.

The fact that $m/a^{13/2}$ is approximately the same for Mimas, Enceladus, Tethys and Dione or, equivalently, that a plot of log a against log m for these satellites can be roughly fitted by a straight line with slope corresponding to p = 0 is only weak evidence in favour of the tidal hypothesis. For if Mimas has always had an orbital radius greater than that of Janus then the initial values of $m/a^{13/2}$ could not have been very much different from the present values.

The important work of A.T. Sinclair, R.J. Greenberg and others on capture into resonance has shown that the observed two-body resonances in Saturn's satellite system could be the result of orbital evolution. If $<\dot{n}>$ is finite and certain stability conditions are satisfied, then capture into resonance is not only possible but also probable. The calculated probabilities of capture do not, however, depend on the evolutionary mechanism and so, although the results support the hypothesis that the resonance are the result of orbital evolution, they do not support the tidal hypothesis in particular. All three resonances in the Saturnian system, however, have been shown to be stable under the action of tidal forces and this in itself is evidence in favour of the tidal hypothesis.

COMMISSION 8: POSITIONAL ASTRONOMY
(ASTRONOMIE DE POSITION)

Report of Meetings, 25, 28 August, 1 September 1976

PRESIDENT: G. van Herk   SECRETARIES: B. L. Klock, H. J. Fogh Olsen, R. H. Tucker

Business Matters

I. REPORT
The President's Report was approved without discussion. Copies will be distributed to the Members as soon as they become available.

II. OFFICERS AND NEW MEMBERS
President: Tucker, Vice-President: Høg, Organizing Committee: Anguita, Débarbat, Fricke, Klock, Marcus, Nemiro, Schombert, Tavastsherna, Teleki, Van Herk, Yasuda;
New Members: Bem, Chernega, Chollet, Dravkikh, Gay, Grudler, Gubanov, Gulyaev, Johnston, Kharin, Lederle, Manrique, Raimond.

III. BY-LAWS
The proposed By-Laws, written by Tucker, were translated into French. These were extended following a suggestion by Tucker, adopted and translated back into English. They will be circulated to all Members.

Reviews on current Problems

I. STATUS REPORTS
a. W. Fricke dealt with the situation of the work on the FK5 (July 1976). There will be an improvement of the individual positions and p.m. and in the system of the FK4 and an extension to stars as faint as 9th magnitude. The team (Bien, du Mont, Gliese, Lederle, Strobel and Walter) at Heidelberg has collected information from all material available, where especially the Pulkovo- and U.S. Naval Observatories are thanked for their cooperation. Billaud's work on astrolabe results is mentioned. No new evidence of large systematic errors in the FK4 have become known since the report of 1973. A list of catalogues was circulated (and can be obtained on request) for the purpose of having its completeness checked. Further material is urgently requested. New catalogues with differential observations can still be incorporated if they are received before June 1978; absolute observations should not come in later than December 1977. The fainter stars are selected from a master-catalogue (AGK3R, SRS and PZT programs combined). Standards are set up for the older observations to fulfill. Six or more catalogue positions are available for 2297 stars north of -5°, and three or more for 2411 stars south of -5°. It is not yet sure whether all these older observations are free from magnitude errors. The inclusion of catalogues prior to 1900 cannot be considered (inadequate observational technique, magnitude equations, unknown polar motion). Dieckvoss asks about the date of the Equinox. This will be answered in the Joint Session with Commissions 4, 19, 31 and 40. Kharin: Will the 1977 observations of the Sun and Planets be included? Reply: No, they will come too late.
b. J.L. Schombert reported on the SRS program. The final results have been reported by Abbadia, Bordeaux, Nikolaiev, Perth (Hamburg), San Fernando, Tokyo, San Juan and Pulkovo (α). By the end of 1976 the following observatories expect to finish their reductions: Bucharest, Cape (Herstmonceux), Pulkovo (δ) and Washington. The SRS will be reduced to an improved FK4 system. The $\Delta\alpha_\delta \cos\delta$ differences with the

FK4, obtained at Perth, El Leoncito and Santiago-Pulkovo show, from +30° to -90°
a very satisfactory resemblance. Fricke, Gliese and Høg discussed the system to be
used in the reduction.

c. H. Yasuda reviewed the NPZT program. The final results from Abbadia and
Bordeaux have been received (the last ones having an accuracy of $\pm 0^s0086$ (reduced
to the Equator) and $\pm 0\rlap{.}''203$); the expected data of delivery from the other observatories range from 1977 till 1981. A list was given, showing the distribution of NPZT
stars with some observational history for every hour in RA and 5 declination zones.

d. Y.S. Yatskiv gave an account of the status of Astrometry in the USSR. He
presented papers from: Polozhentsev, Meridian Astronomy in the USSR; Lengauer, Corrections for Chromatic Refraction when employing photoelectric and other methods
for determining Star Positions; Nefed'eva, The new Tables of Astronomical Refraction; Polozhentsev, Stellar data and Computing Facilities at the Pulkovo Observatory; Orelskaya, On the present State of observations of Asteroids selected for improving the Star Catalogue Positions; Nemiro, Piljugina, Tavastsherna and Shishkina,
Catalogue of Absolute Right Ascensions of 1023 Bright and Faint Stars (Pu-58);
Ivanov, Kirjan and Kirjan, The Two-coordinate Photoelectric Micrometer.

New instruments, reduction techniques and methods of observation were described. Six observatories continue the compilation of catalogues.

	all abs. cat.	N x10³
AOE, Kazan MC	6 2	40
GAISH, Moscow MC (2)	18 1	92
Golosseevo VC	5 1	62
Kharkov MC	8 -	44
Kiev MC	7 -	49
Nikolaev MC VC PI	13 5	191
Odessa MC	16 -	73
Pulkovo MC VC PI HMC A PVC	33 8	485
Tashkent MC	13 2	95

The total number of catalogues produced
in the last 35 years is given in the
table. The mean errors for the best ob-
servations are now $\epsilon_\alpha \cos\delta = \pm 0^s011$ and
$\epsilon_\delta = \pm 0\rlap{.}''25$.
Gliese: Out of the 144 catalogues I have
received, 96 came from the USSR. Fricke:
Will the method to observe declinations
at a station near the Earth's equator
(Murry and Kreinin) be executed?-It will.
Van Herk asked about the work done on
Spitsbergen. Yatskiv: Preliminary results

are already obtained. The maximum period of continuous observing has been 52 hours
by three persons. Anguita: Is Pulkovo's PVC working now? Yatskiv: The instrument is
now under investigation in the workshop.

II. INSTRUMENTATION AND RESULTS

a. E. Høg discussed the instrumentation, the working and the accuracies of
his slit micrometer, used by the Hamburg expedition to Perth. The results are most
promising. Klock: How accurate has Mars been observed? Høg: A mean error of $\pm 0\rlap{.}''20$
was obtained. Dieckvoss: How faint will you eventually come? Høg: With the use of
a tracker instead of a fixed slitsystem, one could probably go 2-3 magnitudes
fainter, provided the proper accompanying equipment is involved.

Høg offered a copy of Lindegren's summary on the work done on planets with a
multislit micrometer. The displacement of the photocentre relative to the geometrical disk centre is found from formulae and is largely confirmed by the Hamburg-
Perth observations. Some displacements (Venus to Jupiter) have to be determined ex-
perimentally at different phase angles. Accidental mean errors of one observation
of $\pm 0\rlap{.}''18$ ($\Delta\alpha\cos\delta$) and $\pm 0\rlap{.}''30$ ($\Delta\delta$) are reported.

b. H.J. Fogh Olsen showed that the $\Delta\delta_\delta$ between the recently finished MC catalogue at Brorfelde and the AGK3 has an unexplained jump of the order of $0\rlap{.}''3$ at $\delta =$
40°. These $\Delta\delta_\delta$ depend on the AGK2 as well as p.m. have been applied. The only physical explanation seems to be the change of measuring machines at 40° for the AGK2
plates, a possibility strongly denied by Dieckvoss. Fricke: Use the latest AGK3 edition. Lacroute: Compare your work with the Strasbourg edition. Anguita: How many
fields of optical counterparts of radio sources, and how many stars per field do
you observe? Fogh Olsen: 41 fields are observed with an average of 15 stars.

c. E. Høg discussed the differences Perth-FK4 for 1200 stars. No differences
depending on clamp position were found. There is a good agreement with the results
obtained at Santiago. Gliese: How absolute is the azimuth determination? Høg: The

azimuth is relative to FK4 stars on an arc of 140° of the meridian.

    d. B.L. Klock reports on the status of the Northern TC Division at Washington. The 6" has been much changed since 1973: a new objective, a new glass circle, a circle scanner, inductosyns rebuilt (read-out 0.''05, pointing within 1"). The ATC has been temporarily equipped with a visual micrometer and has been moved to Arizona.

    e. I. Pakvor read Mitič and Pakvor's paper on the use of vacuum tubes to observe meridian marks. The vacuum is at the moment (reply to Jackson) a few tens of mm, the m.e. of one reading is 0^{s}005. The required tolerances for the stability of glass sealings and the tube itself could be fulfilled.

    f. Pakvor read S. Sadžakov's paper on the differences in Dec for stars in common to the AGK3 and the General Catalogue Latitude Stars. Systematic errors in AGK3 are small.

    g.F. Chollet (i.l. of Débarbat) explained the set up of the filter needed to bring the excessive amount of sunlight down to that of the full moon to observe the Sun with the astrolabe. The observations are made now E and W of the meridian with satisfactory results; the necessary constants are taken from the night observations.

    h. C.A. Murray discussed the need for a continuation of classical MC work even with all the rumours of space astrometry around. There will remain a great need to compare results of old and new techniques. MC observations will also be badly needed to help find reliable positions of radio objects. In this context the next fitted in:

    i. K.C. Blackwell gave the results of an investigation of the movement of the Equinox. The Greenwich solar observations and Morrison's discussion of lunar occultations seem to indicate a movement which can be expressed as a quadratic function of time. To settle this matter: continuous observations of the members of the solar system, including the Sun, are urgently requested. Klock replies to Murray that the Washington observations of Sun and planets will be available in about 18 months. Walter replies to Débarbat that six fundamental stars are radio sources.

    j. At the request of Klock, the President commented on the need to make notes on which component of double stars were observed: p, f, n, s, br or fa. This habit seems to have gone lost through the fargoing automation.

III. WORKING GROUPS

    a. G. Teleki reported on the work done on Refraction. The new Pulkovo Refraction Tables (5th ed.), a collaboration of Soviet astronomers and the Working group, is now being prepared. An averaged atmosphere is taken for the whole globe with regional characteristics; asymmetry of the atmosphere is taken into account. Only data for cloudless nights are used. Chromatic refraction will be calculated from spectrophotometric gradients. B-V values are the best substitutes; astrometric catalogues should include these values. The chromatic characteristics for the detector must be given. A simplified version of the refraction corrections is given by Nefed'eva, based on the Soviet Standard Atmosphere Gost-73. Kolchinskij's use of the geodetic methods of determining terrestrial refractions has led to a value at z=90° (288.2K, 1013.25 mb at the surface) of 33' 5.''5, much lower than what theories claim. Harin and Fukaya are cited to what can be obtained from the methods to correct refraction from U and L culminations. The model of the northern hemisphere is from Sugawa and Kikuchi, based on data given for each 10th degree of latitude and longitude.Regional- as well as time variations are reported. The mean average asymmetry in NS direction is twice that in the EW direction.

    b. G. Billaud dealt with the Astrolabe Contributions to FK5. Results from different catalogues cannot be compared easily as they depend on three unknowns (chaining constants). A systematic sinusoidal effect in the $\Delta\alpha_\alpha$ and $\Delta\delta_\alpha$ could not be explained and was corrected for. The comparison of the coordinates from two catalogues gives equations of condition where one of the required unknowns has to be chosen, after which the differences can be expressed in the same scale. In RA a translation, in Dec a translation and a rotation can occur when the solved unknowns show small errors. From 18 catalogues 1139 $\Delta\alpha$ corrections to the FK4 (m.e. 4ms) were computed, and 943 $\Delta\delta$ corrections (m.e. 0.''07). In both hemispheres some important effects with respect to the FK4 (up to 10 ms and 0.''15) were found. Fricke complimented Billaud warmly for the excellent job he has done. Requieme was reluctant to correct for

effects from unknown sources. Eichhorn commented on the completely different outlook
of these systematic differences compared to what one is used to see from MC work
and asked Fricke how these results will find their way into FK5. Fricke: the results
will be used for individual corrections and to some extent for the improvement of
the FK4 system.

Resolutions

1. Commission 8 approves the new program (proposed by the Institute of Theoretical Astronomy in Leningrad) concerning observations of the selected minor planets
for solving the problems of fundamental astrometry and asks the observatories having
appropriate equipment to take part in the observations according for this program.
The results should be presented in such a manner that later improved reference star
positions can be taken into account.

2. Commission 8 approves the initiative of the Kiev University Observatory
which has begun compilation of the general catalogue of the Bright Stars (BS) and
asks the observatories which have taken part in the observations of these stars to
send their results to the Kiev Observatory. In addition, it is agreed that the Kiev
University Observatory shall make the data available to other establishments working
on the same program.

3. Commission 8 supports the plans of the Hamburg-Bergedorf Observatory for a
fourfold coverage of the northern hemisphere on overlapping plates.

4. Commissions 8 and 24 endorse the collaboration between Denmark and the United
Kingdom for operating the automated Carlsberg Meridian Circle on a good observing
site and support the proposed observing program for this and other automated meridian circles for solving the problems of fundamental astrometry and providing positions of reference stars and of stars of astrophysical interest.

5. Commissions 8 and 24 together, noting the great progress which may become
possible by the use of space astrometry in defining the reference system, in trigo-
nometrical parallaxes, and in proper motions, support strongly the study and possibly the realization as quickly as possible of astrometric observations from space.
This must not affect the planning of ground-based programmes before the accuracy,
reliability and long-term continuity of space astrometry have been assured.

6. In accordance with Resolution No 5 of Commission 8 adopted at the XVth
General Assembly at Sydney, Commission 8 resolves that:

a. All SRS observations of each star for each night should be made available
by the observatories together with the final positions derived from them.

b. The observations of the FK4 stars made with the SRS observations should be
made available in the same way.

c. The SRS observations should be reduced to an improved FK4 system which is to
be derived jointly by the Pulkovo Observatory and the U.S. Naval Observatory in consultation with the Astronomisches Rechen-Institut on the basis of absolute and/or
semi-absolute observations near the epoch 1970. This system shall be called
SRS Preliminary System. The relation between the SRS Preliminary System and the
FK4 has to be specified explicitly.

d. A complete SRS catalogue should be preserved at the U.S. Naval Observatory
and at the Pulkovo Observatory. This should be made available on request. In addition to the mean values for each star, this catalogue should include all individual
observations with the weights and the systematic corrections which were used in
forming the mean values.

e. The AGK3R and SRS catalogues should be made available in the system of FK5.

f. In order to ensure that the AGK3R and SRS will provide an adequate reference
system in the future the observations of SRS must be repeated and it is strongly
recommended that the AGK3R be also reobserved.

7. Commission 8 recommends the application of Kreinin and Murri's method of
determining absolute declinations at a station situated near the Earth's equator.

8. Noting the great improvements of accuracy and efficiency of observations by
means of photographic and photo-electric techniques already obtained, and to be

expected, Commission 8 encourages observers to employ these new techniques.

9. For the refinement of astrometric measurements for the establishment of a highly accurate fundamental system Commission 8 agrees that the development and application of optical astrometric instruments (transit circles, astrolabes, et cetera), radio interferometers and space telescopes are necessary.

COMMISSION 9: INSTRUMENTS AND TECHNIQUES (INSTRUMENTS ET TECHNIQUES)
Report of Meetings August 25, 26, 27, 31, 1976

PRESIDENT: A. B. Meinel

The 8 scientific sessions of Commission 9 were organized on the basis of
Working Groups by the Chairman of each group. In addition, two Joint Discussion
sessions were held with Commission 12 and organized by R. G. Giovanelli. The
business session elected J. Ring as President and E. H. Richardson as Vice
President. The Commission voted to drop inactive members as determined from
results of a poll of all the membership. Fourteen new members were added. The
five Working Groups were continued, the new President to review chairmanship and
composition of these groups. Discussion of a new Working Group on Auxiliary
Instrumentation showed that the rapid advances in this field and the wide inter-
change of instrumentation on major world telescopes may make such a group desirable. The President and Comite d' Organization are to deal with this matter.
A draft resolution was prepared by the Working Group on Data Acquisition and
Processing and endorsed to the General Assembly: Whereas astronomers are rapidly
moving toward use of separate auxiliary on-line instrumentation on many telescopes
around the world, the question of interface standards is of major importance.
Commission 9 therefore urges adoption of international computer language standards
proposed by IEEE (HPIB language) and IEC (CAMAC language) for instrumentation
data handling interfacing. The Working Group also requests that it study whether
definition of a universal real-time language for on-line instrumentation is
possible and/or desirable.

WORKING GROUP ON LARGE TELESCOPES
The session of the Working Group on Large Telescopes was organized by J.
Rosch. I. M. Kopylov described the 6 meter telescope and showed early photographs
taken at prime focus with a Maksutov field corrector. Image diameters of 0".8 were
obtained. A limiting magnitude of 25 was reached at F/4 in 45m exposure on 103a-O
plates. G. Cayrel described the joint French-Canadian-U.S. 3.7m telescope being
erected on Mauna Kea, Hawaii. E. J. Wampler described performance of the 4m
Anglo-Australian telescope at Siding Spring Australia. A. B. Meinel presented
progress on the Multiple Mirror Telescope (MMT), a joint project of Smithsonian
Astrophysical Observatory and University of Arizona, showing optical test results
for the six 1.7m mirrors indicating 90% of the light within 0".5. C. Kuehne
described the Zeiss 0.75m alt-azimuth telescope and associated control system.
A supplemental session devoted exclusively to the Soviet 6m telescope was
arranged by J. Rosch at which the Chief Engineer for the project, B. K.
Joannisiani described the telescope, its construction details and performance. He
also described the problems of making, polishing and mounting the 42-ton
borosilicate (Pyrex) glass mirror. In spite of the short notice for this meeting,
it was well attended.

WORKING GROUP ON PHOTOGRAPHIC PROBLEMS
Mr. William C. Miller was elected Chairman of this new Working Group during a
meeting of Commission 9 in Sidney, 1973. Mr. Miller retired from the Hale
Observatories on July 1, 1975 and requested to be relieved from this office.
Richard M. West was appointed chairman of the Working Group. Secretary is J. L.
Heudier.
Two sessions of the Working Group were held during the General Assembly meet-
ings on August 25, 1976. They were attended by approximately 130 astronomers.
During the Working Group business session, the terms of reference for this new
Working Group were established by the adoption of the following resolution: "The
IAU Working Group on Photographic Problems, having met during the XVIth General
Assembly in Grenoble, and recognizing the continued importance of photographic
detection techniques in current astronomical research, considers as its principal

aim to support astronomers using this technique, in particular by: promoting a rapid and efficient dissemination of information about astronomical photography in the widest sense, e.g. sensitization, calibration, exposure, processing, storage, copying, and extraction of data from exposed plates; encouraging manufacturers to develop and market new, improved emulsions for use in astronomy, while assuring the continued production of other emulsions of astronomical merit, and; striving to facilitate the international delivery of astronomical plates and films. (It was tacitly assumed that the area of electronographic emulsions belongs to the activities of the Group). In the same session, the following Organizing Committee was elected by acclamation: R. Cannon (UK), O. Dokuchaeva (URSS), J. Grygar (Czechoslovakia), K. Henize (USA), J. L. Heudier (France), D. Latham (USA), A. Smith (USA), B. Takase (Japan), R. West (Denmark). As special consultants to the Working Group were elected: Wm. C. Miller (USA), A. Millikan (USA). Chairman for the period until 1979 is Richard M. West. The Secretary of the Working Group is Jean-Louis Heudier. The scientific sessions were divided into three major topics.

Emulsions and sensitization techniques
A. Millikan: "A review of photographic image detection theory as applied to astronomy." D. Latham: "The detective performance of Kodak spectroscopic plates." A. G. Smith: "Comparative hypersensitization of Kodak astronomical plates with nitrogen and hydrogen: sensitometry and sky tests." O. Zichova and J. Grygar: "Sensitization of plates for the Ondrejov 2m telescope."

Processing and storage of photographic plates
A. Millikan: "Optimal processing of large photographic plates." W. van Altena: "Envelopes and environmental conditions for the archival storage of exposed photographic plates."

Examples of practical work
H. C. Arp: "Photography of faint surface brightness regions in galaxies." J. L. Heudier: "The photographic laboratory of the INAG Schmidt telescope." A. G. D. Philip: "Classification of objective prism spectra on IIIa-J plates." I. R. King: "Automatic measurement of photographic plates."

A supplemental working session was held, at which time it was decided that a bibliography covering publications (after 1970) of interest to astronomers using photographic techniques should be compiled and prepared for distribution early 1977. It appeared desirable to convene a meeting of European astronomers in this field some time in 1977 and the chairman was asked to take the necessary steps towards such a meeting. The question of availability of electronographic emulsions was discussed in view of recent difficulties (some manufacturers having discontinued the production) and a report will be prepared and distributed.

In view of the large interest in astronomical photography, it has been decided that no formal membership of the Working Group will be established other than a steering committee. The circular letters and reports which are issued by the Working Group will, however, be sent to all astronomers and institutes who so wish , and a distribution list is maintained and updated by the secretary. Requests should be addressed to: Dr. Jean Louis HEUDIER, U.E.R. Observatoire de Nice, F-06300 NICE - Le Mont Gros, France, or the chairman: Richard M. WEST, ESO Sky Atlas Laboratory, c/o CERN, CH-1211 GENEVA 23, Switzerland.

WORKING GROUP ON AUTOMATION AND DATA REDUCTION
Two scientific sessions were organized by E. J. Wampler. Papers were pre-sented by P. B. Boyce, M. S. Ewing, B. M. Lasker and T. B. McCord. There was some general discussion of the possibilities and merits of standardization of computer languages and interfaces.

It is clear that since the last General Assembly, the use of mini-computers in observatories has become routine. Sixteen bit machines are most commonly used and the low cost of the central processors has permitted many users to build powerful systems. Experience has shown that to realize the full capability of the mini-computers, it is necessary to assemble systems with extensive peripherals such as disk-packs, tape drives, plotters, etc. In 1976 the cost of a minisystem, together with its peripherals, runs between 45 and 70 thousand American dollars. These machines are being used to automate telescopes and instruments, collect and reduce data, etc. We are moving towards a situation where complete automation of ground-based observation is feasible for many observatories. A meeting held at MIT in April 1975 discussed the situation current at that time.[1]

More recently, microprocessors have been introduced to fulfill special functions and in some cases to act as backup to a larger mini-computer. The chief limitation at present with microprocessors is that they are difficult to program. They are not a replacement for minicomputers. It was pointed out that the minimum minicomputer configurations are often a poor investment because of the difficulty of programming without such peripheral devices as a fast printer plus magnetic tape and disc.

Generally, the reliability of the electronic part of a computer system is very good, although the electromechanical parts are more troublesome and some early versions of microprocessor kits have had reliability problems.

The working group discussed the use of FORTH as a possible common language in astronomy. Although it is crash-prone and expensive, and there may be some problems of proprietary rights, those who have used it are generally enthusiastic about it. The group agreed that they would study "whether definition of a universal real-time executive programming language for on-line instrument interfacing is possible and/or desirable."

There was a discussion of the merit of a standard computer interfacing scheme such as CAMAC. CAMAC has now been adopted by some observatories, and should make it easier to move complex computer aided instruments from one telescope or observatory to another.

The WG recommends that Commission 9 and the IAU draw attention to the existence of international standards for equipment interfacing, and urges that observatories make use of these standards for new developments where appropriate.

WORKING GROUP ON INFRARED TECHNIQUES

A scientific session was organized by J. Ring. The first part of the session was devoted to a review of the new ground-based facilities for infrared Astronomy which were either in construction or planned. The requirements for an infrared telescope or flux-collector were outlined by J. Ring as follows: 1) The site must be as free as possible from absorption by water-vapour and from "sky-noise." These requirements seemed to imply a high-altitude site. 2) The collecting area should be as large as possible since for some types of observation in the infrared, the observing time was inversely proportional to the square of the area. 3) The structure should be thermally "clean." This was a problem with some attempts to convert existing optical telescopes. 4) Since daytime use was often possible in the infrared, good tracking and pointing accuracy was important.

He then briefly listed several categories of telescopes which were in the course of construction or modification, including old and new optical telescopes with infrared capability, new systems planned from the outset with a joint infrared/optical role, and specifically infrared telescopes or flux-collectors. He paid particular attention to the three new telescopes being constructed for Mauna Kea, the joint Canada/France/Hawaii telescope which was designed with infrared use in mind, the NASA 3m infrared telescope which was to have high image quality and the U.K. 3.8m infrared flux-collector which would have only 2-3 arcsecond images but which provided a large, "clean" collecting aperture very cost-effectively. He also mentioned the infrared alternative role of the M.M.T., whose optical performance had been described in an earlier session by A. B. Meinel. W. E. Brunk gave further details of the NASA telescope and L. Luud briefly outlined plans in the U.S.S.R. for infrared astronomy. Good progress was being made in the near I.R. and long-wave detectors were being investigated. It was likely that good infrared sites were available in the U.S.S.R.

The second part of the session was devoted to papers on new infrared spectrometric techniques including Michelson and Fabry-Perot interferometers, an image tube echelle spectrograph and a heterodyne spectrometer.

J. W. Brault described two Fourier transform spectrometers which had been built at K.P.N.O. and showed examples of visible and infrared solar spectra obtained with one of them. (A fuller description of these instruments was given in the joint discussion between Commissions 9 and 12).

J. E. Beckman described an interferometer constructed jointly, by groups at Meudon, E.S.A. Noordwijk and L.P.S.P. France. It was intended for high resolution (R $\sim$ 2.10^4) spectroscopy of interstellar emission lines in the middle infrared. This range, especially between 18μm and 100μm, requires the use of an airborne telescope, viz. NASA's 92cm Kuiper Airborne Infrared Observatory, and the instrument design is suited to its bent Cassegrain focus. The two-beam interferometer uses conventional optics, plane mirrors and dielectric beamsplitters. Its moving mirror employs a 15cm drive table travelling at speeds calculated to produce an audio output on the detector, centered at 80Hz. Speed, position and direction are monitored and used to control the continuous motion and take discrete data, via an HP5525 magnetically split auxiliary laser interferometer. Bandwidth is limited by helium-cooled 10% optical passband filters, in series with an electrical analog filter in the detector signal chain. The detector itself is a "3-parts" bolometer, of rugged construction , giving N.E.P. between 2 and 3x10^{-14} watt Hz$^{-\frac{1}{2}}$ under operating conditions.

The system operates on-line to an HP2100A computer, which controls the motion, accepts data, and yields spectra by fast Fourier transform after automatic phase correction. For astronomical work, spectra taken off source in times of order 1 minute are subtracted from relevant on-source spectra. This system has the advantage that for lines where the theoretical prediction is susceptible to errors of order 1%, the method can distinguish terrestrial from interstellar lines without ambiguity. In a recent observing run using the aircraft, lines of SIII at 18.71μm, and OIII at 88.35μm were detected from the Orion Nebula. The former was resolved, yielding a half-power velocity width of 28 km.sec^{-1}, in good agreement with radio recombination line data. In addition new standards of precision were set for atmospheric emission line frequencies within the 10% passbands, including those of the isotopic species of water HDO, and H$_2$^{18}O.

E. R. Wohlman described several interesting, high-resolution Fabry-Perot spectrometers which are in operation and under development in C. H. Townes' group at the University of California, Berkeley, for use in the 5μm, 10μm and 20μm windows. Observations have been made during the past several years with a coude-mounted tandem scanning Fabry-Perot spectrometer in series with a cooled 1% bandwidth circular variable filter. The system has a resolving power of $\lambda/\Delta\lambda \sim 10^4$ and has been used between 3μm and 13μm to observe molecular absorptions in cool stars and planets and fine-structure line emission from HII regions. The NeII 12.8μm emission from the galactic centre has been resolved both spectrally and spatially.

A single scanning Fabry-Perot interferometer in series with a cooled grating has been used in the 20μm window. The 18.7μm SIII line has been detected in emission from several sources. This spectrometer is also used at coude. A Cassegrain spectrometer designed for observations in the 10μm window is near completion. This system, also of resolving power $\sim$10^4, employs a single cooled Fabry-Perot interferometer in series with a cooled grating. Cooling the Fabry-Perot plates and mounting the system at the Cassegrain focus will result in an improvement in sensitivity of about an order of magnitude above that of the tandem Fabry-Perot.

The principal advantage of the Fabry-Perot spectrometer is that the narrow bandwidth reduces sky noise well below the level of random photon noise at all wavelengths in the infrared. In addition, the Fabry-Perot spectrometer can be used to obtain a spectrum of an arbitrarily narrow spectral interval. For this reason, the Fabry-Perot is a particularly sensitive instrument for the study of individual emission lines.

T. R. Gull described an echelle spectrograph which has been designed for use at Cassegrain focus of the four-meter telescope at K.P.N.O. The optical design of the spectrograph is optimized with consideration of the properties of the large

telescope to achieve peak performance for high-dispersion spectroscopy of trailed
stellar objects and untrailed nebular objects with seeing-limited angular resolution.

The echelle spectrograph is proving to be very successful with over 1300
echellograms logged as of mid-July, 1976, less than two years after the first
testing on the telescope. Examples of its performance are many, but include the
following studies: 1) H and K line profiles in supergiants. 2) Interstellar
diffuse bands. 3) Interstellar atomic and molecular lines. 4) First spectra
of the x-ray flare A0620-00. 5) Spectra of several comets, including Comet
D'Arrest. 6) Rotation and limb darkening of the outer planets. 7) Spectral
variations of planetary satellites such as Io and Triton. 8) Planetary nebulae
and their nuclei. 9) Multi-slit line mapping of planetary nebulae and HII
regions. 10) First high dispersion, seeing-limited multi-slit spectra of supernova remnants. 11) Nuclei of ordinary galaxies. 12) Nuclei of Seyfert galaxies, and 13) The brightest quasi-stellar objects.

The echelle is proving to be a most useful observing tool when the extra
dimension of spatial resolution is added. Its use, however, will be limited until routine data reduction techniques are developed by the supporting national
facilities.

A. Betz pointed out that when resolving powers greater than 10^5 are required
in the 10μm region of the infrared, heterodyne spectroscopy becomes an attractive
alternative to conventional high resolution techniques. For the study of the
shapes of individual line profiles of CO_2 in the atmospheres of Mars and Venus, a
heterodyne spectrometer has been constructed at the University of California,
Berkeley, with the following characteristics. The infrared signal received
through the telescope is combined with 1 milliwatt of the coherent output power
of a stabilized CO_2 laser serving as the local oscillator. The laser can
oscillate on any one of a number of selectable vibrational-rotational transitions
of CO_2 in the $00^01 - (10^00, 02^00)_I$ band. The combined beams are focussed on a
high speed mercury-cadmium telluride photodiode cooled to 77 Kelvin, which pro-
duces a difference frequency spectrum in the radio-frequency range of 0 to 1500
MHz. After amplification, a 200 MHz segment of this intermediate frequency
spectrum is converted into the band of 50 to 250 MHz by a single-sideband (SSB)
mixer and then directed into a multi-channel RF filter bank, similar to the types
used in microwave spectral line receivers. The oscillator frequency for the SSB
mixer is chosen to center the desired line profile in the filter bank, and it is
tracked in frequency to accommodate changes in source Doppler shift during the
course of the observation. The filter bank analyzes the 50 to 250 MHz spectrum
simultaneously into 40 independent channels of 5 MHz ($1.7 \times 10^{-4} cm^{-1}$) each. The
40 detected power outputs are synchronously demodulated at the telescope beam
switching frequency of 150 Hz and multiplexed into a computer for on-line
integration and analysis. After each integration cycle on the source, usually 4
minutes, a blackbody calibrator is set in the infrared signal beam. The sensitivity is measured to be 2.7×10^{-16} watts for a signal-to-noise ratio of 1 after
a 1-second integration in each 5 MHz channel. However, the detection of only one
source polarization and the loss of half the signal due to sky-chopping can be
viewed as effectively degrading this sensitivity by an additional factor of 4.
With an input infrared frequency of about 3×10^3Hz, a 5 MHz channel bandwidth
implies a resolving power of 6×10^6. Some of the more recent results obtained
with this instrument may be found in "Astrophys. J. (Lett.) 208, L141-148"; while
a more general introduction to heterodyne detection at infrared frequencies is
contained in the review of Blaney, which is published in "Space Science Reviews,
17, 691-702."

Summarising the discussion, Prof. Ring predicted very considerable advances
in ground-based infrared astronomy when the new telescopes were followed by the
powerful spectrometric tools which had been described and by the newer detectors
which were emerging rapidly at present. These systems would be particularly needed to follow-up the discovery of new types of source by infrared space observations such as those proposed with the L.S.T. and the NASA/Dutch Infrared

Astronomy Satellite survey.

WORKING GROUP ON PHOTOELECTRONIC IMAGE DEVICES

Owing to the occurrance of IAU Colloquium 40 on Photoelectronic Detectors with Linear Response in Meudon immediately following the General Assembly, the scientific session of the Working Group was limited to a series of seven invited review papers, covering the field of photoelectronic image detectors. These papers were designed to review the types of devices currently available and under development and typical use, and were directed toward the average user rather than the image tube specialists.

ELECTRONOGRAPHIC IMAGE DETECTORS, D. McMullan

The characteristics of the three classes of electronographic image tubes - the "classical" Camera Electronique, the "valve" type, and the "barrier membrane" type - were discussed. Ease of operation by the ordinary astronomer was identified as the most important requirement although this can only be obtained with some sacrifice in performance. Electronographic cameras in actual astronomical use were briefly described and the problem of commercial availability considered.

IMAGE INTENSIFIERS, W. A. Baum

The development of image intensifiers dates back about 40 years, and they have been the subject of various earlier reviews, including those by W. K. Ford (Annual Rev. Astron. Astroph. 6, 1, 1966), W. C. Livingston (Annual Rev. Astron. Astroph. 11, 137, 1973), and E. J. Wampler (Methods in Exp. Phys. 12, 237, 1974). A complete list of users would include astronomers at nearly all well-known observatories of the world.

Four astronomical applications of image intensifiers appear likely to continue or to expand: (1) intensified sky photography in the near-infrared, (2) intensified photographic spectroscopy where simplicity and low cost are important, (3) image photon-counting utilizing phosphor persistence in an image intensifier as a temporary memory, and (4) image photon counting behind an intensifier using signal integration external to it.

THE PRINCETON SEC CAMERA PROJECT, M. Schwarzschild

The main engineering effort of the project aims at present at putting a SEC tube with 5 cm x 5 cm cathode and target into operation. The present smaller SEC tube has been used recently for obtaining direct images at Mt. Palomar (Morton and Williams) and at Cerro Tololo (Williams and Schwarzschild). The photometric exploitation of these images (each effectively with 350 x 350 pixels of 50 x 50μm) is in progress.

SIT SYSTEMS, J. A. Westphal and J. Krishan

SIT tubes have been in use for several years, at a number of observatories, for direct imaging, two-dimensional photometry, sky-subtraction spectroscopy, and acquisition and guiding. Their advantages include linearity over a wide range, cheapness and ready availability, a full two-dimensional format, ruggedness, large dynamic range, and simplicity and reliability. Sufficient observing experience is now available that the tubes are reasonably well understood and used routinely. Some disadvantages are the necessity for cooling to reduce thermal noise and spectral response limited to 8500 Å by the 520 photocathode and 4000 Å by the fiber optics input (although UV transmitting tubes are now available on special order). A detailed report is available in the proceedings of the IAU Symposium on detectors (Meudon, Sept. 10-13, 1976).

DIODE ARRAY DETECTORS, Robert G. Tull

The silicon photodiode has high quantum efficiency (>50%) for visible and near-infrared radiation, with a useful spectral range from 0.39 to 1.1μm. Its response is linear when used in a charge storage mode. It is also an efficient detector of sub-atomic particles and has been used as the electron multiplier in experimental photomultiplier tubes. Integrated circuit linear and area arrays of silicon photodiodes, complete with integral shift-register readout circuits and operating in the charge-storage mode, became commercially available (Reticon Corp., Sunnyvale, CA, U.S.A.) about 5 years ago.

This paper will review existing astronomically useful systems which make use of silicon diode arrays, primarily self-scanned Reticon arrays, both as direct photon detectors and as intensified detectors. Silicon and SIT vidicons, CCD's, and CID's, are excluded from this review.

Linear arrays as direct photon detectors are used in spectroscopy at the Universities of Texas, British Columbia, Arizona, Toronto, and California (L.A.), and at Lund Observatory. Electron Bombarded Silicon (EBS) arrays exist at the University of California (San Diego) and University of Texas. C. Coleman and B. Morgan in Great Britain have developed EBS arrays used behind a Spectracon tube. Arrays optically coupled to image intensifier phosphors are in use at the Universities of Wisconsin and Michigan, Hale Observatories, and Kitt Peak. Magnetographs using Reticon arrays (not intensified) have succeeded at Lockheed Solar Observatory and Kitt Peak. Non-intensified area arrays have been used at Vancouver for photometry, and at University of Texas for telescope autoguiding.

CCD AND CID IMAGE SYSTEMS, W. C. Livingston

Compared to Reticon-type arrays the charge coupled devices (CCD) and charge injection devices (CID) are well adapted to large 2-dimensional formats and low-light-level operation. Noise sources are dark current (which is eliminated by cooling with LN^2), output amplifier noise (which depends on video capacity Cu and transister channel-resistance), and surface/bulk-state trapping noise.

The CCD has the smallest value of Cu and promises a threshold detectivity of a few electrons. Its main disadvantage is related to imperfect charge transfer efficiency which leads to a loss of MTF and various photometric problems. The CID advantages include random access and non-destructive readout. The latter allows the averaging of many reads to reduce the effects of amplifier noise. The disadvantage is a large value of Cu leading to relatively high amplifier noise.

A Fairchild CCD has been used successfully by D. Wilkinson and E. Loh for nebular surface photometry (BAAS 8, 350, 1976). B. Oke, at Palomar, has experimented with a Texas Instrument 100 x 160 CCD for stellar spectroscopy. B. Smith has used a TI 400 x 400 element array for planetary imaging and arrays up to 800 x 800 are planned for the Jupiter Orbiter Probe. A general purpose CID camera employing a GE 100 x 100 array has been developed by R. A. Kens at K.P.N.O. under the astronomical guidance of R. Lynds, J. Harvey, and M. Belton. Preliminary tests for narrow band imaging on the 4m telescope are encouraging.

MICROCHANNEL PLATES AND THEIR APPLICATION TO PHOTON-COUNTING IMAGE SYSTEMS, M. Lampton

The use of microchannel plate electron multipliers has led to the development of a variety of electrooptical image intensifiers and sensors. A microchannel plate (MCP), is a planar array of several million parallel cylindrical-channel electron multipliers. Such a device offers an electron gain of 10^4 or more, and can be cascaded for yet higher electron gains. MCP's are sensitive to low energy electrons, and can also sense UV, EUV, and x-ray radiation directly. Thus they are useful in sealed visible-light image tubes and as demountable sensors in space astronomy.

The Working Group on Seeing and Site Testing was disbanded.

REFERENCES
    1. Telescope Automation. Editors Huguenin and McCord. Copies obtainable from Remote Sensing Laboratory, Rm. 24-422, Cambridge, Mass. 02139.

COMMISSION 10: SOLAR ACTIVITY (ACTIVITÉ SOLAIRE)

ACTING PRESIDENT: G. Newkirk, Jr.   SECRETARY: A. Bruzek

Scientific Sessions

Stellar and Solar Structure (Jointly organized by Commissions 10, 12, and 35)
See Commission 35 Report.

How Can Flares Be Understood? Commission 10 organized a three-hour discussion
on selected problems associated with the occurrence and interpretation of solar
flares. A four-member panel, consisting of E. R. Priest, D. M. Rust, P. A. Sturrock, and H. Zirin had a principal role in the discussion, and the meeting was
chaired by Z. Svestka. A detailed report will be published in SOLAR PHYSICS. Specific contributions included:

PART 1: MAGNETIC CONFIGURATIONS AND INSTABILITIES IN FLARES

Z. Svestka: Introductory Talk; M. J. Martres: The Relation of
Flares to "Newly Emerging Flux" and "Evolving Magnetic Structures";
A. B. Severny: How Flares Can Be Understood; M. Pick: Relationship
between Type III and Microwave Radio Bursts and the Role of Magnetic
Configuration; M. R. Kundu: The Location and Size of Microwave
Bursts; S. I. Syrovatskii: Basic Questions in Our Understanding of
Flares; D. S. Spicer: The Thermal and Non-Thermal Flare: A Result
of Non-Linear Threshold Phenomena; J. Reyvaerts, E. R. Priest and
D. M. Rust: An Emerging Flux Model for Solar Flares.

PART 2: LOCATION OF THE PRIMARY FLARE SITE AND ENERGY TRANSFER IN FLARES

J. C. Brown: Introductory Talk; G. E. Brueckner: The Prime Energy
Release and Flare Development; J. A. Vorpahl: Comments Regarding
Release and Transfer in Solar Flares; K. G. Widing: Multiple
Loop Activations and Continuous Energy Release in a Solar Flare;
J. C. Henoux and Y. Nakagawa: Location of the Primary Flare Site
and Energy Transfer in Flares; A. B. Severny: Comments on Salyut-4
Observations of Active Regions on the Sun; H. W. Dodson-Prince: The
Early and Late Loops in Flares; R. Falciani: Photometric Studies of
the Starting Phase of Flares; H.S. Hudson: Effects of Electrons
versus Protons in the Solar Atmosphere; S. A. Colgate: Thermal
Effects in Flares; P. A. Sturrock: An Overview of the Energy-Flow
Problem.

Small Scale Solar Magnetic Structure (Jointly sponsored by Commission 10 and
12). Proceedings will appear in Highlights of Astronomy with F. L. Deubner and
J. O. Stenflo, editors.

Reports on Results of Skylab and OSO-8 (Jointly sponsored by Commissions 10,
12, and 44)

SKYLAB FLARE RESULTS - S. Kahler

Over 200 solar flares were observed on Skylab with enough data to follow the
evolution of a large fraction of those flares. In addition to active region flares,
flarelike phenomena were seen in X-ray bright points and in X-ray filament cavities.
The basic flare component observed by all groups is the loop, which is delineated

by magnetic flux tubes. Loops range in size from several arc seconds to several
arc minutes in length. The NRL spectroheliograph data show that heating appears
to take place at or near the tops of loop structures.

The flare of June 15 has been analyzed by the AS&E and NRL groups who find
evidence of continued heating well after the end of the impulsive microwave burst.

One consistent result obtained by all groups is that the calculated conductive
cooling times for flares are consistently shorter than the observed cooling times.
This may indicate that additional heating during the decay phase is required.

Attempts to find a soft X-ray signature of impulsive microwave bursts have not
been successful. The relationship between the impulsive phase of the flare and the
heating of flares is uncertain at this time.

SKYLAB CORONAL ACTIVITY RESULTS - R. MacQueen

Coronal activity observed with the Skylab telescope complement has provided
a wealth of new results. Significant transient activity---coronal brightness
changes at least equal to the diffuse X-ray coronal brightness over areas exceeding
10^{19} cm^2---has been seen in more than 40 events by Webb, Krieger and Rust, employing
results from the American Science & Engineering Company soft X-ray telescope. These
changes are manifest in long lived (3-40 hour) brightness increases of the general
size scale of $H\alpha$ filaments following $H\alpha$ filament disappearances. These observations often imply there is a rather long-lived heating phase following the most
dramatic initial phase of coronal ejection observed most readily in the outer corona
in white light. Corollary evidence for this post-ejection heating phase in also
found in EUV observations from the Naval Research Laboratory spectroheliograph,
which show excellent temporal correlation with gradual rise and fall events long
observed at radio wavelengths.

As noted above, the ejection of coronal material itself--typically 10^{15} - 10^{16}
grams total--has been observed in over 100 cases with the white light coronagraph
supplied by the High Altitude Observatory. These events, associated both with flare
and eruptive prominence events seen on the solar disk, have velocities ranging from
~ 100 to > 1200 km/sec and thus involve kinetic energies in the range 10^{30} - 10^{32}
ergs. The morphology and kinematics of a number of events have been examined in
some detail, and evidence based upon the distribution of material when a number of
events, and the metric wavelength radio association in a single event, presented for
the dominence of magnetic forces. These transient events represent major perturbations to the local solar corona and dramatically modify the coronal evolution; however, the interplanetary (1 AU) signature of transient coronal events is yet indef-
inite. Their role in modifying the interplanetary medium--through electron density
and/or magnetic fluctuations--is under study as is the potential role of transients
in modulating the cosmic ray flux.

SKYLAB ACTIVE REGION RESULTS - E. M. Reeves

The sun is known to be highly structured in both quiet regions and in active
centers, with previous space observations characterizing active regions as areas of
increased temperature and density compared to the surrounding quiet sun, and with
the ultraviolet line and X-ray band intensities increasing with the strength of
the observed magnetic field. The improved spatial resolution of the ATM instruments
in the ultraviolet and X-ray domains permitted detailed studies to be made of the
complex three-dimensional loop structures which form the dominant active region morphological features at higher temperatures of the transition region and corona.
Models of active region loops suggest a cooler inner core surrounded by cylindrical
sheaths forming the transition to the surrounding coronal temperatures. Although
the intensity along the loop can be quite constant, models indicate higher tempera-

tures toward the top of the loop, and indicate structural scales that would be resolved only in the hotter outer regions (T$\geq 10^{6}$°K). Formulations can also be derived to relate the size of the loop to the isothermal extent and the necessary energy input under assumptions of radiation or conduction loss mechanisms. The nature
of the necessary energy input processes required to achieve the observed loop stability
is not as yet understood.

The appearance of the loops provides strong evidence for inferring the dominant
role of the magnetic field in shaping and stabilizing the loop structures. Coronal
magnetic field extrapolations from observed photospheric fields have been undertaken
by several research groups and can provide quite good comparisons with observed solar structures on a global scale, including large loop structures interconnecting
active regions. Over limited regions of the sun comprising a single active region
complex the field extrapolations can be used to infer field strengths and coronal
pressure, and frequently provide good comparisons with the observed details. There
is some evidence for force-free currents flowing in loops, which would be oppositely
directed in loops on either side of the connecting line, and also evidence that some
loops cannot be reproduced by potential field extrapolations. There is quite good
evidence of magnetic field line reconnection above the photosphere from changes in
active region loop structure observed in both the ultraviolet and X-ray regions.
Loop patterns show definite rearrangements on time scales inconsistent with significant motions of the photospheric footpoint. Although there remains some uncertainty
about the unambiguous interpretation of the observations, magnetic field reconnection appers to offer the most direct interpretation.

Observations of bright points in the soft X-ray and extreme ultraviolet regions
indicate that they can also be represented as loop structure embedded in the chromospheric network. Although the bright points are generally small (approximately 20
arcsec) with lifetimes of about 8 hours, there appears to be continuous size and lifetime distributions over into the domain of the more classically recognized active
regions. For most of these low-lying features the density is about 4 times the
surrounding area but the overlying coronal temperature is comparable to that of
the quiet sun.

Studies of sunspots in the extreme ultraviolet frequently reveal that they
form the footpoint of many active region loops extending out to connect in the
surrounding plage. When viewed from above in radiation in the range 5 x 10^{4} -
6 x 10^{5}K they appear very bright compared to the surrounding plage.

OSO-8 RESULTS - R. Bonnet

The afternoon session was devoted to the description of the results obtained
by the instruments on OSO-8, both pointed toward the sun and spinning with the
spacecraft's wheel, viewing the sun once per orbit. Results must be regarded as
preliminary. This session included a brief description of the results obtained by
the Crimean Astrophysical Observatory onboard the Soviet station Salyut 4.

The American pointed instrument on OSO-8, a high resolution UV spectrometer,
was described by E. C. Bruner, Prinicipal Investigator. It consists of a small
telescope, 1.8 m equivalent focal length, followed by a single channel Ebert-Fastie
spectrometer covering the range between 120 and 200 nm, with a spectral resolution
of $\simeq$ 20 mA. The instrument is commanded through a small incorporated computer.
Such a high spectral resolution is ideal to study flows of matter and velocity
fields from the photosphere to the top of the transition region. Most observations
were made with an angular resolution of 20 arcsec.

Observation of waves with periods of 300 sec and smaller were described. Emphasis was given to the SiII lines at 181.7 nm formed at $\simeq 15000^\circ K$. Intensity fluctuations of periods 240 sec and 300 sec and velocity fluctuations of 300 sec were observed over the Quiet Sun while the same observations made over a sunspot lead to periods of 180 sec for both the intensity and velocity fluctuations. Over a plage only velocity fluctuations appear with a period of $\simeq 90$ sec and an amplitude of 2.8 km/sec. However, no power appears at the corresponding frequency in power spectra averaged over several orbits made of this same plage. The same results hold for the C IV line at 154.8 nm formed in the transition region.

A red shift is measured over bright regions in the network and simultaneous observations in Si II and the Ca II line at 854.2 nm made at Sac Peak give a good correlation between red shifts and regions of strong megnetic fields. Puzzling systematic blue shifts were observed at the limb, and high resolution profiles of the Lyman-α line show that the solar line is red shifted with respect to the geo-coronal absorption line.

Intensity transients were observed in the C IV line which preceded by 5 min a redshift of the line and an increase in its width. A series of questions and an active discussion ensued the presentation of these results which concentrated on the significance of intensity enhancements in terms of temperature fluctuations and the low amplitude of the steady flow observed over active regions where velocities as high as 20 km/sec are expected.
S. Jordan reported on observations of the 90 sec oscillations and computations of the mechanical energy carried by waves of periods smaller than 100 sec assuming that the magnetic field is equal to zero. From the lack of 90 sec oscillations in the Si II line, he concludes that radiative losses exceed the mechanical heating.

The French instrument on OSO-8 was described by P. Lemaire. This instrument is also a high resolution telescope giving a spatial resolution of 2 arcsec followed by a 6 channel spectrometer centered on 6 strong chromospheric lines: the H and K resonance lines of Ca II, the resonance lines of Mg II in the near ultra-violet, and the Lyman-α and Lyman-β lines of hydrogen. In addition, two lines formed in the transition region O VI 103.2 km and Si III, 120.6 km can be observed nearly simultaneously with the 6 main lines. The spectral resolution achieved with this instrument is also 20 mA. The telescope secondary mirror can be moved to generate small rasters on the disk of up to a sqaure of 64 arcsec with steps of one arcsec.

Many results were presented on intensity and velocity oscillations in the lower chromosphere where periods of 175 sec were measured in Ca II, and Mg II. The red and blue peaks of the Mg II lines oscillate in nearly exact opposition. Apart from instrumental effects, no oscillation could be detected in Lyman-α and Lyman-β. Examples were shown of high resolution profiles made in different regions of the network, plages and sunspots. An apparent systematic asymmetry appears in the intensity of the blue and red peaks of Lyman-α showing a higher blue peak and Lyman-β a higher red peak on the average. The profiles of Ca II and Mg II in different regions of prominences show strong asymmetries and vary over areas of one arcsec.

R. C. Catura described the X-ray heliometer in the OSO-8 wheel (Lockheed Research Laboratories) which operates in the energy range 1.5 - 14 KeV with a spatial resolution of 2 arcmin.

R. Novick described the University of Columbia wheel instruments which make spectra of the whole sun between 2 and 7 KeV, measure polarization, and can also observe stars in the direction parallel to the spin axis of the spacecraft. Preliminary results on the Crab Nebula show strong polarization effects.

A. Severny briefly described the solar instruments onboard Salyut 4: (1) a crossed dispersion spectrometer operating between 97 nm and 140 nm with a spectral resolution of .315A fed by a telescope which was able to resolve ± 2 arcsec, which recorded ≃ 600 spectra of active regions, flares, prominences; (2) a grazing incidence spectrometer operating between 74 and 97 nm with no spatial resolution.

Administrative Sessions

Report of the business meetings of Commission 10 (26 August and 1 September 1976) and on the meeting of the Organizing Committee (30 August 1976).

At the beginning of the first meeting, G. Newkirk (Acting President) welcomed Mme. d'Azambuja and honoured--audience standing--the late president of Commission 10, Prof. Kiepenheuer, and L. de Feiter, late member of the Organizing Committee.

(1) Election of Officers and New Organizing Committee. The Acting President proposed a slate suggested by the present Organizing Committee and noted that a main task of the members of the O. C. is the preparation of the Commission section of the Report on Astronomy. Kundu and Pick proposed that "Energetic Particles" should be taken care of in related sections of the report by Smerd, Sturrock and Wilcox. The proposed slate was accepted unanimously.

Président: G. Newkirk, Jr.
Vice-Président: V. Bumba
Comité d'Organisation: R. J. Bray, T. Hirayama, V. A. Krat, J.-L. Leroy,
                        S. F. Smerd, N. V. Steshenko, M. Stix, P. Sturrock,
                        J. Wilcox, C. Zwaan.

(2) Working Groups

    Working Group on International Programs (Chairman, P. Simon). Members:
        L. Dezso, M. Dryer, W. Kreplin, R. P. Lin, J. Rush, D. M. Rust,
        A. Shapley, H. Tanaka.

    Working Group on Solar Maximum Year (Chairman, D. Rust). Members: Still
        to be established from a list of candidates proposed by the Commission.

    Flare Build-Up Study (Chairman: Z. Svestka). Members: G. Brueckner,
        A. Bruzek, J. Harvey, M. Pick, E. Priest, D. Rust, P. Simon, S. Smerd,
        P. Sturrock, S. Syrovatskii, J. Vorpahl.

    Solar Mass Ejections (Chairman, E. Tandberg-Hanssen). Members: To be
        established from a list of candidates proposed by the members of
        Commission 10.

(3) Working Group Reports.

(3.1) WG International Programs. P. Simon reported that the chief responsibilities of the working group are the continuity of synoptic observations, organization of new international data programs, and the collecting of data over long periods for later research requiring them. Simon noted several problems: (a) flare patrol and publication (flare reporting), (b) the sunspot program of the Royal Greenwich Observatory, (c) the coronal observations (which were not discussed further).

The chairman had sent a memo on "Flare Patrol and Report of Subflares" to the members of Commission 10 in April 1975 and submitted another memo, "The Flare Survey Programme", to the Commission meeting. He noted that H. W. Dodson-Prince has prepared a new booklet with recommendations for flare observations.

Discussion following the WG report resulted in Recommendations or Resolutions regarding Flare Reporting, the SOON network, and the Royal Greenwich Photoheliographic Results.

(3.2) <u>WG Solar Maximum Year (SMY)</u>. H. Zirin reported that the WG was formed at the Sydney IAU with the following membership: Athay, Bruzek, Bappu, de Jager, Dunn, C. Jordan, R. Noyes, Smerd, Stepanov, Svestka, Tanaka, Zirker, Rösch. The objective of SMY is to formulate a cooperative study of the sun by combined observations of ground based observatories and space equipment during the coming maximum. A tentative program has been developed in the "Prospectus for the Solar Maximum Year". Zirin wishes to resign as chairman because of other commitments.

Discussion culminated in the action on the SMY and the re-establishment of the Working Group.

(3.3) <u>WG Flare Build-Up Study (FBS)</u>. Z. Svestka reported that the FBS is a common undertaking of magnetospheric and solar physicists with the objective of investigating the common processes which may occur in solar flares and in the terrestrial magnetosphere. A workshop was held last Fall (1975) in Falmouth (Mass.) and the proceedings are in print. Svestka proposed three recommendations originatint from the Workshop concerning: (1) organization of future workshops (2) observations to be coordinated by FBS and SMY during the Solar Maximum Mission, (3) support of a balanced program in solar research. The President suggested and the WG chairman agreed that only the third recommendation be brought before Commission 10.

(3.4) <u>WG Mass Ejection</u>. Tandberg-Hanssen reported on the history of the WG, which started as the Spray Patrol Project. He noted the expanded capability of modern observations to study many aspects of mass ejections from the sun and proposed that the WG be reconstituted to reflect this change.

(3.5) <u>WG Ad hoc Group on Solar Nomenclature</u>. Bruzek reported that the ad hoc group preparing the Solar Glossary was established on the initiative of the late President of Commission 10, Prof. Kiepenheuer. The group comprises the following: Beckers, Bruzek, Dodson-Prince, Durrant, Fokker, Harvey, Howard, C. Jordan, Koutchmy, Martres, Patel, Roxburgh, Svalgaard, Tandberg-Hanssen. The Glossary gives extended definitions and/or quantitative descriptions of solar terms and phenomena supported by illustrations and will be published in the Astrophysical and Space Science Library (Reidel/Dordrecht).

(4) <u>Membership</u>. (a) Commission 10: the President submitted a list of scientists who have applied for membership with the Commission and all were accepted. (b) SCOSTEP Executive Committee: Svestka wants to retire as IAU representative. Smerd was proposed and adopted as his successor. (c) SCOSTEP MONSEE Committee: It was proposed and accepted to replace Jaeger (Potsdam) by Bumba; Tanaka remains the other representative.

(5) Duties of Commission 10 and 12. The President raised the question of whether the division of responsibility between the two solar commissions should be re-examined. Smerd proposed that a joint WG (10, 12, and 40) should be established to bring a recommendation to the next General Assembly. The proposal was accepted and the President was requested to pursue the matter.

Recommendation 1

Flare Reporting - Considering the essential role of flare patrols in current
solar research and in the study of solar terrestrial physics and the growing interest in possible long-term variation in the level of solar activity and noting the
increasing gap in 24 hour coverage of the sun, Commission 10 recommends that the
solar community support a continued, viable program for flare patrol and flare
reporting during the next solar cycle.

Recommendation 2

Support of Solar Research - Reflecting upon the continuing contribution of
research at ground based observatories, plasma laboratories, and theoretical institutes on the nature of solar flares;

Recognizing that large gaps in solar observations will occur between individual
space missions, and being deeply concerned about the many recent abandonments of
ground based programs due to lack of funds; Commission 10

Recommends: that the various national funding agencies support a balanced
program of space and ground based solar research with appropriate recognition of
the vital role played by ground based observatories, theoretical institutes, and
plasma physics laboratories.

Resolution 1

Royal Greenwich Observatory Heliograph - The Royal Greenwich Observatory has
served the solar physics community for one hundred years by the production of high
quality, homogeneous, photoheliographic observations. Commission 10 of the IAU
expresses its gratitude to the RGO for this valuable service to the scientific
community and notes with regret that the RGO will terminate its photoheliographic
program at the end of 1976.

Recognizing the need of ensuring the continuation of this long series of homogeneous reports performed during one century and noting the capability and interest
of the Debrecen Observatory to continue such a program, Commission 10 encourages
the Debrecen Observatory to undertake the following responsibilities:

- To carry out direct photoheliographic observations at Debrecen and,

- To organize cooperation between other observatories willing to contribute
  to such a project

- With the assistance of the Greenwich Observatory to ensure a homogeneous
  continuity of the gathering, reduction and publication of such data

- To ensure the archiving of the original photographs and this access to
  interested scientists from around the world.

Resolution 2

SOON SYSTEM - Commission 10 is pleased to note the development of the Solar
Optical Observing Network (SOON) and thanks the U. S. Air Force for their generous
offer to make the solar data gathered available to the international scientific
community. Commission 10 hopes that the representatives of the U. S. Air Force
and the World Data Centers will consult frequently in the near future to assure
that continuity will be maintained between the new data to be generated by the
SOON system and the older data of the international flare reporting system.

Other Actions

H. Newton – Commission 10 of the IAU, noting the 100th anniversary of the photoheliograph program at the Royal Greenwich Observatory, sends cordial greetings to Mr. Harold Newton, whose study of sunspots for so many years was a major contribution to the long enduring solar program of the Royal Greenwich Observatory.

Mass Ejections – Recent ground-based and space-borne observations have revealed many unsuspected properties of mass ejections from the sun. Progress in the study of these phenomena will be vastly aided as a variety of observational methods are coordinated and information on particular events exchanged among interested scientific parties.

Accordingly, Commission 10 establishes a Working Group on Mass Ejections with E. Tandberg-Hanssen as Chairman and suggested members as: J. Kleczek, M. McCabe, R. MacQueen, J. Parkinson, R. Stewart, and a representative from Commission 10 to be established by negotiation between the Chairman of this Working Group and that Commission President.

The _charge_ to the Working Group is to:

1) Organize a system for the appropriate reporting of mass ejections to the World Data Centers

2) Define an appropriate nomenclature for mass ejections from the sun, and,

3) Encourage participation in collaborative programs for the observation, reporting, and interpretation of solar mass ejections,

4) Report these findings to the Commission at the next General Assembly.

Commission 10 recommends that the World Data Centers receive observations pertinent to mass ejections from the sun and disseminate this information through their normal channels.

Solar Maximum Year – In view of the importance of coordinated ground and space observations of the forthcoming maximum of solar activity and realizing the potential interest in establishing a program for a Solar Maximum Year on the part of Commission 10 and other organizations of the ICSU organizations for the purpose of developing a detailed plan of the program for the SMY for submission to the IAU Executive Committee at its first meeting in August 1977.

The charge to the Working Group for SMY is as follows:

To develop a detailed plan for the SMY to include:

• its scientific objectives

• the observational and theoretical work required to meet those objectives

• a plan for the coordination of the observing programs of such satellites as: solar maximum mission; international sun/earth explorer; intercosmos series; the proposed Japanese solar satellite; with ground based observatories as well as with observations which are part of the International Magnetospheric Study in order to organize the SMY to make best use of known space opportunities

- a plan for the implementation of the SMY including specifically the role of the IAU, the scope of the overall effort, and the collaborative roles expected from the various national committees, national scientific agencies, etc.

- a proposal for the structure of a permanent Steering Committee for SMY

- a specific timetable for the implementation of the SMY

- a survey of the likely cooperating observatories and spacecraft missions and their scientific contributions as proposed.

The Working Group shall prepare a draft plan for the SMY for submission to the IAU Executive Committee for consideration for adoption by the IAU at the time of its August 1977 meeting.

Rust agreed to serve as Chairman of this Working Group with the responsibility of setting up the Working Group and carrying out the above charge with their cooperation. Suggestions for membership in the Group include: I. Axford, J. Brown, A. Bruzek, L. Fisk, K. Frost, A. Gabriel, T. Holzer, C. Jordan, R. MacQueen, R. Manka, P. Simon, W. Stepanov, Z. Svestka, K. Tanaka, H. Zirin, R. Michard, E. J. Smith.

Proposals for Colloquia and Seminars - The Commission received a proposal from E. Jensen that an IAU sponsored colloquium on "Prominences and Their Solar Environment" be held in Oslo, 8-12 August 1977. The proposed Organizing Committee is to include Orrall (Chairman), Leroy, Morozhenko, and Engvold. After discussion, the Commission suggested that the objectives of the Colloquium be sharpened to take advantage of the timely emergence of such new observations as those now available from the Skylab and that the Organizing Committee be broadened to include a larger disciplinary distribution. Proposed additions to be considered should include: radio physics, Kundu; EUV, Moe; MHD, Anzer; EUV, Withbroe; coronal manifestations, MacQueen; EUV, Sheeley. A postponement of the colloquium to August 1978 should be considered. The President of the Commission and the proposers agreed to coordinate further actions in keeping with these recommendations.

COMMISSION 12: RADIATION AND STRUCTURE OF THE SOLAR ATMOSPHERE
(RADIATION ET STRUCTURE DE L'ATMOSPHÈRE SOLAIRE)

Report of Meetings, 25, 26, 28, 31 August and 1 September 1976

PRESIDENT: R. G. Giovanelli.

25 August 1976 - Business Meeting

I.   ORGANIZING COMMITTEE
     The Commission elected the following Organizing Committee:
     .President: M. K. V. Bappu.
     Vice-President: Y. Uchida.
     Organizing Committee: J. M. Beckers, G. E. Brueckner, A. N. Cox,
S. R. Gopasyuk, C. Jordan, W. Mattig, V. M. Sobelev, P. Souffrin,
R. G. Giovanelli (ex officio).

II. NEW MEMBERS
     New members of the Commission were endorsed as follows: J. E. Beckman,
N. Bel, A. Bhatnagar, V. Bumba, C. Cannon, F.-L. Deubner, L. Dezsö, E. Fossat,
E. M. Frazier, J. W. Harvey, S. Jordan, F. Kneer, S. Koutchmy, A. Kubičela,
M. Kuperus, W. C. Livingston, S. McKenna-Lawler, M. Marik, R. Mewe, D. Mihalas,
M. Semel, A. Skumanich, G. Stellmacher, V. E. Stepanov, J. O. Stenflo, K. Tanaka,
W. Unno, G. Ya. Vasil'eva, M. Vukičević-Karabin, H. Wöhl, S. M. Youssef, A. Zelenka.

III. REPORT OF THE COMMISSION
     The problems associated with preparing the Commission's Report for 1976-79
were discussed. While the Report for 1973-76 attempted to be an account of all
activities with comprehensive references, it was generally considered that a
complete coverage of all work with full references would be impossible in the
future. Dr. Bappu said that he would prefer to make the next Report a critical
review of outstanding work and trends, with only a limited bibliography, but some
members of the Commission considered the full reference list to be valuable, even
if only a limited text were possible. The new President and Committee undertook
to consider the matter further.

IV. WORKING GROUP ON ECLIPSES
     The Working Group (G. Newkirk) reported as follows: "Since the last General
Assembly the Working Group (1 member) has restricted its activities to placing
parties interested in observing the total eclipses of June 1974 and October 1976
in direct contact with Mr. A. Driver (CSIRO), Australian Coordinator, and
Mr. R. La Count (NSF).
     The Group recommends that it should continue in existence so as to facilitate
contact between scientists and coordinators in the countries involved in future
eclipses."
     The Commission expressed its gratitude to the eclipse coordinators for their
efforts on behalf of the astronomical community. G. Newkirk agreed to continue
as head of the Working Group.

V.   SYMPOSIA
     The Commission agreed that it would be timely to hold a symposium in 1979
dealing with stellar chromospheres ("what have we learned in the past 25 years")
and their relation to the solar chromosphere.

It was also suggested that a suitable topic for a future symposium would be "small scale velocity fields and the interpretation of line profiles."

VI. RESOLUTIONS
The Commission supported unanimously the following resolution by Commission 14: "The International Astronomical Union rates highly the activities of the United States National Bureau of Standards in the compilation and critical evaluation of atomic and molecular data, and considers these activities essential to the advancement of astronomy."

VII. FUTURE OF THE SOLAR COMMISSIONS
The Commission agreed to a proposal by G. Newkirk, President of Commission 10, that a joint Working Group be established to discuss the possible amalgamation of Commissions 10 and 12. The Working Group will make its recommendations at the next General Assembly.

Scientific Sessions

The scientific sessions took the form of symposia on selected subjects of current interest, several being joint meetings with other Commissions. In general, the discussions were based on some 2-4 invited talks, with some very short presentations restricted to the subject matter of the symposium. On the whole the meetings were extremely successful, though there was a good deal of support for one session to be devoted to general papers during the next Assembly.

Two half-day sessions were held jointly with Commissions 10 and 35 on "Stellar and Solar Structure," the first on magnetic fields, dynamics and convection, the second on large-scale motions and oscillations. An account of these sessions appears in the Commission 35 report.

A Joint Discussion on "Small-Scale Structure of Solar Magnetic Fields" is described in Highlights of Astronomy.

A full day was devoted to "The Interpretation of Atmospheric Structure in the Presence of Inhomogeneities." A. Skumanich discussed multi-dimensional geometrical radiative transfer effects in spectral line formation. U. Frisch detailed the difficulties involved in incorporating "microturbulence" into a physically realistic study of line formation. F. Kneer followed by reviewing the radiative transfer calculations which have attempted to infer the macroscopic velocity field structure of the solar atmosphere. R. F. Stein emphasized briefly the importance of solving simultaneously the radiative transfer equation coupled to the pertinent aerodynamics. A. Omont discussed the problems encountered in atomic physics which limit our understanding of the redistribution function. R. W. Milkey then presented the radiative transfer effects due to angle and frequency redistribution at a photon scattering event. R. G. Athay discussed the types of observations and diagnostic analyses one can make which enable the structure of the solar atmosphere to be determined better. Emphasis was placed on the need for simultaneous observations of several spectral lines. R. G. Giovanelli emphasized the strong time-dependence of the inhomogeneities involved by presenting a movie of the solar atmosphere taken simultaneously in CaK and Hα.

A further full-day session was devoted jointly with Commissions 10 and 34 to a "Report on Skylab and OSO-8". This is described in the Commission 10 report.

A joint meeting on "Advances in High Spectral and Spatial Resolution" was held under the sponsorship of Commissions 9 and 12. S. P. Worden, Sacramento Peak Observatory, spoke on solar and stellar speckle interferometry. Speckle technique was reviewed and new observations were given on umbral dots, indicating that their

size is approximately 1/4 arc-sec. P. Lena, Meudon Observatory, described a
successful attempt to resolve IR sources using speckle at 2.0 μm wavelengths; the
seeing-disk of the 4-m Mayall telescope was rapidly scanned by a slit aperture.
J. Brault, Kitt Peak National Observatory, summarized the advantages of Fourier
transform spectroscopy for visible wavelengths. He stressed that even for the
same resolution figure as a grating instrument, the resulting FTS line profile is
considerably freer of scattered light and instrumental smear effects. First
results with the KPNO 1-m FTS system for spot spectra were presented.

J. Ring gave a progress report on the Imperial College visible light FTS
instrument, which is in use for stellar observations with the Isaac Newton
telescope. Finally, F. L. Deubner and W. Mattig showed a movie of solar
granulation taken on the spectrostratoscope balloon flight in 1975. The extended
time of good seeing obtained from the stratosphere platform enabled one clearly to
see "exploding" granules, a phenomenon which they demonstrated to be a common
occurrence.

Session on Solar Abundances, 1 September 1976.

A session on solar abundances, organized and chaired by E. A. Müller, resulted
in a new composite list of photospheric abundances (Table I) which can serve as a
standard reference, together with abundances in the corona (Table II) and the solar
wind (Tables III, IV).
The session was divided into three parts dealing with these regions separately.
Most time was devoted to photospheric abundances, where different groups of elements were discussed separately. The grouping of the elements was made according
to common problems characteristic to the determination of their abundances. It was
encouraged to include in the discussion the results from sunspot spectra, molecular
bands, chromospheric lines, prominence lines, as well as isotopic ratios, if known.
Summaries of the various reports are given below and in Table I are listed the
recommended values of the elemental abundances. The photospheric abundances are given on the standard scale $\log \varepsilon$ (H) = 12.00. Uncertainties are included if they are
given by the authors.
It should be noted that while the plans for this session were in progress,
three compilations of solar abundances both for the photosphere and the corona
appeared in the literature, or are ready to be published, namely Hauge, O. and
Engvold, O. (1976, in press); Ross, J. E. and Aller, L. H. (1976, Science 191,
1223); and Withbroe, G. L. (1976, Solar Physics, in press).

I. PHOTOSPHERIC ABUNDANCES

Y. CHMIELEWSKI, (Observatoire de Genève): "The Abundances of the Light Elements,
He, Li, Be, and B".

The common characteristics in the determination of the light elements (He
through B) are the following:
(1) These elements are represented in the solar spectrum by only very few
lines which are usually the resonance lines and which, in most instances, are more
or less seriously blended. This fact requires in all cases an abundance determination by the method of spectral synthesis. The number of observational constraints
can be increased adequately by using center-to-limb observations which (a) help
disentangle the blends, and (b) provide a powerful check on the relevance of the
physics used in the derivation of the abundances. The effects of blending and the
possibility of departures from LTE, especially in the ionization equilibria, have
to be carefully investigated.
(2) The atomic structure of these elements is simple thus giving the advantage
that non-LTE calculations can be performed with model atoms that are at the same
time simple and realistic. Generally, good f-values and reasonably accurate cross-
sections are available.

(3) The lines being faint or of only moderate strength, the computation of their line profiles is not much affected by uncertainties in the damping parameters. Moreover, inasmuch as these elements are very light, thermal broadening dominates and the resulting abundances are relatively insensitive to microturbulence.

HELIUM is somewhat different from the other light elements since its lines can only be observed in the upper chromosphere or in prominences. The results of Hirayama (1971) are in agreement with some of the cosmic ray and solar wind data. A detailed study of the photospheric LITHIUM feature at $\lambda 6707.8$A (Brault and Müller, 1975; Müller et al. 1975) yielded an abundance similar to that of GMA. The photospheric and the recent sunspot results (Stellmacher and Wiehr, 1971) are in agreement. A thorough re-investigation of the BERYLLIUM lines (Chmielewski et al.,1975) led to no appreciable change in the Be abundance value because the effects due to the additional continuous opacity in the region λ 3000 - 4000 A compensate the non-LTE effects on the ionization equilibrium. BORON has been definitely identified in the solar spectrum by Kohl et al. (1976) who give an abundance result which is no longer an upper limit. ISOTOPIC RATIOS have been reported for H (Beckers, 1975), He (Hall 1975), and Li (Müller et al., 1975) as follows:

$$^2\text{H}/^1\text{H} < 2.5 \times 10^{-7}, \quad ^3\text{He}/^4\text{He} = (4\pm2) \times 10^{-4}, \quad ^6\text{Li}/^7\text{Li} \lesssim 0.01.$$

References

Beckers, J.M., 1975: Astrophys. J. Letters 195, L43.
Chmielewski, Y. and Müller, E.A., 1975: Astron. Astrophys. 42, 37.
Hall, D.N.B., 1975: Astrophys. J. 197, 509.
Hirayama, T. 1971: Solar Phys. 19, 384.
Kohl, J.L., Parkinson, W.H., and Withbroe, G.L., 1976:
    submitted to Astrophys. J.
Müller, E.A., Peytremann, E. and de la Reza, R., 1975: Solar Phys. 41, 53.
Stellmacher, G., and Wiehr, E., 1971: Solar Phys. 21, 97.

D. L. LAMBERT (University of Texas): "The photospheric C, N, and O abundances".

Improved C, N, and O abundances have been derived from the most reliable atomic and molecular indicators. Five model solar atmospheres have been examined, tests against continuum observations showing two of them to be superior. These are the model given by Holweger and Müller, (1974) and an unpublished model by R. Allen. Atomic and molecular lines give consistent abundances when analysed with either model. Other models (e.g. the Harvard-Smithsonian reference atmosphere) do not provide such consistency.

Primary abundance indicators are recognized. They are the [CI] 8727 A line, the CH A-X system, the C_2 Swan bands, the NI lines and the [OI] 6300 and 6363 A lines. These provide the recommended abundances listed in Table I. An uncertainty of ± 0.10 dex is suggested and includes possible NLTE effects and the neglect of the granulation. (This paper will be submitted for publication in Monthly Notices of the Royal Astronomical Society).

References

Holweger, H. and Müller, E.A., 1974: Solar Phys. 39, 19.

D. L. LAMBERT (University of Texas):"Photospheric Abundances: Sodium through Calcium."

An interim revision of photospheric abundances for elements with atomic numbers from sodium through to calcium was presented. Attention was drawn to critical improvements and deficiencies in the f-values. A comparison with meteoritic abundan-

ces shows excellent agreement (to ± 0.1 dex) except for chlorine (volatile in meteorites) and argon (undetectable in the photospheric spectrum).

SODIUM. The reference analysis is Holweger's (1971). Recent pumping experiments involving two dye lasers give radiative lifetimes in excellent agreement with the
coulomb approximation f-values used in the solar analyses. Independent discussions
of the NaD line wings (Blackwell et al., 1972;Worral, 1973) used the refined line
broadening calculations of Lewis et al., (1971) and obtained an abundance consistent with the value given in Table I, derived from 8 weak lines.
MAGNESIUM. The abundance is most reliably extracted from MgII lines for which the
f-values (MgII is isoelectronic with NaI) are reliable. Many weak lines of MgI are
available but the presence of severe configuration interaction renders a simple
theoretical calculation of the f-values uncertain. Recently, Froese Fischer,(1975),
has published results of a MCHF calculation. These provide f-values for 21 weak
solar lines and an abundance log ε (Mg) = 7.60 ± 0.20 which is consistent with the
MgII analysis. The large rms error is attributed to the f-values.
ALUMINIUM. The abundance is not very well determined. Although good lines exist,
the theoretical f-values are often uncertain owing to severe cancellation in the
radial integral. Other lines are strongly broadened and the line broadening coefficients are uncertain. There are no measurements of f-values for lines in the solar
line list; one may anticipate that lifetime measurements by pulsed dye lasers will
be reported shortly. The present abundance is taken from Lambert and Warner (1968).
SILICON. The abundance is based on the discussion by Holweger, (1973) who used
weak lines with measured f-values. His line list did not overlap with that of Lambert and Warner, (1968), who selected lines with reliable theoretical f-values.
When the two line lists are compared for a common model atmosphere, the Si abundances are consistent.
PHOSPHORUS. About 10 weak PI lines from the 4s-4p array and theoretical f-values
provide the abundance listed in Table I.
SULPHUR. The abundance comes from a recalculation using a sample of weak SI lines
(see Lambert and Warner, 1968) and the forbidden lines λ7725 and λ10821 (see Swings
et al., 1969).
CHLORINE. The abundance is provided by Hall and Noyes, (1969), who identified HCl
vibration-rotation lines in infrared sunspot spectra.
ARGON. No photospheric lines.
POTASSIUM. The abundance is based upon the resonance lines (a NLTE analysis is reported by de la Reza and Müller, (1975), and about 6 weak excited lines. The coulomb approximation f-values (except for cases of severe cancellation) are confirmed
by several accurate measurements.
CALCIUM. The reference analysis is Holweger's (1972). The Ca abundance can be derived from the CaI intercombination resonance transition at 6572 A, weak excited
CaI lines, the CaI autoionizing transitions, the [CaII] lines at 7323 A and weak
excited CaII lines. The most reliable indicators are the [CaII] and CaII lines.
The CaI lines give a consistent result and this sets a good limit on possible departures from LTE in the Ca-Ca$^+$ ionization equilibrium.

References

Blackwell, Kirby, and Smith, 1972: M.N.R.A.S. 160, 189.
Fischer, F., 1975: Can. J. Phys. 53, 184 and 338.
Hall, D., and Noyes, R.W., 1969: Ap. Letters, 4, 143.
Holweger, H.,1971: Astron.Astrophys. 10, 128.
Holweger, H., 1972: Solar Phys. 25, 14.
Holweger, H., 1973: Astron. Astrophys. 26, 275.
Lambert, D.L., and Warner, B., 1968: M.N.R.A.S. 138, 181 and 213.
Lewis, E.L., McNamara, L.F., and Michels, H.H., 1971: Phys. Rev. A3, 1939.
de la Reza, R., and Müller, E.A., 1975: Solar Phys. 43, 15.
Swings, J.-P., Lambert, D.L., and Grevesse, N., 1969: Solar Phys. 6, 3.
Worrall, G., 1973: Astron. Astrophys. 29, 37.

N. GREVESSE (Institut d'Astrophysique, Université de Liège):
"The Iron Group Elements (Sc through Ni)".

Between 1960 (when Goldberg, Müller and Aller, 1960 (GMA), published their
detailed analysis of the abundances of the elements in the solar photosphere) and
1969 a great number of new determinations of the photospheric abundances of these
elements were published. Although using more refined techniques, now applicable
because of the rapid evolution of the computational facilities, using new solar
data, i.e. equivalent widths and profiles, of better quality, and using new extensive sets of transition probabilities, mainly issued from the National Bureau of
Standards, the new results confirmed GMA's values.

The old iron abundance problem has disappeared completely. When one compared
photospheric values with those obtained from meteorites or the corona, the photospheric iron abundance (GMA) was systematically lower, by a factor of 10, than the
others. Other members of the iron group showed similar but smaller discrepancies.

The explanation appeared in 1969; the differences are not real but are due only
to systematic errors in the transition probabilities used to derive abundances. In
the last 7 years many papers have been published relating either to oscillator
strengths of iron-group elements or to revisions of solar abundances of these elements. Special efforts have been made by the spectroscopy groups at Aarhus, AFCRL,
Caltech, Harvard, Kiel, NBS, Oxford and elsewhere to determine these oscillator
strengths using refined modern techniques. We shall look at the case of iron in
some detail, and then pass through the other elements rather rapidly.

Since 1969, when it was shown for the first time by Garz and Koch (1969) that
the absolute scale of the available gf-values for Fe I was in error by a factor
of ~ 10, 25 papers giving new gf-values for Fe I and Fe II lines have been published. At first some disagreement persisted between absolute scales proposed by
different laboratories. More recent results, such as those of Bridges and Kornblith
(1974), May, Richter and Wichelmann (1974) and the Oxford group (Blackwell et al.,
1975), show an agreement which is within 0.10 dex in the log gf (in many cases the
agreement is even better), but minor discrepancies remain to be explained, especially for faint lines. From these new sets of transition probabilities, new solar photospheric abundances have been derived and it is rather surprizing to see
that the results appear to scatter from ~ 7.0 to ~ 7.65. Most of the scatter can be
reduced if one uses faint lines for which the gf-values have been measured recently.
Thus, most of the scatter in the results is probably due to uncertainties in microturbulence and damping. For this reason we shall retain from among the recent results only those based on faint lines. They are all in agreement with a rather high
iron abundance. The recommended value is A_{Fe} = 7.50. It takes into account the
above-mentioned results and the result J.P. Swings and myself obtained in
1969 (Grevesse and Swings, 1969) from a study of forbidden Fe II lines whose transition probabilities are known with high accuracy.

Let us now turn to the other elements. As we have already said, they did not
attract the same attention as iron. Nevertheless, new sets of gf-values have been
obtained during recent years and, for most of them, the initial discrepancy photosphere - meteorites was explained and removed. The adopted photospheric values
of the iron group elements are given in Table I; they are in excellent agreement
with the meteoritic results.

In concluding I wish to make the following remarks and suggestions:
(1) The techniques used to measure transition probabilites are long, expensive
    and time consuming. But, we still need a lot of accurate transition probabili-
    ties for lines of neutral and, particularly, once-ionized iron group elements.
    As these measurements are difficult, even extremely difficult for faint lines
    which are those we need for abundance determinations, we suggest that theore-
    ticians help us to supply the data we need. Biémont (1974) and Huber and
    Sandeman (1976) have shown that relatively accurate transition probabilities
    can be computed using semi-empirical methods for neutral iron group elements.
(2) With a few rare exceptions all the photospheric abundances rely upon high qua-
    lity spectra obtained at the center of the disk. It would be extremely inte-

resting to base the determinations on high quality spectra going from the center up to the limb ($\cos\theta \simeq 0.1$). We hope that Brault's interferometer at Kitt Peak will fill this gap soon.

(3) All the results quoted here were obtained under the assumption of local thermodynamic equilibrium (L.T.E). Athay and Lites (1972) have investigated the effects of departure from L.T.E. on a number of iron lines. They showed that low excitation levels (χ_{exc} < 2.5 eV) are underpopulated by a factor 3 in the high layers where the $_{exc}$ cores of the lines originating from these levels are formed. But Holweger (1973) and Smith (1974) showed that these departures are extremely sensitive to the collisional cross-sections used and that the non-LTE effects on the abundances are very small (an increase of the abundance of 0.03 to 0.04 dex). Furthermore, no variation in abundance is found when going from the neutral to the once-ionized element, or when examining the abundances derived from lines originating in quite different levels of excitation potential. Therefore, non-LTE effects may be neglected as far as the iron group elements in the solar photosphere are concerned.

(4) Recent work by Carrol et al. (1976) on Fe H, Cr H, and Mn H suggests that Fe H could be present in the spectra of the photosphere and of sunspots, thus opening a new field for abundance determinations of these elements. On the basis of their statistical analysis the temptation is great to conclude on their presence, but complete analyses of these complex spectra is needed before the question can be settled.

(5) Extensive abundance determinations of the iron group elements in sunspots have not yet been performed. The only result I know of is that of Van Paradijs (1975) who made a curve of growth analysis of one sunspot spectrum and obtained A (Fe) = 7.45 ± 0.32. In prominences, the abundances of Sc, Ti and Fe have been determined recently by Yakovkin et al. (1975). Their results, A (Sc) = 3.2, A (Ti) = 4.94 and A (Fe) = 7.40 also agree with the recent photospheric results.

Athay, R.G. and Lites, B.W.,1972: Astrophys. J. 176, 809.
Biémont, E., 1974: Solar Phys. 38, 15; Solar Phys. 39, 305;
    see also Solar Phys. 44, 269, 1975; Astrophys. Letters 17, 127, 1976.
Blackwell, D.E., Ibbetson, P.A. and Petford, A.D., 1975: Mon. Not. R. astr.
    Soc. 171, 195.
Bridges, J.M. and Kornblith, R.L., 1974: Astrophys. J. 192, 793
Carroll, P.K., McCormack, P. and O'Connor, S., 1976: Astrophys. J. 208, 903;
    see also Carroll, P.K. and McCormack, P. 1972; Astrophys. J. 177, L73.
Garz, T. and Koch, M., 1969: Astron. Astrophys. 2, 274.
Goldberg, L., Müller, E.A. and Aller, L.H., 1960: Astrophys. J. Suppl. Ser. 5, 1.
Grevesse, N. and Swings, J.P., 1969: Astron. Astrophys. 2, 28.
Holweger, H., 1973: Solar Phys. 30, 35.
Huber, M.C.E. and Sandeman, R.J., 1976: 8th annual conference of the European
    Group for Atomic Spectroscopy, Oxford, July 13-16, paper 5.
May, M., Richter, J. and Wichelmann, J., 1974: Astron. Astrophys. Suppl. 18, 405.
Smith, M.A., 1974: Astrophys. J. 190, 481.
Van Paradijs, J.,1975: Astron Astrophys. 44, 395.
Yakovkin, N.A., Zeldina, M.Y. and Rakhubovsky, A.S., 1975: Soviet Astron. 19,203.

TABLE I. - ELEMENTAL ABUNDANCES IN THE PHOTOSPHERE

Z. El.	Log A	Z.	El.	Log A	Z.	El.	Log A
1 H	12.0	26	Fe	7.5	59	Pr	0.8
2 He	10.8 ± 0.1	27	Co	5.0	60	Nd	1.2
3 Li	1.0 ± 0.1	28	Ni	6.3	62	Sm	0.7
4 Be	1.15 ± 0.20	29	Cu	4.2	63	Eu	0.7
5 B	2.6 ± 0.3	30	Zn	4.4	64	Gd	1.1
6 C	8.67 ± 0.10	31	Ga	2.8	66	Dy	1.1
7 N	7.96 ± 0.10	32	Ge	3.4	68	Er	0.8
8 O	8.90 ± 0.10	37	Rb	2.6	69	Tm	0.3
9 F	4.56 ± 0.33	38	Sr	2.9	70	Yb	0.8
10 Ne	−	39	Y	2.1	71	Lu	0.8
11 Na	6.29	40	Zr	2.8	72	Hf	0.9
12 Mg	7.56	41	Nb	2.0	74	W	0.8
13 Al	6.40	42	Mo	2.2	75	Re	<- 0.3
14 Si	7.60	44	Ru	1.9	76	Os	0.8
15 P	5.45	45	Rh	1.5	77	Ir	2.2
16 S	7.25	46	Pd	1.5	78	Pt	1.8
17 Cl	5.5	47	Ag	0.9	79	Au	0.1
18 Ar	−	48	Cd	2.0	80	Hg	< 2.1
19 K	5.14 ± 0.10	49	In	1.7	81	Tl	0.9
20 Ca	6.32	50	Sn	2.0	82	Pb	1.9
21 Sc	3.1	51	Sb	1.0	83	Bi	< 1.9
22 Ti	5.0	55	Cs	< 2.1	90	Th	0.2
23 V	4.0	56	Ba	2.1	92	U	< 0.6
24 Cr	5.7	57	La	1.1			
25 Mn	5.4	58	Ce	1.6			

O. HAUGE (Institut of Theoretical Astrophysics, University of Oslo):
"The Abundances of all Elements Heavier than Nickel: Cu through U."

The abundances of these elements are mainly determined from analyses of photo-spheric spectra. In a few cases sunspot spectra have been studied (In, Cs, Tl). Coronal lines (Cu), chromospheric lines (Sr, Ba) and prominence lines (Sr, Ba) have also been used.

Abundances published by different authors are commonly in good agreement, par-ticularly for elements present in the photospheric spectrum with lines of medium strength. When studying strong lines, saturation effects and damping may make the derived result more uncertain. In strong lines the contribution to a line comes from higher levels in the solar atmosphere where different atmospheric models are in some disagreement. When faint lines are studied, the derived abundance depend critically on the setting of the continuum level.

Abundances are mostly derived on the assumption of LTE; non-LTE calculations should surely be undertaken, particularly when resonance lines are studied.

The oscillator strengths of lines from some elements are supposed to be well known. Independent determinations undertaken with different methods give results in close agreement. But for many elements, particularly among the RE's, the amount of data is very limited.

Many authors have recently given valuable contributions to the solar abundance table. Here we only refer to the work of two groups. J.E. Ross and L.H. Aller have recalculated photospheric abundances for a large number of elements using improved f-values and subsequently used the HSRA model atmosphere in order to put the determinations in a more uniform system. T. Andersen and his group in Aarhus have succeeded in measuring life-times by beam-foil techniques of energy levels of 11 heavy elements where new f-values were very desirable. These investi-gations have resulted in a revision of solar abundances by up to a factor of 10 (as in the case of Sm and W). The recommended abundance values are listed in Table I.

II. THE CORONAL ABUNDANCES

CAROLE JORDAN (Department of Theoretical Physics, University of Oxford)

The present summary gives a critically-evaluated list of abundances derived from EUV emission lines formed in the transition region and inner corona. Abundances derived from X-ray spectra present rather different problems and have been recently discussed by Parkinson (1976).

Early analyses of EUV spectra used whole-sun intensities and simple approximations regarding the region of line formation and the relative populations of levels in the ground term and in metastable terms. Also, improved excitation cross-sections have become available. The early work therefore has not been included although it would be possible to re-analyse some of the data.

The abundances given in Table II result from analyses from spectra of regions of the quiet sun (Burton et al. 1971, Dupree, 1972), and a whole sun spectrum in the far EUV (Malinovsky and Heroux, 1973) since quiet sun spectra are not yet published for this spectral region. Some re-analysis of the data published by the above authors has been made. In particular iteration of the region of line formation has been performed, and the writer's choice of atomic data has been used.

The following comments can be made regarding the abundances given in Table I. The high abundance of carbon relative to oxygen given by Withbroe for the EUV lines is weighted heavily by early analyses and is not, in the writer's opinion, borne out by later more sophisticated treatments. The oxygen abundance is not well determined from EUV data, since the lines lie in a region where the shape of the emission-measure distribution is changing rapidly with temperature. The neon to magnesium ratio from EUV lines is between 1 and 2, rather than neon being less abun-dant than magnesium. Silicon is taken as standard at 7.65 for the present work.

Withbroe's tabulations show a systematic difference in the abundance of iron as determined from EUV lines and visible region forbidden lines. This is not, however, a systematic difference between heights or regions in the solar atmosphere. The writer considers that the difference could lie in a current underestimate of populations of metastable levels in the iron ions.

TABLE II. - ELEMENTAL ABUNDANCES IN THE CORONA

Element	Photosphere	EUV	EUV
	(Withbroe, 1976)	(Withbroe 1976)	(This work)
C	8.54	8.76	8.44
N	8.06	8.08	7.81
O	8.84	8.63	8.84
Ne	-	7.60	7.95
Mg	7.54	7.65	7.75
Si	7.65	7.67	7.65
S	7.21	7.21	7.11
Fe	7.48	7.72	7.65
Ni	6.28	6.45	6.25

References

Burton, W.M., Jordan, C., Ridgeley, A. and Wilson, R., 1971, Phil. Trans. Roy.
      Soc. Lond. A. 270, 81.
Dupree, A.K., 1972, Astrophys. J., 178, 527.
Malinovsky, M. and Heroux, L., 1973, Astrophys. J., 181, 1009.
Parkinson, J.H., 1976, to be published.
Withbroe, G.L., 1976, Sol. Phys. (In Press).

III. ELEMENTAL ABUNDANCES IN THE SOLAR WIND

PETER BOCHSLER and JOHANNES GEISS (Physikalisches Institut, Universität Bern)

1. Introduction

So far, three essentially-different methods have been used for investigating abundances in the solar wind:
-   Energy/charge spectrometers have been used for abundances of several elements in times of low solar wind temperature;
-   The foil collection technique has provided data on isotopic and elemental abundances of He, Ne and Ar;
-   Trapped solar wind gases in lunar soil (and meteorites) have given information on the isotopic composition of several additional elements (H, C, N, Kr, Xe). In addition, some general limits on the abundances of these elements in the solar wind can be derived.

The observed time variations in the solar wind composition indicate that fractionation occurs in the source region and that average solar wind abundances do not have to be identical with photospheric abundances. Some models have been developed about ion fractionation in the solar wind acceleration region, but there exists no theory yet which would allow a straightforward calculation of photospheric abundances from solar wind abundances. However, average solar wind abundances are seen to agree with photospheric abundances within a factor of about two, even for elements with very different characteristics such as ionization potential, mass or mass/charge. Therefore, it is concluded that the relative abundances of isotopes in the solar wind are a good approximation to photospheric abundances.

2. Elemental abundances in the solar wind

TABLE III : ELEMENTAL ABUNDANCES IN THE SOLAR WIND.

	Value	Method	Reference
He/H	0.04 ± 0.01	E/Q-Analyzers 1962-1975	Ogilvie and Hirshberg (1974)
O/H	$(5 \pm 2) \cdot 10^{-4}$)		(Bame et al. (1975)
Fe/H	$(5 \pm 3) \cdot 10^{-5}$) -	E/Q-Analyzers 1969-1972	((Grünwaldt (1976)
Si/H	$(7.6 \pm 3) \cdot 10^{-5}$)		((Bame et al. (1975)
He/Ne	530 ± 70)	Foil collection	(Geiss et al. (1972)
Ne/Ar	41 ± 10) -	1969-1972	(Cerutti (1974)

Comments: The He/H ratio is varying with time. Ogilvie and Hirshberg (1974) have reviewed values measured during a whole solar cycle and discussed correlations of this ratio and its fluctuation with the velocity of the solar wind, phase in the solar cycle, temperature etc. The value of 0.04 is a good average of He/H in the solar wind over a whole solar cycle but it is most probably lower than the value in the source region. The Los Alamos group (Bame et al., 1975) has published O/H, Si/H and Fe/H ratios in the solar wind for several selected periods from 1969 to 1971. The authors judge the experimental errors of their results to be approximately a factor of two. Grünwaldt (1976) has given the average O/H and Fe/H ratio for a period of two days of exceptionally-quiet solar wind conditions. His values are compatible with those of Bame et al. (1975). He/Ne and Ne/Ar ratios have been obtained by means of the foil collection technique.

The errors in Table III indicate the range of fluctuations in a sample which is not complete. Therefore the true elemental ratios in the solar wind could in a few cases be beyond the given limits of error.

Investigations in the lunar soil have shown that krypton and xenon are present in the solar wind, but due to the unknown trapping and storing conditions it is difficult to give reliable abundance values. The same holds for C and N. The C/N ratio in the trapped gas is 1.5 (Kerridge, 1975).

A comparison of the solar wind abundance values given in Table III with spectroscopic data shows that the deviations are not larger than a factor of about two.

3. Isotopic abundances

Comments: No deuterium has been found in solar wind trapped in lunar soil. The given upper limit is determined by the contamination of lunar fines with terrestrial hydrogen and by spallation-produced deuterium.

The ^{4}He/^{3}He ratio fluctuates in the solar wind (Bame et al., 1975; Geiss et al., 1972; Grünwaldt, 1976) reflecting fractionation processes in the corona. Therefore, the average solar wind ratio given above could be lower than the ratio in the outer convective zone, perhaps by as much as 10 to 30 %.

The isotopic composition of solar neon is significantly different from terrestrial neon, while no deviation is found between the solar and the terrestrial ^{36}Ar/^{38}Ar ratio.

Trapped solar wind carbon has within the limits of errors the same isotopic composition as terrestrial carbonates. Nitrogen behaves strangely: There are strong variations observed in lunar fines and there is an indication that the nitrogen composition in the solar wind has been changing with time.

Solar wind krypton is only slightly fractionated relative to terrestrial krypton. Xenon, however, is linearly fractionated relative to terrestrial atmospheric xenon with a fractionation factor of 3.6 % per mass unit, the heavier isotopes being depleted. Furthermore, it appears that in terrestrial xenon a component rich in the lightest and heaviest isotopes is slightly depleted (Lewis et al., 1976). Terrestrial xenon has been augmented by ^{129}Xe from the decay of ^{129}I taking place after the gas-grain separation in the planetary nebula.

TABLE IV : ISOTOPIC ABUNDANCES IN THE SOLAR WIND.

	Value	Method	Reference
D/H	$< 3 \cdot 10^{-6}$	Lunar soil	Epstein and Taylor (1972)
^{4}He/^{3}He	2350 ± 120	Foil collection)	Geiss et al. (1972)
^{20}Ne/^{22}Ne	13.7 ± 0.3	Foil collection)	
^{22}Ne/^{21}Ne	31 ± 4	Foil collection)	Filleux et al. (1977)
^{36}Ar/^{38}Ar	$(5.3 \pm 0.3$	Foil collection	Cerutti (1974)
	$(5.33 \pm 0.03$	Lunar Soil	Eberhardt et al. (1972)
C	(1.5 ± 1.5) % δ ^{13}C	Lunar Soil	Epstein and Taylor (1972)
N	(3 ± 7) % δ^{15}N	Lunar Soil	Kerridge (1975)

4. Conclusions

The importance of the investigation of abundances in the solar wind lies in the unique possibility to obtain precise values on the isotopic composition of some elements in the outer convective zone of the sun. These data are relevant for the understanding of the history of the solar system and the universe; they are placing limits on models for the synthesis of elements.

It is yet difficult to derive precise solar element abundances from solar-wind measurements because of the fractionation processes in the corona. However, the noble-gas data from the solar wind provide even now some valuable evidence on elemental abundances in the sun.

References

Bame S.J., Asbridge J.R., Feldman W.C. Montgomery M.D., and Kearney P.D., (1975), Solar Physics 43, 463.
Cerutti H., (1974), Ph.D. Thesis, University of Bern.
Eberhardt P., Geiss J., Graf H., Grögler N., Mendia M.D., Mörgeli M., Schwaller H. and Stettler A., (1972), Geochim. Cosmochim. Acta. Suppl. 3, Vol. 2, 1821.
Epstein S. and Taylor H.P. Jr., (1972), Geochim. Cosmochim. Acta, Suppl. 2, Vol. 2, 1421.
Filleux Ch., Bühler F., Cerutti H., Eberhardt P. and Geiss J., (1977), in preparation.
Geiss J., Bühler F., Cerutti H., Eberhardt P. and Filleux Ch., (1972), Sec. 14 of Apollo 16 Prel. Sci. Report, NASA SP-315.
Grünwaldt H., (1976), to be published in Space Res. XVI.
Lewis R.S., Srinivasan B. and Anders E., (1976), Science 190, 1251.
Kerridge J.F., (1975), Science 188, 162.
Ogilvie K.W. and Hirshberg J., (1974), J. Geophys. Res. 79, 4595.

COMMISSION 14: FUNDAMENTAL SPECTROSCOPIC DATA
(DONNÉES SPECTROSCOPIQUES FONDAMENTALES)

Report of Meeting, 25 August 1976

PRESIDENT: R. H. Garstang

The President thanked the Chairmen of the Working Groups for their part in
the preparation of the Report of the Commission.

Membership of the Commission. The President reported the deaths of four
members of the Commission (W. R. Hindmarsh, T. A. Littlefield, B. Rosen and E. W.
Salpeter), the replacement of L. J. Kieffer by E. C. Beaty as a Consultant and
the nomination of I. Martinson as Consultant. He reported the nomination of E.
Trefftz as President and J. G. Phillips as Vice President for the next three
years. R. H. Garstang, W. Lochte-Holtgreven, S. L. Mandelsh'tam, A. H. Gabriel,
R. W. Nicholls, S. Sahal and W. L. Weise were elected as the Organizing Committee
for 1976-1979.

There was a brief discussion of the Report of the Commission. Concern was
expressed on the limited distribution which the Report receives in relation to
the work involved in its preparation.

Resolution. It was moved by M. J. Seaton, seconded by A. H. Cook, and after
minor amendments, carried unanimously, that "The International Astronomical Union
highly values the activities of the United States National Bureau of Standards
in the compilation and critical evaluation of atomic and molecular data, and
considers these activities essential for the advancement of astronomy." (Following the meeting the President obtained the agreement of Commissions 12 and 29 to
co-sponsor the Resolution for submission to the General Assembly.)

The President presented a brief report by K. M. Baird supplementing his
printed Report. Much work is in progress on extending frequency measurements to
the visible part of the spectrum. There seem to be many alternatives more
accurate than the present caesium standard but much remains to be done before a
practical alternative can be proven. However, it should soon be practical to
redefine the meter in terms of the caesium frequency standard and the adopted
value of the velocity of the light. There have been new measurements by G.
Guelachvili of 87 absolute wave numbers in HCℓ and HF spectra and by C. Freed
and co-workers of absolute frequency values for a very large number of transitions
in CO_2 isotope lasers.

The President reported work by G. H. C. Freeman (Consultant to the
Commission), who now has a Michelson interferometer in the vacuum ultraviolet.
Problems of polishing and coating of magnesium fluoride discs have been over-
come. The interferometer is being used to study the shape of the Xe I 147 nm
line under conditions where the width of the line is less than 500 fm as a
prelude to measuring its wavelength.

W. Huebner reported on his opacity library, set up so that the user can
obtain opacities of mixtures of his choice.

Report of Meeting, 26 August 1976

PRESIDENT: R. H. Garstang   SECRETARY: M. C. E. Huber

ATOMIC DATA FOR ASTROPHYSICS

     W. C. Martin presented information supplementing the Commission Report.
He referred to Edlén's recent review (Beam Foil Spectroscopy, Eds., I. A. Sellin
and D. J. Pegg, Plenum, New York, 1976, Vol. 1, p. 1) which has a bibliography
of 300 references on the spectra of atoms from helium to nickel. A new NBS
bibliography will be published soon (L. Hagan, NBS Special Publication 363,
Supplement 1, 1976) covering July 1971 to June 1975. A compilation on Cr I to
Cr XXIV will be published in 1977 (J. Sugar and C. H. Corliss, J. Phys. Chem.
Ref. Data) and work on a compilation for manganese is in progress. A compilation
on O I is now available (C. E. Moore, NSRDS-NBS 3, Section 7, 1976). Reports
from several laboratories were received too late or were unintentionally omitted
from the list in the Draft Report. Perhaps of most direct astronomical interest
in this regard is work on Fe X-XIII and Fe XVIII-XXIII in the Astrophysics Research Div., Culham Laboratory, Abingdon, Oxon, England (reported by B. C.
Fawcett). Space does not permit giving detailed references to individual papers
which have been published.

     K. Widing presented a paper by C. Moore-Sitterly and himself, in which they
reported that extensions of known C I series account for many lines in solar
limb and flare spectra. Over 300 additional lines of Si I have also been
identified in limb spectra. Further laboratory studies on Fe I, Sn I and Pb I
spectra are in progress and work on Ge I has been completed. G. D. Sandlin has
prepared a list of 100 forbidden lines in the solar corona and transition zone
in the ultraviolet.

     M. J. Seaton spoke on progress and prospects in the computer calculations
of atomic data. He drew attention to very general computer programs which have
been developed at Queen's University Belfast (QUB) and at University College
London (UCL). These programs are in use at a number of other institutes. Other
groups active in the calculation of atomic data include those at ETH Zurich,
Observatoire de Paris (Meudon); Observatoire de Nice; University of Maryland;
Goddard Space Flight Center; Lebedev Institute (Moscow); Leningrad State
University; Louisiana State University; JILA, Boulder, Colorado; IBM, San Jose,
California; MPI Munich; Physics Institute, Riga, Latvia. Data calculated include
atomic structures and energy levels, radiative transition data (bound-bound,
bound-free and free-free) and cross sections for electron collisions with atoms
and ions. The most accurate formulation of the collision problem is "close-
coupling plus correlation terms" (CC + CT). The groups at QUB and UCL have
shown that this formulation also provides a powerful technique for the calculation
of radiative data. The distorted wave (DW) and Coulomb-Born (CB) methods are
less accurate but are economic in computing resonances and should be satisfactory
for electron collisions with more highly ionized systems. However, in some cases
the errors in DW calculations turn out to be larger than might have been expected
-- for example, in the 2s-2p transition in N V the error is nearly a factor of
two (due to strong coupling between the n = 3 states).

     Seaton mentioned the accurate measurements recently made for 1s-2s and
1s-2p excitation in H (J. F. Williams): the results are in excellent agreement
with (CC + CT) calculations. For the isoelectronic case of He^+ 1s-2s, there is
however, a puzzling disagreement between calculations and experiment. Sophisticated calculations for Be^+ 2s-2p give results about 15% larger than those of
recent measurements; the difference may be within the accuracy of the absolute
calibration of the experiment. The QUB and UCL groups have made (CC + CT)
calculations for excitation of atoms and ions with outer $2p^q$ electrons.

Although there is little direct comparison experimental data, consistency checks suggest that the calculated results are correct to within about 5%. Calculated photodetachment cross sections for alkali negative ions are in good agreement with measurements made using lasers. Accurate photoionization cross sections have been calculated for Be, Mg, C, N, O, Ne and Aℓ. Similar techniques have been used to calculate large numbers of f values for O I and Mg I. In summary one can say that (i) good progress has been made for systems up to Ne in the periodic table, (ii) similar work for systems up to Ca should be possible with existing techniques, (iii) for heavier systems some work is being attempted but real progress may require the use of new-generation computers and (iv) for highly ionized systems the DW and DB methods can be used but further work is required to establish the exact range of validity of these approximations.

Replying to A. Dalgarno, Seaton said that it was difficult with present techniques to calculate cross sections for simultaneous ionization and excitation. He also pointed out that difficulties may arise in calculations involving highly excited states, because, there, one ought to consider the coupling with many states. A. Burgess commented that ionization cross sections calculated by the Exchange Classical Impact Parameter method agree quite well with plasma measurements, in fact, better than the results obtained with Coulomb–Born calculations.

Presenting a review by himself and D. R. Flower, H. Nussbaumer remarked that the calculation of ionization rates is at present in an uneasy situation in that the Coulomb–Born calculations disagree with the Exchange Classical Impact Parameter (ECIP) method. While satisfactory agreement between the Coulomb–Born calculations and crossed electron–ion beam experiments is found at high energies, the astrophysically important region is mostly that close to threshold. It appears that close to threshold the Coulomb–Born method rather systematically overestimates the cross sections. Statistically the crossed beam experiments are better represented by the ECIP method, however, the individual error bars are rather large. Ionization cross sections deduced from time dependent plasma experiments seem to favour the ECIP results. In a review on dielectronic recombination theory just completed, Seaton and Storey find that Burgess's general formula for recombination is usually accurate to better than 30%. The best available method for calculating collision strengths is now the Close Coupling method (CC). One expects the Distorted Wave method (DW) to be applicable to ionized systems with accuracy increasing with the degree of ionization and it is much more economical than CC. It is not possible to give safe rules as to when DW may be sufficient and we have to learn from accumulated experience, although it appears that DW works for strong transitions even in lowly ionized systems, as for example in C III. Based on the same bound state functions, collision strengths calculated in the DW approximation and by the CC method agree to better than 10%, even close to threshold. Collision strengths for $2s^2 - 2s3\ell$ transitions are also needed. But we must not conclude that the DW method will be sufficient for these transitions because it was good enough for transitions within the n = 2 complex. This reservation is based upon experience gained from N V, where DW and CC collision strengths for 2s – 2p and 2s – 3d agree well but where they disagree for 2s – 3s and 2s – 3p. The discrepancies are due to collisional coupling between the 3ℓ terms. We may expect similar effects on the less ionized C III. Doing a CC calculation is in itself no guarantee of good results, one also requires a good representation of the bound states. To represent the ground state $2s^2$ ^{1}S in C III one must allow for configuration interaction with $2p^2$ ^{1}S. The CC collision strength for the $2s^2$ ^{1}S – 2s2p ^{1}P transition differs by a factor of two between the two-configuration calculation and the result allowing for $2s^2 - 2p^2$ interaction. The same interaction also has an important effect on the $2s^2$ ^{1}S – 2s2p ^{3}P transition probability.

The collision strengths for that intercombination transition are strongly
energy dependent because of the presence of resonances converging on $2s2p$ $^1P^o$.
Although the atomic physicist may be interested mostly in the complicated resonance structure, the result needed to interpret observations is the collision
strength averaged over the Maxwellian energy distribution. For the solar case
the resonance contribution is approximately equal to the direct contribution.
Collision strengths for the fine structure transition $2s2p$ $^3P^o_J$ - $^3P^o_{J'}$ are also
strongly influenced by resonance contributions. A further illustration of the
need for investigating resonances is the B sequence. For transitions between
levels of the configurations $2s^22p$ and $2s2p^2$ the collision strengths for
$^2P^o_{1/2}$ - $^2P^o_{3/2}$, $^2P^o_J$ - $^4P_{J'}$, ^4P_J - $^4P_{J'}$, ^4P - ^2P are expected to show resonance
structures resulting in important contributions to the total collision strengths.
Spectral lines from the B sequence are of considerable astrophysical interest.
The sun provides an example. For the N III $\lambda\lambda991$, 686 lines an intensity ratio
$I(^2P^o$ - $^2D)/I(^2P^o$ - $^2P)$ of about 3 is observed for quiet regions and of about
7 for active regions. Based on DW calculations one finds from the observed
ratios $T_e \approx 40000°K$ for quiet and $T_e \approx 26000°K$ for active regions, thus a lower
temperature for active regions. Both these temperatures are considerably below
the temperature for which the fractional abundance of N^{+2} attains its maximum
value which is approximately 80000°K. This is certainly an interesting result
concerning the solar transition region, but should be viewed with caution until
the applicability of the DW approximation in this case has been confirmed.
Nussbaumer stressed again the importance of resonance contributions to the
collision strengths for forbidden and intercombination transitions. An exact
treatment of them is laborious, but verifications are needed for various cases
to ensure that approximate methods may be employed. Collision strengths for
strong transitions in ionized atoms can probably be safely calculated by the
DW approximation. To establish the lower ionization limit for the validity of
DW one needs a CC calculation for each isoelectronic sequence. Weak transitions
to nearly degenerate terms may in any case demand a CC calculation although
this may not be feasible in practice.

D. R. Flower cited further support for the accuracy of Nussbaumer's C III
data: computations recently made by Hibbert yield a transition probability for
the 1908 Å intercombination line that lies within 10% of the value derived by
Nussbaumer. In reply to C. Jordan, Nussbaumer confirmed that no close-coupling
nor distorted-wave calculations involving more than three configurations have
been published for C III. However, the inclusion of 24 rather than three
configurations resulted in a change of 8 to 10% for the transition probabilities
of the 977 and 1176 Å lines. The collision strengths for these strong transitions might be expected to vary like the oscillator strength; this had, indeed,
been confirmed by recent calculations at University College London. Thus, he
had confidence in the collision strengths he used, since they had been scaled
by comparing oscillator strengths from 3 and 24 configurations. Seaton noted
that cross sections for collisions $2s \rightarrow 2p$ of Be^+, calculated in the close-
coupling approximation, had changed only 5% if n = 3 configurations were
included. This makes it probable that the inclusion of n = 3 states would make
little difference to the C III collision strengths for transitions between n = 2
states.

S. Sahal-Bréchot and V. Bommier described a method for determining magnetic
fields in solar prominences by use of the Hanle effect in the D_3 line of He
(3^3D-2^3P). Radiation in this line is polarized because of its anisotropic
radiative excitation from the photosphere. Owing to Zeeman coherences, the
degree of polarization and the direction of linear polarization are sensitive
to the strength and direction of magnetic fields. Calculations have been made
for fields up to 15 Gauss (i.e. in the range that is most sensitive and also
free of level crossings). Comparisons between these calculations and solar

observations showed that the method is feasible. The determination will be
complete when the D_3 line is observed simultaneously with an additional line.

G. Brueckner discussed the need for improved atomic data for the
interpretation of the ultraviolet solar spectrum. Improved line identification
compilations of high-temperature lines (180-600 Å), $10^6 < T_e < 6 \times 10^6$ are
necessary to identify several hundred flare lines in this region which have been
detected but remain unidentified. The chromospheric spectrum 1100 to 1800 Å
shows approximately 4000 emission lines, of which 2000 still remain unidentified.
Many of these mostly weak lines are enhanced in sunspots. From their appearance
it is likely that these lines are formed in a temperature regime $4000° < T <
10000°K$. Complete laboratory line lists of Fe I, Mg I, Co I, Cu I, Ni I and Mn I,
including high level transitions, are needed. In addition, improved ultraviolet
laboratory spectra of the second and third spectra of all elements found in the
sun are necessary. From the comparison with the visible chromospheric flash
spectrum one would expect ultraviolet lines of the less abundant elements like
V I, V II, As I, As II, Sr II, La II, Y II, Ba II, Sc II, Zr II, and Mn II to be
present in the sun. Laboratory UV spectra of these elements are needed.

Brueckner then considered the wavelength reference system, which needs vast
improvement for all prominent solar lines in the 180 to 1800 Å region. Presently
the mean deviation of solar from laboratory wavelength is $\Delta\lambda/\lambda = 8.7 \times 10^{-5}$
for 50 lines in the 274-4100 Å region. In order to carry out meaningful Doppler
measurements in the solar spectrum, laboratory standards for the prominent solar
lines need to be known with an accuracy of $\Delta\lambda/\lambda = 1 \times 10^{-6}$. This measurement
accuracy has been achieved with C I standard lines in the 1274-1459 Å region, but
other areas of the 1100 to 1800 Å solar spectrum do not contain enough standard
lines. A list of low-temperature, unblended solar lines covering this area of
the spectrum has been compiled; approximately half of them need new laboratory
wavelength measurements. In addition, standard wavelengths with an accuracy of
$\Delta\lambda/\lambda = 10^{-6}$ of the most important transition zone lines in the solar spectrum
$1100 < \lambda < 1800$ Å are needed. Brueckner concluded by emphasizing the importance
of atomic cross sections and transition probabilities. By using combinations
of allowed and spin-forbidden transitions, the ultraviolet solar spectrum has
great advantages over the visible spectrum when carrying out refined diagnostic
work on the solar chromosphere. Improved electron excitation cross sections,
ionization cross sections and f values are needed for C I, Si I, O I, Si II and
Fe II, especially for the intersystem lines. Numerous new observations of
density-sensitive line ratios in the transition zone have been made recently,
using C III, Si III, O IV and O V. Conflicting electron densities result if the
presently available atomic parameters are used. The errors of these parameters
must be decreased below ±10%. Only then can meaningful density values be
derived.

In discussion Seaton pointed out that the identification of transitions
involving high-lying states can frequently be hampered by series perturbations.
Jordan said that because the intensity ratio of the lines $\lambda\lambda 1908$ to 1176 was
determined at the limb, $\lambda 1176$ would be expected to be optically thick. This
should influence the results derived from the intensity ratio. E. M. Reeves
reminded the audience that photoelectric observations had shown considerable
variations with time in the intensity of many of the lines used for diagnostics;
consequently, time-dependent ionization-equilibrium calculations might be expected
to shed more light upon many currently contradictory results. Replying to
Garstang, Jordan and A. K. Dupree thought that the variation of N_e and T_e over
the region where the C III lines are formed would not explain the C III discrepancy. Nussbaumer reiterated his opinion that the atomic data are now accurate
enough to make the C III problem a question of interpretation -- possibly including time dependences -- rather than of atomic data.

In commenting on the oscillator strength section of the Commission 14
Report R. H. Garstang pointed out that references had been given in full only
if they did not appear in Astronomy and Astrophysics Abstracts. He also presented some extracts from a report on work in the U.S.S.R. which had been
received too late for inclusion in the printed Report. He drew attention to the
Russian work on wavelengths and transition probabilities in the hydrogen, helium
and lithium isoelectronic sequences, to several important papers on ultraviolet
spectroscopy of highly charged ions produced in vacuum spark and laser plasmas,
and to lifetime and oscillator strength measurements in many elements. Up-dating
his Report, Garstang drew attention to continuing work on improved furnace
measurements of Fe I by Blackwell, beam-foil measurements by Andersen and
colleagues on ions such as Fe III and Ti III, recent high-accuracy beam-foil
lifetime measurements by Curtis and others, combined hook and absorption
measurements on Cr I and Ni I by Huber and Sandeman, calculations on O I by
Saraph using the frozen-cores approximation, and work by Grant on relativistic
intensity calculations.

E. Trefftz commented on her report, and pointed out that many items had been
omitted because of lack of space. She drew attention to recent work by E. W.
Smith (NBS Boulder) on calculations of molecular line broadening by neutral particles. K. T. Tang worked on H - H_2 collisions. The anisotropic part of the
potential is now reasonably well established. There are still discrepancies in
the collisional calculations. J. Schäfer (MPI Munich) used a carefully calculated potential of H_2 - H_2 by W. Meyer (Univ. of Mainz) to do close coupling
calculations. He finds large probabilities for the rotational excitation of both
H_2 molecules. Chemical reactions are sometimes restricted to head-on collisions.
Manz (TU Munich) suggests generalizing a one-dimensional calculation to three
dimensions by statistical methods.

J. W. Liebert described his work on the blue white dwarf suspect Feige 7,
which has been found to have a rich optical spectrum, and variable circular
polarization with a period of 2.2 hours. The mean longitudinal field is estimated to be 5 million Gauss at peak polarization. The spectrum fits Zeeman
patterns of hydrogen and He I in the presence of mean homogeneous fields of about
20 million Gauss. The star provides the first confirmation of the theoretical
spectra of hydrogen and helium in such high fields, inaccessible to laboratory
measurements. The period must be due to rotation, and the blue continuum indicates that it is the hottest of the known magnetic degenerate stars. The
comparable intensity of H and He I lines may be unique in white dwarf stars. The
star must have a helium dominated atmosphere. Liebert showed spectrophotometric
observations of the magnetic stars GD 229, G 240-72 and G 195-19.

G. Wegner remarked that the magnetic field in the peculiar white dwarf
BPM 25114, suggested by Bessel and Wickramasinghe, has been confirmed with
circular polarization in the hydrogen lines, and Hγ line profiles indicating a
field of 10^7 Gauss.

Report of Meeting, 31 August 1976

PRESIDENT: R. H. Garstang   SECRETARY: J. B. Tatum

MOLECULAR DATA FOR ASTROPHYSICS

R. W. Nicholls described some recent work on molecular spectra. For
diagnostic applications in astrophysics the principal molecular data needed are
(a) wavelengths (for the location of energy levels and transitions between them,
(b) intensities (for the determination of transition probability data), and (c)
cross sections and rate constants (for the definitive assessment of energetic
processes). We have been principally concerned with (a) and (b), contrary to

popular belief reliable data in both areas for molecules and band systems of
astrophysical interest are quite fragmentary. This is clear from a careful
assessment of the data compilations (Données Spectroscopiques of Rosen and of
Barrow, and more particuarly in Suchard's two recent works Spectroscopic Data I
(Heteronuclear Molecules - parts A and B) (Plenum Press, New York, 1975),
Spectroscopic Data II (Homonuclear Molecules) (Plenum Press, New York, 1976).
Many of the early analyses of molecular spectra were not extensive enough nor
were made with sufficient precision to be useful in the high resolution computer
synthesis of molecular spectra. The situation is far worse for transition
probability data, as indicated in a review chapter on the subject in the 1977
Annual Review of Astronomy and Astrophysics by R. W. Nicholls. Nicholls illustrated his remarks by discussing work in his own laboratory on excitation, wavelength analysis, intensity measurement and theory (including computer simulation)
of (mainly diatomic) spectra of astrophysical, aeronomical and atmospheric
importance. Molecules currently under study are O_2, C_2, CN, CℓO, ScO, and YO.
Examples of work on O_2 and CℓO were presented.

L. E. Snyder reviewed microwave molecular spectra. As of August 1976, 40
molecular species had been identified in the interstellar clouds. The two most
complex, dimethyl ether and ethyl alcohol, are isomers with nine atoms each; 29
of the molecules contain one or more carbon atoms, the other 11 contain no carbon
atom. The identification of interstellar X-ogen as HCO^+ (formyl ion), suggested
by Klemperer, has been confirmed. In 1975 the isotope $H^{13}CO^+$ was found in space
and subsequently the group of R. C. Woods reported laboratory measurements of both
$H^{12}CO^+$ and $H^{13}CO^+$ which were in excellent agreement with the interstellar measurements. Later Woods' group measured DCO^+ which then was found to have remarkably
high intensity in several galactic molecular clouds. The identification work on
interstellar HNC (hydrogen isocyanide) has been completed. As a result of a suggestion of G. Herzberg, D. Buhl and Snyder searched for and found HNC in the interstellar clouds. Recently three different laboratory groups were successful in
measuring HNC and its isotopes and $HN^{13}C$ was detected in space. Laboratory measurements have confirmed the identification of interstellar N_2H^+. This molecule
was discovered accidentally by Turner, tentatively identified by Green et al. on the
basis of molecular computations, and measured in the laboratory by Saykally et al.

Snyder surveyed progress on the chemistry of interstellar molecules.
Theoretical models utilizing ion-molecule chemical formation schemes have shown
HCO^+ to be a keystone molecule for the formation of other interstellar molecules.
In agreement with model predictions, observations have shown HCO^+ to be abundant
and widespread throughout the galactic molecular clouds and N_2H^+ to be anti-
correlated with CO and HCO^+ in Orion. DCO^+ was found to be enhanced in cool
clouds in agreement with chemical fractionation models. The dark cloud L134
produced a spectrum of HNC which (a) is of higher resolution than any currently
available laboratory spectrum and (b) has approximately the same intensity as HCN.
Ion-molecule formation theory predicts that the reaction H_2CN^+ + e forms either
HCN + H or HNC + H. Hence the dark cloud L134 may represent the first observed.
case where the branching ratio (HNC/HCN) is approximately unity. The formyl
radical, HCO, has been studied extensively in the laboratory and often advocated
as an interstellar molecule. All searches for HCO have been unsuccessful until
very recently when the J = 3/2-1/2, F = 2-1 transition at 86,670.65 MHz was found.
In agreement with the models of W. D. Langer, the formyl radical was not detected
in the densest molecular clouds but rather in clouds which appear to be of intermediate density. The interstellar measurements indicate that typically the number
density ratio HCO^+/HCO is 2-5 except for W51 where the ratio is 7-17.

Concluding, Snyder stressed the need for continuing fundamental microwave
laboratory measurements on small refractory molecules (such as TiO and VO). On
the basis of the discovery of large interstellar molecules he suggested that

microwave measurements of small optically active molecules could become important
for future polarization studies. Unidentified interstellar microwave lines
continue to be detected: U86.76 in Sgr B2 interferes with $H^{13}CO^+$ and may be
CH_3C_3N (methylcyanoacetylene); U90.146 could be COH^+; and several identified
lines were found as the result of observations which began as a search for the
gauche isomer of ethyl alcohol. Reaction intermediates such as NH_2^+, CH_3CO^+,
H_3CO^+ and H_2CN^+ are important in ion-molecule formation schemes but there is
little or nothing known about their microwave spectral properties.

In the ensuing discussion E. Trefftz asked whether there is any explanation
of the fractionation of hydrogen and deuterium in the molecules HCO^+ and DCO^+.
Snyder replied that the degree of fractionation depends on the electron density,
the density of hydrogen molecules and the temperature. The detectability of
deuterium is greater at lower temperatures. It is possible that in the dark
clouds we are observing a primordial deuterium abundance. Responding to questions
by P. K. Carroll and A. Dalgarno, Snyder indicated that metallic hydrides are
likely candidates for detection in space, and that it is surprising that OD has
not been seen, it would be worthwhile searching for OD again in the dark clouds.

A. H. Delsemme discussed molecules in comets. Until recently only water ice
could reasonably be trusted as a major constituent of the volatile fraction of a
cometary nucleus. Now for the first time radio astronomical observations of
comets have yielded results. They reconfirmed water as a parent molecule. CH_3CN
was identified by Ulich and Conklin, HCN by Huebner et al., and several unidentified lines were observed. OH and CH, known in the optical range, were also detected via the splitting of the ground-state doublet. Brightness profiles across
the coma yield decay times of the observed molecules and of their unobserved
parents. A better knowledge of molecular absorption cross sections for dissociation and ionization in the ultraviolet should allow parent identifications. Pure
photochemistry can be observed in the cometary exosphere. Therefore observed
large velocities can result from the energy balance of the dissociations and
should be used for the identification of parent molecules from their dissociation
mechanisms. High-dispersion spectra have confirmed the usual assumption of a
small expansion velocity for C_2 and CN. Charge-exchange reactions play a fundamental role in the collisional zone of the inner coma, reshuffling many molecular
species; in particular, the role of H_3O^+ must predominate, but its spectrum is
unknown. The observation and detailed analysis of a laboratory spectrum of H_2O^+
by Lew and Heiber has led to the first identification of a cometary ion since
1942. The detailed analysis was a prerequisite for predicting the very low
temperature spectrum observed in comets.

Delsemme emphasized that with the increasing availability of vacuum UV spectra of comets more complete data are needed on diatomic and polyatomic molecules.
The resonance lines of H, C and O have now been repeatedly observed (Comets
Kohoutek and West) from satellites and rockets; in Comet West, the first negative
system of CO^+, the bands of CS from 2500 to 2700 Å, and CN^+ near 3200 Å have also
been observed. Surprisingly, a large production rate of C I in its ^1D state is
also mentioned (Feldman et al., and Smith et al.), its origin is unknown yet,
although it could probably be a mechanism analogous to the excitation of the red
forbidden line of CO_2. In the laboratory, Meinel has obtained C_2^+ in absorption,
and Herzberg, probably NH_2^+. Radiative lifetimes of cometary ions and molecules
in their excited levels are much in demand. Most oscillator strength data are
still poor for the simplest molecules: UV photoabsorption data, including ionization or dissociation rates and collisional excitation cross sections, are still
missing or insufficient to compute space lifetimes of most molecules, ions and
radicals; more accurate data on the solar flux in the extreme UV are also badly
needed to predict the branching ratios of parent molecules into ionized or
dissociated fragments.

A. Dalgarno discussed the role of molecular processes in theories of evolution of interstellar clouds, star formation and the star-gas interaction. The main heat loss in a diffuse cloud arises from excitation of the $^2P_{3/2}$ fine structure level of C^+. Collisions of C^+ and H_2 in the zero and first rotational levels require more detailed study. In a dense cloud, rotational excitations of interstellar molecules control the heat loss. Cross sections are needed also for diagnostic purposes. Much progress has been made in developing procedures for the solution of the scattering problem but less progress has been made in developing methods for calculating potential energy surfaces to the chemical accuracy required. Nevertheless due mainly to Sheldon Green, data are available for the atoms colliding with N_2, CO, N_2H^+, $HC\ell$ and H_2CO. The main collision partner however, is H_2 and only limited information is available. Cross sections of collision-induced molecular hyperfine and fine structure occur in theories of astrophysical masers but are available only as arbitrary estimates. Charge transfer of highly stripped ions with atomic hydrogen modifies the ionization structure of a gas ionized by cosmic rays or X-rays. Work has been carried out on the charge transfer of C III, C IV, N III and Si III. Charge transfer with helium also needs exploration. Radiative charge transfer rates have been calculated. Dalgarno drew attention to a major compilation by W. Huntress on ion molecule reactions in interstellar chemistry. Little information is available at temperatures appropriate to interstellar clouds, and in some cases the exothermicity of critical reactions has not been established. Of major significance is the rate coefficient and its temperature dependence for the radiative association of C^+ and H_2. The rate coefficient for C^+ and H has been determined now to be 3×10^{-17} cm^3 sec^{-1} at 100°K. Dissociative recombination of CH^+ is still entirely uncertain as are the branching ratios for polyatomic ions. Photoionization and photodissociation cross sections of interstellar molecules are critical but little progress has been achieved except for the case of OH where an unattenuated interstellar rate of about 10^{-10} s^{-1} seems now to be established.

In discussion B. Zuckerman pointed out that vibrationally excited transitions have been observed in cyanoacetylene, and A. B. Underhill referred to observations of CS in Comet West. In response to a question by Snyder, Dalgarno agreed that probably only about one-fifth of the observed interstellar molecules have so far been understood in terms of ion-molecule reactions; in reply to Trefftz he indicated that H_3O^+ should be observable probably in the infrared or radio regions, but that the triplet transitions of CO would be hard to observe. P. D. Feldman reported that in Comet West the singlet Fourth Positive system of CO was observed but the triplet Cameron bands were not present. The President drew attention to work by R. McCarroll on charge exchange involving multiply charged ions, and by A. J. Sauval on calculations of the molecular equilibrium in cool stars; lack of time prevented detailed presentations of these contributions.

COMMISSION 15: PHYSICAL STUDY OF COMETS, MINOR PLANETS AND METEORITES
(L'ETUDE PHYSIQUE DES COMETES, DES PETITES PLANETES ET DES METEORITES).

Report of Meetings, 26, 27 August and 1 September 1976

PRESIDENT: A. H. Delsemme   SECRETARY: J. Rahe

26 August 1976

REVIEWS ON TRANSIENT PHENOMENA IN COMETS

This session was dedicated to Karl Wurm's memory.

1. The Neutral Coma - L. M. Shul'man (in absentia)

The large abundance of H and OH in the neutral comas is generally accepted as evidence of water ice as a major constituent of the nucleus, although recent arguments suggest CO_2 as another major constituent (Biermann and Diercksen, Delsemme and Combi). Two other parent molecules have at last been detected, namely CH_3CN and HCN, whereas the observed radicals C_2 and C_3 suggest complicated unsaturated molecules, whose origin is still controversial (Stief, Shul'man, Kaimakov, Cherednichenko). After the discovery of the Lyman alpha halo, a number of models of the hydrogen atmosphere have been developed. The proper radiation transfer was studied and applied by Keller. The dynamics of the neutral gas in comas has been extensively studied by different workers (Wallis, Shul'man, Mendis and co-workers). Recent work agrees on some general features. A collision-dominated flow takes place in the inner part of the coma (typically 10^4 km). Photochemical reactions give a large contribution to the energy balance of the coma. Heating from gas-dust interaction is much smaller than that from photo-dissociation in a water-dominated coma (Wallis). A multi-component hydrodynamic model including an extended source of water (icy halo) is described by Ip and Mendis. Many reactions, including charge-exchange reactions can take place in the dense part of the coma and therefore can reshuffle the neutrals (Akin, Oppenheimer).

2. The Cometary Ionosphere - D. A. Mendis

The physical structure and chemical composition of the cometary ionosphere were discussed. The relative importance of radiative and collisional ionization processes were evaluated, and it was shown that ionization by an energetic flux of electrons discharging from the tail through the inner coma, may be 1 to 2 orders of magnitude more efficient than photoionization. The importance of ion-molecule reactions in determining the chemical structure of the ionosphere was stressed and a detailed ionospheric model computation for a H_2O dominated comet containing some CO presented. The dominance of two hitherto unobserved ionic species H_3O^+ and HCC^- in the inner ionosphere was noted.

3. Dust in Cometary Comas and Tails - Z. Sekanina

Results of studies of the distribution of light in cometary dust tails were summarized. The emission rate of dust, the particle-size distribution function and the ejection velocity of dust particles were established for a few comets with

the use of the Finson-Probstein technique. Attention is drawn to the expulsion
of the relatively large particles from comets; these are comparable in size with
meteoroids that would produce faint meteors. Infrared and polarimetric methods
were also discussed.

4. Review of Plasma Tails - J. C. Brandt

Transient phenomena in comet tails were reviewed from the observational
viewpoint with emphasis on the interpretation of apparent motions in the tail, the
capture of magnetic field from the solar wind, currents flowing in the tails, and
production of the ionized species. Our knowledge of most of these areas is fragmentary, but different lines of evidence appear to indicate magnetic fields
greater than $\approx 100_\gamma$ and currents $\sim 10^8$ amperes in the tail. Definite results will
require space missions to comets.

ADMINISTRATIVE MEETING

1. Election of Officers for 1976-1979

These elections were the first taking place within the new mandate of
Commission 15, resulting from the extension of its terms of reference to the
Minor Planets and Meteorites. For this reason they had to be very formal. In
order to give all members a chance to vote, the following procedure was used:

1. Nominations were called by mail, and the slate of nominees that had been duly
seconded, were proposed by mail to all Commission members.
2. The single-transferable vote system and the majority vote rule was used.
Sixty-one members have voted, out of a total of seventy. Professor N. Richter
has been unanimously elected President at the first ballot with 61 votes, or 100%
of the votes expressed. The Vice-Presidential first ballot yielded: B. D. Donn
33 votes; C. R. Chapman 14 votes; B. J. Levin 7 votes; O. V. Dobrovolskii 5 votes;
F. Miller 1 vote; G. Wetherill 1 vote. The second ballot gave B. D. Donn 39 votes;
C. R. Chapman 19 votes. The following members of the organizing Committee were
also elected, each with more than 90% of the votes expressed by mail: Anders,
Chapman, Delsemme, Dobrovolskii, Gehrels each 69 votes; Yavnel 68 votes; Roemer
66 votes; Rahe 65 votes; Arpigny 64 votes; since Arpigny had sent his resignation
for health reasons, his mandate was vacant and D. Morrison was unanimously
elected for this mandate at the Grenoble Meeting. Since the President, the Vice-
President and the other members of the organizing Committee had been elected by
mail with more than 2/3 of the votes, the rules were waived and all officers
were unanimously elected by applause at the Grenoble meeting.

2. By-Laws of the Commission

a. The following motion has been carried: The members of the organizing
Committee should not generally serve more than two consecutive terms; this rule
does not apply to the Retiring President; other exceptions should be duly
approved by a majority vote of the other members of the organizing Committee. The
terms of the present officers will be counted from the Sydney 1973 Meeting.
b. After discussion, the motion, seconded by mail, for the Constitution of
several permanent working groups to separate comets, minor planets, and meteorites
within the Commission, has been lost.

3. Ad-Hoc Committees of the Commission

1. Committee for Cometary Archives (October 1974)
Chairman: A. H. Delsemme; Members: J. Rahe; H. L. Giclas.

2. Committee on Cometary Observations and Experiments in Space (August 1976)
Chairman: M. Greenberg; Members: J. E. Blamont, A. H. Delsemme, B. Donn, H. U.
Keller.
3. Committee on Minor Planet Observations from Space (September 1976)
Chairman: D. L. Matson; Members: C. R. Chapman, A. Dollfus, L. Kresak, D.
Morrison, G. Wetherill, K. and I. van Houten.
4. Committee on Cometary Spectra (September 1976)
Chairman: B. Donn; Members: C. Arpigny, A. H. Delsemme, G. Herbig, P. Wehinger.
   The mandate of the first of these committees has been confirmed and the
last three committees have been created by motions unanimously carried on August
26 (1 and 2) and September 1 (3 and 4).

4. Report of the Committee for Cometary Archives

   This Committee was constituted in October 1974, upon recommendation of the
participants of IAU Colloquium No. 25, and immediate approval by the attending
members of the Organizing Committee of Commission 15. A. H. Delsemme was elected
chairman, J. Rahe and H. L. Giclas, members. The Committee has established
targets and priorities. As a first step, it will try to publish a limited edition
of Bobrovnikoff's notebooks, which are a precious source book on cometary data
in the 19th and early 20th century. It proposes also to establish a catalogue,
in a format easy to transfer on IBM cards, concerning a limited number of bright
comets of the 20th century, beginning with 1908 III and 1910 II. The catalogue
would record the location of photographs and spectra, with dates, size and scale
of plates, dispersion and range of spectra, assessment of spatial and spectral
resolving power, type of guiding, name of observer. The idea, at least at an
early stage, is to concentrate the information on comets for which cross-references
are likely to be useful. The mandate of this Committee has been extended
indefinitely.

5. Resolutions

   a) A resolution concerning priorities in comet radio observations is carried
and transmitted to the Executive Committee for further action. It reads:
   The transitory nature and unpredictable appearance of most comets prevent
astronomers from scheduling time on large radio telescopes. In order to exploit
the potential for unique comet data as indicated by recent radio observations,
the XVIth General Assembly of the International Astronomical Union strongly
recommends the priority allocation of time on short notice for comet observations
on large centimeter and millimeter wavelength telescopes.
   b) A resolution concerning the importance of searches for cometary antitails
is carried to be brought to the attention of observers. It reads:
   Although the association of meteor streams with short-period comets is well
established, direct information concerning separation of large dust particles
from the nuclei of short-period comets, available in principle through observations
of antitails, is lacking. Because of the special geometrical circumstances
required and the general faintness of short-period comets, the opportunities for
appropriate observations are rare. Commission 15 therefore calls to the special
attention of observers with access to fast, wide-field telescopes the importance
of searches for antitails.

6. New Members

   New members elected: A. Brecher, C. R. Chapman, C. Cristescu, E. Everhart,
A. Eviatar, E. I. Gerard, L. Grossman, S. Grudzinska, H. F. Haupt, T. V. Johnson,
H. P. Larson, D. L. Matson, D. A. Mendis, B. Milet, E. Moore, J. S. Neff,
L. E. Snyder, K. Tomita, H. J. Schober, M. Wallis, J. T. Wasson, P. Wehinger,
S. Wyckoff, B. H. Zellner. The Commission now counts 94 members.

27 August 1976

SESSION OF SHORT PAPERS ON RECENT RESULTS, WITH EMPHASIS ON COMET WEST 1975n.

1. Production Rates in Comet West, derived from Rocket Spectra - P. D. Feldman and W. H. Brune (in absentia).

Ultraviolet spectra of Comet West were obtained by an Aerobee rocket launched 1976 March 5.49. The principal emission features were lines of C, O and C^+ and bands of OH, CO, CO^+ and CO_2^+. Estimates of the production rates, in units of 10^{28} s^{-1}, are OH: 96; C: 31; CO: 42; O: 110. The CI ($^1D-^1P^0$) line at 1931 A indicates that a large fraction (possibly one third) of the carbon is produced at the metastable ^1D state.

2. Low-Resolution Photoelectric Spectrophotometry of the Inner Coma of Comet West - J. S. Neff and Dean A. Ketelsen.

Absolute flux measurements were obtained at 23 wavelength points separated by 123.2 A with a 130 A bandpass between 3100-5800 A on seven nights. The continuous albedo was found to be wavelength independent. Five continuous flux wavelengths were found to vary as $r^{-3.30} \Delta^{-2}$ with no suggestion of a phase angle dependence. Thus if scattering is important in producing the continuous spectrum the scattering in this wavelength range is colorless and diffuse. The specific intensities of NH, CN and C_2 bands were found to vary as r^{-8} to 9.4 for r>2 AU. For r<1.0 the value of n in r^{-n} was between 1.5 and 4. The low values of n for r<1 are probably due to temporarily enhanced production of molecules due to the exposure of fresh surfaces when the nucleus fragmented near perihelion.

3. Polarization Observations of Comet West - S. Iobe, K. Saito, K. Tomita, and H. Maehara.

Contour maps of Comet West, showing the polarization intensity and angle were obtained from four photographs in the visible taken by setting the polarizer in front of the focus plane. We found that the light is depolarized in a region adjacent to the core and extending sunwards.

4. Motions of the Plasma Tail of Comet Kohoutek - K. Jockers.

Photographs of Comet Kohoutek 1973 XII are being collected from all over the world to make an atlas of the plasma tail of this comet. It will show the comet several times per each 24 hours and will therefore allow the study of the detailed kinematics of the plasma tail. Preliminary data from this atlas were shown and discussed.

5. A UV Spectrum of Comet West - A. M. Smith, R. C. Bohlin, T. P. Stecher.

Exposure of 1/2 to 32 seconds were made on Comet West with an objective grating spectrograph on an Aerobee rocket. The spectral resolution was 3 A and the field of view was 11^0. Atomic lines identified are CI($^3P-^3P0$), Si II, and CI($^1D-^1P0$). Molecules identified are CS, OH, NH and CN. Molecular ions are CO^+, CN^+ and CO_2^+. An absolute calibration was made and we intend to publish absolute intensities for the observed features and continuum.

6. CO^+ Intensity Profiles in Comet West - P. A. Wehinger and S. Wyckoff.

Calibrated spectrograms of Comet West (1975n) have been measured with a PDS microdensitometer as a function of distance in the sunward and tailward directions for CO^+ and H_2O^+ features. Unwidened image tube spectrograms (126 A mm^{-1}, scale perpendicular to dispersion 151 arc sec mm^{-1}) were obtained 1976 March 10-11 when

Comet West was at r = 0.47 a.u. and Δ = 0.94 a.u. The linear extent of the slit at the distance of the comet was ~4 x 10^5 km. Profiles of the relative intensity as a function of distance (log I vs. log ρ) were obtained. Further measurements and analysis are in progress.

7. Ionic Brightness Profiles in Comet West - M. Combi and A. H. Delsemme.

Spectra of Comet West were obtained with the slit aligned along the radius vector to the sun on March 7, 8, 11, 15, 17, 22, 23 and April 3, 5, 7 and 9, 1976. The dispersion was 30 A/mm from 3800 to 4700 A. Brightness profiles of a space resolution better than 4" (about 2500 km) are being obtained from CO^+ (3-0, 2-0, 1-0), N_2^+ (0-0) and possibly CH^+ (1-0) for all dates of March. A density map of each spectrum is being made, using the PDS microdensitometer at KPNO. The instrumental and atmospheric profiles are being removed from the cometary spectra, using the spectrum of one of Oke's standard stars taken before each cometary spectrum.

8. Radio Observations of OH in Comet West - L. E. Snyder, J. C. Webber, R. M. Crutcher, and G. W. Swenson, Jr.

The main lines of OH at 1667 MHz and 1665 MHz have been observed in Comet West (1975n) during post-perihelion passage (9 March through 7 April, 1976) using the 120-ft (37-m) radio telescope of the University of Illinois. Channel widths of 3.13 kHz (0.56 km s^{-1}) and 7.10 kHz (1.28 km s^{-1}) were used. The 1667 MHz line was observed in emission with time-varying intensity. The 1665 MHz line was initially in emission and became stronger than the 1667 MHz line but later changed to weak absorption. Multiple velocity components and daily changes in the emission profile were observed but a zero velocity (with respect to the rest frame of the comet) component was always present. Our results generally tend to support the model of ultraviolet pumping by the sun which was proposed earlier to explain the OH observations of Comet Kohoutek (1973f). For the 17-24 March period, we found an OH production rate $Q=2.2x10^{29}$ s^{-1} which is in rather good agreement with the ultraviolet result of $Q=9.6x10^{29}$ s^{-1} found on 5.5 March by Feldman and Brune (1976 Ap.J. Letters, submitted). For a more complete discussion, see Ap.J. Letters (1976) 209, L49.

9. OH Observations of Comets at 18 cm - E. Gerard, F. Biraud, J. Crovisier, I. Kazes and B. Milet.

Results of observations conducted at Nançay for Comets Kohoutek and West combined with those we obtained at Dwingeloo for Comet Kobayashi-Berger-Milon strongly support the model suggested by Biraud et al. (1974), Astron. Astroph. 34, 163. From March 25 to March 30, 1976 Comet West was tracked not only at the position of the nucleus but also 3'5 east and 3'5 west of it. The measured brightness temperatures of the comet at the latter two positions are both equal to 56%±20% of the nucleus brightness temperature. Therefore Comet West is not a point source when observed with the Nançay radio telescope whose beamwidth is 3'5. The extended source seems to be at least an order of magnitude wider than that observed in optical spectra.

10. A Search for Radio Frequency Emission From CH in Comet West 1975n - E. Churchwell, J. Rahe, H. Keller.

A search was conducted for two hyperfine transitions (F=1-1) and (F=0-1) at 9 cm, of the ground state Λ-doublet of CH in Comet West 1975n, using the 100-m telescope of the Max-Planck-Institute fur Radioastronomie in Bonn/FGR. The observations were carried out on 1976 March 18 at r=0.72 AU and Δ=1.01 AU when the comet had already passed perihelion. No lines of CH were detected above the noise level, but from the observed noise an upper limit on the mean column density, N(CH) in the $^2\Pi_{1/2}$, J = $\frac{1}{2}$ state could be derived. With a full line width at half-maximum intensity of Δv = 5 km s^{-1} and a peak-to-peak antenna temperature ΔT_L(p-p $\approx$ 0.05K, it was found N(CH) $\leq$ 1.1 x 10^{14} molecules cm^{-2}.

11. Splitting of Comet West (1975n) - Z. Sekanina

The theory explaining the relative motions of comet fragments in terms of the differential nongravitional forces gives a very satisfactory representation of observations of the nuclei of Comet West until the beginning of June 1976. Systematic deviations, on the order of a few arcsec, in late June and July are apparently due to a normal component of the velocity of separation, amounting to a fraction of 1 m/sec.

12. Microwave Continuum Emission from Comet West - R. W. Hobbs, J. C. Brandt, and S. P. Maran.

We have detected 3.71 cm radiation from Comet West using the interferometer at the National Radio Astronomy Observatory. On 5 March 1976 the comet was unresolved (<1.8") at a flux of 0.040 flux density units (10^{-26} watts/m^2/Hz). These observations are consistent with an emitting region diameter less than 1100 km and effective temperature greater than 330°K ($\pm 25\%$). On 4 March we failed to detect any cometary emission at the level of .010 flux density units.
We interpret this emission as originating thermally in the icy grain halo proposed by Delsemme, but with increased number of particles. The variability makes possible the solution of these problems easily consistent with the theory of icy grain halos, particularly since the nucleus of Comet West probably split on 5 March.

13. Chemistry of the Inner Coma, A Progress Report - W. F. Huebner.

The composition of the inner coma is modeled assuming that about 30 chemical species composed of H, C, and O undergo reactions. Ionization and dissociation by solar radiation and over 100 forward and reverse reactions between atoms, molecules and ions are considered in the kinetics. Vaporization from a simple H_2O - CO_2 nucleus provides the initial composition of the gas near the surface.

14. Franck-Condon Factors for the Interpretation of Comet Kohoutek's Spectrum of H_2O^+ - B. Petropoulos and R. Botter.

Franck-Condon factors have been calculated for H_2O^+, by the use of the Rydberg-Klein-Rees-Cooley method; they are in good agreement with experimental results, and they will be used for the interpretation of the spectrum of Comet Kohoutek.

1 September 1976

JOINT MEETING OF COMMISSIONS 15, 20 AND 22. RELATIONSHIPS BETWEEN COMETS, MINOR PLANETS, METEORITES, AND METEOROIDS.

This meeting was intended to cover some of the highlights of IAU Colloquium No. 39, that had taken place two weeks before, in Lyon, France; here the emphasis has been put on the astronomical relationships and the field of meteoritics has rather been neglected.

1. Asteroids: Their Relationships to Meteorites and Comets - C. R. Chapman.

Observations and interpretations of the physical properties of asteroids have been revolutionized in the last few years. We now have data on spectral albedos and diameters for hundreds of asteroids. In general, many meteorite types are found to be represented among the main belt asteroids, although parent bodies for the ordinary chondrites and apparently chondritic Earth-approaching objects

are rare or absent in the main belt. Bias-corrected statistics have been assembled
on the distribution of the major compositional types as a function of diameter
and semi-major axes. Syntheses of these data into plausible scenarios for the
geochemical, collisional, and orbital evolution of asteroids are in a preliminary
stage. But it is likely that the asteroids, perhaps along with comets, played
an especially important role in early chapters of planetary history and that they
hold unique clues to fundamental early solar system processes.

2. Comets, Minor Planets and Meteorites: Orbits and Relationships -G. Wetherill.

Ceplecha has identified about 2/3 of analyzed fireballs with meteorite
classes (ordinary and carbonaceous), based on ablation and fragmentation. Orbits
of fireballs show that most were first earth-crossing with aphelion near ~4 A.U.,
a property shared by ordinary chondrites, indicated by distribution of radiants
and time of fall (Simonenko, Wetherill). Some short-period comets share this
property with asteroid fragments accelerated in the 2:1 Kirkwood Gap. Scholl
and Froeschle have now shown the 5:2 gap to be possibly more effective. Irons
and achondrites are most naturally associated with the innermost asteroid belt,
accelerated by non-linear interaction of Mars perturbations and the 7:5 resonance
of Williams. Earth-impacting fragments are certainly derived from Apollo and
Amor objects, but identification with known meteorite types continues to present
problems.

3. Cometary Meteoroids - P. M. Millman

This review covers the mass range of interplanetary particles from a few
kilograms down to 10^{-12} grams. Dynamical evidence, supported by physical data,
suggests that the great bulk of this material encountered by the earth is of
cometary origin and has an integrated mass peak near 10^{-4} to 10^{-6} grams. Several
teams of scientists are now active in the collection and study of small particles
of extra-terrestrial origin at the low-mass end of the above range near 10^{-8} or
10^{-10} grams. Quantitative studies of the chemical composition of both the large
and small meteoroids by four distinct experimental techniques suggest that the
relative abundances of some 10 or 12 common elements correspond closely to those
of the carbonaceous chondrites type I, in other words, to very primitive un-
differentiated material.

4. The Chemical Nature of the Cometary Nucleus - A. H. Delsemme.

Cometary dust reflects the infrared spectrum of silicates, whose vaporization
can explain metallic lines seen in spectra of sun-grazing comets. The gas-to-dust
ratio R observed in Comets Arend-Roland and Bennett is 3 to 9 hundred times as
small as that predicted from solar abundances. However, if we call "primitive"
(=R_0) the solar ratio after excluding free hydrogen and helium, then R≈R_0 in
Bennett whereas R≈1/3 R_0 in Arend-Roland. The Lyman α halo clearly comes from that
hydrogen that was originally bound in molecules like H_2O, HCN and CH_3CN. Of
course the observed R's are only production-rate ratios; since they are both close
to R_0, we assume that both comets were a rather homogeneous mixture of frozen
gases and dust, so that the dust was dragged away almost in proportion to the gas
production rate.
Another clue is the C/O ratio. If we assume that the volatile fraction of
comets comes from the solar nebula, this C/O "primitive" ratio must be higher than
the solar, because of the removal of some oxygen by the condensation of the silicates. Deduced from Ross and Aller's (1976) abundances, it (nominally) is 0.88,
although it could be as low as 0.5 and, interestingly, higher than 1. Recent U.V.
rocket spectra give C/O = 0.23 for comet Kohoutek and 0.28 for comet West, low
enough in both cases to be probably outside the error bars of the "primitive"
ratio. Models suggest that 70% of the carbon was not condensed because it was in
CH_4 (or 95%, if still in CO); therefore R(new comets) would rather be 1/2 or 1/3

R_O. Dust accretion into larger grains, more difficult to drag away, leading eventually to the building-up of a "crust", could explain the apparently larger R of a rather old comet like Bennett; these views are confirmed by Encke's infrared continuum. All this is consistent with a primitive condensation temperature larger than 55°K (or 45° K if CO) whereas it must be lower than 120°K in order to condense CO_2. Indeed, recent results (Delsemme and Combi, Ap.J. Letters, 1 November 1976) suggest that CO_2 is another major constituent of cometary snows, bringing to four (with H_2O, HCN and CH_3CN) the number of parent molecules reasonably well identified. Surprisingly, thermal equilibrium models of the solar nebula are not ruled out to explain these four molecules as major constituents of the condensable fraction of this nebula between 50° and 120°K (Delsemme and Rud 1976), although several other hypotheses must still be explored.

5. The Significance of Cometary Nuclei - F. L. Whipple.

Observation unambiguously supports the theory that cometary nuclei are a low-temperature condensate and agglomerate, formed presumably in the outer primitive solar nebula. Comets are the logical building materials for Uranus, Neptune and several major satellites. Some asteroids may be defunct cometary cores. Comets undoubtedly contributed to the volatiles of the terrestrial planets. To Earth they may have added a major fraction of the life-giving elements.
Because of the vital role played by comets in the formation of the solar system, they deserve intensive study by all possible methods, especially by unmanned space missions.

6. NASA's Cometary Science Program - B. D. Donn

NASA's primary effort in cometary research is expected to be a flyby mission in the 1980's. The tentative plan is a proposed new start in 1982 for dual spacecraft launched from the space shuttle in 1985. One spacecraft intercepts Comet Halley and the other, first comet Giacobini-Zinner and then Comet Borrelly. In anticipation of a comet mission in about a decade, a program is being initiated for instrument development with emphasis on neutral and ion mass spectrometers, dust composition analysis and imaging devices. A comprehensive cometary research program consisting of observational, theoretical and laboratory investigations forms a second part of NASA's Cometary Science Program.

FINAL ADMINISTRATIVE MEETING

Several items of unfinished business were taken care of during a final administrative meeting that took place on September 1, 1976. Most of the final decisions have been reported earlier for clarity. The final list of Consultants of Commission 15 was also established. They are: M. F. A'Hearn, V. A. Bronshten, Ed Bowell, L. R. Burlaga, A. I. Ershkovich, W. K. Hartmann, C. F. Lillie, R. Ong, M. Pérez de Tejada, F. Scaltriti, M. Shimizu, V. Zappala.

COMMISSION 16: PHYSICAL STUDY OF PLANETS AND SATELLITES (ÉTUDE PHYSIQUE DES PLANÈTES ET DES SATELLITES)

Report of Meetings, 27, 31 August and 1 September 1976

PRESIDENT: C. H. Mayer

27 August 1976

JOINT DISCUSSION WITH COMMISSION 17: SPACE MISSIONS TO THE MOON AND PLANETS
Chairmen: F. El Baz, E. Anders

N. Ness: Magnetic Field of Mercury.
S. K. Runcorn: Magnetic Field of the Moon.
J. A. van Allen: Magnetospheres of Jupiter and Saturn.
M. Marov: Results From Venera 9 and 10-Surface and Atmosphere of Venus.
M. J. S. Belton: Cloud Patterns-Waves and Convection in the Venus Atmosphere-Results From Mariner 10.
W. K. Hartmann: Bombardment Histories for Mercury, Mars, and the Moon.
T. Gehrels: Jupiter Atmosphere-Results From Pioneer 10 and 11.
B. A. Smith: Preliminary Results From the Viking I Mars Orbiter.
S. I. Rasool: Preliminary Results From the Viking I Mars Lander.
S. I. Rasool: Future U. S. Planetary Space Missions.

31 August 1976

ADMINISTRATIVE SESSION
The proposal to the scientific unions for an International Solar System Program and the form of its recommendation to the ICSU by COSPAR were reviewed by C. de Jager with additional comments by S. K. Runcorn and A. Dollfus. The consensus of the following discussion, which included Presidents and representatives of other interested Commissions (4, 7, 17, 22) and of the Planetary Data Centers and the IAU Working Group on Numerical Data, was to endorse participation by interested Commissions of the IAU.

Proposed joint resolutions by Commissions 4 and 16: (1) to organize a Joint Working Group to study and report recommendations on the cartographic coordinates and rotational elements of the planets and satellites, and (2) on the Physical Ephemeris of Mars were discussed and approved for presentation to the Commission. A resolution proposed by M. E. Davies to define a new coordinate system for Mercury was discussed and considered appropriate for study and action by the proposed new Joint Working Group.

1 September 1976

I. COMMISSION BUSINESS
The following nominations for officers and organizing committee and a list of 27 proposed new members and one new consultant were presented to the Commission.
President: T. C. Owen
Vice Presidents: B. A. Smith, V. G. Teifel'
Organizing Committee: M. J. S. Belton, D. Gautier, J. E. Guest, C. H. Mayer, S. Miyamoto, D. Morrison, C. Sagan

Topics discussed included the organization of the Commission, the possibility of maintenance of a current list of physical and orbital elements of planets
and satellites, and improvements in the organization of the General Assemblies.

RESOLUTIONS

The following resolutions were approved by the Commission and submitted for
approval by the IAU Executive Committee.

1.   Joint Resolution of Commissions 4 and 16 on Cartographic Coordinates and
Rotational Elements of the Planets and Satellites. (Adopted by Commissions 4 and 16)

Commissions 4 and 16 noting that
(a)   confusion exists regarding the present rotational elements of some of
      the planets
(b)   extensive amounts of new data from radar observations and by direct
      imaging from spacecraft have made cartography of the surfaces of the
      Moon, Mercury, Venus, and Mars a reality
(c)   there will be an extension of these techniques to the mapping of larger
      satellites of Jupiter and Saturn in the near future

assert that
(a)   to avoid a proliferation of inconsistent cartographic and rotational
      systems, there is a need to define the rotational elements of the planets
      and satellites on a systematic basis and to relate the new cartographic
      coordinates rigorously to the rotational elements.

and therefore recommend that
(1)   Commission 4 (Ephemerides) and Commission 16 (Physical Study of Planets
and Satellites) establish a Joint Working Group to study the cartographic
coordinates and rotational elements of the planets and satellites and to
report recommendations thereon at the next general assembly of the IAU.

2.   Joint Resolution of Commissions 4 and 16 on The Physical Ephemeris of Mars.
(Adopted by Commissions 4 and 16)

Considering that recent new determinations of the rotational elements of Mars
indicate the need for a revision of the elements currently adopted in the
physical ephemeris of Mars, and that a new approach to the definition of the
origin of areographic longitudes appears useful (G. de Vaucouleurs, M. E.
Davies and F. M. Sturms, Jr., J. Geophys, Res. 78, 4395, 1973), Commissions 4
and 16 recommend

(1)   that the tie between the new and current physical ephemeris of Mars be
firmly established by appropriate comparisons between ground-based and Mariner
coordinate systems, and
(2)   that new elements and a new definition of the origin of the areographic
longitudes consistent with the results of (1) above and the definitions adopted
previously (IAU Trans. XVB, 107, 1973) be incorporated in the physical
ephemeris of Mars as soon as deemed practicable in the judgement of the
cognizant Directors of the National Ephemerides Offices.

II.   SCIENTIFIC SESSION

INVITED REVIEWS

R. Smoluchowski: The Interiors of the Outer Planets.

G. Pettengill: Radar Studies of Planets and Satellites.
D. Campbell, G. Pettengill: Arecibo Radar Maps of Venus.
R. Goldstein: Goldstone Radar Mars of Venus.
D. Morrison: Recent Research on Planetary Satellites.
A. Betz: Infrared Heterodyne Spectroscopy of Planetary Atmospheres.
G. Hunt: The Lower Atmosphere of Jupiter.
D. Gautier: The Upper Atmospheres of the Outer Planets.

SHORT REPORTS

A. Dollfus: Report on the IAU PLanetary Photographic Data Center at Meudon.
W. A. Baum: Report on the IAU Planetary Research Center at the Lowell
             Observatory.
B. Andrew: Longitude Dependence of Mars Radio Emission.
P. Wehinger, S. Wyckoff: Io's Extended Sodium Cloud Torus.
W. Irvine: Saturn's Rings.
C. Macris, B. Petropoulos: Seasonal Variations of the Pressure in the
             Martian Atmosphere.
F. Johnson: Solar System Formation.

COMMISSION 17: THE MOON (LA LUNE)

Report of Meetings, 28 August - 1 September, 1976

PRESIDENT: S.K.Runcorn   SECRETARY: J.A.O'Keefe

BUSINESS MEETING
        The business meeting of Commission 17 was called to order at 0900 on Thursday
26 August, in Room 14, ENS d'Electrotechnique of the University of Grenoble,
Professor S.K.Runcorn in the chair.
        A resolution in support of a lunar polar orbiter was read by O'Keefe and
passed;   see below.
        Moutsoulas then brought forward the question of defining the position of a
lunar crater.   Up to the present, the only practical way to define the position
of most lunar craters has been as the center of the rim, since only the rim has
been visible.   Orbital photography, however, would now permit the use of other
points, such as a central peak, or a central point on the crater floor.   For the
sake of continuity with the older observations, it is important that we should
continue to use the center of the rim, even though the points so defined are high
above the level of the lunar surface near the crater.   A resolution was therefore
passed making the center of the rim the reference point for selenographic
measurement;   for the text see below.
        The chairman then drew attention to the International Solar System Decade,
proposed by COSPAR and ICSU, to be modeled on the IGY and the IQSY.   The purpose
is to coordinate studies of the solar system with chemists, physicists, mathe-
maticians and others, as well as with ground-based observatories.   Special
emphasis seems appropriate on the use of new infra-red detectors.   W.A.Baum
commented that voluntary programs of this kind are not sufficient;   telescopes
must be dedicated to the program.   A joint meeting on Tuesday, 31 August, was
announced.
        The chairman then announced the names of E.Anders for the next president of
Commission 17, and K.P.Florensky for vice-president;   these were accepted without
objection.

RESOLUTIONS
        The following resolutions were passed without objection:
1.   The International Astronomical Union noting that improved values of the
second harmonic are critical to theories of the lunar interior and noting further
that the free librations of the moon provide an important clue to its past (impact
or volcanic) urges that subsatellites be put in polar orbit round the Moon, to
determine its gravitational field, especially for the low harmonics, as precisely
as possible.
2.   We recommend that the long-established practice of referring lunar crater
coordinates to the center of the crater at the mean elevation of the rim be
continued, noting that such coordinates pertain, in general, to positions elevated
above the immediate surroundings.

SCIENTIFIC SESSIONS
        A half-day of discussion was devoted to Working Group No. 1, "Figure et
rotation de la Lune".   The session was held at 1400 on 25 August in Room P1 of
the building Sciences Physiques of the University of Grenoble.   The general theme
of the discussion was the need for more precise theoretical developments to meet
the needs of modern observations.   The speakers and their titles were as follows:

1.   Mr.L.V.Morrison, Greenwich Observatory, "Comparison of occultation observations with lunar theory".
2.   Dr.J.Kovalevsky, Observatoire de Meuden, "Developments in analytical theory".
3.   Dr.T.C.van Flandern, U.S.Naval Observatory, "Comparison of occultation observations with integrations".
4.   Dr.V.K.Abalakin, Institute of Theoretical Astronomy, Leningrad, "Estimated accuracy for lunar parameters from laser ranging".
5.   Dr.J.G.Williams, Jet Propulsion Laboratory, NASA, "Results from laser ranging".
6.   Dr.D.H.Eckhardt, USAF Cambridge Research Laboratories, "Developments in analytical theory".
7.   Professor A.H.Cook, Cavendish Laboratory, "Theory of lunar librations".
8.   Dr.L.A.Shimerman, Defense Mapping Agency, St.Louis, "The expanding Apollo control system".
9.   Dr.J.G.Williams, Jet Propulsion Laboratory, NASA, "Results from lunar laser ranging", second part.
10. Dr.C.C.Counselman, III, Massachusetts Institute of Technology, "Results from Very Long Range Baseline Interferometry".

After the business session, on Thursday, 26 August, a scientific session was held on a General Survey of Recent Developments in Lunar Research. The speakers and their titles were as follows:
1.   Professor S.K.Runcorn, University of Newcastle, "Physics of the Moon".
2.   Professor E.Anders, University of Chicago, "Chemistry of the Moon".
3.   Dr.G.Turner, University of Sheffield, "Ages of the Moon".
4.   Professor C.P.Sonnett, University of Arizona, "Solar wind induction in the Moon, and its internal electrical conductivity".
5.   Dr.T.Johnson, Jet Propulsion Laboratory, NASA, "Lunar Polar Orbiter".
6.   Dr.S.Asaad and Dr.J.S.Mikhail (presented by Dr.Asaad), Helwan Observatory, "Report on the lunar work at Kotamia, 1973-1976".

In the afternoon of the same day, Working Group No. 2, "Physics, chemistry and geology of the Moon" met at 1400 in Room D-1, Faculte des Sciences Sociales, for a General Discussion of Processes Involving Surface Features. By correspondence, the Working Group had been discussing the reliability of the impact theory for the origin of most lunar craters. Copies of the correspondence on this subject were made available. The speakers and their titles were as follows:
1.   Dr.R.J.Pike, U.S.Geological Survey, "Crater form and cratering process: diagnostic tests from multivariate statistics".
2.   Professor T.Gold, Cornell University, "The exogenic view of the lunar surface".
3.   Dr.J.A.Bastin, Queen Mary College, "The liquefaction hypothesis: the origin and distribution of marial craters".
4.   Dr.J.Green, University of California at Long Beach, "Lunar volcanism at all scales".
5.   Dr.J.A.O'Keefe, Goddard Space Flight Center, NASA, "External vs. internal causes of lunar craters: a summary".
6.   Dr.J.Iriyama, Chuba Institute of Technology, Japan, "Lunar chronology and evolution inferred from the radiometric age data of Apollo and Luna rocks and soils", and, with Dr.M.Honda, University of Tokyo (presented by Dr.Iriyama) "Movement process of the lunar surface part and the cosmic ray exposure age of the lunar materials."
7.   Mr.E.A.Whitaker, University of Arizona, "New approaches to lunar cratering statistics".
8.   Dr.C.R.Chapman, Planetary Research Institute, Tucson, Arizona, "Origin of sub-kilometer diameter craters".

On Friday, 27 August, at 0900 in the Weil amphitheater of the University of Grenoble, a joint discussion was held with Commission 16 on Space Missions to the Moon and Planets, for which see the report of Commission 16.

REPORT OF THE WORKING GROUP. "Figure and rotation of the Moon" 1974-1976 by
Dr.T.Weimer (chairman)

Le Groupe de travail "Figure et rotation de la Lune" compte en 1976 20
membres; 12 d'entre eux appartiennent a l'U.A.I.; 8 sont membres consultants et
ne font pas partie de l'U.A.I.

Entre janvier 1973 et février 1976, 5 bulletins bibliographiques ont été
publiés, donnant les références d'environ 160 articles (60 sur la figure, carto-
graphie, etc..., 20 sur la rotation, 80 sur le mouvement orbital, le champ gravi-
fique, etc...). Ce sont là des chiffres équivalents à ceux de la période 1970-73.

Il est inutile de revenir en détail sur les travaux et recherches faits dans
les divers instituts; les bulletins bibliographiques tiennent lieu de résumé.
Nous nous bornerons à énumérer ci dessous ce qui n'y a pas trouvé place:

-La "Defense Mapping Agency", Saint-Louis (U.S.A.) continue ses travaux de
sélénodésie en utilisant surtout les documents obtenus par Apollo 14-17. Voici
les sujets des recherches en cours:

1° Relation of Radio Transmitters to Laser Retroreflectors at Apollo 14-15
landing sites;

2° Development of an Apollo Selenodetic System;

3° Development of a new earthbased telescopic Selenodesic System;

4° Lunar positional Reference System (1974) completed.

-K.Koziel à Cracovie (Pologne) a montré que la valeur de f (0,633) déduite
des observations héliométriques concorde dans la limite des erreurs, avec celle
obtenue par laser (0,642). Il cherche à déterminer la valeur de la libration
arbitraire d'apres 10.000 observations héliométriques s'étendant de 1841 a 1945.

-A l'Observatoire de Kiev, Gavilov et ses collaborateurs Kisliuk et Duna

1° etablissent un système de positions sélénographiques de cratères en
comparant les differents catalogues;

2° determinent les hauteurs absolues de 960 points d'après les profils
obtenus à l'astrographe de Kiev (D = 40cm, F = 550cm);

3° font d'étude du profil près du méridien 250°W d'après des clichés de
"Zond 8" et confirment l'existence de la dépression découverte par l'altimètre
laser de Apollo 15-16;

4° comparent les altitudes obtenues à partir de la Terre (catalogue séléno-
désique de Kiev) avec celles déduites des observations spatiales;

5° concident, après étude des divers catalogues, que 40 à 100 points suffi-
sent pour définir un système fondamental de références.

-De nombreux travaux sur les problèmes de sélénodésie et le système fondamental de référence ont été faits à l'Institut de Recherche spatiale de l'Académie
des Sciences de l'U.R.S.S. (Moscou) par A.A.Gurshtein et ses collaborateurs; ils
seront mentionnés en détail dans le bulletin bibliographique no. 14.

Durant la période 1973-76 il n'y a pas eu de découvertes nouvelles sensation-
nelles. L'heure est à l'exploitation plus complète des observations et au per-
fectionnement des méthodes d'observation et des théories (détails dans le bulletin
bibliographique). Actuellement les méthodes classiques d'observations (visuelles
ou photographiques) continuent à coexister avec les techniques les plus modernes
telles que photographie depuis les vaisseaux spatiaux, utilisation de cellules
photoélectriques ou de laser. Mais il est vraisemblable que dans quelques années
les nouvelles méthodes supplanteront les anciennes et fourniront la forme, la
libration et même l'orbite de la Lune avec une précision si grande que d'autres
domaines en tireront bénéfice. Dès à présent les mesures par laser des distances
Terre-Lune ont donné d'excellents résultats pour la rotation de la Terre et les
mouvements de son pôle (Observatoire de MacDonald, Texas). On envisage même d'
étudier le mouvement des plaques techniques par ces mêmes méthodes (Bender,Silver-
berg). Cela implique la création de nouveaux centres d'observation laser.
Effectivement, en plus des stations existant aux U.S.A. et en U.R.S.S., d'autres
sont en cours d'édification en Australie, en France, au Japon, à Hawai.

Report of Ad Hoc Working Party of Commission 17, IAU, on Transient Lunar Events
by Dr.J.E.Geake

The membership comprised A.Dollfus, Chairman, Paris Observatory, Meudon, France,
W.E.Brunk, NASA HQ Washington DC, USA, M.E.Davies, Rand Corpn., Santa Monica, Cal.
USA, F.El-Baz, Smithsonian Inst., Washington, DC, USA, J.E.Geake, Secretary,
UMIST, Manchester, UK, S.K.Runcorn, University of Newcastle, UK and E.A.Whitaker,
University of Arizona, Tucson, Arizona, USA.

Terms of Reference: This ad hoc working party has been set up by S.K.Runcorn,
retiring President of Commission 17, to discuss transient lunar events (TLE's),
and any action that should be taken to study them.

Definition: TLE's are any temporary changes observed to occur on the Moon.
Those reported include obscuration or blurring of surface details, and brightness
changes, which are sometimes coloured.   They are usually 10-100 km across, and
last from a few minutes to an hour or so;   some point flashes have also been
reported.

The present situation was discussed, and may be summarised as follows:

TLE's reported: The first systematic list of 579 TLE's was compiled by Miss B.M.
Middlehurst and others in 1968, from historical and modern records;   it was
extended by Patrick Moore in 1971 to include a further 134 reports, with an
attempt to weight them as regards reliability.   An up-to-date catalogue of over
1400 reports is in preparation by Mrs.W.S.Cameron.
    Most of the recent reports are from amateur observers, with a wide range of
facilities and experience, and it seems probable that many of the effects seen are
really caused by atmospheric effects, instrumental aberrations or eye fatigue.
However, a few observations have been made by experienced and skeptical
astronomers and cannot be so easily dismissed;   but these are nearly all visual
observations (although usually confirmed by other observers), and there is a
serious lack of quantitative instrumental records of TLE's.   Probably the most
convincing visual observation was by Greenacre & Barr in 1963, using a 24 inch
telescope at Flagstaff;   they saw sparkling red light in Aristarchus, and this was
confirmed by other observers using the 69 inch telescope at Perkins Observatory.
    Probably the strongest piece of recorded instrumental evidence is the spectrum
obtained by Kozyrev in 1958, which he ascribed to gas emission from the central
peak of Alphonsus.   Dollfus and Kuiper have both examined the original negative
very critically, and separately concluded that it could not be faulted, and that
the event was real.   Kozyrev later obtained other transient spectra for
Alphonsus and Aristarchus.
    The only other strong piece of evidence for the reality of TLE's is their
correlation in time with lunar perigee, as discovered by Middlehurst - before
the same correlation was reported for Apollo seismic events by Ewing et al.. This
apparent connection with seismic events, which are themselves not in doubt and
which seem very likely at least to disturb surface dust, strengthens the
probability that at least some TLE's are real.
    Correlation of TLE's in location, with mare rims, was discovered by Middle-
hurst and Moore, and independently by Mrs.Cameron.   This correlation seems
plausible in view of the deeply fractured nature of the rock in these regions.
    Correlations of TLE's in time and location must nevertheless be treated with
caution, in view of the possibility of selective observation, once the supposed
correlations were announced.   For this reason, Middlehurst has considered it best
to ignore all recent observations.
    El-Baz pointed out that from the geological point of view there should be some
TLE's associated with motions of the lunar surface.   Most common among these is
the collapse of crater walls that are steeper than the angle of repose.   This

happens most commonly within young or 'Copernician-age' craters.   Another type
of motion must occur along fractures and graben rilles.   It would be logical
for this type of vertical motion to occur during times of high moonquake activity
(it must be added that the majority of these fractures are concentrated on or
near the <u>rims</u> of the lunar maria).   Both types of motion would be expected to
cause a temporary dust cloud that might be big enough to see from Earth, and
could cause the temporary obscuration described by some TLE observers.

<u>Observational attempts to confirm TLE's</u>
    The obvious need is for more instrumental evidence of TLE's, and several
attempts have been made to use a network of amateur observers, who are willing
to monitor the Moon continuously and to alert well-equipped professional
observatories when events are reported.   Organised attempts to confirm TLE
observations were reported by those present, as follows:
(1) Brunk described a program supported by NASA and carried out at the Corralitos
    Observatory of Northwestern University, in New Mexico, USA.   It used a 24-
    inch reflector, with an image orthicon, a choice of filters, and a zoom lens
    for image scale selection.   The TV output was used, rather than direct
    observation, and the display could be photographed at any time.   The
    observers were linked by amateur radio to Argus Astro-Net amateur observers
    at at least 10 locations; both the amateurs' reports and the Corralitos
    findings were tape-recorded.   Reports were also received from amateurs of
    the Moon-Blink network, who used alternating colour filters to look for
    lunar colour changes.
    The observer in charge at Corralitos was Justus Dunlap.   Dr.J.A.Hynek was
    in charge of the program, under Brunk as NASA monitor until 1971, and R.
    Bryson thereafter.   This program was run for 7 years (1965-1972);   98
    amateur reports of TLE's were received and it was possible to check 39 of
    them.   Not one of them was confirmed.
    It was pointed out that the Corralitos observers were mostly local students,
    who were neither professionals nor astronomers.   Brunck agreed to circulate
    copies of the final report of this program (NASA CR-147888) to those present.
    Both the Astro-Net and the Moon-Blink networks are now inoperative.
(2) Miss Middlehurst organised the Lunar International Observers' Network (LION)
    from MSC (now JSC) Houston, with NASA support.   This operated throughout the
    period covered by the Apollo 8 - 14 missions.   No results are known.
(3) Mrs.Cameron (at GSFC) collects amateur and other observations of TLE's, and
    has organised a TLE observing program for the Association of Lunar and
    Planetary Observers (ALPO).   This is still active.
(4) H.Ford in Dundee, Scotland, Director of the BAA Lunar Section, is organising
    a network in the UK, to alert Dr.R.Maddison of Keele University who will
    check reported TLE's with his 18" reflector, which is equipped with a
    spectrograph.
(5) Davies reported a JPL program of visual observations using a 16" reflector on
    Table Mountain.   No reporting network was used.   No TLE's were observed.
(6) Dollfus reported that he was alerted when a dark-side TLE was reported in
    1969, during the Apollo 11 mission.   He was at the Pic du Midi Observatory,
    and the seeing conditions were excellent.   He used the 1 m telescope in
    coronograph mode, to suppress the light scattered by the illuminated crescent,
    giving unprecedented observational conditions.   He immediately observed the
    site of the reported TLE (the wall of Aristarchus), and at first saw nothing
    unusual;   then he saw some flashes.   He was doubtful as to whether they were
    real, and took a rest, after which he again saw no flashes at first, but did
    see some after a minute or so.   He concluded that the flashes he saw were all
    due to eye fatigue.   He saw no events that he regarded as real during 1½
    hours of observation.
(7) Brunk reported a NASA-supported program using a telescope at Port Tobacco,
    Maryland;   the seeing conditions there were poor, and the observers were not
    astronomers.   This observatory took part in the Moon-Blink program, which

operated for 1½ years from 1964, with Mrs. Cameron as Technical Monitor.   5
events were reported from Port Tobacco: 3 were reported from a similar
instrument at Huntsville Alabama, and 2 from one in Edinburg Texas; both
of these instruments were run by professional astronomers.
The Moon-Blink alert network (of 12 east coast amateur and professional
observatories) was only in operation for the last of the events reported
from Port Tobacco (on Nov 15 1965); only 6 of these observatories had clear
sky, and of these 4 saw nothing unusual but 2 confirmed an anomaly at the
reported lunar site.   The Moon was not up at Corralitos.

Relevant NASA observations involving spacecraft, existing or planned.

(1) Davies reported on the Lunar Polar Orbiter: the only relevant instruments
   will be the Spectro-Stereo Scanner, and possibly the Electron Reflection
   Experiment.   The most directly applicable will be the Spectro-Stereo
   Scanner: this will scan a 40 km width, swept along by spacecraft motion,
   with barely overlapping 1° apart tracks on successive polar orbits.   The
   resolution will be ~ 1 km, and there will be 6 filter bands.   It might
   detect patches of obscuration, but it takes about 360 orbits of nearly 2
   hours each to cover the whole Moon, so the chance of happening to catch a
   TLE lasting a few minutes, and of a few 10's of km diameter, is very small.
   It would not detect flashes of emitted light, as the data processor would
   regard these as interference, and ignore them.   Also, the scanner is only
   to be switched on for a small proportion of the time, because the data-
   recording capacity (required when the LPO is behind the Moon) is required
   for other experiments.   This limitation does not apply to the near-side
   part of each orbit, when communication is direct.   However, the whole thing
   will probably only be switched on for part of the time, to economise in
   earth-based communication, space-tracking and data-handling facilities.
   In short, the chance of the LPO detecting a TLE is probably negligible.
   It would not be realistic to suggest to NASA any change in the LPO program
   of observations, in order to look for TLE's.   The present mood in the US,
   as regards funding, is such that the future of the whole LPO program is in
   some doubt.

(2) The other most relevant NASA experiments are the various Apollo surface
   (ALSEP) and orbiting (CSM) instruments.   Of the surface instruments, the
   mass spectrometer gas analyser (Apollo 16 only) and the Cold Cathode Gauge
   (CCG) showed transient gas (e.g. H_2O and NH_4), but this was thought to be of
   spacecraft origin.   These instruments are not now working.   3 of the
   Passive Seismic Experiments (PSE) are still in operation (at the Apollo 15,
   16 and 17 sites) and are still able to locate moonquakes as regards site and
   depth; correlation of these in time and location with independent evidence
   of TLE's would be of the greatest interest.   However, these instruments are
   now only interrogated occasionally (when their accumulation of recorded data
   is acquired) so the discovery of correlation in time with observed TLE's will
   be difficult.   Runcorn suggested that it might be possible to monitor the
   most likely times of activity (i.e. perigee) more continuously, together with
   Earth-based visual observation of the 12 or 13 most likely sites.

(3) Instruments used on board the CSM's while in orbit included α-particle
   detectors.   The results were described by Hodges (Commission 17, Aug. 26
   1976) and by him and Gorenstein (7th Lunar Science Conf., Houston, March 76).
   The instrument detected:

   (a) He^{++} from solar wind and solar-ionised surface emission

   (b) ^{222}Rn (by $^{222}Rn \xrightarrow{\alpha} {}^{218}Po \xrightarrow{\alpha,\beta^-,\beta^-} {}^{214}Po \xrightarrow{\alpha,\beta^-,\beta^-} {}^{206}Pb)$

Estimates of the amounts of those gases released per year from the surface,
and from the lunar atmosphere, were given.   This evidence is relevant to
the possible mechanisms of TLE's, but did not (and could not) give
information about separate events.

(4)   There were 3 reports by Apollo astronauts, in lunar orbit, of flashes on the
        surface.   Mattingly (Apollo 16) saw one on the Earth-lit dark side of the
        Moon;   it appeared to come from below his horizon, so he could not identify
        the site.   It might have been a cosmic-ray induced eye flash, but he differs
        from other astronauts in not usually seeing these.   On Apollo 17 Schmitt
        saw a flash in Grimaldi, and Evans saw one on the eastern rim of the
        Orientale basin.   These astronauts were all, by nature and training,
        skeptical of such events, but they at least agreed that these observations
        were unexplained;   they are discussed in the Apollo 16 and 17 Preliminary
        Science Reports (under 'Visual Observations', by El-Baz as co-author with
        the Astronauts).
        The Apollo 11 astronauts, while in lunar orbit, were alerted to look at
        Aristarchus on their next pass, because observers on Earth were reporting
        activity there;   they did so and reported that the NW wall of the crater
        appeared to be unusually bright.   At the same time, two astronomers in
        Bochum, Germany, observed the same effect.

Possible future NASA-supported observations

Brunk and Davies agreed that there was at present no chance of obtaining
NASA support either for a further Earth-based TLE observing program, or for
the special use for this purpose of existing or planned spacecraft equipment.
Even a recommendation in this direction from a group of internationally
recognised astronomers of the IAU would probably be ineffective.
The only approach likely to succeed might be for some distinguished
astronomer dedicated to this work, and able to convince others of its
importance, to apply to NASA for a general research grant for lunar studies.
Dollfus pointed out that, as Kozgrev's spectra were probably the most
convincing evidence to date, it was important that any observer willing to
make systematic and extensive spectroscopic observations should be
encouraged and supported.   Mrs.Cameron at GSFC has indeed been doing this
kind of work;   she has now photographed over 300 spectra for 20 suspected
TLE sites, and has found a possible anomaly on one of them, in the form of
an extra absorption line at 4908A in Plato.
A search of Lunar Orbiter photographs, and of Apollo CSM photographs from
orbit, might yield evidence of events, especially obscurations.   So far,
no studies of this nature have been initiated, and this might be well worth
doing.   Anyone willing to undertake it should be supported, perhaps through
the Lunar Science Institute.
The present NASA thinking tends to be dominated by geologists who have little
interest in the physical processes implied by TLE's, or even by moonquakes,
which they only study in order to elucidate the structure and composition
of the lunar interior.

Action to be taken

It was agreed that some of the TLE's reported were probably real, but that
no further progress could be made without convincing quantitative instrumental
evidence.   Unambiguous results, using photography, photometry, and especially
spectrophotometry, would be of great interest, and of major importance in
increasing our understanding of the physics of the lunar surface.
        It was decided that the only immediate action possible was to ask for IAU
support for a resolution drawing attention to the importance of this area of
investigation.   It was therefore agreed to place a proposal as follows before
the General Assembly on September 2 1976, from Commission 17:

        "IAU Commission No. 17 considers that lunar transient phenomena, that
        have been reported by many experienced astronomers, warrant further
        quantitative observational and theoretical studies, especially in view of
        their suggested correlation with moonquakes and the discovery of releases
        of argon and radon from the lunar interior." (This was passed by the Assembly)

COMMISSION 19: ROTATION OF THE EARTH
(ROTATION DE LA TERRE)

Report of Meetings, 26, 27 and 31 August 1976

PRESIDENT: C. Sugawa   SECRETARY: H. Abraham

<u>26 August 1976</u>

FIRST SESSION, WITH COMMISSIONS 4, 8, 31 AND 40

<u>Advances in Techniques for the Determination of the Rotation of the Earth</u>

        The President opened the meeting and invited R. Anderle to report on polar
motion determined by Doppler satellite observations. The speaker said that the
standard error in pole positions from 48 hours of Doppler observations was only
7 cm. However, because of uncertainties in the gravity field the standard deviation was 60 cm. This gave a standard error, for a 5-day mean based on two satellites, of 25 cm.

        P.L. Bender spoke about determination of the Earth's rotation by lunar laser
ranging. The accuracy for a measurement of UTO in 3 hours was 0.5 ms, or even
0.2 ms, but gaps of several days could occur, especially near new moon. Comparisons indicated that BIH smoothing did not remove real variations and that most
causes of LLR scatter were known. His personal opinion was that LLR would give
the long term values and LAGEOS would interpolate.

        E.M. Gaposchkin spoke on laser ranging to LAGEOS. Artificial satellites
had no fundamental longitude reference and so could not determine UT but could
be excellent for precise metrology of the Earth. Preliminary LAGEOS results gave
a standard error of 4 1/2 cm for a single pass and showed great future potential.

        J.D. Mulholland then spoke on the EROLD programme which had been arranged
to ascertain whether, initially, five lunar laser ranging stations in four countries
could carry out sustained operations successfully. There were good prospects of
operating by 1977.

        Another technique was presented by C. Counselman who explained determination
of the Earth's rotation by VLBI. This was shown to be a powerful means of determining the Love numbers for earth tides, polar motion, universal time, clock
comparisons, precession and nutation and the coordinates of celestial sources.

        In contrast with the usual VLBI systems, B. Elsmore then discussed the poten-
tialities of measuring UT1 with connected element radio interferometers. Time
could be obtained with a radio interferometer built for that purpose, or could be
obtained in the course of synthesis mapping. The probable error in UT1 from a
simple 12-hour observation with the 5-km telescope, was ± 4.0 ms.

        K.J. Johnston described the determination of UT1 using a 35-km baseline.
Three antennas, 2.7 km apart, were connected by cable, and a 45-ft dish was
situated 35 km from these. This array, which had a baseline of 220,000 wavelengths,
formed an interferometer that measures phase difference (whereas VLBI measures
delays) and its accuracy was expected to reach about 0".01.

SECOND SESSION, WITH COMMISSION 31

IPMS Problems (I)

       This being a joint meeting the President asked the President of Commission
31, H. Enslin to take the chair and H. Fliegel to be the Secretary. The following
is a condensation of the Secretary's detailed report.

       B. Guinot explained the MEDOC experiment. The work would be similar to that
by DMA but with non-identical networks and a different model of forces. The
experiment would attempt to reveal effects due to different reduction programmes
and would promote an international scientific polar motion service based on
satellite techniques. DMA would collaborate.

       A.R. Robbins then gave his report, "A Future International Earth Rotation
Service". The speaker summarised correspondence with members of IAG Special Study
Group 1.04 and other interested colleagues; this referred to the roles of the
ILS, IPMS and BIH and to lunar and satellite laser ranging and VLBI. Correspondents
had agreed almost unanimously that it was not yet possible to determine the form
of a new service and that the IPMS and ILS were essential at present. He also
listed details about the three new systems that will need to be estimated for
future debate.

       Evidence which supported the IPMS was presented by Wm. Markowitz in a paper,
Comparison of ILS, IPMS, BIH and Doppler polar motions with theoretical. For the
past eight years the IPMS results had given the best agreement with the model
and their annual and Chandler components had been remarkably stable. However, the
ILS curve showed large deviations and has a non-periodic component. Markowitz
suggested that ILS observations should be continued for some time to permit comparisons of secular motions.

       In the ensuing discussion BIH, IPMS, ILS and Doppler results were compared
as to agreement, stability, significance and resolution of the 12 and 14-monthly
spectral peaks. Members considered basic objectives and the urgency for involvement
in new techniques. A particular problem was whether new equipment should be sited
for close relationship to existing stations or for best results in the future.

                              27 August 1976

THIRD SESSION

Observations and Reference System

       The need for care in selecting sites for astrometric observations was emphas-
ised by G. Teleki since there could be refraction anomalies of $0\rlap{.}''01$ even at small
zenith distances. The nightly variations of latitude observed at Kitab and Ukiah
had opposite signs, and both sites were near to regions where the atmospheric
density and radio-refractivity fields were perturbed.

       R. Verbeiren discussed the reduction of results obtained by the ILS stations
during the past 75 years. Faulty declination corrections affected the results
when complete groups were not observed; also changes had occurred in the nutation
elements, in the constant of aberration and in the micrometer screw calibrations.
Rounding off errors reached at least $0\rlap{.}''03$. A correction of $-\rlap{.}''0080 \pm 0\rlap{.}''004$ for
the nutation in obliquity had been obtained from the ten pairs of stars that had
been observed by the ILS continuously.

R. d'E. Atkinson spoke about the Earth's axes of rotation and figure.
He showed that observations of zenith distance can give completely the pole of
figure on the celestial sphere; hence the displacement of the pole of date from
the adopted pole of figure is known on the Earth. The Eulerian pole, however,
moved on the surface of a fixed cone and, unlike the pole of date, could not
be obtained from a single observation.

D. Djurović then discussed the stability of the reference system for observing the rotation of the Earth. In the case of many stations their series of observations did not conform to a certain single formula, nor did many of the classical
instruments reproduce the same errors from year to year. There was also correlation
between the time and latitude residuals, especially for some of the stations.

FOURTH SESSION

Physical Interpretation

L. Randić spoke on the influence of the rotation axis on declinations.
Results indicated that the Earth was not rotating as a rigid body with a unique
axis. To find the true nature of the rotation we should compare the declinations
of equatorial stars determined from the northern and southern hemispheres.

E.P. Fedorov and Ya. S. Yatskiv emphasised that the CIO was attached to the
zenith of the five international latitude stations; it was not fixed to them.
Results presented showed how variations occur in the observed axes between these
stations; thus there was no sense in considering the CIO as a fixed origin of the
polar coordinates.

Ya. S. Yatskiv then reviewed evidence on models of the Chandler wobble.
Up-to-date estimates of the parameters agreed well but fluctuations were suggested
by the instantaneous values. Suggested causes of the wobble were discussed,
including atmospheric and seismic excitations. Finally, problems of the nearly
diurnal wobble were reviewed. Observations had shown both prograde and retrograde
components.

K. Yokoyama presented a paper by Y. Wako and himself on nutation terms
derived from time and latitude observations. The equation for determining the polar
position and UT1 from time observations was modified to include a new term.
This τ-term described the effect that nutation errors have on the longitude and
is analogous to the z-term for latitude observations. It would help in evaluating
the nutation amplitudes.

P. Brosche discussed tidal friction of the oceans at present and in the past.
Average values of the torque and the energy dissipation exerted by the M_2 tide
were presented on the basis of 4° models of the present and Permian oceans.
The oceanic tidal friction was probably low at the critical stage when the moon
was close to the Earth.

At the request of the President, G. Winkler described the tiltmeter around
the USNO PZT. This was a system of plastic tubes containing mercury and mounted on
invar. It indicated earth tides but could be affected by temeprature changes
or very heavy rain.

FIFTH SESSION

Administrative Meeting

The report of the IPMS was presented by the Director, S. Yumi. He reviewed
the status of reports not only from the ILS and IPMS latitude observations but
also from the IPMS time observations. The latter had been calculated both with
and without the τ-term. Weights had been estimated both from the internal and
the external consistency. Weight from latitude was about twice that from time.
Cooperative observations for time and latitude would soon be made with PZTs on
the $39°8'$ parallel at Mizusawa, Kitab, Cagliari, Washington and Ukiah. Doppler
satellite observations had been made at Mizusawa and it was reported that Carloforte (or Cagliari), Gaithersburg and Ukiah would be equipped as soon as possible.

The report of the BIH was presented by the Director, B. Guinot. He discussed developments and also the demands for promptness, stability, accuracy and
revisions.The BIH origin for coordinates of the pole had been adjusted to the CIO
in 1968 but discrepancies between the BIH and ILS results had rapidly appeared.
On the introduction of satellite data in 1972 uncertainties in coordinates had
been reduced by more than 50% and UT1 had been improved also. The BIH would take
part in the EROLD campaign (lunar distances) and in the MEDOC experiment (Doppler
observations of artificial satellites). The Rapid Service had been improved since
1976 by the inclusion of USSR data and would continue to be needed for predictions
of DUT1. Before 1984 IAU recommendations would be needed as to corrections for
the deflection of the vertical, the effect of zonal tides on UT1, the forced
diurnal nutation and nutation for a rigid Earth.

Votes of thanks to Yumi and Guinot were passed with acclamation.

S. Yumi next presented the report of the Working Group on the Pole Co-
ordinates.Original data had been collected at Mizusawa, except for missing Batavia
data for 1935. The conversion to machine readable form had been done mainly at
Mizusawa and should be completed in a year. Carloforte data for 41 years had been
punched in Cagliari. Further discussion by the Working Group was needed on the
method of reduction, a definition of each station's mean latitude, and on certain
errors. Completion was expected by mid-1977, funds permitting. Financial support
was gratefully acknowledged. The President thanked Yumi for the report.

H. M. Smith, Chairman of the BIH Directing Board, spoke about revising
the statutes of the BIH. After discussions on the BIH pole, the IPMS pole and
the pole of figure a committee was chosen to draft a resolution.

The meeting stood in memory of Dr J. Witkowski, former member of
Commission 19, who died on 25 May 1976.

It was agreed that the names of the following members should be submitted to the Executive Committee for future office.

President: R. O. Vicente
Vice-President: P. E. G. Pâquet
Organizing Committee: H.J. Abraham, P.L. Bender, B. Elsmore, H. Enslin,
K. Lambeck, G. Teleki, C. Sugawa, Ya. S. Yatskiv, G. Winkler, S. Yumi (ex officio).

31 August 1976

SIXTH SESSION, WITH COMMISSION 31

IPMS Problems (II)

E. Proverbio spoke on the status of Cagliari/Carloforte. At Carloforte living
conditions and logistic and legislative problems hindered recruitment and
scientific development; consequently an observatory was to be set up near Cagliari.
Bureaucratic delays had retarded construction until 1975 but observations had
been made at Cagliari University with the 80 mm VZT. Despite the latitude offset
and provisional screw value the mean external error of a pair per night was
± 0".48. (The 100 mm VZT at Carloforte gave ± 0".28). Instruments for the new site
would include the 80 mm VZT, a Danjon astrolabe, the old Mizusawa PZT, a Doppler
station and a laser station. It would seem useful to transfer the Carloforte VZT
there too, to compare the old and new techniques.

S. Yumi reported on the 39°8' ILS stations meeting that was held in Cagliari
on 20 and 21 August 1976. It had been small but successful. The meeting considered
that the ILS observations, extending from 1899, had the advantage of being free
from star position errors, could reveal local effects and warranted further investigation. It was tentatively agreed that ILS observers should be exchanged in a
4-year programme to ascertain relative personal equations. With reference to the
prospective chain of PZT stations on 39°8' it was also tentatively agreed that there
should be a unified star list, observing programme and reduction method, and
weekly airmail transmissions of data on punched tape. Lisbon Observatory proposed
to set up a time and latitude station on 39°8' very shortly.

G.A. Wilkins reported briefly on IAU Colloquium N°35 on "The Compilation,
Critical Evaluation and Distribution of Stellar Data" which had just been held
at Strasbourg. (See report of meetings of Commission 5 for further details of the
Colloquium and of the recommendations that have been based on its discussions).

H. Fliegel spoke on timing and polar motion as applied to space research.
He explained that Doppler shift measurements showed the rate of change of range, and
this fluctuated because of the Earth's rotation. The phase depended upon UT1 and
the longitude of the station. Conversely, the observations could determine this.
A frequency error of 10^{-12} would correspond to 5 metres in longitude. The BIH Rapid
Service give sufficient accuracy for present programmes.

The following resolutions, each in English and French, were adopted by the
joint meeting.

The International Astronomical Union

recognizing
that the activities of the International Polar Motion Service and of the Bureau
International de l'Heure are complementary, and that they both make essential
contributions towards the determination and understanding of the motion of the
pole, and

recognizing
that the new laser and radio techniques will make an important contribution to the
study of polar motion but that it is at present too early to determine the form of
a new service based on these techniques, and

noting
with satisfaction that the International Polar Motion Service multi-station
derivation of polar motion has attained the precision needed to resolve long-standing
problems,

recommends
that the International Polar Motion Service continue to operate in its present
form, and that the Scientific Council of the International Polar Motion Service
and the Directing Board of the Bureau International de l'Heure jointly keep under
continuous review the possibility of the utilisation of modern techniques on a
permanent basis, and

urges
that the international and national agencies concerned continue their support of
the Central Bureau of the International Polar Motion Service and of each cooperating
observatory.

The International Astronomical Union, in accordance with previous resolutions of
the International Astronomical Union and of the International Union of Geodesy
and Geophysics,

recommends
that the five ILS observatories (or their replacements) be equipped with photographic
zenith tubes and Doppler satellite tracking equipment, and that their existing visual
zenith telescopes be not phased out until there has been a sufficient overlap of
simultaneous observations.

        At the conclusion of the meeting the President expressed his appreciation
of the work done by members of the Commission and a vote of thanks to the President
was carried with acclamation.

COMMISSION 20: POSITIONS AND MOTIONS OF MINOR PLANETS, COMETS, AND SATELLITES
(POSITIONS ET MOUVEMENTS DES PETITES PLANETES, DES COMETES ET DES SATELLITES)

Report of Meetings, 25 and 30 August, and 1 September 1976

PRESIDENT: L. Kresák.   SECRETARY: E. Roemer.

25 August 1976

FIRST ADMINISTRATIVE SESSION
    The President welcomed members of the Commission and announced the appointment
of H. Scholl and C. Froeschlé as interpreters and of E. Roemer as secretary. All
present stood in silent respect as the names of members and former members deceased
during the triennium were read: G. A. Chebotarev, G. M. Clemence, S. Herrick, H. G.
Hertz, H. M. Jeffers, G. P. Kuiper, E. Rabe, and G. Van Biesbroeck.

    It was announced that, for lack of a second proof, the few corrections that
had been suggested by members to the Report of the Commission could not be made.
The most notable among these concerned the observational program on minor planets
and comets with the 33-cm astrograph at Perth.

    Unanimous approval was given to the proposed list of new members of the Commis-
sion: V. K. Abalakin, A. Borsenberger-Bec, J. A. Burns, O. Calame, M. A. Dirikis,
D. W. Dunham, R. Dvorak, F. A. Franklin, C. Froeschlé, B. Garfinkel, A. C. Gilmore,
T. Kiang, Y. Kozai, J. D. Mulholland, P. E. Nacozy, D. Pascu, E. M. Pittich, H.
Scholl, P. J. Shelus, G. E. Taylor, R. M. West, and D. K. Yeomans. Accepted as
consultants were: N. A. Belyaev, N. S. Chernykh, E. Helin, H. Rickman, V. A. Shor,
S. Vaghi, V. Zappalà, and K. Ziolkowski.

    The proposed new officers of the Commission, B. G. Marsden, President, and G.
Sitarski, Vice-President, and members of the Organizing Committee, F. K. Edmondson,
P. Herget, E. I. Kazimirchak-Polonskaya, L. Kresák, W. H. Robertson, J. Schubart,
together with the chairmen (ex officio) of all Working Groups, were then accepted
unanimously. [S. Ferraz-Mello, E. Roemer, and G. E. Taylor were confirmed as
Working Group chairmen at the second administrative session of the Commission.]

    A proposal to the IAU Finance Committee for renewal of the subvention toward
the support of the Minor Planet Center at the Cincinnati Observatory also received
unanimous support. [The amount of S Fr. 6220 for the triennium 1977-1979 subsequently was approved on the recommendation of the Finance Committee.]

    Marsden then reviewed proposals for changes in the form and contents of the
annual ephemeris volume for minor planets that is prepared at the Institute for
Theoretical Astronomy, Leningrad. Suggestions from a number of sources had been
discussed by the Organizing Committee in consultation with Abalakin. These included (1) A tabulation of osculating elements and absolute magnitudes $B(1,0)$ for
each object annually, (2) An augmented list of bright minor planets and special
objects for which extended ephemerides to the improved accuracy of 0.01 and 0.1
should be given, and (3) Revision of the precepts for calculation of ephemeris
magnitudes, in that both phase and opposition effects should be included in the
extended ephemerides, but they should continue to be excluded in calculation of
ordinary opposition ephemerides. Comments were invited so that a comprehensive
recommendation could be presented for consideration at the appropriate time.

Roemer read a resolution proposed by herself and E. I. Kazimirchak-Polonskaya urging follow-up astrometric observations of recently discovered and particularly interesting minor planets, comets, and satellites over the longest possible arc and calling special attention to the need for observations with large telescopes.

Formal consideration of these and other recommendations and resolutions, none of which was expected to require joint action with another Commission or was to be referred explicitly to the General Assembly, was deferred to the second administrative session.

FIRST SCIENTIFIC SESSION ON MINOR PLANETS

Scientific reports were then presented as follows:

(1) T. Gehrels and C. J. van Houten: Palomar-Leiden Survey of the Various Populations of Minor Planets. Gehrels described the observational programs with the Palomar Schmidt telescope to search for special populations of minor planets, to study light variation of faint asteroids, and to improve population statistics generally. Van Houten reported on analyses of the photographic plates at Leiden, particularly regarding measurements of positions for use in computation of orbits, and photometric calibration of the plates of the two Trojan search programs. It is unclear whether the very considerable work involved in determination of positions and orbits for the 1300 normal asteroids observed in the 1971 and 1973 Palomar-Leiden Trojan surveys would be justified by the improved knowledge of asteroid statistics that might result.

(2) V. I. Orel'skaya: Present Status of the Observations of Selected Minor Planets for Improving the Constants of Star Catalogues. (Read by P. Herget.) Nearly 23,000 precise observations of 10 bright minor planets for use in improving constants of star catalogues have been collected by the Institute for Theoretical Astronomy from 21 cooperating observatories in the years 1949-1975. A proposal is made for a new project involving observations of 20 selected objects, including 10 new ones chosen to obtain even coverage over the declination band to ±30°, during the years 1974-1990. Some 35 observatories have agreed to cooperate in the new program.

(3) T. Gehrels: Photographic Photometry of Faint Asteroids and the Ephemeris Magnitudes. A new list of absolute magnitudes, $\underline{B}(1,0)$, verified by photoelectric observations to magnitude 14-15, is in preparation. New observations are planned to obtain standardized photometric data for recently numbered minor planets. The brightness surge at phase angles less than about 7° (opposition effect) has now been well studied in six objects. There is some evidence that the phase effect differs between bright and faint asteroids, and that it may be correlated with compositional type. Retention of the coefficient 0.023 mag/degree for ephemeris purposes is recommended. Use of the symbol $\underline{g}$ for absolute magnitude should be dropped.

30 August 1976

SCIENTIFIC SESSION ON COMETS

The session was organized and presided over by E. Roemer, chairman of the Working Group on Orbits and Ephemerides of Comets. Short reports on observational topics were presented by the following speakers:

(1) Y. Kozai (with Ko. Tomita): Comet Patrol Observations in Japan. Factors that have contributed to the great success with comet discoveries of amateur astronomers in Japan include a favorable longitude, without large population centers for a long distance to the east, widespread availability of suitable small telescopes, great public interest encouraged through popular journals and by many groups for amateur astronomers, and also by the Astronomical Society of Japan, which has awarded since 1907 a medal to any Japanese astronomer who discovers a new celestial object. Time on a number of the main instruments of the Tokyo

Astronomical Observatory is available for observation of comets, but light pollution from industrial areas near the Okayama and Dodaira Stations has seriously hindered recent work. The Tokyo Astronomical Observatory acts as a Central Bureau in Japan, coordinating confirmation of possible new discoveries.

(2) T. Gehrels: Discoveries of Faint Comets with the Palomar Schmidt Telescope. Five comets have been found on plates taken with the Palomar Schmidt in the continuing work on minor planets. Two, of magnitude 15-16, were found immediately upon inspection of the plates. Three others, of magnitude 18-19, were found by careful blink examination. One comet has unusually large perihelion distance, and four, including the accidentally rediscovered P/Swift 1, renamed P/Swift-Gehrels, are of short period.

(3) E. Roemer: Observations of Faint Comets with Large Reflectors. Because of the large scale and limited field of typical large reflecting telescopes (10"/mm and 35', respectively, for the two large telescopes of the University of Arizona), the position of an object to be observed must be fairly well known, and the calculated motion must be compensated during the exposure. The "nuclear" (m_2) magnitudes are appropriate for determination of the required exposure. Use of the Astrographic Catalogue, or of field transfers, is necessary in astrometric reductions because of the small field. Semiaccurate positions, accurate to $0\overset{.}{.}1$, can be scaled by careful use of a reseau.

(4) R. M. West (with H.-E. Schuster): Discovery and Observation of Comets with the ESO 1-m Schmidt Telescope. Plates taken in the survey of the sky from -90° to -20° provide a unique opportunity for discovery of minor planets and comets. The plates are first checked visually at La Silla, then at the Sky Atlas Laboratory in Geneva, whither they are transported weekly by air. Negative glass copies are sent as soon as possible to the Uppsala Observatory, where the plates are again searched for special objects. The triple search is believed to be thorough enough that no exceptional asteroid trails or comets brighter than 18th magnitude have escaped detection. Six comets have been found on the 729 plates obtained in the interval 1973 to mid-1976, three of them being new; two more were already known, and no confirmation could be obtained for the other. Discovery rates for new comets are approximately 1 comet/year, or 1 comet/6000 sq. degrees, and twice these rates for all detected comets.

(5) J. Hers: Comet Observations in South Africa. With concentration of the South African observational facilities at Sutherland, J. A. Bruwer, who continues to use the 10-inch Franklin-Adams camera, is now the only observer of comets in South Africa on a professional basis. Hers uses his own 20-cm Celestron reflector to obtain 15-20 precise positions a year. J. C. Bennett, Pretoria, continues to use a 20-cm telescope in search for new comets.

(6) E. Høg: Reference Stars in the Southern Hemisphere. The Perth 70 catalog of positions of 24,900 stars, compiled by the expedition of the Hamburg Observatory to Perth as a part of the SRS International Program, is now available in both machine-readable and printed form.

Reporting on theoretical investigations were the following:
(1) M. Bielicki: Present Status of the Catalogue of Orbits of One-Apparition Comets. (Read by E. Roemer.) Almost all of about 50,000 original observations of 165 comets observed in the interval 1900-1950 have been collected in Warsaw. Some 15,000 observations of comets observed between 1800 and 1900 have been collected in Czechoslovakia, through the cooperation of the Astronomical Institute of the Slovak Academy of Sciences. Reduction and analysis of observations is in progress. Most of the necessary computer programs for calculation of gravitational perturbations and determination of Newtonian osculating elements near perihelion are completed.

(2) B. G. Marsden: Recent Results on Short-Period Comets. The discovery of
six new short-periodic comets in the year 1975, resulting mostly from patrols with
Schmidt telescopes, is quite remarkable--and presented a number of problems with
bungled discovery reports and marginally adequate follow-up. If not all short-
period comets are to be observed at each return, how is a selection to be made when
one cannot fully anticipate the flares of P/Tuttle-Giacobini-Kresák, the selection
of targets for space probes, or the need for highly accurate ephemerides of close-
approaching P/d'Arrest for attempted radar observations? P/Perrine-Mrkos appears,
after erratic behavior in 1968 and unsuccessful searches in 1975, to have joined
the ranks of lost comets of more than one appearance.

(3) E. I. Kazimirchak-Polonskaya: On the Capture of Comets and on the Evolution of Cometary Orbits. The classical theory for the capture of comets by Jupiter
was criticized. The capture of comets by Neptune to Jupiter was presented instead
as a complicated mechanism that acts in many steps lasting thousands and millions
of years. The laws were described for the capture and for the evolution of orbits
of observed and fictitious comets which belong to different planetary families or
come from interstellar space.

(4) Z. Sekanina: Orbits and Photometric Behavior of Comets of Large Perihelion
Distance. A brief review was presented of recent work by Marsden and Sekanina on
the concentration in original $1/a$ determined from comets of large perihelion distance.
Discovery of more such comets for study of possible concentrations in $1/a$ corresponding to heliocentric distances less than that of the Oort Cloud is greatly
desired. The brightness of some large-q comets, including 1956 I and 1959 X, seems
to change in a peculiar fashion. Comet Lovas, 1974 c (q = 3.0 AU), faded by some
5 magnitudes soon after discovery but then brightened again. Observations in 1976
indicate that the magnitude is still following an inverse square law.

(5) E. Everhart: Evolution of Cometary Orbits. Some 3×10^8 comet orbits
have now been calculated in an effort to simulate the capture process of comets
from near-parabolic to short-period orbits. Of recent test calculations involving
more than 12,000 orbits with perihelion near the orbit of Neptune ($21 < q < 34$ AU),
only 18 led to capture, the remainder to ejection. The capture efficiency of Neptune
is thus very low, 18/12,230. Capture efficiencies calculated similarly were 40/69
for Uranus, 229/500 for Saturn, and 92/229 for Jupiter, giving a final capture
efficiency of 1/6000 for observable ($q < 6$ AU) short-period comets of the Jupiter
family from those in near-parabolic orbits with perihelion distance near the orbit
of Neptune.

SCIENTIFIC SESSION ON SATELLITES

The session was organized and chaired by B. Morando, chairman of the Working
Group on Satellites. Reports on topics concerned with observations and reductions
were given as follows:
(1) E. Roemer: Follow-up Observations of Newly Discovered Faint Satellites.
Searches for discovery of new satellites are normally made near opposition to
facilitate identification of candidate objects. Significant fading occurs rapidly
thereafter as quadrature is approached. It is necessary to verify quickly through
additional observations whether a candidate object is a satellite or an asteroid,
and then the nature of the orbit, whether direct or retrograde. Until the orbit
can be fairly well determined, there is observational loss in limiting magnitude
through the trailed images that result from inaccurately compensated motion. Loss
of observations also will occur because of difficulties in obtaining time on short
notice with large, suitably equipped telescopes. Weather, bright moon, and temporary unobservability arising from proximity of a faint satellite to a bright planet
also will cause gaps. Coordination is essential between those who may be planning
searches for discovery and those whose cooperation is needed to secure sufficient
follow-up observations.

(2) J.-E. Arlot: Problems Related to the Reduction of Observations of
Planetary Satellites. The differential position of two satellites can be measured
more precisely than can the relative position of satellite and primary, but O-C's
of both satellites enter into the differential position. Positions measured with
respect to the planet permit separation of the O-C's for each satellite and are to
be preferred in spite of difficulties in getting a satisfactorily measurable image
of the planet. Good determination of orientation and scale results if catalogue
stars are measured and plate constant methods of reduction are used. Limb darkening introduces a serious difficulty in defining the limb of the planet, the observed
phase defect being in general larger than that calculated geometrically. A phase
defect error displaces the apparent position of the planet but can be partially
compensated by inclusion of a longitude error of the planet as an unknown.

(3) S. Ferraz-Mello (M. Tsuchida): New Comparisons Between Sampson's Theory
and Observations. Comparison of observations of the Galilean satellites made in
the interval 1912-1925 with Sampson's Tables confirms earlier conclusions that the
concept of the time scale (UT) of Sampson's theory is poor. Residuals are apparent
for the observations 1914-1925, and have the same sense as found previously, but
the indicated rate of increase with time is less than found from analysis of earlier
observations made with instruments of focal length only 6-7 m.

Reports on theoretical topics were then given as follows:
(1) Y. Kozai: On the Motion of the Satellites of Saturn. The work reported
was in extension of that published in 1957, in which observations made at the U.S.
Naval Observatory in the interval 1928-1947 of the five inner satellites of Saturn
were used to improve the elements published by G. Struve in 1930 and 1933. Some
errors have been found and corrected, the theory has been extended, and new values
for the masses of three satellites and for J_2 and J_4 of Saturn have been derived
(in press, Publ. Astron. Soc. Japan). The source of a discrepancy between the
value found for J_2 and that recommended for adoption by Commission 4 was not immediately clear.

(2) K. Aksnes (with F. A. Franklin): Analyses of the Mutual Phenomena of the
Galilean Satellites. Final results from analysis of 91 light curves of mutual phenomena of the Galilean satellites in 1973 were presented. Good values were derived
for the radii of JII, JIII, and JIV, and for corrections to the orbital constants
of Sampson's theory, decreasing the errors to 100 km, an order of magnitude improvement over Sampson's theory. (Astron. J. 81, 464, 1976)

1 September 1976

SECOND SCIENTIFIC SESSION ON MINOR PLANETS
    The following scientific reports were presented in a session chaired by
President Kresák:
    (1) B. G. Marsden (with contributions from Yu. V. Batrakov): The Lost Minor
Planets and the Critical List. Several categories of the Critical List of minor
planets were reviewed. These included (1) Those minor planets observed in only one
definite opposition, almost all of which are hopelessly lost, (2) Those of unclear
recent observational status, needing new investigations of orbits, (3) Those of
satisfactory observational status, but requiring orbit improvements (some 300 objects
are involved; only the worst cases were listed), (4) Objects with orbits expected to
be satisfactory but not observed within the past ten years, (5) Planets observed in
fewer than four oppositions but with recently improved orbits, (6) Objects with
recently improved orbits (and in some cases new observations) that qualify for
removal from the Critical List, and (7) Recently numbered objects that have not yet
been observed in four oppositions. Some objects listed could already be removed
from the Critical List on the basis of observations made just before the General
Assembly. Particular attention was called to minor planets on the Critical List
that have favorable oppositions in the near future. The usefulness of reporting

negative searches was recognized, but observers should state the dimensions of the field searched, the ephemeris used, and the magnitude limit.

(2) J. Schubart: Present Status of Mass Determination of Minor Planets. Masses have been determined for the three largest minor planets, by Schubart for Ceres and Pallas, from their mutual perturbations, and by the late H. G. Hertz for Vesta, through perturbations in the orbit of (197) Arete during the periodic close approaches, the most recent of which occurred late in 1975. Hertz had continued to collect observations of Arete, and the number available has more than doubled from the set used in the 1968 solution. Observations of Arete in early 1977, referred to a known fundamental reference system, are of particular importance to an improved solution. A comment by J. G. Williams called attention to the possibility of obtaining additional information on minor planet masses from spacecraft tracking data.

(3) L. Kresák: Passages of Minor Planets and Comets Near the Earth. Close approaches of long- and short-period comets to within 0.15 AU of the earth during the last 300 years were analyzed to conclude, because the rate of close approaches does not increase with decrease of the intrinsic brightness, that there are probably no intrinsically very faint comets at all. Close-approach data for minor planets is obviously incomplete, most particularly because of frequently unfavorable observing conditions for discovery of Apollo-type objects. It is suggested that only a few objects observed presently as asteroids are likely to be of cometary origin. Even so, the large disproportion between the frequency of asteroidal objects and active comets permits the assumption that comets do leave small extinct nuclei of considerable lifetime.

(4) O. Møller and L. K. Kristensen: The (51) Nemausa Project. Møller reviewed the project begun by P. Naur with aims similar to those of the new project proposed by the ITA, Leningrad. Some 2200 observations of (51) Nemausa have been collected in the interval 1946-1974, many of them fully documented by information about reference stars, dependences, etc. A preliminary investigation is underway to ascertain whether the desired goal of an accuracy of $0\overset{..}{.}03$ in the correction to the equator point, $\Delta\delta_0$, can be achieved. Kristensen called attention to a special opportunity during the 1976-1977 opposition of (51) Nemausa to tie the reference star system rigorously to the FK4, independent of errors in star positions, through observations of the minor planet against exactly the same star field on both 1976 October 25.8 and 1977 January 12.7 UT.

(5) G. E. Taylor: Occultations by Minor Planets and Satellites. Observations of occultations of stars by minor planets and satellites can give (1) Evidence for presence of atmospheres, (2) Data concerning sizes and shapes of the occulting bodies, (3) Positions of very high intrinsic accuracy, though realization may have to await improvement of star positions, and (4) Information on double and multiple star systems. Predictions, even when limited to the largest objects, face still unresolved difficulties because of (1) Errors in star positions and unavailability of positions in usable form for stars fainter than about 10th magnitude, (2) Frequently inadequate ephemerides of the occulting objects, being of highly variable accuracy for minor planets, and involving for satellites uncertainties also in the position of the planet, and (3) Impossibility of determining adequate data for the guidance of observers until improved positions of the minor planet relative to the star field can be obtained, which not infrequently is no more than a few days before the event. D. W. Dunham and J. D. Mulholland described the program of G. F. Benedict D. S. Evans, and P. J. Shelus to predict occultations by use of a high-speed micro-densitometer to scan Palomar Sky Survey glass plates for stars along the ephemeris path of solar system objects.

SECOND ADMINISTRATIVE SESSION
    With the President in the chair, the proposed membership of Working Groups,

as well as formation of a new Working Group on Prediction of Occultations, were
approved as follows:

Orbits and Ephemerides of Comets: M. P. Candy, A. C. Gilmore, E. I. Kazimirchak-
Polonskaya, L. Kresák, B. G. Marsden, E. Roemer (chm.), G. Sitarski, Ko. Tomita,
and R. M. West.

Satellites: K. Aksnes, Yu. V. Batrakov, A. Borsenberger-Bec, C. Cristescu, S. Ferraz-
Mello (chm.), Y. Kozai, J. H. Lieske, B. Morando, J. D. Mulholland, D. Pascu, E.
Roemer, J. L. Sagnier, V. A. Shor, and A. T. Sinclair.

Prediction of Occultations: K. Aksnes, J. A. Burns, D. W. Dunham, P. Herget, A.
Klemola, L. Kohoutek, W. H. Robertson, P. J. Shelus, V. A. Shor, and G. E. Taylor
(chm.).

   Establishment of an ad hoc committee to advise on the naming of minor planets,
particularly those receiving numbers near 2000, also was approved, with the following membership: V. K. Abalakin, P. Herget (chm.), L. Kohoutek, B. G. Marsden, J.
Schubart, C. J. van Houten, and P. Wild.

   Several matters in the area of international cooperation that had been brought
to the attention of the Commission already had been dealt with through one or
another of the Working Groups. The cooperative observing program on minor planets
described by Orel'skaya is the subject of a proposed resolution (see below).

   A proposal, which had been sent to the IAU Executive Committee, for an IAU
Symposium in honor of Professor Y. Hagihara on the occasion of his eightieth birthday was described by Marsden. Subject of the proposed Symposium, to be sponsored
jointly by Commissions 4, 7, and 20, and to be held in or near Tokyo in May 1978,
is: "Dynamical Astronomy: Theory and Application." Topics to be covered include,
but are not restricted to the following: the three-body problem, planetary theory,
preparation of planetary ephemerides, the Hirayama families of minor planets, the
motions of satellites, and studies of the nongravitational forces acting on comets.

   The President reported that a Colloquium on "Dynamics of Comets and Meteors",
to be held in Bratislava under the joint sponsorship of Commissions 20 and 22 in
1980, is under consideration. Ferraz-Mello mentioned the possibility of a colloquium in Brazil, either in conjunction with the IAU General Assembly in 1979, or
in 1980.

   Kazimirchak-Polonskaya presented the following statement, which was warmly
received by members of the Commission:
   "All researchers working on the motion of comets and minor planets at the
Institute for Theoretical Astronomy express their deep gratitude to Miss Professor
Elizabeth Roemer for her observations, which are exceptional in their precision,
for her selfless and unselfish labor.
   "We wish Professor Roemer further achievements in her most useful activity.
   "We are also extremely grateful to other observers, wishing them success in
their important observational work."

   Two resolutions in support of observational programs, the first sponsored by
E. Roemer and E. I. Kazimirchak-Polonskaya, and the second presented at the request
of V. I. Orel'skaya, received the approval of Commission 20:

Resolution 1

   Commission 20 commends the efforts that have led recently to the discovery
with powerful wide-field instruments of objects of unusual interest, including
MINOR PLANETS of the Apollo and Amor types, of unusually high orbital inclination

or eccentricity, or in motion commensurable with that of Jupiter; COMETS, both new
ones of short period and long-period objects of great perihelion distance; and two
new SATELLITES of Jupiter. At the same time, Commission 20 calls attention to the
urgent need for, and great importance of, follow-up astrometric observations of
these and other objects over the longest possible arc. Many of these objects are
faint and the use of large telescopes with efficient detectors capable of good
astrometric precision is required. Such observations are essential to the determination of reliable orbital elements upon which both future observations and
studies of the dynamical evolution of these bodies can be based.

La Commission 20 reconnait les efforts qui ont abouti récemment à la découverte,
à l'aide d'instruments appropriés à grand champ, d'objets d'un intérêt particulier,
comprenant les petites planètes du type Apollo et Amor, ainsi que des objets ayant
une grande inclinaison orbitale ou excentricité, ou ayant un mouvement commensurable
avec Jupiter; des comètes nouvelles à courtes periodes et des objets à grandes
periodes ayant une grande distance perihélique; enfin deux nouveaux satellites de
Jupiter. Au même temps, la Commission 20 attire l'attention sur un but urgent et
de très grande importance, la poursuite des observations astrométriques de ces
objets et d'autres sur un arc le plus long possible. La plupart de ces objets sont
faibles et nécessitent l'usage de télescopes puissants avec des techniques permet-
tant une bonne précision astrométrique. Les observations sont essentielles pour la
détermination des éléments orbitaux sur lesquels seront basées les observations
futures et l'étude de l'évolution dynamique de ces corps.

Resolution 2

Commission 20 endorses the following resolution of Commission 8:
"Commission 8 approves the new program (proposed by the Institute for Theoretical Astronomy in Leningrad) concerning observations of the selected minor planets
for solving the problems of fundamental astrometry and asks the observatories having
appropriate equipment to take part in the observations according to this program."

La Commission 20 soutient la résolution suivante de la Commission 8:
"La Commission 8 approuve le nouveau programme (proposé par l'Institut d'astronomie théorique à Leningrad) concernant l'observation de petites planètes choisies,
afin de résoudre les problèmes fondamentaux de l'astrométrie, et demande les observatoires ayant les instruments appropriés de prendre part aux observations selon ce
programme."

Recommendation

The following recommendation concerning the contents of the annual volume of
Minor Planet Ephemerides prepared at the Institute for Theoretical Astronomy,
Leningrad, was approved:

In order to make the annual volume of Minor Planet Ephemerides more suited to
the current requirements of observers, the Institute for Theoretical Astronomy is
requested to give consideration to the following suggestions and the possibility
of introducing them into the 1979 edition:
   (a) That osculating elements be derived for all numbered minor planets (except
for those that are hopelessly lost) for all epochs when the Julian Date divided by
200 leaves remainder 0.5 and that each annual volume tabulate that set which corresponds to the last of these epochs falling during the corresponding year; these
elements should be given with the increased precision used for the new elements
introduced at the beginning of, e.g., the 1976 edition.
   (b) That the lists of bright (say, to opposition magnitude 12.5) and unusual
planets for which special ephemerides are prepared be extended and that the precision of these ephemerides be increased to 0.01 in right ascension, and $0!1$ in
declination.

(c) That the magnitudes in special ephemerides include the opposition effect with the phase according to the precepts set out by Gehrels; in the ordinary ephemerides, where the magnitude is given only for the fourth tabulated date, no allowance should be made for phase or opposition effects, and in these same ephemerides the precision of the mean anomaly should be increased to 0.1; for absolute magnitudes the symbol $\underline{B}(1,0)$, rather than $\underline{g}$, should be used.

President Kresák closed the meeting with an expression to the Commission of the enjoyment and honor he had felt in serving the Commission in succession to such outstanding former Presidents as Arend, Herget, Chebotarev, and Edmondson.

JOINT MEETING OF COMMISSIONS 15, 20, and 22

The report of the Joint Meeting of Commissions 15, 20, and 22, in which highlights of IAU Colloquium 39 were overviewed, is included in the Proceedings of Commission 15.

COMMISSIONS 21: LIGHT OF THE NIGHT SKY (LUMIÈRE DU CIEL NOCTURNE)

Report of Meetings, 26, 30 and 31 August 1976

PRESIDENT: J. L. Weinberg.   SECRETARY: J. G. Sparrow.

During the Assembly, the activities of Commission 21 (Light of the Night Sky) included a half day joint meeting with Commission 22 (Meteors and Interplanetary Dust), two 90-minute scientific sessions and a business meeting.

26 August 1976

JOINT MEETING

Results on Interplanetary Dust and Zodiacal Light obtained from the Pioneer 10 and 11 Deep Space Probes and the Helios 1 and 2 close-in Solar Probes

The following communications were presented during the meeting, in order of their presentation:

J. L. Weinberg: Summary of Pioneer 10/11 Results, Zodiacal Light and In situ. A new methodology was described which should facilitate the processing of the large amount of data obtained on zodiacal light and background starlight. Weinberg also summarized results from the two discrete particle detectors for R. K. Soberman, who was unable to attend.

C. Leinert: Summary of Helios 1/2 Results, Zodiacal Light. The performance of the zodiacal light photometers on both Helios probes was discussed, and preliminary results were presented based on 18 months of data received from Helios 1.

E. Grün (H. Fechtig, J. Kissel and P. Gammelin): First Results of the Micrometeoroid Experiment on Board Helios 1. Results support the findings of the Pioneer 8 and 9 dust experiments that small particles are leaving the solar system on hyperbolic orbits.

H. Fechtig: Dust Fluxes near 1 AU and in the Earth-Moon System. HEOS 2 dust experiment results were reported, including the interpretation of three categories of dust fluxes, namely random particles, fairly clustered particles (groups) and heavily clustered particles (swarms).

A. N. Simonenko and T. N. Nazarova: Space Probe Fluxes and Ground-Based Meteor Fluxes, Recent USSR Results. Read by title only.

P. M. Millman: A Comparison of Meteor Data and In situ Results. Examination of the various size ranges corresponding to different methods of observation.

R. H. Giese: The Compatibility between Optical and Mechanical Results and Theoretical Interpretations. Evidence was presented for the suggestion that the brightness and polarization of zodiacal light can be explained by scattering from fluffy, absorbing particles in the size range from a few microns to some hundreds of microns.

Brief overviews of the preceding papers and of other recent significant results were then given by F. L. Whipple, P. M. Millman and J. M. Greenberg as part of a panel discussion which concluded the joint meeting.

<u>30 August 1976</u>

SCIENTIFIC PRESENTATIONS
        Four invited papers and nine contributed papers were included in the scientific
sessions.
        G. Weill: Recent Studies of the Airglow, a Review. Results were summarized
from some 170 papers published during the preceding three years.
        N. B. Divari: Recent USSR Studies of the Light of the Night Sky. Read by
title only.
        R. Dumont: Zodiacal Light off the Ecliptic. Results were presented based
on measurements made by the author in Tenerife between 1964 and 1975 (see, also,
<u>Astron. Astrophys. 51</u>, 393, 1976).
        E. Pitz (A. Schulz, C. Leinert and H. Link): Ultraviolet Observations of
Zodiacal Light with the Astro 7 Rocket Experiment. Observations at elongations
from 30 to 50 degrees essentially support the deviations of the color of zodiacal
light in the ultraviolet from that of the sun as published by C. F. Lillie (<u>NASA SP-
310</u>, 1972).
        P. J. Edwards (G. R. Stanley and G. Neilson): Mid-Latitude Night Skylight
Fluctuations of Geomagnetic and Geoelectric Origin. Mid-latitude auroral light
fluctuations at 391 nm and 558 nm were studied using a cross correlation technique.
        A. S. Asaad: Recent Zodiacal Light Studies at the Helwan Observatory.
        Ph. L. Lamy (for M. Maucherat, P. Cruvellier, J. Maucherat, M. Hanus, M. Renard
and J. P. Thouvenin): Absolute UV Photometry of the Zodiacal Light from the ELZ
Instrument aboard the D2B-AURA Satellite. Preliminary results on the upper limits
to the zodiacal light brightness at three bands in the ultraviolet are not in
agreement with the results of Lillie or with those presented here by Pitz.
        Ph. L. Lamy and J. M. Perrin: The Ultra-violet Brightness of the Zodiacal
Light. Results are presented on a parametric scattering model of the zodiacal
cloud that is being used to interpret results from the D2B experiment.
        D. H. Morgan (K. Nandy and G. I. Thompson): The Ultraviolet Galactic Back-
ground from TD-1. Satellite Observations. Measurements of zodiacal light were
also discussed. These observations should assist in resolving the question of the
ultraviolet 'excess' mentioned earlier (<u>Monthly Notices Roy. Astron. Soc.</u>, in press).
        H. Tanabe: Studies of Background Starlight. A discussion was given of the
Tokyo starcounting program and the comparison of these data with the background
starlight results obtained from the Pioneer 10 and 11 Jupiter probes.
        K. Mattila: A Photoelectric Method for Measuring the Integrated Starlight.
Earlier methods used to obtain information on the surface brightness of background
starlight were discussed, and a new method was suggested based on comparative
measurements by two photometers having large and small fields of view, respectively.
        K. Mattila: The Extragalactic Component of the Light of the Night Sky. A
new method which utilizes the screening effect of a high latitude dark nebula gives
9 ± 3 $S_{10}(B)$ for the extragalactic light at 4000A (<u>Astron. Astrophys.</u> 47, 77, 1976).
        G. Schwehm: Temperature Distribution of the Interplanetary Dust Grains close
to the Sun. These calculations take account of more recent data on particle
properties than had previously been used.

<u>31 August 1976</u>

ADMINISTRATIVE SESSION
1. <u>Commission Membership</u>

        The membership accepted the recommendation of the Commission officers as to the
proposed new officers of Commission 21 during the next triennium:
        President: R. Dumont
        Vice President: H. Tanabe
        Organizing Committee: A. S. Asaad, R. H. Giese, C. Leinert, Yu. L. Trutse,
G. Weill, J. L. Weinberg, R. D. Wolstencroft.

The following new members were welcomed into Commission 21:   A. C. Levasseur, T. Mukai, J. G. Sparrow.

2.   Commission Name

It was agreed that the French title of Commission 21 should be changed to Lumiere du Ciel Nocturne, to make it more consistent with the English title and with the current emphasis of the Commission. This change will be requested of the General Secretary.

3.   Relationship to Commission 22

A similarity in part of the Reports of Commissions 21 and 22 led the General Secretary to suggest a merger of the Commissions. This similarity was primarily a consequence of the joint meeting on interplanetary dust (IAU Colloq. 31) and the inclusion of combined zodiacal light/dust experiments on the Pioneer 10/11 and Helios 1/2 probes. Techniques of the two Commissions for providing data on interplanetary dust are quite different, although complementary, and there is practically no overlap in membership due to the differences in these techniques and to the other disciplines included in each Commission. In separate meetings, the officers and membership of the two Commissions have argued strongly against a merger at this time, although joint studies and colloquia on interplanetary dust will be continued.

4.   Working Group

Reactivation of the background starlight working group was agreed upon following the suggestion by Weinberg. This group is to consist of J. L. Weinberg, H. Tanabe, R. D. Wolstencroft, H. C. van de Hulst and M. Maucherat, subject to agreement by the last two people listed. The task of the working group is to survey what data are available on background starlight and the methods that can be used to separate the components; it will make recommendations to Commission 21 at the next General Assembly.

5.   Other Topics

After some discussion it was decided that a newsletter to acquaint members with the availability of data from space experiments and ground based measurements would be beneficial. The incoming President volunteered to produce such a newsletter at intervals not greater than yearly from information received from members.

Meetings of interest to members of the Commission were listed, in particular the next General Assembly in Montreal, August 14-23, 1979. It was proposed to hold a Zodiacal Light and Interplanetary Medium Conference in Ottawa either before or after the General Assembly, cosponsored by IAU Commissions 21, 22, (15?) and the COSPAR Cosmic Dust Panel. The next COSPAR meeting is scheduled for Israel in June 1977.

After an interesting exchange on technical matters not particularly relevant to the business meeting, R. Dumont, the incoming President, thanked Dr. J. L. Weinberg for his extensive and enthusiastic efforts on behalf of the Commission during the past three years.

COMMISSION 22: METEORS AND INTERPLANETARY DUST
(METEORES ET LA POUSSIERE INTERPLANETAIRE)

Report of Meetings, 28 August - 1 September 1976

PRESIDENT: B.A. Lindblad   SECRETARY: C.S.L. Keay

ADMINISTRATIVE SESSION

I. GENERAL

The business meeting was called to order at 10.00 on 28 August 1976. Approximately 30 members and guests were present. The President reported with regret the deaths of two commission members: Professor Astapovich of the USSR and Professor Olivier of USA. The meeting stood in silence for a few moments in their memory.

The appointments by the Executive Committee of I. Halliday as the new President of Commission 22 and of W.G. Elford as the new Vice-President, were announced. New members of the Commission, approved by the Executive Committee, were noted as follows: O.I. Belkovich, D.E. Brownlee, H. Fechtig, R.H. Giese, G.A. Harvey, D.W. Hughes, Z. Kvíz, V. Padevět, V. Porubčan and K. Tomita; bringing the current membership to 69. The President announced that the following have been asked to serve for the 1976-79 term as consulting members: V. Benyukh, E. Grün, K.B. Hindley, F. Hörz, E.P. Mazets, J.W. Rhee, J. Trulsen and D.K. Yeomans.

The following members were elected to the Organizing Committee for the coming term: H. Fechtig, C.L. Hemenway, K.N. Kramer, B.A. Lindblad, B.A. McIntosh, Z. Sekanina, A.N. Simonenko and J. Štohl.

The President reported on correspondence between the General Secretary and the Presidents of Commissions 21 and 22 relating to the subject area of Commission 21. A suggestion had been previously put forward that Commission 21 should merge with Commission 22. The correspondence had been circulated to members of the organizing committees of both commissions. There was unanimous agreement that a merger would serve no useful purpose because the observational techniques of the two commissions are so different. In a joint letter the Presidents of Commissions 21 and 22 had informed the General Secretary of this viewpoint. After some discussion at which the President of Commission 21, J. Weinberg, participated, the decision by the Presidents not to consider a merger was put to the vote and carried unanimously.

The report of the Commission had been previously circulated in manuscript to all members. The President mentioned that the subject of tectites had inadvertently been omitted from the report. The Commission gave formal approval to the report which will be printed in Trans. IAU, Vol. XVI A Part I, and distributed to all commission members.

A short discussion followed concerning the committee, established by the Meteoritical Society, on meteorite nomenclature. It was agreed that the President of Commission 22 should continue to serve as a consultant to this committee.

Following a brief discussion of the report of the committee on Radar
Observations of Meteor Rates and Radiants and Anomalies at the Base of the
Thermosphere, it was agreed that the committee be expanded by the addition
of the following new members: O.I. Belkovich, I. Halliday and J. Jones.

The Vice-President, I. Halliday, expressed the thanks of the entire
Commission to the retiring President for his work on behalf of Commission 22
during the past three years, particularly for the preparation of the Report of
the Commission and for arranging the sessions at the General Assembly.

I. Halliday issued an invitation to a symposium (or colloquium) on the
subject of meteors and interplanetary dust to be held in Ottawa immediately
before, or after, the 1979 General Assembly in Montreal. It was hoped that
the meeting would be co-sponsored by IAU Commission 21 and the COSPAR
Cosmic Dust Panel. The majority of those present favoured a meeting to
follow the General Assembly in 1979.

II. RESOLUTIONS
The following resolutions were proposed and carried without dissent; a
preliminary text of Resolutions 1 and 2 had been circulated to members of
the Commission prior to the business meeting.

1. Commission 22 recommends that a cooperative long-term program of
radar observations of meteor showers from several points on the earth be
initiated with the aim of studying meteor stream cross-sections and longitudinal structure.

2. Commission 22 recommends that a meteor data centre in Lund,
Sweden be established for the collection of meteor observations by radio and
photographic techniques. The Commission expresses the hope that sufficient
financial support will be forthcoming to ensure the operation of such a data
centre.

3. Commission 22 notes with interest the programme for Fireball Photography initiated by the B.A.A. and the operation of an International Centre for
Meteor Observations for the collection of visual meteor data. The Commission
appreciates the value of these programmes and endorses their continuing
operation. The commission expresses the hope that sufficient financial support
will be forthcoming to ensure continued operation.

4. Commission 22 approves the Progress Report by the "Committee on
Radar Observations of Meteor Rates and Radiants, and Anomalies at the Base
of the Thermosphere" established at the XV IAU Congress. The Committee is
requested to continue its efforts to develop a widely acceptable design for an
economical Meteor Rate-measuring Radar System.

5. Commission 22 wishes to call to the attention of observers with
access to fast, wide-field telescopes the importance of searches for comet
anti-tails. The Commission notes that the association of meteor streams with
short-period comets is well established from other studies, however, the
observation of anti-tails provides in a more direct way evidence concerning
the separation of large dust particles from the nucleii of short-period comets.
Owing to special geometrical circumstances and the general faintness of short-
period comets, this information is generally lacking.

SCIENTIFIC SESSIONS

25 August 1976

A half day of discussion was devoted to the topic "The Future of Photographic Meteor Programs". The focus in this session was on photographic fireball networks and meteorite recovery projects. It was noted with regret that the Prairie network fireball program had been terminated. The opinion was expressed by several speakers, that the fireball orbits obtained in the Prairie network program were of an exceptional accuracy and that as many orbits as possible should be fully reduced and published. It was noted that several new photographic fireball networks are presently being put into operation. The wide interest in this field was demonstrated by the reports presented at the meeting. The technical and personnel problems of operating a fireball network over an extended period of time were discussed. It was recognized that a fatigue problem could arise if no meteorite was recovered early in a program.

The following reports were presented:
F. Whipple: Observational Problems in Meteor Photography.
F. Whipple and R. McCrosky: Some Results of the Prairie Network Project.
I. Halliday: MORP - The Meteorite Observation and Recovery Project.
K. Hindley: The British Fireball Network.
J. Kiko: Reliability and Efficiency of the European Meteorite Camera Network: The German Part.
Z. Ceplecha: The European Fireball Network, Some General Remarks.
V. Porubčan: The Czechoslovakian Fireball Network: The Slovakian Part.
A.N. Simonenko and V.V. Fedynsky: The Soviet Bolide Network. (Read by O.I. Belkovich).
P.B. Babadzhanov: Combined Photographic and Radar Observations of Meteors at the Dushanbe Astrophysical Institute.
P. Millman: Electronic Image Intensification in Meteor Photography.

26 August 1976

This session was a joint meeting with Commission 21. Its purpose was to present recent results on Interplanetary Dust and Zodiacal Light obtained from the Pioneer 10 and 11 Deep Space Probes and the Helios 1 and 2 close-in Solar Probes. The following communications were presented:
J.L. Weinberg and R.K. Soberman: Summary of Pioneer 10/11 Results, Zodiacal Light and In Situ Measurements.
C. Leinert: Summary of Helios 1/2 Results, Zodiacal Light Measurements.
E. Grün: Summary of Helios 1/2 Results, In Situ Measurements.
H. Fechtig: Dust Fluxes near 1 AU and in the Earth-Moon System.
P.M. Millman: A Comparison of Meteor Data and In Situ Results.
R.H. Giese: The Compatibility between Optical and Mechanical (In Situ) Results and Theoretical Interpretations.

After these presentations a summary discussion followed with contributions from J.M. Greenberg, P.M. Millman and F.L. Whipple. The reader is referred to the minutes of the sessions of Commission 21 for further information.

28 August 1976

A half day of discussion was devoted to visual and radar techniques. Short papers were presented at this session as follows:

C.S.L. Keay: On Radar Observations of Meteor Rates and Radiants, and Anomalies at the Base of the Thermosphere: Report of IAU/IAGA Joint Committee.

J. Jones: On the Determination of the Meteor Radiant Distributions from Single Station Observations.

B.L. Kashcheev and V.Z. Nechitajlenko: Automatical Radar System for the Study of the Cosmical Nature of Meteors and Their Penetration into the Earth's Atmosphere. (Paper read by O.I. Belkovich).

K. Hindley: An International Centre for Amateur Meteor Observations.

F. Link: On the Presence of Cosmic Dust in the Upper Atmosphere.

B.A. Lindblad: Meteor Radar Rates and Solar Activity.

P.B. Babadzhanov: Winds in the Upper Atmosphere.

K. Tomita: Japanese Meteor Radio Program.

V. Padevĕt: Effective Dynamic Cross Section of a Meteor.

1 September 1976

This session was a joint meeting with Commissions 15 and 20. The meeting highlighted some of the results presented at the IAU Colloquium No. 39, Lyon 17-20 August 1976. The meeting was organized by A.H. Delsemme. The reader is referred to the minutes of the sessions of Commission 15 for further information.

COMMISSION 24: ASTROMETRIE PHOTOGRAPHIQUE (PHOTOMETRIC ASTROMETRY)

Report of Meetings, 28 August 1976

PRESIDENT: P. Lacroute   SECRETARY: M. Creze

First Meeting: 9 H 15

The president announced that, as a result of the postal ballot of the members
of the commission, the following names were proposed for the new officers and the
organising committee, for approval by the general assembly: President, C.A. Murray;
Vice-President, H.K. Eichhorn; C.O., De Vegt Ch., Fredrick L.W., Gliese W., Harrington R.S., Lacroute P., Potters M.J., van Altena W.

The following were confirmed as new members and consultants of the commission:
New members: Firneis M.G., Gatewood G., Ianna F.A., Hershey J.L., Hill G., Jones
B.F., Jones D.H.P., Latypov A.A., Nicholson W., Onegina A.B., Podobed V.V.,
Robertson W.H., Valbousquet A.
Consultants: Cudworth K., Hanson R.B., Kanaev I.I.

DISCUSSION ON THE REDUCTION OF RELATIVE TRIGONOMETRIC PARALLAXES TO ABSOLUTE VALUES

W.F. van Altena, E.D. Hoffleit and H.A. Smith, Yale University Observatory,
presented by W.F. van Altena

Abstract

Systematic differences in trigonometric parallaxes between Allegheny Observatory and Yale Observatory, between Allegheny Observatory and McCormick Observatory
and between the Cape Observatory and Yale Observatory have been investigated for
stars common to each pair. The differences found correlated with right ascension,
naturally suggesting some sort of annual influence. It is proposed that these
differences are related to differences in the annual temperature cycle between
observatories, possibly through the mechanism of temperature dependent decentering
of the telescope objectives. A dependence upon spectral type was also discovered
in the differences between the relative parallaxes from Allegheny and from Yale.
Further work is needed to clarify the nature of these systematic effects and to
insure that they do not significantly bias available trigonometric parallaxes.

It is proposed that a new parallax catalogue be constructed at Yale after a
thorough statistical analysis of all available trigonometric parallaxes has been
made. We solicit suggestions and recommendations from interested users.

C. TURON, M. CREZE, PRESENTED BY M. CREZE
Given a sample of trigonometric parallaxes, it is assumed that:
1) - There is the possibility of defining, from photometric or spectroscopic
observations, a number of subsets, in each of which all stars may be assumed to
have about the same unknown absolute magnitudes.
2) - There exists a partition of the sample such that in each part the systematic errors of trigonometric parallaxes may be considered to be identical.

Then a method is proposed by which all kinds of suspected systematic errors
in trigonometric parallaxes may be derived together with absolute magnitude
calibrations. The method provides absolute systematic errors, and not only

systematic differences between catalogues.

Details will be published soon in Astronomy and Astrophysics.

PROBLEM OF THE HYADES
     In a short report, W.J. Luyten announced many new measures of plates on the
Hyades and Pleiades fields and new proper motions.

Hanson reported on the proper motions in the Hyades.

Afterwards some remarks on the importance of the parallaxes of the Hyades were
made.

DISCUSSION
     P. Connes: Do members of the commission believe that an instrument able to
measure relative parallaxes free from any instrumental effect and with quite
negligible random errors, could provide an important contribution to the problem of
trigonometric parallaxes?

     K. Aa. Strand: A special astrometric facility has been designed by a study
group at NASA-Ames this past summer, with the aim of obtaining accuracies of the
order of one tenth milliarc second.

     W. van Altena: Due to the importance of systematic errors in the study of trig-
onometric parallaxes it is very important that observations be made with several
different instruments. I would strongly urge you to develop an instrument to determine very accurate relative parallaxes.

<u>Second Meeting: 11 H 15</u>

IMPROVEMENTS IN PHOTOGRAPHIC ASTROMETRY

     A.R. Klemola

     Second-epoch photography for the proper motion program continues with the com-
pletion of 685 of the total of 1246 fields lying at declination -20° and northward.
The plate measurements, devoted to those fields lying outside the zone of avoidance,
have been completed for a total of 190 fields at +25°, +30°, and +35°. Reductions
for proper motions, positions, and photometry have been carried out for a group of
overlapping plates around the north galactic pole. The average annual rate of
measurement comes to about 150 fields per year.

     Reductions based on galaxies as a reference frame call for the use of a model,
which includes additionally the cubic terms x^3, x^2y, xy^2 in x and y^3, x^2y, xy^2 in y,
for the satisfactory reduction of residuals locally over the plates. Moreover, we
find that the coefficient of the radial distortion term has changed significantly
between the first and second epoch plates and must be included in the reduction for
proper motions. It is found that on a 6° x 6° field there are possible local systematic residuals over small areas of one or two degrees which reach two microns
or more. This represents a limit to local systematic accuracy of about 0.5 to 1.0
arcsec for proper motions as evidenced from a comparison of overlapping plates
where images are of poorer average quality.

     External comparisons of a preliminary nature have been made with the AGK3 for
the sample of overlapping fields around the north galactic pole. It is clear that
a strong magnitude equation in one of the proper motion sources exists when differences AGK3-Lick are formed separately for stars brighter than and fainter than magnitude 9. These differences are well in excess of the amount expected for the
correction to precession in this part of the sky. The study of these differences

in proper motions continues with the more extensive data available for the +22° to +38° zone.

The sky brightness at Mt. Hamilton, now increasing at a rate of 5% per year, is becoming an important limitation for long exposures with the astrograph. The need for eventual transfer of the astrograph to a dar-sky site is indicated.

DISCUSSION

C.A. Murray: Are the proper motion differences in the overlapping regions more in R.A. or in declinations?

A.R. Klemola: It is the same in both coordinates.

S. Vasilevskis: Do the systematic errors mentioned by you affect mean proper motions of stars distributed over the whole plate?

A.R. Klemola: The differences in proper motions include the effects of precession for this point of the sky. Therefore the large differences in Lick and AGK3 represent errors of systematic nature as well as precession.

CORRECTION OF ERRORS IN THE LOWER PART OF AGK2 NEW REDUCTIONS

P. Lacroute and A. Valbousquet

Summary

Paper to be published in Astronomy and Astrophysics.

Some difficulties in the reduction of AGK2 below +20° by the overlap method suggest the necessity of a new reduction. The authors have established that there are important systematic errors in the measures of AGK2, of about 0".1 on the average, but often as large as 0".30. These errors, which are different for different zones of declination, are the cause of the difficulties in the resolution and give systematic errors in the mean results, which sometimes amount to 0".10.

After correction of the data the systematic errors are progressively eliminated.

Taking into account these difficulties, a new solution of the AGK2 has been obtained by a method giving more emphasis to the reference stars and less to the comparison between the stars. That slightly improves the agreement with the reference stars but diminishes the agreement between the plates. A check shows that in the ne solution the systematic differences between the photographic positions and the reference positions are not significant.

Remark by W. Dieckvoss: In 1960, in Hamburg, we took the "screw-error" of Bonn's machine (decl. +7°5; +5°0; +2°5; 0° ; -2°5) at face value. The discussion of the 'field errors" showed no indication of residual screw error (now determined to be as large as 0".16, original curve at Bonn, 0".4 . . . 0".5). This gives rise to larger mean errors without affecting the overall system (FK4).

ASTROMETRIC POTENTIALITIES OF A NEW FOUR-FOLD COVERAGE OF THE NORTHERN HEMISPHERE

Chr. de Vegt, presented by W. Dieckvoss

During the Perth Symposium the project of a new fourfold coverage of the northern hemisphere including a completely new reduction of the whole AGK2 plate material was discussed (1,2). In the meantime detailed results from the new zone astrograph have confirmed the high accuracy of the instrument (3).

Expected Results from 4-Fold Coverage in Comparison with AGK2/3 Data

Catalogue		Epoch(s)	M.E. of Pos.	P.E. of P.M.	Source Cat.
AGK3	(1)	1960	0".18 - 0".26[+]	about 0".01/a	(1)-(2)
AGK2	(2)	1930	0.15 - 0.22[+]		
4-fold coverage	(3)	1975 adopted epoch	0".07		
AGK2N AGK2 - new reduction	(4)	1930	0".14	0".004/a	(3)-(4)
Carte du Ciel final reduction	(5)	1891 - 1950	0".14 - 0".20	0".002 - 0".003/a	(3)-(5)

[+]dependent on magnitude, see (3).

An important step in the formation of the new catalogue is the complete re-measurement of the AGK2 plates including both exposures. Magnitude dependent errors can then be reduced and recently found discrepancies in the Bonn-zone of the AGK2 (4) could be fully removed.

The suggested catalogue project will provide a homogeneous system of positions and proper motions up to m_V = 12-13. Therefore a high-density 2nd-order reference system is obtained and a large number of faint reference stars for a final reduction of the CdC will then be available.

From the remeasurement of the AGK2 plates the present AGK2/3 solution can be improved considerably.

With a view to the suggested ESA space mission (5) the new catalogue will provide a homogeneous net of stellar positions for a detailed comparison of groundbased and spaceborn astrometric data.

References

(1) de Vegt, Chr., IAU Symp. 61, 209 (1974).
(2) de Vegt, Chr., Ebner, H., MNRAS 167, 169 (1974).
(3) de Vegt, Chr., Mitt. Astron. Gesellsch. 38, 181 (1976).
(4) Lacroute, P., Valbousquet, A. (to be published).
(5) European Space Agency (ESA), Space astrometry, report on the mission definition study, Neuilly (1976).

DISCUSSION

P. Lacroute: Some difficulties may arise with the overlap method if the material has systematic errors on the plates. It is difficult to know to what extent some large observed differences between the simple reduction and the overlap reduction come from the systematic errors in the initial data or from the overlap method.

R.B. Hanson: I agree that good plate material is necessary for the overlap reductions, since the R.M.S. dispersion of the proper motion differences across plate gradient boundaires is equal to that expected by applying the precepts of

Eichhorn and Williams (1963. A.J. <u>68</u>, 221) i.e. 0!'005/year. There is no direct
implication in Hanson's finding of deficiencies in the plate material, only in the
lack of proper overlap reduction.

RESOLUTIONS
    A first resolution presented by Murray endorsing the collaboration between
Denmark and the United Kingdom for operating the automated Carlsberg Meridian
Circle was adopted. This resolution was also submitted to commission 8 and has
been adopted in its final form at the end of the joint meeting of Commissions 8
and 24, on Monday 30 September.

    A second resolution was proposed by van Altena to give support to the compila-
tion of a new general catalogue of trigonometric parallaxes. This last resolution
provoked the following discussion:

    K. Aa. Strand: I am not in favour of having a catalogue produced on the basis
of the paper presented this morning by Dr. van Altena.

    W. van Altena: The present investigation is quite preliminary and is based on
a small amount of material. It is hoped that a complete investigation of all available relative parallaxes will make it possible to isolate the source of the systematic errors. It is unlikely that the wholesale remeasurement of old parallax plates
will result in an overall systematic change in the parallax system of any one observatory. Since the old parallax series were only exposed to obtain a mean reference
frame at about 11th magnitude, there are in general not enough reference stars to
model potential nonlinear field effects in the coordinates.

    W. Gliese: The enormous work of trigonometric parallaxes especially at U.S.
Naval Observatory and the large amount of new photometric distance determinations
of faint red dwarfs make it desirable to compile a third catalogue of nearby stars
within the next three years. The last catalogue, in 1969, was compiled without
applying any corrections at all to the various parallaxes though we did not feel
very happy about this. Therefore, I support van Altena's proposal to establish a
uniform system of trigonometric parallaxes in the near future.

    S. Vassilevskis: I would like to express appreciation for the readiness of
Yale University Observatory to prepare a new catalogue.

    K. Aa. Strand: Instead of a resolution I propose that the commission organize
a working group on the subject.

    W. van Altena: It will be necessary to have close cooperation with all parallax
observatories if reliable corrections are to be determined. Observing and measuring
procedures as well as temperature gradients may be required in order to understand
the systematic nature of the corrections. I plan to work closely with all observa-
ories in the derivation of the corrections and see little value in establishing a
working group since all likely members would be involved in the project at the
appropriate time.

    L.W. Fredrick: At the McCormick Observatory we considered the question of
making a new parallax catalogue and concluded that now is not the time to compile
this catalogue.

    W. van Altena: It is more than 25 years since the closing date of the Jenkins
parallax catalogue and 14 years from that of the supplement. Since that time only
the U.S. Naval Observatory has been a major producer of parallaxes and this situation is not likely to change in the near future. New observing techniques such as
space astrometry and ground based photometric and interferometric instruments are
not likely to have any impact on the field for at least the next 10 years, since

no such operating instruments exist now and the space experiments will only be launched in 1983 at the very earliest. For these reasons, I can see no persuasive argument for delaying a new parallax catalogue when numerous astronomers have expressed a strong desire for one that deals with the problem of the systematic errors.

P. Lacroute: I propose that further discussion on van Altena's resolution be adjourned until the end of the joint meeting of Commissions 8 and 24. At the end of this last session, van Altena's resolution was adopted by a big majority in the following form:

Recognizing the fundamental importance of trigonometric parallaxes for the establishment of the galactic distance scale and the calibration of stellar luminosities, and noting that the Yale general catalogue of trigonometric parallaxes is now more than twenty-five years old, Commission 24 recommends that a new general catalogue be compiled in consultation with interested individuals, and that every effort be made to derive an uniform system of absolute parallaxes.

C.A. Murray read a letter from Professor J. Meurers suggesting that a colloquium on "Modern Astrometry" be held in Vienna in 1978. Commission 24 agreed to sponsor such a colloquium.

COMMISSION 25 : PHOTOMETRY AND POLARIMETRY (PHOTOMÉTRIE ET POLARIMÉTRIE STELLAIRES)

Report of Meeting, 30th August 1976

PRESIDENT : M. Golay

1 - Colour Equations B and V Photometry

By Ivan R. King

The most widely used photometric system, photographic B and V, has serious systematic problems that are not generally recognized. The photographic sensitivity bands do not match well with the corresponding photoelectric sensitivity bands ; and as a result, the photographic systems each have a colour coefficient of the order of 0.1 (B-V) with respect to the photoelectric systems. These colour coeffi-cients are serious, they are poorly known, and they are usually ignored.

Colour coefficients can be calculated theoretically from known flux curves and "known" sensitivity functions ; they predict a strong coefficient for V and a smaller one for B.

The empirical determination of colour coefficients is difficult, and in fact the usual photoelectric calibration sequence does not allow a colour calibration at all. The curve is drawn through the points, and there are no useful residuals. The only useful residuals come from redundant photoelectric observations, where stars of different colour are observed at the same magnitude. Each such group of stars has an unknown zero point, but they should all have the same slope against B-V, which can be found from a least-squares solution over all the groups. I have made such a solution from careful calibrations in SA 51, 57, and 68. My overall conclusion is that in wavelength the photographic B and V stand "outward" of the photoelectric B and V points, each by about 0.10 (B-V), with an uncertainty of ± 0.03 in each coefficient. For V, the sensitivity curves make it obvious that this should be so ; but for B it is hard to understand.

The M dwarfs have a special problem, because of TiO bands that affect the V photographic passband much more than the photoelectric. As a result (confirmed both by calculation and by observational data) the dwarf M's deviate from the linear colour equation by amounts that may exceed a tenth of a magnitude.

Comments

By A.N. Argue :

I would like to make two comments :

1) The (U-B) colours of late type giants are considerably different from the main sequence. So could a term in (U-B) be included in the colour equation to take care of the differences you mentioned between giants and main sequence ?

2) It might be dangerous to derive the colour equation only from the "bright stars" because of the existence of the so-called "photographic Purkinje effects". The relation between brightness and measured density for an in-focus photographic image

is a very complicated function of intensity, wavelength and other factors. For a refractor you also have the chromatic aberration, but the dependence of colour equation on brightness level can also operate for a reflector.

By Willstrop :

Twenty years ago the UBV system was introduced to avoid the systematic errors in m_{pg} caused by the Balmer discontinuity. It seems that we should now try to eliminate the red end of the Vpg response function.

By A.W.J. Cousins :

At Dr. Velghe's request (he is at present in Cape Town) I wish to draw attention to the VRI sequences now available in the Harvard E and F regions and in the Magellanic Clouds. (Mem.R.Ast.Soc. 81, 25 and Mon. Notes Astron.Soc.S. Africa, 35, 70 (1976)). Some hundreds of bright southern stars, a fair number in the equatorial zone, have also been measured, luminosity sequences have been established and the reddening locci determined. Unpublished material can be supplied on request.

2 - Photoelectric magnitude sequences in the yellow and the near infrared

By J. Denoyelle and A.G. Velghe :

In the study of galaxies one of the major problems is the evolutionary process of the disk. Unravelling this evolution requires the collection of many small bits of information. The suggestion I am going to make aims to provide such a bit of information. The problem we want to tackle is : "Are there differences in the large-scale distributions of the objects composing the disk, and if so, are they related to differences in age and evolution ?"

Obviously the first step is to obtain reliable distributions for the disk-components ; therefore, we need space-densities up to very large distances from the sun and also as a function of the distance to the galactic plane. This in turn supposes that the objects to be studied are easily detectable and classified. As Baade discovered, the main contributors of the disk-population are red giants, the M-type giants in particular.

Thanks to the effort and the impulse of J.J. Nassau, classification criteria have been established by him and his associates, for stars of the type M, C and S, based on the aspect of objective prism spectra in the near infrared : the wavelengths region extends from 6800 to 8800 A. These classification techniques were summarised by Nassau and Velghe (1964).

As has been shown already by Blanco (1965), it is possible to adopt natural groups for stars later than M1 :
the group M2-M4   : TiO 7054A : strong/7589A : moderate/8432A : absent
        M5-M6.5   :   7589A : strong/8432A : visible/V07900A: absent
        M7 and later: V07900A : present.
It seems however rather hazardous to separate stars with types M5 and later in two different groups, because the spectra of many of these stars vary within this range during a cycle of light variation. Moreover, statistical studies have indicated that the space distributions are very similar within the range of the group M2-M4 (early type M stars) and the group M5-M10 (late type M stars). When surveys

of M-stars are made, a most interesting byproduct is the detection of S stars and Carbon stars. The more of these stars are found the better the question can be answered how these objects are distributed and what their affinity is with one or another population group.

From the infrared objective prism spectra a sufficient number of M stars is obtained, so that reliable information on their space-distribution can be derived by the method of star counts. Moreover, the problem of interstellar absorption can be solved, provided the apparent distributions are available in two magnitude systems, for instance in yellow (Johnson's V) and infrared (I). For the details of this method, the reader is referred to the paper by Velghe (1972).

It is evident that the precision of this true distribution depends on how good the observed distributions are : this means that the statistical irregularities must be small and the errors in the magnitude scales negligible. With the large Schmidt telescopes now available, the statistical conditions can be matched for objects up to an infrared magnitude of say m_i = 14.0 at least. Due to the colour indices of these cool stars the yellow magnitude limit lies around 16.5 or 17. Of course the only reasonable way to determine the magnitudes of hundreds of stars is by photographic photometry, which in turn requires the availability of photoelectric standard sequences.

As stated in the very beginning, this investigation concerns the disk-population as a whole and therefore a larger project should be programmed. In our view, spectral surveys in the near infrared should be undertaken in a large number of regions, well distributed in galactic longitude, as well in the galactic plane as at different latitudes up to intermediate heights. As the frequency of these objects diminishes quite rapidly with increasing latitudes, one should concentrate first on the galactic plane and the lower latitudes. At the center of these regions, a sequence of about 20 intermediate to late type stars should be measured photo-electrically in the yellow and the infrared. It might be possible to use stars already identified as late type stars.

These sequences are needed as standards for photographic photometry and should be set up very carefully : requirements are principally that they go deep enough in the yellow magnitude (V = 16.5 at least), as to allow the determination of colour-indices for the fainter stars ; secondly the effective wavelength of the photoelectric and the photographic magnitude systems should correspond as closely as possible to avoid the annoying problem of the colour-equation.

At first sight this seems a non-realistic programme, as it sounds like the special plan of selected areas. However, the situation is far more comfortable : 1) the number of stars per region is quite low ; 2) some areas can be included, that are used for other investigations, such as Harvard Standard Regions, Mc Cormick Areas ; 3) the fast photometric techniques now currently used in connection with telescopes of the one meter size, allow a rapid execution. We estimate that the whole project could be terminated within a couple of years. The only condition is that at least some members of this Commission support this proposition and are willing to contribute to such a programme by making some observations.

Bibliography

J.J. Nassau and A.G. Velghe   1964, ApJ 139, 190
A.G. Velghe   1956, Astron.J. 61, 241 (communic. Obs. Roy. Belgique
                               No 104, 1956)
A.G. Velghe   1972, Symposium on the New Astronomy, Bloemfonteim

South Africa (= Communic. Obs. Roy. Belgique
Series A No 25)

Comments

By P.S. The

1) It is indeed of great importance to study the space density of M giants as function of galactic longitude and latitude, since this information can be used for the
study of the evolution of our Galaxy.

2) The study of the space density of the M giants was started long ago by the late
Dr. Nassau. Astronomers contributing to this study are Van Albada, McCuskey, Blaauw,
Westerlund, Hidajat, Mavridis, Vluming and Thé (McCarthy adds : Blanco, Cameron,
Seyfert, Mac Rae, Allers, Sanduleak, Pesh, McConnell, Philips, Stephenson). An
excellent review of the work done up to 1970 was published by Mavridis in "Structure and Evolution of Galaxies", Reidel Publ. Co., 1971. Up till now the situation
of the programme in the galacitc plane is that, except 2, all 15 luminosity function fields (LF-fields) of Dr. McCuskey has been studied for the space density of
M giants.

3) In order to avoid all kinds of difficulties in combined photoelectric photographic photometry I measure directly the M giants photoelectrically in Kron's R,
I-system. This is only possible if we are studying fields fairly far off the
direction of the galactic centre, and off the galactic plane, in which the number
of M giants is not large.

4) To expedite the work I would like to propose to combine forces.

Reply to the comments of Dr. P.S. The

After the private discussions I had during this assembly, I certainly agree to
coordinate first the available information.
We do not propose the R-system because of the poor matching of the photographic and the photoelectric system.

3 - A catalogue of photometric sequences

By A.N. Argne and E.W. Miller : Supplement No 2

An account will be published in the Proceedings of I.A.U. Colloquium No 35 :
"The compilation Critical Evaluation and Distribution of Stellar Data", Strasbourg,
France, 1976 August 19-21.

I trust no further account is needed for the report of Commission 25, but
please let me know if you require an abstract.

4 - Discussion of V magnitude for uvby system

By Ch. R. Tolbert

Since about 1970, we have had a set of standard stars available for the uvby
system (Crawford and Barnes, A.J. 1970). Unfortunately no homogeneous standard
magnitudes are available for these stars. I would like to urge that this commission

recommends the establishment of Y and/or Johnson V magnitudes for the standard stars now being used in the uvby system.

Comments

By A.W.J. Cousins

We have never experienced any problems in reproducing V magnitudes (with different filters, photocathodes, refractors and reflectors) and they correlate quite well with the Strömgren y (except for late type stars). I am not aware what problems Dr. Tolbert has encountered.

By F. Rufener

It should be noted that the V magnitudes that we determine in the Geneva Photometry (UBV B1 B2 V1 G) is equivalent to the one in the UBV system of Johnson-Morgan.

The second catalogue of stars measured in the Geneva Observatory Photometric System which is in press (Astron. Astrophys. Suppl. 26) already contains 206 stars on the 319 which are shown as standard stars in the tables II and III of Crawford and Barnes (AJ, 75, 978). The remaining 113 stars from these lists are still being observed. About thirty of them already have correct V magnitude estimations in the Geneva System.

5 - Magnitude in the photometric catalogue of Geneva

By F. Rufener

After the accurate measurement of the six colour indices, an apparent V magnitude has been determined for the stars of the photometric catalogue of the Geneva Observatory.

Rufener and Maeder (1971, 1973) have described the method utilized for the creation of a standard stars sequence which constitute the basis of this V magnitude scale. Prepared independently of the Johnson and Morgan scale, the zero of this new scale has been fixed in such a way that they correspond to each other. Through an examination of the correlations between these two scales, it has been found that, for most of the uses, the two definitions were identical and therefore the V magnitudes of the two systems interchangeable. However, we always have isolated the determinations done in Geneva and we have never mixed or averaged our measurements to those of the UBV system.

The determination of an independent V magnitude has been done, for each measurement of star, in seven colours. A weight related to its quality has been attributed to each of these determinations, independently of the weight given to the colour indices. For each star, we have calculated a weighted mean of the V magnitude and a corresponding standard deviation. We have, under publication (Rufener, 1976), a catalogue of 4670 stars containing, in addition to the six colours of the Geneva photometry, a V magnitude accompanied by a standard deviation (σ_V) and by the number (Q) of the good measurements which have defined this mean. The σ_V calculated for the values of $Q \leqslant 5$ is statistically not significant ; however it indicates clearly that an anomaly of stability may occur when $\sigma_V \geqslant 0.020$. When the value of $\sigma_V \leqslant 0.015$, this simply indicates the good agreement of the determinations, but is not a proof of great stability or great precision. On the other hand, when Q > 5, σ_V becomes significant. Its typical value for stable stars, easy to measure and which have often been observed, is σ_V = 0.011.

In Table 1, we have defined the criteria that were chosen to characterize A) the cases where the star is considered as variable or microvariable quasi certain and, B) the cases where this conviction is still very probable, say two chances out of three.

We have at present a compilation of 35'000 measurements in seven colours related to more than 7'000 stars and the observations are carried out in the Northern sky as well as in the Southern sky.

Bibliography

Rufener, F., Maeder, A. 1971, Publ. Obs. Genève, Serie A. Fasc. 78.
Rufener, F., Maeder, A. 1973, in B. Hauck and B. Westerlund (eds.),
                         Problems of Calibration of Absolute Magnitudes and
                         Temperature of Stars, IAU Symp. No 54, D. Reidel Publ. Co.,
                         Dordrecht, p. 298.
Rufener, F.   1976, Astron. Astrophys. Suppl. 26, 3, in press.

TABLE 1

CRITERIA OF VARIABILITY OR OF MICROVARIABILITY

Number of measurements Q	Standard deviation for the V magnitude ($\sigma_V.10^3$)	
	Criteria A	Criteria B
6 - 10	$\geqslant$ 25	$\geqslant$ 19 < 25
11 - 20	$\geqslant$ 19	$\geqslant$ 15 < 19
> 20	$\geqslant$ 15	

6 - A note on photometry of faint stars

By H.R. Butcher ; paper presented by B.J. Bok

Members of IAU Commission 25 may be interested in a recent experiment comparing stellar photometry from photographic plates with measurements from electronic camera exposures.

The experiment compared electrographic measures of stars in NGC 1866, reported by Walker (M.N.R.A.S. 169, 199, 1974), and photographic measurements of the same stars on Kodak emulsions type IIIaJ and 127-04. The electrographic data were obtained with a Spectracon type electronic camera, and consist of multiple exposures in the V band of the UBV system, taken with the 1.5 m reflector on Cerro Tololo. The photographic material consists of one 45 min., sky limited IIIaJ plate (with a GG 385 filter) and one 65 min., sky limited 127-04 plate (with a GG 495 filter). These plates were obtained by the writer with the 4-m reflector on Cerro Tololo.

Because the methods of reduction of the two sets of data were different, they should be described briefly : Walker used a density-recording microphotometer to make a single scan across each stellar image. He then measured the area under each profile, and converted this area to a magnitude (cf. ApJ, 161, 835, 1970). The writer, on the other hand, has used a digital microphotometer to record a square raster pattern for a part of each of his plates, the interval between points being ten microns. Each data point has been subsequently converted to an intensity reading, using spot calibrations recorded on each plate simultaneously with the stellar exposures. Next, the centre of each image was found (to an accuracy of 2-3 microns) and the intensity readings were then plotted as a function of distance from this centre. A standard stellar profile was then fitted to this plot of data points, and the resulting calculated "volume" under this profile was converted to a magnitude. Measurements from both the IIIaJ and 127-04 exposures were combined finally to derive a V magnitude, using an equation of the form $V = m(127) + a$. $m(IIIaJ) - m(127) + b$, where a and b are constants whose values were determined by a least-squares fit to Walker's measurements.

It has been found that many of Walker's stars are not single, and that agreement between the photographic and electrographic magnitudes is worse for the blended images. There is also poorer agreement for stars closer than 2.5 arcmin to the centre of the cluster, an effect probably due to an increased background of faint stars.

The attached graph shows a comparison between the photographic and electrographic V magnitudes for single, unblended images farther than 2.5 arcmin from the centre of the cluster. It would appear that over a range of at least four magnitudes, the photographic results compare very favorably with the electronic camera data. Any non-linearity in the photographic magnitude scale is clearly only a few percent over the whole range, if there is in fact any non-linearity at all.

For the range of magnitudes $18.5 < V < 22.5$, Walker estimates his average probable error to be ± 0.06 magnitude. For the restricted group of stars the scatter of the points is less than this estimate ! A similar plot comparing all unblended images of Walker's stars (25 objects) yields a scatter of ± 0.06 mag (p.e.), and one for all the stars faint enough to have unsaturated images ($m > 19$, 55 stars total), blended or not, yields ± 0.08 mag (p.e.). The real accuracy of the derived magnitudes, of course, depends as much on precisely what is measured, and on how it is measured, as on the information content of the plates, a point noted by Walker as well. But it does appear that the photographic results compare remarkably well with the electronic camera data.

There are several factors which are believed to have contributed to the success of this experiment. First, of course, is the high quality and fine grain of the IIIaJ and 127 emulsions. Next is the recording of spot calibrations on the same plates as were used for the stellar exposures. And finally, credit must go to the Photometric Data Systems microphotometer employed in the reductions, which has close to one micron positional accuracy and can record densities up to ~ 5.0.

The actual reductions reported here were performed at the Kitt Peak National Observatory, Tucson, with the Kitt Peak PDS microphotometer, CDC 6400 computer, and interactive picture analysis facility. The necessary computer programmes were written by E. O'Neil, P. Scott, D. Wells and especially C.R. Lynds. Similar efforts, using slightly different methods, have recently been reported by Kinman (Observatory, 95, 280, 1975 : paper read by Clube).

7 - The ESO-SRC Southern Sky Survey

By R.M. West and R.D. Cannon

In response to many requests from IAU participants, we would like to summarise the current state of the ESO-SRC Southern Sky Survey.

This is the main survey to be carried out by the ESO 1 m Schmidt telescope in Chile and the UK 1.2 m Schmidt in Australia. The ESO Schmidt has now almost completed the preliminary ESO(B) "quick blue" Survey, on Eastman Kodak IIaO emulsion, and film and glass copies have been widely distributed. It will soon commence the deep red survey, using Eastman Kodak 127-04 emulsion. Meanwhile the UK Schmidt is well advanced with the deep blue survey on Eastman Kodak IIIaJ emulsion, and has taken very nearly half of the required plates. There are 606 fields in each survey, covering the sky from -17o declination to the south pole.

The ESO-SRC Atlas, consisting of film copies of blue and red copies of 606 fields (in all 1212 films), is being made at the ESO Sky Atlas Laboratory in Geneva. The first shipment, of about 50 SRC IIIaJ photographs, is now ready for despatch. Some examples can be seen at the Royal Observatory Edinburgh display in the Science Library.

The price for the 1212 films is now (1976) Swiss francs 16'480.-. Orders may be placed with : ESO Sky Atlas Laboratory, c/o CERN, CH-1211 Geneva 23, Switzerland, to which further enquiries may also be addressed.

Comments

By Plaut in answer to Gehrels question concerning field corrections to photographic photometry on plates taken with the Palomar Schmidt.

On the average the field correction of the Palomar Schmidt can be neglected. It's varying from one plate to the other. Iris photometry on Schmidt plates is only possible at bad seeing. High accuracy is not possible at all : mean error of a single measurement = ± 0$\overset{m}{.}$15 to 0$\overset{m}{.}$20 (external m.e.).

By M.F. McCarthy

It is most interesting to have these remarks on field errors by Dr. Plaut. Recently Dr. Kunz and colleagues at Berkely have been investigating magnitude and colour errors and the relation between photoelectric on photographic systems. Here at Grenoble Prof. Luyten, in reply to a question concerning distance errors of positions on plates taken with the Palomar Schmidt, indicated that because of the differential methods employed in the automatic engine used by him and because of the numerous standard stars employed as reference points over the whole field, distance errors are negligible. Perhaps by careful measures with the PDS or Joyse-Lobel equipment sufficient photometric standards could be established over the entire Schmidt plate and differential photometry be carried out by reference to these. One would still, in low latitude fields in the Milky Way, have problems of crowding and from background nebulosity especially for faint objects. Accurate Schmidt photometry has been a problem since the time of Schmidt magnificent invention.

8 - Ultraviolet photometric systems

By E. Peytremann

Several replies have been received in response to a questionary on photometric systems in the ultraviolet. The characteristics of each system are summarised hereafter, together with one reference to a published article if applicable.

Wavelength units are in nanometers : 1 nm = 10 Å.

1) Wisconsin Experiment Package : Satellite OAO - 2 (NASA)

1a) Spectrometry. Scans from 360 to 185 nm (width = 2.2 nm) and from 185 to 116 nm (width = 1.2 nm), steps of 1.0 nm. Calibration is absolute (pre-flight) and the relative accuracy is about ± 0ᵐ06. Ref. : Code et al., 1970, ApJ, 161, 377. Data have been published by Code and Meade, 1976, Wisconsin Astrophysics No 30, "Atlas of UV Stellar Spectra".

1b) Photometry. Twelve bands with following wavelengths and widths in parenthesis : 425.2 nm (86), 331.7 (52), 298.5 (41), 294.5 (43), 246.2 (38), 238.6 (36), 203.5 (49), 191.3 (26), 167.9 (26), 155.4 (24), 143.0 (24), 133.3 (20). Relative accuracy is about 0ᵐ04. Absolute accuracy is 35 % from 110 to 130 nm, 21 % from 130 to 180 nm, 8.5 % from 180 to 330 nm, 4 % from 330 to 810 nm. A catalogue of about 500 stars is in preparation. Data also available on magnetic tape or microfilm from National Space Science Data Center (NASA, Goddard Space Flight Center, Greenbelt, Md., USA).

2) Celescope   Satellite OAO - 2 (NASA)
From bands with following wavelengths and widths : U1, 258.2 (55.0) ; U2, 230.8 (90.0) ; U3, 162.1 (32.5) ; U4, 153.7 (62.5 nm.
In-flight calibration : about 0ᵐ2 for U2 and 0ᵐ3 for U3. U1 and U4 are unreliable. Data have been published for about 5000 stars by Davis et al., 1973, "Celescope Catalogue of Ultraviolet Stellar Observations", Smithsonian Press, Washington, D.C., USA). Data are available on tape from National Space Science Data Center, for US residents, and from World Data Center A for Rockets and Satellites, Code 601, Goddard Space Flight Center (Greenbelt, Md., USA) for non-US residents.

3) Ultraviolet Sky Survey Telescope
   Satellite : ESRO-TD1 UK/Belgium Experiment : S2/68
Spectrophotometry. 3 channels from 137.2 to 174.8 nm, 177.1 to 214.7 nm, and 216.9 to 254.5 nm. About 22 steps/channel and resolution of about 3.6 nm. In addition there is one wide-band channel at 274 nm and width about 32 nm. Absolute accuracy is about 20 % and relative accuracy is about 10 %. For calibration, the reference is : Humphries et al., 1976, Astron. Astrophys., 49, 389. A catalogue of about 1400 stars is to be published by European Space Agency : "The UV Bright Star Spectrophotometry Catalogue". Further information about available data (magnetic tape, etc.) should be requested from Prof. A. Monfils (Institut d'Astrophysique, Liège, Belgique).

4) Ultraviolet experiment, Astronomical Netherlands Satellite (ANS)
Six bands with following wavelengths and widths : 154.9 (14.9), 154.5 (5.0), 179.9 (4.9), 220.0 (20.0), 249.3 (15.0), 329.4 (10.1) nm. Absolute and relative accuracy are respectively 20 % and 10 %. Reference : Van Duinen et al., 1975, Astron. Astrophys., 39, 159. A catalogue with data for about 6000 stars will be published. Data can be made available on special request. Requests should be addressed to Dr. R.J. van Duinen, Laboratory for Space Research, P.O. 800, Groningen, Netherlands.

5) Ultraviolet extended sources : French satellite D2B
There are three experiments of the Laboratoire d'Astronomie Spatiale, Marseille, France. Mainly devoted to extended sources photometry (Milky Way, zodiacal light), it can also observe bright stars.

5.1) Anti-Sun instrument for ecliptic plane objects (within 4°) and with 4 bands : 310 nm (30), 255.0 (30), 160.0 (30), 85.0 (10).

5.2) Whole sky survey in the following 4 bands : 310 (30), 220 (30), 160 (30) and 121.6 (30).

5.3) UBV photometry of stars near ecliptic plane ($m_v \leqslant 8$) : 380 (200), 400 (150), 430 (100).

Absolute calibration should allow accuracy better than 20 %. Ref. : Presentation of zodiacal light instrument aboard the D2B astronomical satellite : IAU Colloquium No 31, 1975, Heidelberg, W. Germany. Requests for additional information should be addressed to : Dr. P. Cruvellier, Laboratoire d'Astronomie Spatiale, Traverse du Siphon, 13012 Marseille, France.

6) Ultraviolet Photometry from Balloons (experiments from Observatoire de Genève, Switzerland.

6.1) 6 channels at : 213 (12.5), 227 (12.5), 248 (12.5, 262 (14), 286 (15), and 304 (14) nm. 12 stars have been observed. Average altitude : 38 km.

6.2) 8 channels at : 205.2 (3.8), 218.4 (3.5), 271.8 (5.5), 295.4 (5.6), 307.4 (6.8), 336.6 (4.6), 402 (36), and 520 (40) nm. 14 stars members of Pleiades cluster have been observed so far. Average altitude : 40 km.

Results published so far concern only the analysis of absorption by Ozone. Ref. : Rigaud et al., 1975, Ann. de Géophysique, 31, 455. Further information can be obtained from Prof. M. Golay, Observatoire de Genève, CH-1290 Sauverny, Switzerland.

COMMISSION 26:   DOUBLE STARS (ETOILES DOUBLES)

Report of Meetings, 25 and 27 August 1976

PRESIDENT: S. L. Lippincott   SECRETARY: O. G. Franz

First Session

President Lippincott called for a moment of silence to honor the memory of
Commission members and double star researchers lost by death since the 1973 meeting:
G. van Biesbroeck, W. H. van den Bos, J. Hopmann, H. M. Jeffers, R. Jonckheere,
G. F. G. Knipe, C. P. Olivier, and F. Zagar.

I. ADMINISTRATIVE MATTERS

At the recommendation of the President, the Commission endorsed as new members:
H. Abt, E. H. Geyer, J. L. Hershey, K. D. Rakos, and A. Poveda. Deleted from mem-
bership were S. S. Kumar, R. E. Nather, and U. Güntzel-Lingner, because of inactivity
in double star research. The Commission endorsed A. Batten, M. Fracastoro,
R. S. Harrington, S. L. Lippincott, and C. E. Worley for the Organizing Committee,
P. Muller for Président, and O. G. Franz for Vice President.

The President invited discussion on a definition of the Commission's scientific
objectives: "The Commission on Double Stars has for its objectives the organization
and promotion of all research pertaining to stellar systems for which two or more
components are astrometrically observable."

After Worley, Dunham, and Batten commented on the difficulty of finding a
precise definition and on the need to exercise judgment on what is of interest and
importance, the Commission adopted the proposed statement of scientific objectives.

Concerning proposed working rules, Worley questioned the outgoing president
automatically becoming a member of the Organizing Committee. Batten and Fredrick
defended such a procedure as common practice to provide continuity.

The Commission then adopted, by a vote of 7:0 with three abstentions, the
following three of six articles of the "Proposed internal rules and regulations for
the appointment of officers and Organizing Committee members of Commission 26, IAU":

The Board (President and Vice President and O.C.) comprises one member
per six or fraction thereof members of the commission (ex.: 7 for 38).
The appointment of a member normally covers two General Assemblies (two
terms) of the Union. The Board is changed at each General Assembly according
to the following rules:

1) The outgoing president shall be an ex-officio member of the
    O.C. in excess of the number defined above, for one term.

2) The vice president normally becomes president.

3) Among those to be replaced are: a) those who have completed
    the second term; b) those who wish to retire after one term;
    c) those who leave the commission or die during a term.

No action was taken on articles 4-6 dealing specifically with the designation

of Organizing Committee members and Candidates for vice president, pending further
review by and proposals from Commission members.

II.   INDEX CATALOG AND CATALOG OF OBSERVATIONS

Worley reviewed the status of the catalogs. At present the Index Catalog
contains 70,295 entries, an increase of 6,058 since 1963; this gain does not yet
include new lists by Couteau and Luyten. The Observation Catalog, at 1976.5, contained 301,995 entries, a growth of 93,123 since the catalog's transfer to the
U. S. Naval Observatory. Among these new entries are 46,085 pre-1927 northern-
hemisphere observations not previously contained in the catalog. Their inclusion
is a project undertaken: 1. to make these measures available to astronomers not
having access to the original publications; 2. to preserve the data, since few
copies of the original publications exist and some are deteriorating, posing the
danger that valuable data may be lost. An estimated 30,000 old observations remain
to be processed, for an ultimate total of 76,000.

The use of the catalogs is expanding rapidly, from about 15 to a current 45
to 50 requests per year.

Access to documentation is a problem. When data were transmitted from
E. Doolittle to the Lick Observatory, the original cards with handwritten entries
were passed along. However, when responsibility for the catalogs was transferred
to the U. S. Naval Observatory, these documents remained at the Lick Observatory
and are not readily available for cross-checking, verifications, identification
checks, elimination of duplications, etc.

Addressing this problem, the following resolution was unanimously adopted:
"Commission 26 on Double Stars, recognizing the value and importance of documental
material now located at the Lick Observatory and pertaining to the Double Star
Index and Observation Catalogs, urges that these documents be provided or made
readily accessible, in originals or copies, to U. S. Naval Observatory personnel
responsible for the maintenance of these Catalogs."

III. TERMINOLOGY CONCERNING DYNAMICAL PARALLAXES

Dommanget reviewed the history of dynamical parallaxes. He proposed that the
terms "orbital" and "non-orbital" dynamical parallaxes be used to distinguish a
parallax based on known orbital elements from one computed on the basis of observed
motion insufficient to define an orbit, to alert the user to the difference in
accuracy of these two types of dynamical parallaxes (Dommanget, J. 1976, Ciel et
Terre 92, No. 2).

Worley and Franz questioned the need to consider dynamical parallaxes, since
they provide no independent information. At best they can serve to resolve ambiguities and to provide a check on the consistency of dynamical and physical
characteristics of binaries, e.g., whether a given pair is or cannot be composed
of main-sequence stars.

IV. LUNAR OCCULTATION OBSERVATIONS

D. S. Evans discussed the detection and measurement of close double stars
from observation of the occultation of stars by the moon, warning particularly
against over-interpretation of the recorded diffraction patterns. Illustrating
the capabilities of the technique, he described observations with four tesescopes
at the University of Texas of the occultation of β Sco on 8 July 1976, which led
to the positive identification of five component stars. Combination of these
results with observations at other sites could yield true angular separations and
position angles. While occultations thus provide distinct possibilities for the
discovery and study of close pairs, one should not expect results on specific
(spectroscopic) binaries. Dunham remarked that the July 1976 and September 1975
occultations of β Sco were indeed observed at other sites. He also called

attention to his list of double stars in the zodiacal zone, particularly of new
discoveries. Dunham and G. Taylor hope to expand their predictions of the occultations of stars by minor planets; observations of such events should increase the
accuracy and number of stellar-diameter determinations, yield diameters of asteroids,
and lead to the discovery of more double stars.

V. CIRCULAIRE D'INFORMATION - P. MULLER

La Circulaire d'Information a pour origine une idée venue en même temps et
indépendamment au Prof. W. Rabe et à moi-même; on trouvera dans la première le texte
intégral de la proposition de W. Rabe dont l'essentiel concordait avec mon projet.
La Circulaire a été diffusée à partir de 1954, et la Commission a pris à Dublin
(1955) une résolution qui approuvait la formule adoptée et me confiait la charge
de l'éditer jusqu'à nouvel ordre.

La liste d'envoi comprend d'abord les membres de la Commission; en outre, j'ai
retenu un certain nombre d'astronomes non membres et d'établissements divers en
raison de leur intérêt pour les étoiles doubles. Quelques omissions ont pu per-
sister un certain temps avant d'être réparées et je prie les intéressés de m'excuser
ici pour ces anomalies. Le Secrétariat de l'U.A.I. recevait 11 exemplaires, mais
se contentera de 2 à partir de cette année. Au cours des années, j'ai reçu de nombreuses demandes soit d'astronomes, soit d'institutions qui désiraient ce service
et je les ai ajoutés sur ma liste, sauf dans quelques cas où le demandeur se trom-
pait évidemment sur la nature de la publication. La liste d'envoi compte actuellement (juillet 1976), 75 adresses, et la Circulaire paraît à dates fixes (mars,
juillet et novembre), avec parfois un numéro bis pour des compléments.

Je tiens à souligner, comme je l'ai fait déjà en 1970 à Brighton, que la
Circulaire est à mes yeux en document provisoire, où l'on trouve des données four-
nies avant publication, avec toutes les possibilités d'erreurs matérielles (liées
surtout à la rapidité de la composition) et même de corrections d'auteur ultérieures
que cela comporte; il convient donc de s'imposer la vérification de ces données lors
de leur publication définitive qui doit toujours suivre. Je précise par ailleurs
que la Circulaire a toujours été composée et diffusée par les soins de mon personnel
et des services généraux de mon établissement.

A la Circulaire de mars dernier était joint un bref questionnaire destiné à
recueillir les suggestions des usagers et à contrôler les adresses. Je remercie
tous ceux qui ont répondu soit par cette voie, soit personnellement comme je le
leur proposais également. Dans l'ensemble, la formule et le contenu paraissent
approuvés, et le désir général est de voir la Circulaire continuer telle qu'elle
est. Une seule suggestion précise m'a été faite (Belgrade), celle de publier des
références bibliographiques notamment de séries d'observations. Il me semble,
après réflexion, que dans l'esprit de la Circulaire la référence normale et certainement utile serait l'annonce de publications prochaines. J'invite donc les
observateurs à bien vouloir, s'ils l'acceptent, m'informer au moment de la remise
de leur manuscript ou de la correction des épreuves, avec les indications utiles:
instrument, époque et nombre des mesures, journal ou revue et date probable de
la parution.

En bref, comme la Circulaire est d'abord l'oeuvre de ceux qui m'en fournissent
la matière, bien plus que la mienne, je les remercie tous pour en avoir assuré le
succès et j'espère pouvoir compter encore sur eux dans l'avenir.

Second Session

VI. THE STATE OF DOUBLE STAR ASTRONOMY IN SOUTH AFRICA - J. HERS

At the Lamont-Hussey Observatory, all double star observations ceased with the
departure of F. Holden. The 27-inch refractor was later dismantled, and the objective shipped back to the U.S.A. Today only the dome remains.

The Republic Observatory's 26½-inch refractor remained in intermittent use for planetary photography until 1973; but in the absence of suitable observers, the double star program was never resumed and no observations have been made since 1971. In official quarters a very low priority was assigned to double stars, and no definite decision appears to have been taken on the ultimate future of the telescope. However, the likelihood of the telescope's being moved to another site appears very small. To house the 10-meter tube, a very large, expensive dome would be needed, and funds would be far better employed keeping the instrument in operation at the present site rather than to develop a new site. While it is true that the present site is now entirely surrounded by the city, this has relatively little effect on an instrument of such long focal length. The recent promulgation of smokeless zones has actually tended to improve conditions. As far as seeing associated with atmospheric turbulence is concerned, comparisons of recent observing reports with those of the 1920's show no noticeable change. An excellent instrument remains therefore available, waiting to be used. It was recently suggested that local amateurs might use it to observe visual double stars, but no one has yet come forward, and it seems unlikely that anything will happen. On the other hand, it is probable that facilities would not be refused to observers from elsewhere, as long as this did not involve extra expenditure. It would be of the greatest value to southern double star astronomy if the 26½-inch refractor could be put back into regular use as soon as possible.

VII.   BRIEF SCIENTIFIC COMMUNICATIONS

a) Rakos reviewed the program of area scanner observations of visual double stars on the UBV and the Strömgren systems now in progress in Vienna. Several thousand observations of the combined light, magnitude differences, and relative positions of the components of about 250 pairs with separations of less than 1 arcsec to about 7 arcsec have been obtained in Vienna, Chile and Hawaii. Final data reduction is being completed.

b) Strand reported on new mass determinations for white dwarfs. Analysis of the orbital motion of Stein 2051 = Gl75-34 shows the red-dwarf component to have a 20-year perturbation. Of the two possible orbital solutions, the less plausible one indicates the presence of a "dark" companion of 0.02 solar masses. The computed masses of the visible red-dwarf and white dwarf components would be 0.22 and 0.48 solar masses, respectively. The second solution assumes the red-dwarf to be a close binary of $\rho \sim 0.7$ arcsec and $\triangle$ m = 0.5 mag and yields 0.18 and 0.16 solar masses for the red-dwarf and its close companion, respectively. The mass of the white-dwarf component becomes 0.68 solar masses.

Also studied was the system G107-69, 70, whose fainter component was found to be a close binary. Some photographs obtained at the U. S. Naval Observatory showed the binary image sufficiently elongated to permit measures of the relative positions of its components, yielding a provisional orbit of P = 16.6 years and a = 0.67 arcsec. Photometry and spectrophotometry indicate both components to be nearly identical, late degenerate stars of 0.9 solar masses each, based on τ = 0.085 arcsec. These investigations raised from three to six the number of known white-dwarf masses.

c) Franz reported on his discovery of the variability of the carbon-star component of ADS 14338. The observed, seemingly irregular brightness changes of at least 1.5 mag in B and V can be represented by a combination of two regular variations with periods of 87 and 364 days, suggesting that the carbon star is probably an unresolved binary with two variable components.

COMMISSION 27: VARIABLE STARS (ETOILES VARIABLES)

REPORT OF MEETINGS, 27, 28, 31 August 1976

PRESIDENT: M.W. Feast   SECRETARY: L.N. Mavridis

27 August 1976

SCIENTIFIC MEETING
    Commission 27 was responsible for part of the organization of the Joint
Meeting (Commissions 25, 27, 29, 35, 36 and 42) on "Observational Evidence of
the Heterogeneity of the Stellar Surface" (half-day).   Speakers were:
D.J. Mullan, M. Hack, D.S. Evans, D.M. Popper and J.W. Harvey.

28 August 1976

SCIENTIFIC MEETING
    This meeting was devoted to the subject "Infrared and Radio Observations of
Variable Stars".   Speakers were: C.C. Wu, M.W. Feast, T. Gehrels (who read a
paper by G.V. Coyne), A. Winnberg and J. Morris.

ADMINISTRATIVE MEETING
    The composition of the new Organizing Committee for submission to the IAU
Executive Committee was decided by an election.   Also a list of new members of
the Commission was approved.
    Following the announcement by the President that Professor P.F. Chugainov
had resigned as Chairman of the Working Group on Flare Stars the Commission
appointed Professor L.N. Mavridis as new Chairman of this Working Group.
    There was some discussion of a circular letter from the Moscow Variable
Star Bureau asking for suggestions for the preparation of the Fourth Edition of
the General Catalogue of Variable Stars.   A motion for the continuation of the
annual subvension from the IAU to the Moscow Variable Star Bureau for their work
on the General Catalogue of Variable Stars was proposed by Professor R.E.
Gershberg, seconded and carried unanimously.
    Professor G. Herbig proposed that the editors of Astronomy and Astrophysics
Abstracts be asked to cross reference individual variable stars by name (as in
the old Jahresbericht).   This motion was seconded and carried unanimously.
    Dr B.C. Marsden gave a brief account of the problems connected with the
inclusion of announcements concerning variable stars in the Telegrams and the
Circulars of the Central Bureau for Astronomical Telegrams.   It was decided to
encourage people interested in U Gem stars to contact Dr Marsden so that
information on these stars could be forwarded to them without delay.
    The President informed the Commission about the present state of the
organization of the IAU Colloquium No. 72 "The Interaction of Variable Stars
with their Environment" to be held in Bamberg (FRG) in September 1977.   The
possibility that an IAU colloquium on "Duplicity and its Consequences amongst
Variable Stars" be held in New Zealand was discussed.   It was noted that the
IAU Executive proposed instead that this topic be placed on the programme of the
IAU regional meeting planned for New Zealand.
    Following a suggestion by Professor L. Plaut the Commissions decided to
urge the co-ordinators of co-operative variable star observations to ensure that
the corresponding results are published.

31 August 1976

SCIENTIFIC MEETING
    This meeting was devoted to the subject "Pulsation of Giant Stars in the
Instability Strips (Theory and Observations)".   Speakers were: P. Demarque,
B. Madore, A.N. Cox, R. Stobie, J.R. Lesh, D.S. Evans, W. Dziembowski, N. Nikolov.

COMMISSION 28: GALAXIES (GALAXIES)

Report of Meetings, 25, 26, 30 and 31 August 1976

PRESIDENT: E. B. Holmberg   SECRETARY: B. E. Westerlund

There were four sessions, of which the first dealt mainly with business matters, the second was a joint meeting with Commission 34 on Interstellar Matter in
External Galaxies, the third was devoted to the Structure and Evolution of the
Magellanic Clouds, and the fourth was an all-day joint discussion with Commissions
47 and 48 on Clusters of Galaxies, Cosmology and Intergalactic Matter.

25 August 1976

BUSINESS MEETING
Members stood in silence for one minute in memoriam of R. Minkowski and
F. Zwicky.

I. NEW OFFICERS AND MEMBERS
The Commission unanimously approved the election of the new President, B. E.
Markarian, and Vice-President, B. E. Westerlund. The President proposed and the
Commission unanimously approved, the list of names proposed for the new Organizing
Committee. It was agreed that the Secretary should be selected by the new President.
A large number of proposals for new members of the Commission had been received. The Organizing Committee had screened the list, and presented the names
of those they regarded qualified to the Commission for consideration. As a general rule was accepted that only those who have been members of the IAU for three
years where eligible as Commission members. It was also recommended that the Organizing Committee consider the problem of non-active members. Following the discussion the Commission members approved unanimously the list of new members presented.

II. COMMISSION RULES
It was decided that no further rules are needed at present. A discussion
followed concerning the form of meetings during IAU General Assemblies and the
future form of Commission reports. It is expected that the Executive Committee
of the IAU will present new instructions regarding both items.

III. PROPOSED RESOLUTION
Following a proposal by Dr. G. de Vaucouleurs the Commission decided to recommend to the Union "that in the standard correction of extragalactic redshifts
for solar motion with respect to the Local Group $\Delta V = 300 \cos A$, the definition
of the solar apex be changed from $l^1 = 57°$, $b^1 = 0°$ to $l = 90°$, $b = 0°$, but that
no change be made in the conventional value of the solar velocity $V_\odot = 300$ km s^{-1}".

IV. WORKING GROUPS
The Commission discussed the need of a working group for extragalactic studies
from space and decided that a group should be formed.

V. CATALOGUES OF GALAXIES
The President read a request from Dr. Vorontsov-Velyaminov in which he asks
that the numbers in his Morphological Catalogue be used.

VI. SCIENTIFIC PAPERS
The following short papers were presented during a brief session following

the business meeting:
    S. van den Bergh: M 104
    W. Baum and R. Florentin Nielsen: Old and Young Photons.
    R. S. Dixon: Galaxy Catalogues on Magnetic Tape.
    L. Bottinelli, L. Gouguenheim: HI in Markarian Galaxies.
    J. Stock and H. Alvarez: The Status of the Venezuelan Schmidt Camera.
    A. de Vaucouleurs, G. de Vaucouleurs and H. Corwin: The Second Reference
        Catalogue of Bright Galaxies.
    G. Reaves: Dwarf Galaxies in the Virgo Cluster.
    W. G. Tifft: Mean Redshifts in the Perseus Cluster.
    J. P. Vigier: The Influence of the Local Supergalactic Cluster on Radial
        Velocities.
    B. M. Lewis: Orbits of satellite galaxies.

26 August 1976

SCIENTIFIC MEETING
    A report on this meeting will be found under Commission 34. The following
papers were presented:
    G. Courtès: HII Studies of Local Group Galaxies.
    B. L. Webster: Interstellar Abundances in External Galaxies.
    M. Peimbert: The Helium Problem.
    R. Sancisi: Galactic Warps: Observations.
    A. Toomre: Interpretation.
    H. van Woerden: Gas Content of Early-Type Galaxies.

30 August 1976

SCIENTIFIC MEETING
    In the session on The Structure and Evolution of the Magellanic Clouds the
following papers were presented:
    G. de Vaucouleurs: Structure, Composition and Dynamics of the Magellanic
        Clouds and other Late-type Barred Spirals.
    A. Ardeberg: The Evolution of the Large Magellanic Cloud.
    Th. Schmidt-Kaler: The Spiral Structure, the Dynamics and the Activity
        Centre of the Large Magellanic Cloud.
    M. T. Brück: The Structure of the Small Magellanic Cloud.
    D. S. Mathewson: The Magellanic System.
    R. J. Dufour (read by B. E. Westerlund): The Structure and Chemical Composi-
        tion of Gaseous Nebulae in the Magellanic Clouds.
    B. M. Lasker: Supernova Remnants and H II Regions in the Magellanic Clouds.

ABSTRACTS OF THE PAPERS:

G. de Vaucouleurs: A review was given of the progress in our understanding
of the Structure, Composition and Dynamics of the Magellanic Clouds and other
late-type barred spirals during the past 25 years.

Reference was made to review papers by Buscombe, Gascoigne and de Vaucouleurs
(Australian J. of Science Suppl. 1954), by de Vaucouleurs and Freeman (Vistas in
Astronomy, vol. 14, 1972), to several IAU Symposia and to the bibliography pub-
lished in 1972 by the Carter Observatory, N. Z.

Ardeberg: The clusters in the LMC deserve special interest as they provide
favourable means for improvements of evolutionary tracks of massive stars. The
well-known distance, low reddening and ample number of young clusters are attractive. Large-scale studies of LMC clusters were made already by Westerlund (1961).
Recently, substantial amounts of photographic photometry were made by Robertson
(1974) and by Flower and Hodge (1975). Fundamental discussions were drawn up on
core-helium-burning stages of evolution.

Whereas the colour distribution of field and cluster stars may be similar,
such a similarity can hardly be expected for colour-magnitude diagrams. Thus,
it should be a major concern to discriminate against contamination of cluster
colour-magnitude diagrams by field stars.

For Robertson's data restriction to stars in his central fields markedly
changes the population of the core-helium-burning region. Noticeable are reductions in the number of red giants as well as their luminosity range and in the
number of stars above the main sequence. It is suggested that restriction would
increase the power of these excellent data. Generally, the only practical reme-
dies seem to be restriction of cluster radii observed and/or observations of wide
areas around each cluster for reduction of the effects of field stars.

The picture given by restricted Robertson data is good evidence for the over-
all validity for LMC stars of accepted evolutionary tracks for high-mass stars.
Reference may be made to Iben (1967) and to Stothers and Chin (1976) and to Lamb,
Iben and Howard (1976).

The first large-scale study of LMC structure was that of de Vaucouleurs and
Freeman (1972). Through data compilation from direct photography, surface photometry and star counts they out-lined what they named "an extended spiral structure". Spectacular is a γ-shaped arm structure extending from the Bar. The
methods for structure delineation chosen by de Vaucouleurs and Freeman are maybe
not the very sharpest. A wide variety of populations are mixed in a way hard to
predict and disentangle.

Few objects can challenge the super-giant stars in terms of delineation
power for young features. The surveys of Sanduleak (1970) and Fehrenbach and
Duflot (1970, 1973) gave remarkable possibilities for large-scale studies of
such stars. Sanduleak selected nearly 1300 LMC members from small-dispersion
objective-prism plates. For evolutionary studies a severe restriction is the
spectral-type bias caused by the small dispersion. Fehrenbach and Duflot used
radial velocity as a membership criterion and thereby avoided spectral-type
bias. Nearly 600 LMC members were found.

Using mainly the catalogues of Fehrenbach and Duflot an extensive observational programme was carried out at ESO (Ardeberg et al., 1972) including MK
classification, radial-velocity determinations and photoelectric UBV photometry.
This way more than 400 stars could be classified as definite members of the LMC.

In an attempt to delineate the young structure of the LMC Ardeberg (1976)
made use of primarily these ESO data and secondly of LMC members given by
Sanduleak. Whereas the ESO data should be entitled to the absolute preference,
the richness of the Sanduleak data make it an excellent back-up material. The
picture emanating is one of well-defined concentrations of super-giant stars
with surface densities considerably higher than that of the general field. The
few super-giant stars seen within the Bar area are probably to a high degree
superposed. The Bar is not a place of significant, recent star formation. The
contrary is true for the 30 Dor complex, marked by heavy star formation, which
is also true for three other major concentrations of super-giant stars.

The structure defined by super-giant stars is in excellent agreement with that from far-ultraviolet photographs (Watts, 1972). It is also in good agreement with that defined by associations, HII regions and large HI complexes. Thus, the extreme population I seems to belong to the concentrations mentioned. These concentrations do not present any convincing evidence of spiral structure.

Comparison with structural features given by de Vaucouleurs and Freeman shows that except for the 30 Dor complex only the γ wings are conformed. However, from the picture given by well-defined population I objects it seems hard to see any convincing connection between these wings and the Bar, especially for the South-west feature.

Using the ESO data plus his own photometry of Sanduleak stars Isserstedt (1975) examined the structure of the LMC. The resulting distribution is rather similar to the distribution of super-giant stars referred to above. However, Isserstedt concludes that super-giant stars are distributed in longish filaments with widths increasing with increasing intrinsic colour. Further, he states that "Eine klar gegliederte zweiarmige Spirale ist allerdings nicht erkennbar". Thus, Isserstedt finds no indication of a two-armed spiral structure.

Schmidt-Kaler and Isserstedt reanalyzed the material just used by Isserstedt plus some other objects. The main tracers used are supergiant stars and HII regions. However, only super-giant stars with $(U-B) \leq -0\overset{m}{.}6$ are taken into account. Schmidt-Kaler and Isserstedt find a clear spiral pattern with two main arms emanating from the 30 Dor complex and completely unrelated to the Bar. It may be noted that stars with $(U-B)_0 \leq -0\overset{m}{.}60$ formed one of the groups earlier studied by Isserstedt.

With the strongly mass-dependent evolution of super-giant stars the colour is a doubtful age parameter. The colour limit adopted by Schmidt-Kaler and Isserstedt may not be extremely clear-cut. Accepting the data selected by Schmidt-Kaler and Isserstedt one finds that they connect into spiral pattern features which may otherwise be regarded as individual concentrations of objects from the extreme population I. It is emphasized that the two major gaps in the outer of the two main "arms" are situated in its inner part and at the prolongation of the Bar, whereas the inner "arm" passes right through the Bar or in front of it. As the density-wave theory is incompatible with the structure proposed, Schmidt-Kaler and Isserstedt suggest an ejection theory.

Martin et al.(1976) deal with the structure of the LMC defined by super-giant stars. Using objective prism plates of fairly high dispersion they classify all stars mentioned above and previously not classified at comparable dispersion. A structure of major concentrations is found, which is very similar to that discussed above. Martin et al. find that their data fit the γ wings of de Vaucouleurs and Freeman, whereas the other concentrations do not correspond very well to the structure drawn by these authors. Martin et al. also state that the spiral structure suggested by Schmidt-Kaler and Isserstedt "seems to be out of question".

A modest number of papers have been specifically concerned with the large-scale evolution of the LMC. Payne-Gaposhkin used Cepheids of ages up to 10^8 years. She suggested that during the past 10^8 years star formation slowly advanced along the direction of the Bar. There may well be such a tendency. For the time being there are no other reliable observational data adequately covering this time interval.

Hodge (1973) used young clusters for a study of the recent evolutionary history of the LMC. He covered a time interval of 14×10^6 years and concluded that clusters form in space-time cells with dimensions 1 kpc and 10^6 years, well isolated in space and time. Identification of turn-off points with the brightest stars and (especially) assignment of single cluster ages are procedures which may not be fully compatible with Hodge's estimate of the uncertainty in a single cluster age, being 10^6 years. This in turn throws some doubt on the mentioned space-time cells.

From the ESO data mentioned earlier the large-scale star formation in the LMC was studied over the past 2×10^7 years (Ardeberg, 1976). From the positions

in the $\log L/L_\odot$; $\log T_e$ diagram and corresponding isochrones ages of super-giant
stars were determined. The resulting intrinsic "birth function" shows that
during the time interval studied star formation was essentially a one-event feature, occurring around 8×10^6 years ago. Local deviations from the over-all
birth function seem to be of low significance. If for the total area studied by
Hodge, cluster birth rates are derived, a birth function is found which is in good
agreement with that defined by super-giant stars. It may be concluded that this
evolutionary picture is hardly compatible with spiral-structure type generation
of stars. It is, however, in good agreement with Biermann's (1976) models of
galaxy evolution.

Schmidt-Kaler: If the structure of the LMC is investigated by means of the
best spiral tracers a clear spiral pattern emerges (embedded in an underlying
E5 galaxy) corresponding to an $Sc(-c^+)$ III-IVp galaxy, the peculiarity being the
deviation from the usual diametral symmetry of the two main spiral arms. This was
demonstrated by an overlay of the distributions of: HII regions, blue supergiants,
dark clouds and interstellar reddenings, OB-associations, supernova remnants and
WR-stars etc. Ardeberg's supergiants with ages up to 10^7 yrs are lined up along
exactly the same spiral arms, the older supergiants show (as in our Galaxy) a
broader distribution smearing out the sharp spiral features. The HII-regions with
diameters of 6 pc or more are the best spiral tracers (just as the giant HII
regions in our Galaxy) while the smaller ones are scattered due to age and selection effects. Near 30 Doradus, the centre of the spiral features, dark clouds
delineate the spiral structure best (just as in M31).

The structure is also well visible in Fig. 6 of Martin et al. (Astron.
Astrophys. 51, 1976, 31). It appears that de Vaucouleurs' γ-structure is based
only on B7-A9 supergiants and bright cepheids, but it should be stressed that in
our Galaxy spiral structure is well outlined only by the youngest objects of the
(extreme) population I.

The spiral structure extends over 6.1 kpc almost identical to that of the
comparable ScII-III galaxy M33.

The history of past star formation activity can be studied by means of the
width and distance from the axis of the spiral arms which correlate well to each
other and to the stellar age. Thus, the average age of WR and SNR is estimated
at 3 and $5.5\cdot10^6$ yrs, resp. Two bursts of star formation can be recognized:
the last one culminated $8\cdot10^6$ yrs ago, and centered on 30 Dor and the actual spiral structure; an earlier one culminated $60\cdot10^6$ yrs ago and centered on the bar.

The spiral features and the magnetic fields are focussed on the unique super-
giant HII complex 30 Doradus. It plays the same exceptional role among all HII
regions of the LMC as SgrA among those of our Galaxy, and shows the essential
properties of a galactic nucleus with the semistellar object HD 38268 at the very
centre. Evidence for its activity is given by the peculiar, nearly radial filaments of the nebula, its expansion velocity of about 50 km s^{-1} and the corresponding mass loss of 0.05 $M_\odot$/yr, and the variability of HD 38268 (O + WN, with
the extreme luminosity $M_v = -11$) on short time scales.

Discussions of the kinematics of the LMC yields a geometric line of nodes
at p = 168° with i = 33°, and a kinematic line of nodes at p = 188°, the difference being due to a transversal motion of 150 km s^{-1} parallel to the Magellanic
stream implying bound motion of the LMC around the Galaxy.

Detailed papers appeared in: HII Regions and Related Topics (ed. T. L. Wilson
and D. Downes), Springer Lecture Notes 1975, p. 484; Astrophys. Space Sci. 41,
1976, pp 139 and 357; Third Europ. Meeting IAU (Tbilissi 1976), in press; Astron.
Astrophys. in press.

Mrs Brück: Descriptions of the geometrical structure of the SMC by de Vaucouleurs and Freeman (1972) and of the stellar content of its various features
by Westerlund (1972) have already been the subjects of review articles.

In this paper, the structure of the Cloud and its evolution are studied by
considering the distribution of objects of various ages. The youngest objects,
the HII regions (1 to 2×10^6 years) clearly outline the wing and bar (Davies,
Elliott and Meaburn 1975). Somewhat older B stars (Sanduleak 1975) also occupy
these regions and correlate with HI contours.

Star clusters (Hodge and Wright 1974; Brück 1975 and 1976) when divided by colour and age show the youngest them (10^6 to 10^7 years) in the bar and wing, small open clusters (10^7 to 10^8 years) throughout the disk while red globular clusters (estimated age 10^9 years) congregate in the SW region of the Cloud. Plots of Cepheid distributions by age (Payne-Gaposchkin and Gaposchkin 1966) show a similar tendency for younger objects (4×10^7 years) to favour the bar and wing whose outlines become blurred as one reaches older objects (4×10^8 years). This may indicate the date of possible close encounter (3 to 5×10^8 years ago) of the Magellanic Clouds with the Galaxy, though Mathewson and Schwarz (1976) have recently sounded a note of caution as regards the encounter hypothesis. By contrast the planetary nebulae which are classified as older disk objects in our Galaxy, appear to line up along the bar of the SMC (Westerlund 1968).

Interstellar matter in the SMC derived from galaxy counts (Hodge 1974, Mac Gillivray 1975) and a dark nebulae survey (Hodge 1974b) show elliptical contours and also individual patches of obscuration in the wing and in what appears to be a counter tide or wing in the diametrically opposite region.

The halo of the SMC has been traced on UK 1.2 m Schmidt telescope plates reaching B magnitude 21^m5 in fields to the east of the main body of the Cloud. Using the COSMOS measuring machine in its star-counting mode, equidensity contours show that the halo extends in an apparently circular fashion to a distance of $4°4$ from the centre. This compares with the effective radius of $1°5$ of the elliptical contours of integrated light and of star clusters. The "Shapley wing" protrudes from the circular halo a further $2°$.

Review articles:
de Vaucouleurs, G., and Freeman, K.C., 1972 Vistas in Astronomy, vol. 14, 184.
Westerlund, B.E., 1972 Proceedings of the First European Astronomical Meeting, Vol. 3, 39.

References:
Brück, M.T., 1975, Mon. Not. Roy. astr. Soc., 173, 327.
Brück, M.T., 1976, Occasional Reports of the Roy. Obs. Edin. No. 1
Davies, R.D., Elliott, K.H., and Meaburn, J., 1976, Mem. R. astr. Soc., 81, 89 (Communications from the Roy. Obs. Edin. 212)
Hodge, P.W., 1974a, Astrophys. J., 192, 21.
Hodge, P.W., 1974b, Publ. Astron. Soc. Pacific, 86, 263.
Hodge, P.W. and Wright, F.W., 1974, Astron. J.,19, 860; 29, 858.
Mathewson, D. S. and Schwartz, M.P., 1976, Mon. Not. Roy. astr. Soc., 176, 47P.
MacGillivray, H.T., 1975, Mon. Not. Roy. astr. Soc., 170, 241.
Payne-Gaposchkin, C. and Gaposchkin, S., 1966, Smithsonian Contributions to Astrophysics Vol. 9.
Sanduleak, N., 1975, Astron. and Astrophys. 39, 481.
Westerlund, B., 1968, I.A.U. Symposium 34, 23.

Mathewson: The entire Magellanic System has been surveyed by Mathewson, Schwarz, and Murray for neutral hydrogen with the 18-m telescope at the Parkes Observatory, CSIRO, down to a detection limit of 2×10^{19} atoms cm^{-2} and with a velocity resolution of 7 km s^{-1}. In addition HI observations of the Inter-Cloud region have been made with the 64-m telescope at Parkes with a velocity resolution of 4 km s^{-1} using a 15' of arc grid spacing. Detailed maps are presented showing the intensity and velocity structure of the neutral hydrogen over the whole system. For the first time details are known of the transverse velocity component of the Magellanic Clouds from work on the Magellanic Stream (Mathewson 1976; Mathewson and Schwarz 1976). The effects of this transverse motion on the observed radial velocities of the Magellanic System have been removed from the maps.

The results show:
(1) a severe warping of the disk of the LMC
(2) that doubt is cast upon the existence of the three expanding shells of gas in the SMC which Hindman (1967) suggested were responsible for the double

HI profiles in the SMC. This double velocity structure is probably caused by a warped outer part of the SMC shadowing the central disk in a similar fashion to that found by Rogstad et al. (1976) in M33.

(3) that the Inter-Cloud region is comprised of the outer disk of the LMC, a very extended outer disk of the SMC, and gas in a bridge between the two galaxies.· This bridge gas shows the same velocity characteristics as the gas in the SMC, i.e. two components separated by 40 km s^{-1}; although this separation becomes progressively less with distance from the SMC until they merge.

Mathewson and Ford have found a stellar component of the bridge gas from the tip of the Wing at 2^h15^m to 3^h15^m. The radial velocities of these OB stars differ from that of the gas by 40 km s^{-1} which is different to the young stars in the LMC and SMC whose radial velocities are in close agreement with the gas. This indicates that the gas in the bridge is slowing down at the rate of 40 km s^{-1} every few times 10^6 years. The bridge must be a very young feature as already suggested by Westerlund. It also implies that the bridge is not a tidal feature due to the gravitational interaction of the LMC/SMC which would have occurred more than 10^8 years ago.

The bridge has probably just formed from gas in the outer parts of the SMC and this density "wave" in the disk of the SMC has produced stars in a similar manner to that in a more conventional spiral galaxy.

Galaxy counts along the bridge show that the dust to gas ratio is similar to that in the main body of the SMC. There is a displacement of the dust from the main ridgeline of the gas and it is suggested that radiation pressure of the light from the SMC and LMC is driving the dust outwards from the bridge as here it is only loosely coupled to the gas.

References:
Hindman, J.V., 1967, Aust. J. Phys., 20, 147.
Mathewson, D.S., 1976, R.G.O. Bull., No. 182.
Mathewson, D.S., and Schwarz, P. 1976, Mon. Not. Roy. astr. Soc., 176, 47P.
Rogstad, D.H., Wright, M.C.H., and Lockhart, I.A., 1976, Astrophys. J., 204, 703.

Dufour: The Magellanic Clouds constitute the best systems to study the structure and chemical composition of gaseous nebulae in external galaxies. During 1974-1975 accurate spectrophotometric observations of H II regions in both Clouds were published by the Peimberts, Aller and his coworkers, and Dufour, who found that the interstellar gas in the Clouds was metal deficient, particularly in the SMC. In addition, the observations of the Peimberts and Dufour suggest that the He/H ratio is lower in the Clouds also. Chemical abundancies calculated from these observations are summarized below.

Object	log H	log He	log N	log O	log Ne	log S	log Ar
Solar Neighborhood	12.00	11.00	7.64	8.81	8.06	7.4	7.3
LMC	12.00	10.94	6.95	8.49	7.80	7.2	6.6
SMC	12.00	10.91	6.50	8.00	7.30	7.1	6.5

Similar modern observations of planetary nebulae in the Clouds have been published recently by Osmer, Webster, and Dufour and Killen. The new data confirm Feast's discovery that the planetary nebulae in the SMC are generally of low excitation, while those in the LMC cover a range of excitation similar to planetaries in the Galaxy. Abundances derived from these data are rather heterogeneous, but suggest that N/H for the planetaries in the Clouds is comparable to values found in galactic planetary nebulae, while O/H is similar to the values in the H II regions of each Cloud. Osmer found that He/H was

~ 50 % overabundanct in most of the planetaries he studied in both Clouds.

Spectrophotometric observations of supernova remnants in the two Clouds is
still in its infancy, but new observations of LMC SNR Henize 49 by Dufour confirm
the 1973 results of Osterbrock and Dufour.   More observations of gaseous nebulae
in the Magellanic Clouds are warranted, particularly of planetary nebulae and
supernova remnants, in order to develop a more complete understanding of the chemical composition and evolution of the Clouds and of galaxies in general.

Lasker: The prime focus of the Cerro Tololo 4-meter telescope was used to
assemble an atlas of the known supernova remnants (SNR) and the other nebulae in
the Magellanic Clouds. A sample of these objects are published in Bull. R.G.O.
(in press).

The stratification of [O III] λ 5007 with respect to [S II] λλ 6716, 6731
is discussed for the SNR's, N186 D and N206. The [O III] tends to lie outside
the [S II] by about 3pc, and an explanation is sought in terms of shock flow
through a region of density fluctuations.

A group of nebulae which are not known to be SNR's (i.e., do not have nonthermal radio emission) have [S II] strengths of the same order as Hα. These
nebulae seem to have expansion velocities of ~ 30 km/sec, and it is argued that
the ionization is radiative and that the additional heating required to explain
the strong [S II] is furnished by shocks. From the data it is impossible to determine whether these shocks are driven by supernova blasts or by stellar winds.

<u>31 August 1976</u>

SCIENTIFIC MEETING

As a Joint Discussion between Commissions 28, 47 and 48 this will be reported
in Highlights of the IAU, Vol. IV.

<u>Reports of Working Groups</u>

I. GALAXY PHOTOMETRY AND SPECTROPHOTOMETRY

The Group convened on September 1 with Dr H. D. Ables as the Chairman.

Dr M. Capaccioli was nominated as the new Chairman and elected unanimously
by the members present. It was agreed that new members to the Working Group may
be added by proposals to the Chairman.

The following papers were read:
J. Kormendy: Tidal Distension of Giant Elliptical Galaxies.
J. Kormendy: On the Universality of the Central Surface Brightness
     $B(0) = 21.65$ mag arsec^{-2} in Exponential Disks.
J.E.Solheim and G. de Vaucouleurs: Progress Report on Intermediate Band
     Photometry of Galaxies.
G. de Vaucouleurs and M. Capaccioli: Standard Photometric Profile of NGC 3379.
G. de Vaucouleurs, A. de Vaucouleurs and H. Corwin:Total magnitudes and
     Colors of Galaxies from Multiaperture Photometry.
R. Barbon, Benacchio and M. Capaccioli: Geometrical Parameters for NGC
     3379/84.
S. van den Bergh: Photometry of Galaxies in the Hydra I Cluster.
F. Bertola and di Tullio: Sizes of Galaxies and Their Morphological Types.
R.G. Bingham: Electronographic Polarimetry Applied to M 82, NGC 4594 and
     NGC 1569.
H.D. Ables and P.G. Ables: B,V, Photometry of the WLM Galaxy.

II. THE MAGELLANIC CLOUDS

The Group convened on August 27 with B.E. Westerlund as Chairman.

Following the resignation of Drs J. Graham and D. Thackeray and the new
elections the Group now consists of R.D. Davies, G. de Vaucouleurs, M.W.Feast,
Ch. Fehrenbach, K. G. Freeman, S.C.B. Gascoigne, B.M. Lasker, B.Y. Mills
V.C. Reddish, A. Toomre, S. van den Bergh, and B.E. Westerlund.

Westerlund was reelected Chairman.
The following papers were read:
M. Azzopardi and J. Vigneau: The Structure of the Small Magellanic Cloud
   as Shown by the Supergiants.
A. Florsch: Some Arguments in Favour of a Great Depth of the Small
   Magellanic Cloud.
R.D. Davies: The Structure of the Clouds - Correlation between Optical and
   Radio Data.
A.G. Davis Philip: A Search for Stellar Members of the Magellanic Stream.
S.C.B. Gascoigne (read by D.S. Mathewson): NGC 2209.
B. Olander, H.B. Richer and B.E. Westerlund:The Carbon Stars in the Large
   Magellanic Cloud.

III. INTERNAL MOTIONS IN GALAXIES

The Working Group met at two successive sessions on the afternoon of August
27, 1976. The Chairman of the W.G. acted as chairman of the sessions. In a brief
introduction the Chairman recalled the circumstances that had let to the formation
of the W.G., during the Sydney Assembly.
Business matters were discussed at the start of the second session. It
was agreed that the W.G. should continue to exist; P. Pişmiş was reelected chairman for a period of three years. It was suggested and agreed that a "subgroup"
be formed to consider and propose a sample list of galaxies to be studied by
means of all possible methods of data acquisition: radial velocities by optical-
from stars and gas - by the 21 cm HI line, by radio recombination lines and possibly by molecular lines; distribution of luminosity by multicolor photometry (the
latter to be carried out by the W.G. in Photometry of Galaxies) radio continuum
data etc.
It is hardly necessary to emphasize that the results of such an endeavour
will be extremely helpful in the study of the dynamics and the evolution of
galaxies.
The following persons agreed to take part in this subgroup: Brosche,
Cappaccioli, de Vaucouleurs (G), Einasto, Gouguenheim and Huchtmeier. The list of
galaxies, representing all galaxy types will be discussed by this "Working Subgroup". The final list agreed upon will be distributed to persons on the mailing
list of the W.G. on the Internal Motions in Galaxies or to anyone who expresses
the desire to receive it. Suggestion as to the galaxies desirable to be included
in the list will be welcome.
The W.G. will continue its program of distributing abstracts of papers in
the process of publication.
The following papers on current work on the Velocity Fields in Galaxies were
presented:
G. de Vaucouleurs, A. de Vaucouleurs and W.D. Pence: The Velocity Field of
   NGC 253 from Fabry-Pérot Interferometry.
S.M. Simkin: The Rotation Axis of 3C33 and Cyg A.
V.C. Rubin: Stellar Motions in Galaxies. The Barred Spiral NGC 3351 and the
   SO Galaxy 3115.
M-H. Ulrich: Improved Optical Observations of the Outflow of Gas from the
   Nucleus of the Spiral Galaxy NGC 253".
H.M. Johnson: Recent Fabry-Pérot Observations of NGC 5128.
W. Huchtmeier: Recent H I Work of Sc Galaxies carried out at Effelsberg.
S. Faber: The Rotation Curve, Mass and Mass-to-light Ratio of NGC 4594.
P. Pişmis and L. Maupome: Remarks on Our Present Knowledge of Masses of
   Galaxies.
M. Capaccioli: Velocity Fields of the Early type Galaxies NGC 128, NGC 4125
   and NGC 3998 from Absorption and Emission lines.

IV. EXTRAGALACTIC SURVEYS FROM SPACE

A meeting was held on August 26 with H. Arp as Acting Chairman. About 40 astronomers participated. Three papers were presented:

G. Courtés: Three Steps to Extragalactic Surveys.

M. Capaccioli: Deep Photographic Survey from Spacelab.

K. Henize: Studies of an Allreflecting Schmidt Telescope.

The Working Group was then formed with R. Barbon as the Chairman and H. Arp as Vice-Chairman.

"This Working Group feels that a program of space astronomy for extragalactic and galactic research requires a variety of telescopes with varying capabilities and complementary roles. In addition to large telescopes with high resolution and great light-gathering power, it is also necessary to have optically fast wide-angle telescopes capable of surveying large areas of the sky in UV, visible and near IR wavelengths to limiting magnitudes and surface brightnesses fainter than can be achieved by ground based survey telescopes. Such telescopes can:

1) study extended objects with very faint surface brightnesses which are virtually impossible to reach with telescopes of large focal ratios, 2) explore for very faint galaxies and clusters of galaxies and 3) provide, in short time, UV and IR data of large numbers of objects over large sky areas of use to the entire astronomical community.

We therefore recommend that the space agencies of the IAU member nations give high priority to survey telescopes for space astronomy. In support of this recommendation it is the intention of this Working Group to promote discussions of the scientific objectives of such telescopes and of their impact on telescope design, and to aid in the coordination of study efforts by the various groups of astronomers interested in such telescopes and the data produced by them".

COMMISSION 29: STELLAR SPECTRA (SPECTRES STELLAIRES)

Report of Meeting 25, 31 August 1976

PRESIDENT: A. A. Boyarchuk   SECRETARY: J. P. Swings

First Session, 25 August 1976: Business Meeting

    The following resolutions were adopted:
Resolution submitted to the General Secretary by Commissions 12, 14 and 29
(proposed by R. Garstang):
    " The IAU highly values the activity of the United States National Bureau
of Standards in the compilation and critical evaluation of atomic and molecular
data, and considers these activities essential for the advancement of astronomy."
    Resolution proposed by J. Pasachoff:
    " Because the resonance lines of ionized magnesium at 2803 and 2795 A
are being increasingly studied in the Sun and stars, we propose that these lines
be denoted by their wavelengths rather than by any letter."
    New Officers and Organizing Committee:

President: M. Hack, Vice-President: W. K. Bonsack, Organizing Committee:Y.Andrillat,
A. A. Boyarchuk, C. O. R. Jaschek, J. Jugaku, R. J. Kovachev, D. C. Morton,
J. P. Swings, K. O. Wright.

New Members: J. Boulon, C. R. Cowley, V. Doazan, M. R. Fernandes, L. Goldberg,
D. L. Lambert, D. D. Locanthi, H. M. Maitzen, B. F. Peery, M. Plavec, D. Reimers,
Th. Snow, J. Tech.O. Vilhu, R. Viotti, J. M. Vreux, P. Wehinger, W. W. Weiss,
R. F. Wing, S. Wyckoff.
    The following members have resigned: H. A. Brück
    Deceased members: H. Kienle, S. V. Rublev
    Future Symposia and Colloquia: HR diagram (Symposium, endorsed by Com-
missions 29, 35, 36, 37, 45). Organizing Committee: A. G. Davis Philip,
P. Demarque, S. Strom, S. van den Bergh, K. Kodaira, I. Appenzeller, B.Paczynski,
N. Walborn. To be held in Washington D. C.,U.S.A., November 2-5, 1977.
    Turbulence in Stellar Atmospheres and Implications (Symposium, endorsed by
Commissions 29, 36.) Organizing Committee: D. F. Gray, M. Marlborough, J. Linsky.
To be held in Canada, in August 1979, on the occasion of Olin C. Wilson 70th
birthday.
    Colloquium on Photometry of Emission Line Stars to be held probably in
Hvar (Yugoslavia) in October 1977. It is recommended that Commission 29 will
sponsor this meeting only if spectroscopy of emission line stars is included.
Also it is recommended to reduce the number of topics. The chairman of the
Organizing Committee is J. Grygar.
    Colloquium on Mass-loss and Evolution of O-type stars to be held in
Western Canada in September 1977. The chairman of the Organizing Committee
is P. Conti.
    . The President opened the discussion about the future form of the Com-
mission Report. He reported that about 600 papers were written that are related to Commission 29; it is impossible to summarize all this work in a few
pages. Studies of spectra of variable stars, binaries, radial velocity
measurements, etc. are often included in the reports of Commission 29 as well as
in those of Commissions 27 and 42. It was therefore suggested that Com-

211

mission 29 should be concerned with the physical interpretation of spectra.
M. Hack will write a letter to members of the commission asking for i) name
of commission (e.g. opportunity to change it in "interpretation of stellar
spectra") and ii) recommendation concerning the report.

Suggestions for "Newsletters": Requests are made that there should exist
a circulation of some sort of newsletter, like the reports written by members
of the organizing committee to the President of the Commission.This point should
be included in the letter of M. Hack mentioned above.

Report on working groups: 1) Be stars. This working group will be continued since there are many people actively working in this field, as reported
by R. Herman. 2) UV spectra. It provides a good link between commissions 29
and 44 for interpretation of data. Main work: the Joint Discussion of September 1,
1976 (report by L. Houziaux). 3) Line standards. G. Cayrel reports that she
obtained no collaboration on this matter and therefore resigns. Only working
groups 1) and 2) will remain.

Second Session, 31 August 1976: Scientific Meeting

The following short paper or reports of work in progress were presented:

A. B. Underhill: Identification in the ultraviolet spectra of O and B stars.
J. B. Swings: 8000-11000 A spectra of emission-line stars (in collaboration
with Y. Andrillat, L. Houziaux and T. M. Vreux).
L. Gratton: A spectrophotometric investigation of K-giants (in collaboration with
S. Gaudenzi, A. Giangrande, R. Nesci and C. Rossi).
M. Delcroix: Determination of atmospheric parameters of hot stars.
L. Luud: The infrared hydrogen emission lines in stellar spectra (in collaboration
with M. Ilmas).

Other Activities of the Commission

One Joint Meeting was organized by the President with the cooperation of
Presidents of Commissions 25, 27, 35, 36 and 42 on "Observational evidence
of the heterogeneity of the stellar surfaces", August 27, 1976, (see Highlights
of Astronomy, Vol.4.,1977).
D.J. Mullan: The heterogeneity of the solar atmosphere.
M. Hack: The heterogeneity of surfaces of magnetic Ap stars.
D. S. Evans: An analysis of the slow light variability of BY Draconis.
D. M. Popper: Star spots on AR Lac type stars.
J. M. Harvey, C. R. Lynds and S. P. Worden: Direct observations of the
heterogeneity of supergiant discs.
R. E. Gershberg: On the spottedness and magnetic field of T Tau-type stars.

One Joint Meeting was organized in cooperation with Commission 36, 45
on "Classification criteria for non-normal stars" (August 28, 1976).

Two Joint Discussions were held with the participation of Commission 29:
"Stellar atmospheres as indicator and factor of stellar evolution" with
Commissions 35, 36, 44 (August 30, 1976) and "Impact of ultraviolet obser-
vations on spectral classification" with Commissions 44, 45, (September 1, 1976).

COMMISSION 30: RADIAL VELOCITIES (VITESSES RADIALES)

Report of Meetings, 1976 August 26

PRESIDENT: R.F. Griffin   SECRETARY: A.G. Davis Philip

The President welcomed those present at the meeting. He reported that Professor Heard, who had been expecting to attend and to present a paper, had been prevented from travelling as he was convalescing after a heart attack. The President's proposal that he should write a letter of greetings to Professor Heard in the name of the Commission, expressing the Commission's regret at his enforced absence and its best wishes for his continued recovery, was carried with acclamation. *

After some discussion - principally between the President and Professor de Vaucouleurs - the following changes in the composition of the Commission were agreed for the ensuing triennium:

President:	A.H. Batten
Vice-President:	M. Duflot
Organising Committee:	J.F. Heard*, A.G. Davis Philip, R.F. Griffin
Resigned:	A. Blaauw, R.F. Garrison, L. Gratton, E.K. Kharadze
Elected to membership:	C.T. Bolton, G. Hill, M. Imbert, E. Maurice, O.A. Mel'nikov, L. Prévot, E. Rebeirot

The Commission voted to adopt one rule: that no member should serve more than three consecutive terms on the Organising Committee unless he becomes the President or Vice-President; the retiring President should normally serve one term on the Organising Committee after his term of office.

Dr. Batten presented the report of the Joint Working Group set up in 1973 by Commissions 30 and 40 to study the problem of confusion between the two different uses of the symbol V_r to denote radial velocity.

The Commission unanimously agreed to forward as a resolution to the General Assembly the principal recommendation in the report, viz: "The practice of calling the quantity $c\Delta\nu/\nu_o$ a radial velocity, and denoting it by the symbol V_r, is confusing in extragalactic applications and should be discontinued. Astronomers who find it convenient to publish results in the form $c\Delta\nu/\nu_o$ should clearly indicate that they have done so. A new symbol might be coined for this quantity. We suggest V_ν". The Working Group, having completed its task, was discharged.

On the initiative of Professor de Vaucouleurs, a further resolution (also considered by Commissions 28 and 40) was approved for submission to the General Assembly, as follows: "That in the standard correction of extragalactic redshifts for solar motion with respect to the Local Group $\Delta V = 300 \cos A$, the definition of the solar apex be changed from $l^I = 57°$, $b^I = 0°$ to $l = 90°$, $b = 0°$, but that no change be made in the conventional value of the solar velocity $V_\odot = 300$ km s^{-1}."

The role of the Commission was discussed. Mme. Barbier spoke on "Catalogues of radial velocity" and Mme. Duflot on "La publication des vitesses radiales". The Commis-

---

*Unfortunately Professor Heard died on 1976 October 5 after a further heart attack.

sion agreed to set up a new Working Group

to advise on the information which should be given in publishing new radial-velocity data, and

to advise on the procedure to be followed in finding the mean velocity of a constant-velocity star for which there is more than one value in the literature.

Mme. Barbier was appointed as Chairman of the Working Group, and Dr. Batten, Mme. Duflot, Professor Evans and Dr. Griffin as members.

The following scientific topics were discussed:

Standard stars
Bolton (for Heard): report of continuing observations of the standard stars adopted in 1973.
Maurice: Concerning a possible zero-point error in Evans'southern standards.

Standard wavelengths
Batten: Report of recent work at Victoria.

Radial-velocity spectrometers
Mayor; Wright

Reduction of photographic spectrograms using a PDS microdensitometer
Bolton

Automated reduction of Fabry-Pérot interferograms of nebulae and galaxies
de Vaucouleurs

Radial velocities in the galactic-pole fields
Hilditch; Upgren; Griffin

Radial-velocity measurements with a 60-cm objective prism
Fehrenbach

Measurement of globular-cluster radial velocities with a pressure-scanned Fabry-Pérot
Shawl

COMMISSION 31: TIME (L'HEURE)

Report of Meetings 25, 26, 27, 28, 30 and 31 August, 1976

PRESIDENT: H. Enslin   SECRETARY: H. F. Fliegel

25 August 1976

Participants stood in silence in memoriam of A. Gougenheim, M.R. Madwar, J. Verbaandert, and F. Zagar; the latter, late President of Commission 31 during the term 1968 - 1970.

ADMINISTRATIVE MATTERS; INTERNATIONAL COOPERATION

The Commission unanimously approved the nomination of A. Orte as new President, and of S. Iijima as new Vice-President. The new Organizing Committee members were approved. There were no objections against the names of several new members to the Commission who had applied, or been proposed, for membership. The Commission took note of the resignation of P. Bakulin, H.U. Sandig, B. Sternberk, and M.M. Thomson.

The President presented a list of "Subjects to be considered by Commission 31". Following comments, he agreed to present, during a later session, an amended version. It was approved that the list be published in the Proceedings and thus brought to a wider public.

B. Guinot, BIH Director, discussed general considerations on the generation of TAI and the operation of the UTC system. He noted that time comparison improvements would be desirable to improve the establishment of TAI. Details of the BIH's work are given in its Annual Reports and also in the Report of the Commission. It was unanimously agreed to include a note of appreciation for the BIH's work in the Proceedings of the Commission (see final item in this Report).

J. Terrien, BIPM Director, gave an account of the resolutions concerning TAI and UTC passed by the 15th CGPM (1975), which has also recommended that the value of 299792458 m/s be used for the speed of light. In J. Terrien's opinion, this value will remain unchanged.

The BIH Directing Board's Chairman, H.M. Smith, presented a paper "Proposed Revision of the Terms of Reference of the BIH", and commented that a new resolution on the BIH's work has been made necessary by the BIH's additional responsibilities. He did not believe that the BIH should be made responsible for reporting matters of legal time. It was decided to discuss those parts of the draft which refer to the determination of the Earth's rotation, during a joint session with Commission 19.

Wm. Markowitz reported on the 1974 CCDS meeting. He reminded of the recommendation adopted by Commissions 4 and 31 in Sydney that TAI be changed by 32 s to bring ET and TAI in close accord. The CCDS discussed this proposal, but referred it back to the IAU. It was noted that a proposal that the change should not be made was now before the IAU.

H.M. Smith reviewed the activities of CCIR Study Group 7. At the 1976 meetings of Study Group 7, it was agreed that the Interim Working Party 7/1 continue with extended tasks: collation of users' requirements and study of digital time codes. A new Interim Working Party, 7/3, was created which is concerned with the reduction of mutual interference in the frequency bands allocated for time signal and standard frequency emissions. The speaker pointed out that the IAU should react to CCIR Opinion 36-1 which invited the IAU to consider Universal Time as an angular measure.

The President outlined briefly Draft Resolutions Nos. 1 - 3, distributed in advance of the General Assembly. The Commission approved an additional meeting on 26 August to discuss draft resolutions, thereby providing directions for the drafting committee chaired by H.M. Smith.

26 August 1976

Two joint meetings were held with Commissions 4 and 8 (see Report of Commission 4); one joint meeting with Commissions 4, 8, 19, and 40; and one joint meeting with Commission 19 (see Report of Commission 19).

27 August 1976

ADMINISTRATIVE MATTERS
The following names were agreed upon for consultantship during the term 1977 to 1979; C.O. Alley, J.A. Barnes, G. Becker, C.C. Costain, R. Lake, S. Leschiutta, P. Morgan, P. Mourilhe Silva, A.R. Robbins, J.M. Steele, J. Terrien.
The amended list of the "Subjects to be considered by Commission 31" was unanimously approved after slight changes suggested by Commission members (see last page in this Report).
A.M. Sinzi reported about his many efforts aimed at the establishment of a worldwide office responsible for collecting and providing information about legal and daylight saving times. As a result, the International Hydrographic Bureau has acted as such an office and has published, once or twice a year, such information in the monthly Hydrographic Bulletin, but quite incompletely, as yet.

DISCUSSION OF DRAFT RESOLUTIONS
Draft Resolution No. 1, use of GMT and UT, was presented in 2 versions. Main points of argument concerned the questions as to whether or not it should be recognized that GMT be used in the sense of UT1 in almanacs, and that UT be used in place of UT0, UT1, UT2 and UTC in cases where the distinction between them is not necessary. There were also proposals that the notation UT should imply UT1 exclusively.
When discussing Draft Resolution No. 3, designations and notations for time concepts and time scales, it appeared that "time concept" is interpreted most differently: therefore, it was agreed to drop that description and all clauses referring to time concepts.
Discussion of Draft Resolution No. 2, adjustment of the TAI frequency, was confined to the importance of the uniformity and accuracy of TAI in consideration of the new dynamical time scale, and the date to be recommended for the adjustment. It was noted that the introduction of the steering of TAI could be postponed to some later date after the frequency adjustment was made.
A draft reply to CCIR Opinion 36-1 was presented as Draft Resolution No. 4. No final decision was taken as it was intended to consider the drafts, after revision, during later meetings to be held jointly with Commission 4.

28 August 1976

Two meetings were held, the first jointly with Commission 4.

DISCUSSION AND VOTE ON RESOLUTIONS
The President read the revised draft of Resolution No. 2. Discussion related to the way of explaining the problem and also the necessity of adjusting the frequency of TAI as the latter would cause great inconvenience for users. The decision was deferred to a later meeting, subject to an amended draft.
The final draft of the BIH Statutes was introduced and unanimously approved by Commission 31 (see Resolution No. 7 of the General Assembly).

SCIENTIFIC PRESENTATIONS
A.M. Sinzi reviewed the prediction, observation and reduction of lunar occultations, and the subsequent evaluation of ET from the data. He demonstrated the results of analyses of ET - TAI from occultation data extending over 15 years. Scattering of up to ±0.4 s of the yearly mean values of ET - TAI may be due to the ephemeris used.

L.V. Morrison presented a paper "Comparison of ET(Moon) and ET(Mercury) for the Period 1677 - 1973". The UT of the observation of internal contacts for the transit of Mercury was compared with the ET calculated from the motions theories of Mercury and the Earth. The resulting ΔT's were compared with other ΔT's derived from lunar observations; the tidal acceleration of the Moon thereby found was $-26'' \pm 2''$ cy^{-2}.

C.O. Alley reported an atomic clock General Relativity experiment using a slow flying aircraft for 15 h at 10^4 m on 5 independent flights, and 2 interchangeable sets of atomic clocks. Typical flight data: measured, $+47 \pm 1.5$ ns; calculated, $+47.1 \pm 0.25$ ns $= +52.8$ ns (gravitational potential) $- 5.7$ ns (relative velocity).

J.D. Williams gave an account of relativity corrections as used in lunar laser ranging modelling. The corrections for significant periodic effects to be applied to the earth-based clocks to convert them to a solar system barycentric frame can be classified in yearly (1658 μs), monthly (1.6 μs), and daily (2.0 μs x cos Φ) terms.

The first lecture, 2nd meeting, delivered by G. Winkler on "Time Synchronization via Satellites", outlined methods for determining geometric and ionopheric corrections and reviewed various types of satellites used for clock comparisons, with estimates of accuracy: 0.1 μs actually reached, 0.1 ns perhaps possible with new systems.

P. Morgan demonstrated phase comparison results made in Australia between a ground oscillator and satellite signals on one frequency with residuals varying from 0.8 to 1.5 μs. Agreement of transportable clock versus satellite time transfer between USNO and Australia has been better than 1 μs.

M. Granveaud described the algorithm used by the BIH since 1973 June, for computing TAI from data of - at present - about 80 clocks, the weights of which are determined iteratively. Provision is taken in order to preserve the uniformity of TAI.

J. Azoubib explained the principles which underly the steering of an atomic time scale, by the use of the data of primary standards to improve both its accuracy and long term stability. The steering of TAI will cause no difficulties for the BIH.

30 August 1976

There were two joint meetings with Commission 4.

SCIENTIFIC PRESENTATION

G. Becker reviewed the characteristics of 6 primary caesium standards, 3 of them being operational. The uncertainty of the operating standards is estimated to be about 1×10^{-13}. Improvement to about 2×10^{-14} is envisaged.

DISCUSSION AND VOTE ON RESOLUTIONS

Resolution No. 2, as given below, was adopted after some discussion as to how the frequency adjustment of TAI should be formulated.

The President of Commission 4, R.L. Duncombe, took the chair, for a vote by Commission 4 members only, on the proposal that the new dynamical time scale should be specified with respect to TAI on 1977 January 1, not 1958 January 1, as provided for in Recommendation 5(a), Joint Report of the Working Groups of Commission 4. After some discussion, the proposal was approved.

The President of Commission 31 resumed the meeting by introducing Draft Resolution No. 4. An extended discussion took place as to whether or not the statement expressed in the clause 3(a) was true, and if so, whether it was necessary to include it in the Resolution. Regarding Resolution No. 3, there was strong debating about the proposal that, in order to be logically consistent with other definitions, the notations TU0, TU1, TU2 should be adopted and not UT0, UT1, UT2. The counter opinion was that this would increase confusion. Discussion of Resolution No. 1 concentrated on the point concerning retention or deletion of the clause concerning discontinuation of GMT. Resolutions Nos. 1, 3, and 4 were adopted as given below.

31 August 1976

A joint meeting was held with Commission 19 (see Commission 19 Report).

Resolutions

RESOLUTION NO. 1 BY COMMISSIONS 4 AND 31

Considering
the desirability of a clarification of the use of Greenwich Mean Time (GMT) and Universal Time (UT),

Notice
(a) that GMT and UT are used in the sense of UTC for Statutory, communications, civil use and other purposes in which maximum precision of timing is integer seconds,
(b) that GMT and UT continue to be used in the sense of UT1 as the independent argument of almanacs for astronomical navigation and surveying,

Recognize
that UT may be used in the place of UT0, UT1, UT2 and UTC in cases where the distinction between them is not needed,

Urge
that GMT be replaced by the appropriate designations, and

Recommend
that the unambiguous notations UT0, UT1, UT2 and UTC be used in all scientific publications whenever it is necessary to distinguish between them.

RESOLUTION NO. 2 BY COMMISSIONS 4 AND 31

Considering
(a) that the IAU has adopted for the dynamics of the solar system a new time scale based on the second of the International System of Units (SI),
(b) that the new time scale is closely related to International Atomic Time (TAI) and that high uniformity and accuracy in TAI are desired, and
(c) that it has been established by reference to improved primary standards that the present duration of the scale interval of TAI differs from the SI second at sea level by $(10 \pm 2) \times 10^{-13}$ s,

Recommend
that a single step adjustment of $+10 \times 10^{-13}$ s be made at $00^h00^m00^s$ TAI of 1 January 1977 to bring the duration of the scale interval of TAI into close agreement with the SI second at sea level, and that thereafter the uniformity and accuracy of TAI shall be maintained.

RESOLUTION NO. 3 BY COMMISSIONS 4 AND 31

Considering
(a) that various time scales are in current use which are based, for example, upon the rotation of the Earth or upon quantum transitions,
(b) that agreed designations of the time scales are desirable,
(c) that the time differences between time scales must be expressed unambiguously,
(d) the recommendations of the CGPM and CCIR,

Recognizing
that the designations of the time scales are also used as symbols for the expression of time instants read from the respective scales,

Recommend
(1) that the following notations be used in all languages,
(2) that the following rules be applied.

Time Scales

1. Atomic Time
TAI (International Atomic Time) is the time scale established by the BIH on
the basis of atomic clock data supplied by cooperating institutions;
TA(i) is an atomic time scale established by the institution "i".

2. Universal Time
UTO(i) is the mean solar time counted from midnight of the origin of longitudes
obtained from direct astronomical observation at the observatory "i";
UT1(i) is UTO(i) corrected for the effect of the polar motion at the observatory "i";
UT2(i) is UT1(i) corrected for the effect of the seasonal variation as published by the BIH, of the Earth's rotation.
The specification of the observatory may be omitted from UT1 and UT2, if it
can be inferred unambiguously from the accompanying text. In the case of the BIH,
this specification is usually omitted.

3. Coordinated Universal Time
UTC ist the time scale maintained by the BIH which forms the basis of a coordinated dissemination of time signals and standard frequencies and has been recommended by the CGPM to be used as the basis of civil time. The UTC scale corresponds
exactly in rate with TAI and differs from it by an integer number of seconds. It is
adjusted by the insertion or deletion of seconds to ensure approximate agreement
with UT1.
UTC(i) is a time scale realized by the institution "i" and adjusted to maintain
agreement with UTC.
Where there is any possibility of misunderstanding, the designation UTC(BIH)
should be applied.

Note. UT may be used to designate a time scale related to the diurnal rotation
of the Earth in cases where the distinction between UTO, UT1, UT2 or UTC ist not
needed.

RESOLUTION NO. 4 BY COMMISSIONS 4 AND 31

1. Take notice of CCIR Opinion 36-1 on Time Scales which invites the IAU to
consider whether the UT scale could be considered henceforth as an angular measure
and should be differentiated accordingly,

2. Agree that Universal Time may be considered as an angular measure of the
rotation of the Earth, but

3. Are of the opinion
(a) that a useful time scale is generated by any process which enables dates
to be assigned to events,
(b) that the designation Universal Time as a time measure is firmly established
and of such great convenience in astronomy, geodesy, astronomical navigation and
related applications that it would not be desirable to attempt to change this practice and that considering the need to express differences between other time scales,
it is necessary to retain the existing designations in hours, minutes and seconds.

RESOLUTION NO. 5 BY COMMISSIONS 19 AND 31

This was adopted by the General Assembly specifically (Resolution No. 7).

Subjects to be considered by Commission 31

1. Fundamentals
(a) Time concepts
(b) Time units, time scale unit intervals and multiples
(c) Definition and nomenclature of time scales
(d) Legal aspects of time

2. Investigations about dynamical time scales, atomic time scales, and Coordinated Universal Time

3. Time determination
(a) Clocks and frequency standards
(b) Auxiliary equipment
(c) Overall review of the determination and coordination of Universal Time
(in cooperation with Commission 19, which has the major responsibility in the determination of rotational time)

4. Time dissemination and synchronization
(a) Time signals, standard frequencies, time codes
(b) Methods of precise time dissemination
(c) Time coordination and synchronization

5. Applications of time
(a) Applications of time and frequencies to astronomy, space research, Earth sciences, and navigation
(b) Information about time and frequencies for users

6. Relativistic effects on time measurements

7. Cooperation with international organizations concerned with time

Acknowledgement

Commission 31 of the IAU wishes to extend its sincerest gratitude to the Director of the BIH, Dr. B. Guinot, and his staff, in appreciation of the effective work of the BIH and the rapid publication of results.

COMMISSION 33: STRUCTURE AND DYNAMICS OF THE GALACTIC SYSTEM
(STRUCTURE ET DYNAMIQUE DU SYSTEME GALACTIQUE)

Report of Meetings 25, 26, 27, 30 and 31 August 1976

PRESIDENT: L. Perek
VICE-PRESIDENT: F.J. Kerr

25 August 1976

Joint Discussion 1. Galactic Structure in the Direction of the Polar Caps.
(Included in the Highlights of Astronomy 4.)

26 August 1976

Business Meeting of Commission 33. The new officers and Organizing Committee members were duly elected. Guidelines and rules of operation of the Commission as printed in IAU Trans. XIV B, p. 200, were deemed sufficient. It was agreed that the tradition of detailed Commission reports should be continued. The co-options of new members were endorsed and additional new members were elected. Proposals for new Symposia were discussed.

W.P. Bidelman reported about the main results of Colloquium No. 35, The Compilation, Critical Evaluation and Distribution of Stellar Data.

D.S. Matthewson presented a paper on the Magellanic Stream and discussed the possibility of its tidal origin.

27 August 1976

A. Large-scale Distribution of Stars and Total Mass

Invited speakers were:
    S. van den Bergh: The Bulge and Disk of the Galaxy.
    M. Schmidt: The Halo of the Galaxy.
    J. Einasto: The Mass of the Galaxy.

The speakers discussed the role of dust in various types of galaxies, the mass of the visible part of the halo of the Galaxy and the total mass of the Galaxy. Also under discussion was the relation of population II to the hypergalaxy formed by the Galaxy and its several neighbouring stellar systems.

B. Informal Presentations

R.H. Miller showed a film of a three-dimensional stellar system as it evolved from an initial stage of a spherical shape in rotation.

T.A. Agekian analyzed the role of the gradient of directions of motion along the normal of the trajectory in the meridional plane of a stellar system.

W.J. Luyten reported on a survey by G. Hill of proper motions of 43,000 stars in and around the Hyades and Pleiades clusters.

A.R. Upgren, investigating the dynamics of young and old stellar populations from dK2-M2 stars, found outward motions of young disk stars.

30 August 1976

Large-scale Distribution of Interstellar Matter
Joint Meeting with Commission 34

Invited speakers were:
    W.B. Burton: Distribution of CO and the General Morphology of Hydrogen.
    J.P. Puget: Information from Gamma-ray Studies.
    L. Hart: Recombination-line Observations of Ionized Hydrogen.
    W.W. Roberts: The Distribution in the Context of the Density Wave Theory.

The common theme of the meeting was the discussion of new discoveries in a region at 5-7 kpc from the centre. The carbon monoxide, in contradistinction to neutral hydrogen, seems to be confined to distances below 8 kpc in a rather clumpy distribution. The gamma-rays and the ionized hydrogen show most prominent peaks at 5-6 kpc. All these effects can be understood within the frame of the density-wave theory which predicts that the spiral shock-front would be strongest in that part of the Galaxy.

31 August 1976

Non-circular Motions in the Galaxy
Joint Meeting with Commission 34

Invited speakers were:
    B.J. Robinson: Motions of Molecular Clouds in the Nuclear Disk.
    P.G. Mezger: Radial Velocities of the Ionized Gas in the Centre of the
        Galaxy (presented by T. Pauls).
    E.R. Wollman, T.R. Geballe, J.H. Lacy, C.H. Townes, D.M. Rank: Spectral
        and Spatial Resolution of the 12.8 mu Ne II Emission from the
        Galactic Centre (presented by E.R. Wollman).
    S.V.M. Clube: The Galaxy as an Expanding Spiral.
    F.J. Kerr: Large-scale Motions of H I in the Galaxy.
    M.S. Roberts: H I Motions in the Andromeda.
    G. Monnet: Analysis of the Expansion Along the Minor Axis in M 31, M 33
        and the Galaxy.
    J.H. Oort: A Dark Arm in M 31 with a Large Radial Motion.

There are many recent observations of significant non-circular motions in the Galaxy as well as in the Andromeda nebula. Their interpretation by Clube as a general expansion, mainly supported by a discussion of proper motions from the Lick Pilot Survey, was opposed by Kerr and Oort who quoted important evidence that the local standard of rest has a zero radial velocity with regard to the galactic centre, such as a prominent absorption feature at Sgr A or the symmetry of radial velocities of distant planetary nebulae.

The success of the above meetings is due to the efforts of the authors of invited papers and of the contributors to discussion. We record here our appreciation to M.F. McCarthy S.J. and his organizing committee for the preparation of the excellent programme of the Joint Discussion, and to F.J. Kerr and J. Einasto for the organization of the meetings on Large-scale Distribution of Interstellar Matter and of Stars and Total Mass respectively.

COMMISSION 34: INTERSTELLAR MATTER AND PLANETARY NEBULAE
(MATIERE INTERSTELLAIRE ET NEBULEUSES PLANETAIRES)

Report of Meetings, 25 - 31 August 1976

PRESIDENT: Hugo van Woerden.
VICE-PRESIDENT: George B. Field.

Business Session, 27 August

The President spoke briefly in commemoration of two outstanding deceased members: S.B. Pikel'ner and R. Minkowski. Pikel'ner's major contributions include work on the galactic halo, on the two-phase model of the interstellar gas, and several books; he was President of the Commission in 1964-67. Minkowski is noted for his work on planetary nebulae, supernovae and supernova remnants, and the identification of radio sources; also he supervised the famous Palomar Observatory Sky Survey. Members stood in silence for a minute in memoriam.

Membership

The membership list being submitted to the General Secretary contains over 300 names. It was composed of: the membership approved by the 1973 General Assembly, minus a few withdrawals, plus additions proposed by members of the Organizing Committee and by National Committees of Astronomy, plus applications made by individual IAU Members (mostly during the current Assembly); the applications have not been screened. President and Organizing Committee consider the present list too long to be practical, and suggest a reduction by asking members to reapply, say, every six years.

Following considerable discussion, the Commission authorized its new President to arrange for a reduction in membership, after proper consultation with the Organizing Committee and with Commission members.

Organizing Committee

The Commission agreed to nominate the following Organizing Committee (OC) for 1976-79: G.B. Field, President; V. Radhakrishnan, Vice-President; L.A. Higgs, G.S. Khromov, J. Lequeux, B.T. Lynds, M. Morimoto, D.C. Morton, M. Peimbert, H. van Woerden, B. Zuckerman.

Scientific Sessions during current Assembly

The program for these sessions, which consists mainly of invited reviews, was drawn up after consultation with the OC and with Presidents of other Commissions. One session of $1\frac{1}{2}$ hours is being devoted to selected short contributions; a few others are presented in the appropriate review sessions. The selection of these contributions from some 35 submitted was made by the OC.

Members expressed their appreciation of the size, quality and general setup of the program. However, they would prefer more time for contributed papers, perhaps 4-5 hours in total.

The reports of sessions will contain summaries of the invited reviews and (briefly) of relevant discussion; contributed papers will be mentioned by title.

Symposia and Colloquia

Dr. Terzian summarized plans for IAU Symposium No.76, "Planetary Nebulae", to be held at Ithaca (NY, USA) in June 1977.

The President reviewed plans for various meetings now under consideration: an IAU Colloquium on Photometry of Emission-Line Objects (Hvar, Jugoslavia, 1977 or 1978); an IAU Symposium on Large-Scale Characteristics of the Galaxy, to be sponsored by Commissions 33, 34 and 40 (Maryland, 1978 or 1979), an IAU/URSI Symposium on Interstellar Molecules, to be sponsored by Commissions 34 and 40 (Ottawa, 1979). The Commission strongly endorsed both proposed symposia.

Designation of Planetary Nebulae

Following a proposal by Drs. Perek and Kohoutek, the Commission adopted the following resolution:

"In order to keep up a unified system of designations of planetary nebulae, consistent with the "Catalogue of Galactic Planetary Nebulae" (Perek, Kohoutek, 1967) new designations will be approved by the Organizing Committee of Commission 34 and will be published. Dr. L. Kohoutek will prepare the first list of new designations since the appearance of the Catalogue and will transmit it to the President of Commission 34 before the General Assembly of 1979."

Bibliography of Non-Stellar Objects

Drs. A. Acker and M.C. Lortet reported on the efforts of the Strasbourg Centre for Stellar Data in compiling and distributing information pertaining to a Bibliography of Non-Stellar Objects. The Commission took note of these efforts and suggested that Drs. Acker and Lortet prepare a brief report, for possible inclusion in the IAU Information Bulletin after consultation with the President of Commission 34.

Scientific Sessions

The following pages give brief summaries of the invited reviews and of the discussion; contributed papers are mentioned by title only. The reviews will be published in extenso in a book, "Topics in Interstellar Matter", by Reidel.

Session 1: The Hot Interstellar Gas Phase

Joint Meeting of Commissions 44 and 34,
held 25 August 1976. Chairman: G.B. Field;
Secretary: E.B. Jenkins.

OBSERVATIONS of O VI, Edward B. Jenkins.

A large proportion of the early-type stars observed by the Copernicus satellite show evidence of O VI ions in the lines of sight. The absorption profiles are always broader than normal interstellar lines, which suggests the ions are formed by collisions in a very hot gas ($T \sim 4 \times 10^5$ K), rather than from nonthermal processes on ordinary, cool gas, such as interactions with cosmic rays or X-rays. The lack of conspicuous amounts of N V and S IV indicates there is relatively little gas at $T < 2 \times 10^5$ K, where the radiative cooling times are short.

Present evidence seems to disfavour a close relationship between the hot gas and the particular stars observed; practically no correlation between stellar and gas radial velocities is present, and the velocities and column densities of O VI seem unrelated to the stars' spectral types. While the amount of O VI is correlated with a star's distance, there is a large scatter in average density from one line of sight to the next. The measured densities are consistent with a density $n(O\ VI) = 2.8 \times 10^{-8}$ cm^{-3} in the galactic plane and an exponential dropoff with a scale height of about 300 pc. A conspicuous overabundance of O VI is seen near the Vela supernova remnant. Aside from that, there seems to be no correspondence between the distribution of coronal gas and other noteworthy regional properties, such as local galactic structure.

O VI profiles with larger column densities generally have a larger velocity spread, which is consistent with our viewing a blend from several coronal-gas regions, each with its own random velocity of about 25 km s^{-1} and $N(O\ VI) \sim 10^{13}$ cm^{-2}. If these regions have a pressure not much greater than the general interstellar gas, they must occupy at least a few percent of the total volume of space in our region of the Galaxy.

OBSERVATIONS OF THE SOFT X-RAY BACKGROUND, Donald P. Cox.

During the last several years, the study of celestial soft X-rays ($\sim 1/4$ keV) has shown that they are very likely galactic, diffuse in origin, and thermal.

To date, the Wisconsin group has mapped the surface brightness of 60% of the sky. Several things have been noted (e.g. W. Sanders, thesis): (1) There are interesting correlations between some bright areas and HI column-density features. (2) There is a general but imprecise anticorrelation between surface brightness and N_{HI}, but the dimmer regions do not show the spectral hardening characteristic of absorption. (3) The surface brightness tends to correlate with measured interstellar O VI absorption column density for distant stars at medium and high latitudes.

The limited amount of spectral information available from proportional counters has led to the following conclusions (e.g. P. Burstein, thesis): (1) No single-temperature interstellar plasma has a spectrum consistent with the observed pulse-height distributions, even with various spectral changes possible from absorption. (2) Combination models can be made to fit with two or more temperatures. For two-component models the higher temperature must be at least 2×10^6 K and the lower one in the range $(2\ to\ 8) \times 10^5$ K. It is not yet clear whether the distribution function in temperature is essentially bimodal. (3) The two-component models can be subjected to O VI column-density and pressure constraints; in this case, the easiest accommodation occurs with the lower temperature around 6×10^5 K. The pressures of both components are then found to be at least 10^{-12} dyne$\cdot$cm^{-2} or $P/k = 2nT \sim 10^4$cm^{-3}K. (Burstein, Borken, Kraushaar, and Sanders, Ap.J., in press)

A recent reference that covers these matters is Williamson, F.O., Sanders, W.T., Kraushaar, W.L., McCammon, D., Borken, R., and Bunner, A.N. 1974, Ap.J. (Letters), 193, L133.

NATURE AND ORIGIN OF THE HOT GAS, C.F. McKee.

Supernova explosions in a cloudy interstellar medium (ISM) produce a large volume of hot gas (HIM) in the disk of the galaxy with $T \gtrsim 3 \times 10^5$ K (Cox and Smith, 1974, Ap.J.(Letters) 189, L105; McKee and Ostriker, 1975, Bull.Amer.Astr.Soc. 7, 419). The evolution of supernova remnants (SNR) in the resulting 3-component medium (cold clouds, warm clouds of 10^4 K, and HIM) is altered by evaporation of the clouds, which injects a significant amount of mass into the hot interior of the SNR (McKee and Ostriker, 1975; L. Cowie, 1976, Thesis, Harvard). The onset of the dense shell phase of SNR expansion is hastened by the higher internal density resulting from evaporation and by the enhanced cooling in the conductive interfaces between the clouds and the HIM.

The density and temperature of the HIM are determined by energy balance (SN input = radiative loss) and mass balance (cloud evaporation rate = dense shell formation rate), with typical values being $(n,T) = (10^{-2.5}$ cm^{-3}, $10^{5.7}$ K$)$ (McKee and Ostriker, 1975). The resulting pressure $P/k = 10^{3.5}$ K cm^{-3} agrees with 21-cm and H_2 determinations. The predicted mean density of O VI in the HIM is $\sim 10^{-7}$ cm^{-3}, significantly greater than observed; but much of the O VI should have too high a velocity dispersion to be readily detectable. Soft X-ray emission arises in large

SNR (R ~ 170 pc) near the cooling point, where the pressure is higher than average; the predicted intensity agrees with observation.

NATURE AND ORIGIN OF THE HOT GAS, R. McCray.

No summary available.

SUMMARY OF COMMENTS AND DISCUSSION

One participant (unidentified) questioned whether the exceptionally low abundance of O VI toward the star γ Cas places constraints on our interpretation of the filling factor of the hot gas. Jenkins replied that the known irregularatiry in the distribution of coronal material makes a general interpretation from a single measurement untrustworthy, but the γ Cas measurement makes the notion that we are immersed in a coronal region less tenable. A comment was made that the softest X-ray emission was more isotropic than the background at higher energies; McKee and Cox suggested this may imply we are inside a region of hot gas (perhaps too hot to produce O VI toward the very nearest stars). Jenkins raised two possible objections to this viewpoint: one being the observation of backscattering of solar Lα emission by local neutral hydrogen, and the other being a suggestion by Parker that the constancy of the cosmic-ray flux with time (based on meteoritic evidence) would preclude our being recently enveloped by a bubble of coronal material.
Field asked about the preponderance of negative velocities seen in an early survey of O VI. Jenkins replied that the laboratory wavelengths of the transitions are poorly known, but that a reconsideration of presently available data seemed to favour an upward revision of all the older velocities by about 6 km s^{-1}. McCray emphasized that better laboratory measurements of wavelengths still need to be made, since the lack or presence of negative velocities is crucial to deciding between circumstellar or general interstellar origins of the gas.
Radhakrishnan asked McKee what was a typical radius for a supernova remnant when breakup occurred. McKee replied this radius should be on the order of 170 pc. Jenkins questioned McCray on the alteration of shell geometry by a motion of the star with respect to the surrounding material. McCray said this problem has been considered, and he briefly showed a diagram of the situation.
The session chairman encouraged McKee and McCray to engage in a direct debate on their opposing viewpoints on the origin of the coronal gas. During this confrontation McKee questioned why we do not see more compelling observational evidence for the widespread presence of distinct shells around the hotter, more luminous stars. He suggested that dense blobs of cool material inside the stellar wind region may quench the formation of the shell through losses by conduction. McCray suggested we should investigate whether or not such density inhomogeneities are swept out by the stellar wind, or do they remain inside the cavity?
Van der Laan expressed a general caution that if we thought of the filling factor of hot gas not being very much less than unity, we might have trouble understanding the establishment of a "grand design" in galactic structure.

Session 2/3: Interstellar Molecules and Dust

Joint Meeting of Commissions 34, 40 and 44,
held 25 August 1976. Chairmen: B.J. Robinson
and A.D. Code; Secretary: J.M. Greenberg.

OBSERVATIONS OF MOLECULAR CLOUDS, B. Zuckerman.

Since the 1973 IAU meeting about 15 new interstellar molecules have been detected, and a number of unidentified lines have been identified both astronomically and in the laboratory. HCO^+, HNC and N_2H^+ are the most interesting of these. Various observations suggest that only 10-30% of the interstellar carbon is contained in CO.

I will discuss giant molecular clouds seen near bright HII regions and other clouds not associated with luminous stars. Many detailed studies exist, for both kinds of cloud. For example, M17 has been observed by Lada and W49 by Mufson and Liszt. Dark clouds not associated with bright stars have been studied, for example, by Evans and Kutner, Blair and Encrenaz in molecular lines, by Strom and co-workers in the infrared, and by Gilmore and Brown in the radio continuum.

Questions of the dominant motions in these clouds were considered by Liszt and co-workers, Zuckerman and Evans and Milman who discussed the relative virtues of cloud models dominated by systematic or turbulent motions. The absence of self-reversed ^{12}CO profiles was used as an argument in favour of systematic motions with $v \alpha r$. Recently self-reversed lines have been observed in about a half-dozen clouds suggesting that, at least in such clouds, systematic motions are not large and/or not monotonic. Loren and Snell in a recent paper argue that certain asymmetries in the ^{12}CO and ^{13}CO profiles suggest that four clouds showing self-reversals are free-fall collapsing according to a $v \alpha r^{-1/2}$ law. There are, however, various theoretical and observational problems with this interpretation, particularly when H_2CO absorption profiles are compared with the CO profiles.

Very recent results on the Orion molecular cloud concern both the very large and very small scale. Kutner, Evans and Tucker have made maps in ^{12}CO, ^{13}CO and H_2CO (2-mm emission) which extend from the Kleinmann-Low Nebula in the south to NGC 1977, suggesting that this nebula is ionization-bounded by the molecular cloud. Various optical, infrared, and radio measurements could be made near the interface of HII region and molecular cloud. The H_2CO map shows several fragments of the molecular cloud separated by ~ 1 pc; each fragment contains ~ $10^3 M_\odot$. Based mainly on observations of M17, Elmergreen and Lada suggest fragmentation on a scale of ~ 20 pc, comparable to the separation of subgroups in OB-associations noted by Blaauw.

At the Kleinmann-Low Nebula Zuckerman, Kuiper and Kuiper have found ^{12}CO emission over a total velocity range of at least 150 km/s. This localized high-velocity emission (extent $\leq 1'$) is probably associated with pre-, rather than post-, main-sequence objects and probably represents extensive mass outflow rather than inflow. The implied outflow of mass (~ $10^{-3} M_\odot/yr$) and momentum is, however, quite large if it is to be supplied by a cluster of young stars.

Discussion   Burton: Can one eliminate side-lobe contributions to the broad CO profiles? Zuckerman: The broad wings disappear away from the K-L Nebula.

ISOTOPIC COMPOSITION OF INTERSTELLAR CLOUDS, C.H. Townes.

The observation of microwave spectra of molecules in interstellar clouds allows separation and detection of the lines of isotopes of many of the more common elements. Comparison of intensities of isotopic lines shows that the relative isotopic abundances for C, O, S, N, and Si are generally rather similar to those found on Earth. However, there are interesting and provocative differences.

Special conditions of opacity, of cloud structure, of excitation, or of chemical fractionation of isotopes can confuse and make very uncertain the determination of isotopic abundances from the intensities and shapes of molecular lines. Opacity high enough to make relative abundances of isotopes difficult to determine with precision is commonly encountered, and so is chemical fractionation at least in the case of the hydrogen isotopes. The extent of resulting uncertainties is still unclear. However, careful selection and interpretation of measurements and comparison of different spectra seem to allow the determination of relative isotopic abundances to a useful precision and degree of certainty. The results indicate that the $^{12}C/^{13}C$ ratio is generally about 45, one-half that found on Earth, but is as low as about 20 in the Sgr A and Sgr B clouds, and as high as about 80 in some other clouds. Such apparent variations are probably real, and do not depend simply on distance of the cloud from the Galactic Center, as might be expected if interstellar clouds at a given distance are intermingled. The $^{17}O/^{18}O$ abundance ratio is slightly greater in interstellar clouds than on Earth. Both deuterium and ^{15}N are substantially depleted in the Sagittarius clouds as compared with most other parts of the Galaxy. This provides some evidence that deuterium in the Galaxy is a relict of events other than stellar activity.

Most of these results fit rather well current views of the nucleosynthesis and evolving stellar history of the Galaxy. However, the variation in isotopic ratios seems to show that the large molecular clouds, of mass $10^5 - 10^6$ $M_\odot$, retain their integrity for approximately 10^9 years or longer, i.e. comparable with the galactic lifetime. The Earth may have been formed in a cloud which was somewhat poorer than average in ^{13}C relative to ^{12}C. The long lifetime of massive clouds is not surprising except in view of possible gravitational collapse, for which important details of the dynamics involved are obscure.

Discussion   Van den Bout: Comparison of 2-cm absorption and 2-mm emission by H_2CO gives a higher $^{12}C/^{13}C$ ratio than do other molecular lines - Townes: The discrepancy may not be real. - Watson: The 3757 Å line of CH^+ in ζ Oph gives a terrestrial $^{12}C/^{13}C$ ratio - Townes: The optical observations are restricted to tenuous rather than dark clouds; this may make a difference in the abundance ratios.

THEORY OF INTERSTELLAR CLOUDS AND THE FORMATION AND EXCITATION OF MOLECULAR HYDROGEN, A. Dalgarno.

The observational data from the Copernicus satellite on the relative abundances of atomic and molecular hydrogen are generally consistent with a theory that postulates an equilibrium between formation of H_2 on grain surfaces and destruction by fluorescent dissociation induced by the interstellar radiation field.

H_2 is detected in excited rotational levels. The rotational populations can be explained by a combination of ultraviolet pumping and excitation during the formation process. Application of the theory requires the solution of the vibratrional-rotational radiative cascading problem. Cascade tables have been constructed and used to infer the densities, the temperatures and the ultraviolet radiation-field intensities for several clouds. The derived densities range from 10 to 1000 cm^{-3} and the gas pressures from 10^3 to well over 10^4 cm^{-3} K; there is little evidence for a uniform cloud pressure supported by an intercloud medium. In some of the clouds the derived radiation field is unusually large, suggesting that the cloud is close to the parent star and presumably physically associated with it.

There is also clear observational evidence for clouds that are sheets 0.01 pc thick with densities between 100 cm^{-3} and 1000 cm^{-3}, produced presumably by shock waves assosiated with expanding HII regions or old supernova remnants. Stellar winds of early-type stars may also produce thin, dense circumstellar shells which will contain rotationally excited H_2.

The Copernicus data also reveal the presence of HD in amounts which show that there must be a source of HD in addition to grain formation, which is probably the reaction sequence $H^+ + D \rightarrow H + D^+$, $D^+ + H_2 \rightarrow H^+ + HD$. From the measured abundance of HD, the proton density can be derived and from it the ionizing flux within the cloud. Ionizing fluxes ranging from 10^{-17} to 10^{-15} sec^{-1} for different clouds have been derived. Although the variation in the fluxes is probably real, the estimates are sensitive to the adopted models. Ionizing fluxes can also be derived from the observed abundances of OH. For ζ Oph the value is 1.6×10^{-17} sec^{-1} which, if correct, excludes the possibility of low-energy cosmic-ray ionization in the cloud.

Emission lines of the 1-0 band of H_2 have been detected recently in Orion and in NGC 7027. Emission from higher vibrational levels was not detected and the origin of the excitation is uncertain. Whether it is ultraviolet pumping or collision excitation, densities of order at least 10^6 cm^{-3} appear to be required.

Controversy persists over the heating due to the grain photoelectric effect, but it seems likely that it is the major source in diffuse clouds. Other sources are cosmic rays, H_2 photo-dissociation, and possibly H_2 formation. If H_2 formation is a heat source, hydrogen atom recombination during cloud collapse may significantly affect the thermal structure of the cloud. The hydrogen atoms are assumed to exist in excess of their equilibrium abundance, because in the evolution of a dense cloud the time scale for molecular-hydrogen formation may be long compared to the cloud lifetime.

Cooling is affected by the formation of molecules, which may generate instabilities leading to fragmentation and to the onset of gravitational collapse. Radiative trapping tends to suppress the instability.

FORMATION OF INTERSTELLAR MOLECULES, W.D. Watson.

Recent and forthcoming reviews of this topic include those by Dalgarno and Black (Rep. Prog. Phys., in press), Herbst and Klemperer (Phys. Today 29, 32), and Watson (Rev. Mod. Phys., Oct 1976).

Until about 1973, formation of most interstellar molecules was thought to occur on surfaces of interstellar grains. Neither a precise understanding of the relevant processes, nor clear predictions that can be tested, are available due to the uncertainties in the surface physics. Consideration of reactions between positive ions and molecules, which began in 1973, has shown that many small interstellar molecules can be produced readily by gas phase reactions. For some molecules (e.g. HD, CH^+, HCO^+, DCO^+, N_2H^+, CCH, HNC) this may be the only adequate process. The key to achieving adequate ionization rates is the efficient transfer of ionization of hydrogen and helium by cosmic rays to less abundant elements. In dense interstellar clouds, HCO^+ is a "cornerstone" of the ion-molecule scheme. Its unambiguous identification with the X-ogen line during the last year provides strong support for ion-molecule reactions in interstellar clouds.

The production of OH and H_2O begins with $O + H^+ \rightarrow O^+ + H$, for which the cross-section is uncertain, and continues via OH^+ and H_3O^+. Initial detections of OH in diffuse clouds during the past year at wavelengths near 3080 A and 1220 A are in excellent agreement with abundance predictions based on these reactions.

Reactions on grain surfaces should produce OH and NH at roughly the same rate, except for the difference in the abundances of oxygen and nitrogen. The improved upper limit (NH/OH) < 1/100 obtained this year toward Omicron Persei is inconsistent with formation on grain surfaces having the expected properties.

In dense clouds, the ionization of H_2 and He by cosmic rays initiates the ion-molecule reactions. H_2^+ reacts immediately to produce H_3^+ ($H_2^+ + H_2 \rightarrow H_3^+ + H$), of which most probably leads to HCO^+ and N_2H^+ in collisions with CO and N_2. At higher densities, the reaction $CO + N_2H^+ \rightarrow HCO^+ + N_2$ preferentially depletes N_2H^+. This phenomenon apparently is observed in maps of N_2H^+, HCO^+ and SO_2 across the Orion Nebula. The observed variation is interpreted to indicate an upper limit for water, $[H_2O]/[CO] \gtrsim 1/6$.

Ionized helium is perhaps more important than hydrogen in the ion-molecule scheme. The reaction of He^+ with H and H_2 is extremely slow, so that most He^+ is used to produce reactive atoms and ions: $He^+ + (CO, N_2,...) \rightarrow He + (C^+ + O, N^+ + N, ...)$. The ions react with molecular hydrogen and the atoms with H_3^+ to produce hydrides (e.g. H_2O, NH_3(?), CH_2). Charge-exchange with metal atoms may be significant for the neutralization of ions.

Quantitative predictions can also be made for certain small molecules containing more than one "heavy" atom. The most direct process for producing HCN predicts $[HCN/NH_3] < 0.6$, a result only marginally consistent with observation. Unambiguous identification of the HNC microwave line which is comparable in strength to that of HCN is especially interesting for studies of reactions. A gas-phase formation is strongly indicated. Despite its wide occurrence, even in relatively diffuse regions, formaldehyde's formation is still a problem. CCH is also expected to occur widely – a prediction that has been verified by observation. Specific formation schemes in the gas phase for more complex molecules are not available.

The discussion centred on the importance of activation energies in ion-molecule reactions.

THE NATURE OF DUST GRAINS, P.G. Martin.

The formation and destruction of grains, dust in HII regions and molecule formation are omitted.

The gas-to-dust ratio is 100, from 21-cm, X-ray and Lyman-α observations, consistent with cosmic abundances and gas depletion. R is 3.3, with variable-extinction and colour-difference methods agreeing, even in Orion. Dust changes from place to place are nevertheless seen in variations of ultraviolet extinction, of λ_{max} and λ_c, and of the ratio of ice (3.1μ) to silicate (10μ) extinction.

The albedo and phase function (g) are used in diverse problems, yet are still not well known. Brightness profiles of dark nebulae might give g. Interstellar circular polarization is consistent with an albedo ~ 0.7. The UV region is uncertain.

A pinwheel hypothesis is proposed to achieve grain alignment in a weak field. The 10μ polarization shows silicates can be aligned. UV polarization measurements at the 2200 Å feature are needed.

Much laboratory work has yielded optical constants of silicates, carbonates and sulphates needed for IR identification. Normal rocks have too high an intrinsic band strength to explain the 10μ polarization spectrum. The graphite interpretation of 2200 Å absorption is not firm.

Diffuse bands are well-correlated with eachother. The lack of a λ 4430 blue emission wing and of polarization structure at $\lambda\lambda$ 4430, 5780, 6284 rules out a normal-sized grain explanation; the λ 5780 asymmetry is consistent with small (unaligned) grains. Lab work has also eliminated the H^-, C^- or O^- alternatives.

Electromagnetic scattering by spheroids or even 'bricks' can now be computed.

Temperature fluctuations of small grains, important for mantle accretion and molecule formation, are discussed.

The study of far-infrared (100μ-1mm) thermal emission from grains represents a new frontier.

Discussion   Isobe: The ratio R of total to selective extinction is longitude-dependent. - R.W. Wilson: Both $A_V/E(B-V)$ and the ratio of ultraviolet and visual extinction are strongly variable.

Contributed paper: Ultraviolet extinction observations by ANS in ρ Ophiuchi and Orion, by D.P. Gilra.

Session 4/5: Interstellar Matter in External Galaxies

Joint Meeting of Commissions 34 and 28;   26 August 1976
Chairmen: H. van Woerden and E.B. Holmberg;
Secretaries: W.B. Burton and M. Peimbert.

H II STUDIES OF LOCAL-GROUP GALAXIES, G. Courtès.

The galaxies of the Local Group showing H II regions have the great advantage of being close enough to provide, sometimes more easily than in our Galaxy: the fine morphology and distribution of those H II regions; their unambiguous positions compared with those of stars and H I contours; the precise shape of spiral patterns and their true galactocentric distances.

Recent observations using high-sensitivity monochromatic detection and new optical designs have given very efficient means for understanding of galactic and extragalactic structures.

Efforts have been made to obtain standard sizes of H II regions, in order to resolve extragalactic distance-scale problems.

Studies of star formation at the front of spiral features rich in H II regions have been discussed in relation with the kinematics of the gas as well as with stellar distribution and evolution.

Several new kinds of H II regions have been distinguished, up to the very extended diffuse emissions of spiral arms, disk and central regions.

Results from high-sensitivity and high-resolution spectrography, and from far-UV space astronomy, give a first general explanation of the different modes of excitation of these emission phenomena.

Discussion   Pecker: In M31 and M33 the sizes of H II regions are well-defined functions of galactocentric radius R; hence they can be used as distance indicators. – Courtès: In fact, the sizes are similar in M31 and M33.– De Vaucouleurs: The range of sizes of H II regions is much smaller than that of galaxies.– Van den Bergh: Why do the shapes of H II regions in M33 vary with R, with ring shapes in the outer parts? – Courtès: I don't know.

INTERSTELLAR ABUNDANCES IN EXTERNAL GALAXIES, B. Louise Webster.

It is now well established that the relation between the ionization level of an H II region and its galactocentric distance which was seen in some spiral galaxies by Aller and then by Searle, is a consequence of a gradient in the abundances of oxygen and nitrogen relative to hydrogen. Searle's original interpretation has been confirmed by further model computations and by detailed observations of the emission-line intensities, especially those giving the electron temperatures. There is also an indication of a second effect, a variation in the energy distribution of the radiation ionizing the H II regions with galactocentric distance. It has been suggested that the maximum mass that may form a star depends on the heavy-element abundance and that massive hot stars cannot form in the metal-rich, inner regions of galaxies. Alternatively, it has been proposed that more dust grains modify the radiation field by preferentially absorbing the higher-frequency photons.

Work is going on at present to apply these interpretations of H II regions to the wider problem of evolution within different types of external galaxies. There is a close relation between the morphological type of a galaxy and the excitation properties of its H II regions. The trend is from generally low excitations in Sbc supergiant galaxies, through strong gradients in later-type spirals and giant galaxies, higher excitation and lower abundances with no gradients in LMC-like galaxies, to very low abundances in dwarf and very-low-surface-brightness systems.

THE HELIUM PROBLEM, M. Peimbert.

(a) Our Galaxy and the Magellanic Clouds

Observations of H II regions and planetary nebulae suggest that the pregalactic helium to hydrogen abundance ratio is $N(He)/N(H) = 0.074 \pm 0.006$. There is no observational evidence for a smaller pregalactic ratio. Infrared and radio observations of the galactic center indicate that $N(He^+)/N(H^+) = 0.085$ and that $N(He)/N(H)$ could very well be twice as high. The pregalactic helium abundance can be used to determine the increase of the helium abundance by mass, ΔY, for different objects in our Galaxy, to choose appropriate models of stellar evolution for metal-poor objects, and to derive accurate ages for metal poor globular clusters.

A ratio of helium to heavy-element enrichment by mass of $\Delta Y/\Delta Z \sim 3 \pm 1$ has been found from observations of H II regions and planetary nebulae. This ratio is larger by at least a factor five than the one predicted by standard models of stellar evolution. Possible solutions to this discrepancy are mentioned.

In the interstellar medium of the solar neighbourhood it has been found that $\Delta Y + \Delta Z \sim 0.1$. This value can be used to estimate the total energy emitted by the Galaxy in its lifetime.

(b) Cosmology

Observations indicate that the pregalactic helium abundance derived from our Galaxy and the Magellanic Clouds holds for objects as distant as 100 Mpc. Moreover the most recent results from quasars are not in contradiction with this value. Therefore it seems that there is a general process which is responsible for a pre-galactic production of N(He)/N(H) ~ 0.07.

This pregalactic helium abundance coupled with the standard big-bang model implies an open Universe. A similar result is derived from the deuterium abundance. Under the adoption of a big-bang model with nonzero lepton numbers the pregalactic Y value provides us with a restriction on the model which is independent of the deuterium restriction.

Discussion of reviews by Webster and Peimbert   Mrs. Collin emphasized the difficulties in deriving the O/N ratio from certain lines. - Pecker: Heavy elements may be expelled from the Galaxy while helium is retained. This would invalidate Peimbert's assumptions in the derivation of ΔY/ΔZ. - Anonymous: Nitrogen may be a factor four underabundant in the LMC compared to the solar vicinity; or the solar oxygen abundance may be overestimated. ʹ- Peimbert: The comparison Magellanic Clouds - solar neighbourhood is based on several H II regions, not on the Sun alone.

WARPED HYDROGEN DISKS IN GALAXIES, R. Sancisi.

Observations of nearby edge-on galaxies in the 21-cm hydrogen line with the Westerbork Synthesis Radio Telescope have revealed a significant bending of the outer gas layer in four out of five systems. Three of these systems (NGC 5907, 4565 and 4244) are fairly isolated in the sky. The most pronounced and unambiguous warp is found in NGC 5907, where the deviation of the HI layer from the optical plane is about 20 percent of the radius. NGC 4565 and 4244 have a smaller distortion. NGC 4631 has a large warp but also a close and disturbed companion, NGC 4656. NGC 891 has neither.

The origin and survival of these warps are not understood. Tidal explanations do not seem to be adequate in at least three cases.

Discussion   M.A. Gordon: Can you say where the inflection originates? - Sancisi: Usually at the edge of the optical image. - Field: If the point of inflection were on the line of sight to the main body of the galaxy you would not notice it. - Verschuur: Several years ago I suggested that the high-velocity clouds might be explained by a warp in our Galaxy. - M.S. Roberts: Very deep plates of M31 by Arp and Wilson show that this galaxy looks like an integral sign as does NGC 5907.

WARPED HYDROGEN LAYERS: THEORY, Alar Toomre.

The warped neutral-hydrogen layers of the seemingly isolated edge-on spirals NGC 5907 and 4565 just reported by Sancisi constitute a fascinating dynamical puzzle, exactly as he said. To this theorist, who with Hunter (1969, Ap.J., 155, 747) concluded that such integral-sign shapes could not long endure, they are indeed a source of embarrassment: Unlike the warps of our Galaxy or those of NGC 3190 and 3628, they cannot even hypothetically be attributed to any visible companions - nor, unlike with the already serious analogous claims made for M31, M33 and M83, can one remotely claim these data to be geometrically ambiguous.

This talk contained no new answers. It mostly just reviewed such past work as the recognition already by Kahn and Woltjer (1959, Ap.J., 130, 705) that bent shapes are very prone to distort and ultimately destroy themselves via differential precession, or the couterclaim by Lynden-Bell (1965, M.N., 129, 299) that at least in highly-flattened Maclaurin spheroids and some closely related (if perhaps artificial) models one can find genuine modes of bending which precess indefinitely without any deformation. It also touched on the pros and cons of Kahn and Woltjer's old suggestion of distortion due to intergalactic winds, and the suggestion by Rogstad et al. (1976, Ap.J., 204, 703) that the blame might instead lie in some fairly recent but massive infall of gas into those galaxies. Most likely, this reviewer feels, Hunter and Toomre were simply mistaken in concluding that long-lived, discrete modes of bending are impossible in realistic disks - but he certainly remains at a loss to know just where or why such a mistake may have occurred.

Discussion   Mrs. Pismis: Attempts at interpreting galactic warps will depend on
their assumed age: 10^9 or 10^{10} years? - Oort: The external material may still be
reminiscent of the conditions at the time of formation of the galaxy. - Toomre:
A tilted halo is possible, but dynamical friction will make the material find the
plane in a relatively short time. - De Vaucouleurs: In several galaxies tidal interaction must be responsible for distortions. Also observations are compatible with
the idea that our Galaxy is tidally affected by the Large Magellanic Cloud. -
Toomre: Tidal interaction is not dismissed, but the conditions for the warp in our
Galaxy make a tidal interpretation very unlikely.

THE GAS CONTENT OF EARLY-TYPE GALAXIES, Hugo van Woerden.

The amount of gas present in elliptical (E) and lenticular (L, or SO) galaxies
has been a matter of controversy for some time. While HI in ellipticals remained
undetected despite persistent efforts, Balkowski et al. (1972, A & A 21, 303) found
a gas fraction M_H/M_t in lenticulars similar to that in early spirals (Sa-Sb).
However, Gallagher, Faber and Balick (1975, Ap.J. 202, 7) conclude that normal SO's
are gas-poor and fit the general run of M_H/M_t (or $\overline{M_H/L_B}$, where L_B = blue luminosity)
with morphological type, although peculiar and interacting systems are often gas-
rich. The lack of gas in E and SO galaxies is difficult to understand, unless the
gas shed by evolving stars is removed by galactic winds or by intergalactic ram
pressure.
     Recent unpublished results have changed the picture considerably. Observations
by Knapp, Faber and Gallagher at Green Bank, confirmed by Knapp and Kerr at Arecibo,
give the first reliable detection of hydrogen in an elliptical: NGC 4278 has M_H =
7×10^8 $\odot$, or M_H/L_B = 0.05. Krumm and Salpeter at Arecibo, and Knapp et al. at
Green Bank, have detected several SO's with M_H/L between 0.01 and 0.1. Van Woerden,
Mebold, Goss, Siegman and Hawarden (1976, Proc. Astr. Soc. Australia, in press),
having measured some fifty SO and SO/a galaxies at Parkes, find hydrogen in at least
sixteen, with M_H/L_B ratios ranging from 0.1 to 2, a value high even for a magellanic
irregular. Deep IIIa-J plates taken with the Siding Spring Schmidt lead to revised
classifications for many galaxies. However, gas-richness clearly does not correlate
with optical morphology: the highest M_H/L_B ratios occur in NGC 1512, a distorted
SBO with faint tidal arms, interacting with a blue(!) EO; in NGC 6902, an SO/a with
a bright lens surrounded by well-developed but faint spiral arms; and NGC 1533, a
true lenticular. The cause of the great spread (a factor 100) in M_H/L among SO's
remains unclear; and why are some gas-rich systems devoid of spiral structure?

Discussion   Mrs. Faber: The detection of hydrogen in NGC 4278 is not in disagreement with previous upper limits. The new Parkes results suggest that southern SO's
are richer in gas than northern ones. - De Vaucouleurs: Accurate galaxy types are
required; the classifications of many southern galaxies are poor. - Van Woerden:
However, no reclassification explains the high M_H/L_B ratios mentioned above.

Contributed Paper: Observations by ANS of dust in galaxies, by C.-C. Wu.

Session 6: New Results in Interstellar Physics

Meeting of Commission 34, held 27 August 1976..
Chairman: H. van Woerden.

The following contributed papers were presented:

The temperature distribution of neutral hydrogen at high galactic latitudes, by
Y. Terzian.

A Nançay H I absorption survey towards extragalactic radio sources, by J. Crovisier, I. Kazès and D. Aubry.

Molecular hydrogen in the Orion Nebula, by U. Fink.

CH spectra of Sgr B 2, by B. Andrew.

An experimental study of the 10μ band of cosmic dust, by C. Friedemann.

Spectral indices of background radiation between 38 and 408 MHz, by I. Milogradov-Turin.

Detection of S III (18.7μ) and O III (88.4μ) infrared-line emission from the Orion Nebula, by J.P. Baluteau.

Laboratory studies of Mg TBP and the diffuse interstellar lines, by F.M. Johnson.

Session 7/8: Interaction of Stars and Interstellar Medium

Meeting of Commission 34, held 28 August 1976.
Chairmen: G.S. Khromov and M. Peimbert; Secretary: H.M. Johnson.

COMPACT HII REGIONS, P.A. Shaver.

Most if not all compact HII regions have associated massive molecular clouds, observable at least in the lines of CO. These molecular clouds produce a great deal of visual extinction, so the optically visible HII regions must be on the near sides of the associated clouds, on the average. Recent studies of molecular and recombination lines show that the latter are significantly blue-shifted relative to the former on average, which is to be expected if the HII regions are comprised of ionized gas streaming away from the molecular clouds, and being replenished by them.

It is now known that HII regions emit almost all of their energy in the infrared. The 20μm/6cm flux ratio is particularly large for many of the smallest, most dense HII regions. An extreme case is the source Oph 4, a compact HII region in the Ophiuchus dark cloud; it is invisible, a weak radio and 2μm source surrounded by extended far-infrared radiation and carbon recombination-line emission, probably excited by an early B-star embedded in the dark cloud. Such radio and infrared studies of dark clouds appear very promising for the study of compact HII regions and early star formation.

The association of compact HII regions with maser sources is now well-established. 80 percent of Type-I OH masers coincide with compact HII regions within a few arcsec, and 20-50 percent of compact HII regions have associated Type-I OH-maser sources. H_2O sources are also frequently associated with these sources. However these maser sources do not appear to be associated with larger HII regions: the OH-maser phenomenon seems to disappear when the compact HII region has expanded to about 15000 A.U.

Radio recombination-line studies of compact HII regions have advanced recently with the use of large synthesis radio telescopes; it is possible now to study the velocities and physical properties of individual compact components of HII complexes. Another important development has been the discovery of extended carbon recombination-line regions in dark clouds such as that in Ophiuchus; these probably indicate the presence of embedded early B-stars, and open a new window on star formation.

A tentative evolutionary scheme for HII regions is beginning to emerge. Initially the HII regions are extremely compact and dense, optically obscured, and have a large near-infrared flux and associated OH/H_2O maser sources. As they evolve, the maser phenomenon disappears and the HII regions become visible with decreasing reddening and near-infrared flux. Old HII regions eventually melt together into a diffuse complex, the infrared flux is comparatively weak, and the associated molecular cloud may have disappeared altogether.

DUST IN HII REGIONS, Syuzo Isobe.

There is much direct and indirect evidence for dust in HII regions: (1) The continuum light scattered by dust grains (O'Dell and Hubbard 1965, Ap.J. 142, 591), (2) Thermal radiation from dust grains at infrared wavelengths (Ney and Allen 1969, Ap.J. 155, L193), (3) Abnormal values of $R \equiv A_V/E(B-V)$ in the direction of some HII regions, (4) The three-micron and ten-micron absorption features (Gillett and Forrest 1973, Ap.J. 179, 483), (5) Abnormal helium abundance in some HII regions (Peimbert and Castero 1969, Bol. Obs. Tonantzintla 5, 3), and so on.

O'Dell et al. have found dust deficient in the inner parts of HII regions, especially of the Orion Nebula, but rather overabundant in the outer parts. Consideration of the effect of an isotopic scattering by dust grains changes these conclusions.

Münch and Persson (1971, Ap.J. 165, 241) found that dust is well mixed with gas in the Orion Nebula. The similarity of intensity distributions of hydrogen recombination lines and infrared radiation appears to confirm this. However, calculation of the expected infrared radiation indicates dust depletion by a factor 10 to 100 (Harper et al. 1976, Ap.J. 205, 136).

Graphite and silicate grains are stable even in HII regions, but will be blown out from the central parts (< 0.1 pc) by radiation pressure. On the other hand, gas expansion from adjacent molecular clouds will supply dust to HII regions.

Churchwell et al. (1974, A & A 32, 283) find from radio recombination lines that only three out of 39 HII regions have abnormally low helium abundances. Since the main cause of the observed helium deficiency in HII regions is the destruction of helium-ionizing photons by dust grains (Mezger et al. 1974), there must be few dust grains in many HII regions. Although radio data indicate a normal helium abundance in the Orion Nebula, optical data show depletion of helium in the central region (Peimbert and Castero 1969).

These contradictions are resolved by consideration of the globule model proposed by Dopita et al. (1975, Ap. Space Sci. 34, 91). Optical observations are weighted to the high-density region, where dust grains just ejected from globules destruct the helium-ionizing photon; by the radio technique we look through the dust-free regions.

From a statistical analysis of intensity fluctuations in the central region of the Orion Nebula Tamura (1976, Sci. Report Tohoku Univ.) finds a total number of 400 condensations and sizes of order 0.001 pc, consistent with earlier results of Dyson.

We conclude that the observations summarized in this review can be explained by introducing globules with dust envelopes in the HII regions.

UNUSUAL MOTIONS IN HII REGIONS, John Meaburn.

One of the most interesting, and still unexplained, phenomena associated with HII regions are the huge volumes of ionized gas emitting split lines. Large-scale line splitting has been found by Wilson, Münch, Flather and Coffeen (1959) in M42, by Deharveng and Maucherat in the Orion and Carina Nebulae, by Smith in the Rosette, and by Meaburn and collaborators in M42, M16, M8, M17 and NGC 2264.

In M42, the linear scale of this type of splitting increases from the dense (10^4 to 10^3 e/cm^3) core to the outer regions. Moreover, all the split regions in this nebula have been shown (by Elliott and Meaburn and Dopita, Isobe and Meaburn) to be over the dark areas surrounded by bright rims produced by ionization fronts eating into the adjacent neutral masses.

In M16, M17, M8 and the Doradus Nebula the nebular lines are doubled to both negative and positive velocities (a) with respect to their positions when bright and single and (b) with respect to the mean motion of the adjacent neutral masses of HI and molecules, out of which the nebulae are forming. In these nebulae the splitting again occurs, exclusively, over the obvious dark neutral intrusions, away from their surrounding bright rims; over these rims the bright nebular lines are always single. In M17 up to five components appear in several adjacent fields. Most curious, though,

is the apparent association in this nebula of the splitting in the [OIII] and CO
lines.

In the Large Magellanic Cloud, the blue supergiants and HII regions appear to
delineate a spiral galaxy, with the Doradus Nebula as its nucleus. In this nebula
the huge volume (50 pc size), surrounded by bright rims, which is emitting split
lines also contains a massive group of young stars. The H and K absorption lines in
their spectra, as measured by Feast, are centred on the most positive velocity component of the doubled nebular line. We suggest that this may imply a state of
implosion: a neutral shell of 50 pc diameter, with an ionization front on its inner
boundary, could be falling in towards the ionizing stars.

The high velocity dispersions, > 100 km/s (not splitting), found in small
($\lesssim$ 0.1 pc), dense (10^4 to 10^5 e/cm^3) ionized and neutral concentrations (Herbig-Haro
objects?) are distinctly separate phenomena. These motions have been found by Gull
(échelle spectrograph), by Münch and Taylor (two-etalon Fabry-Pérot), and by Glushkov
and his associates (conventional grating spectrograph) in a variety of HII regions.

Discussion   The interpretation of velocities in photographic interferograms is
difficult in the presence of discontinuities of background brightness. - T. Wilson:
I question the assumptions of spatial symmetry in the 30 Dor region.

Contributed paper: High-velocity structures in the Orion Nebula, by T.R. Gull.

PLANETARY NEBULAE, G.S. Khromov.

While the theory of the radiation of planetary nebulae (PN) appears satisfactory,
the quality of observations often leaves much to be desired. Even in close doublets,
relative line intensities may be seriously in error, probably owing to erroneous
estimates of the continuum. These errors affect the determinations of electron density.

The radio and optical images of PN are much alike. More surprises may come from
emission in the 21-cm and molecular lines.

Infrared observations suggest that dust is well mixed with the gas; in the
younger planetaries the dust may be hotter and the infrared emission is stronger.

Determinations of the physical parameters of PN are hampered by incomplete and
unreliable spectrophotometric data, and by structural peculiarities. Computer models
have so far been too primitive. Both density and ionization must vary with radius,
and are interdependent; hence, shell models are required.

Studies of small-scale condensations have centred on the comet-like structures
in NGC 7293. The physical conditions there are probably close to those in the main
body of this nebula, but spectra must resolve this matter.

Since the publication of the Perek-Kohoutek Catalogue, few misidentifications
and new discoveries have been reported; we may actually know most of the galactic
planetary nebulae. Shklovskij's method of relative distance determination, based on
visible disks, appears reliable but absolute calibration remains a problem.

The evolution of PN shells is observed over 4 or 5 orders of magnitude in
density. However, densities above 10^8 cm^{-3} are not observed and the evolution of PN
progenitors (out of old red giants?) remains speculative. Age estimates based on a
constant rate of expansion are clearly too primitive. The increase of the ratio of
the HeII λ 4686 to Hβ with time does not necessarily imply an evolutionary rise in
temperature of the PN nucleus.

Discussion   Peimbert: The scale of the comet tails in NGC 7293 is 10^{16} cm. -
Anonymous: While in NGC 7027 the infrared and HII maps agree, in other PN the IR
region is the smaller. - Seaton: In NGC 7027, hot and cool dust have quite different
distributions.

Contributed papers: Rocket-UV spectra of planetary and diffuse nebulae, by T.P.
Stecher. Binary nuclei in planetary nebulae, by Mrs. A. Acker.

CIRCUMSTELLAR MASERS, Lewis E. Snyder.

OH, H_2O and SiO are the molecules found in maser emission from circumstellar shells. SiO is the newest circumstellar maser. The SiO maser has small source size, shows some evidence for circular polarization, has time-varying intensities and nonthermal populations. Several pumping theories have been developed to explain the nonthermal populations and observed SiO intensities.

At present there are 42 known SiO maser sources. Ori A is the only SiO maser source which is also associated with a molecular-cloud region. Most of the other SiO sources readily can be associated with known late-type stars.

Van Blerkom and Auer noted that the high excitation energies required for SiO require the inversion to occur close to the stellar photosphere (perhaps within a few stellar radii), while the OH maser is formed in a region more than 100 stellar radii distant. Their solution of the SiO radiative transfer problem for a spherical shell gives an asymmetric emission doublet; for a rotating disk, it is an emission triplet. The distinctive triplet structure in the velocity pattern of the SiO maser lines observed in both VY CMa and NML Cyg suggests that these stars may be young objects still surrounded by "solar nebulae"; this conclusion supports Herbig's model for VY CMa.

Reid and Dickinson have used ground-vibrational state SiO data to argue quite convincingly that the true stellar velocity of long-period variable stars is not given by the optical absorption lines but instead lies between the optical absorption and emission lines near the midpoint of the OH radial velocity pattern.

The SiO emission in Orion has the doublet structure typical of the spherical shell surrounding a late-type star; the velocities match that of the H_2O maser emission. Results of a recent VLBI observation show the strongest SiO maser source in Orion to be greater than 4 A.U., which is consistent with the size of the corresponding H_2O feature.

There is a high degree of correlation, perhaps 80% or more, between OH, H_2O and SiO circumstellar maser stars. The failure to detect SiO masers in known OH/H_2O/IR sources associated with HII regions and molecular cloud regions (other than Orion) could be due to beam dilution and/or phase effects. If not, the absence of SiO maser emission would serve as an observational indication that particular OH/H_2O/IR sources are not associated with late-type stars and hence may be much younger, protostellar regions.

Discussion   Buscombe: Have the searches for SiO masers been limited to known OH and H_2O sources, or extended to lists of other cool red giants with extended atmospheres, especially semi-regular and long-period variables? In my Third General Catalogue of MK Spectral Classifications (soon to appear) are hundreds of such stars, with positions for epoch 2000 and some UBV photometry at maximum light.

Contributed paper: Circumstellar CO and HCN, by B. Zuckerman.

Session 9: Large-Scale Distribution of Interstellar Material

Joint Meeting of Commissions 33 and 34, held 30 August 1976.
Chairman: F.J. Kerr.

This session is reported by Commission 33.

Session 10/11: Non-Circular Motions in the Galaxy

Joint Meeting of Commissions 33 and 34, held 31 August 1976.
Chairmen: H. van Woerden and L. Perek.

This session is reported by Commission 33.

COMMISSION 35: STELLAR CONSTITUTION (CONSTITUTION DES ETOILES)

Report of Meetings, 25, 26 and 27 August 1976

PRESIDENT: L. Mestel   SECRETARY: R. C. Smith

27 August 1976

BUSINESS MEETING
    The following nominations were approved.

    PRESIDENT: B. Paczynski (Astronomical Observatory, Warsaw)

    VICE-PRESIDENT: R. J. Tayler (Astronomy Centre, University of Sussex)

    ORGANIZING COMMITTEE: D. J. Faulkner, P. Giannone, C. Hayashi, I. Iben,
                           R. Kippenhahn, A. G. Massevich, G. Ruben, J.-P. Zahn.

    Thirty-three new Commission members were elected. The Commission Report was
formally adopted. It was agreed that the Report should continue to review in depth
a limited number of topics rather than attempt to cover the whole field. It is
hoped that with the new format agreed with the publishers, reprints of the next
Report will be available before the next General Assembly.

25 August 1976

SOLAR AND STELLAR MAGNETISM
(Joint Meeting with Commissions 10 and 12; Chairman Professor T. G. Cowling)

1.   Observations of Stellar Magnetism (Dr W. K. Bonsack)

    This review concerns itself only with observations of effects in circularly
polarized light which can be interpreted as the Zeeman effect. It is emphasized
that the numerical values obtained for stellar magnetic fields depend strongly on
the technique used. The classical photographic method assumes that the wavelength
difference between line components in the two states of circular polarization can be
interpreted as the Zeeman shift due to the component of the field along the line of
sight (effective field), but this interpretation neglects such effects as incipient
resolution of the individual Zeeman components and saturation in the line profiles.
In addition, different lines, and different parts of the individual line profiles,
are formed at different depths, and the depth-dependence of the field is not known.
Individual line shifts are small, and are strongly affected by photographic grain.
At least 25 lines per plate must be measured for reliable results, and numerical
agreement between photographic results and photoelectric measurements depending
on a single line should not be expected.

    Among the magnetic A- and B-type stars, the largest field observed is still the
34-kG (surface) field found by Babcock in HD 215441. Resolved or partially resolved
Zeeman patterns have permitted measurement of the total surface field in a few other
stars (G.W. Preston, Ap.J. 164, 309, 1971) yielding values up to 17 kG. Interpretation of resolved and unresolved patterns has permitted derivation of the magnetic field geometry for β CrB (S. and R. Wolff, Ap.J. 160, 1049, 1970) and for 53
Cam (J. Huchra, Ap.J. 174, 435, 1972); in both cases a dipole field offset from
the centre of the star represented the data reasonably well. The hottest known
main-sequence magnetic star is the Bp helium variable HR 7129, in which R. and S.
Wolff (Ap.J. 203, 171, 1976) found an effective field varying between +7 and -5kG;
the effective temperature is near 20000 K. At the other extreme is HD 101065
(Przybylski's star), which has a constant field of +2 kG (S. Wolff and W. Hagen
Pub.A.S.P. 88, 119, 1976) and probably has the temperature of an early F-type star.

The majority of the magnetic stars which have been investigated in sufficient detail have been found to be periodic, including Babcock's prototype irregular variable, 78 Vir (G. W. Preston, Ap.J. 158, 243, 1969). Well established periods exist as short as three days. The upper limit to the periods cannot be easily established, but a monotonic field variation in γ Equ was interpreted as a part of a periodic variation of possibly 75 years duration by W. Bonsack and C. Pilachowski (Ap.J. 190, 327, 1974). Photoelectric observations (J. Landstreet, et al. Ap.J. 201, 624, 1975) indicated that fields in excess of 1 kG do not occur in stars with rotation periods shorter than approximately three days. There is no evidence that the character of the magnetic variation, or of the accompanying spectrum and light variations, change with period in the observed range; this suggests that the period is the rotation period in all cases. It remains possible that aperiodic phenomena do occur in these stars; sometimes these are superimposed on periodic effects (e.g. HR 2727, W. Bonsack, Ap.J. 1976, in press), and in other cases there may in fact be no discernable periods.

In summary, in the upper main sequence, magnetic fields are found in Ap stars of the Si and Cr-Eu-Sr types, providing that they rotate sufficiently slowly, and in at least some helium-variable Bp stars. Fields do not occur in Mn-Hg type Ap stars, Am stars, or normal A and B stars.

Magnetism in white dwarfs was detected by means of broad band polarization in the continuum in the DC (continuous spectrum) star Grw+70°8247 by J. Kemp et al. (Ap.J. 161, L77, 1970) following the theoretical prediction of the effect by Kemp (Ap.J. 162, 169, 1970). The measurements were interpreted as representing a field of 1×10^7 gauss. Several additional magnetic DC white dwarfs have since been found. Attempts to detect the (quadratic) Zeeman effect in the hydrogen lines of DA stars yielded only an upper limit of 5×10^4 gauss (J. Elias and J. Greenstein, Pub.A.S.P. 86, 957, 1974; J. Angel and J. Landstreet, Ap.J. 160, L147, 1970) until Angel, et al. (Ap.J. 194, L47, 1974) recognized resolved patterns of the quadratic Zeeman effect in the Balmer lines of GD90. The data were interpreted in terms of a field of 1×10^7 gauss. The systematics of white dwarf magnetic fields are unclear, except that the fields are quite rare, and usually constant in time.

In the discussion, Dr A. Severny, summarizing observational work in the USSR, claimed that their photoelectric method measured essentially the longitudinal component of the field, whereas the photographic method mainly used in the US measures the total field; but Dr Bonsack argued that this is an oversimplification. Dr Severny reported on the occasional appearance of weak fields in bright stars such as Vega and Sirius A; and of fields of several hundreds of gauss in two Am-stars.

2.   The Solar Magnetic Field - Observation and Theory (Dr N. O. Weiss)

Both observations and theory were recently discussed at IAU Symposium No. 71 (Basic Mechnisms of Solar Activity, ed. Bumba and Kleczek; see especially the review by Stix). The basic features of the solar cycle are well known: the 11 year sunspot cycle is really a 22 year magnetic cycle, dominated by strong toroidal fields, of opposite sign in the two hemispheres, whose development is summarised in the butterfly diagram. Other stars with convective envelopes show similar behaviour, as measured by variations in Ca K emission. So the sun should be regarded as a model, unique in that we can observe its field in some detail. Magnetograph measurements show active regions and weaker, irregular polar fields. The latter reverse around sunspot maximum but they behave erratically and there are periods when both poles show fields of the same sign.

Babcock's phenomenological model of the solar cycle had flux ropes which were drawn out by differential rotation. Leighton added turbulent diffusion, caused by supergranules, and, with plausible assumptions, reproduced the butterfly diagram and the correct period. Yoshimura (Ap.J.Supp. 29, 467, 1975) has recently obtained and solved the equations for a more elaborate model.

The same equations emerge from recent developments in kinematic dynamo theory. The theory of turbulent dynamos (mean field electrodynamics) has been developed principally by Krause, Rädler and Steenbeck, starting from Parker's concept of cyclonic eddies. The mean magnetic field $\underline{B}$ satisfies the equation

$$\frac{\partial \underline{B}}{\partial t} - \text{curl } (\underline{u} \times \underline{B}) = \text{curl } (\alpha \underline{B}) + \eta \nabla^2 \underline{B},$$

where $\underline{u}$ is the velocity and η is the total (molecular and turbulent) diffusivity. The poloidal field is generated from the toroidal field through the "α-effect", and the parameter α depends on a lack of isotropy in the turbulence. More specifically, for turbulence that is not mirror-symmetric, α is proportional to the average value of $\underline{u}.\text{curl }\underline{u}$ (the helicity). A reversed toroidal field is then produced by differential rotation. Many oscillatory dynamo models have now been computed, reproducing the butterfly diagram with remarkable fidelity.

Despite this success, mean field dynamos have been criticized, especially by Piddington (see Cowling, Nature, 255, 189, 1975). The observed fields form discrete flux tubes instead of a turbulent mess; moreover, opposed fields can only be eliminated by invoking some dynamically driven reconnection process. The theory also depends on averaging over the smaller of two distinct scales, and on first order smoothing, neither of which is valid for the sun. Piddington has therefore revived the magnetic oscillator theory, while Layzer, Rosner & Doyle have proposed a hybrid model. Nevertheless, mean field dynamos are now generally accepted as reasonable illustrative models of the solar cycle, and emphasis has switched to the construction of nonlinear dynamos, including the effect of forces exerted by the magnetic field.

The next stage must be to relate dynamos to the hydrodynamics of convection in the sun. There are several scales of cellular convection, affected differently by rotation, and it is necessary to compute the resulting distributions of angular velocity and helicity. Flux ropes are formed and react back on the motion. There must be a balance between the tendency of convection to pump flux downwards and that of buoyant flux ropes to float upwards. Other problems which can probably be reconciled with dynamo theory are the erratic behaviour of the polar fields (which must be indicators, rather than essential features, of the dynamo process); the rigid rotation of the sector structure and coronal holes; and the remarkable secular variation of the solar cycle itself (Eddy, Science, 192, 1189, 1976). We now have a crude understanding of solar magnetic fields but proper, detailed models have yet to be constructed.

3.   Stellar Dynamos (Dr W. Deinzer)

Turbulent $\alpha\omega$- and α^2-dynamos. Applying the theory of the turbulent dynamo (see P. H. Roberts and M. Stix, The Turbulent Dynamo, NCAR, 1971) to a differentially rotating star, an axisymmetric magnetic field

$$\underline{B} = (0,0,B) + \nabla \times (0,0,A)$$

is produced according to the equations

$$\frac{\partial B}{\partial t} - \eta \nabla^2 B = \hat{Q}_B(\omega, \alpha) A$$

$$\frac{\partial A}{\partial t} - \eta \nabla^2 A = \hat{Q}_A(\alpha) B$$

Here η is the magnetic diffusivity, $\hat{Q}_A$ and $\hat{Q}_B$ are linear operators, ω is the angular velocity and α is a quantity describing (to lowest approximation) the inductive action of the turbulent medium. For vanishing α the dissipation of poloidal flux cannot be compensated and the dynamo breaks down. If the inductive action of the differential rotation is much stronger than the α-effect, the latter may be neglected in Q_B and the so-called $\alpha\omega$-dynamo is considered. If the matter is in solid-body-rotation Q_B as well as Q_A depend on α only and the so-called α^2-dynamo is working. A similar formalism can be set up if there is no axial symmetry.

Application to stars. (a) Fully convective stars in the Hayashi-phase. No convective envelopes as the seat of dynamos for an outer magnetic field seem to exist in magnetic stars. Hence even dynamo theory has to rely on "fossil" fields in this case, fossil in the sense that the fields were produced by a dynamo acting during the convective pre-main-sequence phase. An α^2-dynamo in fully convective stars is excited, if (P. H. Roberts, Phil.Trans.Roy.Soc. A 272, 663, 1972) $P_\alpha = |\alpha R/\eta| \gtrsim 8$. M. Schüssler (Astron.Astrophys. 38, 263, 1975) has shown that these fields survive when the stars evolve towards the main-sequence.

(b) Convective cores in main-sequence stars. Both types of dynamo have been applied. E. H. Levy and W. K. Rose (Astrophys.J. 193, 419, 1974) found excited $\alpha\omega$-dynamos for

$$P_{\alpha\omega} = P_\alpha \cdot \left| \frac{\Delta\omega . R^2}{\eta} \right| \gtrsim 10^3$$

$\Delta\omega$ is a measure for the variation of angular velocity due to differential rotation.

A. Pähler (Diplomarbeit, Göttingen 1976) found excited α^2-dynamos for $P_\alpha \gtrsim \pi$ (slightly smaller than the value found by F. Krause and M. Steenbeck, (Z.Naturforsch, 22a, 671, 1967), due to different boundary conditions). Obviously an α^2-dynamo is more easily excited in a convective core than in a convective envelope, because the field is enclosed completely in the finite volume of the core. The high electrical conductivity of the envelopes prevents the field from penetrating to the outside during the main-sequence life-time.

Excitation conditions. To find out whether the values requires for P_α and $P_{\alpha\omega}$ are feasible in the stellar context, formulae for α and η must be available. Using values presented by Krause (Habilitationsschrift, Jena, 1967),H. Köhler (Astron. Astrophys. 25, 467, 1973) was unable to reproduce the solar cycle; only after reducing α by three orders of magnitude a realistic solar cycle was obtained. Considering a random distribution of sound waves W. Deinzer (in preparation, 1976) found a value for α which in the limit of slow rotation is

$$|\alpha| = u^2 \frac{2\pi}{k} \frac{1}{\eta} \frac{\frac{\omega}{ck}}{1+(\frac{c}{\eta k})^2} \quad \text{for} \quad \frac{\omega}{ck} \gg 1$$

(u velocity amplitude, c velocity of sound, k characteristic wave number). This formula gives for $\eta \backsim u . \ell$ (the turbulent magnetic diffusivity) and $c \gg \eta k$

$$|\alpha| \backsim 2\pi (\frac{u}{c})^3 \omega\ell = (\frac{u}{c})^3 \alpha_{\text{Krause}}$$

which gives the reduced value required by Köhler, if $u \backsim 1$ km/sec is assumed. Evaluating the dynamo number, $P_\alpha = 2\pi(u/c)^2 (\omega R/c)$ is obtained. Assuming still $u \backsim 1$ km/sec it is easily seen that only for low temperatures $T \lesssim 10^4$K, P_α is of order unity as required for excitation. Hence only the dynamo in the Hayashi-phase

seems to be excited, whereas for dynamos possible in convective cores the production of magnetic flux is much smaller than the energy dissipated by Ohmic losses
- at least if the inductive action of the turbulent medium is based on sound waves.

4.   Stellar Magnetism - Theory (Dr L. Mestel)

    Most of the points raised are summarised in Section 4 of the Commission 35
Report (Transactions Vol.A). Particular emphasis was laid on: (1) the distinction
between a field maintained by contemporary dynamo action, and a slowly decaying
"fossil" field, that may nevertheless be a relic of a field built up by dynamo
action in an earlier phase of the star's life; (2) the necessity for a complete
dynamo theory to include the dynamical back-reaction of the magnetic forces on the
driving motions, as well as the condition that Ohmic flux destruction be balanced
by kinematic flux replenishment; (3) the relevance of hydromagnetic stability
theory in restricting the allowed field topologies; (4) the possible role of
rotationally-driven circulation in explaining observed anti-correlation between
surface field and rotation rate, and as a possible reason for the apparent preference
for high angles of obliquity between rotation and magnetic axes.

26 August 1976

LARGE SCALE MOTIONS IN THE SOLAR ATMOSPHERE
(Joint Meeting of Commissions 10, 12 and 35; Chairman Dr M. Schwarzschild)

1.   Recent Observations on Large Scale Solar Oscillations (Dr H. A. Hill)

    In the fall of 1973 the observational program at SCLERA obtained its first
evidence for whole-body oscillations of the sun. In the last several years considerable additional evidence for whole-body oscillations has been obtained by the
group. At the present time, 20 modes have been identified with periods between 68
minutes and 6 minutes.

    The statistical significance of the observed acoustic spectrum has often been
questioned, and in some cases, claims have been made that the reported oscillations
can be interpreted as noise. These analyses and conclusions, however, are considered
inapplicable to the brightness measurements at SCLERA due to invalid statistical
tests. A thorough treatment of the statistics shows that several tests strongly
indicate that the reported observations are indeed statistically significant. The
most striking evidence is that two modes remain coherent in phase over a duration
of 40 days.

    The brightness amplitudes of the oscillations observed at SCLERA are signifi-
cantly larger than those inferred from either velocity or temperature measurements.
The brightness and velocity amplitudes are in good agreement, however, after
incorporating a proper theoretical treatment of acoustic wave propagation in the
sun's outer layer. In fact, the resolution of this "paradox" has shown that the
commonly used boundary conditions for pulsation theory in the sun are considerably
in error.

    With the incorporation of the acoustic wave theory, the temperature amplitudes
inferred from the SCLERA measurements remain larger than those inferred using
brightness changes in spectral lines. The theory used to interpret the spectral
line observations, however, is applicable only for a non-oscillating system. It has
been possible to show that the sensitivity of spectral lines to short-term temperature
changes was in the past overestimated by an order of magnitude due to misapplication
of static theory. This removes the above apparent temperature discrepancy and
seriously impairs this technique in the study of small-amplitude oscillations.

In summary, a very strong case can be made for the existence of whole-body oscillations in the sun. Much progress has also been made in understanding the physical processes leading to their detection. The observed normal mode spectrum is quite rich and should be possible to study in considerable detail.

2.   Observations of the Line-of-Sight Velocity (Dr E. Fossat)

The sodium optical resonance device of the Nice group has been used by G.Grec and E.Fossat to measure the line-of-sight velocity averaged across the whole solar disk, with a sensitivity of about 1 ms^{-1}, for periods ranging between 5 and 90 minutes. The 5-min oscillation is clearly visible, with a r.m.s. amplitude of 2 ms^{-1}, but it is shown that the low frequency spectrum can be interpreted as pure atmospheric noise, due to fluctuations of inhomogeneous transparency in front of the rotating solar disk. A very simple atmospheric model can be deduced from the numerical results.

By comparison, an evaluation of what can be the atmospheric noise in the solar diameter measurements of the SCLERA group has been made. This noise comes from fluctuations in the differential refraction. A measurement of air-mass fluctuations has been made in Kitt Peak by E.Fossat and J.Harvey. With the only assumption that the same atmospheric model is still valid (because transparency and air-mass fluctuations are probably driven by the same physical processes, mainly convection), the noise induced by this result is exactly consistent with the quantitative SCLERA results. Furthermore, it is shown that the peaks in the SCLERA spectrum are consistent with statistical fluctuations around a continuous spectrum, which confirms the preceding interpretation.

It is concluded that up to now, no long period solar oscillation has been observed with periods between 5 and 90 minutes.

3.   Observations of Oscillations of the Entire Sun (Drs V. Kotov, A.Severny and
      T. Tsap)

The existence of a stable 160^m period, previously found by Severny et al. (1976) has been confirmed by further observations during 1975 and 1976. We have also suspicions of the same periodicity in the solar luminosity and magnetic field.

The periodicities found near 147^m and 171^m are not definitely confirmed. It seems encouraging that periods found here ($\simeq$ 147^m-149^m, 160^m, 171^m) coincide approximately with theoretical periods (147^m, 159^m, 171^m) of normal modes of the sun's vibrations found by Christensen-Dalsgaard and Gough (1976). The possible existence of oscillations with 147^m and 171^m periods in our data might indicate non-radial (ℓ=2) oscillations; if so, this is evidence in favour of the improved standard model of the solar interior, with a large abundance of heavy elements (Z=0.04). However, because the 147^m and 171^m periodicities, unlike the 160^m, are not persistent in phase, their nature is still not quite clear.

It would be premature to speculate about the physical nature, radial or non-radial, of either the 160^m or the less certain 147^m and 171^m oscillations, all of which may prove to be decisive for the solar interior problem.

4.   Spatially Resolved Observations of Solar Global Relations (Dr F-L. Deubner)

Periods of solar pulsations found by Hill, Severny, and by Brookes and their collaborators were compiled in a frequency diagram and compared with the theoretically predicted frequencies of some low-order radial and non-radial global oscillations.

The comparison shows that unless the observations yield very high resolution in frequency and sufficient spatial resolution to distinguish between various modes with similar periods, it is impossible to deduce meaningful information about the solar interior from these low-frequency pulsations.

The brightness of Uranus and Neptune have been monitored with a double-beam photometer at ESO, La Silla. No periodic fluctuations of the solar luminosity in excess of 3×10^{-3} mag. have been detected in the low-frequency domain.

Spatially resolved observations obtained with a photoelectric lambdameter at the Capri observatory show low-frequency power equally distributed over a range of wavelengths from 25 Mm to about half the solar diameter, indistinguishable from convective motions.

On the other hand, the well resolved trapped acoustic modes of the 5-min oscillations appearing in the same data, present themselves as the most powerful tool presently available to probe the solar interior. The differences between the observed and the predicted modes should readily serve to improve the current models of the vertical structure of the solar convection zone as well as of the deeper layers which are affected by low-order non-radial globular pulsations.

The close coincidence of the predicted and observed non-radial f-modes seems to indicate that there is nothing basically wrong with the standard solar model.

5.   CI 5380 as a Temperature Indicator and a Search for Global Oscillations
     (Dr W. Livingston)

The high excitation, weak Fraunhofer line CI 5380.3 $\overset{o}{\text{A}}$ is shown to originate within the same photospheric layers as the sun's continuous radiation. By monitoring the central depth of the CI line relative to the local continuum, in unfocussed sunlight, we may follow temperature, and hence luminosity changes of the whole sun as a function of time. The technique is practically insensitive to telluric absorption effects and instrumental spectral response. A power spectrum analysis of 100 hours of observations reveals no dominate oscillation of period P rising above the 3σ uncertainty limit of 0.4K for $5^m \leq P \leq 60^m$. For power at $2^h 40^m$ the limit is 1.0K, for $5^h 20^m$, 2.0K. The day-to-day fluctuation for a three month period is 0.85K, rms (which corresponds to 0.06% in the solar constant, or 0.0006 mag. in luminosity.

6.   Theory of Solar Oscillations (Dr D. O. Gough)

Whether the sun is simultaneously oscillating in many modes is not yet resolved. Professor Hill has presented a case which makes it look likely, but his arguments have not been widely accepted. To the eye, the averaged spectrum does show a qualitative discontinuity at about the acoustic cut-off frequency of the solar atmosphere, and I should emphasise that the eye provides a very sensitive, though occasionally unreliable, statistical test. Oscillations are presumably apparent in the raw data, since it was by visual inspection of the oblateness record that led Hill to the discovery. The most convincing evidence is perhaps the long term phase coherence, but we must await more thorough analysis before that is assured. The failure of Livingston, Milkey and Slaughter to observe oscillations in temperature does not contradict Hill's claim, because the sensitivity of their method to detect oscillations may be too low.

Fossat has pointed out that under certain statistical tests Hill's data is consistent with noise. Care must be taken when interpreting this remark, for it does not follow from this that Hill's data contains only noise. Moreover, what one means by noise must be made quite explicit before debate continues. An ensemble of apparently randomly excited discrete oscillators is in some contexts considered noise, and may be a good description of the sun. But such noise is not devoid of information, and analysis of it may tell us as much about the solar interior. More important is Fossat's claim that the source of the noise is the terrestrial atmosphere. This was argued from some knowledge of only the temporal behaviour of atmospheric refractive index fluctuations and must be substantiated by measurements of spatial coherence.

The three years of phase coherence of the $2^h 40^m$ oscillations reported by Kotov, Severny and Tsap makes the interpretation as large scale solar vibrations most plausible. I should point out that Brookes, Isaak and van der Raay at Birmingham, who are unable to be at this meeting, have independently observed oscillations almost in phase with the Russian results, in an Na line of light integrated from the entire solar disk. Worden and Simon's interesting analysis shows that supergranulation may be an important factor contaminating a power spectrum of the data, but it is not an explanation of the $2^h 40^m$ oscillation since the superposed epoch analysis filters out phase incoherent signal. Moreover, the phenomenon influences the Russian and UK measurements differently, and so would be unlikely to lead to phase agreement.

Solar oscillations have important theoretical implications. For example, they provide a potential tool for diagnosing the internal structure of the sun. Since the amplitudes are so low linear theory is adequate for computing the frequencies, and the existence of the oscillations does not upset the hydrostatic and energy balance deep inside the sun. It is important to realise that frequency information alone is insufficient because the spectra are too dense to enable one to identify the modes. It is essential to measure spatial structure so that the appropriate spherical harmonics characterizing the oscillations can be determined. The data will be useful only if phases are maintained for many periods, as in the case of the $2^h 40^m$ oscillation, so that accurate frequency measurement is possible: the periods form almost a harmonic sequence and it is only the deviations from such a sequence that contain detailed information about the sun's structure. The internal solar rotation might be inferred from rotational splitting of degenerate non-radial modes.

It is also important to try to understand the mechanism that excites the oscillations. Random excitation by the turbulence in the convection zone seems most likely, but a thorough analysis has not been made. Goldreich and Keeley have estimated amplitudes ignoring the reaction of the oscillations on the convection, and have obtained values much too low, both for the oscillations discussed here and for the five minute oscillations. The apparent concentration of power into a single g mode, which seems to be implied by the Doppler data, suggests resonant driving by the p modes.

Oscillations provide a potential mechanism for transporting angular momentum. If the solar core is as quiescent as current folklore tells, dissipation in most places is so weak that coupling between the core with the surface by the modes discussed here takes place on a time-scale longer than the solar age. But more rapid dissipation of long period gravity waves generated by the large cells at the base of the convection zone does seem likely. Moreover, many of the modes may be absorbed in critical layers, which would lead to severe inhomogeneities in the distribution of angular velocity, and consequent meridional circulation and material mixing.

Dissipation of oscillations may be a significant factor in the energy balance of the solar atmosphere. Indeed it may be responsible for the latitude dependence of the limb darkening function observed by Hill and his collaborators in their study of the solar oblateness. If this is what is responsible for Dicke's oblateness data, then the 12.2^d precession of the oblateness reported recently by Dicke is presumably a beating between two rotationally split quadrupole modes. This provides an integral measure of the solar rotation which, whatever the modes that are presumed to be split, implies that the interior mean angular velocity exceeds maximum surface value. On the other hand, this interpretation also renders it most unlikely that the oblateness of the solar gravitational field is sufficient to have a significant influence on planetary orbits.

7.   Large-Scale Solar Motions (Dr H. Wöhl)

The review began with a short description of several theoretical ideas to explain the solar differential, rotation and the problems connected with them. The main section of the review contained the observational aspects of large-scale solar motions. The well known motions of sunspots detected by tracer techniques were mentioned. The main result is that the plasma and the solar fine structures show different rotation velocities at the equator and that the amount of the differential rotation is also quite different. Almost no differential rotation was found for long lived magnetic features and coronal holes.

The motions of Ca^+-mottles of the quiet sun which have been studied since 1974 rather extensively by E. H. Schröter and the author at the Swiss station of the Göttingen Observatory were described in more detail. The most important result is the detection of a giant circulation cell pattern within the equator belt in early summer 1975. Several attempts to search for a similar giant circulation pattern of four antisymmetric cells crossing the equator in photospheric layers by using Doppler shift measurements were unsuccessful.

8.   Variations in Solar Activity and Rotation (Dr J. Eddy)

Although the early part of the sunspot number curve is not well documented, the 17th century shows a major anomaly (the Maunder minimum) when no spots were observed on the N. hemisphere of the sun for 43 years and no spots at all for up to 10 years. The 11-year cycle was also absent, a fact that was a matter of contemporary concern. The rarity of aurorae, lack of eclipse sightings of the corona and C^{14} data all agree in showing that this was a period of low activity.

Records of aurorae, coronal sightings and C^{14} data go back to about 5000 B.C. and all show variations in activity. There was a maximum in about 1200 A.D. and there were at least 5 periods of essentially zero activity. There is however no sign of a long term periodicity.

Studies of sunspot drawings by Scheiner and Hevelius suggest that the rotation period of the sun was about 1 day shorter during the Maunder minimum than it was beforehand, when the rotation rate was essentially equal to the present-day rate.

27 August 1976

SCIENTIFIC MEETING

1.   Solar Neutrinos (Dr M. J. Newman)

While work continues towards the development of alternative and independent techniques, observational evidence concerning the flux of neutrinos from the sun at present is offered only by the ^{37}Cl experiment of Raymond Davis and his collaborators at the Brookhaven National Laboratory. This is in a sense unfortunate, for the energy of the copious flux of neutrinos believed to be produced by the basic proton-proton reaction is below threshold for the ^{37}Cl detector, and the high-energy neutrinos from the PP II and PP III completions (^7Be and ^8B decays) which produce the bulk of the counts on ^{37}Cl predicted by standard solar models can be quenched in a variety of ways. A definite test of the essential correctness of our understanding of solar structure and energy generation can only be provided by an experiment of increased sensitivity, and preferably one sensitive to the presence or absence of the low-energy neutrino flux.

Although the Brookhaven experiment is currently indicating a counting rate on ^{37}Cl of about 1.5 ± 0.5 SNU (one Solar Neutrino Unit = 10^{-36} capture per second per target atom) and standard solar models predict about 5 ± 1.5 SNU, the extreme

temperature sensitivity of the branches leading to the troublesome high-energy
neutrinos has provided theorists with several possible escape routes. None,
however, have proved entirely convincing.

Straightforward attempts at reducing the central temperature, such as appealing
to significant differential rotation, large interior magnetic fields, or a reduced
opacity coefficient, encounter difficulties: oblateness measurements constrain the
possibilities for differential rotation, the magnetic fields required are improbably
large, errors in the opacity calculation, if present, are more likely in the opposite
direction from that required. The opacity could be reduced if the surface composition was not representative of the interior metallicity, but the differential
required is considerable, and problems exist for achieving the required surface
contamination. Non-radiative energy transport has been suggested as a possibility
for depressing the mean temperature gradient, but known processes are inadequate.

The central temperature could also be depressed by increasing the rate of
energy generation possible at a given temperature. The ^{3}He-mixing approach achieves
low neutrino-counting rates at present by transporting large amounts of fragile
^{3}He into hot regions, providing a transient energy source; but it has not been shown
that the required mixing has occurred. Non-nuclear sources of energy have been considered, but are perhaps not very convincing. It has been suggested that the
proton-proton rate, which dominates the energy generation in standard solar models,
may be seriously in error, but this is unlikely.

The neutrino-counting rate for the ^{37}Cl experiment can be depressed while
actually increasing the central temperature, if the distribution function for mutual
ion velocities in the solar interior departs from the Maxwellian in the sense that
particles farther out on the tail are preferentially depleted. That condition,
however, has not been shown to hold.

Hoyle has found low counting rates in his solar models with a cosmological
convective core, and Prentice has found similar models in his study of possibilities
for the origin of the solar system. Several authors have found low counting rates
for solar models in which the gravitational constant is a function of position.

Many other suggestions have been made in the literature. In summary, one must
say that all of the low-neutrino-counting-rate models which have been offered to
date are implausible to greater or lesser extent, and the solar neutrino problem
remains serious. A conclusive test of our ideas concerning solar structure, however,
awaits further experimental developments.

2. Periodic Full Amplitude Calculations for Double Mode Cepheids (Drs A. N. Cox
   and S. W. Hodson)

The dozen or so double mode or beat Cepheids have masses between one and two
solar masses as determined by matching the ratios of the observed two or three
periods (2-6 days) with results from the radial linear nonadiabatic theory. The
mass of a typical one of these variables, U Tr A (2.568 and 1.825 days), is 1.2 $M_{\odot}$,
less than half the evolutionary mass at the T_e necessary to be in the pulsational
instability strip. Theoretical nonlinear results of Stellingwerf previously indicated that this double mode behaviour might be possible at very cool temperatures,
now however thought to be beyond the red edge. Nonlinear initial value studies of
purely radiative 1.6 $M_{\odot}$ models at the predicted 5800 K, or even cooler at 5600 K did
not show the mixed mode behaviour, leaving in question the required double mode
conditions. Cool 1.2 $M_{\odot}$ models also give only pure mode pulsations. New Los Alamos
periodic nonlinear solutions at 1.6 $M_{\odot}$ using the same method as Stellingwerf give
model stability for both the fundamental and first overtone modes, confirming the
initial value hydrodynamic studies. Low masses, such as these stars seem to have,
should not evolve to the pulsational instability strip according to evolution

calculations. If they were more massive during earlier evolution, the mass loss
could not exceed ten per cent if there is direct evolution from the red giant tip
region to the pulsational instability strip. Maybe these double mode Cepheids are
mode switching at the transition line in the instability strip, but where they are
in the instability strip their anomalous masses need further explanation.

3.   Convection in RR Lyrae Models (Dr R. G. Deupree - read by Dr A. N. Cox)

RR Lyrae models including convection have been computed by nonlinear numerical
integration of the conservation equations in two spatial dimensions and time. Convection arises naturally from unbalanced buoyancy forces in convectively unstable
regions and hence no phenomenological theory of convection is required. A red
edge of the instability strip is found with this approach. Convection need be
important only in the hydrogen ionization region for pulsation to be quenched.
The quenching is achieved by time dependent convection altering the thermal structure
in such a way so as to negate the traditional driving mechanisms. The red edge
becomes slightly bluer as the helium abundance is decreased, making the width of
the instability strip a sensitive indicator of helium abundance. A value of $Y \sim 0.3$
for globular cluster RR Lyrae stars is indicated.

4.   Excitation of Pulsation in Stellar Convection Zones (Dr W. Unno)

Dynamical as well as thermodynamical coupling between the convection and the
pulsation are evaluated by use of the mixing length treatment for the time dependent
convection. Dynamical coupling occurs through the turbulent pressure and the eddy
viscosity. It can be as important as the thermodynamical coupling which includes
the κ-mechanism and the Cowling-Spiegel-Souffrin mechanism. The convective flux is
shown to have destabilizing effects especially for the g-mode oscillations.

5.   Helium Abundance Estimation of Disk K-Dwarfs from their Position in an
      Empirical (log T_{eff}, M_{bol}) Diagram (Drs M. N. Perrin, G. Cayrel de Strobel
      and P. M. Hejlesen)

There is some evidence that the observational width of the main sequence
limited to disk stars (-0.62 < $[Fe/H]_{\odot}$ < +0.50, effective temperatures lower than
5500°K) is not well explained by   a variation of the metal/hydrogen ratio alone.
It seems necessary to assume the existence in the stars of a simultaneous variation
of the helium content with the metal content (roughly $\Delta Y \simeq (5 \pm 3)\Delta Z$), which acts
in the opposite way to the metal content on the position of the observational ZAMS.

6.   On the Evolution of Massive Stars Through the Core Carbon-Burning Phase
      (Drs S.A.Lamb, I. Iben Jr., and W. M. Howard)

Complete stellar models of Population I composition (X = 0.7, Z = 0.02) and of
mass 15 $M_{\odot}$ and 25 $M_{\odot}$ are evolved from the main-sequence phase into the shell carbon-
burning phase. By using a fine grid of mass shells throughout the evolution, and a
full reaction network for the carbon-burning phase, care is taken to follow detailed
changes in the chemical composition and in the physical parameters of the models.

Both the 15 $M_{\odot}$ and 25 $M_{\odot}$ models ignite helium as blue supergiants and remain
blue supergiants through most of the core helium-burning phase. The 15 $M_{\odot}$ model
ignites carbon as a red supergiant, whereas the 25 $M_{\odot}$ ignites carbon well to the
blue of the red supergiant branch. We conclude that the longest period Cepheids
(P $\gtrsim$ 126 days) are massive stars ($\gtrsim$ 18 $M_{\odot}$) in the core carbon-burning phase of their
evolution.

A comparison of the times spent by our 15 $M_{\odot}$ stellar model first as a blue,
then as a yellow, and finally as a red supergiant with the supergiant statistics
presented by Humphreys and Wildey indicates that the majority of red supergiants

cannot be massive stars. We suggest that between 3/5 and 3/4 of the red supergiants
are stars of intermediate mass ($1 \lesssim M/M_\odot \lesssim 9$) with helium- and hydrogen-burning
shells above a carbon-oxygen core. For such stars to coexist with more massive
stars currently burning carbon it is required that, in a typical OB association,
star formation should have taken place over a period of time comparable to the
main-sequence lifetime of a 7.7 $M_\odot$ star (i.e., $\sim 3 \times 10^7$ years).

We find that the envelope ratio of ^{14}N to ^{12}C is enhanced by a factor of 5 over
the initial envelope ratio of both the 15 $M_\odot$ star, and the 25 $M_\odot$ star, and this
material could eventually make a significant contribution to the enhancement of
^{14}N in the interstellar medium.

7. Turbulent Diffusion in Stars (Dr E. Schatzman)

The depletion of lithium in giants seems to result from the destruction of
lithium on the main sequence prior to the formation of the giants. Lithium is
carried by turbulent diffusion towards the regions where it is burned. The measurement of the abundance of lithium in giants gives the possibility of determining the
rate of transport. It turns out that a turbulent diffusion coefficient $D = Re^* \nu$,
where ν is a kinematic viscosity, and Re^* a constant, a sort of Reynolds number,
explains the depletion of lithium on the main sequence. The depletion of lithium
in the Sun is compatible with the loss of angular momentum of the Sun. The same
value of Re^* suggests an explanation of the (V sin i) distribution function for
main sequence stars and for giant stars. The value of Re^* is around 180. A
possible explanation of the number Re^* is the following. Differential rotation
can induce a turbulence which has the tendency to establish near solid body rotation.
The decrease in the gradient of angular velocity stops the turbulence. Transport
processes are then due to microscopic viscosity. Loss of angular momentum re-
establishes a larger gradient of angular velocity, which regenerates turbulence.
Assuming that the star is always marginally unstable with respect to turbulence gives
the possibility of relating Re^* to the critical Reynolds number for the onset of
turbulence, $Re^* = (1/9)Re(critical)$. It leads to Re^* of the order of 170.

8. Diffusion and a Lower Limit to the Mixed Mass in the Solar Envelope
(Dr G. Michaud)

If one wishes to assume that the heavy elements on the solar surface were
accreted by the sun, one must remember that either the zone immediately below the
surface convection zone is stable enough for diffusion to be important or other
transport processes (i.e. turbulence, meridional circulation), more efficient than
diffusion, will tend to homogenize the star. Diffusion is the slowest of the transport processes and will become important as soon as the other transport processes
become inoperative. Diffusion theory then allows one to determine the minimum
mass of the convection zone, if transport processes at the bottom of the convection
zone are not to influence the abundances in the convection zone. If diffusion
time scales, θ, are shorter than, or of the order of, the life of the star,
diffusion will modify the abundances in the convection zone. The diffusion theory
gives $\theta \sim 2 \times 10^{11} \Delta M^{0.535}$ (years), with ΔM in units of $M_\odot$. The mass in the
convection zone for which diffusion does not modify surface abundances in the sun
by more than a factor of two is around $3 \times 10^{-3} M_\odot$. It is larger than the mass
assumed in most, but not all, of the accretion models, which are so ruled out.

9. Astrophysical Opacity Library at Los Alamos Scientific Laboratory
(Drs W.F. Huebner, A. L. Merts, N.H. Magee Jr., and M.F. Argo)

The main contents of the astrophysical elements opacity library is composed
of equation of state data, radiative Rosseland mean opacity, Planck mean opacity,
electron conduction opacity, total (combined radiative and electron conduction)
Rosseland opacity, and 2000 values of the frequency dependent extinction coefficients

in equally spaced intervals of $\Delta u = 0.01$ from $u \equiv h\nu/kT = 0$ to 20. Among available auxiliary quantities are the degeneracy parameter (η), the free electron density, the number of free electrons per atom (N_f), the mass density, and the plasma cutoff frequency.

The basic library contains the following 20 elements: H, He, C, N, O, Ne, Na, Mg, Al, Si, P, S, Cl, Ar, Ca, Ti, Cr, Mn, Fe, and Ni. Since the library is intended to produce opacities for astrophysical mixtures by matching the electron pressures for each constituent, molecules are by necessity excluded. To match the electron pressures, a standard grid is set up for all elements as a function of temperature, T, and degeneracy parameter, η. The temperature range is from $kT = 1eV$ (≈ 11600 K) to 10 keV with ten approximately logarithmically spaced grid points for each decade. A few additional points are supplied between 10 and 100 keV. The η grid goes in steps of 1 from $\eta = -25$ to -1, and then in increasing steps up to 500. The T-η grid is, however, only piece-wise rectangular: the area of very negative η (non-degenerate, very low density) and very high T is of little interest. The area of high degeneracy ($\eta > 0$) at low T is only considered as long as the Coulomb inter-action between ions (with effective charge N_f), $(N_f e)^2/(r\ kT) \lesssim 1$, where r is the mean separation between ions. This limits approximately the density range from $\rho \approx 10^{-12}$ to $10^{-3} g/cm^3$ at $kT \approx 1$ eV, and $\rho \approx 10^{-3}$ to 10^7 g/cm^3 at $kT \approx 10$ keV.

The mixing program requires as input the chemical element composition (atomic number abundance, atomic weight abundance, or log weight abundance on a scale on which H is 12), and the temperature-density grid. The library contains the improvements given by Magee, Merts and Huebner (Astrophys.J. 196, 617, 1975). Additional important improvements include the use of non-hydrogenic bound-free cross sections and effects due to level broadening. Progress is being made on a program consistent with the opacity library to calculate molecular effects.

COMMISSION 36: THE THEORY OF STELLAR ATMOSPHERES (LA THEORIE DES ATMOSPHERES STELLAIRES)

Report of Meetings, 25 and 26 August 1976

PRESIDENT: R. Cayrel   SECRETARY: F. Praderie

Business Meeting, 25 August 1976

Mr. N. Heidmann was appointed acting secretary in the absence of F. Praderie.

I. ELECTION OF PRESIDENT AND VICE-PRESIDENT
    The first topic was the election of the new President and the new Vice-President. Dr. D. Mihalas was elected new President with 15 votes for and none against. Dr. G. Traving was elected new Vice-President with 17 votes for and none against. These two votes were taken by secret ballot. Before the second vote a short discussion took place on the fact that Dr. V.V. Sobolev had not accepted to be candidate for the vice-presidency.

II. ELECTION OF ORGANIZING COMMITTEE
    After a discussion and taking into account proposals from the floor a positive vote was taken on the following list of names for the new Organizing Committee: Hearn, A.G. ; Jefferies, J.T.; Kodaira, K.; Kuhi, L.V.; Marlborough, J.M.; Pagel, B.E.J.; Praderie F.; Saper, A.A.; Vardye, M.

III. ADMISSION OF NEW MEMBERS
    The President proposes a list of 24 names for new members of the Commission. Five other names are proposed from the floor. A final list of 29 names is accepted, subject to the condition, for those not yet member of the Union, of becoming member of the Union at the sixteenth General Assembly. The list reads as follows: Altrock R.C. (U.S.), Bell R.A. (U.S.), Blanco (I), Boesgaard A. (U.S.), Carbon D. (U.S.), Castor J.I. (U.S.), Cowley C.R. (U.S.), Davis C.G. (U.S.), Evangelinis E. (Gr.), Foy R. (F), Gail H. (D), Gray D.F. (Can.) Gustafsson B. (S), Heidmann N. (F), Hekela J. (Gr.), Holweger H. (D), Kodaira J. (J), Kolesov A.K. (USSR), Lambert D.L. (U.S.), Linsky J.L. (U.S.); Matsumoto M. (J), Mukai S. (J), Pasinetti L. (I), Reimers D. (D), Seldmayr E. (D), Thompson R.I. (U.S.), Tsuji J. (J), Wehrse R. (D), Wickramasinghe C. (U.K.).

IV. SUPPORT OF THE COMMISSION FOR IAU SYMPOSIUM
    The Commission decides to give support to a symposium on "Turbulence and mass loss in stellar atmospheres" to be held in connection with the XVII th General Assembly in Canada in 1979. This symposium results from joining two separate proposals from D.F. Gray and J.M. Marlborough from one side and from J.L. Linsky from another side. This symposium could be given in honour of O.C. Wilson for his outstanding work on chromospheric H and K emission in stars. The Commission proposes Gray, Marlborough and Linsky as members of the Organizing Committee of the symposium.
    The Commission also gives support to a project of symposium on the HR diagram linked to the 100th anniversary of Russel (proposed by A.G.D. Philip) to be held in 1977 in the U.S.
    A discussion takes place on other possible topics for symposia or colloquia of interest for Commission 36. It is decided that the new President will circulate a letter among the members of the Commission relative to this topic.
    The session is closed at 10:30 a.m.

Scientific Sessions: August 26, 1976

The scientific sessions, initially planned for August 27 morning, took place on August 26 afternoon in order to avoid an overlapping with the scientific meeting of Comission 44 on ultraviolet stellar astronomy: "stellar mass loss and coronae".

The first session was devoted to "Competitive line broadening" and included the following contributed papers:

(1) Empirical evidence for competition between microturbulence and collisional broadening at the first turn-off of the solar and stellar curves of growth of FeI by R. Foy.

(2) Recent theoretical works on collisional broadening of spectral line by N. Feautrier.

(3) Fourier analysis of line profiles for turbulence and rotation, by D. Gray.

The second session was devoted to the topic "how to deal with molecules in stellar atmospheres" and included the following contributed papers:

(1) Effects of molecular absorption on the temperature profile of cool stellar atmospheres by G. Gustafsson.

(2) LTE or non LTE for molecules in stellar atmospheres, by D.L. Lambert.

(3) On the treatment of atomic and molecular opacities by H.R. Johnson.

The Commission 36 has also taken part in several joint meetings and in the joint discussion number 5 on "stellar atmospheres as indicator and factor of stellar evolution".

COMMISSION 37:   STAR CLUSTERS AND ASSOCIATIONS
(AMAS STELLAIRES ET ASSOCIATIONS)

Report of Meetings, 27, 30, and 31 August 1976

PRESIDENT: I. R. King   SECRETARY: R. Wielen

27 August 1976

BUSINESS SESSION
     Because of scheduling constraints, it was necessary to precede this day's
scientific meeting with a brief business session. The slate of new officers was
approved, and likewise the list of proposed new members of the Commission. Since
the list of proposed new members would undoubtedly include individuals who were
not yet being elected to I.A.U. membership, it was agreed that such persons would
be given the status of consultants to the Commission, for the 3-year period that
ends with the next I.A.U. General Assembly. This accomplishes the purposes of the
Commission without disturbing the orderly procedures of I.A.U. membership.

SCIENTIFIC SESSION
     All of the scientific sessions of the Commission were conducted in an informal
way. A number of individuals had been invited by the President to present reviews
of areas that are of interest to the Commission. Each review was followed by an
extended discussion period, during which participants also had an opportunity to
deliver very brief reports.

     The reviews presented at this session were as follows:
     G. Clark: X-ray Sources in Globular Clusters.
     G. Larsson-Leander: Stellar Associations.
     W. F. van Altena: Astrometry in Clusters and Associations.

30 August 1976

SCIENTIFIC SESSIONS
     The following reviews were presented:
     A. F. J. Moffat: Young Clusters.
     T. G. Hawarden: Older Open Clusters.
     P. R. Demarque: HR Diagrams and Stellar Evolution.
     G. D. Illingworth: Southern-Hemisphere Programs.
     R. Wielen: Dynamics of Star Clusters.
     W. E. Harris: Distribution of Globular Clusters in the Galaxy (presented by
I. R. King).

31 August 1976

BUSINESS SESSION
     The Report of the Commission was accepted without comment. The President
stated that additions and corrections to the Report would be included in the
present report (see below).

     The Commission agreed to endorse two proposed I.A.U. symposia:
     (a) "The HR Diagram," proposed by A. G. D. Philip for October 1977, possibly
at Albany, N. Y., U.S.A.
     (b) "Star Clusters," proposed by S. van den Bergh for a time just prior to
the 1979 General Assembly. This symposium would be held in Toronto, Canada.

Dr. B. Balázs reported on the status of the <u>Catalogue of Star Clusters and Associations</u>. The First Supplement to the second edition is in press and will hopefully appear in the Spring of 1977, unless it is further delayed by financial problems. The Commission agreed to request that the I.A.U. Executive Committee grant a subsidy of $1000 (or as close to that amount as possible) toward the expenses of this publication.

Dr. Balázs reminded the Commission of the importance of communication of results, and the Commission thereupon adopted the following resolution:

"Recognizing the importance of the <u>Catalogue of Star Clusters and Associations</u> to all astronomers, Commission 37 urges all observers to send copies of their papers directly to the Editors of the <u>Catalogue</u>, B. Balázs or J. Ruprecht (open clusters and associations), or R. E. White (globular clusters)."

There was some discussion of the future form of the <u>Catalogue</u>, especially in view of its increasing size. It was agreed that the third edition should be pre-pared in machine-readable form and that it should be made available through the Stellar Data Center in Strasbourg.

Dr. A. G. D. Philip described a collection of globular-cluster HR diagrams published by himself, Cullen, and White (Dudley Obs. Report No. 11).

The President called attention to an appeal by Dr. B. V. Kukarkin for pre-prints and reprints for use in preparing the second edition of his book "Globular Star Clusters."

The Commission discussed the question of nomenclature of star clusters. There was general agreement that the traditional system of nomenclature has become unwieldy and that a new system should be devised. Although it is clear that a new nomenclature ought to be based on some regular coordinate system, it is not obvious which coordinates should be used, or what sort of notation. A Working Group on Nomenclature was appointed; its members are A. F. J. Moffat (Chairman), B. Balázs, J. Ruprecht, W. L. Sanders, and R. E. White.

SCIENTIFIC SESSION
    One review was presented:
    R. E. White: Globular-Cluster HR Diagrams.

Appendix

CORRECTIONS AND ADDITIONS TO COMMISSION REPORT
    (Since the submission of the Commission Report, a number of errors have been called to the attention of the President, along with some additions that should be made. The President wishes especially to express his apologies to the Basel Observatory, whose contribution was omitted through an oversight on his part.)

W. E. Harris (Yale) has redetermined distances of 111 globular clusters and discussed their distribution in the Galaxy (Astron. J., in press). R. F. Garrison (Toronto) is determining integrated spectral classes and Morgan metallicity classes for 60 globular clusters. G. Alcaino (ESO) has determined preliminary photometric distances to 43 faint globular clusters south of $\delta = -22°$, and he has completed a general study of morphological data for globular clusters in the Galaxy.

A sixth globular-cluster X-ray source has been found, through the discovery by Liller (Harvard) of an obscured globular cluster of high central concentration, at the position of an observed X-ray source.

In the following supplements to the tables in the Report, "c" identifies corrections to previous items, while "a" means that the item is an addition.

1. Supplement to Table 1 (Associations)

a   Ara   R1   Herbst, Havlen (York, ESO) UBVRI, sp.
a   CMa   R1   Herbst, Racine (York, Toronto) UBV, sp.
c   Cyg   T1   observer is Gieseking
a   Mon   R2   Herbst, Racine (York, Toronto) UBV, sp.

2. Supplement to Table 2 (Open Clusters)

a   188   Helfer (Rochester) UBVIyz
a   559   Grubissich RGU (Astron. Astrophys. Suppl., in press)
a   637   Grubissich RGU (Astron. Astrophys. Suppl., in press)
a   1027   Helfer (Rochester) UBVIyz
a   1502   Hagen-Harris (Yale) sp., r.v.'s
a   1528   Helfer (Rochester) UBVIyz
a   1647   Helfer (Rochester) UBVIyz
a   1778   Helfer (Rochester) UBVIyz
a   2169   Hagen-Harris (Yale) r.v.'s
a   2323   Hagen-Harris (Yale) sp., r.v.'s
a   2421   MacConnell (Mérida) UBV, E(B-V), Be stars
a   2422   Hagen-Harris (Yale) sp., r.v.'s
a   2423   Hassan (Cairo, Basel) UBV
a   2546   Hagen-Harris (Yale) sp., r.v.'s
a   2670   Hagen-Harris (Yale) sp.
c   2682   Pulkovo: Kadla, Frolov p.m., UBV, c-m
a   2910   Topaktas (Istanbul, Basel) UBV
a   2925   Topaktas (Istanbul, Basel) UBV
a   3114   Hagen-Harris (Yale) sp., r.v.'s
a   3532   Hagen-Harris (Yale) sp., r.v.'s
a   3766   Hagen-Harris (Yale) r.v.'s
a   Yilmaz (Istanbul, Basel) RGU
a   5168   Fenkart (Basel) RGU
a   5316   Hagen-Harris (Yale) sp.
a   5617   Hagen-Harris (Yale) sp.
a   5822   Hagen-Harris (Yale) sp.
a   6031   Topaktas (Istanbul, Basel) RGU
a   6383   Hagen-Harris (Yale) sp.
a   6405   Hagen-Harris (Yale) sp., r.v.'s
a   6416   Hagen-Harris (Yale) sp.
a   6425   Hagen-Harris (Yale) sp.
a   6494   Hagen-Harris (Yale) sp.
a   6633   Hagen-Harris (Yale) sp.
a   6673   Helfer (Rochester) UBVIyz
a   6705   Helfer (Rochester) UBVIyz
a   6755   Helfer (Rochester) UBVIyz
a   6882/5   Helfer (Rochester) UBVIyz
a   7062   Helfer (Rochester) UBVIyz
a   7063   Helfer (Rochester) UBVIyz
a   7209   Helfer (Rochester) UBVIyz
c   7243   Pulkovo p.m. by Koroleva
c   7788   Pulkovo: Frolov p.m., UBV, c-m
c   7789   omit Pulkovo reference
c   7790   Pulkovo: Frolov p.m., UBV, c-m
a   IC 1805   Helfer (Rochester) UBVIyz
a   IC 2395   Hagen-Harris (Yale) sp.
a   IC 4725   Helfer (Rochester) UBVIyz

2.   <u>Supplement to Table 2 (Open Clusters) [continued]</u>

a	IC 4756	Helfer (Rochester) <u>UBVIyz</u>
a	Ba 11a	Grubissich <u>RGU</u> (<u>Astron. Astrophys. Suppl</u>. 11, 283, 1973)
a	Ba 16	Fenkart (Basel) <u>RGU</u>
a	Be 6	Spaenhauer (Basel) <u>RGU</u>
a	Be 58	Frolov (Pulkovo) p.m., <u>UBV</u>, c-m
a	Cr 140	Hagen-Harris (Yale) sp., r.v.'s
a	Cr 272	Fenkart (Basel) <u>RGU</u>
a	Cr 299	Topaktas (Istanbul, Basel) <u>RGU</u>
a	H 10	Topaktas (Istanbul, Basel) <u>RGU</u>
a	Lyngå 6	Lyngå says really a cluster (<u>Astron. Astrophys</u>., in press)
a	Mel 71	Hassan (Cairo, Basel) <u>UBV</u>
a	Ru 79	Topaktas (Istanbul, Basel) <u>RGU</u>
a	Ru 82	Topaktas (Istanbul, Basel) <u>RGU</u>
a	Stock 2	Hagen-Harris (Yale) sp., r.v.'s
a		Helfer (Rochester) <u>UBVIyz</u>
a	Stock 16	Fenkart (Basel) <u>RGU</u>
a	Tr 2	Hagen-Harris (Yale) sp., r.v.'s
a		Helfer (Rochester) <u>UBVIyz</u>
a	Tr 10	Hagen-Harris (Yale) sp., r.v.'s
a	Anon Feinstein	Hagen-Harris (Yale) sp., r.v.'s
a	Anon	Frolov (Pulkovo) p.m., <u>UBV</u>, c-m

3.   <u>Supplement to Table 3 (Globular Clusters)</u>

a	362	Alcaíno (ESO) <u>UBV</u>, c-m
a	1904(M79)	Alcaíno (ESO) <u>UBV</u>, c-m
a	3201	Alcaíno (ESO) <u>UBV</u>, c-m
c	4590(M68)	Alcaíno (ESO) <u>UBV</u>, c-m
c	5024(M53)	Panova (Pulkovo) p.m., <u>BV</u>, c-m
c	5272(M3)	Kadla <u>et al</u>. (Pulkovo) distr. of RG, <u>UBV</u>, surf. phot.
a	5904(M5)	Bashtova (Pulkovo) p.m., <u>BV</u>, c-m
a		Kadla <u>et al</u>. (Pulkovo) distr. of RG, <u>BV</u>, surf. phot.
a	6205(M13)	Kadla <u>et al</u>. (Pulkovo) distr. of RG, <u>UBV</u>, surf. phot.
c	6341(M92)	Kadla <u>et al</u>. (Pulkovo) distr. of RG, <u>UBV</u>, surf. phot.
a		Cudworth (Yerkes) vel. disp. from p.m.
a	6397	Alcaíno (ESO), <u>UBV</u>, c-m
a	6656(M22)	Alcaíno (ESO), <u>UBV</u>, c-m
a	7078(M15)	Kadla <u>et al</u>. (Pulkovo) distr. of RG, <u>UBV</u>, surf. phot.
		Panova (Pulkovo) p.m.
		Cudworth (Yerkes) vel. disp. from p.m.
a	7089(M2)	Kadla <u>et al</u>. (Pulkovo) <u>UBV</u>

COMMISSION 38: EXCHANGE OF ASTRONOMERS
(ECHANGE DES ASTRONOMES)

Report of Meetings, 26 and 30 August 1976

PRESIDENT: P. M. Routly   VICE-PRESIDENT: D. A. MacRae

Present at both meetings were the President and Vice-President, P. M. Routly and D. A. MacRae, all six members of the Organizing Committee, J. Delhaye, G. S. Khromov, A. Reiz, J. Sahade, F. G. Smith, and F. B. Wood, and some fifteen general members and interested participants.

The President began by giving a verbal report of the procedures and criteria that were followed in administering the Exchange of Astronomers Program since the XVth General Assembly in Sydney. While unanimously approved by the Commission members, the President pointed out the continuous need to justify the existence of the Program, to re-affirm its objectives, and to devise new ways in which the Program might be improved in the future.

Accordingly, after full discussion and debate, the Commission recommended for adoption the following restatement of Program Aims, Minimum Requirements, Guide-lines, and Application/Selection Procedures.

Exchange of Astronomers Program

I. BASIC AIMS
    To award grants to qualified individuals to enable them to visit institutions abroad for the purpose of working on problems which they could not pursue advantageously in their own countries. It is hoped, in particular, that each visitor have ample time and opportunity to interact with the intellectual life of the Host Institution so that scientific and cultural benefit is derived on both sides. It is a specific objective of the program that astronomy in the home country be enriched after the applicant returns.

II. MINIMUM REQUIREMENTS
    (1) Each candidate must submit a curriculum vitae showing that he/she is professionally qualified and must submit a viable plan of scholarly activity to be carried out during the proposed visit.

    (2) Candidates may be faculty/staff members, post-doctoral fellows, or graduate students at any recognized educational/research institution or observatory. All candidates must have excellent records and must have made permanent and professional commitments to astronomy.

    (3) All visits must be formally agreed to by the Directors of the Home and Host Institutions involved. Such endorsements must confirm that the proposed plan of study is a reasonable one and will be of benefit to astronomy.

    (4) All visits must consist of a stay of at least 3 months at a single Host Institution. Stop-overs at other Institutions en route may be permitted.

(5)   Each applicant must give details, with supporting documentation, of
      funds currently available to him to finance his proposed visit.   In
      particular, the applicant must state what other applications he has
      submitted in efforts to obtain support from other sources and must
      indicate the status of such applications.   In the event that an
      applicant receives funds from another source which may be used, in
      whole or part, for the same proposed visit, he is required to revise
      his application or make a refund to the I.A.U.

(6)   Each recipient is required to submit a brief report to the President
      of Commission 38 after the conclusion of his visit.   Acknowledgement
      of support from the I.A.U.'s Exchange of Astronomers Program should
      also be made in any published paper resulting from a visit.

III. GUIDELINES
(1)   Grants are made within limitations imposed by the budget allocated
      to the Commission by the Executive Committee of the Union.

(2)   The amount of a grant will be governed by the cost of a single re-
      turn economy air fare between the Home and Host Institutions and
      normally is to be used by the applicant for such travel.   In excep-
      tional cases, and with prior Commission approval, the funds can in-
      stead be used wholly or in part for subsistence costs during the
      visit.

(3)   Grants to attend symposia, summer schools, conferences, society
      meetings, etc. are outside the scope of the program.

(4)   Grants will not be made for the sole purpose of obtaining observa-
      tional data; other benefits to astronomy must also be realized during
      the visit.

(5)   An individual should normally not expect to receive an I.A.U. award
      for a second visit.

(6). The Program is normally designed to support the work of young astro-
      nomers, but established astronomers who can benefit from the Program
      may also participate.

(7)   Some grants may be awarded on the basis of one-way fare.   Examples
      are visits of long duration - a year or more - or cases where highly
      qualified graduate students apply for funds to go abroad to begin
      graduate studies at an institution where they have been formally
      accepted.

(8)   It is to be emphasized that all recipients should return to their
      Home Institutions or Home Countries upon the completion of their
      visits.

IV. APPLICATION AND SELECTION PROCEDURES
(1)   Each applicant must formally submit his request for a grant in the
      form of a letter to the President of Commission 38.   The information
      supplied in this one document should be complete and detailed as it
      will be used to judge whether the proposal is in conformity with the
      aims of the Program, whether the minimum initial requirements are
      being met, and whether the guidelines will permit a favorable deci-
      sion.   Any special circumstances must be carefully set forth.

(2) It is the applicant's responsibility to arrange for two confidential
     letters of endorsement from senior officials of the Home and Host
     Institutions to be sent without delay directly to the President of
     Commission 38. The letters should, as well, confirm statements made
     by the applicant concerning his financial status and availability of
     support from sources other than the Exchange of Astronomers Program.

(3) Upon receipt by the President of a complete application, a copy will
     be sent to the Vice-President. These two officers of the Commission
     will jointly decide on the disposition of the application. Experi-
     ence shows that the majority of applications can be dealt with effec-
     tively in this manner.

(4) If the application has unusual features rendering a decision diffi-
     cult, if a significant deviation from the guidelines arises, if there
     is disagreement between the two officers, or for any other reason,
     the application will be distributed to members of the Organizing
     Committee for their opinions before a final decision is reached. In
     case of an appeal against an unfavorable decision, the Organizing
     Committee will be asked to adjudicate. In this way, the guidelines
     will be continuously under review and improved in the light of
     experience.

(5) When a favorable decision has been reached, the President will for-
     ward copies of the application and all pertinent documents to the
     General Secretary and will recommend that a grant be made in a
     specified amount.

(6) At six or eight month intervals during the triennium, the President
     will issue interim Reports on the Program to members of the Organiz-
     ing Committee of the Commission and to the General Secretary for the
     information of the Executive Committee of the Union. The members of
     the Organizing Committee will be asked to comment on the operation of
     the Program on each such occasion.

(7) Brief summaries of Commission activities will also be distributed to
     general members of the Commission during the intervals between
     General Assemblies.

Another question which was discussed thoroughly was whether the Exchange of
Astronomers Program should continue to be operated as in the past or whether the
Program's objectives would be better served by a small ad-hoc committee of the
Executive Committee.

Because of the desirability for the widest possible exposure to, and participation by, the international astronomical community, it was felt unanimously that
the Exchange of Astronomers Program should continue to be operated by a full
Commission, similar in structure to the other Commissions of the Union, and consisting of a President, Vice-President, Organizing Committee, and General Membership. Among the reasons in favor of a General Membership were the following -
to provide a pool of potential members for the Organizing Committee, to optimize
the opportunity for professional contact and the flow of information involving,
especially, the smaller and/or less well developed countries and, finally, to
provide a mechanism for interested members of the Union to participate directly
in the affairs of the Commission.

It was therefore recommended that Commission 38 continue to operate the
Exchange of Astronomers Program and that the composition of this Commission over
the next three years be as follows:

President: D. A. MacRae (Canada)

Vice-President: J. Delhaye (France)

Organizing Committee: M. K. V. Bappu (India), G. S. Khromov (USSR), A. Reiz (Denmark), P. M. Routly (USA), J. Sahade (Argentina), F. G. Smith (U.K.), F. B. Wood (USA).

General Membership: G. Abetti (Italy), A. W. Alsabti (Iraq), B. J. Bok (USA), H. F. Haupt (Austria), G. Keller (USA), V. Kourganoff (France), M. Marik (Hungary), J. M. Mohr (Czechoslovakia), S. Myamoto (Japan), S. E. Okoye (Nigeria), A. Opolski (Poland), T. L. Page (USA), G. Righini (Italy). S. Rosseland (Norway), G. Ruben (G.D.R.), R. H. Stoy (U.K.), P. Swings (Belgium), G. Teleki (Yugoslovia), C. R. Tolbert (USA), J. P. Wild (Australia), S. P. Wyatt, Jr. (USA).

It was also suggested that the General Secretary write the National Committees of each of the adhering countries of the Union not represented in the proposed membership above to ask whether they would like to be represented and, if so, by whom.

COMMISSION 40: RADIO ASTRONOMY
(RADIO ASTRONOMIE)

Report of Meetings, 25, 26, 27, 28 and 30 August and 1 September 1976

ACTING PRESIDENT: H. van der Laan   SECRETARY: R. G. Strom

JOINT DISCUSSIONS

On 25 August Commission 40 cosponsored a meeting with Commissions 34 and 44 on Interstellar Molecules and Dust.

On the afternoon of 26 August a joint discussion was held with Commissions 4, 8, 19 and 31 on New Techniques for the Determination of the Rotation of the Earth.

Commissions 40 and 48 held a joint discussion on Physics of Radio Sources and Quasars on 27 August.

These activities are outlined in one of the reports of the cosponsoring Commissions.

<u>28 August 1976</u>

In the unfortunate absence of the President, Yu. N. Parijskij, the Vice President took the chair. The following business was transacted.

I. REPORT OF THE ACTING PRESIDENT
    Commission activities since the Sydney General Assembly were briefly summarized. The Acting President pointed out the changing character of Commission 40, particularly the fact that radio studies now embrace nearly all fields of astrophysics. In view of this, each Commission 40 member should also be a member of at least one other Commission where his purely astronomical interests can be represented.

II. ELECTION OF THE ORGANIZING COMMITTEE
    The list of nominations for Commission Officers and Organizing Committee was adopted without change:
    President: H. van der Laan
    Vice President: G. Swarup
    Organizing Committee: Blum, Fanti, McLean, Mezger, Moffet, H. P. Palmer, Parijskij, Robinson, Zuckerman.

III. RESOLUTION CONCERNING SYSTEM III FOR JUPITER
    H. P. Palmer summarized a resolution on behalf of Bozyan, Riddle, Seidelmann and others, which was adopted by the Commission: 'RESOLVED, that the provisional rotation period adopted for Jupiter's System III (1957.0) longitude measure, being inadequate for current use, be replaced by a new System III measure for which the sidereal rotation rate of Jupiter is $870\overset{\circ}{.}536$ per Ephemeris day. The epoch shall be 1965 Jan. 1 $0^h0^m0^s$ ET, the longitude at epoch of the central meridian, as observed from Earth, shall be $217\overset{\circ}{.}595$, and the system shall be called System III (1965).'

IV. RESOLUTION CONCERNING THE STANDARD OF REST
    Heidmann proposed, on behalf of De Vaucouleurs, a resolution which was adopted by the Commission: 'Commission 28, 30 and 40 RECOMMEND, that in the standard correction of extragalactic redshifts for solar motion with respect to the Local Group $\Delta V = 300 \cos A$, the definition of the solar apex be changed from $\ell^I = 57^o$, $b^I = 0^o$ to $\ell^{II} = 90^o$, $b^{II} = 0^o$, but that no change be made in the conventional value of the

solar velocity $V_{\odot} = 300$ km s^{-1}.'

V. RECOMMENDATION ON THE DEFINITION OF RADIAL VELOCITY

Menon outlined the conclusions of the joint working group set up by Commissions 30 and 40 in Sydney. There was considerable opposition to the introduction of a new symbol, v_ν, to define the fictional velocity $c\Delta\nu/\nu_o$. Therefore the original recommendations were modified, to be considered further by Commission 30:

(1) The practice of calling the quantity $c\Delta\nu/\nu_0$ a radial velocity, and denoting it by the symbol v_r is confusing in extragalactic applications and should be discontinued.

(2) Astronomers who insist on publishing results in the form $c\Delta\nu/\nu_0$ should clearly indicate that they have done so.

VI. PROPOSALS RECEIVED FROM OTHER COMMISSIONS

Note was taken of the following resolutions, for which there was general support in our Commission:

(1) Commission 8 urges the continued support and coordinated long term planning of meridian astronomy. (2) Commission 10 strongly recommends that ground based solar research not be neglected in favor of satellite-borne observations. (3) Hagen of Commission 10 requests coordination of solar flare reporting.

VII. REPORTS ON ASTRONOMY

Westerhout proposed a vote of thanks to Parijskij and his colleagues for the extensive bibliography they prepared for the Commission Report in Volume XVIA of the Transactions. This proposal met unanimous acclaim. After some discussion it was decided that account should be taken of the wide ranging character of Commission 40 in preparing future Reports. These could be more efficiently prepared in close consultation with other Commissions.

VIII. IUCAF NEWS

The discussion was dominated by the upcoming (1979) World Administrative Radio Conference (WARC), by interference from satellites and aircraft and by the need for protecting spectral lines.

A working group chaired by Robinson will look into the priorities for protection of frequency bands for molecular lines and report to IUCAF before 1 November 1976. It will also study protection for the 106-116 GHz band which contains many lines (especially CO), and consider whether the protected 130-140 GHz band should be exchanged for this. Barret was asked to look into protection for ammonia lines and submit a report to IUCAF

Westerhout summarized the activities of IUCAF for the remainder of 1976. These include rewriting proposals, contacting appropriate people and calling for cooperation in obtaining the required frequency protection.

Findlay proposed the following resolution which was passed by the Commission: 'The International Astronomical Union, CONSIDERING (a) that the World Administrative Radio Conference (WARC) to be held in 1979 will study the technical requirements and frequency allocations for all radio services, including the use of radio frequencies for scientific research purposes; (b) that the decisions of WARC-1979 can be expected to remain in force for about 20 years; (c) that the deliberations and recommendations of the Inter-Union Commission on Frequency Allocations for radio astronomy and space science (IUCAF) are the appropriate means of indicating the requirements for radio frequencies for research purposes; (d) that individual national administrations are now making preparations for WARC-1979; RESOLVES to encourage IUCAF, 1. to undertake in a timely manner the deliberations and studies required to determine the needs of radio scientists for the use of the radio spectrum; 2. to bring its recommendations to the attention of the members of URSI, IAU and COSPAR so that those bodies may comment on them; 3. to invite national

TABLE 1. MILLIMETER RECEIVERS (from the review by P. Zimmermann)

Device	Ref. (see text)	Frequency Range (GHz)	Instantaneous Bandwidth (MHz)	Gain, G (dB); Conversion Loss, L (dB)	Noise Temp. (K)	Operating Temp. (K)	Remarks
Maser	(a)	29-35	30-60	G=30	30±5*	4.2	
	(b)	35	20	G=20	-	-	
	(c)	85-90	140	G~3-5	<100*	4.2	upper frequency limit ~100 GHz due to lack of high frequency pumps, and problems of circulator design
Parametric Amplifiers	(d)	33.6	900	G=15	144†	300	
	(e)	46	200	G=22	40†	18	
	(f)	60	200	G=18	605*	300	
	(g)	~60	800	G=14	850*	315	
Josephson Paramp	(h)	33	3400	G=15	220†	4.2	20±10 K from the paramp alone; saturation problem; advantage of zero bias operation and low pump power required
Josephson Mixer	(i)	36	50	{L=5, G=1.35}	210-54*	4.2	pump power in the range of microwatts; bandwidth restricted by IF-amplifier
	(j)	220-325	100	L=12.3	71*	4.2	
InSb Mixer	(k)	115	4	L~10	250†	4.2	pump power only ~0.1 µW; bandwidth restricted by element response time; expected to be used up to 500 GHz
		230	4	L~10	300†	4.2	
Schottky-Barrier Mixer		85	100	L=4.6	420*	300	limiting factor is diode quality; good up to >200 GHz with present diodes
	(l)	115	100	L=5.5	500*	300	
		115	100	L=5.8	300*	77	

*single side band
†double side band

administrations to include the IUCAF recommendations, as appropriate, in the documents they will prepare for the CCIR and WARC-1979.'

Howard proposed a resolution which was passed by the Commission and subsequently by the General Assembly of the Union. (See Resolution No. 8.)

IX. FUTURE SYMPOSIA AND COLLOQUIA
International conferences being planned for the next three years were called
to the attention of Commission members. These include symposia and colloquia being
arranged in conjunction with the 1979 IAU General Assembly, a colloquium which will
precede the 1978 URSI meeting in Helsinki and a conference on normal galaxies to
be held in the summer of 1977. As these plans take shape Commission members will
be informed through the IAU Bulletins or by circular letter.

X. MEMBERSHIP
Nominations for membership were reviewed by the Organizing Committee. A total
of 64 new members were elected. Their names are incorporated in the current membership list.

XI. OTHER BUSINESS
Sullivan described a project in which he is collecting information bearing on
the history of radio astronomy. He is particularly interested in original source
material and appealed for help in locating any he may not be aware of.

Dixon announced that the master catalogue of radio sources continues.

30 August 1976

Two scientific sessions were held on the afternoon of 30 August.

I. TECHNICAL DEVELOPMENTS IN MILLIMETER ASTRONOMY
The meeting, chaired by M. A. Gordon, heard of recent developments in millimeter wave receivers, and the status of a number of millimeter telescope projects.

Millimeter Receivers (P. Zimmermann)

For radio astronomical and in particular spectroscopic measurements in the
range 30-300 GHz receiver types employed are masers, parametric amplifiers, Joseph-
son junction devices, InSb-bulk mixers and Schottky-barrier mixers. Existing receivers which have been reported in the literature and their salient features are
listed in Table 1. (References to Table 1 are: (a) Cardiasmenos et al. June 1976,
IEEE-MTT Int. Microw. Symp.; (b) Zagatin et al. 1976, Radio Eng. Electr. Phys.,
12, 501; (c) Kolberg and Lewin. 1976, IEEE Trans. Microw. Theory Techn., 24, in
press; (d) Cohn et al. 1969, IEEE G-MTT Int.Microw. Symp. Dallas; (e) Edrich.1973,
Int. Microw. Symp. Dig. No. III-5; (f) Stover et al. 1973, IEEE Int. Solid State
Circuits Conf. Dig. Philadelphia; (g) Whelehan et al. 1973, Microwave J., 16 (11),
35; (h) Chiao and Parrish. 1976, J. Appl. Phys., 47, 2639; (i) Taur et al. 1974,
IEEE Trans. Microw. Theory Techn., 22, 1005; (j) Edrich. June 1976, IEEE-MTT Int.
Microw. Symp. (late paper); (k) Phillips and Jefferts. 1974, IEEE Trans. Microw.
Theory Techn., 22, 1290; (l) Kerr. 1975, IEEE Trans. Microw. Theory Techn., 23,
781.) Up to the present the Schottky-barrier mixer is the most widely employed
receiver type, although increasing emphasis is now being placed on other cryogenic
receivers.

Millimeter Telescope Projects

Reports on recent projects are summarized below. Each contribution is identified by the university or institute involved, and features of the antennas are
listed in Table 2.

TABLE 2. RECENT MILLIMETER TELESCOPE PROJECTS

Group (country)	Site (altitude)	Diameter	RMS accuracy	Status	Other Information
NRAO (U.S.A.)	(undecided)	25 m	.075 mm	early planning	in radome
MPG (Germany) &	South Spain (3300 m)	30 m	<.1 mm	} under design	joint MPG/INAG institute in Grenoble; early 1980's; French project is four element synthesis array
INAG (France)	S. of Grenoble (2550 m)	4 x 10 m	.1 mm		
Caltech (U.S.A.)	Owens Valley (1200 m)	10 m	.07 mm	built	several 10 m dishes for interferometer planned
U. of Mass. (U.S.A.)	5 College Obs. (306 m)	13.7 m	.11 mm	built	in radome
CSIRO (Australia)	Parkes (392 m)	16.7 m	.27 mm	operating	center of 64 m dish
	"	37 m	.8 mm	"	" " "
	Epping	4 m	.1 mm	built	2 m dish being built for interferometry
	Tidbinbilla	64 m	-	operating	NASA Deep Space Station; has operated at 13 mm
Chalmers (Sweden)	Onsala (14 m)	20.1 m	.2 mm	operating	in radome
Bell Labs (U.S.A.)	New Jersey (90 m)	7 m	.1 mm	built	off-axis parabolic surface
U. C. Berkeley (U.S.A.)	Hat Creek (1050 m)	2 x 6 m	(1 mm)*	operating	synthesis array, 2 elements

*quoted wavelength limit

National Radio Astronomy Observatory (B. E. Turner) The plan is to erect a homologous dish in a dome with a slit. At 0.8 mm the telescope would have an 8" arc
beam and a 15% aperture efficiency. The total cost would be $8 million and though
no site has been chosen, Hawaii is a possibility.

Max-Planck-Institut für Radioastronomie (J. W. M. Baars) This part of the Franco-
German project (CNRS-MPG, see below) will place a homologous dish on a mountain
55 km from Granada. A joint institute for operating the two observatories is planned in Grenoble. The estimated cost of the 30 m telescope alone is DM 17 million.

California Institute of Technology (M. S. Ewing) In a project headed by R. B.
Leighton a prototype 10.4m f/0.4 reflector has been built whose aluminum honeycomb
panels have been machined in place on the backup structure and surfaced with sheet
aluminum. An improved mirror is under construction and a reflector with 15 μm
errors appears possible. Three improved reflectors will be used for interferometry
in Owens Valley, while a specially treated antenna is to be built for high altitude submillimeter observations.

Institut National d'Astronomie et de Geophysique (E. J. Blum) The proposed synthesis array consisting of four dishes on a T-shaped baseline 2 km (EW) x 1 km (NS)
will operate between 22 GHz and 150 GHz. It will be possible to synthesize a field
of 1'arc in less than a week with 3"arc resolution, and the maximum resolution will
be 0".5 arc (at 115 GHz). Studies have begun to develop cooled mixer R.F. heads
and an oversized waveguide transmission system linking the antennas. This project
is part of a French-German proposal (CNRS-MPG) for a millimeter facility.

University of Massachusetts (W. M. Irvine) Initial tests of the telescope system
are planned for the autumn of 1976. The 1024 channel digital autocorrelator has a
bandwidth of 25-30 MHz (50-60 MHz for 512 channels with bandwidth doubling). The
computer system, based on a Mod Comp IV/25, enables almost complete data reduction
at the telescope. Receivers include a ruby travelling wave maser (f = 20-25 GHz,
Δf = 30 MHz, T_s = 50-100 K), with a rutile TWM (85-95 GHz, 140 MHz, 100 K), state
of the art mixers at 2 and 3 mm and a 115 GHz rutile maser planned.

Australian projects, CSIRO (B. J. Robinson) The 16.7 m reflector, using cooled
mixer receivers, has surveyed CS (J=1→0, 48 GHz) emission in the southern Milky
Way, and searched for SiO masers at 43 GHz. The 4 m dish will survey CO, HCN,
HCO^+, etc., in the Galaxy and Magellanic Clouds. An 80-120 GHz cooled mixer receiver and 512 channel "electro-acoustic spectrograph" are under development. An interferometer (4m and 2m dishes) with baselines up to 100 m is under construction.

Chalmers University (B. Höglund) The 20 m Cassegrain dish has a tracking accuracy
of 2" to 3" arc. At present there is a 100 K maser available for the 21 to 25 GHz
band, with a 29 to 35 GHz maser planned for the near future. An 88 to 115 GHz
uncooled mixer will soon be available and there are plans for a cooled mixer.
Spectrometers are being developed for line observations.

Bell Laboratories (R. W. Wilson) The reflector is a 7 m circular section of a
16 m diameter parabolic surface. This permits the operation of an offset Cassegrain
system with no aperture blockage. The initial astronomical receiver will cover 70
to 150 GHz and have two 256 channel spectrometers of 0.25 and 1 MHz resolution.

University of California at Berkeley (J. Welch) The interferometer is designed
for operation between 1 mm and 15 mm. The two dishes can be moved along a 'T'
shaped rail line 300 m EW by 200 m NS. For the 11 mm to 15 mm wavelength range
the sensitivity after a full synthesis is 10 mJy.

II. INTERSTELLAR RADIO SPECTROSCOPY
    During this session, chaired by T. L. Wilson, the following papers were pre-

sented:
   Radio Recombination Lines (M.J. Seaton)
   Molecular Line Excitation: The Density and Mass of Molecular Hydrogen Clouds
in the Galaxy (P. Solomon)
   Radial Gradients in Isotopes and Elements (M. Walmsley)
   Galactic OH towards extragalactic radio sources (I. Kazès)
   Radio recombination lines at 300 MHz (A. Parrish)
   Spatial extent and gas motions in the ρ Oph. dark cloud (P. Myers)
   Interstellar HC_5N in a dark cloud (J. MacLead)
   Recombination line broadening (P. Encrenaz)

1 September 1976

   Two scientific sessions and an extra business meeting were held on Wednesday,
1 September.

I. MORNING SCIENTIFIC SESSION
   In the meeting, chaired by H. van der Laan, the following papers were presented:

Hard- and Software

   Progress report on the VLA (D. S. Heeschen)
   Data processing for synthesis arrays (W. N. Brouw)
   Theoretical analysis of CLEAN (U. J. Schwarz)

Nearby Extragalactic Systems

   Introductory review (W. K. Huchtmeier)
   Line and continuum studies of edge-on galaxies (R. Sancisi)
   Nearby spirals (J. H. Oort)
   CO in external galaxies (B. M. Zuckerman)
   Neutral hydrogen in the far south of M31 (R. D. Davies)
   Gas motions in the M81/M82 group from HI observations (L. Weliachew)
   HI observations of NGC3077/M81 and NGC4038/39 (J. M. van der Hulst)
   HI warps in M31 and M33 (D. Emerson)
   Warps and bar-like disturbances in galaxies (A. Bosma)
   The peculiar galaxy NGC3718 in HI (U. J. Schwarz)
   Radio continuum studies of IC342 (R. Wielebinski)
   The radial distribution of various constituents in M31, M33 and the Galaxy
(E. M. Berkhuijsen)
   Radio maps of NGC3310 and 3079 at 2.7 and 8.1 GHz (E. R. Seaquist)

II. AFTERNOON SCIENTIFIC SESSION
   The chairman, J. R. Shakeshaft, introduced the following papers:
   The Stereo I experiment for type I radio bursts (J. L. Bougeret)
   The proper motion of the Crab pulsar (S. Wyckoff)
   The origin of Loop I and its relation to the local gas (H. Weaver)
   Stimulated recombination line emission from M82 (P. A. Shaver)

Radio Galaxies, their Nuclei and Quasars

   Preliminary results from the 151 MHz synthesis array at Cambridge (P.F. Scott)
   Fifteen southern radio galaxies mapped with the Fleurs Synthesis Radio Telescope (W. N. Christiansen)
   Brightness distributions from VLBI data, using closure phase (P. Wilkinson
and A. Readhead)
   High resolution observations of M81, M82 and the galactic center (K. Kellermann)

NRAO 150, the smallest angular size radio source; and other VLBI results
(D. Shaffer)

VLBI observations of the radio nuclei in extended radio sources (I. Pauliny-
Toth)

The rôle of central components in extended radio sources (T. K. Menon)

The complete polarization properties of compact radio sources (R. S. Booth)

The spectra of radio source variations at meter - wavelengths (W. Erickson)

The radio spectral evolution of some quasars (W. Dent)

Appeal for flux density data on 3C454.3 (J. Wardle)

Optical spectra of radio galaxies - Lick spectrophotometric observations
(R. Costero)

III. EXTRA BUSINESS MEETING

About sixty members were present when the chairman, H. van der Laan, opened
the meeting. One point was on the agenda, the reorientation of Commission 40. The
chairman stated his case for changes in the Commission's activities and its relation to other Commissions of the IAU (this and the ensuing discussion are summarized
here as succintly as possible).

The Chairman's Statement

(a) Commission 40 is both a very large and a very active component of the
IAU. Rumors of its abolition are so unrealistic that they must be dismissed as fri-
volous.

(b) Radio astronomy is now a mature constituent of modern astronomy regarded
in its spectral dimension. As such radio astronomy touches upon virtually every
subject in astrophysics and observes practically every category of astronomical
object. This implies that those who use radio telescopes, i.e. usually members of
Commission 40, normally have interests in one or more other Commissions whose activities and goals are circumscribed by astrophysical themes or astronomical objects.
All Commission 40 members must therefore be emphatically encouraged to join and
participate actively in the Commission(s) where their astronomical interests are
developed.

(c) Radio astronomical observations using established techniques should be
integrated and analysed in the best astrophysical context. Such research should
therefore be reported in the astronomically most appropriate commissions and those
commissions should in their meetings and written reports take cognizance of and
fully profit from these investigations. Commission 40 should discontinue publishing
all and sundry subjects in its triennial Reports in Astronomy and instead encourage
and assist other commissions to integrate radio astronomical results in their reviews.

(d) Radio observatories are very vulnerable to terrestrial and space interference. Radio astronomers must be constantly vigilant to protect essential frequency intervals of the radio spectrum. They can do so effectively only by a concerted global effort. For this effort the activities of Commission 40 are indispensable. Commission G of URSI and Commission 40 of the IAU together can provide the
network of relations which IUCAF requires for frequency protection.

(e) Astronomical applications of radio techniques continue to be developed. In
the R & D stage of new devices and methods Commission 40 provides the best context
for astronomers to exchange and compare experience in this area of astronomical
techniques.

(f) Radio astronomers frequently engage in joint programs (e.g. monitoring
variable sources) and complex multinational cooperative arrangements. Commission 40
can provide organizational and moral support.

Under (b) and (c) the chairman's views of how Commission 40 ought to reorient
are stated, while under (d), (e) and (f) important reasons for continued Commission
40 activities are given.

Discussion

Several members proposed that an additional reason for Commission 40's continued existence is the 'bond of kinship' among radio astronomers. There is no justification for disregarding history and disowning the enjoyment of sharing in this
development and communicating in a common jargon, even though this concerns a great
variety of astronomical interests. Some members appear not to be a member of any
other Commission at all, but most of those in attendance agreed that Commission 40
cannot be expected to cover even a fraction of the astronomical research carried
out by radio astronomical techniques. The meeting therefore strongly supported
the chairman's proposal that each member join other Commissions to represent his
interests as well. For solar astronomers especially this appears to be of vital
importance if their research is to gain the attention it deserves in solar physics.

Commission 40 members who are on the IAU Executive Committee are encouraged
to make it crystal clear in the Executive that Commission 40 remains a strong and
active Commission, which sees many excellent reasons for its existence and which
has no intention whatsoever of being abolished. A motion to this effect was passed
by acclamation. The meeting ended with a discussion concerning the Commission's
activities at General Assemblies. Many members said they enjoyed the meetings which
were organized in Grenoble on the spot and which dealt with many subjects, mainly
extragalactic, but where due to the common language and the general supposition
that techniques used were already understood, a very great information rate was a-
chieved. These members felt that in the future such meetings, of many short contributions, should be held again. Others thought it a pity that beautiful results obtained at radio observatories should in their presentation be confined to Commission
40 meetings and strongly urged that these results be aired at the meetings of other
Commissions. The conclusion was that in the future the Organizing Committee of the
Commission should, in preparation for the General Assembly, organize as many joint
discussions and multiple Commission meetings as is feasible, to assure a proper
forum for radio investigations. The chairman appealed to Commission members to
support its Organizing Committee by responding positively when requested to contribute. He promised to do what is necessary to organize Commission 40 work in preparation for the 1979 General Assembly in the spirit of this discussion.

COMMISSION 41: HISTORY OF ASTRONOMY (HISTOIRE DE L'ASTRONOMIE)

Report of Meetings, 30, 31 August 1976

PRESIDENT: O. Gingerich   SECRETARY: M. Hoskin

30 August 1976

I. BUSINESS SESSION

Members stood for a few moments to honor the memory of the late Dr. Per
Collinder, and then elected J. Dobrzycki (Poland) as President, M. A. Hoskin (UK)
as Vice-President, and S. M. R. Ansari (India), W. Hartner (GFR), P. G. Kulikovsky
(USSR) and O. Gingerich (USA) as members of the Organizing Committee together with
O. Pedersen (Denmark) as representative of International Union of the History and
Philosophy of Science. The president expressed his gratification that a large
number of former consulting members of Commission 41 had now been nominated for full
membership of the IAU by their respective national committees: Z. Horsky, H. D.
Howse, J. North, R. Taton, V. Thoren, and J. P. Verdet. Other new members of the
Commission include D. Heggie, G. Jackisch, G. S. Khromov, K. Lang, J. Merleau-Ponty,
J. Rybka, and W. Sullivan. It was then resolved that "The members of Commission 41
wish to thank our Soviet Colleagues for their regular publication of *Bibliography
of Books and Papers Published in the History of Astronomy*, and urge the continued
publication of this valuable aid to research", and that "The members of Commission
41 recommend to the General Secretary that this *Bibliography* be listed as an
official IAU Publication".

M. A. Hoskin then reported on the progress of the *General History of Astronomy*. Contracts had been signed with Cambridge University Press for publication of
the work in four volumes of about 600 pages each. The volumes would appear separately, and it was hoped volume 3 would be ready for the printer early in 1978;
authors for all four volumes were now being commissioned. The presence of every
volume editor at Grenoble would allow the editorial board to meet and settle the
remaining questions of general strategy.

II. OBSERVATIONS IN ANCIENT AND MEDIEVAL ASTRONOMY.

R. R. Newton, in a series of books and articles, has compared the available
reports of various astronomical observations, from the eighth century B. C. to the
thirteenth century A. D., with the results expected on the basis of modern astronomy. He came to distrust the authenticity of many of the alleged observational
data from Ptolemy's *Almagest*, reaching the general conclusion that Ptolemy was the
fudger of observations *par excellence*. In a review paper on the role of observations in ancient and medieval astronomy, W. Hartner agreed that some of
Ptolemy's "observations" were actually calculations, but explained that it was
historically anachronistic to label the Alexandrian astronomer's work fraudulent.
Hartner then presented a far-ranging discussion of the relationship of observation
to theory in the astronomy of Greek Antiquity, China, Islam, and the Renaissance.

K. P. Moesgaard described a specific and technical study of "Hipparchus's
Solar Theory Derived from Lunar Eclipse Observations." To save the lunar eclipse
phenonomena, Ptolemy by and large adopted the entire Hipparchian model machinery
which, after three hundred years, still worked perfectly. Ptolemy stuck to an
apparent mess of really good and very faulty parameters handed down from his predecessors, but one cannot simply characterize his behaviour as fudging. He may well
have had good reasons for his attempt at preserving the continuity with earlier
astronomy.

Hartner's contribution will be published in full in the February 1977
Journal for the History of Astronomy (JHA), and an extended summary of Moesgaard's
paper appears in *JHA* 7, 216-17, 1976.

Shorter papers were given by F. Link on the role of ancient observations in

establishing the solar cycle, and by R. Movahed on recent excavations of the old
Maragha Observatory in Iran.

III. RETIRING PRESIDENTIAL ADDRESS

O. Gingerich reviewed some recent research on Copernicus and in particular
his interpretation of the Tycho documents he had discovered in the Vatican Library
in 1973. The most interesting of these manuscript pages from 1578 (illustrated in
the IAU *Highlights in Astronomy* 1973) shows a proto-Tychonic system in which the
Sun carries Mercury and Venus around the central Earth, but in which the superior
planets retain Ptolemaic epicycles. Tycho did not formulate his own system until
five years later, and his hesitation resulted from his belief in the crystalline
spheres; had he placed Mars in orbit about the Sun, it would have cut the orbit
of the Sun about the fixed Earth, yielding a mechanically unacceptable system.
By 1583 Tycho was willing to adopt this arrangement and abandon the crystal spheres.
N. Swerdlow had suggested on the basis of a page in the so-called "Uppsala
notebook" that Copernicus at one time considered a geocentric Tychonic system.
Gingerich argued that the parallel evolution of Tycho's own cosmology supports the
view that Copernicus was driven to a heliocentric system in order to retain
simultaneously the crystalline spheres and the unity of an orbital system surrounding the Sun.

IV. PRESERVATION OF TWENTIETH-CENTURY ASTRONOMY

This session, occupying the whole Monday afternoon, was devoted to the
question of what records of contemporary astronomy should be preserved for future
historians. In the opening paper. M. A. Hoskin showed by example how complete a
record of the "intellectual biography" of a problem in astronomy could be available
from as recently as the 1920s. The two traditional interests of the historian--
intellectual biography and the impact of instrumental changes--would always be of
central importance, but today other questions had gained in significance, and
sources must be preserved to enable the historian to investigate these questions
also. They included: the action of grant-giving agencies, the dynamics of particular observatory groups, the intellectual networks on an international scale, the
identification of indicators of scientific activity, and the development of instrumentation.

The records on which answers might be based included written documents and
"oral history." Hoskin briefly illustrated some of the dangers of oral history,
before directing his attention to problems of preserving written records. Documents occupy space and space is expensive; they are of use only if the historian can
locate them, and the preparation of catalogues is expensive; much modern paper is
made from highly-acidic wood pulp and will need costly lamination if it is to
survive many decades, together with preservation in stable conditions with relative
humidity low enough to prevent the multiplication of spores; and historians must be
given access to documents under (expensive) supervision. The preservation of
archives is therefore a costly business, and historians must bear this in mind
when advocating the nature, and the scale, of records to be kept.

Commenting, S. Weart described the experiences of the staff of the Center
for History and Philosophy of Physics of the American Institute of Physics in the
collection of oral history, and discussed some of the problems involved in preserving the history of very large (governmental) scientific organizations. S. M.
R. Ansari outlined the contrasting situation in India, where the tendency was
towards autonomous observatories; and O. Pedersen announced the deposit of the
Hertzsprung papers in the Institute for History of Science at Aarhus University.

Two senior astronomers then discussed problems that had arisen in their
own experience. B. Strömgren listed a number of possible research problems in
history of astronomy in the last half-century, and emphasized the special
importance of establishing the input from physics into a given astronomical
problem area. W. H. McCrea drew attention to the influence of referees' reports
and, more generally, to the impact of editorial decisions.

31 August 1976

V. MEGALITHIC ASTRONOMY: FACT OR SPECULATION?

The Tuesday morning session drew an audience so large that some were unable even to find standing room in the lecture hall. D. C. Heggie opened with a review paper on "good" and "bad" evidence, well illustrated with colored slides of megalithic sites in Britain. What is difficult to decide about almost all orientations is whether they were deliberately incorporated by the people who built the monuments, or whether they occur quite by chance; and, if they were deliberate, why was it done? Heggie's analysis opened with structural considerations. For example, some types of orientation are impractical. Tomb passages are one, and those involving the centers of stone circles are another, for the centers are not often marked. Turning to astronomical evidence, he pointed out that even when one knows of megalithic orientations to the astronomically significant points on the horizon, there remains the possibility that they occur by chance. This is what is meant by "bad" evidence: it is evidence quite consistent with the view that the megalith-builders did not deliberately incorporate astronomical orientations. For a long time, almost all of the evidence on megalithic astronmy was bad evidence, and thanks are due mainly to A. Thom for the fact that this situation has changed.

Even if one is persuaded that megalithic orientations were deliberately astronomical, it remains to be seen what purpose they served. The most popular theory for solar lines is a calendrical one. The difficulty with this explanation is the abundance of solstitial lines; for the very reason that solstitial lines are easy to set up, they are difficult to use for the determination of the time of the solstice. On the other hand, one would *guess* that solstitial lines would play a central role if the purpose of the orientations was religious. Lunar orientations are usually discussed in the context of eclipse predictions. Thom's method requires the use of orientations of comparatively high accuracy, in order that solar perturbations could have been detected. However, the absence of entirely satisfactory evidence for accurate orientations combine with astronomical and archaeological difficulties to make this theory unattractive. Even Hawkins's method (or Hoyle's modification of it) is unnecessarily involved. (To say that Stonehenge was a computer is to say in essence that it was used to facilitate the the counting.)

An extensive summary of Heggie's paper is found in *JHA* 7, 220–222, 1976.

In his commentary, O. Pedersen remarked that so little is known of Stone Age religion that to call the monuments religious rather than astronomical is to beg the question. He himself held on the present evidence that the builders of simple monuments had *some* astronomical ideas in mind, even if these were subordinate to (say) religious ones. Pedersen also warned against the transfer of credibility (as when D. G. Kendall's support for the megalithic yard had swung sentiment in favor of Thom's *astronomical* thesis, despite Kendall's own disclaimer).

J. Dobrzycki, in examining the difficulties of a hypothetical megalithic observer from the viewpoint of modern astronomical knowledge, cast doubt on the reasonability of highly accurate alignments, for example, for lunar extrema. In the discussion from the floor, F. Biraud, appealed for support for a campaign to protect megalithic remains in France from the destruction now going on. The session concluded with an account by J. Eddy of his studies of the solstitial and stellar alignments of Big Horn Medicine Wheel in Wyoming and newly investigated similar sites of the Plains Indians in Canada and the USA. Eddy's beautifully illustrated and fascinatingly recounted report were enthusiastically received by the large audience.

VI. RARE BOOKS IN ASTRONOMICAL LIBRARIES

In the final session, held jointly with Commission 5 (Documentation), a number of speakers outlined the history and holdings of the principal collections

of rare books now in observatory libraries. An introductory paper by M. Conti-
Grassi (and read by G. Feuillebois) described her census of 16th-century
books in observatory collections. Contributions were given concerning the Paris
Observatory (G. Feuillebois), the Crawford Collection at the Royal Observatory
in Edinburgh (M. Smyth and E. Forbes, read by M. Smyth), the Uppsala University
Observatory (N. Olander), and the Spanish Naval Observatory in San Fernando
(D. Almorza, read by J. Benavente). In addition, O. Gingerich reported on his
survey of Copernicus *De revolutionibus*; he has located 230 copies of both the
1543 Nuremburg edition and the 1566 Basel edition, and has personally examined
approximately 180 copies of each edition.

VII. CONSULTING MEMBERS OF COMMISSION 41
      This commission differs from most in the comparatively large number of
consulting members, almost all of whom are professional historians of science
working on the history of astronomy or instrumentation. Those currently
appointed are: A. Aaboe, J. A. Bennett, M. L. Righini Bonelli, B. R. Goldstein,
D. B. Herrmann, E. S. Kennedy, D. A. King, H.-G. Körber, P. Kunitzsch,
H. Labat, F. Maddison, Y. Maeyama, K. P. Moesgaard, J. Needham, N. I. Nevskaja,
D. Pingree, E. Poulle, D. J. de Solla Price, T. Przypkowski, B. A. Rosenfeld,
G. Rosinska, H. Sandblad, Z. Sokolovskaya, F. R. Stephenson, N. Swerdlow,
R. Taton, G. J. Toomer, B. E. Tumanjan, A. Van Helden, J. Vernet Ginés,
B. L. van der Waerden, D. W. Waters, S. Weart, R. S. Westman, D. T. Whiteside,
and C. Wilson.

COMMISSION 42: CLOSE BINARY STARS (ÉTOILES BINAIRES SERRÉES)

Report of the Meetings August 25, 28 and September 1, 1976

PRESIDENT: T. J. Herczeg   SECRETARY: D. B. Wood

Commission 42 convened for two business meetings and two scientific sessions. In the following, the minutes of the business meetings and a summary of the scientific meetings are presented, with the abstracts of some of the papers read.

Business Meeting I - August 25, 1976

The President called the meeting to order at 0910, and outlined the following schedule of events which was agreed to by the attendees: (1) Triannual history; (2) nomination and election of officers and new members; (3) future meetings; (4) coordinated programs, information systems; (5) scientific priorities.

1. Triannual History

Herczeg indicated that the Report of the Commission did exist, but that it was unavailable for the commission members to comment on. The work on X-ray binaries attracted particular interest during the last three years. Three major developments were Symposium No. 73 (Cambridge), the Greenbelt conference, and the X-ray-optical coordinated observing programs.

The group extended a vote of confidence and thanks to Dr. Larsson-Leander for his very important "Bibliography and Program Notes" and also to Dr. Szeidl for reporting much of the current work on eclipsing binaries in the IBVS of Commission 27.

2. Nomination and Election of Officers and New Members

The Organizing Committee (OC) has proposed for the 1976-79 triennium for President,
        G. Larsson-Leander;
for Vice President,
        B. Warner.
The OC further proposed the addition of Dr. Kondo and Dr. Whelan to the Committee while Herczeg reported that Dr. F. B. Wood intended to resign from his membership in the OC to make room for younger astronomers.

In addition, the National Committees have proposed Dr. Kruszewski (Poland) and Dr. Cherepashchuk (USSR); further, Herczeg proposed Dr. Sinvhal. It was found that there was good reason to add Cherepashchuk to represent research in the Soviet Union; since research in Poland is already represented by Smak, an addition from this country may not be necessary; Sinvhal would add to the representation of the rapidly growing astronomical work in India.

To Dr. Wood, the group expressed their gratitude for his service which dates to the beginning to Commission 42.

Since Plavec and Popper both are at UCLA, they each offered to resign to make for room on the OC. Popper prevailed in tendering his resignation, and the group expressed their thanks for his service.

Plavec expressed his concern for the lack of a "classical photometrist" on the OC. Herczeg pointed out that four members of the proposed OC do photometric work on a fairly regular basis.

The slate was then unanimously accepted by the Commission members.

Note that the final Membership of the OC is as follows:
        Batten,* Charepashchuk, Fracastoro, Gyldenkerne,
        Herczeg, Kitamura, Kondo, Plavec, Sinvhal, Smak,
        van den Heuvel, Whelan.

F. B. Wood expressed his concern that the Commission go on record that National Committee recommendations will always be taken under advisement by the Commission but that we are in no way bound to accept them.

The following new members to the Commission were proposed:

    By the OC: Anderson, Bath, Brownlee, Faulkner, Gursky, Geyer,
    Hazlehurst, Hilditch, Krezminski, Lucy, Scarfe, Seggewiss.

    By the respective National Committees: Breinhorst, Budding,
    Harmanec, Rovithis, Semeniuk, Schoeffel.

    By Commission Members: Chambliss, Rahe, van'tVeer, Ziólkowsky.

Smak expressed his view that it would be important to know more about the proposed new members in advance of the meeting. The following formal proposal, as a reminder, was approved unanimously:

    "For further additions to the Commission, it is desired that the
    OC be ready to support each proposed member".

Herczeg read a letter of resignation from Commission 42 from Dr. A. J. W. Cousins, as he ceased to be active in the field of eclipsing binaries.

The new Commission members were then all admitted as a group.

3. Future Meetings

Possible future meetings were discussed. IAU-Colloquium No. 42, to be held in Bamberg (September 1977), already approved by the Executive Committee, was announced. Specific attention has been paid to a proposal by Czech astronomers (conference on emission line objects) and by astronomers from New Zealand (conference on duplicity among intrinsic variable stars). Decision about possible co-sponsoring by Commission 42 was postponed until the 2nd business meeting. However, the following formal proposal by Plavec was approved unanimously:

    "Commission 42 proposes and is ready to sponsor a conference on
    close binaries, patterned after the highly successful Parkesville
    or Cambridge conferences, to be held in 1979 in conjunction with
    the IAU General Assembly".

---

*Dr. Batten's name was omitted inadvertently from the original list, and has been added here as approved at Session II.

4a. Coordinated Programs

X-ray and optical coordinated programs for Cyg X-1, Her X-1, 3U0900-40, and 3U1700-37 were discussed. Bolton commented on the problems of scheduling satellite time for the X-ray observations. Kondo suggested extending the list to include Cyg X-2 and other objects.

The first business meeting ended at 1035.

Business Meeting II - September 1, 1976

The President called the meeting to order at 0905. It was pointed out that those who wish to become Commission members should contact Larsson-Leander.

4a. Coordinated Programs (continued)

Kondo and Bolton reported on their recent efforts concerning cooperative X-ray and optical programs. The UK-5 satellite will be available for such programs, and efforts are being made to use SAS-3 and OSO-8.

Herczeg expressed interest in "revitalizing" the two pertinent subcommittees: Coordinated Program Subcommittee (Glydenkerne), and Extraterrestrial Observations (Kondo).

It was suggested to retain all four objects, Cyg X-1, Her X-1, 3U0900-40 and 3U1700-37, on the coordinated X-ray and optical program. If other stars are of interest, the Subcommittee on Extraterrestrial Observations should initiate the effort, through the Subcommittee on Coordinated Programs.

The Commission recommended coordinated optical programs on RX Cas (for two years, as proposed by Dr. Martynow acting as coordinator) and on VV Cep (proposed to the OC by Dr. A. Galatola).

Dr. Gibson reported on the coordinated radio and optical program on RS CVn variables. At the present, he is organizing a temporary working group, which will probably eventually be headed by D. Hill.

The above set of coordinated programs was unanimously approved.

There was some discussion concerning earlier programs for Y Cyg (O'Connell) and for AR Cas (Herczeg).

3. Future Meetings (continued)

The OC suggested postponing final position on the conferences put forward on emission line objects and duplicity, as mentioned in Session I under pt. 3. In the ensuing discussion the opinion prevailed that the Commission saw itself unable to co-sponsor these meetings at the present time.

It was proposed, and carried, to support (jointly sponsor) an upcoming colloquium in Victoria B.C. in the fall of 1977 on Of and OB stars, as proposed to the Executive Committee by Drs. Swings, Conti and others.

There was more discussion regarding a symposium on close binaries in conjunction with the next IAU General Assembly. (See Session I, pt. 3.) Cooperations should be sought from Commissions 29, 35 and 44. A likely site could be Toronto, Calgary or Montreal. Smak proposed that we formally propose very strong support of such a symposium. It was unanimously voted to do so, and a nucleus of an organizing committee was named, consisting of Plavec, Batten and Larsson-Leander.

There was some discussion regarding the inclusion of radio astronomers. Bolton noted that Commission 40 was planning a meeting on radio stars, including binaries.

4b. Information Systems

Herczeg introduced the topic of information sources, namely data repositories, catalogs, and published data. For binary star researchers, the basic catalogs are the Moscow General Catalog of Variable Stars, Batten's Catalog of Spectroscopic Binaries, and the Finding List for Observers of Eclipsing Binaries.

Batten reported on the status of the 7th edition of his catalog. He noted the problem of obtaining V magnitudes for variables stars. It is clear that observers should be encouraged to at least provide the V magnitude of their comparison stars. F. B. Wood reported on his card catalog of eclipsing binary publications, and asked if another edition of the Finding List would be useful. The Commission voted unanimously to support the publication of such a revised Finding List.

Data repositories are maintained by the RAS and by Odessa Observatory, but there is no catalog of what systems are in these repositories. Apparently little data is actually in them (less than 50 systems); observers should be encouraged to send their data. It was proposed that Commission 42 should contact the Editor of the IBVS to publish a list of systems in the repositories.

5. Priorities of Research

Basically, the priorities have not changed since the last General Assembly. Herczeg proposed that we support the same priorities, but with increased emphasis on X-ray and radio studies, and dropping reference to work with the intensity interferometer. The Commission members voted to accept that proposal.

The business meeting was adjourned at 1050.

Scientific Sessions

The main scientific session of Commission 42 was held on August 28 under the title "Current Trends in Binary Star Research" and continued on September 1 under the title "Short Scientific Meeting". The following summary covers both sessions. The intention was to give in these meetings a fair cross-section of the present work on close binaries, with the possible exception of the X-ray binaries (which formed the topic of a Joint Discussion with Commissions 44 and 48).

Seven review papers (invited) and 16 short reports were presented; for 13 of these contributions the abstracts are given below.

Among the invited papers were J. Smak, disk-structures and the outbursts of dwarf novae; M. Plavec dealt with mass transfer and Be stars; D. M. Popper reviewed binaries with H and K emission; A. H. Batten presented recent spectroscopic studies of circumstellar matter considering, in particular, the systems U Cep and RZ Oph. Anne P. Cowley reviewed the VV Cep-type systems, with emphasis on the recent eclipse of AZ Cas; she pointed out that all these systems exhibit high eccentricities which circumstance may play an important role in their particular evolution. (Two further papers, by K. O. Wright and R. Faraggiana, dealt with the system VV Cep itself, now approaching eclipse.) D. B. Wood reviewed the light-curve synthesis methods in calculating eclipsing binary orbits and Y. Kondo reported on his and G. McCluskey's studies of mass flow in early type close binaries, based on UV observations.

G. S. Mumford gave a short discussion of the period of U Gem and J. Ziólkowski presented his interpretation of the relationship between Algol-type systems and

X-ray binaries. J. Anderson commented on the accuracy of mass and radius estimates from binary systems; T. B. Horak reported on a five-colour photometry of HO Tel and RW CrA done in collaboration with C. J. van Houten and J. Grygar.

D. Y. Martynov gave a detailed description of the remarkable system RX Cas, proposed and accepted for a coordinated program. This 32-day eclipsing system consists of two evolved components (G3III + A5IIIe) with strongly variable light curve and many indications of a strange spectroscopic behaviour. In a short report, Anne P. Cowley informed the Commission that the X-ray source 3U1809+50 has quite recently been identified with the (irregular) variable AM Her which also was found to be a short period spectroscopic binary.

Abstracts of Papers

The following abstracts are arranged in a topical order: observational techniques, radio astronomy of binary stars, early-type systems and binaries with emission lines, special systems, numerical solutions. For reasons of limited space, in a few cases the abstracts have been slightly shortened here.

The Systematics of Visible-band Linear Polarization for Close Binaries (R. H. Koch, U. Pennsylvania).

If cool contact, flare, and degenerate systems are excluded from consideration, there now exist some 1500 filtered and almost 1000 unfiltered linear polarization measures distributed among 63 close binaries. With this quantity of data, it is possible to assess the incidence of intrinsic polarization as a function of binary evolution.

No single stellar characteristic--e.g. rotation, mass, mass ratio--is sufficient to predict if a binary will or will not be intrinsically polarized. It is convenient, however, to parameterize the polarization in terms of the separation between the photospheres of the component stars.

Among unevolved binaries, intrinsic polarization is rare, presumably indicating the lack of efficient scattering envelopes. At present, the only convincing exception to this generality is U Oph, which is known to be photometrically complicated. For evolved binaries, intrinsic polarization is widespread. It is uncommon only if the two stars are very close together in linear units and this is partly due to the finite precision of present measures. Typically because of stream, disk, shell, and envelope scattering--but not because of photospheric scattering--polarization is common among relatively wide pairs.

There exist about 1000 filtered measures for some 20 contact pairs between B5 and K1. Most of these data are not yet analyzed, but certain systems show variations of the orientation of the electric vector reminiscent of the calculations by Collins and Buerger.

Variable Polarization in U Cephei (V. Piirola, Helsinki).

Large increase in linear polarization of U Cep was observed during the primary minimum on 1975 Sept. 8 and confirmed during several minima from Sept. 1975 to April 1976. Typically, the polarization increased from 0.10 percent after first contact to 0.80-1.08 percent near second contact (beginning of totality), then decreased towards mid-eclipse close to 0.10 percent and increased again near third contact (end of totality) to 0.80-1.04 percent. The maximum value of polarization varied from cycle to cycle and decreased towards the spring 1976 to about 0.50 per cent. Sometimes the increase near second contact was small. Mean polarization outside eclipses was 0.17 $\pm$ 0.03 percent. Position angle was close to 95° except near second contact where it changed to about 60°.

The changes in the polarization of U Cep could be explained by circumstellar matter surrounding the primary. Polarimetric observations correlate with the spectroscopic observations and give further evidence and information about changes in mass transfer and amount of circumstellar matter in U Cep.

Microwave Emission from RS CVn Binaries (D. M. Gibson, Manchester).

Surveys of some 30 known RS CVn binaries at centimeter wavelengths with tele-scopes at the NRAO and Jodrell Bank have resulted in the detections of AR Lac, UX Ari, RT Lac, HR 1099, RZ Eri, and λ And, and the probable detection of PW Her. Subsequent monitoring of these systems at the NRAO, Jodrell Bank, Cambridge, and Algonquin Radio Observatory have shown them to be variable on a timescale of $\sim$1 day with luminosities as large as 6×10^{17} ergs s^{-1} Hz^{-1} at 8085 MHz. The observed behaviour is consistent with recurring outbursts where large outbursts are expo-nentially less probable than small ones. Statistical considerations make it proba-ble that the RS CVn systems as a class are the first optically-selected class of stellar microwave radio-emitters.

Analyses of ten large ($L_{8085 MHz} > 6\times10^{16}$ ergs s^{-1} Hz^{-1}) outbursts from AR Lac, UX Ari, RT Lac, and HR 1099 permit the unique determination of the source parameters during extremes of the observed behaviour. The evolution of the source spectrum during an outburst from optically thick to optically thin and the presence of circular polarization clearly indicate the radiation mechanism is gyrosynchro-tron emission from a power law distribution ($\Gamma \sim 2$) of mildly relativistic elec-trons in magnetic fields $B_{\perp} \sim 30$ Gauss. These properties allow independent calcu-lations of the brightness temperature $T_B \sim 10^{10}$K, and, thus, the source size $R \sim 10^{11}$ cm. The total energy in fields and particles during such an outburst can be as large as 10^{36} ergs.

Radio Emission from AG Pegasi (P. C. Gregory and S. Kwok, U. British Columbia, E. R. Seaquist, David Dunlap Obs.)

Radio emission from the symbiotic nova AG Pegasi has been detected at 2.8 and 3.7 cm and an upper limit obtained at 11 cm. The observations are interpreted as free-free emission from an ionized nebula formed by continuous mass ejection from the WN6 star. Our observations indicate a mass loss rate of 10^{-6} M$_{\odot}$/yr and a total mass for the nebula of 7×10^{-5} M$_{\odot}$.

Recent Ultraviolet Observations of the Mass Flow in Close Binaries (Y. Kondo, NASA Johnson Space Center, G. E. McCluskey, Lehigh U.)

The ultraviolet spectrum of the eclipsing binary UW CMa (07f + 0-B) has been observed with the Copernicus Princeton University Telescope Spectrometer in the wavelength region 950-1560 Å at a resolution corresponding to 0.2 Å. These obser-vations were obtained near phases 0.25 and 0.75 to investigate the presence or the absence of the orbital Doppler shifts in the spectral features. A number of stellar and interstellar lines appear in the spectrum. The following lines showing P-Cygni characteristics have been observed: C III (977, 1175 Å), S IV (1062, 1072 Å), P V (1117 Å), Fe III + Si IV? (1122 Å), P V + Si IV? (1128 Å), N V (1238, 1424 Å), Si IV (1393, 1402 Å), and C IV (1549 Å). The centers of the absorption components of the P-Cygni lines yield radial velocities of from -200 to -800 km/s while the peaks of the emission components are shifted by +400 to +800 km/s. These veloci-ties are significantly larger than the projected orbital velocity of about 200 km/s

and indicate that gas motions in the system are occurring. A mass-loss rate of
about 3×10^{-6} solar mass per year is estimated. A few photospheric absorption
lines from the O7f component are also present. Analysis of the data shows that
the high temperature gas giving rise to the P-Cygni features in the far ultraviolet
spectrum is located primarily around the entire binary system and does not share in
the motion of either component. The effects of radiation pressure on the Roche
(Jacobian) equipotential surfaces and in generating a stellar wind are discussed
and the spectral findings are compared with ultraviolet observations of β Lyrae.

Massive Contact Systems (Kam-Ching Leung, U. Nebraska).

    In recent years more realistic models employing the Roche model have been
developed. As a result of these new approaches many systems which we believed to
be in contact have been proven to be either semi-detached or detached systems. At
the same time, many true contact systems were found. Those systems identified were
almost entirely among low mass stars--W UMa type systems.

    We have chosen twelve promising candidates in our initial study of massive
contact binaries. All the systems in this study were analyzed with the Wilson and
Devinney programs. To start with, we always assumed a system that was detached and
let the photometric solution evolve its final least square solution. It was found
that nine out of twelve systems selected were contact binaries. In the cases
where radial velocity curves or other information are available the absolute dimensions of the systems are calculated. It was discovered that there were two groups
of contact binaries: zero-age contact and evolved contact. The contact systems
V701 Sco and BH Cen are members of very young clusters. They are found to be zero-
age by the evidences of their ages and radii. The formation of these systems must
be the consequence of star fission under critical angular momentum: The angular
momentum is not large enough to allow the system to become detached, and not small
enough to result in a single star. The systems 29 CMa, AO Cas, V729 Cyg, V1010
Cyg, and V1073 Cyg were found to be, and AU Pup and V535 Ara were suspected to be,
evolved contact binaries.

    The other three systems investigated, BF Aur, μ' Sco, and V Pup, proved to be
semi-detached systems only.

    The three contact systems reported by others, SV Cen (Wilson and Starr),
V382 Cyg and RZ Pys (Devinney) were found to be evolved contacts.

A Search for Variables Amongst Early Type Spectroscopic Binaries and Be Stars
(R. W. Hilditch, St. Andrews, and G. Hill, Dominion Aph. Obs.)

    A search was carried out for variables amongst bright, early type spectrosco-
pic binaries, Be stars and mass-losing OB supergiants. From the first three years
(1971-73) of this continuing survey, a total of 1808 observations of 42 programme
stars have been obtained on the DAO photometric system. Eleven new variables have
been discovered including, notably, the O-type spectroscopic binaries 14 Cephei
and DH Cephei (NGC 7380-2) and the enigmatic objects HD 187399 and HD 190467.
These results will appear in the DAO publications.

Close Binaries with H and K Emission (Daniel M. Popper, UCLA).

    Properties of the components of detached main-sequence and subgiant close bi-
naries with H and K emission outside of eclipse are compared with those of detached binaries of similar mass without the emission. Distributions in the mass-
color, mass-radius, and HR diagrams lead to the hypothesis that the emission

binaries represent a later stage of evolution, towards which the non-emission systems are evolving. The instabilities in periods and in the light curves, characteristic of the emission systems, appear to develop as the more massive component reaches the end of core hydrogen burning and obtains a convective envelope. While it is not necessary to invoke mass exchange in some of the detached emission systems, in others a mild amount of mass exchange, perhaps through an enhanced stellar wind, could lead to their present properties. In some cases considerations of the time scale for evolution may require mass loss from the system.

The Structure of VV Cephei System (K. O. Wright, Dominion Aph. Obs.)

Since VV Cephei will go into eclipse in November 1976, the Victoria observations of the red region of the spectrum obtained between 1956 and 1976 have been studied in order to determine the parameters of the system. Radial velocities of the M-type star have been obtained from fifteen well-defined lines in the region 6322 to 6663 A measured on 123 plates. A period of $7430\overset{d}{.}5$ was adopted and the data fitted very well a velocity curve with elements e = 0.345, ω = 59°, V_0 = -20.2 km s^{-1}, K_1 = 19.4 km s^{-1}, T = J.D. 2,438,461.0. The date of mid eclipse computed from these data agrees with the 1957 observed value within twenty days. The orbit of the secondary B-type component of the system was derived from measures of the emission Hα line that was assumed to be formed by an envelope surrounding the secondary star. After subtracting the intensity profile of α Orionis (which represents the M-type component of VV Cephei adequately for this purpose) from the observed intensity profile of the VV Cephei spectrum, an emission profile with symmetrical wings and half-width $\sim$5 Å was obtained; the centre of this line was assumed to represent the velocity of the secondary star. Values of K_2 = 19.1 km s^{-1} and V_0 = -18.5 km s^{-1} were found to represent these observations when the other elements found for the primary star were adopted. The difference in V_0 may be explained by the necessity to measure M-type lines close to Hα rather than over the extended range used for the definitive orbit. The ratio of the semi-amplitudes, K_1 and K_2 combined with an inclination of 77° adopted by Hutchings and Wright gives masses of 19.7 and 20.0 $M_\odot$ for the M and B star respectively.

The additional absorption features of the Hα profile have been measured relative to the emission line; i.e. they are assumed to be produced by matter between the B star and the observer. The principal absorptions can be explained in terms of a mass of gas flowing from the M star, presumably through the Lagrange point and around the B star. The presence of two strong absorptions differing in velocity by 50 km s^{-1} at secondary eclipse makes this model plausible. An additional emission line with velocity $\sim$-60 km s^{-1} relative to the B-type star appears at most phases after secondary eclipse; it may be produced by interaction between the gas flowing around the secondary star and the stream coming from the primary star.

The chromospheric spectrum, produced by absorption of the outer envelope of the M star as the B star goes behind the primary, was observed in mid 1975, a year and a half before first contact of the eclipse, which indicates a very extended atmosphere for the M star. This spectrum was well developed by August 1976, particularly for low excitation lines of Ca I, Ca II, Fe I, Ti I, Ti II, Mn I, Ni I, etc.

On the Present Eclipse of VV Cephei (R. Faraggiana, Trieste).

High dispersion spectra of VV Cep have been taken at the Haute Provence Observatory since 1967. A region of the spectrum very sensitive to atmospheric eclipses is around the Ca I resonance line λ 4227. Comparing the spectra taken at different epochs, we noticed that sharp cores are becoming visible first for the lines of ions; among them the strong line of Sc II was appearing first. Next the lines due to neutral elements strengthened and in July 1976 strong Fe I and Cr I chromospheric lines became visible. Among the emission lines, starting from December 1974, Fe II 4233 is becoming steadily weaker in comparison with λ 4243; this diminution of intensity may be explained if the Fe II lines are formed in the vicinity of the companion.

An Eccentric Close Binary Model for the X Persei System (H. F. Henrichs and E. P. J. van den Heuvel, Amsterdam).

A model for the X-ray source 3U0352+30 connected with the 6th mag 0 9.5V pe star X Per is proposed. It is suggested that a $\sim$ 1.5 $M_\odot$ neutron star pulsating with a $13^m.9$ period is moving in $22^h.4$ around a relatively normal 0 9.5V star (M $\sim$ 20 $M_\odot$) in an inclined, slightly eccentric orbit (a $\simeq$ 11.2 $R_\odot$; i $\lesssim 53^o$; e $\simeq$ 0.1) giving rise to a $22^h.4$ modulation in the X-ray flux, produced by capture of stellar wind material from the main star, which slightly underfills its critical lobe at periastron. The apsidal motion of the elliptic orbit may explain the 581^d period observed in the wavelength shifts of the higher Balmer absorption lines. A full paper will be published in Astronomy and Astrophysics.

Physically Accurate Models of Eclipsing Binary Stars (D. B. Wood, NASA Goddard Space Flight Center).

In the 1960's and early 1970's, the synthesis approach grew rapidly, with many workers entering the arena with more and more models; such people as Biermann and Thomas, Cochran, Doughty and Mochnacki, Hill and Hutchings, Lucy, Nagy, Rucinski, Whalen and Moss, Wilson and Devinney, Van Landingham, and others.

At this point, let me stress the importance of using a filter-defined photometric system. In fact, all these models are basically monochromatic, so it is important that observations are made in a narrow or intermediate band system such as the Strömgren uvby.

Most of these synthetic models are based on the Roche model, and many are designed specifically to deal with the contact and over-contact binaries. The models are quite demanding of computer resources. The wood model is designed with practical application in mind. Certain compromises in the physics are made, and approximations introduced, to make the model useful on moderately sized computing facilities and budgets. The basic unique feature of this model which permits it to be run at least an order of magnitude faster than other synthesis models is the use of a triaxial ellipsoid star shape. This permits the use of many closed analytical expressions, and permits the integrations over the apparent disks to be done with Gauss-type numerical quadrature.

All of the models designed so far are geometrically symmetric (E.g. ellipsoids or Roche surfaces) and thermally relaxed, so that they are also photometrically symmetric. The only asymmetry which exists is that due to orbital eccentricity. Obviously many observed light curves are not symmetric, as evidenced by the sine terms in the old rectification process. We are now ready for the next stage of complication in our synthetic models--that of magnetic and hydrodynamic forces. The physical models need to be developed to quantitatively account for the observed "perturbations" of the present generation of synthesis models.

We are now in the "Copernical Revolution" in eclipsing binary research where we no longer need to use more and more epicycles (i.e. rectification) to describe the relationship between the stars themselves and what we observe from our distant vantage point.

Information Resolution in the Context of Close Binary Photometry (E. Budding and H. Al-Naimiy, Manchester).

Photometric observations of certain eclipsing binary stars have been investigated using recently developed frequency domain techniques. The combination of both minima in the analysis results in a distinct methodological improvement over the single-minimum method discussed hitherto. This improvement has two aspects: (1) increased accuracy of the determined elements, (2) agreement of the results of the two-minimum method with the single-minimum method provides a criterion whereby the self-consistency of the underlying model with its representation of the light curve between minima by a cosine series may be assessed. Such self-consistent solutions may be further improved by the inclusion of "photometric perturbations".

Additionally, an investigation of the infrared light curve of Algol (β Per) was also reported upon.

COMMISSION 44: ASTRONOMICAL OBSERVATIONS FROM OUTSIDE THE TERRESTRIAL ATMOSPHERE (OBSERVATIONS ASTRONOMIQUES AU-DEHORS DE L'ATMOSPHERE TERRESTRE)

Report of Meetings, 25, 27, 28 and 31 August 1976

PRESIDENT: A. D. Code   SECRETARY: R. J. Davis

25 August 1976

BUSINESS MEETING

The President opened the meeting by reviewing the report of the Commission and initiated discussion of the role of Commission 44. R. Bonnet presented a prepared report outlining the important contributions that the Commission, as a multidisciplinary forum, can make, particularly in the climate of ever increasing international cooperation in space research. The President reported that Commission 44 was a co-sponsor of the Joint Discussion 2 (X-Ray Binaries and Compact Objects), Joint Discussion 5 (Stellar Atmospheres as Indicator and Factor of Stellar Evolution), and Joint Discussion 7 (Impact of Ultraviolet Observations on Spectral Classification). Business items requiring action on the part of the Commission membership were reviewed and action deferred until the business meeting scheduled for 28 August in order to permit adequate time for discussion and study of items.

SCIENTIFIC SESSION

E. Jenkins organized a session on the hot interstellar gas component in which reviews of the observational data on the soft X-ray background and of the ultraviolet O IV lines were presented followed by discussions of the theoretical interpretations. These presentations were co-sponsored jointly with Commission 34 on Interstellar Matter. A complete account of these sessions is contained in the Commission 34 report.

27 August 1976

SCIENTIFIC SESSION

This meeting on Stellar Mass Loss, Coronae and Winds was organized and chaired by C. de Jager. The invited review papers are as follows:

Observed Mass Loss of Individual Stars

H. Lamers - Survey of observations of mass loss from ultraviolet observations.
J. Hutchings - Survey of ground based observations providing evidence of
                mass loss.
Y. Kondo - Observations of MgII resonance lines.

UV and X-ray Observations Related to Stellar Chromospheres and Coronae

A. Dupree - Review of UV experiments for late type stars.
C. Jordan - Comparison of Procyon emission measures with the Sun.
R. Mewe - X-rays from stellar coronae.

Theoretical Aspects of Stellar Coronae, Winds and Mass Loss

T. Hearn - Review of theories on formation of stellar coronae and of mass loss.
J. Cassinelli and L. Hartmann - Predictions of Infrared flux from hot
                coronae of early type stars.

28 August 1976

BUSINESS MEETING
    I.   The Commission members approved the list of proposed members of the
Organizing Committee for submission to the IAU Executive Committee.

    II.   Nominations for Commission 44 Vice President were proposed and R. J. Van
Duinen was elected by written ballot. His name was therefore submitted to the IAU
Executive Committee.

    III.   Proposals for new members of Commission 44 having been previously
solicited were approved and it was agreed to submit that list to the IAU General
Secretary.

    IV.   The working group appointed at the first business meeting to consider
the question of dissemination of information on space flight opportunities and
space research activities presented their report. It was agreed that a
subcommittee be appointed to publish a Commission 44 newsletter. Y. Kondo was
appointed Chairman of the subcommittee on Commission 44 newsletter and
J. D. Rosendhal, NASA Headquarters, K. Henize, NASA Johnson Space Flight Center
and E. Peytremann, ESA, volunteered to serve on the subcommittee. It was hoped
that the chairman could recruit additional representatives to broaden the base of
input.

    Two to three newsletters per year are envisioned. Initially the newsletter
would be distributed directly to a reasonably small mailing list that would include
all members of Commission 44 and Presidents of all IAU Commissions. It was also
suggested that the subcommittee approach a journal such as Nature or Science in
regard to publishing concise summaries of the newsletter. It was agreed that the
newsletter would perform an extremely useful function not otherwise provided, as
follows:

    a.   It would include only space research opportunities of direct
          interest to astronomers.

    b.   It would have a small mailing list not completely duplicated within
          the large general mailing lists maintained by the various
          national space agencies.

    c.   It would summarize advanced planning, pre-launch announcements of
          opportunity, and post-launch status announcements of opportunity
          and give names and addresses to be contacted for further information

    d.   It would announce pertinent meetings and symposia.

    V.   The President reported on discussions with Commission 38 (Exchange of
Astronomers) on the proposal that IAU funds might be made available to young
astronomers wishing to pursue space research when funds for the necessary travel
were not available from any agency. The need for such grants has been considered
by Commission 38.

    VI.   The Commission voted to co-sponsor the IAU Symposium on O stars planned
for 12-16 September 1977 in Vancouver B.C., Canada.

    VII.   The President summarized the Commission 44 special discussion on Absolute
Spectrophotometric Calibration of Stellar Fluxes in the Vacuum Ultraviolet, held on
27 August 1976. The Commission members requested that a full report of the
discussion appear in this report. A paper prepared by R. C. Bohlin, G. J. Strongylis
and F. Beeckmans is found as Appendix A of this report.

SCIENTIFIC SESSION
     A comprehensive review of NASA flight programs in astronomy was presented by
J. D. Rosendhal. The following is a summary of this presentation.

NASA FUTURE FLIGHT PROGRAMS
     1.   Dr. Rosendhal began his presentation with a summary of the status of
approved programs:

          a.   IUE (International Ultraviolet Explorer)

          Descriptive material on this high dispersion UV telescope is readily
available from NASA. Space craft integration has been completed with all flight
subsystems and components with the exception of the Inertial Reference Assembly
which will be delivered in December 1976. Flight scientific instrumentation is
complete except for the spectrograph camera system and the flight fine error
sensors. Launch is presently scheduled for the 4th quarter of calendar year 1977.
The pacing item is delivery of the spectrograph camera system. Rosendhal presented
a summary of approved guest investigator programs: NASA: 47 U.S. programs, ESA: 53
programs, SRC: 45 programs. It is estimated that obtaining observations for the
proposals accepted by NASA will require the first 9 months of NASA's share of the
observing time. An additional Announcement of Opportunities will be issued by
NASA at about the time of launch, for observations beyond the initial 9-month
period.

          b.   HEAO (High Energy Astronomical Observatory)

          HEAO-A is on schedule for launch in April 1977, spacecraft and experiments
have been mated and systems tests are underway. B is on schedule for launch in
June 1978; experiment fabrication is now essentially complete for launch. C is
planned for 1979. The world's largest facility for calibration of X-ray
telescopes is now in operation at Marshall Space Flight Center. Discussions are
currently underway regarding the nature and details of Guest Investigator programs
for the HEAO missions. Because it is a pointed instrument, HEAO-B may lend itself
more readily to use by Guest Investigators than the other missions which are
scanning satellites.

          c.   Solar Maximum Missions

          This mission is the sole approved astrophysics new start in the Fy 1977
NASA budget. Wavelength coverage of instruments ranges between gamma rays and
visible wavelengths. Emphasis will be on high spatial and temporal resolution and
correlated observations. An Announcement of Opportunity for guest investigators
will take place during the summer of 1977. Discussions regarding arrangements for
coordination of observations with other satellites (especially ISEE-C) and exchange
of data between PIs are now underway.

     2.   Advanced Planning - Explorers

          a.   IRAS (Infrared Sky Survey): The phase B study was completed in
May 1976. This mission is a strong candidate for an Fy 1977 start. It will be a
cooperative mission with the Dutch and British.

          b.   High Energy Astrophysical Transient (X and γ-ray) Explorer: For
observing transient X-ray sources and for locating the sources of γ-ray bursts.

          c.   Soft X-ray and extreme UV explorer (6 - 950 Å Survey). Both the
British and Germans have indicated an interest in cooperating in this effort. An
additional study of a scout-class scanning EUV sky survey mission has also been
initiated.

     d.   UV photometric-polarimetric explorer.

     e.   Cosmic background explorer

Funds are insufficient for implementing all of these missions. At the completion of the mission definition studies it is planned to have an inter-disciplinary panel review the results and assign priorities to various missions.

     3.   Space Telescope (ST)

Current characteristics and performance specifications are:

Aperture: 2.4 meters   limiting m_v ~ 27th mag.
System f/number: f/24   0.1 arcsec angular resolution
Weight: 9318 kg λ response: 1200 A thru 1 mm

Provision is made for up to 5 scientific instruments and detectors. The five candidate scientific instruments for the first payload are: f/24 field camera (SEC orthicon), Faint Object Spectrograph (photon counting detector), IR photometer (bolometer), Faint Object Camera (instrument and associated photon counting detector may be supplied by ESA), Astrometer (part of fine guidance system). The actual instrument complement will be selected by evaluation of proposals received in response to an AO. Assuming Fy 1978 new start approval, and Announcement of Opportunity for Focal plane instruments is scheduled for 1st quarter, calendar year 1977 and launch in 4th quarter 1983. No new funding was approved for Fy 1977. Studies are being carried on with remaining Fy 1976 and transition period money. Particular emphasis has been placed on detector development. The Fy 1977 Congressional Appropriations Bill provided that, if the President's Fy 1978 budget contains the ST as a new start, NASA will be permitted to issue an early Request for Proposals for the Optical Telescope Assembly (OTA).

     4.   Space Shuttle and Support System Module (SSM) and Spacelab.

Orbital flight tests of the shuttle will begin in 1979. The first flight of the spacelab manned module provided by ESA (Spacelab 1) will take place in the 3rd quarter of 1980 and the first flight of the pallet-only version of spacelab (Spacelab 2) is currently planned for the 4th quarter. There will be some very limited opportunities for scientific experiments on the Orbital Flight Tests. Egnineering verification of the Spacelab is the primary objective of Spacelabs 1 & 2 but it is anticipated that there will be opportunities for a substantial scientific program as well. Emphasis will be on atmospheric physics on Spacelab 1 and astrophysics on Spacelab 2. Responsibility for Spacelab 3 (1st quarter of 1981) has been assigned to the office of appplications. Proposals for Spacelab 1 have been received and evaluated and a preliminary payload is now being selected. AO's for the Orbital Flight Tests and for Spacelab 2 will be issued in September 1976. According to current plans there will be a total of 18 flights of the shuttle by mid-1982 and 50 by the end of 1987.

Funds for Spacelabs 1, 2 and 3 are included in the approved Fy 1977 budget. An augmentation has been requested in the Fy 1978 as a first step in building towards a possible level-off-effort program. It is anticipated that discipline oriented AO's will be issued within the next year followed by yearly announcements of new opportunities. Early emphasis will be on interdisciplinary payloads and on smaller PI class instruments. According to current plans emphasis will eventually shift to discipline dedicated flights and the use of larger facility class instruments. Relative balance between PI and facility instruments will depend upon the discipline involved. Facility instruments which have been studied include a meter-class general purpose UV telescope, a cryogenically cooled infrared telescope, a meter-class UV/optical solar telescope, a hard X-ray imaging solar telescope, and solar EUV, XUV and soft X-ray telescopes.

The session on scientific results from recent spacecraft was chaired by R. Bonnet. The presentations were as follows:

P. Wesselius - Ultraviolet Photometry Experiment, Astronomical
              Netherlands Satellite.
M. Marov - Venus probes, Venera 9 and 10.
P. Cruvellier - Ultraviolet observations from D2B Aura Satellite
J. P. Delaboudiniere - Solar Spectroheliograms from D2B.
G. Brueckner - High spectral and spatial resolution solar rocket
               measurements.

31 August 1976

SCIENTIFIC SESSIONS
    Reports on solar observations from Sky Lab and from OSO-8 were presented in an all day session held jointly with Commission 10 (Solar Activity) and Commission 12 (Radiation and Structure of the Solar Atmosphere). The Sky Lab presentations were organized by R. MacQueen and the OSO-8 presentations by R. Bonnet. Details of the program are to be found in the report of Commission 10.

Appendix A

A COMPARISON OF ABSOLUTE FLUX MEASUREMENTS OF STARS IN THE ULTRAVIOLET

R. C. BOHLIN[*], G. J. STRONGYLIS[*], AND F. BEECKMANS[†]
[*]GODDARD SPACE FLIGHT CENTER   [†]INSTITUTE d'ASTROPHYSIQUE

Measurements of the absolute spectral-energy distribution of stars have a long history, with most ground based efforts concentrated on the star Vega (see review by Oke and Schild 1970 and Hayes and Latham 1975). The accuracy of the Vega calibration is about 5 percent from 3300 to 10800 Å. In the rocket ultraviolet, the precision is worse, but the maximum difference between several modern measurements is now only 35 percent. The problem of defining a network of known standard stars is best broken into two parts. First, what are the relative fluxes between a single standard star and a larger set of stellar spectra, as measured by a photometric spectrometer? Second, what is the absolute flux of a single standard star? The ultraviolet standard chosen is η UMa B3V, because it is the best measured star with substantial flux near and shortward of Lα.

To investigate the question of what sets of available data are from photometric instruments, individual scans are compared in Fig. 1 with scans of the same star obtained by OAO-2 (Code and Meade 1976). To compute the flux ratios shown, the data of higher resolution were averaged over the bandpass of the lower resolution instrument. The values labeled Bohlin are revisions downward by about 10 percent of the Bohlin, et al. (1974) calibration. The revisions (Bohlin and Strongylis 1976) are necessary, because better O_2 cross sections and air extinction coefficients were discovered in the literature. The extinction of air over a 73m path in the laboratory was needed to obtain the absolute calibration of the flight detector from a standard NBS tungsten lamp. The three stars from TD1-S2/68 appear in Humphries, et al. (1976) and the ANS data is from Wu (1975). The independent measurement of the flux from γ Ori by Hessberg, et al. (1975) are in disagreement with the data discussed here with ratios to OAO-2 from 0.05 at 1250 Å to 1.43 at 2200 Å. A similar problem exists with the fluxes of Evans (1972).

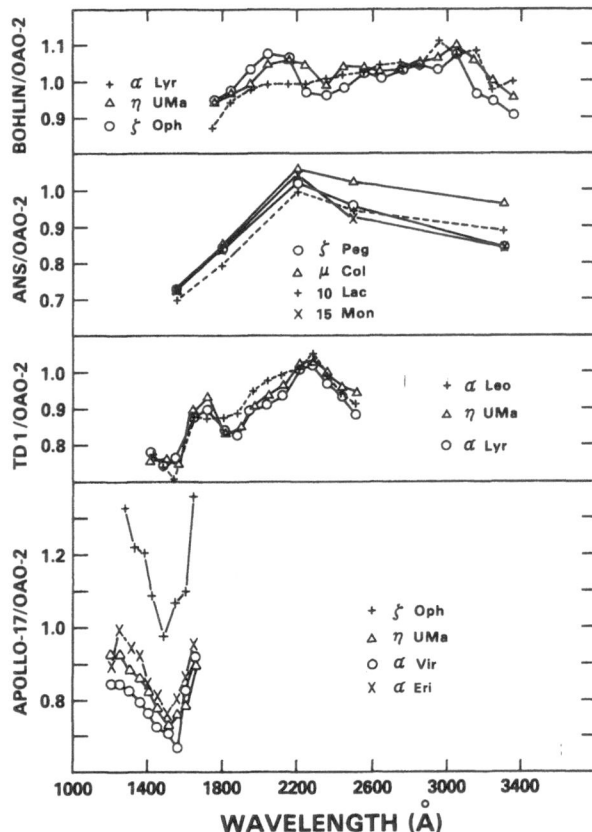

Fig. 1.   The ratios of absolute fluxes of four independent measurements to the final
fluxes derived from the OAO-2 satellite.

        The deviation of the points in Fig. 1 from unity represents the difference
between the absolute fluxes of OAO-2 and the other data. The typical spread in
the ratios of ±3 percent for each experiment is a measure of the reproducibility
of the data relative to OAO-2. The Apollo 17 spectrum of ζ Oph (Henry, et al.
1975) has an uncertain background correction and should not be considered (Henry,
private communication). The ratios for the other three Apollo 17 stars shown and
a fourth star α Gru have a mean scatter of about ±5 percent. Averages of the mean
TD1/OAO-2 ratios for 25 stars are given in Table 1 and have a 1σ mean error of ±1
percent. Consistent ratios from star to star do not prove that an instrument is
photometric, but any errors, therefore, must be common to all five experiments.

        Assuming that all five sets of data are actually photometric and that Fig. 1
and Table 1 represent the differences between their calibrations, the determination
of the absolute flux from one star will permit corrections of all five sets of
photometry to a common absolute scale. Various measurements of the flux of the
primary standard η UMa are shown in Fig. 2. The results of Stecher (1968) assume
that the response of sodium salicylate is flat and are shown as averaged values
over 100 Å bandpasses. Code and Meade (1976) have normalized the long wavelength
OAO-2 data to the ground based fluxes of the Hayes and Latham (1975) scale. The
maximum difference among the data shown in Fig. 2 is 35 percent near 1500 Å.
Longward of 1700 Å, most values agree to about ±5 percent.

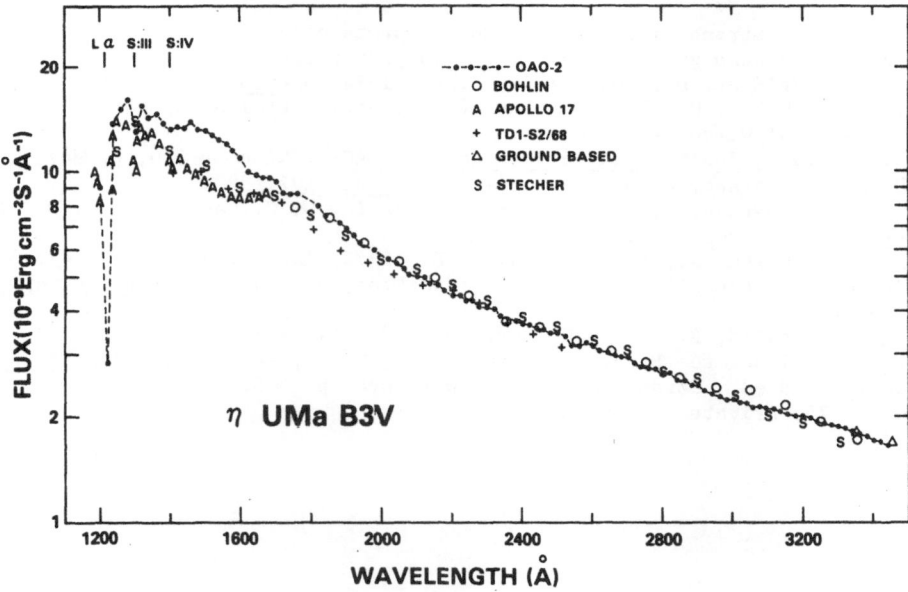

Fig. 2. The absolute flux of η UMa from various experiments.

More measurements of absolute flux values are needed, particularly in the 912 to 1700 Å range. For the present, the best estimate of the UV flux of η UMa might be obtained by using a model stellar atmosphere as an interpolating device to fit the points shown in Fig. 2 (Bohlin and Strongylis 1977). For effective temperatures near 1700°K, neither line-blanketing nor non-LTE effects in the models are a serious problem.

Table 1

Mean Ratio of TD1/OAO-2 for 25 Stars

λ (Å)	Ratio	λ(Å)	Ratio
1400	.789	2050	.978
1450	.797	2100	.984
1500	.753	2150	1.005
1550	.734	2200	1.011
1600	.794	2250	1.015
1650	.907	2300	1.008
1700	.920	2350	.975
1800	.859	2400	.965
1850	.868	2450	.951
1900	.871	2500	.934
1950	.906	2540	.949
2000	.948		

References

Bohlin, R. C. and Strongylis, G. J. 1976, in preparation.
Bohlin, R. C. and Strongylis, G. J. 1977, in preparation.
Bohlin, R. C., Frimout, D., and Lillie, C. F. 1974, Astron. Astrophys. 30, 127.
Code, A. D. and Meade, M. 1976, "Atlas of Ultraviolet Stellar Spectra",
     University of Wisconsin preprint.
Evans, D. C. 1972, The Scientific Results from OAO-2, NASA SP-310, p. 347.
Hayes, D. S. and Latham, D. W. 1975, Astrophys. J. 197, 593.
Henry, R. C., Weinstein, A., Feldman, P. D., Fastie, W. G., and Moos, H. W. 1975,
     Astrophys. J. 201, 613.
Hessberg, H. Niekerke, J., and Stephan, K. H. 1975, Astron. Astrophys. 42, 395.
Humphries, C. M., Jamer, C., Malaise, D., and Wroe, H. 1976, Astron. Astrophys.
     49, 389.
Oke, J. B. and Schild, R. E. 1970, Astrophys. J. 161, 1015.
Stecher, T. P. 1968, see Hess, W. N. and Meade, G. D., An Introduction to Space
     Science, 2nd ed., Gordon and Breach, New York, p. 825.
Wu, C. C. 1975, private communication.

COMMISSION 45: SPECTRAL CLASSIFICATION AND MULTIBAND COLOUR INDICES
(CLASSIFICATIONS SPECTRALES ET INDICES DE COULEUR A PLUSIEURS BANDES)

Report of Meetings, 26, 27, 28 and 31 August 1976

PRESIDENT: C. Jaschek.   SECRETARY: N Houk.

26 August 1976

I. SCIENTIFIC MEETING
    P.C. Keenan (U.S.A.) discussed his revised spectral classification for the
cooler stars, based on the original MK system with the addition of abundance indices
for stars which differ from the sun in composition. The system is defined and illustrated in a forthcoming photographic atlas.

    G. Cayrel, R. Foy and F. Spite (France) all briefly spoke on the spectral
classification of metal deficient stars. There seems to be a tendency to classify
such stars too early and to underestimate their luminosity. Also, because of lack
of standards, the classifications vary between observatories; it is suggested that
giving a general indication of the abnormality is sufficient.

    Garrison (Canada) reported that an MK – UBV study (with W.A. Hiltner) of OB
stars south of $\delta = -20°$ and brighter than B = 10 has been completed and is in preparation for publication. A second program of classifying very homogeneously all
stars in both hemispheres brighter than V = 4^m75 is under way, using the new forthcoming Atlas of Morgan and Keenan.

    E.E. Mendoza V (Mexico) reported that Hα and OI ($\lambda7774$Å) photometry of over
100 main-sequence A stars shows that this photometric system neatly separates the
metallic-line (Am) stars from normal A stars.

    J.R. Mould (United Kingdom) discussed abundance effects on the classification
of four M dwarfs. The strength of the $\gamma(0,0)$ band of TiO at a given temperature is
a coarse abundance indication. The CaH bands provide a clear-cut separation of
giants and dwarfs but abundance also has an effect.

    A.M. Hubert-Delplace (France) reported that she and H. Hubert will publish a
photographic spectral atlas of 40 northern hemisphere Be stars (77Å/mm) showing
typical spectral variations. The stars were selected from about 200 emission-line
stars observed in a program by Mrs. Herman et al. since 1955.

    Wing (U.S.A.) reported the identification of the feature observed near 9900Å
in M dwarfs and S stars as FeH, as suggested by L. Nordh, although stellar observations of higher resolution are needed for confirmation.

    Th. Schmidt-Kaler (Federal Republic of Germany) discussed a pilot project on
classification of stars of type F2 and earlier from Hamburg Schmidt objective-prism
plates (600Å/mm on Hα). The measurements, reductions, and classification procedure
are all carried out on a mini-computer, with 2000-3000 stars per plate being analyzed. Under best conditions it is possible to get two-dimensional types with an
accuracy approaching that of MK classification.

    H. Richer (Canada) reported on work with Olander and Westerlund on carbon stars
in the Large Magellanic Cloud identified by Westerlund. So far VRI photometry for
103 objects and slit spectra of 20 have been obtained, in the range $11^m9 < I < 13^m7$.
The spectra indicate a large range in the abundances of C^{13} in these stars.

Golay (Switzerland) pointed out that because of its accuracy, homogeneity and the large number of stars measured, the Geneva ($UBVB_1B_2V_1G$) system is well suited for sorting stars of similar colors into photometric "boxes." Stars in the same box are nearly the same also in spectral type, Hβ index, Copenhagen photometry, UV and IR photometry, and absolute magnitude. Since the range of absolute magnitudes of stars in the same "box" is within .1 mag, a calibration can be made to obtain distance moduli of stars and clusters from such Geneva photometry.

Bidelman (U.S.A.) reported that an extension list of shell stars will appear as an appendix to IAU Symposium No. 70 on Be and Shell Stars. Also in the lists are some of the more notable P Cyg stars, a few eclipsing binaries and some early-type stars involved in nebulosity. Bidelman also noted that he plans to make available in the near future numerous hitherto unpublished spectral classifications by the late G.P. Kuiper of faint high-proper-motion stars.

27 August 1976

II. ADMINISTRATIVE MEETING

Activity Report

The President reported about the Commission activities since the last Assembly. The main points are the following:

a - Meetings organized: IAU Symp. No. 72 on "Abundance Effects in Classification" held on July 8-11, 1975, at Lausanne. The meeting was dedicated to Dr. W.W. Morgan; the proceedings are being published by P. Keenan and B. Hauck at Reidel.

b - Meetings co-sponsored: IAU Colloquium No. 32, "Physics of Ap Stars," at Vienna on 8-10 September 1975; the meeting on "Multicolor Photometry and the Theoretical HR Diagram" organized by D. Philip, at Albany on October 25-27, 1974; IAU Colloquium No. 35 on "The Compilation, Critical Evaluation and Distribution of Stellar Data " at Strasbourg on August 19-21, 1976. The Commission has also co-sponsored the Joint Discussions on "The Impact of Ultraviolet Observations on Spectral Classification" and on "The Galactic Structure in the Direction of the Polar Cap" held during the present Assembly.

c - Circulars: Four circulars were mailed to all Commission Members and about ten to the Organizing Committee Members.

d - The Commission report was divided in parts, each one written by a different specialist. Due to an unfortunate shortening of Dr. Bidelman's report, it became almost useless and circulation of an unabridged version is recommended. In the ensuing discussion the usefulness of the Commission report was questioned. It was finally suggested that an effort should be made next time to circulate the report before the Assembly and that members should be provided with a more detailed report than the one that is printed by the IAU.

e - Grant: Upon the recommendation of the Commission, the IAU made a grant of $1000 (U.S.) to Dr. P. Keenan for the publication of his "Atlas of Spectra of Stars of Types Later than G0."

f - The President of the Working Group reported on its activities. (See Commission report).

Questions Submitted to Vote

The members accepted by vote:

a - to co-sponsor the meeting on the "HR diagram" on the hundredth anniversary of H.N. Russell, as proposed by Philip for October 1977. The Commission representative will be van den Bergh.

b - to co-sponsor the meeting on the "Photometry of Emission-Line Objects" as proposed by Grygar, to be held at Hvar in October 1977. The Commission representative will be Cester.

c - the following two rules for the Organization of the Working Group: 1 - The com-

position of the WG is to be examined at each Assembly. 2 - The chairman of the
WG is proposed by the group and is appointed by the Commission for three years.
d - the proposed new organizing Committee (confirmation of previous mail ballot):
B. Hauck (Switzerland), President; A. Slettebak (USA), Vice-President; Ardeberg
(Sweden); Bartaya (USSR); Cowley (USA); Jaschek (France); Keenan (USA); Kharadze
(USSR); Mendoza (Mexico); Straizys (USSR).
e - the proposal of the new IAU members (confirmation of previous mail ballot):
Albers H. (USA) and Claria J. (Venezuela).
f - the new Commission members: Albers H. (USA), Bell R.A. (USA), Crampton D. (Can-
ada), Hill P.W. (UK), Lutz J. (USA), Maeder A. (Switzerland), Maehara (Japan),
Morguleff N. (France), Osborn W. (Venezuela), Pasinetti L.E. (Italy), Wesselius
P.R. (Netherlands).
g - the new consultant members: Levato H. (Argentina), Mead M. (USA).
h - the composition of the Working Group on Spectroscopic and Photometric Data:
Jaschek C. (Chairman), Barbier M., Bidelman W.P., Dluzneyskaya O., Hauck B.,
Houk N., McCarthy M., Mead J., Nandy K., Philip D.

28 August 1976

III. JOINT MEETING OF COMMISSIONS 29, 36 AND 45 ON "CLASSIFICATION CRITERIA FOR
NON-NORMAL STARS"
    A.P. Cowley (U.S.A.) and H. Houk (U.S.A.) summarized the way non-normal stars
are classified in the Michigan Spectral Catalogue. The technique involves using
features in normal MK standards of more than one spectral type to give a rough
indication of the degree as well as the type of the peculiarity. For example, a
K2 III with strong CN similar to that found in a supergiant might be designated
K2 III CN Ib. The method has the advantage of not requiring use of numerous non-
normal standards, but for some types of peculiarity and for some strong cases, a
verbal remark is required to adequately describe the spectrum. Volume 2 of the
Michigan Spectral Catalogue, containing over 30 000 HD stars from $\delta = -53.0°$ to
$-40.0°$, will probably be available around September 1977, but no orders should be
placed until an announcement of availability is made.

    Pagel (United Kingdom) spoke on two subjects. Regarding the classification of
extreme metal-weak stars, he noted that work by Spite shows that the necessary information is present in 80Å/mm spectra, but not easy to extract. In visual classification, the absolute strength of the Balmer lines is a good temperature indicator only for Teff > 5200. Comparison of metal lines then gives a useful indication
of line weakening. Luminosity is indicated by $\lambda 4173$, but only well enough to distinguish two classes. Within these limits the classification by Bond gives very
reasonable results and led him to discover subgiant CH stars.

    Secondly Pagel discussed the Wilson Bappu effect. He has developed a simple
scaling law which supports the hypothesis of optically thick Doppler broadening for
the full width at half maximum of Ca^+K2 and Mg^+K2 and provides a theoretical basis
for metallicity affecting Mv(K). Whether this effect exists seems to depend on
which calibration is adopted for Mv(K), an unsettled question; one must be able to
allow for errors in the reduction to absolute parallax.

    Mrs. Herman (France) reported on progress in the classification of Be stars.

31 August 1976

IV. JOINT MEETING OF COMMISSIONS 25 AND 45
    A.G.D. Philip (U.S.A.) summarized his analysis of the Hauck-Mermilliod Cata-
logue of homogeneous four-color data. Estimates of the values of the astrophysical
parameters Mv, θe, log g, and $[Fe/_H]$ as well as for the color excess E b-y were
made for 5183 stars of spectral type O through F. The probable errors for the calculated values for Mv, θe, log g, and $[Fe/_H]$ were ±0.4, ±0.01, ±0.13 and ±0.1
respectively.

H.U. Nørgaard-Nielsen (Denmark) spoke about the application of trigonometric parallaxes for absolute luminosity calibration of photometric systems. He demonstrated that it is very difficult to avoid introducing systematic errors into absolute magnitude calibrations caused by accidental errors in the trigonometric parallaxes. Nørgaard-Nielsen thinks that the widely used corrections by Lutz and Kelker contain fundamental errors.

A. Maeder (Switzerland) discussed problems of photometry and stellar structure in relation to basic calibrations. The relation between a) the deficiency of stars near A7-F0 (MS gap), b) the location of the cool edge of Am and δ Scuti stars, and c) the difference in photometric effects of rotation and the appearance of efficient convection in outer stellar layers is emphasized. The effect on the basic Teff calibration, and the influence of the overshooting from convective cores on age and mass calibrations were also pointed out. A similar discontinuity is located at B6-B7.

G. Cayrel (France) gave a report (co-authors Foy, Hardorp and Perrin) on a determination of the metal content of the Hyades. A careful comparison made between the results obtained using different photometric systems ($[Fe/_H] = + 0.25 \pm 0.08$) and those by detailed analysis ($[Fe/_H] = + 0.08 \pm 0.15$) shows that a small discrepancy still exists. As a byproduct HD 76151, proposed by Hardorp as a solar-type standard, was analyzed in detail. Results: Tef = 5600, log g = 4.44; $[Fe/_H] = - 0.06 \pm 0.15$.

COMMISSION 46: TEACHING OF ASTRONOMY (ENSEIGNEMENT DE L'ASTRONOMIE)

Report of Meetings held in Grenoble

PRESIDENT: D. McNally.   SECRETARIES: J. M. Pasachoff, D. Wentzel.

Session I, 25 August 1976

I. REPORT OF COMMISSION
    The President's report of the activities of Commission 46 for the period 1973
- Jan. 1976 was approved.

II. NATIONAL REPORTS
    The form of the National reports was approved. In a discussion of the form
of the report, it was suggested that reports should be brief and take account of
the reports previously published. New developments should however be treated in
extenso. The report should cover university, school and public education in
astronomy and related subjects.

III. MEMBERSHIP

A. Organising Committee

    By virtue of the constitution of Commission 46 E. A. Müller and T. Swihart
leave the organising committee, H. E. Jørgensen also retired. They are replaced
by W. Buscombe, L. Mavridis, B. F. Peery and A. Riguelet-Kaswalder. The organising committee for the period 1976-1979 is given at the head of this report.

B. Members of the Commission

    During the period 1973-76, N. P. Grushinsky (USSR), J. Riihimaa, (Finland)
resigned, S. Torres-Peimbert resigned as national representative of Mexico, but
remained a member of the Commission. W. Buscombe (U.S.A.) and S. E. Okoye
(Nigeria) joined the Commission. The following new members were proposed and
approved:
    Canada: J. E. Kennedy,
    France: M. Gerbaldi, L. Bottinelli,
    GDR: H. Zimmermann,
    Spain: Catala-Poch, M$^{\underline{a}}$. A.,
    U.K: V. Barocas, D. Clarke, D. R. Fawell, H. G. Miles,
    U.S.A: J. M. Pasachoff, B. F. Peery, R. R. Robbins,
    U.S.S.R: V. V. Porfir'ev,
    Yugoslavia: B. M. Ševarlić.

C. Consulting Members

    The list of consulting members of the Commission was reviewed. The list of
consulting members is considered to lapse at each General Assembly and a list
drawn afresh. All retiring consulting members are eligible for reappointment.
The list proposed and adopted for the period 1976-1979 is given at the head of
this report.
    In a discussion on membership the following points emerged.
    (i) Each country adhering to the Union had the right to nominate a National
Representative to Commission 46. Care should be taken to ensure that National

Representatives had an intimate connection with astronomical education and a direct connection with some aspect of such education. National Representatives had the responsibility of maintaining contact between Commission 46 and its projects and national astronomical education.

(ii) In order to maintain the vitality and range of interest of Commission 46 additional membership was necessary. Additional members should be selected with reference to the special expertise they can offer the Commission.

(iii) With membership now at 68 undue proliferation of members should be avoided - in particular a policy of deletion and retirement of inactive members should be actively followed.

IV. RELATIONS WITH ICSU, UNESCO, COSTED AND OTHER BODIES

The President reported on relations between Commission 46 and ICSU, UNESCO, and COSTED. The content of his remarks may be found in the report of Commission 46, IAU Transactions XVI.

The President was kept informed of the work of other bodies such as Task Group for Education in Astronomy (TGEA) in the United States, the Royal Astronomical Society's Education Committee in the U.K., the activities of the International Society of Planetarium Educators among others. One notes the work done by these bodies e.g. the introduction of astronomy in the National Parks by TGEA, the preparation of teaching units by the RAS and the exchange of ideas for keeping the public informed of planetarium visits by ISPE. There is one resource which needs stressing and that is the local astronomical society. These are a valuable resource in that young people could first get to know a little about astronomy in the company of enthusiastic amateurs - this is a large resource in many countries and one which we should capitalise more heavily.

Session II, 27 August 1976

V. VISITING PROFESSORS' PROJECT

In a written report M. Rigutti recommended the termination of the Visiting Professors' Project in the absence of tangible financial support. Whilst sympathising with Rigutti's analysis, the Commission decided to keep the project alive to the extent that the purpose of the project should be advertised in the Information Bulletin, that persons interested in giving lectures, meeting with astronomers etc., in developing countries should send their names to Rigutti who would inform the appropriate National Representative who could then take action to make the necessary arrangements. This form of the project should be reviewed in 1979.

VI. EXCHANGE OF EQUIPMENT WORKING GROUP (JOINTLY WITH COMMISSION 9)

L. Houziaux reported that it had proved very difficult to find institutions with equipment to donate or lend to this project and impossible to find acceptable arrangements to provide for transport and installation. With the agreement of the Presidents of Commission 9 and 46 it was agreed to terminate this project.

VII. BOOK PROJECT

L. Houziaux reported a situation similar to that for Exchange of Equipment existed. While suggesting joint consultation with Commission 5, he felt there was no extensive pressure for its continuation. It was clear from the ensuing discussion that many members of the commission felt that there was still a need for the Book project. The following suggestions were made.

(a) That the IAU should endeavour to donate some copies of its publications to developing countries even though this involved the IAU in some expense.

(b) That publishers generally might make available some copies of books free or at a low charge, though the copyright difficulties and financial stringencies were recognised. But it was pointed out that even a few copies would help.

(c) That Commission 46 might prepare a select list of textbooks with a view to indicating those texts most advantageous for instructional purposes.

(d) That duplicate journals might be sent to a corresponding institution in a developing country.

Professor Houziaux undertook to look into these matters with a view to seeing what could be acheived in practical terms.

VIII. INTERNATIONAL SCHOOLS FOR YOUNG ASTRONOMERS (ISYA)

The details of all ISYA held in the period 1973/76 have been given in the report of the Commission in IAU Transactions XVI.   There was an extensive debate on the nature, character and future of the schools.   The following salient features emerged.

(a) The geographical range of the schools was considerable: Asia, 2 (India, Indonesia); Europe, 3 (Greece, Italy, U.K); South America, 2 (Argentine).

(b) Although the schools were originally held under joint IAU/UNESCO auspices, the Commission accepted a suggestion by the Executive Committee that UNESCO should now be dropped from the title of the schools in view of the fact that UNESCO now make no direct financial contribution in support of the schools.   It was recognised that some advantage did exist through association with UNESCO and that some indirect financial support was obtained.

(c) It was concluded that sites for schools should be chosen with due regard to effectiveness of instruction in astronomy and support for existing astronomical communities.

(d) The Executive Committee had insisted that, before IAU finance could be made available for future schools, better rules than presently existed, for the operation of the schools, would be needed.   Draft proposals were rejected by the Commission on the grounds that the freedom of action of the Secretary of the ISYA would be fettered unnecessarily.   A committee (President, Houziaux, Wentzel) was constituted to draft more acceptable rules and guidelines.   The new rules and guide lines were presented at a later meeting and accepted.   The Rules and Guidelines for the operation of the ISYA are appended to this report.

(e) Commission 46 wishes to record:

(i) that it regards the ISYA as the project of the Commission which has highest priority.

(ii) its warm appreciation and approval for the work of Kleczek in organising the ISYA.

Session III, 30 August 1976

IX. ASTRONOMICAL EDUCATIONAL MATERIAL (AEM)

The next update of AEM for 1979 will be prepared by:

B. F. Peery assisted by R. R. Robbins : English Language Material.

E. V. Kononovich assisted by C. Iwaniszewska et al : Slavic Language Material.

L. Mavridis assisted by A. E. Jørgensen : Material in all other languages.

It was agreed that the collators of AEM should have the assistance of a further person in addition to the help expected from National Representatives in the form of information on new material.

In a discussion of AEM the following points emerged:

(a) The cost of producing and distribution of an update of AEM for a single language group is of the order of $300. This cost is borne by the collator's institution, the IAU having no funds for this purpose.

(b) While it would be desireable from many points of view to give clear recommendations in favour of a selected list of material, it was not only invidious but impractical for the collators to undertake further selection of material recommended.   However, it was agreed to classify material more carefully in relation to its use in teaching (noting that many reviews of textbooks made little reference to pedagogic value) particularly in regard to level and degree of sophistication.

(c) Evaluation of text books at upper school/university level in major international and semi international languages should have high priority in (b).
(d) The Commission should seek recommendations for translations and encourage
publishers to produce the texts so translated.

X. PROJECT CONTRATYPE
     M. Gerbaldi reported that the trial period for ordering material (56 2 × 2
slides) of the Project held at the Institut d'Astrophysique in Paris had resulted
in 32 orders - 18 of these orders were from European Astronomers 8 from
Universities and 6 from teachers.   The first six slides of the material of the
Project held in Moscow were received by the Commission but with no assessment of
demand.
     It was clear that the aim of the Project to get good teaching materials into
the hands of teachers, particularly in the developing countries had failed.   The
following steps to remedy this situation were proposed.
     (a) Dr. West of ESO and Dr. Littmann Director of the Hansen Planetarium (a
non-profit making educational enterprise) have expressed interest in marketing the
Project Material.   It was agreed to explore with Dr. West and Dr. Littmann how
the Project Material might be marketed.
     (b) It was further agreed that certain other bodies e.g., the Royal Astronom-
ical Society, might be approached to see if they might act as further outlets for
the material.
     (c) The need for documentation of the material was also pressed.   This
question will be actively addressed by Gerbaldi and Wentzel.   Gerbaldi already
has documentation for a set of slides on gravitation.
     (d) It was noted that profound difficulties existed in making known the
existence of this material in the developing countries.   Once the future of the
Project was settled National Representatives would be required to take steps to
locally advertise the existence of the project.
     E. E. Proverbio briefly reported on a scheme with integrated illustrative
material now being used in Italian Schools.   The material was expressly designed
for young children.

XI. A NEWSLETTER FOR COMMISSION 46?
     The President pointed out that he receives a great deal of correspondence
from members of the Commission.   In that correspondence there is a great deal of
material of interest to the entire Commission.   Because it would be burdensome to
circulate all such correspondence to all members of the Commission it is a pity
that such items may be referred to only briefly in a report.   There is a need to
broaden the exchange of ideas between Commission Members rather than use a
presidential filter.   The Education Committees of other Scientific Unions had
found it useful to establish a newsletter.   These ranged in style from a 4000
circulation, 16-page quarterly of IUPAC to the four-page newsletter of IUB.
     The President proposed that the Commission might experiment with an 8-page
Newsletter published twice a year.   The President volunteered to serve as editor
to get the Newsletter established.   This suggestion was accepted by the
Commission.   D. Wentzel offered to produce and distribute the Newsletter from
Maryland.   Some funding might be available from the IAU to help defray the costs
of production and distribution.   It is hoped to present the first issue of the
Newsletter in January 1977.

XII. FUTURE MEETINGS OF THE COMMISSION
     It was agreed that a whole day should be given to a discussion of Teaching
Methods for Astronomy at University Level.   Professors Climenhaga and Ovenden
were asked to consider the organisation and format of such a meeting during the
Montreal General Assembly.   Such a meeting would be additional to the meeting of
the Commission with Canadian Schoolteachers.

Session IV, 30 August 1976

THE TEACHING OF ASTRONOMY
    This session divided naturally into two parts - the first on the teaching of
astronomy specialists and the second on the teaching of astronomy to non-
specialists.
    McNally referred to the importance of specialist courses for astronomers in
view of the way astronomy was developing. Astronomy is both a pure and an applied
science in its own right and not a branch of some other discipline. However, in
its role as applied science, astronomy must indent heavily on other sciences and
it was essential to ensure that desireable developments in the teaching of these
sciences did not prove detrimental to instruction required for Astronomy. A clear
example is the trend towards pure, and away from, applied mathematics but there
were areas of physics where a growing neglect in the curriculum could be detrimental to astronomy teaching. Rigutti stressed the need to make astronomy available
to students before postgraduate level in order that they should appreciate the
role played by astronomy in our cultural heritage. Iwaniszewska outlined the
astronomy programme in Polish Universities for astronomy specialists and teachers
of astronomy. Okoye urged that an international centre be established to support
training and continuing study for under-developed countries. He also raised a
number of thought provoking issues regarding the role of astronomy in relation to
education in the developing countries. Kourganoff stressed that astronomy was an
integrating science and astronomers should be trained with this in mind and not as
an adjunct of physics - it was not recognised by practitioners of other sciences
that the problems of astronomy required astronomical solutions. He favoured an
early acquaintanceship with astronomy and certainly no later than the early years
at university. A lively discussion ensued clearly indicating the need for a more
extended meeting on these topics.
    Haupt (Austria) reported on a survey of 18 year olds aimed at assessing how
much astronomy had been retained from their courses at school. The results were
very encouraging e.g. 45% of the sample could correctly discriminate between
astronomy and astrology. Robbins described an astronomy programme devised at the
University of Texas for teacher training through active participation in experiments designed for classroom use. Friedmann reported on the methods being used
in the GDR to train teachers for the astronomy section of the new school science
curriculum. The courses take two years and are backed by summer schools designed
to promote confidence in presenting astronomy. Sandqvist reported on the
apprenticeship scheme operating in Sweden where astronomy is offered as an option
for 9th grade students. This scheme is designed to give career guidance.
Participation in the scheme, while demanding on the institution, had been found to
be rewarding.

Extra Session

    The tele-recording of the Royal Astronomical Society's first Young Person's
Lecture" - "Not Seeing is Believing" by D. Sciama was shown. This tele-recording
of 50 minutes duration is available in colour from Dr. A. Crilly, Open University,
Alexandra Palace, Wood Green, London N.

2 September 1976

MEETING WITH FRENCH SCHOOL TEACHERS
    About 150 teachers and 50 astronomers attended.

Information was given on:
1) the new programmes for physics teaching in French schools,
2) the pedagogical and audiovisual material available,
3) the teaching of astronomy at the same level in different countries:
   Belgium, Canada (Quebec), Poland and Switzerland.
Discussions were held on:
1) The training and permanent recycling of teachers for astronomy.
2) Astronomy Clubs.
3) The teaching of Astronomy during "free" activity sessions.
4) The importance of Astronomy at secondary level:
   a) in the first cycle (11-14 years) it helps the children to understand their
      surroundings.
   b) in the second cycle (14-18 years) the universe can be taken as a good
      laboratory.

Resolutions were adopted concerning:

   The introduction of astronomy in physics programmes.
   The permanent recycling of teachers.
   The organisation of a Summer School in 1977.
   The development of audio-visual material.

   An exciting keynote address was given by Dr. Bonnet on "Astronomy in the era
of investigation from space".
   Particular thanks are due to Professor Omnes, president of the "Commission
Laganique" and Professor Schatzman vice-president of the French Physical Society
who both co-sponsored the meeting, Professors Gie and Lena who chaired the sessions,
the "Inspecteurs Generaux" Delarue and Guinier who attended the meeting, Mme. Coq,
Drs. de la Cotardière and Delacote who chaired the working groups, and Professors
Houziaux, Iwaniszewska, Landry, Lena, de Loore, Maeder and Terlon who made important
contributions.   The meeting received financial support from the French National
Committee of Astronomy and from the Ministry of Education.   Their support is
gratefully acknowledged as is the unstinting help and support of Lucelle Bottinelli
and Michele Gerbaldi with the organisation of the Meeting.
                     L. Gougenheim - Local Organiser.

        RULES AND GUIDELINES FOR INTERNATIONAL SCHOOLS FOR YOUNG ASTRONOMERS

A. RULES:
     1) The President, Vice-President(s), Secretary of the ISYA and Past President
will constitute a sub-committee of Commission 46 to regulate the organisation of
the ISYA.   The Secretary of the ISYA will be the Secretary to the sub-committee
and will keep the sub-committee informed of all proposed ISYA, the details of
their organisation and evaluation.   The President of the Commission will inform
the General Secretary of the Union of the programme for each proposed School only
when approved by the sub-committee.
     2) The Secretary of the ISYA at each General Assembly will ask the opinion of
the Commission for possible venues, topics, lecturers for ISYA in the succeeding 3
year period.
     3) The ISYA sub-committee will meet at least once during each General Assembly
following discussion in the full Commission to discuss the suggestions, order of
priorities and discuss the evaluation of previous schools.   A report on immediate
past schools and plans for future schools should be submitted to the IAU Executive
Committee by the President.
B. GUIDELINES:
     1) The choice of venue, topic(s), lecturers and general calibre of students
for each ISYA should be carefully considered with respect to local conditions and

goals, e.g. a school may be held to support astronomy at an embryonic stage, or to support a growing astronomical activity, or to explain and demonstrate techniques.

The goals of each school should be carefully stated in advance. Topics should be directly appropriate to local conditions and lecturers should be chosen with due regard to the contribution they may be expected to make towards the realisation of the goals of the school.

2) The students should have an adequate academic background to profit from the school (normally a person with a degree in physical or related sciences from an established university or institution of similar standing or, a person undergoing training in an established observatory).   Students of inadequate academic background are unlikely to profit, and should be discouraged from attending the schools, as they may hinder the progress of the majority of the better qualified students.

As far as practicable, the local organiser of the school must inform the Secretary of the ISYA and the ISYA sub-committee of the qualifications of the proposed students. The sub-committee may at its discretion ask for the removal of any proposed student who is believed to be inadequately qualified.

3) The Secretary of the ISYA, one year after each school, should collect such information as is necessary to evaluate the school.

COMMISSION 47:   COSMOLOGY   (COSMOLOGIE)

Report of Meetings   25, 26, 28, 30 and 31 August 1976

PRESIDENT: M.S. Longair.

25 August 1976

BUSINESS MEETING

        The President described Commission activities since the last General
Assembly.   The Commission had cosponsored IAU Symposium No.74 "Radio Astronomy
and Cosmology" jointly with Commission No 40 (Radio Astronomy).   Over 150
invited participants from outside Cambridge attended the Symposium for which the
hosts were the members of the Radio Astronomy Group, Cavendish Laboratory,
Cambridge.The emphasis of the symposium was upon those aspects in which studies
of discrete radio sources can provide cosmological information.   All observational
evidence available at the time of the symposium was surveyed in depth by the
astronomers who had made the observations.   The proceedings which will be published
shortly by D. Reidel and Co. and for which the editor is D.L. Jauncey will provide
a complete survey of how radio astronomical observations can contribute to
cosmological problems.

        Certain members of the Organising Committee had been very helpful in the
preparation of the Draft Report by writing short review papers which covered a
wide range of cosmological topics.   The President warmly thanked Drs Zeldovich.
Novikov, Nariai, Maccallum and Ne'eman for their efforts.

        The following elections were made for the period 1976 to 1979:

President   I.D. Novikov

Vice-President   G.O. Abell

Organising Committee   G. Dautcourt,   G de Vaucouleurs,   S. Hayakawa,   K.I. Kellermann,
M.S. Longair, M. Rowan-Robinson and K.S. Thorne.

        A long list of proposed new members of the Commission was discussed and with
the addition of other names, proposed and seconded by members of the Commission,
these proposed new members were elected Commission members.   Their names are included in the full list of Commission members which will be found elsewhere in this
volume.
        The Commission discussed the proposal originating with J. Einasto that
an IAU Symposium be cosponsored by the Commission jointly with Commission No 28
(Galaxies) entitled "The Large Scale Structure of the Universe" to be held
in Tallinn, Estonia, USSR in September 1977.   The proposal was enthusiastically
supported by the Commission members who urge the Executive Committee to view the
proposal favourably.

<u>26 August 1976</u>

SCIENTIFIC SESSION   <u>Unconventional and Pathological World Models</u>

    The following papers were read:

Y. Ne'eman: Review of Unconventional and Pathological World Models.

B. Carr: Cold and Tepid Universes and Primordial Stars.

V.L. Ginzburg: Black Hole Physics and Fundamental Length.

I. Segal: Chronometric Redshift Theory.

A. Lausberg: Distance Functions for Inertial Interactions in Friedmann Universes.

A. Sapar: The Fundamental Role of Planck Units in Cosmology.

I. Roxburgh: Dirac Cosmology.

J. Barnothy: FIB Cosmology.

<u>28 August 1976</u>

SCIENTIFIC SESSION   <u>Classical Cosmological Tests</u>

    The following papers were read:

A.S. Webster: The Large Scale Structure of the Universe.

V. Rubin: Criteria for Tests of Isotropy.

M.S. Longair: Survey of Scientific Result of IAU Symposium No 74 "Radio Astronomy
                and Cosmology".

G. Abell: Mean Matter Density of the Universe as a Cosmological Test

G. de Vaucouleurs: The Distance Scale and H_0.

B. Tinsley: Counts of Galaxies and Related Topics.

K. Matilla: The Background Optical Radiation

J. Solheim: Ghost Images.

W. Tifft: On the Continuity of Redshifts.

<u>30 August 1976</u>

SCIENTIFIC SESSION   <u>Galaxy Formation</u>

B. Jones: Review of the Current Status of Theories of Galaxy Formation

M. Anile: Isotropy of the Microwave Background Radiation in Turbulent Cosmologies.

R. Gott: Evolution of the Spectrum of Primordial Density Fluctuations.

R. Larson: The Collapse and Formation of Galaxies.

I. Novikov: Numerical Calculations of the Non-linear Stages of Galaxy Formation.

There followed a general discussion led by B. Jones during which short contributions were made by B. Lewis,   T. Gold, M.J. Rees and I King.

31 August 1976

Joint Discussion No 4   Clusters of Galaxies,   Cosmology and Intergalactic Matter

This Joint Discussion was sponsored jointly by Commissions 28 (Galaxies), 47 (Cosmology) and 48 (High Energy Astrophysics). The papers presented at this Joint Discussion will be published in the Volume "Highlights of Astronomy Vol 4". The proceedings are edited by M.S. Longair and J.M. Riley. The programme was as follows:

A. Oemler: The Galaxy Content of Clusters (including a contribution by N. Bahcall on the Structure of Clusters of Galaxies).

S. White: The Dynamical Evolution of Clusters of Galaxies.

R. Gott: Groups of Galaxies.

M. Kalinkov: The Existence of Higher-Order Clusters of Galaxies.

There were also short contributions from J. Dawe, K. Mattila and N. Vidal.

L. Culhane: X-rays from Clusters of Galaxies.

M.S. Longair: Recent Radio Observations of a Complete Sample of Clusters.

D. Harris: Survey of the Radio Properties of Clusters of Galaxies.

S. Lea: Hot Gas in Clusters of Galaxies.

S. Gull: The Microwave Background Radiation in the Direction of Clusters of Galaxies.

There was a short contribution from P.Gorenstein.

COMMISSION 48: HIGH ENERGY ASTROPHYSICS (ASTROPHYSIQUE DE GRANDE ÉNERGIE)

PRESIDENT: M. J. Rees

BUSINESS MEETING

   At a business meeting held on 25th August, the following new organising committee was proposed for Commission 48: I. S. Shklovski (President), F. Pacini (Vice-President), J. Audouze, J. L. Culhane, K. I. Kellermann, L. M. Ozernoi, E. N. Parker, M. J. Rees, J. Shaham. A list of proposed new members of the Commission was also approved.

   Two general topics of concern to the Commission were briefly discussed: (i) the possible need for rationalisation of X-ray source nomenclature, and (ii) the desirability of coordinating and exchanging information about transient phenomena recorded in various wavebands.

SCIENTIFIC SESSIONS

25 August 1976

"Physics of dense matter"

(with commission 35)

G. Baym (NORDITA and Illinois)	An overview of neutron star structure; latest developments on equation of state and pion condensation.
J. Shaham (Jerusalem)	Astrophysical consequences of vortex line pinning, wobble, starquakes etc.
N. Holloway (Sussex)	Properties of the surface layers of neutron stars and their relevance to pulsar theories.

   (Short contributions were presented by S. Tsuruta (Munich) and K.Brecher (MIT).

26 August 1976

Joint discussion II

"X-ray binaries and compact objects"

(also sponsored by commissions 42 and 44)

G. Clark (MIT)	X-ray bursts
P. J. N. Davison (London)	Periods in X-ray sources
J. Grindlay (Harvard)	Globular cluster sources
A. P. Willmore (Birmingham)	Transient X-ray sources
J. Hutchings (DAO Canada)	Optical observations of X-ray sources
Y. Avni (Weizmann Inst.)	Mass estimates

E. P. van den Heuvel (Amsterdam)   Evolution of X-ray binaries
Y. Kondo (NASA)   Report on coordinated observations
R. McCray (JILA)   Accretion flows in X-ray binaries

(Short theoretical contributions were presented by F. Lamb (Illinois)
R. N. Henriksen (Ontario) and L. Maraschi (Milan).

27 August 1976

"Physics of radio sources and quasars"

(with commission 40)

G. Miley (Leiden)   Radio structure of extended sources
M. S. Longair (Cambridge)   Statistical properties of radio sources
M. Cohen (Caltech)   Compact (VBLI) structure in quasars and
                                   galactic nuclei.
D. De Young (NRAO)   Theories of double radio sources
A. Boksenberg (London)   Spectra of quasars and related objects
E. M. Burbidge (La Jolla)   Optical spectra of quasars
J. J. Perry (Munich)   Radiation pressure effects in quasars
S. Colgate (Colorado)   Theories of galactic nuclei involving
                                   multiple supernovae
T. Gold (Cornell)   Present status of stellar collision theory
V. L. Ginzburg & L. M. Ozernoi
   (Moscow)   Theories of galactic nuclei

30 August 1976

"CNO isotopes in astrophysics"

R. N. Clayton (Chicago)   Nucleosynthetic oxygen anomalies in meteorites.
                                   Secular variation of the $^{15}N/^{14}N$ ratio in the
                                   solar wind.
C. Rolfs (Münster)   New cross section measurements relevant to
                                   the CNO nucleosynthesis.
R. Caughlan (Montana)   CNO bi-cycles.
P. Demarque (Yale)   Peculiar CNO abundances in evolved globular
                                   clusters
D. Dearborn (Cambridge)   Isotopic abundances in cool stars.
S. Starrfield (Tempe, Arizona)   Nova outbursts and hot CNO cycles.
G. Wannier (Amherst)   Isotopic abundances in dense interstellar
                                   clouds.
J. Encrenaz (Meudon)   Isotopic abundances in dense interstellar
                                   clouds.
W. Truran (Illinois)   Nucleosynthesis of CNO isotopes.
J. Lequeux (Meudon)   CNO isotopes and chemical evolution of galaxies.
W. Watson (Illinois)   Isotope fractionation in interstellar molecules.
G. Steigman (Yale)   Further comments on fractionation processes
                                   in the interstellar medium.

<u>31 August 1976</u>

Joint discussion IV

"Clusters of Galaxies, Cosmology and Intergalactic Matter"

(also sponsored by commissions 28 and 47)

A. Oemler (Yale)
: The luminosity function structure and galaxy content of clusters.

J. R. Gott (Princeton)
: The formation of clusters of galaxies.

S. White (Cambridge)
: The dynamical evolution of clusters by galaxies.

M. Kalinkov (Bulgaria)
: The existence of high-order clusters of galaxies.

J. L. Culhane (London)
: X-rays from clusters of galaxies.

M. S. Longair (Cambridge)
: Statistical properties of radio sources in clusters.

D. Harris (Dwingeloo)
: Radio trail sources in clusters.

S. Lea (NASA/Ames)
: Hot gas in clusters of galaxies.

S. Gull (Cambridge)
: The microwave background in the direction of clusters of galaxies.

(Additional short contributions were presented by J. A. Dawe (Edinburgh) and N. Vidal (Tel Aviv).

<u>1 September 1976</u>

"Supernovae"

D. Branch (Oklahoma)
: Supernova observations

L. Rosino (Asiago)
: Supernova observations

G. Lasher (I.B.M.)
: Supernova light curves

L. Culhane (Mullard, London)
: X-rays from supernova remnants

R. Chevalier (Arizona)
: Supernova remnants

S. Colgate (Colorado)
: Supernovae and quasars

G. Tammann (Basel, Switzerland)
: Supernova statistics

B. Tinsley (Yale)
: Masses of supernova progenitors

Z. Barkat (Jerusalem)
: Evolution of supernova progenitors

J. W. Truran (Illinois)
: Explosive nucleosynthesis

K. Sato (Kyoto)
: Supernova mechanisms

D. K. Nadyozhin (Moscow)
: Gravitational collapse, weak interactions and supernova outbursts

R. Epstein (Harvard and NORDITA)
: Gravitational collapse and supernova mechanisms

The full proceedings of Joint Discussions II and IV, edited by E. van den Heuvel and J. M. Riley respectively, will appear, according to custom, in "Highlights of Astronomy". Special arrangements are being made to publish the proceedings of the sessions on "CNO isotopes" and "Supernovae", in two volumes to be edited by J. Audouze and D. N. Schramm.

COMMISSION 49: THE INTERPLANETARY PLASMA AND THE HELIOSPHERE
(PLASMA INTERPLANETAIRE ET L'HELIOSPHERE)

Report of Meetings, 25 and 26 August 1976

PRESIDENT: W.I. Axford.   VICE PRESIDENT: A. Hewish.

The first meeting of Commission 49 was held during the General Assembly on 25 and 26 August 1976. A business meeting was held on the 26 August and new officers were elected as follows:
President: A. Hewish, Vice President: H.J. Fahr, Organizing Committee: S. Grdzielski, A.Z. Dolginov, J.L. Bertaux, S. Cuperman, G. Thomas, W.A. Coles.

It has been agreed that the IAU should co-sponsor the SCOSTEP Symposium on Travelling Interplanetary Phenomena (in memory of the late Dr. L.D. de Feiter), which will be held in Tel Aviv prior to the 20th COSPAR Plenary Meeting in June 1977. It was agreed to seek IAU co-sponsorship of "Solar Wind IV", a specialist meeting on the solar wind which is planned to be held in Germany in September 1978.

25 August 1976

SESSION I: INTERSTELLAR GAS WITHIN THE SOLAR SYSTEM
P. Blum and H.J. Fahr: Neutral gas in the heliosphere
H.J. Fahr: The change of interstellar gas parameters within the solar system
C. Wulf-Mathies: The intensity of the interplanetary helium 584 Å background radiation and the solar line shape at 584 Å
F. Paresce: Resonance absorption cell techniques for the observations of helium in the interstellar wind
C.S. Weller: Present status of and future possibilities for UV measurements of local interstellar gas
J.L. Bertaux: Temporal measurements of neutral hydrogen

SESSION II: INTERPLANETARY SCINTILLATIONS
W.A. Coles: Interplanetary scintillations
N.A. Lotova and I.V. Cheshey: Dispersion analysis of solar wind velocity
S.D. Shawhan, F.T. Erskine, W.M. Cronyn, E.C. Roelof, D.G. Mitchell and B.L. Gotwols: Interplanetary scintillation events associated with solar wind sector boundary crossings at Earth
A. Readhead: Observations of the interstellar medium using interplanetary scintillation

26 August 1976

SESSION III: MISCELLANEOUS TOPICS
M. Dryer, D.S. Intriligator, E.J. Smith, R.S. Steinolfson, J.H. Wolfe and S.T. Wu: Dynamic MHD models of solar-initiated disturbances from the lower corona to 10 AU
H.J. Fahr, H. Ripken and M. Bird: Solar wind expansion from strongly diverging magnetic fields
C.P. Sonett: The moon as a pseudo-scatterer of the interplanetary field

SESSION IV: EXPLORATION OF THE INTERPLANETARY MEDIUM
W.I. Axford: Future possibilities for the exploration of interplanetary space
I. Roxburgh: A solar probe
J.A. Van Allen: Cosmic ray gradient measurements from Pioneers 10 and 11
C. Jordan and D. Bohlin: Solar physics from out-of-ecliptic spacecraft

COMMISSION 50: IDENTIFICATION AND PROTECTION OF EXISTING AND POTENTIAL
OBSERVATORY SITES
(PROTECTION DES SITES D'OBSERVATOIRES EXISTANT ET POTENTIELS)

Report of Meetings, 25 and 30 August 1976

PRESIDENT: M.F. Walker.   SECRETARY: P.J. Treanor.

25 August 1976

I. ORGANIZATION
    The Commission elected, subject to the approval of the Executive Committee of
the Union, the following new officers: President, R. Cayrel; Vice-President,
F.G. Smith. The proposed membership of the Organizing Committee was approved
by the membership.

II. REVIEW OF CURRENT SITE TESTING PROGRAMS AND EFFORTS TO PROTECT POTENTIAL SITES
    The following communications were presented:
    F. Sanchez: The Work of the Joint Astronomical Site Survey in the Canary
Islands, 1974 - 1975.
    F.G. Smith: The British Northern Hemisphere Site Testing Program.
    L. Barreto: Observing Sites in Brazil.
    G. diTullio: The Italian Site Testing Program.
    P.J. Treanor: The Vatican Site Testing Program.
    J. Osorio: Observing Sites in Portugal.
    B.M. Lewis: Observing Sites in New Zealand.
    W. Mattig: JOSO Site Testing Results.
    The British Program, conducted by the Royal Observatory Edinburgh and coor-
dinated by B. McInnes, and the Joint Survey directed by F. Sanchez, indicate that
good observing conditions exist at Mauna Kea (Hawaii), Izaña (Tenerife), and
Fuente Nueva (La Palma)at each of which about 70% of the dark hours are usable and
about 60% are photometric. The number of clear hours is somewhat less at Madeira,
especially in springtime, the yearly average being about 50% useable and 40% photometric. Fuente Nueva is outstanding for good seeing, about 40% of the useable
hours having seeing ≤ 1". The British Science Research Council has decided that
La Palma is the preferred site for the proposed Northern Hemisphere Observatory.
A similar conclusion has been reached by JOSO (Joint Organization for Solar Research),
 who have concluded that the summit of Pico de Teide (Tenerife), and Roque de los
Muchachos (La Palma), are the best solar sites within a radius of 3000 km of
western Europe. In Spain, a commission has been established with F. Sanchez as
President to draft legislation to protect observatory sites on Spanish territory
from such adverse factors as light pollution, road construction and air traffic;
the IAU, through Commission 50, will be asked to provide criteria and detailed
recommendations.
    The Italian site testing program was designed to select a site for the 3.5 m
national telescope to be located within Italy. Starting in 1960, preliminary tests
were made at St. Barthelemy (Valle d'Aosta), Pescopagano (Basilicata), Gravina
(Puglia), and M.S. Venere (Sicilia). The best conditions were found at Pescopagano.
From July 1972 - December 1973, more detailed observations were made on the mountain
Toppo di Castelgrande (altitude 1285 m) located a few kilometers southwest of
Pescopagano. During this period, the site had average seeing of ≤ 2" (as measured
by the Polaris trail method) on 65% of the nights, and fair meteorological condi-
tions.
    Owing to the artificial brightness of the night sky on the Italian peninsula,

the search for a new site for the Vatican Schmidt has been concentrated on the
island of Sardegna. Considerations of turbulence and access rule out the Gennargentu
Mountains and Mount Limbara, while the city of Caglieri (population 350,000) creates
a zone of avoidance 60 km in radius. Observations on various high plateaus at
about 1000 m altitude reveal the risk of excessive crest cloud and mist. However,
Polaris trail and visual observations indicate that seeing adequate for Schmidt
camera work, dark sky, and easy access are obtainable below the inversion layer
at altitudes of about 600 m. The most promising such region found to date is in
central Sardegna in the hills east of Oristano, but more on-site observations of
local meteorological conditions are required as the island has a complicated cli-
matic pattern.

Site testing in Portugal has been concerned with the selection of a site within
Portugal for the new 76 cm reflector of the University of Porto. From a preliminary study of meteorological records, the mountain of Serra Amarela (1300 m altitude, 30 km from the sea) in the north of Portugal has been selected for on-site
study. No results have as yet been obtained.

In June, 1975, the National Committee for Astronomy in New Zealand established
a working party to encourage and coordinate site testing activities in New Zealand.
Present activities include: construction of instruments to measure sky brightness,
seeing, extinction, cloud cover and other meteorological data, and the operation
of these instruments at established sites such as the Mount John Observatory
(λ = 190° W, ϕ = 44° S, altitude 1029 m) and at various possible sites such as
Black Birch (λ = 186° W, ϕ = 41° S, altitude 1400 m). These studies will emphasize seeing measured by recording star trails of Sigma Octantis with polar star
trail telescopes of the Lick design, and observations of cloud cover.

III. REVIEW OF CURRENT PROGRAMS TO PROTECT EXISTING OBSERVATORY SITES
The following communications were presented:
R.E. White: Protection of Observatories near Tucson, Arizona.
M.F. Walker: Proposed Light Control Legislation to Protect Lick Observatory.
E.E. Mendoza: Protection of the Site at San Pedro Martir, Baja California.
Z. Suemoto: Protection of Observatory Sites in Japan.

The effect of the lighting control ordinance adopted by the city of Tucson in
1972 (see IAU Inf. Bull. No. 33, 1975) has been to stop the increase of sky illumination with time at Kitt Peak for wavelengths shortward of the λ 4400A cutoff
of the filters required for lamps having >15% of their emergent flux shortward of
that limit; longward of 4400A, the intensity has continued to increase. Pima
County, Arizona, enacted a similar ordinance in May, 1975. This ordinance differs
from the former in the creation of special "dark zones," 40 km in radius, around
the observatory sites at Kitt Peak, Mt. Lemmon (Catalina Mountains) and Mt. Hopkins
(Santa Rita Mountains). Within these zones, lighting is restricted to incandescent
type lamps.

The sky brightness at Mount Hamilton is now about four times the natural intensity and the Lick Observatory is therefore attempting to secure lighting control
legislation in the surrounding Santa Clara County. As a result of presentations
by the Observatory, the Council of Mayors of the cities in Santa Clara County directed the County Association of Public Works Directors to set up a Committee consisting of Public Works Directors, and representatives of the Observatory and the
Pacific Gas and Electric Company to study the effect of lighting on the Observatory
and recommend solutions. The Committee report found that the work of the Observatory is seriously hampered and will ultimately be rendered impossible unless cor-
rective action is taken. The Committee concluded that it is in the public interest
to preserve the scientific capability of the Observatory and to improve lighting
efficiency to save energy and money. They therefore have recommended to the cities
the adoption of a control ordinance similar to the Tucson city ordinance discussed
above, but differing in restricting the types of exterior lighting to incandescent,
clear filtered mercury vapor, and low pressure sodium.

The National Astronomical Observatory at San Pedro Martir, Baja California,

Mexico, located in the San Pedro Martir Forest Reserve and National Park which covers about 20 x 30 km^2, is protected by a law which states that it is in the public interest to conserve and protect the San Pedro Martir Forest in such a way as to insure the normal development of astronomy and similar sciences.

The protection of three observatory sites in Japan has been arranged through informal contacts between the observatory and the local firms or organizations involved, sometimes assisted by prefecture authorities. It is felt that this type of procedure has been and will be the most effective way of dealing with problems of site protection in Japan. The observatories that have been protected in this way are: the Okayama Astrophysical Station of the Tokyo Astronomical Observatory, which was threatened by lights from the Mizushima and Fukuyama refinery and industrial areas and by lights and vibrations from the new super rapid railway system; as a result of the control programs, the sky brightness due to the industrial areas has remained constant since 1970; the Hida Observatory of the Kyoto University, which was threatened by a searchlight on a bowling alley; the International Latitude Observatory at Mizusawa, threatened by the construction of the super rapid railway and a limited access highway.

IV. PRELIMINARY DISCUSSION OF COMMITTEE RECOMMENDATIONS

After a general discussion of possible recommendations which the Committee might make with regard to site identification and protection, it was decided to appoint a working group on recommendations consisting of Walker, Cayrel, F.G. Smith, Mendoza, and Sanchez to consider these matters and report back to the Commission at the next meeting on 30 August.

<u>30 August 1976</u>

I. RESOLUTION

The Commission approved with emendations the following resolution drafted by the working group on recommendations:

The IAU notes with alarm the increasing levels of interference with astronomical observation resulting from artificial illumination of the night sky, radio emission, atmospheric pollution and the operation of aircraft above Observatory sites.

The IAU therefore urgently requests that the responsible civil authorities take action to preserve existing and planned Observatories from such interference. To this end, the IAU undertakes to provide through Commission 50 information on acceptable levels of interference and possible means of control.

This resolution was submitted for approval and approved by the XVIth General Assembly.

II. COMMISSION RECOMMENDATIONS

On the advice of the working group on recommendations, the Commission decided to make only the following general recommendations at this time:

(1) The Commission considers that the most vital problem is that of preserving those sites known to have a very high quality from adverse conditions of all kinds. The most urgent aspect is to limit artificial illumination to a small fraction of the natural sky brightness, making these sites available for observations of faint objects that cannot be made in any other way. The Commission therefore urges astronomers and civil authorities to give highest priority to this problem.

(2) The Commission recommends that at existing observatories where a more limited range of observations can be made despite considerable levels of light pollution, every effort should be made to prevent these levels from increasing.

(3) The Commission recommends that the power of radio transmitters be limited so as to avoid interference with sensitive electronic detectors. Present experience suggests that to avoid interference with electronic equipment used on optical telescopes, the power flux from radio transmitters should not exceed one millivolt per meter (or 1.6 microwatts per square meter) at the observatory site. (Note that this value supercedes that given in the Report of Commission 50, <u>Trans. IAU XVIa</u>).

(4) The Commission recommends that aircraft routes be planned to avoid the skies over observatory sites, so as to prevent interference by condensation trails. Restriction of this nature would involve flights at altitudes of ⩾10°as seen from the sites, which implies that civil air lanes should be placed at least 60 km from sites of very high quality. An example of the effectiveness of this type of restriction was reported by Cayrel: all air traffic over Haute Provence was banned for the total solar eclipse of 1961 February 15. The sky remained clear, but when air traffic was resumed in the afternoon, clouds were generated by the aircraft trails.

III. WORKING GROUP

Detailed recommendations will require considerably more study than was possible during the course of the General Assembly. For example, the provisions of recommendations (1) and (2), above, will clearly be different for different locations. In order to develop the specific recommendations called for in the resolution, the Commission voted to establish a working group which will, in consultation with all members of the Commission, study these matters in depth and report back to the Commission in about one year. This group will consist of Cayrel, F.G. Smith, Walker, Sanchez, White, and an additional member from the USSR to be appointed by the USSR Academy of Sciences. The working group will draft specific recommendations with regard to the identification of potential sites as well as the protection of these and existing locations.

IV. DISCUSSION OF SPECIFIC SITES

Regarding specific sites, the Commission noted that research on island sites with suitable latitude, airmass conditions, and altitude of peak has narrowed these down to a very small number, including notably Guadaloupe, the Canary Islands, Madeira, Pico, and Hawaii.

Observations by the British observers covering 17 months show that Madeira is a good site, only slightly inferior to the Canary Islands. Osorio emphasized that there is an urgent need for light control legislation on Madeira to prevent the illumination from reaching a harmful level. The Commission resolved to inform the Portugese authorities, through the Secretary General, that the Commission considers the Madeira site of high quality, worthy of protection, and in urgent need of legislation in view of the present light pollution danger.

Less is known about Pico, which is a volcanic cone promising laminar air flow. Osorio reported that the Portugese meteorological authorities are willing to collaborate in improving meteorological data for the peak, and that the Portugese astronomers are willing to assist in astronomical observations in site testing on Pico. The Commission expressed the view, to be communicated to the Portugese authorities, that it would be desirable to initiate a preliminary program of meteorological observations on Pico.

IV. OTHER COMMISSION ACTIVITIES

In addition to the matters discussed above, the Commission decided:

(1) That the next President of Commission 50 shall expand the existing contacts with Commission 42 and the Inter-Union Commission on Frequency Allocations for Radio Astronomy and Space Science, and initiate an exchange of documents.

(2) That closer relationships shall be established with JOSO and the question of expanding the activities of the Commission to include solar astronomy will be explored.

(3) That to fulfill its role as a clearing house of information, the Commission will request site testing groups and institutions active in site protection to send copies of relevant documents to the President. Arrangements will be made by F.G. Smith to set up an archive of this material at Herstmonceux. The Commission will prepare and distribute to Commission members and other interested individuals index lists of available documents and published references.

WORKING GROUP FOR PLANETARY SYSTEM NOMENCLATURE

NOMENCLATURE DU SYSTÈME PLANÉTAIRE

PRESIDENT: P. M. Millman

At the XV General Assembly of the International Astronomical Union in Sydney, Australia, August 21-30, 1973, a new working group was formed, the Working Group for Planetary System Nomenclature (IAU/WGPSN). Unlike most other working groups in the IAU the WGPSN does not report through any commission, or group of commissions, but is responsible only to the Executive Committee of the IAU, and reports directly to this Committee. The WGPSN is charged with formulating and coordinating all topographic nomenclature on the planetary bodies of the solar system and has certain powers of action in the interval between General Assemblies. The establishment of the WGPSN was found advisable because of the recent rapid advance in our knowledge of the topography of the surfaces of planetary bodies, and the necessity of coordinating the approved systems of nomenclature among the different planets and their satellites.

In the period 1973-1976 the WGPSN held three meetings as follows:-

> First Meeting - Ottawa, Canada, June 27 and 28, 1974;
> Second Meeting - Moscow, USSR, July 14 and 18, 1975;
> Third Meeting - Grenoble, France, August 30 and 31, 1976.

The following members of the IAU have served on the WGPSN in the interval noted:-

A. Dollfus	D. Morrison
B.Ju. Levin	T.C. Owen
C.H. Mayer	G.H. Pettengill
D.H. Menzel	S.K. Runcorn
P.M. Millman	B.A. Smith

In addition to those listed above, the following have been members of the various nomenclature task groups, responsible for compiling the detailed material to be presented to the WGPSN:-

K. Aksnes	I.K. Koval
M.S. Bobrov	A.D. Kuz'min
C.R. Chapman	Yu.N. Lipskij
M.E. Davies	M.Ya. Marov
F. El-Baz	H. Masursky
K.P. Florenskij	S. Miyamoto
D. Gautier	A.V. Morozhenko
O.J. Gingerich	C. Sagan
R.M. Goldstein	V.V. Shevchenko
J.E. Guest	V.G. Tejfel'

The nomenclature resolutions passed by the WGPSN, and later approved by the International Astronomical Union, are listed in the following pages.

First Meeting

Resolutions from the First Meeting of the I.A.U. Working Group for Planetary
System Nomenclature, Ottawa, Ontario, June 27 and 28, 1974.

Resolution I

BASIC PRINCIPLES FOR PLANETARY SYSTEM NOMENCLATURE
     (a) Nomenclature is a tool and the first consideration shall be to make it
simple, clear and unambiguous.

     (b) The number of names chosen for each body should be kept to a minimum, and
governed by the anticipated requirements of the scientific community.

     (c) Although there will be exceptions, duplication of the same name on two or
more bodies should be avoided.

     (d) In general, individual names chosen should be single words, and expressed
in the language of origin. Transliteration and pronunciation for various alphabets
should be given, but there will be no translation from one language to another.

     (e) Where possible, consideration should also be given to the traditional
aspects of any nomenclature system, provided that this does not cause confusion.

     (f) Solar system nomenclature shall be international in its choice of names.
Recommendations submitted by I.A.U. National Committees will be considered. Final
approval of any selection is the responsibility of the International Astronomical
Union.

     (g) We must look to the future in general discussions of solar system nomen-
clature and attempt to lay the groundwork for future requirements that will result
from the development of the space program.

Resolution II

LATIN TERMS FOR DIFFERENT TYPES OF FEATURES, TO BE USED IN PLANETARY SYSTEM
NOMENCLATURE
     The following Latin terms, already approved for use on the moon or Mars, are
suitable for use with a nomenclature system on any planet or satellite in the solar
system (plurals are given in brackets):-

Latin Term	Approximate Description
(a) CATENA (Catenae)	a chain or line of craters
(b) CHASMA (Chasmata)	a deep, elongated, steep-sided depression
(c) CRATER (Crateres)	an essentially circular depression
(d) DORSUM (Dorsa)	a ridge
(e) FOSSA (Fossae)	a long, narrow, shallow depression
(f) LABYRINTHUS (Labyrinthi)	a complex of intersecting narrow depressions
(g) MENSA (Mensae)	a flat-topped prominence with cliff-like edges
(h) MONS (Montes)	a mountain
(i) PATERA (Paterae)	an irregular crater, or a complex one with scalloped edges
(j) PLANITIA (Planitiae)	a plain
(k) PLANUM (Plana)	a plateau
(l) RIMA (Rimae)	a fissure
(m) RUPES (Rupes)	a scarp

Latin Term	Approximate Description
(n) THOLUS (Tholi)	a hill
(o) VALLIS (Valles)	a valley
(p) VASTITAS (Vastitates)	an extensive plain

When required, additional Latin terms may be added to this list, but it is recommended that the number of terms used be kept to a minimum. The following terms, already in use on the moon, should be discussed in each case before being used on other planetary bodies:-

(q) LACUS (Lacus)
(r) MARE (Maria)
(s) PALUS (Paludes)
(t) PROMONTORIUM (Promontoria)
(u) SINUS (Sinus)

Resolution III

POSSIBLE NAME CATEGORIES FOR USE IN PLANETARY SYSTEM NOMENCLATURE

Traditionally, the names of distinguished, deceased scientists have generally been used to name craters on the moon and Mars. Although this source can still be used it is obvious, when we examine the future requirements of planetary system nomenclature, and particularly for the case of the other planets and satellites, that we should consider the possibility of using additional name categories.

Recommendations concerning the name categories for any planet and its satellites shall be approved by the Working Group for Planetary System Nomenclature before the individual names are assigned by the Task Group concerned. Task groups shall operate in compliance with Resolution I. It is agreed to prohibit the assignment of names of individuals known primarily

- as religious figures;
- as military leaders, political leaders, and philosophers of the 19th and
  20th centuries.

Some examples of name categories that can, without difficulty, provide several hundred names, and in some cases considerably more, are:-

(a) distinguished, deceased - artists (painters)   Where names of specific individ-
(b) distinguished, deceased - muscians   uals are used the dates of birth
(c) distinguished, deceased - sculptors   and death, and very brief bio-
(d) distinguished, deceased - writers and poets   graphical details, should be
                                                   published.

(e) animals   (j) lakes
(f) birds   (k) minerals
(g) cities   (l) mountains
(h) first names of men and women   (m) rivers
(i) islands   (n) villages

Some examples of name categories capable of providing less than one hundred names are:-

(o) deserts   (s) scientific instruments
(p) fundamental particles   (t) ships of discovery
(q) geographical provinces   (u) the name of the particular planet or
(r) observatories   satellite in various languages

The preceding lists should in no way be considered restrictive.

Eventually, we may have to consider the surface nomenclature for a total of more than thirty different planetary bodies. Hence, the choice of name categories should be made with this in mind.

Resolution IV

SCHEDULES FOR MAP PRODUCTION
The development of lists of names for various bodies in the solar system is an important but time-consuming task that must involve a cooperative effort by representatives of several countries. To avoid decisions hastily made to satisfy contractual deadlines or mission constraints, it is essential that the nomenclature task groups be made aware of these requirements well in advance.

We therefore request the Executive Committee of the I.A.U. to notify those organizations that may be responsible for production of maps of solar-system bodies (e.g. NASA), asking them to inform the IAU/WGPSN of any plans for mapping that will involve deadlines for the availability of names. The IAU/WGPSN should also receive advance notice of any missions that may involve landing sites or areas of reconnaissance requiring special nomenclature.

Resolution V

ADVANCE NOTICE OF MEETINGS
Working Group meetings and Task Group meetings should be scheduled at least six months in advance, if at all possible. When convenient, such meetings might be scheduled in conjunction with international meetings which a majority of members are likely to attend.

Resolution VI

PROCEDURES IN TASK GROUPS
Task-group members unable to attend meetings shall be contacted by the Chairman regarding concurrence in the choice of names. Adequate documentation shall be provided. Lack of response within 45 days (allowed for two-way mail or wire service) shall be regarded as concurrence.

Resolution VII

LUNAR MAPPING
Until the next meeting of the IAU/WGPSN, approximately one year from June, 1974, names of non-scientists shall not be chosen for lunar maps.

Resolution VIII

NOMENCLATURE FOR MERCURY
(a) The classical nomenclature, as used by E.M. Antoniadi, will be adopted for regional names and albedo features but probably not for topographic features.

(b) A maximum of six features will be named for deceased scientists who have made exceptional contributions to the study of this planet.

(c) The craters may be named for birds of the world, or for cities of the world.

(d) Other features may be named for (i) ships of discovery, (ii) names of Mercury or associated with Mercury in various languages, (iii) observatories.

(e) A shorthand notation, similar to that used for Mars, will be sought as a potential means for designating small craters on Mercury.

(f) A Latin term will be chosen for the class of geological feature called in English a "basin".

(g) The following three names, already in provisional use, are approved:-

Kuiper
Caloris
Hun kal

Resolution IX

NOMENCLATURE FOR MARS
    The names Kuiper and Vishniac are approved for craters at the following locations:-

Kuiper, G.P. long. 157   lat. -57
Vishniac, W.   275   -76

Resolution X

NOMENCLATURE FOR MARS
    On the 1:1,000,000 and 1:250,000 series maps of Mars a system is proposed for naming the previously undesignated craters of approximately 5-20 km diameter, and craters 20-100 km in size which have double-letter designations. Names for these craters have been chosen from a list of small towns and villages of the world. Criteria used in compiling the list of names were - (i) names of three or less syllables which are easy to pronounce, (ii) worldwide representation, (iii) names limited to small towns or villages. Two and three syllable names are proposed for 10-20 km undesignated craters and 20-100 km double-lettered craters on the 1:1,000,000 maps; one syllable names are proposed for very small (5-10 km) craters on the 1:250,000 maps. We do not propose at the present time to name the other 6000 double-lettered craters on Mars, but feel that, within the small landing site areas, named craters would be more meaningful as reference points than lettered craters.

Second Meeting

    Resolutions from the Second Meeting of the I.A.U. Working Group for Planetary System Nomenclature, Moscow, USSR, July 14 and 18, 1975.

Resolution I

LUNAR NOMENCLATURE

1. (a) For sheets of the 1:250,000 lunar map series we recommend that where old lettered crater names (Mädler system names) are replaced by approved new names, the old names be printed on the map in brackets under the new names.

    (b) For Edition 3 of the 1:5,000,000 lunar map series we recommend retention of all the lettered crater names now present on Edition 2. Where these lettered craters have been assigned approved new names the old name will be shown in brackets on the map under the new name.

2. For the 1:1,000,000 lunar map series (LAC) we recommend the following sheet names:-

LAC No.	Name	LAC No.	Name
1	Peary	51	Cockcroft
2	Carpenter	52	Joule
3	Anaxagoras	53	Fersman
4	Meton	54	Robertson
5	Petermann	55	Vasco da Gama
6	Schwarzschild	56	Hevelius
7	Karpinskij	57	Kepler
8	Kirkwood	58	Copernicus
9	Brianchon	59	Mare Vaporum
10	Pythagoras	60	Julius Caesar
11	J. Herschel	61	Taruntius
12	Plato	62	Mare Undarum
13	Aristoteles	63	Neper
14	Endymion	64	Babcock
15	Belkovich	65	Ostwald
16	Compton	66	Mendeleev
17	Störmer	67	Mandel'shtam
18	D'Alembert	68	Sharonov
19	Birkhoff	69	Zhukovskij
20	Coulomb	70	Kibalchich
21	Omar Khayyam	71	Michelson
22	Lavoisier	72	Nobel
23	Rümker	73	Riccioli
24	Sinus Iridum	74	Grimaldi
25	Cassini	75	Letronne
26	Eudoxus	76	Montes Riphaeus
27	Geminus	77	Ptolemaeus
28	Gauss	78	Theophilus
29	Fabry	79	Colombo
30	Millikan	80	Langrenus
31	Campbell	81	Ansgarius
32	Chandler	82	Pasteur
33	Schneller	83	Langemak
34	Fowler	84	Dellinger
35	Landau	85	Keeler
36	Lorentz	86	Icarus
37	Russell	87	Korolev
38	Seleucus	88	Vavilov
39	Aristarchus	89	Lucretius
40	Timocharis	90	Lowell
41	Montes Apenninus	91	Eichstadt
42	Mare Serenitatis	92	Byrgius
43	Macrobius	93	Mare Humorum
44	Cleomedes	94	Pitatus
45	Hubble	95	Purbach
46	Joliot	96	Rupes Altai
47	Seyfert	97	Fracastorius
48	Mare Moscoviense	98	Petavius
49	Komarov	99	Humboldt
50	Fitzgerald	100	Hilbert

LAC No.	Name	LAC No.	Name
101	Fermi	121	Apollo
102	Gagarin	122	Brouwer
103	O'Day	123	Rydberg
104	Van de Graaff	124	Phocylides
105	Mohorovicic	125	Schiller
106	Lodygin	126	Clavius
107	Houzeau	127	Hommel
108	Mare Orientale	128	Biela
109	Vallis Inghirami	129	Lyot
110	Schickard	130	Fechner
111	Wilhelm	131	Planck
112	Tycho	132	Hess
113	Maurolycus	133	Minkowski
114	Rheita	134	Fizeau
115	Furnerius	135	Arrhenius
116	Mare Australe	136	Bailly
117	Milne	137	Moretus
118	Jules Verne	138	Manzinus
119	Mare Ingenii	139	Helmholtz
120	Oppenheimer	140	Schrödinger
		141	Minnaert
		142	Zeeman
		143	Hausen
		144	Amundsen

3. We recommend the approval of the following list of crater names as assigned on the moon:-

AL-KHWARIZMI
780–850?
Arab mathematician
7.0 N, 107.0 E

ARTSIMOVICH (Lev A.)
(replaces Diophantus A)
1909–1973
Russian Physicist
27.5 N, 36.5 W

AVERY (Oswald T.)
(replaces Gilbert U)
1877–1955
Canadian Biologist
1.2 S, 81.3 E

BACK (Ernst E.A.)
(replaces Schubert B)
1881–1959
German physicist
1.2 N, 80.6 E

BLACK (Joseph)
(replaces Kästner F)
1728–1799
French chemist
9.0 S, 80.4 E

BOREL (Félix Édouard Émile)
(replaces Le Monnier C)
1871–1956
French mathematician
22.7 N, 26.4 E

CAVENTOU (Joseph Bienaimé)
(replaces La Hire D)
1795–1877
French cehmist
29.8 N, 29.3 W

DALE (Henry Hallett)
1875–1968 (Nobel, 1936)
English Physiologist
9.4 S, 83.0 E

ELMER (Charles W.)
1872–1954
American amateur astronomer
10.0 S, 84.2 E

ESCLANGON (Ernest B.)
(replaces Macrobius L)
1876–1954
French astronomer
21.5 N, 42.0 E

FREDHOLM (Erik Ivar) (replaces Macrobius D)	1866-1927 Swedish mathematician	18.5 N, 46.5 E
GARDNER (Irvine Clifton) (replaces Vitruvius A)	1889-1972 American physicist	17.7 N, 33.7 E
GAST (Paul W.)	1930-1973 American geochemist	Dorsum near 24 N, 9 E
GOLGI (Camillo) (replaces Schiaparelli D)	1843-1926 (Nobel, 1906) Italian cytologist	27.8 N, 59.9 W
GREAVES (William M.H.) (replaces Lick D)	1897-1955 British astronomer	13.2 N, 52.6 E
HELMERT (Friedrich Robert)	1843-1917 German geodicist-astronomer	7.5 S, 87.7 E
ISAEV (Aleksej M.)	1908-1971 Russian rocket engineer	17.5 S, 147.5 E
KROGH (Schack August Steenberg) (replaces Auzout B)	1874-1949 (Nobel, 1920) Danish physiologist	9.7 N, 65.8 E
LANDSTEINER (Karl) (replaces Timocharis F)	1868-1943 (Nobel, 1930) Austrian/American pathologist	31.2 N, 14.9 W
LEBESGUE (Henri Leon)	1875-1941 French mathematician	5.1 S, 89.0 E
LIOUVILLE (Joseph)	1809-1882 French mathematician	4.2 N, 73.7 E
McADIE (Alexander George)	1863-1943 American meteorologist	2.0 N, 92.4 E
McDONALD (W. Johnson) (replaces Carlini B)	1844-1926 American amateur astronomer	30.4 N, 21.0 W
MORLEY (Edward Williams) (replaces MacLaurin R)	1838-1923 American chemist	2.8 S, 64.6 E
NECHO	610-593 B.C. Egyptian geographer	5.0 S, 123.0 E
POMORTSEV (Mikhail M.) (replaces Dubiago P)	1851-1916 Russian rocket scientist	0.7 N, 67.0 E
PUPIN (Michael I.) (replaces Timocharis K)	1858-1935 Yugoslavian/American physicist	23.8 N, 11.0 W
RAMAN (Chandrasekhara V.) (replaces Herodotus D)	1888-1970 (Nobel, 1930) Indian physicist	27.0 N, 55.2 W
RESPIGHI (Lorenzo) (replaces Dubiago C)	1824-1890 Italian astronomer	4.0 N, 72.0 E
RUTHERFORD (Ernest)	1871-1937 (Nobel, 1908) English physicist	10.5 N, 137.0 E
SAMPSON (Ralph Allen)	1866-1939 British astronomer	29.7 N, 16.7 W
SANTOS-DUMONT (Alberto) (replaces Hadley B)	1873-1932 Brazilian pioneer in aeronautics	27.7 N, 4.8 E
SCHEELE (Carl Wilhelm) (replaces Letronne D)	1742-1786 Swedish chemist	9.5 S, 38.0 W

SHERRINGTON (Charles Scott)	1856-1952 (Novel, 1932) English neurophysiologist	11.0 S, 117.5 E
SLOCUM (Frederick)	1873-1944 American astronomer	3.0 S, 89.0 E
STEWARD (John Quincy) (replaces Dubiago Q)	1894-1972 American astronomer	2.2 N, 67.0 E
SWASEY (Ambrose)	1846-1937 American inventor	5.5 S, 89.7 E
SWIFT (Lewis) (replaces Peirce B)	1820-1913 American astronomer	19.4 N, 53.5 E
TALBOT (William Henry Fox)	1800-1877 English chemist	2.3 S, 85.3 E
TOSCANELLI (Paolo dal Pozza) (replaces Aristarchus C)	1397-1482 Italian physician, map-maker	27.9 N, 47.6 W
TOWNLEY (Sidney Dean) (replaces Apollonius G)	1867-1946 American astronomer	3.8 N, 63.2 E
VAN ALBADA (Gale Bruno) (replaces Auzout A)	1912-1972 Dutch astronomer	9.8 N, 64.5 E
VAN VLECK (John Monroe) (replaces Gilbert M)	1833-1912 American mathematician, astronomer	1.4 S, 78.2 E
WARNER (Worcester Reed)	1846-1929 American inventor	3.9 S, 87.3 E
WEIERSTRASS (Karl Theodor W.) (replaces Gilbert N)	1815-1897 German mathematician	0.9 S, 72.3 E
ZASYADKO (Alexander D.)	1779-1837 Russian rocket scientist	3.8 N, 94.0 E

4. We recommend the approval of the following list of first names of men and
    women for use as crater names in restricted areas of the moon:-

Agnes	Ching-te	Isis	Natasha	Soraya
Akis	Christel	Ivan		Stella
Alan	Cleopatra		Osama	Sung-mei
Aloha	Courtney		Osiris	Susan
André		Jehan	Osman	
		Jerik		
Ango	Dag	Jomo	Patricia	Taizo
Ann	Delia	José	Pierre	Thera
Annegrit	Dieter	Julienne	Priscilla	
Ardeshir	Dilip			Vera
Artemis		Karima	Ravi	Verne
	Edith	Kathleen	Reiko	Vladimir
		Kira.	Robert	
Bawa	Fairouz	Krishna	Rocco	Wan-yu
Béla	Felix		Rosa	
Brigitte		Manuel	Rudolf	Yoshi
	Ganau	Marcello	Ruth	Yuri
Carlos		Mary		
Carmen	Harold	Mavis		
Chang-ngo	Ian	Michael	Siegfried	Zahia
Charles	Ina	Monira	Sita	

5. (a) We recommend that the following list of generic Latin terms be used hence-
         forth in assigning names to non-crater features on the moon:-*

Catena	Mons	Rupes
Dorsum	Planum	Sinus
Lacus	Promentorium	Tholus
Mare	Rima	Vallis

    (b) We recommend that the following generic Latin term be retained where
         presently assigned on the moon but not be used in the future assignment
         of names:-†

         Palus

6. In three exceptional cases we recommend the adoption of the following names,
    already in use on the moon:-

         Reiner Gamma
         Mons Hadley Delta
         Mons Maraldi Gamma

                              Resolution II

MERCURY NOMENCLATURE

1. We recommend approval for the following names for features on Mercury, all of
    which have been assigned by the Mercury Nomenclature Task Group to specific
    features:

PLANITIA	RUPES
Borealis	Astrolabe
Budh	Discovery
Odin	Endeavour
Sobkou	Fram
Suisei	Heemskerck
Tir	Hero
	Mirni
MONTES	Santa María
	Victoria
Caloris	Vostock
DORSUM	VALLIS
Antoniadi	Arecibo
Schiaparelli	Goldstone
	Haystack
	Simeiz

---

* It was felt that for the terms Lacus, Mare, Planum, Promontorium, Sinus and
Tholus, there should be a case by case discussion before using them with new
names.

† It was recognized that the term Oceanus was suitable for use in its present
application on the moon but that there was little likelihood of the need for this
term in conjunction with new names.

2. We recommend approval of the following bank of names to be applied as needed to Rupes on Mercury:

Adventure	Gjøa	Persej	Resolution	Zarya
Beagle	Kon Tiki	Pinta	São Gabriel	Zeehaen
Endurance	Niña	Pourquoi-Pas	Vityaz'	

3. We propose approval for the names on the attached map of albedo features observed telescopically. These names can be used for albedo features only, and are provided at this time for use by telescopic observers (see page 332 for map).

4. For the initial 1:5,000,000 topographic maps of the parts of Mercury photographed by Mariner 10, we suggest that map sheets be assigned dual names. The first sheet name will be that of a prominent topographic feature, and the second name, to be placed in parentheses, will be that of an albedo feature as indicated (see page 333 for map).

Resolution III

VENUS NOMENCLATURE

1. It is proposed that two themes, particularly appropriate for Venus, be adopted as a source of names for surface features. The more important of these themes derives from the feminine mystique long associated with Venus, and it is proposed that all craters received feminine names. Basins and extended basin-like plains (i.e. like Mare Imbrium on the moon and Hellas on Mars) are to be given names of goodesses from ancient cultures. Craters larger than about 100 km in diameter (the actual size will depend on the number of large craters found) should be assigned female names drawn from mythology, while craters of smaller diameter would receive feminine first names drawn from all cultures. The lowest crater size limit to be named cannot be specified in advance of observations.

2. The second of these themes derives from the extensive and opaque cloud cover which surrounds the planetary sphere, and which requires the use of radio and other techniques in order to study and map the solid surface. It is, therefore, proposed to assign the names of deceased radio, radar and space scientists to topographic features. These features will be described using the classes and corresponding latin names as defined for the moon and Mars where applicable.

3. An exception to the above themes is proposed in the case of a single prominent surface feature which was among the first to be observed, and which has served to help define the origin of the official IAU system of longitudes for the planet. This feature, some 1000 km in diameter, is located at approximately -25° latitude and 0° longitude. It is proposed that this feature be called Alpha.

Resolution IV

MARS NOMENCLATURE
    We recommend that the following actions be accepted:

1. Add the following terms to describe specific topographic features on Mars:

Chaos   - distinctive areas of broken terrain

Cavus   - hollows, irregular steep-sided depressions usually in
(plural, Cavi)   arrays or clusters

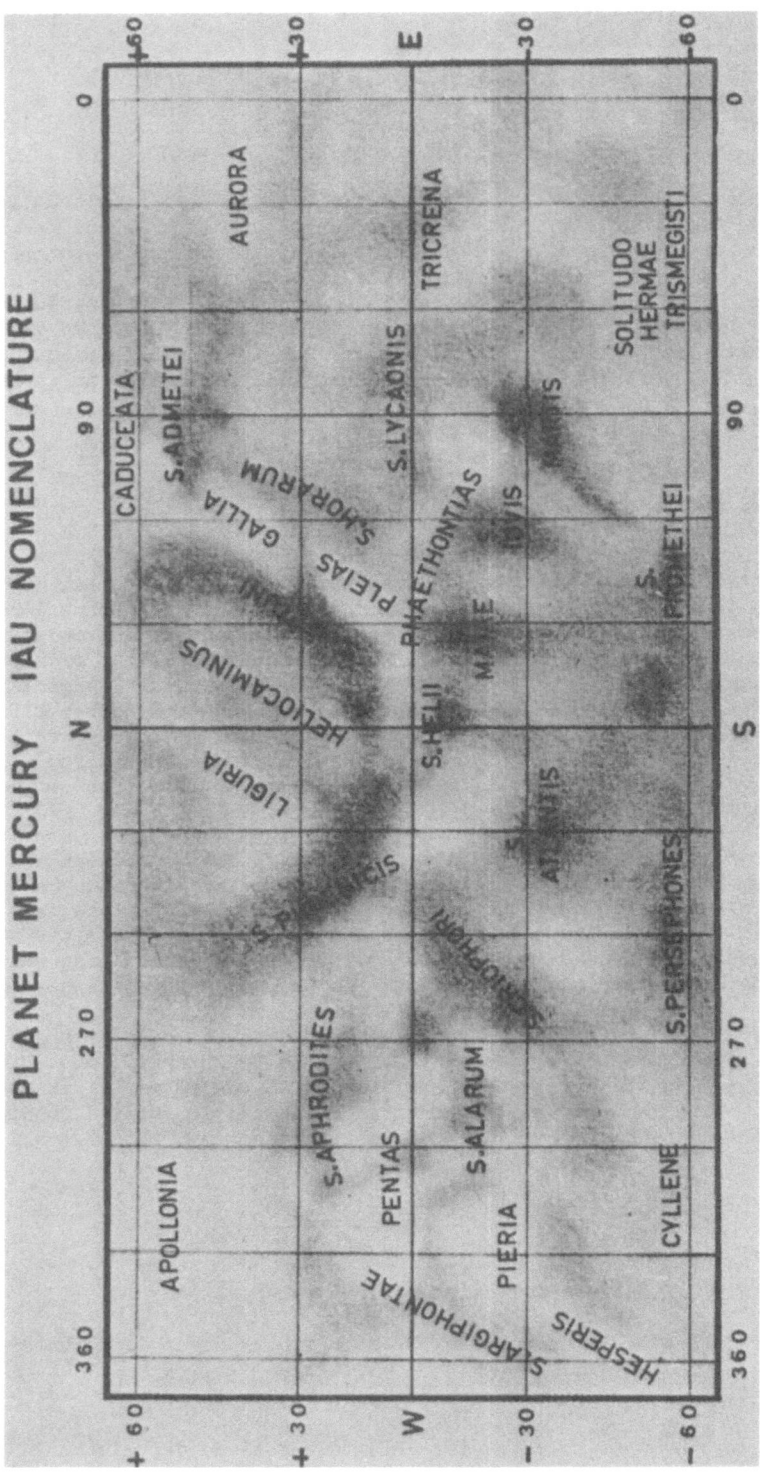

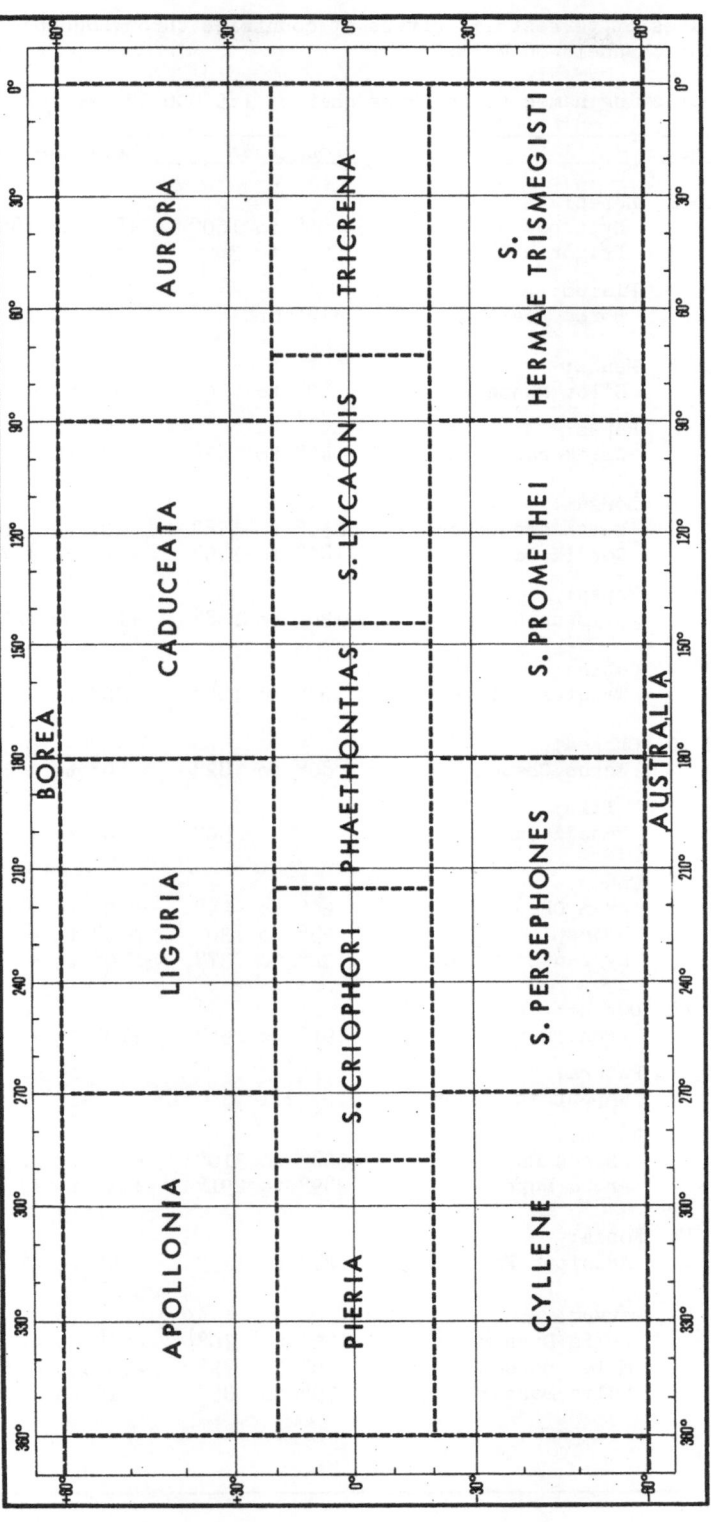

2. Use the names of terrestrial rivers of countries throughout the world for small valleys and channels on Mars.

3. Add the following names to the Mars charts, 1:5,000,000 series:

Sheet		Longitude W	Latitude
MC 1	Rupes:		
	Hyperboreus Rupes	60° to 160°	+80° to +85°
	Frigoris Rupes	20° to 305°	+80°
	Planum:		
	Borea Planum	circumpolar	+80° to +90°
MC 4	Mensa:		
	Siloe Mensae	10° to 20°	+30° to +40°
	Rupes:		
	Nilokeras Rupes	40° to 52°	+32° to +40°
MC 8	Dorsum:		
	Eumenides Dorsum	156° to 158°	0° to +10°
	Gordii Dorsum	142° to 146°	+ 1° to + 7°
	Rupes:		
	Olympus Rupes	130° to 138°	+15° to +23°
MC 9	Catena:		
	Tractus Catena	103° to 104°	+20° to +35°
MC 10	Chasma:		
	Echus Chasma	80° to 82°	0° to + 7°
	Vallis:		
	Nanedi Vallis	50°	0° to + 7°
MC 11	Chaos:		
	Aram Chaos	20° to 24°	0° to + 5°
	Hydaspis Chaos	25° to 30°	+ 1° to + 4°
	Hydraotes Chaos	32° to 37°	0° to + 3°
MC 13	Dorsum:		
	Arena Dorsum	291° to 294°	+10° to +16°
	Patera:		
	Nepenthes Patera	271°	+12°
	Rupes:		
	Phison Rupes	308° to 310°	+26° to +27°
	Arena Rupes	289° to 291°	+11° to +16°
MC 17	Fossa:		
	Aganippe Fossa	128° to 129°	− 4° to −11°
MC 18	Dorsum:		
	Felis Dorsum	63° to 70°	−20° to −27°
	Melas Dorsum	70° to 75°	−15° to −23°
	Solis Dorsum	79° to 85°	−21° to −27°

Sheet		Longitude W	Latitude
MC 19	Chaos:		
	Margaritifer Chaos	17° to 22°	- 7° to -11°
	Fossa:		
	Erythraea Fossa	29° to 32°	-28° to -29°
	Vallis:		
	Samara Valles	18° to 20°	-22° to -26°
	Ladon Valles	27° to 30°	-21° to -24°
	Uzboi Vallis	34°	-26°
MC 26	Fossa:		
	Oceanidum Fossa	27° to 31°	-60° to -63°
	Rupes:		
	Bosporos Rupes	53° to 60°	-39° to -47°
	Ogygis Rupes	53° to 55°	-33° to -35°
	Tholus:		
	Charitum Tholus	41°	-55°
MC 30	Dorsum:		
	Argentea Dorsum	15° to 40°	-76° to -80°
	Brevia Dorsa	280° to 310°	-68° to -73°
	Planum:		
	Australis Planum	circumpolar	-80° to -90°
	Rupes:		
	Promethei Rupes	230° to 315°	-16° to -80°
	Hypernotius Rupes	185° to 250°	-19° to -83°
	Ultimi Rupes	200°	-65° to -77°
	Chasma:		
	Thyles Chasma	230° to 235°	-69° to -73°
	Cavi:		
	Angusti Cavi	60° to 90°	-70° to -84°
	Sisyphi Cavi	340° to 0°	-76° to -82°
	Cavi Novi	320° to 340°	-66° to -70°
	Ultimi Cavi	195° to 210°	-73° to -75°

4. Change Pavonis Patera to Ullyses Patera 121°.5 +3°.

5. The following names for the Viking Landing Site Maps are proposed:

Cydonia Region	Longitude W	Latitude	Country of Origin
Apt	9°.5	+40°.1	France
Arandas	15.1	+42.6	Mexico
Banberg	3.1	+39.9	West Germany
Can	14.6	+48.4	Turkey
Cray	16.2	+44.4	U.K.(England)
Eagle	8.2	+44.0	USA
Eil	9.7	+42.0	Somalia
Elath	13.7	+46.1	Israel
Esk	7.0	+45.4	Australia
Faith	11.9	+43.2	USA
Freedom	9.0	+43.6	USA
Garm	9.2	+48.4	USSR

Cydonia Region	Longitude W	Latitude	Country of Origin
Gol	10°.7	+47°.4	Norway
Hope	10.3	+45.1	Canada
Jen	10.6	+40.1	Nigeria
Laf	5.9	+48.3	Cameroon
Lagarto	8.4	+50.2	Brazil
Land	8.8	+48.4	USA
Lota	11.9	+46.5	Chile
Mohawk	5.4	+43.2	USA
Tem'	9.5	+42.1	USSR
Vils	11.7	+39.2	Austria
Yakima	3.2	+43.3	USA

Chryse Region

	Longitude W	Latitude	Country of Origin
Batoş	29.5	+21.7	Romania
Bled	31.4	+21.8	Yugoslavia
Bok	31.6	+20.8	New Guinea
Cave	35.6	+21.9	New Zealand
Concord	34.1	+16.9	USA
Gold	31.3	+20.2	USA
Hamelin	32.8	+20.4	West Germany
Kin	34.4	+20.4	Japan
Kok	28.1	+15.7	Sarawak
Libertad	29.4	+23.3	Venezuela
Lins	29.8	+15.9	Brazil
Lod	31.6	+21.2	Israel
Luck	36.9	+17.4	USA
Mut	35.7	+22.7	Turkey
Nune	38.6	+17.7	Mozambique
Nutak	30.3	+17.6	Canada
Oraibi	32.4	+17.4	USA
Ore	33.9	+16.9	Congo
Pylos	30.1	+16.9	Greece
Ribe	29.2	+16.6	Denmark
Shawnee	31.5	+22.7	USA
Soochow	28.9	+16.8	China
Sūf	38.2	+16.6	Jordan
Surt	30.6	+17.0	Libya
Taxco	40.0	+20.8	Mexico
Tile	28.6	+17.8	Somalia
Vaux	42.8	+18.1	France
Wahoo	33.7	+23.5	USA
Wabash	33.7	+21.5	USA
Warra	37.5	+20.8	Australia
Yala	36.5	+17.5	Thailand
Yar	39.1	+22.5	USSR
Yat	29.1	+18.3	Niger
Yoro	28.0	+23.0	Honduras
Yuty	34.1	+22.4	Paraguay
Zir	36.6	+18.7	Turkey
Zuni	29.6	+19.6	USA

6. Delete the name Hippalus Tholus 89°W +76° on Sheet MC 1.

7. Change the term for the crater Galileo to Galilaei.

<u>Resolution V</u>

OUTER SOLAR SYSTEM NOMENCLATURE

1. The names of new satellites should follow the traditions established by the
    existing names for satellites in a given system.

2. Within this guideline the discoverer of a new satellite should be free to choose
    the name for his discovery.

3. In the case of Jupiter, there are now many reasons for applying names to the
    satellites that presently only have numbers. Following the rule given in V 1
    above, we propose the following names:-

<u>the innermost satellite</u>

J V   Amalthea

<u>the outer satellites</u>

J VI	Himalia	J VIII	Pasiphae
J VII	Elara	J IX	Sinope
J X	Lysithea	J XI	Carme
J XIII	Leda	J XII	Ananke

The left hand column contains satellites with <u>direct</u> orbits, and names ending
in "a". The right hand column contains satellites with retrograde orbits, and
names ending in "e". The name "Leda" was proposed by the discoverer of J XIII,
Dr. Charles Kowal.*

4. In anticipation of the discovery of surface features on the satellites of
    Jupiter and Saturn, we proposed to use the mythologies of various cultures as
    sources of names for such features. We can use epics from a given mythology to
    provide a natural association for names of various types of features in a given
    area or on a given satellite. We also propose to use words from Esperanto to
    name features on some of the smaller satellites.†

---

* We acknowledge with gratitude our indebtedness to Dr. Jürgan Blunck for his essential assistance in the development of satellite names for Jupiter. We depart from
Dr. Blunck's proposal in two respects: We are not concerned about duplication of
names already assigned to asteroids. Such duplication already exists and has not
led to any confusion. We omit the name "Thebe" from consideration, because it is
too close to "Phoebe", already assigned to a satellite of Saturn.

The tradition of using only numbers for J V - J XII stems from Barnard's discovery
of J V at a time when even J I - IV were formally only recognized by number. This
tradition is now out of favor, and at least two informal name systems are in use
in some places. We also face the possible embarrassment of naming features on an
unnamed object. It therefore seems necessary to name these satellites. Dr. Kowal
has expressed a preference for using only numbers - his choice of "Leda" was
offered if we decided to name all the satellites. Since this was our decision, we
have accepted his suggested name.

†We shall try to adhere to whatever general practice is established by other task
groups in the naming of topographic features within the bank of names we establish
from our sources - e.g., latin names for types of feature: Rupes, Vallis, etc.;
person names for craters: Jason, Thor, Gilgamesh, etc. We must recognize, however, the possibility that entirely new types of topographic features may be
found on the satellites of the outer planets.

5. Because of the existing confusion in the designation of the ring system of Saturn, we propose the following scheme:

   (a) We adopt the traditional divisions known as the A,B and C rings.

   (b) For material inside the inner boundary of ring C, we propose the designation D.

   (c) For material outside the outer boundary of ring A, we propose the designation E.

   (d) If additional divisions are discovered within any of these segments, we propose the use of subscripts such as A_1, A_2, etc.

Action Following Second Meeting

Resolution approved by the WGPSN in October, 1975.

Resolved that the large craters on Mercury be named after great contributors to the humanities and the arts; including (but not limited to): authors of drama, prose and poetry; painters; sculptors; architects; composers; and musicians.

Third Meeting

Resolutions from the Third Meeting of the I.A.U. Working Group for Planetary System Nomenclature, Grenoble, France, August 30 and 31, 1976.

Resolution I

PUBLICATION OF PLANETARY SYSTEM NOMENCLATURE
   We recommend that:-

1. All official actions on nomenclature, approved by the I.A.U. at a General Assembly, be printed in the Transactions of the I.A.U.

2. That there be produced, for wide distribution, a comprehensive publication including all planetary nomenclature resolutions and lists of names officially approved by the I.A.U. up to and including the XVI General Assembly; and that all future approved planetary nomenclature be published in supplementary volumes following each General Assembly.

Resolution II

LUNAR NOMENCLATURE

1. We recommend approval of the following list of 107 names of scientists and 13 Latin words as now assigned and printed on the 1:250,000 lunar map series and marked "provisional".

ABETTI (A.)	1846-1928 Italian astronomer	19.8 N, 27.7 E
ACOSTA (C.) (replaces Langrenus C)	c.1525-c.1594 Spanish; natural history, medicine	5.5 S, 60.0 E
AGRICOLA, MONTES (Georgius)	1494-1555 German earth scientist	LAC 38

AL-BAKRI (A.A.) (replaces Tacquet A)	1010–1094 Spanish/Arabian geographer	14.5 N, 20.3 E
ALDROVANDI, DORSA (Ulisse)	1522–1605 Italian earth scientist	LAC 42
AL-MARRAKUSHI (replaces Langrenus D)	fl. 13th century Muslim astronomer, mathematician	10.2 S, 55.8 E
AL-TUSI	fl. 13th century Muslim astronomer, mathematician	7.0 N, 120.0 E
AMEGHINO (F.) (replaces Apollonius C)	c. 1854–1911 Italian anthropologist	3.4 N, 57.0 E
AMMONIUS (replaces Ptolemaeus A)	c. 517–526 Greek/Egyptian philosopher	8.5 S, 0.8 W
AMONTONS (Guillaume)	1663–1705 French physicist	5.2 S, 46.8 E
AMORIS, SINUS	Latin name meaning love	LAC 43
ANDRONOV (Aleksandr A.)	1901–1952 Russian physicist	22.7 S, 146.0 E
ANDRUSOV, DORSA (Nicolai I.)	1861–1924 Czechoslovakian palaeontologist	LAC 80
ANVILLE (J.B.) (replaces Taruntius G)	1697–1782 French cartographer	2.0 N, 49.5 E
ARDUINO, DORSUM (Giovanni)	1713–1795 Italian earth scientist	LAC 39
ARGAND, DORSA (Emile)	1879–1940 French earth scientist	LAC 39
ARMINSKI (Franciszek)	1789–1848 Polish astronomer	16.4 S, 154.1 E
ASADA (G.) (replaces Taruntius A)	1734–1799 Japanese astronomer	7.5 N, 49.9 E
ASPERITATIS, SINUS (replaces Torricelli R and surrounding mare)	Latin word meaning roughness	LAC 78
ATWOOD (G.) (replaces Langrenus K)	1745–1807 English mathematician, physicist	5.7 S, 57.5 E
AZARA, DORSUM (F. de)	1742–1811 Spanish earth scientist	LAC 42
BALANDIN (A.A.)	1898–1967 Russian chemist	19.0 S, 152.5 E
BANCROFT (W.D.) (replaces Archimedes A)	1867–1953 American chemist	28.0 N, 6.5 W
BARLOW, DORSA (W.)	1845–1936 English earth scientist	LAC 61
BEKETOV (N.N.) (replaces Jansen C)	1827–1911 Russian chemist	16.3 N, 29.2 E
BENEDICT (F.G.)	1870–1957 American chemist, physiologist	4.7 N, 141.7 E

BERGMAN (T.O.)	1735-1784 Swedish chemist, mineralogist, astronomer	7.4 N, 137.5 E
BILHARZ (T.) (replaces Langrenus F)	1825-1862 German anatomist, zoologist	5.5 S, 56.2 E
BINGHAM (H.)	1875-1956 American explorer	8.0 N, 115.0 E
BOBILLIER (E.) (replaces Bessel E)	1798-1840 French geometrist, mechanics	19.5 N, 15.5 E
BOETHIUS (replaces Dubiago U)	480?-524 Greek physicist	5.8 N, 72.3 E
BOMBELLI (R.) (replaces Apollonius T)	1526-1572 Italian algebraist	5.8 N, 56.0 E
BONITATIS, LACUS	Latin name meaning goodness	LAC 43
BOWDITCH (Nathaniel)	1773-1848 American navigator, astronomer, mathematician	25.0 S, 103.0 E
BREWSTER (D.) (replaces Römer L)	1781-1868 Scottish; optics	23.5 N, 34.7 E
BUCHER, DORSUM (W.H.)	1889-1965 Swiss earth scientist	LAC 39
BUCKLAND, DORSUM (W.)	1784-1856 English earth scientist	LAC 41,42
BURNET, DORSA (Thomas)	1635-1715 English earth scientist	LAC 38
CARTAN (E.J.) (replaces Apollonius D)	1869-1951 French mathematician	4.5 N, 59.3 E
CATO, DORSA ('The Censor')	234 B.C. - 149 B.C. Greek founder of geological engineering	LAC 61
CAYEUX, DORSUM (Lucien)	1864-1944 French sedimentary petrographer	LAC 62
CLOOS, DORSUM (Hans)	1885-1951 German earth scientist	LAC 64
CONCORDIAE, SINUS	Latin name meaning harmony	LAC 61
CONDON (Edward U.) (replaces Webb R)	1902-1974 American physicist	2.2 N, 60.3 E
VON COTTA, DORSUM (Carl B.)	1808-1879 German earth scientist	LAC 42
CRILE (G.) (replaces Proclus F)	1864-1943 American surgeon	14.5 N, 46.0 E
CTESIBIUS	ca. 100 B.C. Egyptian physicist	1.5 N, 118.5 E
CUSHMAN, DORSUM (J.A.)	1881-1949 American micropaleontologist	LAC 61
DANA, DORSA (James D.)	1813-1895 American earth scientist	LAC 64

D'ARSONVAL (Jacques Aresne)	1851–1940 French physicist	10.0 S, 124.3 E
DOLORIS, LACUS	Latin word meaning anguish	LAC 41
EPPINGER (H.) (replaces Euclides D)	1879–1946 Czechoslovakian physician	9.5 S, 25.8 W
EWING, DORSA (W. Maurice)	1906–1974 American geophysicist	LAC 75
FABBRONI (Giovanni V.M.) (replaces Vitruvius E)	1752–1822 Italian chemist	18.6 N, 29.3 E
FAHRENHEIT (Gabriel Daniel) (replaces Picard X)	1686–1736 Dutch physicist	13.3 N, 61.6 E
FELICITATIS, LACUS	Latin word meaning happiness	LAC 41
FIDEI, SINUS	Latin word meaning trust	LAC 41
FINSCH (O.F.H.)	1839–1917 German zoologist	23.7 N, 20.7 E
FISCHER (Emil) " (Hans)	1852–1919 German chemist (Nobel, 1902) 1881–1945 German organic chemist(Nobel,1930)	8.2 N, 142.6 E
GAUDII, LACUS	Latin word meaning joy	LAC 42
GEIKIE, DORSA (Archibald)	1835–1924 Scottish geologist	LAC 80
GEISSLER (Heinrich) (replaces Gilbert D)	1814–1879 German physicist	2.3 S, 76.5 E
GLAUBER (Johann Rudolph)	1603/4–1668/70 German chemist	11.7 N, 142.8 E
GRABAU, DORSUM (Amadeus W.)	1870–1946 American earth scientist	LAC 40
GRAVE (Dmitriy A.) " (Ivan P.)	1863–1939 Russian mathematician 1874–1960 Russian; ballistics	17.0 S, 150.0 E
GUETTARD, DORSUM (Jean-Etienne)	1715–1786 French earth scientist	LAC 76
HARDEN (Sir Arthur)	1865–1940 English chemist (Nobel, 1929)	5.5 N, 143.7 E
HARKER, DORSA (Alfred)	1859–1939 English petrologist	LAC 62
HEIM, DORSUM (Albert)	1849–1937 Swiss earth scientist	LAC 40
HERON (also written Hero)	Ca. 100 B.C. Egyptian physicist	1.0 N, 119.6 E
HIEMALIS, LACUS (replaces Menelaus R)	Latin word meaning winter	LAC 60
HIGAZY, DORSUM (Riad)	1919–1967 Arab earth scientist	LAC 40

HONORIS, SINUS	Latin word meaning honor	LAC 60
HUME (David)	1711-1776 Scottish historian and philosopher	4.5 S, 90.5 E
IBN BATTUTA (replaces Goclenius A)	c.1304-c.1368/9 Arabian geographer, traveller	7.0 S, 50.5 E
IBN FIRNAS	A.D. 274/887 Spansih; humanities, technology	7.5 N, 122.5 E
IBN-RUSHD (Averroës) (replaces Cyrillus B)	1126-1198 Muslim philosopher, physician	11.5 S, 21.5 E
KOSBERG (C.A.)	1903-1965 Soviet; aircraft construction	20.2 S, 149.5 E
KUIPER (Gerard P.) (replaces Bonpland E)	1905-1973 American astronomer	10.0 S, 22.5 W
KUNDT (August) (replaces Guericke C)	1839-1894 German physicist	11.5 S, 11.5 W
LANDER (Richard Lemon)	1804-1834 English explorer	15.3 S, 131.5 E
LEAKEY (Louis S.B.) (replaces Censorinus F)	1903-1972 British archaeologist	3.0 S, 37.5 E
LENITATIS, LACUS (replaces Manilius N)	Latin word meaning softness	LAC 60
LINDBERGH (Charles A.) (replaces Messier G)	1902-1974 American aviator	5.3 S, 53.0 E
LISTER, DORSA (Martin)	1638-1712 British stratigrapher	LAC 42
MACMILLAN (William Duncan)	1871-1948 American mathematician	24.2 N, 8.0 W
MOISSAN (Ferdinand F.H.)	1852-1907 (Nobel, 1906) French chemist	5.0 N, 137.4 E
MORO, MONS (A. Lazzaro)	1687-1740 Italian earth scientist	12.0 S, 19.6 W
NAONOBU (Ajima) (replaces Langrenus B)	c. 1732-1798 Japanese mathematician	4.5 S, 57.7 E
NICOL, DORSUM (William)	1768-1851 Scottish physicist	LAC 42
NIGGLI, DORSUM (Paul)	1888-1953 Swiss earth scientist	LAC 38
NOBILI (Leopoldo) (replaces Schubert Y)	1784-1835 Italian physicist	0.5 N, 76.0 E
NORMAN (Robert) (replaces Euclides B)	ca. 1590 English physicist, navigator	12.0 S, 30.5 W
ODII, LACUS	Latin word meaning hatred	LAC 41
OPPEL, DORSUM (Albert)	1831-1865 German palaeontologist	LAC 44,62
OWEN, DORSUM (George)	1552-1613 English earth scientist	LAC 42

PENCK, MONS (Albrecht)	1858–1945 German geographer	10.0 S, 21.5 E
PETIT (Alexis Therese) (replaces Apollonius W)	1791–1820 French physicist	2.5 N, 64.0 E
RANKINE (William John M.)	1820–1872 Scottish physicist, engineer	3.4 S, 71.2 E
RASPLETIN (Aleksandr A.)	1908–1967 Soviet radio engineer	22.5 S, 151.7 E
RICHARDS (Theodore W.)	1868–1928 (Nobel, 1914) American chemist	8.0 N, 140.0 E
RUBEY, DORSA (William W.)	1898–1974 American geologist	LAC 75
SCILLA, DORSUM (Agostino)	1639–1700 Earth scientist	LAC 38
SMIRNOV, DORSA (Sergei S.)	1895–1947 Russian earth scientist	LAC 42
SMITHSON (James) (replaces Taruntius N)	1765–1829 English chemist	2.5 N, 53.5 E
SODDY (Frederick)	1877–1956 (Nobel, 1921) British physicist	0.8 N, 121.5 E
SOLITUDINIS, LACUS	Latin word meaning loneliness	LAC 100
SOMERVILLE (Mary Fairfax) (replaces Langrenus J)	1780–1872 Scottish physicist, mathematician	8.0 S, 65.0 E
SORBY, DORSA (Henry C.)	1826–1908 English chemist	LAC 42
STILLE, DORSA (Hans)	1876–1966 German earth scientist	LAC 40
TERMIER, DORSUM (Pierre)	1859–1930 French geologist	LAC 62
TOLANSKY (Samuel) (replaces Parry A)	1907–1973 English physicist	9.8 S, 16.0 W
VAN BIESBROECK (G.A.) (replaces Krieger B)	1880–1974 American astronomer	28.8 N, 45.6 W
VIVIANI (Vincenzo)	1622–1703 Italian physicist, mathematician	6.0 N, 117.0 E
WHISTON, DORSA (William)	1666–1753 English earth scientist	LAC 38
WINTHROP (John) (replaces Letronne P)	1714–1779 American physicist and astronomer	10.8 S, 44.5 W
WROBLEWSKI (Sigmund von)	1845–1888 Polish physicist	24.0 S, 152.7 E
XENOPHON	434–355 B.C. Athenian historian	22.7 S, 121.9 E
ZÄHRINGER (Josef) (replaces Taruntius E)	1929–1970 German cosmochemist	5.7 N, 40.2 E
ZIRKEL, DORSUM (Ferdinand)	1838–1912 German earth scientist	LAC 40

2. We recommend approval of the following list of 195 names of scientists as a
   bank of names for future use on the moon. Nobel prize winners are starred.

AIKEN, Howard Hathaway	1900-1973	American mathematician
ALBERT, Abraham Adrian	1905-1972	American mathematician
ALBERTUS MAGNUS	c. 1200-1280	German natural scientist
*ALDER, Kurt	1902-1958	German chemist (Nobel, 1950)
ALDEROTTI, Taddeo	1223-c. 1295	Italian physician
ALDINI, Giovanni	1762-1834	Italian physicist
ANUCHIN, Dmitrii N.	1843-1923	Russian geographer, anthropologist
BACCELLI, Guido	1832-1916	Italian physician
BACHMANN, Augustus Q.	1652-1723	German botanist
BAER, Karl Ernst von	1792-1876	Estonian embryologist, biologist
BAILEY, Liberty Hyde, Jr.	1858-1954	American botanist
BAIRD, Spencer Fullerton	1823-1887	American zoologist
BALBIANI, Edouard G.	1823-1899	French biologist
BANKS, Joseph	1743-1820	English botanist
BANTI, Guido	1852-1925	Italian pathologist
*BÁRÁNY, Robert	1876-1936	Austrian physician (Nobel,1914)
*BARKLA, Charles G.	1877-1944	British physicist (Nobel, 1917)
BARTOLOTTI, Gian Giacomo	c. 1470-d. after 1530	Italian physician
BARTON, Benjamin S.	1766-1815	American botanist, zoologist
BASSI, Agostino M.	1773-1856	Italian natural scientist
BATES, Henry W.	1825-1892	English natural scientist
BECCARI, Nello	1883-1957	Italian anatomist
BELLINI, Lorenzo	1643-1704	Italian physiologist
BELON, Pierre	1517-1564	French zoologist, botanist
BENEDETTI, Alessandro	c. 1450-1512	Italian anatomist
BENEDICKS, Carl Axel Fredrik	1875-1958	Swedish metallographer
*BERGIUS, Friedrich	1884-1959	German chemist (Nobel, 1931)
BERKELEY, George	1685-1753	British scientific philosopher
BERNARD, Claude	1813-1878	French physiologist
BEZOUT, Étienne	1739-1783	French mathematician
*BLACKETT, Patrich Maynard S.	1897-1974	English physicist (Nobel, 1948)
BOAS, Franz	1858-1942	German-American anthropologist
BÖHLER, Lorenz	1885-1973	Austrian surgeon
BONNET, Charles	1720-1793	Swiss natural scientist, biologist

BORDET, Jules	1870-1961	Belgian bacteriologist, physiologist
*BORN, Max	1882-1970	German physicist (Nobel, 1954)
BOS, Willem Hendrik van den	1896-1974	Dutch astronomer
*BOSCH, Carl	1874-1940	German chemist (Nobel, 1931)
*BOTHE, Walther	1891-1957	German physicist (Nobel, 1954)
BOULE, Marcellin	1861-1942	French geologist
BOVERI, Theodor	1862-1915	German zoologist, biologist
BOWER, Frederick Orpen	1855-1948	English botanist
BRAHMAGUPTA	598-d. after 665	Indian astronomer
BRANDT, Georg	1694-1768	Swedish chemist, mineralogist
*BRAUN, Ferdinand	1850-1918	German physicist (Nobel, 1909)
BROCA, Pierre Paul	1824-1880	French physician, anthropologist
BRODE, Wallace R.	1900-1974	American chemist
BROOM, Robert	1866-1951	Scottish paleontologist
*BUCHNER, Eduard	1860-1917	German chemist (Nobel, 1907)
BUCKINGHAM, Edgar	1867-1940	American physicist
BUONANNI, Filippo	1638-1725	Italian natural scientist
CAILLETET, Louis Paul	1832-1913	French physicist
CANDOLLE, Augustin-Pyramus de	1778-1841	Swiss botanist
*CARREL, Alexis	1873-1944	French surgeon, physiologist (Nobel, 1912)
CASTILLON, Johann	1704-1791	Italian mathematician
*CHADWICK, James	1891-1974	British physicist (Nobel, 1935)
CHERRIE, George K.	1865-1948	American naturalist
CHEVREUL, Michel E.	1786-1889	French chemist
CLAPEYRON, Benoit-Pierre-Émile	1799-1864	French physicist
COCKERELL, Theodore D.A.	1866-1948	English zoologist
*CORI, Gerty Theresa R.	1896-1957	Czechoslovakian/American biochemist (Nobel, 1947)
CORRENS, Carl	1864-1933	German botanist
CUÉNOT, Lucien	1866-1951	French zoologist
*DALÉN, Nils Gustaf	1869-1937	Swedish engineer (Nobel, 1912)
DE BARY, Anton	1831-1888	German botanist
DE L'HOPITAL, Guilliaume F.A.	1661-1704	French mathematician
*DIELS, Otto Paul Hermann	1876-1954	German chemist (Nobel, 1950)
DIETRICH, Amalie Nelle	1821-1891	German botanist
*DOMAGK, Gerhard	1895-1964	German chemist, pathologist (Nobel, 1939)
DULONG, Pierre Louis	1785-1838	French physicist, chemist

EAST, Edward Murray	1879-1938	American biologist
EDDY, Nathan Browne	1890-1973	American pharmacologist
EKEBERG, Anders Gustaf	1767-1813	Swedish chemist
EMICH, Friedrich Peter	1860-1940	Austrian chemist
ENGLER, Adolf G.H.	1844-1930	German botanist
ERASISTRATUS	b.c. B.C. 304	Greek anatomist
*ERLANGER, Joseph	1874-1965	American physiologist (Nobel, 1944)
FAGON, Guy-Crescent	1638-1718	French botanist
FARLOW, William G.	1844-1919	American botanist
*FIBIGER, Johannes A.G.	1867-1928	Danish pathologist (Nobel,1926)
*FINSEN, Niels Ryberg	1860-1904	Danish physician (Nobel, 1903)
*FLOREY, Howard Walter	1898-1968	British pathologist (Nobel, 1945)
FLÜCKIGER, Otto	1881-1942	Swiss geographer
FLUDD, Robert	1574-1637	English chemist, physician
FORBES, James David	1809-1868	Scottish physicist
FÜCHSEL, Georg C.	1722-1773	German stratigrapher
GALTON, Francis	1822-1911	English anthropologist
*GASSER, Herbert Spencer	1888-1963	American physiologist (Nobel, 1944)
GEGENBAUR, Karl	1826-1903	German anatomist, zoologist
GIBBON, John Heysham, Jr.	1903-1973	American physician
GORE, John Ellard	1845-1910	Irish astronomer
GRASHCHENKOV, Nikolai I.	1901-1965	Soviet neurologist
GRAY, Asa	1810-1888	American botanist
GREW, Nehemiah	1641-1712	English physician, botanist
*GRIGNARD, Victor	1871-1935	French chemist (Nobel, 1912)
GROOTEN, Christian	c. 1530-c. 1603	German cartographer
*GUILLAUME, Charles Édouard	1861-1938	Swiss physicist (Nobel, 1920)
*HABER, Fritz	1868-1934	German/Swiss chemist (Nobel, 1918)
HAECKEL, Ernst H.	1834-1919	German zoologist, biologist
HALES, Stephen	1677-1761	English physiologist, botanist
HARRISON, Ross Granville	1870-1959	American biologist
*HAWORTH, Walter Norman	1883-1950	English chemist (Nobel, 1937)
*HENCH, Philip Showalter	1896-1965	American physician (Nobel,1950)
*HEVESY, George	1885-1966	Hungarian chemist (Nobel, 1943)
*HEYROVSKÝ, Jaroslav	1890-1967	Czechoslovakian chemist (Nobel, 1959)
*HINSHELWOOD, Cyril	1897-1967	English chemist (Nobel, 1956)

HOFMEISTER, Whilhelm F.B.	1824-1877	German botanist
HOOKER, Joseph Dalton	1817-1911	English botanist
*HOUSSAY, Bernardo Alberto	1887-1971	Argentinian physiologist (Nobel, 1947)
HUTCHINSON, John	1884-1972	English botanist
IBN BAJJA	1106-1138	Spanish-Arab astronomer
IBN BUTLAN	c. 1000-1068	Arabian physician
*KARRER, Paul	1889-1971	Russian-Swiss chemist (Nobel, 1937)
*KENDALL, Edward Calvin	1886-1972	American biochemist (Nobel, 1950)
KERR, John	1824-1907	Scottish physicist
*KOCHER, Emil Theodor	1841-1917	Swiss surgeon (Nobel, 1909)
KOHLRAUSCH, Friedrich W.G.	1840-1910	German physicist, chemist
*KOSSEL, Albrecht	1853-1927	German chemist (Nobel, 1910)
*KUHN, Richard	1900-1967	Austrian chemist (Nobel, 1938)
*LAVERAN, Charles Louis Alphonse	1845-1922	French physician (Nobel, 1907)
LEFSCHETZ, Solomon	1884-1972	Russian-American mathematician
LEHRMAN, Daniel S.	1919-1972	American psychologist
*LENARD, Philip Edward Anton	1862-1947	German physicist (Nobel, 1905)
LEUCKART, Karl G.F.R.	1822-1898	German zoologist
*LIPPMAN, Gabriel	1845-1921	French physicist (Nobel, 1908)
LISSAJOUS, Jules Antoine	1822-1880	French physicist
MACLEOD, John J.R.	1876-1935	Scottish physiologist
MALPIGHI, Marcello	1628-1694	Italian biologist, physician
MALUS, Étienne Louis	1775-1812	French physicist
MATHIAS, Émile-Ovide-Joseph	1861-1942	French physicist
MAXIM, Hiram Percy	1869-1936	American inventor
*MEYERHOF, Otto	1884-1951	German physician (Nobel, 1922)
MILLIONSHCHIKOV, Mikhail D.	1913-1973	Soviet applied physicist
*MINOT, George R.	1885-1950	American physician (Nobel,1934)
*MORGAN, Thomas Hunt	1866-1945	American zoologist (Nobel,1933)
MURPHY, Robert Cushman	1887-1973	American zoologist
MUSSCHENBROEK, Petrus van	1692-1761	Dutch physicist
NÄGELI, Karl Wilhelm von	1817-1891	Swiss botanist
NICHOLS, Ernest Fox	1869-1924	American physicist
OSBORN, H. Fairfield	1857-1935	American paleontologist
PECORA, William Thomas	1913-1972	American geologist
PELETIER, Jacques	1517-1582	French mathematician
PEREGRINUS, Peter	fl. c. 1270	French engineer, inventor

PREFFER, Wilhelm F.P.	1845-1920	German botanist, chemist
PLANTE, Gaston	1834-1889	French physicist
PLÜCKER, Julius	1801-1868	German physicist, mathematician
*POWELL, Cecil Frank	1903-1969	English physicist (Nobel, 1950)
*PREGL, Fritz	1869-1930	Austrian chemist (Nobel, 1923)
RABINOWITCH, Eugene	1901-1973	American biophysicist
*RICHARDSON, Owen Williams	1879-1959	English physicist (Nobel, 1928)
*RICHET, Charles Robert	1850-1935	French physiologist (Nobel, 1939)
ROTCH, A. Lawrence	1861-1912	American meteorologist
*ROUS, Peyton	1879-1970	American physician (Nobel,1966)
ROUX, Wilhelm	1850-1924	German zoologist
RUHMKORFF, Heinrich Daniel	1803-1877	German-French inventor
*SABATIER, Paul	1854-1941	French chemist (Nobel, 1912)
SAUVEUR, Joseph	1653-1716	French physicist, mathematician
SCHIMPER, Andreas F.W.	1856-1901	Swiss botanist
SCHWANN, Theodor	1810-1882	German anatomist
SEEBECK, Thomas Johann	1770-1831	German physicist
SIEMIENOWICZ, Kazimierz	fl. c.1650	Polish artillery expert
SIKORSKY, Igor Ivanovich	1889-1972	Russian-American aeronautical engineer
*SPEMANN, Hans	1869-1941	German biologist, zoologist (Nobel, 1935)
*STANLEY, Wendell M.	1904-1971	American biochemist (Nobel, 1946)
*STAUDINGER, Hermann	1881-1965	German chemist (Nobel, 1953)
STEARNS, Carl Leo	1892-1972	American astronomer
*STERN, Otto	1888-1969	German physicist (Nobel, 1943)
STRASBURGER, Eduard Adolf	1844-1912	German botanist
*SVEDBERG, Theodor	1884-1971	Swedish chemist (Nobel, 1926)
*TAMM, Igor	1895-1971	Soviet physicist (Nobel, 1958)
*THEILER, Max	1889-1972	South African bacteriologist (Nobel, 1951)
*TISELIUS, Arne	1902-1971	Swedish biochemist (Nobel,1948)
TORREY, John	1796-1873	American botanist, chemist
TSVET, Mikhail S.	1872-1919	Russian botanist
ULRICH, Edward O.	1857-1944	American paleontologist
VAN BENEDEN, Edouard	1846-1910	Belgian cytologist
VINOGRADSKI, Sergei N.	1856-1953	Russian/French microbiologist
VIRCHOW, Rudolph L.K.	1821-1902	German pathologist
*VIRTANEN, Artturi I.	1895-1973	Finnish biochemist (Nobel,1945)

*VON BAEYER, Adolf	1835–1917	German chemist (Nobel, 1905)
*VON BEHRING, Emil A.	1854–1917	German bacteriologist (Nobel, 1901)
*VON BÉKÉSY, Georg	1899–1972	Hungarian physicist (Nobel, 1961)
VON GOEBEL, Karl I.E.	1855–1932	German botanist
VON HALLER, Albrecht	1708–1777	Swiss physician
VON SACHS, Julius	1832–1897	German botanist
*WAGNER-JAUREGG, Julius	1857–1940	Austrian physician (Nobel,1927)
*WALLACH, Otto	1847–1931	German chemist (Nobel, 1910)
*WARBURG, Otto H.	1883–1970	German biochemist (Nobel, 1931)
WEISMANN, August	1834–1914	German biologist
*WIELAND, Heinrich Otto	1877–1957	German chemist (Nobel, 1927)
*WIEN, Wilhelm	1864–1928	German physicist (Nobel, 1911)
*WILLSTÄTTER, Richard	1872–1942	German chemist (Nobel, 1915)
WINCHELL, Newton Horace	1839–1914	American geologist
*WINDAUS, Adolf O.R.	1876–1959	German chemist (Nobel, 1928)
ZWICKY, Fritz	1898–1974	Swiss astrophysicist

3. We recommend approval of the following list of 20 names of scientists as a bank of names for future use on the moon, and the second biography of one name to be added to the records of a name already approved and assigned.

BIALOBRZESKI (C.)	1878–1953	Polish physicist
BRASIL (V.)	1865–1950	Brazilian biologist
BRONK (D.W.)	1897–1975	American physiologist
DE MORAES (A.)	1916–1970	Brazilian astronomer
HARKHEBI	circa 300 B.C.	Egyptian astronomer
HOPMANN (J.)	1890–1975	Austrian astronomer
KAMIENSKI (M.)	1879–1973	Polish astronomer
KEPINSKI (F.)	1885–1966	Polish astronomer
KOWALCZYK (J.)	1833–1911	Polish astronomer
LINDSAY (E.M.)	1907–1974	Irish astronomer
OLIVIER (C.P.)	1884–1975	American astronomer
PIENKOWSKI (S.)	1883–1953	Polish physicist
PIKEL'NER (S.B.)	–1975	Russian astronomer
POCZOBUTT (O.M.)	1728–1808	Polish astronomer
PRAZMOWSKI (A.)	1821–1885	Polish astronomer
PRISCILLA BOK (F.)	1896–1975	American astronomer
VERTREGT (M.)	1897–1973	Dutch chemist & space scientist
VINOGRADOV (A.P.)	1895–1975	Russian geo-chemist & cosmo-chemist

| WILDT (R.) | 1905-1976 | American astronomer of German birth |
| ZONN (W.) | 1905-1975 | Polish astronomer |

| MINKOWSKI (R.L.B.) | 1895-1976 | American astronomer (second biography to be added to name already approved) |

4. We recommend approval of the following 12 names, assigned to lunar elevations:-

MONS	EULER	LAC Region		39
"	GRUITHUISEN DELTA	"	"	23
"	GRUITHUISEN GAMMA	"	"	23
"	HANSTEEN	"	"	74
"	HERODOTUS	"	"	38
"	MORO	"	"	76
"	PENCK	"	"	78
"	VITRUVIUS	"	"	43
MONTES	AGRICOLA	"	"	38
"	ARCHIMEDES	"	"	41
"	SECCHI	"	"	61
MONS	MARALDI	"	"	43

(correction of Mons Maraldi Gamma)

5. We recommend approval of the following list of 6 lacus names on the moon:-

LACUS	EXCELLENTIAE	35 S, 045 W
"	LUXURIAE	19 N, 175 E
"	OBLIVIONIS	21 S, 169 W
"	SPEI*	43 N, 064 E
"	TEMPORIS	46 N, 055 E
"	TIMORIS	39 S, 028 W

*Previously Lacus Struve

6. We recommend approval for the name of one valley and of the names of 12 crater chains, named after the nearest named crater to facilitate their location and recognition:-

VALLIS	BOHR	LAC Region		55
CATENA	ABULFEDA	"	"	96
"	ARTAMONOV	"	"	46
"	DAVY	"	"	77
"	DZIEWULSKI	"	"	46
"	GREGORY	"	"	65
"	HUMBOLDT	"	"	99
"	KRAFFT	"	"	55

CATENA	KURCHATOV	LAC Region 31
"	LEUSCHNER	" " 71
"	MENDELEEV	" " 66
"	SUMNER	" " 30
"	SYLVESTER	" " 1

7. We recommend that the area bounded approximately by Montes Carpatus (N), Crater Flammarian (E), Mare Cognitum (S) and Crater Kepler (W) be named MARE INSULARUM.

8. We recommend that the term DORSUM, plural DORSA, include curvilinear elevations in the lunar maria (i.e. wrinkle ridges) as well as normal ridges; and that all DORSA continue to be named after geoscientists, as is now the practice (see Resolution II(1) above).

9. We recommend that RIMAE, singular RIMA, long depressions on the moon (straight, arcuate or sinuous) if within a crater, be named after that crater; or otherwise, be named after the nearest named formation, preferable a crater.

Resolution III

MERCURY NOMENCLATURE

1. We recommend the approval of 136 crater names as assigned on Mercury Quadrangles H-6,7,8,11,15 of the 1:5,000,000 Mercury map series.

NAME, with biographical information	Quadrangle	Lat.(°)	Long.(°)	Diam.(km)
ABU NUWAS, (762-810)	H-6	17.5	21	115
AFRICANUS HORTON, (1835-1883)	H-11	-50.5	42	120
AL-JĀHIZ, (d. 869)	H-6	1.5	22	95
AMRU AL-QAYS, (Pre-Islamic)	H-8	13	176	50
ANDAL, (18th century)	H-11	-47	35.5	90
ASVAGOSHA, (1st century)	H-6	11	21	80
BACH, J.S., (1685-1750)	H-15	-69	103	225
BALAGTAS, F., (1788-1862)	H-6	-22	14	100
BALZAC, H. de,(1799-1850)	H-8	11	145	65
BEETHOVEN, L. van, (1770-1827)	H-7	-20	124	625
BELLO, A., (1781-1865)	H-7	-18.5	120.5	150
BERNINI, G.L.,(1598-1680)	H-15	-79.5	136	145
BOCCACCIO, G., (1313-1375)	H-15	-80.5	30	135
BOETHIUS, (480-524)	H-7	- 0.5	74	130
BRAMANTE, (1444-1514)	H-11	-46	62	130
BRUNELLESCHI, F., (1377-1446)	H-6	- 8.5	22.5	140
BYRON, G.G., (1788-1824)	H-6	- 8	33	100
CALLICRATES, (5th century BC)	H-11	-65	32	65
CAMÕES, L.V. de, (1524-1580)	H-15	-70.5	70	70
CARDUCCI, G., (1835-1907)	H-11	-36	90	75

NAME, with biographical information	Quadrangle	Lat.(°)	Long.(°)	Diam.(km)
CERVANTES, M. de, (1547-1616)	H-15	-75	122	200
CHAIKOVSKIJ, P.I., (1840-1893)	H-6	8	50.5	160
CHAO MENG-FU, (d. 1322)	H-15	-87.5	132	150
CHEKHOV, A., (1860-1904)	H-11	-35.5	61.5	180
CHIANG K'UI, (12th century)	H-7	14.5	103	105
CHOPIN, F.F., (1810-1849)	H-15	-64.5	124	100
CHU TA, (c. 1624-c. 1705)	H-7	2.5	106	100
COLERIDGE, S.T., (1772-1834)	H-11	-54.5	66.5	110
COPLEY, J.S., (1738-1815)	H-11	-37.5	85.5	30
DARÍO, R., (1867-1916)	H-11	-26	10	160
DICKENS, C., (1812-1870)	H-15	-73	153	72
DONNE, J., (1572-1631)	H-6	3	14	90
DÜRER, A., (1471-1528)	H-7	22	119.5	190
DVORAK, A., (1841-1904)	H-6	- 9.5	12.5	80
EITOKU, (1543-1590)	H-8	-21.5	157.5	105
EQUIANO, A., (c. 1750-1797)	H-11	-39	31	80
FUTABATEI, S., (1864-1909)	H-7	-15.5	83.5	55
GHIBERTI, L. (1378-1455)	H-11	-48	80	100
GIOTTO, (1266/67?-1337)	H-6	12.5	56	150
GOYA y L.F.J. de, (1746-1828)	H-8	- 6.5	152.5	135
GUIDO D'AREZZO, (c. 995-1050)	H-11	-38	19	50
HANDEL, G.F., (1685-1759)	H-6	4	34	150
HARUNOBU, (1725-1770)	H-7	15.5	141	100
HAYDN, J., (1732-1809)	H-11	-26.5	71.5	230
HESIOD, (c. 800 BC)	H-11	-58	35.5	90
HIROSHIGE, A., (1797-1858)	H-6	-13	27	140
HITOMARO, (7th-8th century)	H-6	-16	16	105
HOLBERG, L., (1684-1754)	H-15	-63.5	61	66
HOMER, (8th or 9th century BC)	H-6	- 1	36.5	320
HORACE, (65-8 BC)	H-15	-68.5	52	48
IBSEN, H.J., (1828-1906)	H-6	-24	36	160
ICTINUS, (2nd half 5th century BC)	H-15	-79	165	110
IMHOTEP, (c. 2680 BC)	H-6	-17.5	37.5	160
JUDAH HA-LEVI, (c. 1075-1141)	H-7	11.5	108	85
KĀLIDĀSĀ, (5th century?)	H-8	-17.5	180	110
KEATS, J., (1795-1821)	H-15	-69.5	154	110
KENKŌ, Y., (14th century)	H-6	-21	16.5	90
KHANSA, (Pre-Islamic)	H-11	-58.5	52	100

NAME, with biographical information	Quadrangle	Lat.(°)	Long.(°)	Diam.(km)
KUROSAWA, K., (18th century)	H-11	-52	23	180
LEOPARDI, G., (1798-1837)	H-15	-73	180	69
LERMONTOV, M. Yu., (1814-1841)	H-6	15.5	48.5	160
LI CH'ING-CHAO, (1081-after 1141)	H-15	-77	73	60
LI PO, (701-762)	H-6	17.5	35	120
LU HSUN, (1881-1936)	H-6	0.5	23.5	95
LYSIPPUS, (4th century BC)	H-7	1.5	133	150
MA CHIH-YUAN, (fl. 1251)	H-11	-59	77	170
MACHAUT, G. de, (c. 1300-1377)	H-7	- 1.5	83	105
MAHLER, G., (1860-1911)	H-6	-19	19	100
MARK TWAIN, (1835-1910)	H-7	-10.5	138.5	140
MARTÍ, J.J., (1853-1895)	H-15	-75.5	164	63
MATISSE, H., (1869-1954)	H-7	-23.5	90	210
MELVILLE, H., (1819-1891)	H-6	22	9.5	135
MENA, J. de, (1411-1456)	H-7	0.5	125	20
MENDES PINTO, F., (c. 1510-1583)	H-11	-61	19	170
MICKIEWICZ, A., (1798-1855)	H-7	23.5	102.5	115
MILTON, J., (1608-1674)	H-8	-25.5	175	175
MISTRAL, G., (1889-1957)	H-6	5	54	100
MOFOLO, T., (1873-1948)	H-11	-37	29	90
MOLIÈRE, (1622-1673)	H-6	16	17.5	140
MOZART, W.A., (1756-1791)	H-8	8	190.5	225
MURASAKI S., (11th century)	H-6	-12	31	125
NAMPEYO, (c. 1860-1942)	H-11	-39.5	50.5	40
NEUMANN, B., (1687-1753)	H-11	-36.5	35	100
OVID, (43 BC-17 AD)	H-15	-69.5	23	40
PETRARCH, (1304-1374)	H-11	-30	26.5	160
PHIDIAS, (active from about 475-430 BC, or later)	H-8	9	150	155
PHILOXENUS, (3rd century BC)	H-7	- 8	112	95
PIGALLE, J.B., (1714-1785)	H-11	-37	10.5	130
PO CHÜ-I, (772-846)	H-8	- 6.5	165.5	60
PO YA, (8th-5th century BC)	H-11	-45.5	21	90
POLYGNOTUS, (c. 500-400 BC)	H-6	0	68.5	130
PROUST, M., (1871-1922)	H-6	20	47	140
PUCCINI, G., (1858-1924)	H-15	-64.5	46	110
PUSHKIN, A.S., (1799-1837)	H-15	-64.5	23	200
RABELAIS, F., (c. 1483-1553)	H-11	-59.5	62.5	130

NAME, with biographical information	Quadrangle	Lat.(°)	Long.(°)	Diam.(km)
RAJNIS, Ya., (1865-1925)	H-7	5	96.5	85
RAMEAU, J.P., (1683-1764)	H-11	-54	38	50
RAPHAEL, (1483-1520)	H-7	-19.5	76.5	350
RENOIR, P.A., (1841-1910)	H-6	-18	52	220
REPIN, I.E., (1844-1930)	H-6	-19	63	95
RILKE, R.M., (1875-1926)	H-11	-45.5	13.5	70
RODIN, A., (1840-1917)	H-6	22	18.5	270
RUBLEV, A., (c. 1380-1430)	H-8	-14.5	157.5	125
RŪDAKĪ, (10th century)	H-6	- 3.5	51.5	120
SADĪ, (c. 1213-1292),	H-15	-77.5	56	60
SCHOENBERG, A., (1874-1951)	H-7	-15.5	136	30
SCHUBERT, F., (1797-1828)	H-11	-42	54.5	160
SCOPAS, (1st half 4th century BC)	H-15	-81	173	95
SEL, S., (11th century)	H-11	-63.5	88.5	130
SHEVCHENKO, T.G., (1814-1861)	H-11	-53	47	130
SINAN, I.A.Al'M., (1489-1588)	H-6	16	30	140
SNORRI, S., (1179-1241)	H-7	- 8.5	83.5	20
SOPHOCLES, (c. 496-406 BC)	H-8	- 6.5	143.5	145
SŌTATSU, (d. 1643)	H-11	-48	19.5	130
SPITTELER, C.F.G., (1845-1924)	H-15	-68	62	66
SULLIVAN, L., (1856-1934)	H-7	-16	87	135
TANSEN,	H-7	4.5	72	25
THĀKUR, R., (1861-1941)	H-6	- 2.5	64	115
THEOPHANES, (14th century)	H-7	- 4	143	50
TINTORETTO, (1518-1594)	H-11	-47.5	24	60
TITIAN, (c. 1487/1490-1576)	H-6	- 3	42.5	115
TOLSTOJ, L.N., (1828-1910)	H-8	-15.0	165	400
TS'AI WEN-CHI, (2nd century AD)	H-6	23.5	22.5	120
TS'AO CHAN, (c. 1715-1763)	H-7	-13	142	110
TSURAYUKI, K.N., (10th century)	H-11	-62	22.5	80
TYAGARAJA, (18th century)	H-8	4	149	100
UNKEI, (13th century)	H-11	-31	62.5	110
VĀLMIKI, (1st century BC)	H-7	-23.5	141.5	220
VAN GOGH, V.W., (1853-1890)	H-15	-76	135	95
VIVALDI, A., (1678-1741)	H-7	14.5	86	210
WAGNER, R., (1813-1883)	H-15	-67.5	114	135
WANG MENG, (1308-1385)	H-7	9.5	104	170
WERGELAND, H.A., (1808-1845)	H-11	-37	56.5	35

NAME, with biographical information	Quadrangle	Lat.(°)	Long.(°)	Diam.(km)
YEATS, W.B., (1865–1939)	H-6	9.5	35	90
YUN SŎN-DO, (1587–1671)	H-15	-72.5	109	61
ZEAMI, (15th century)	H-8	- 2.5	148	125

2. We recommend the approval of the following 6 rupes names as assigned on Mercury:-

	Lat.	Long.
ADVENTURE	-64°	063°
GJÖA	-65	163
POURQUOI-PAS	-58	156
RESOLUTION	-62	052
ZARYA	-42	022
ZEEHAEN	+50	158

3. We recommend the following 9 sheet names for the 1:5,000,000 map series of Mercury:-

Sheet No.	Name	Sheet No.	Name
1	BOREALIS	8	TOLSTOJ
2	VICTORIA	11	DISCOVERY
3	SHAKESPEARE	12	MICHAELANGELO
6	KUIPER	15	BACH
7	BEETHOVEN		

Resolution IV

MARS NOMENCLATURE

1. We recommend approval of the following list of 6 names as assigned to large craters (diameter greater than 100 km) on Mars.

	Lat.	Long.
BYRD	-26°	235°
GUSEV	-14°	184°
LASSWITZ	-09°	222°
LUZIN	-26°	328°.5
SCHÖNER	+20°	309°.5
WIEN	-11°	220°.5

2. We recommend approval of the list of 271 names of small cities and villages, as assigned to small craters (diameter less than 10 km) on Mars.

NEREIDUM MONTES REGION

Name	Country of origin	Coordinates	
Albi	France	34°7	−41°9
Azul	Argentina	42.3	−42.5
Baltisk	USSR	54.5	−42.6
Bozkir	Turkey	31.9	−44.5
Camiri	Bolivia	41.9	−45.1
Choctaw	USA	37.0	−41.5
Cypress	USA	47.0	−47.6
Delta	USA	39.0	−46.3
Dese	Ethiopia	30.3	−45.8
Dessau	GDR	53.0	−43.1
Eger	Hungary	51.8	−48.7
Flora	USA	51.2	−45.0
Gah	Indonesia	32.3	−45.1
Gali	USSR	36.9	−44.1
Gandu	Brazil	47.0	−45.8
Ham	France	32.2	−45.0
Hilo	USA	35.5	−44.8
Ježa	USSR	37.7	−48.8
Kartabo	Guyana	52.3	−41.2
Kakori	India	29.6	−41.9
Kampot	Cambodia	45.4	−42.1
Karpinsk	USSR	31.8	−46.0
Kem'	USSR	32.7	−45.3
Kifrī	Iraq	54.1	−46.0
Kribi	Cameroon	43.4	−43.4
Kushva	USSR	35.2	−44.3
Lemgo	FRG	34.5	−42.9
Luga	USSR	47.2	−44.6
Mafra	Brazil	53.0	−44.4
Maidstone	UK (England)	54.1	−41.9
Nazca	Peru	38.2	−44.0
Ochakov	USSR	31.6	−42.5
Podor	Senegal	43.0	−44.6
Porvoo	Finland	40.6	−43.7
Rengo	Chile	43.5	−43.9
Salaga	Ghana	51.0	−47.6
Sarno	Italy	54.0	−44.7
Satka	USSR	36.7	−43.0
Sauk	USA	32.25	−45.0
Sokol	USSR	40.5	−42.8
Tabou	Ivory Coast	34.8	−45.5
Talsi	USSR	49.1	−41.9
Tara	Ireland	52.7	−44.4
Tarakan	Borneo	30.1	−41.6
Taza	Morocco	45.1	−44.0
Tombe	Sudan	44.4	−42.8
Tōno	Japan	52.2	−45.2
Torsö	Sweden	51.0	−44.7
Turbi	Kenya	51.2	−40.9
Valga	USSR	36.3	−44.6
Vätö	Sweden	53.2	−43.9
Wau	New Guinea	42.4	−45.2
Yungay	Peru	44.6	−44.3

ERYTHRAEUM REGION

Name	Country of Origin	Coordinates	
Alga	USSR	26°5	−24°6
Aspen	USA	22.9	−21.6
Balta	USSR	26.4	−24.1
Bar	USSR	19.3	−25.6
Bend	USA	27.5	−22.6
Bentong	Malaysia	18.9	−22.6
Bigbee	USA	34.6	−25.0
Bison	USA	29.0	−26.6
Blitta	Togo	20.8	−26.3
Bogra	Bangladesh	28.6	−24.4
Boru	USSR	27.7	−24.6
Buta	Zaire	32.2	−23.5
Calbe	GDR	28.7	−25.5
Campos	Brazil	27.6	−22.0
Cartago	Costa Rica	17.8	−23.6
Chapais	Canada	20.4	−22.6
Chekalin	USSR	26.6	−24.5
Cheb	Czechoslovakia	19.3	−24.5
Circle	USA	25.4	−22.4
Cluny	France	27.1	−24.1
Cobalt	USA	26.8	−26.1
Crewe	UK (England)	19.4	−25.2
Deba	Nigeria	17.1	−24.3
Dingo	Australia	17.3	−24.0
Dison	Belgium	16.3	−25.4
Eads	USA	29.8	−28.9
Echt	UK (Scotland)	28.0	−22.2
Edam	Netherlands	19.9	−26.6
Ely	USA	27.1	−23.9
Falun	Sweden	24.4	−24.2
Fastov	USSR	20.2	−25.4
Flat	USA	19.4	−25.8
Gagra	USSR	21.9	−20.9
Galu	Zaire	21.5	−22.3
Gardo	Somalia	24.6	−27.0
Gatico	Chile	20.9	−21.2
Globe	USA	27.1	−24.0
Glazov	USSR	26.4	−20.8
Goba	Ethiopia	20.9	−23.5
Golden	USA	33.3	−22.2
Gori	USSR	28.6	−23.2
Grójec	Poland	30.6	−21.6
Guir	Mali	20.4	−21.8
Harad	Saudi Arabia	27.8	−27.8
Honda	Colombia	16.2	−22.7
Inta	USSR	24.9	−24.6
Irbit	USSR	24.7	−24.6
Jal	USA	28.6	−26.6
Kagul	USSR	18.9	−24.0
Kaj	USSR	29.2	−27.4
Kanab	USA	18.8	−27.6
Kansk	USSR	17.1	−20.8
Kantang	Thailand	17.3	−24.8
Karshi	USSR	19.2	−23.6

ERYTHRAEUM REGION (Cont'd.)

Name	Country of Origin	Coordinates	
Kashira	USSR	18°1	-27°5
Kasimov	USSR	22.8	-25.0
Kimry	USSR	16.2	-20.5
Kirs	USSR	19.3	-26.7
Kirsanov	USSR	25.0	-22.4
Kita	Mali	17.0	-23.1
Kuba	USSR	19.5	-25.6
Lamas	Peru	20.5	-27.4
Lar	Iran	28.8	-26.1
Lebu	Chile	19.4	-20.6
Livny	USSR	28.9	-27.5
Longa	Angola	25.7	-20.9
Loto	Zaire	22.3	-22.2
Lorica	Colombia	28.1	-19.9
Manzī	Burma	27.3	-22.3
Murgoo	Australia	22.3	-24.0
Mila	Algeria	20.6	-27.5
Nan	Thailand	19.8	-27.1
Nardo	Italy	32.7	-27.8
Navan	Ireland	23.2	-26.2
Nepa	USSR	19.5	-25.3
Nitro	USA	23.8	-21.5
Noma	Namibia	24.0	-25.7
Ōmura	Japan	25.0	-25.7
Ostrov	USSR	27.9	-26.9
Plum	USA	18.9	-26.4
Pabo	Uganda	22.9	-27.3
Polotsk	USSR	26.1	-20.1
Rana	Norway	21.6	-26.0
Revda	USSR	28.3	-24.6
Romny	USSR	18.0	-25.7
Ruby	USA	16.9	-25.6
Sandila	India	30.2	-25.9
Sangar	USSR	24.1	-27.9
Say	Niger	29.5	-28.5
Seminole	USA	18.9	-24.5
Shambe	Sudan	30.5	-20.7
Sibu	Malaysia	19.6	-23.3
Sigli	Indonesia	30.6	-20.5
Singa	Sudan	17.2	-22.8
Thule	Denmark (Greenland)	25.5	-23.6
Tak	Thailand	28.45	-26.4
Timaru	New Zealand	22.2	-25.6
Tiwi	Oman	24.6	-27.9
Tura	USSR	21.8	-27.0
Voo	Kenya	19.8	-27.3
Yegros	Paraguay	23.5	-22.5

CHRYSE REGION

Bor	USSR	33°8	+18°5
Chur	USSR	29.3	+17.1
Banff	Canada	31.6	+17.8
Kaup	New Guinea	36.6	+22.5

CHRYSE REGION (Cont'd.)

Name	Country of Origin	Coordinates	
Kipini	Kenya	32°0	+24°8
Naukan	USSR	30.1	+21.3
Trud	USSR	30.7	+17.7

CYDONIA REGION

Name	Country of Origin	Coordinates	
Chom	Tibet	2°2	+38°7
Cruz	Venezuela	1.7	+38.6
Dein	New Guinea	2.4	+38.5
Gaan	Somalia	3.8	+38.9
Gwash	Pakistan	3.1	+39.3
Lutsk	USSR	3.2	+38.9
Wer	India	6.2	+45.9

CAPRI REGION

Name	Country of Origin	Coordinates	
Azusa	USA	40°55	− 5°5
Bahn	Liberia	43.45	− 3.5
Balboa	Panama Canal Zone	34.05	− 3.8
Batoka	Rhodesia	36.8	− 7.6
Berseba	South W. Africa	37.7	− 4.4
Bamba	Zaire	41.7	− 3.5
Butte	USA	39.05	− 5.1
Byske	Sweden	31.1	− 4.95
Camiling	Philippines	38.1	− 0.8
Chimbote	Peru	39.8	− 1.5
Chinju	Republic of Korea	42.3	− 4.6
Conches	France	34.3	− 4.2
Creel	Mexico	38.95	− 6.1
Daet	Phillippines	41.9	− 7.4
Dia-cau	Vietnam	42.8	− 0.3
Glide	USA	43.3	− 8.2
Groves	USA	44.75	− 4.15
Huancayo	Peru	39.85	− 3.7
Innsbruck	Austria	30.1	− 6.5
Kaid	Iraq	44.8	− 4.55
Kholm	Afganistan	42.1	− 7.3
Kong	Ivory Coast	38.7	− 5.45
Locana	Italy	38.3	− 3.45
Lachute	Canada	39.9	− 4.3
Manti	USA	37.7	− 3.65
Mega	Ethiopia	37.05	− 1.5
Misk	Turkey	35.45	− 0.95
Manah	USSR	33.75	− 4.75
Oglala	USA	38.35	− 3.1
Paks	Hungary	42.15	− 7.75
Pinglo	China	36.9	− 3.0
Quorn	Australia	33.75	− 5.55
Rakke	USSR	43.5	− 4.75
Rincon	Netherlands Antilles	43.1	− 8.1
Rypin	Poland	41.05	− 1.25
Sfax	Tunisia	43.5	− 7.75
Sitka	USA	39.35	− 4.3
Spry	USA	38.6	− 3.75

CAPRI REGION (Cont'd.)

Name	Country of Origin	Coordinates	
Stobs	UK (Scotland)	38°4	− 5°0
Tarata	Bolivia	41.35	− 3.8
Timbuktu	Mali	37.6	− 5.6
Tuskegee	USA	36.2	− 2.9
Vaales	Spain	33.1	− 3.95
Wicklow	Iceland	40.7	− 2.0
Windfall	Canada	43.5	− 2.1
Wink	USA	41.5	− 6.6
Žulanka	USSR	42.3	− 2.3

TRITONIS LACUS REGION

Name	Country of Origin	Coordinates	
Bacht	USSR	257°45	+18°9
Basin	USA	253.15	+18.1
Bluff	New Zealand	250.1	+23.7
Brush	USA	248.8	+21.9
Cost	USA	256.1	+15.25
Dank	Oman	253.15	+22.2
Gastre	Argentina	247.6	+24.8
Goff	Somalia	255.2	+23.5
Gulch	Ethiopia	251.15	+16.1
Kaw	French Guiana	255.85	+16.7
Linpu	China	247.0	+18.3
Marbach	Switzerland	249.2	+17.9
Moss	Norway	250.7	+19.4
Naic	Philippines	252.75	+24.7
Phon	Thailand	257.3	+15.8
Porth	UK (Wales)	255.85	+21.45
Troika	USSR	255.0	+17.1
Viana	Brazil	255.25	+19.45

WESTERN CHRYSE REGION

Name	Country of Origin	Coordinates	
Banh	Upper Volta	55°3	+19°9
Belz	USSR	43.2	+21.8
Bise	Okinawa	56.7	+20.6
Blois	France	55.6	+24.1
Bole	Ghana	53.7	+25.5
Broach	India	56.6	+24.1
Cairns	Australia	47.3	+24.0
Changsŏng (Lun-Gu)	Republic of Korea	57.1	+24.0
Chauk	Burma	55.6	+23.9
Chinook	Canada	55.2	+23.0
Chive	Bolivia	55.7	+22.2
Clogh	N. Ireland	47.6	+20.8
Dromore (Lun-Cr)	N. Ireland	49.3	+20.2
Guaymas (Lun-Aw)	Mexico	44.9	+26.0
Jijiga	Ethiopia	53.6	+25.7
Lexington	USA	48.5	+22.2
Naar	Egypt	42.1	+23.2
Nema (Lun-Ds)	USSR	52.0	+21.2
Nif	Yap	56.1	+20.3
Ottumwa (Lun-Fe)	USA	55.5	+25.2
Peixe	Brazil	47.5	+20.6

WESTERN CHRYSE REGION (Cont'd.)

Name	Country of Origin	Coordinates	
Poona (Lun-Du)	India	24°3	+52°0
Puńsk	Poland	41.2	+20.7
Quick	Canada	49.1	+18.4
Rauch	Argentina	57.8	+21.8
Rong	Tibet	45.3	+22.8
Santa Fe (Lun-Br)	USA	47.8	+19.4
Sögel (Lun-Fs)	FRG	54.9	+21.9
Spur	USA	52.0	+22.5
Sucre	Colombia	54.3	+24.2
Tarsus	Turkey	40.2	+23.4
Troy	USA	52.3	+23.7
Valverde (Lun-Fr)	Dominican Republic	55.7	+20.6
Vol'sk	USSR	51.1	+23.5
Weer	Austria	51.5	+20.2
Yorktown	USA	48.5	+23.3
Waspam	Nicaragua	56.5	+20.9
Wassamu (Lin-Ew)	Japan	53.0	+26.1

3. We recommend approval of the following 10 names as assigned to channels on Mars:-

	Latitude	Longitude
ARDA VALLES	-20° to -22°	031° to 033°
BAHRAM VALLIS	western chryse	
CLOTA VALLIS	-25°5 to -26°5	20°5
DRINUS VALLES	-20°5 to -22°5	021°5 to 023°
MAMERS VALLES	+32° to +47°	338° to 348°
MAUMAE VALLIS	western chryse	
OLTIS VALLES	-22°5 to -25°	021° to 022°
RUBICON VALLES	+45°	116°
TAGUS VALLES	-07°	245°
VEDRA VALLES	western chryse	

4. We recommend approval of the following 7 names, as assigned to topographic features on Mars:-

AUREUM CHAOS	MC 19
IANI CHAOS	MC 19
AMENTHES FOSSAE	MC 14
SACRA FOSSA	MC 10
AEOLIS MENSAE	MC 23
NEPENTHES MENSAE	MC 14
XANTHE DORSA	western chryse

5. We recommend approval of the following 2 Latin terms for use on Mars:-

    SCOPULUS (Scopuli) lobate or highly irregular scarps

    SULCUS (Sulci)   a complex area of subparallel furrows and ridges

6. We recommend approval of the following 8 names, as assigned to topographic
    features on Mars:-

	Latitude	Longitude
*FRIGORIS SCOPULUS	+80°	020° to 305°
*HYPERBOREUS SCOPULUS	+80° to +85°	060° to 160°
*NILOKERAS SCOPULUS	+32° to +40°	040° to 052°
*HYPERNOTIUS SCOPULUS	-76° to -80°	230° to 315°
CYANE SULCI	+23° to +28°	126° to 130°
GIGAS SULCI	+08° to +14°	126° to 135°
LUCAS SULCI	+25° to +40°	130° to 148°
SULCI GORDI	+16° to +20°	125° to 127°

*Corrected from RUPES to SCOPULUS

7. We recommend that the name SILOE MENSAE (lat. +30° to +49°, long. 010° to 020°)
    be changed to CYDONIA MENSAE.

In compliance with resolution I (1) of the third meeting of the WGPSN in Grenoble the following Table gives the names of the Apollo landing sites as approved by the IAU at the XVth General Assembly in Sydney.

Table

ASTRONAUT - NAMED LUNAR FEATURES IN THE APOLLO EXPLORATION SITES*

Apollo 11 Landing Site

West: Sharp-rimmed, rayed crater, about 180 m in diameter and 30 m deep. The crater occurs on the western edge of the Apollo 11 landing ellipse, hence the name "West". The LM landed approximately 400 m to the west of said crater.

Apollo 12 Landing Site

A. Crater Cluster names

    1. Crescent: Descriptive name of a row of seven craters arranged in the form of an arch west of the landing site.

    2. Snowman: An arrangement of five craters around the large crater in which Surveyor 3 landed. The geometry resembles the fabled and familiar "snowman" figure.

B. Crater names

    1. Middle Crescent: (Sampling site) Crater in the middle of the afore-mentioned Crescent. Its rim was the farthest stop from the LM on the first EVA.

    2. Head: (Sampling site) Crater that forms the head of the afore-mentioned Snowman pattern.

    3. Bench: (Sampling site) Crater that forms the right arm to the Snowman arrangement; it displays a prominent bench which is indicative of excavation of bedrock beneath the regolith.

    4. Sharp-Apollo: (Sampling site) Sharp-rimmed and bright-rayed crater west of Bench; it is located on the extreme southwestern end of the sampling traverse on the second EVA.

    5. Halo: Small haloed crater on the south rim of the crater in which Surveyor 3 landed.

    6. Surveyor: (Sampling Site) The site of the Surveyor 3 landing.

    7. Block: Last sampling site, small crater within Surveyor crater.

*The designation "-Apollo" has been added to a few names which duplicate those of previously named lunar craters. The designation indicated that the crater is in one of the Apollo landing sites.

Apollo 14 Landing Site

1. Cone: A 350 m crater situated on the western edge of one of the high ridges of the Fra Mauro Formation. The physical location and ejecta of the crater give it a cone-shaped appearance. The south rim of the crater was the farthest stop of the second EVA surface sampling traverse.

2. Triplet: Three craters in a row, "North", "Center", and "South", that served as the first major landmark for landing the craft. The Apollo 14 LM landed west of North Triplet. Samples were also collected from North Triplet.

3. Doublet: Two superposed craters west of the landing point that served as the second major landmark for landing the craft. The Laser Ranging Retro Reflector was deployed on the southeast rim of Doublet crater.

4. Flank: A 30 m crater on the southwestern flank of Cone crater.

5. Old Nameless: Crater with broken rim.

6. Weird: A 40 m unusual cluster of probably two or possibly three craters forming a unique or "weird" shape. A large rock sampled east of this crater has already been named weird rock !

Apollo 15 Landing Site

The following 14 names were carefully selected from an original list of 81 names given to features and craters in the Apollo 15 landing site area.

A. Feature names

1. Apennine Front: The explored foothills of Hadley Delta which is part of the Apennine Mountain range on the eastern rim of Mare Imbrium.

2. North Complex: Complex of hills, craters, scarps and apparent flow fronts to the north of the landing site.

3. South Cluster: A cluster of secondary craters located to the south of the landing site. The western part of the cluster was explored on the second EVA.

4. Plain: A flat mare region on which the LM landed east of Rima Hadley.

5. Terrace: Slight projection of a basalt-mare unit out into Rima Hadley. The farthest sampling point to the west on EVA 3 was in its vicinity.

B. Crater names

1. <u>Bridge</u>: Crater within Rima Hadley whose rim appears to form a bridge across
        the rille. Crater was used as a landmark.

2. <u>Dune</u>: Crater named for a dune-shaped structure on the southeast rim. Dune
        crater was the sampling site of Station number 4.

3. <u>Earthlight</u>: A crater named after an Arthur C. Clarke novel by the same name.
        The crater was described in detail during the second EVA.

4. <u>Elbow</u>: Crater at a part of Rima Hadley resembling a bent elbow. The crater
        was the site of sampling station number 1.

5. <u>Index</u>: A prominent crater near the landing site that served as the major
        landmark for orbital tracking and for LM descent.

6. <u>Last</u>:This crater was supposed to be visited on the last traverse; it became
        the last crater to be approached during descent. The LM landed in its
        vicinity.

7. <u>Rhysling</u>: A crater named for Rhysling, the blind poet of "The Green Hills
        of Earth", a science fiction story by Robert Heinlein. Sampling
        station number 3 is 125 m west-southwest of the crater.

8. <u>Spur</u>: Crater located on a small spur of the Apennine Front. The southern-
        most part of the second EVA traverse was in the vicinity of this crater.

9. <u>St.George</u>: In Jules Verne's "From the Earth to the Moon", the moon-bound crew
        members celebrated a successful launch by drinking a type of wine by the
        name of St. George. This 2.5 km in diameter crater on the Apennine
        Front was the source of soil sample.

<u>Apollo 16 Landing Site</u>

     The following 19 names were chosen from several lists of crew-given
names of features and craters in the landing site area.

A. Feature names

1. <u>Smoky Mountains</u>: Mountainous mass north of the landing site.

2. <u>Stone Mountain</u>: Mountainous mass south of the landing site.

B. Crater names

1. <u>Baby Ray</u>: A small rayed crater atop bright rays of a larger crater.

2. <u>Cinco</u>: A group of five craters ("cinco" is "five" in Spanish) on the foot-
hill of Stone Mountain in the vicinity of the southern-most portion
of the second EVA traverse.

3. <u>Spot</u>: Two overlapping craters that served as a landing landmark. The LM
landed about 100 m north of the craters.

4. <u>End</u>: End for being the last planned stop on the last EVA.

5. <u>Flag</u>: Sampling site of Station number 2 on the rim of Plum.

6. <u>Gator</u>: Large crater named for alligator.

7. <u>Halfway</u>: Crater centrally located between sampling sites 1 and 2.

8. <u>Kiva</u>: Crater named after the Pueblo Indian architectural structure which
is usually round.

9. <u>North Ray</u>: One kilometer diameter bright-rayed crater north of the landing
point. The ejecta and rim of the crater were sampling sites number 11
and 13.

10. <u>Palmetto</u>: A crater, one kilometer in diameter, named after the palm tree
of the same name.

11. <u>Plum</u>: Large crater with Flag on the rim.

12. <u>Ravine</u>: A large irregular depression on the base of Smoky Mountains, part
of which appears much like a ravine, a small narrow, steep-sided
valley.

13. <u>South Ray</u>: One kilometer diameter bright-rayed crater south-southwest
of the landing point. Ray materials from this crater were collected
at several locations in the landing site area.

14. <u>Spook</u>: A crater that is one-half kilometer downrange from the landing spot.
The crater's location represented a hazard and worried the crew
during the preparation and training for the mission. Sampling station
number 2 was on the rime of this crater.

15. <u>Stubby</u>: A crater, one kilometer in diameter, with a stocky and thick pro-
trusion from Stone Mountain. Sampling station 6 is on the north rim
of this crater.

16. Trap: An old, subdued and partly hidden depression that hindered planning a southwesterly traverse to sample and study Baby Ray.

17. Wreck: Sampling station number 8; a relatively old crater whose original features appear disordered and ruined by later events. The crater is located between Stubby and Trap. All three craters were used as landmarks for LM landing.

Apollo 17 Landing Site

The following 29 names were selected from the list of 67 crew-given names in the Apollo 17 landing site area.

A. Feature names

1. Bear Mountain: A cluster of hills that forms the shape of a bear.

2. Family Mountain: Named for the families of the crew members, their associates and families in general.

3. Light Mantle: Fingered mantling deposit, believed to be a landslide.

4. North Massif: Massif is a French term for a large mountain mass, this one north of the landing site.

5. Scarp: A scarp, with mare ridge-like segments, located west of the landing site.

6. Sculptured Hills: Domical hills, surrounding the landing site area, which appear sculptured.

7. South Massif: Mountain mass south of the landing site area.

8. Tortilla Flat: Flat region near the Light Mantle.

9. Wessex Cleft: A cleft in the eastern border of North Massif. Wessex was an ancient Anglican kingdom.

10. Taurus-Littrow Valley: General landing site area

B.   Crater names

1. <u>Bronté</u>: Crater near which the first LRV stop was made; Charlotte Bronté was a 19th century English novelist.

2. <u>Bowen-Apollo</u>:   Crater near the farthest eastern limits of the third EVA traverse.

3. <u>Camelot</u>:   Crater named for the legendary King Arthur of the Round table, Station number 5 was on the rim of this crater.

4. <u>Cochise</u>: Crater named for the American Indian Apache chief.   The crater was studied and described on the third EVA.

5. <u>Emory</u>: Large crater used as a major landmark.   William H. Emory was a member of the Topographical Engineers who explored the American West.

6. <u>Falcon</u>: Small crater on Family Hill used for landmark tracking from orbit prior to lunar landing.   It is also believed to be one of few cinder cones in the Apollo 17 site (previously, F. Crater).

7. <u>Hess-Apollo</u>: Large crater near Mackin-Apollo, named after the geologist H. Hess.

8. <u>Horatio</u>: Crater southwest of Camelot.

9. <u>Lara</u>: Girl's name; sampling site of station number 3.

10. <u>Mackin-Apollo</u>: Large crater used as a landmark, named after J. Hoover Mackin, an American geomorphologist.

11. <u>Nansen-Apollo</u>: Sampling site of station number 21

12. <u>Powell</u>: One of the large craters in the landing site area named for John Wesley Powell, an explorer of the American West.

13. <u>Shakespeare</u>:   Large crater northeast of the landing site named after the English poet and playwright.

14. <u>Sherlock</u>: A crater that was used as a tracking landmark from orbit, also the last (12th) LRV sampling stop was north of the crater.   It is named after Sherlock Holmes, the hero of Sir Arthur Conan Doyle's novels.

15. <u>Shorty</u>: A dark-rimmed crater with relatively short, dark rays; it is named after a character in Richard Brautigan's contemporary novel "Trout Fishing in America."

16.   Steno-Apollo: Sampling site of the first station named after the
          geologist Steno.

17.   Trident: A triplet crater cluster shaped like the three pronged spear
          carried by Neptune to Poseidon, classical mythology's god of the sea.

18.   Van Serg: Sampling station number 9 was on the rim of a small crater.
          It was named after the pseudonym that Prof. Hugh McKinstry, a
          20th century mining geologist used in writing educational satire.

19.   Victory: A large V-shaped depression.

PART 4

ASTRONOMER'S HANDBOOK

ASTRONOMER'S HANDBOOK

The Astronomer's Handbook (*IAU Transactions* XIIC, edited by J.-C. Pecker, *IAU Transactions* XIIIB, edited by L. Perek, *IAU Transactions* XIVB, edited by C. de Jager and A. Jappel, and *IAU Transactions* XVB, edited by G. Contopoulos and A. Jappel) contains documents of permanent importance. Only the parts changed after the XVIth General Assembly are supplemented here.

I. SHORT HISTORY OF THE INTERNATIONAL ASTRONOMICAL UNION

Table 2 of *IAU Transactions* XIIC (see also volumes mentioned above) should be supplemented as follows:

XVIth General Assembly, 1976, Grenoble, Isère, France

Members of the Union, after the General Assembly	3.805
Adhering Countries	48
Participants in the General Assembly (Members 1062)	2.134
Number of Commissions	40
Volumes of Transactions: XVI A (in three parts)	pp. XXI+690
XVI B	pp. X +537
Highlights 1976	pp. VII+754
President: A. Blaauw	1976-1979
General Secretary: Edith A. Müller	1976-1979
Assistant General Secretary: P. A. Wayman	1976-1979
Vice Presidents: D. S. Heeschen	1976-1982
E. K. Kharadze	1976-1982
S. van den Bergh	1976-1982

II. ADMINISTRATION AND FINANCES OF THE UNION

A. ADMINISTRATION

The address of the General Secretary is:

Prof. Edith A. Müller
Observatoire de Genève
CH 1290 Sauverny
Switzerland

The Secretariat, consisting of the Executive Secretary A. Jappel, the Administrative Secretary Mrs. J. Dankova, and a typist, has its office at the Bâtiment des sciences physiques of the University of Lausanne in Dorigny, 1015 Lausanne.

All matters related to Symposia and Colloquia are dealt with by the Assistant General Secretary

Prof. P. A. Wayman
Director, Dunsink Observatory
Castleknock
Co. Dublin
Ireland

B. *FINANCES*

The XVIth General Assembly resolved that the unit of contributions payable by Adhering Countries to the IAU be increased to 1.465.00 Swiss francs for the years 1977, 1978 and 1979.

III. THE INTERNATIONAL ASTRONOMICAL UNION AND OTHER INTERNATIONAL SCIENTIFIC ORGANIZATIONS

A. *THE INTERNATIONAL COUNCIL OF SCIENTIFIC UNIONS*

The terms of office of other international scientific bodies differ from those of the IAU. Information in this respect should be looked up in the latest Yearbook of the ICSU. Only IAU representatives are mentioned in the following paragraphs.

The IAU is represented by the General Secretary, Professor Edith A. Müller, on the General Committee of ICSU.

B. *IAU REPRESENTATIVES ON SCIENTIFIC AND SPECIAL COMMITTEES OF ICSU*

1. *Committee on Space research (COSPAR)* : Edith A. Müller
2. *ICSU Abstracting Board (IAB):* J.-C. Pecker
3. *Federation of Astronomical and Geophysical Services (FAGS):*
                                            H. Enslin and Edith A. Müller
   IAU *representatives in* FAGS *services:*
   a) *Bureau International de l'Heure (BIH):* H. Enslin and H. M. Smith
   b) *International Polar Motion Service (IPMS):* S. Yumi
   c) *International Ursigrams and World Day Service (IUWDS):* F. W. Jäger
   d) *Scientific Ballooning and Radiations Monitoring Organization (SBARMO):* P. Simon

C. *IAU REPRESENTATION IN INTER-UNION COMMITTEES AND IN COMMISSIONS OF ICSU*

1. *Inter-Union Committee on Frequency Allocation for Radio Astronomy and Space Sciences (IUCAF):* G. Westerhout and R. Wielebinski
2. *Special Committee for Solar Terrestrial Physics (SCOSTEP):* Z. Svestka
3. *Inter-Union Commission for Science Teaching (IUCST):* L. N. Houziaux
4. *Committee for Data on Science and Technology (CODATA):* G. A. Wilkins
5. *Inter-Union Commission on Spectroscopy (IUCS):* B. Edlén, J. G. Philips and
                                            M. J. Seaton
6. *Committee on Science and Technology in Developing Countries (COSTED):*
                                            L. N. Houziaux
7. *Scientific Committee on Problems of the Environment (SCOPE):* R. Cayrel and
                                            J.-C. Pecker

D. *IAU REPRESENTATION IN OTHER ORGANIZATIONS*

1. *Fondation Internationale du Pic-du-Midi:* A. Lallemand
2. *Comité Consultatif pour la Définition de la Seconde (CCDS):* Wm. Markowitz
3. *Comité Consultatif pour la Définition du Mètre (CCDM):* A. H. Cook
4. *International Radio Consultative Committee (CCIR):* F. G. Smith and H. M. Smith
5. *European Physical Society:* P. A. Wayman

IV. SERVICES AND FUNCTIONS OF THE IAU

There are no changes as against the corresponding chapter in Transactions Vol. XV B.

V. AND VI. PUBLICATIONS AND SYMPOSIA OF THE INTERNATIONAL ASTRONOMICAL UNION

Reference is made to Information Bulletin No. 36 - June 1976, and No. 37 - January 1977.

VII. EXECUTIVE COMMITTEE 1976-1979

PRESIDENT

Professor Dr. A. Blaauw, Sterrewacht-Huygens Laboratorium, Wassenaarseweg 78, Leiden, 2405, The Netherlands.

VICE-PRESIDENTS

Mr. J. G. Bolton, CSIRO, Division of Radiophysics, P.O.Box 276,Parkes, Epping, N.S.W. 2870, Australia.
Professor Ch. Fehrenbach, Directeur de l'Observatoire de Haute Provence, Saint Michel l'Observatoire, 04300 Forcalquier, France. .
Dr. D. S. Heeschen, National Radio Astronomy Observatory, Edgemont Road, Charlottesville, Virginia 22901, U.S.A.
Professor W. Iwanowska,Institute of Astronomy, Nicolaus Copernicus University, ul. Chopina 12/18, Pl-87-100 Toruń, Poland.
Academician E. K. Kharadze, Director Abastumany Astrophysical Observatory, M.T. Canobili, Abastumani 383762, USSR.
Professor S. van den Bergh, David Dunlap Observatory, University of Toronto, Richmond Hill, Ontario L4C 4Y6, Canada.

GENERAL SECRETARY

Professor Edith A. Müller, Observatoire de Genève, CH-1290 Sauverny, Switzerland.

ASSISTANT GENERAL SECRETARY

Professor P. A. Wayman, Director, Dunsink Observatory, Castleknock, Co. Dublin, Ireland.

ADVISERS TO THE EXECUTIVE COMMITTEE

Professor L. Goldberg, Director, Kitt Peak National Observatory, P.O.Box 26732, Tucson, Arizona 85726, U.S.A. - former President.
Professor Dr. G. Contopoulos, Astronomy Department, Panepistimiopolis, Athens-621, Greece.

VIII. BY-LAWS

Article 12 (a) of the By-laws (IAU Transactions XIV B, p. 273-275, 1971) reads as follows:
"12. (a) Proposals for elections to the President of the Union, six Vice-Presidents, the General Secretary and the Assistant General Secretary are submitted to the General Assembly by the Special Nominating Committee. This consists of the President and past President of the Union, a member proposed by the retiring Executive Committee, and four members elected by the Nominating Committee from among twelve Members proposed by Presidents of Commissions. Other than the President and immediate past President, present and former members of the Executive Committee shall not serve on the Special Nominating Committee".

IX. WORKING RULES

Concerning the rules for scientific meetings reference is made to Transactions Vol. XIV B, 1971 p. 284-288. As to IAU publications (Transactions, Highlights, Symposia) the instructions for preparing camera-ready manuscripts are published in the Information Bulletin No. 36, June 1976.

X. LIST OF COUNTRIES ADHERING TO THE UNION

The following is a list of the 48 countries that adhered to the Union in September 1976, giving also the year of admission, the approximate number of Members, and the Adhering Organizations.

Country	Year	Members	Adhering Organizations
Arab Republic of Egypt	1925	18	Academy of Scientific Research and Technology Dept. Sci. Soc. and Internat. Unions 101, Kasr El Eini Street Cairo
Argentina	1927	31	Comité Nacional de Astronomia Observatorio Astronomico "Felix Aguilar" Av. Benavidez 8175 Oeste 5407 Marquesado, San Juan
Australia	1939	78	Australian Academy of Science P.O.Box 216 Civic Square Canberra, A. C. T. 2608
Austria	1955	24	Oesterreichische Akademie der Wissenschaften Dr. Ignaz Seipel-Platz 2 1010 Vienna

Country	Year	Members	Adhering Organizations
Belgium	1920	61	Académie Royale de Belgique Palais des Académies 1, rue Ducale Bruxelles 1
Brazil	1961	21	Comissao Brasileira de Astronomia (CBA) Conselho Nacional de Pesquisas Avenida Marechal Câmara 350 Rio de Janeiro GB
Bulgaria	1957	17	Bulgarian Academy of Sciences 7th November Street 1 Sofia
Canada	1920	134	National Research Council of Canada International Relations Ottawa K1A OR6
Chile	1947	9	Universidad de Chile Observatorio Astronomico Nacional Casilla 36-D Santiago de Chile
Colombia	1967	2	Academia Colombiana de Ciencias Exactas Fisicas y Naturales, Carrera 8a Apartado Nal. 2584, Bogota 1, D.E.
Cuba	1970	0	Instituto de Astronomia de la Academia de Ciencas de Cuba Calle 174/1722, Cubanacan, Habana
Czechoslovakia	1922	50	Czechoslovak Academy of Sciences Narodni 3 Praha 1 Czechoslovakia
Denmark	1922	31	Videnskabernes Selskab Dantes Plads 5 DK - 1556 Copenhagen V
Finland	1948	16	Académie des Sciences et Lettres Snellmaninkatu 9-11 Helsinki 17
France	1920	290	Académie des Sciences Institut de France 23, Quai de Conti 75, Paris 6e
Germany, D.R.	1951	37	Akademie der Wissenschaften der DDR Otto Nuschkestrasse 22/23 DDR 108 Berlin

Country	Year	Members	Adhering Organizations
Germany, F.R.	1951	199	Rat Westdeutscher Sternwarten Institut für Theoretische Physik und Sternwarte der Universität Neue Universität, Olshausenstrasse 2300 Kiel
Greece	1920	37	Academy of Athens Panepistimiou Str. Athens
Hungary	1947	21	Hungarian Academy of Sciences Roosevelt tér 9 Budapest V
India	1964	63	Indian National Science Academy Bahadur Shah Zafar Marg New Delhi- 110001
Iran	1969	3	Université de Téhéran Institut Géophysique Amirabad Shomali Teheran
Iraq	1976	1	Foundation of Scientific Research P.O.Box 255 Baghdad
Ireland	1947	15	National Committee of Astronomy The Royal Irish Academy 19 Dawson Street Dublin 2
Israel	1954	29	The Israel Academy of Sciences and Humanities 43 Jabotinsky Rd., P.O.B. 4040 Jerusalem 91 040
Italy	1921	141	Consiglio Nazionale delle Ricerche Piazale delle Scienze 7 Rome
Japan	1920	174	Science Council of Japan 22-34 Roppongi, 7 chome Minato-ku, Tokyo 106
Korea (DPR)	1961	3	Academy of Sciences Pyongyang
Korea (Republic)	1973	4	Korean National Astronomical Observatory Yoksam-Dong, Kangnam-ku Seoul, 134-03

Country	Year	Members	Adhering Organizations
Mexico	1921	21	Instituto de Astronomia, UNAM Apartado Postal 70-264 México, D.F.
Netherlands	1922	111	Koninklijke Nederlandse Akademie van Wetenschappen Kloveniersburgwal 29 Amsterdam - C
New Zealand	1964	14	The Royal Society of New Zealand 6 Kelburn Parade P.O.Box 12249 Wellington
Norway	1922	16	Det Norske Videnskaps-Akademi i Oslo Drammensveien 78 Oslo 2
Poland	1922	54	Polska Akademia Nauk Palac Kultury i Nauki Warsaw
Portugal	1924	14	Seccao Portuguesa das Unioes Inter- nacionais Astronomica e Geodésia e Geofisica Praça da Estrela, Lisboa 2
Roumania	1928	20	Roumanian National Committee of Astronomy Astronomical Observatory Str. Cutitul de Argint 5 Bucarest 28
South Africa	1938	12	Council for Scientific and Industrial Research Information and Research Services Pretoria 0001
Spain	1922	30	Comision Nacional de Astronomia Instituto Geografico y Catastral General Ibanez 3 Madrid (3)
Sweden	1925	58	Kungl. Vetenskapsakademien Fack, S-104 05 Stockholm 50
Switzerland	1923	35	Schweizerische Naturforschende Gesellschaft Zentralsekretariat, Laupenstr. 10 3000 Bern
Taiwan	1959	11	Academia Sinica No. 130 Yen Chiu Yuan Road Section 11 Nankang-Taipei

Country	Year	Members	Adhering Organizations
Turkey	1961	30	Türk Astronomi Dernegi University Rasathanesi Beyazit, Istanbul
United Kingdom	1920	339	The Royal Society 6, Carlton House Terrace London, SW1Y 5AG
Uruguay	1970	0	Ministerio de Educacion y Cultura Comité Nacional de Astronomia Casilla de Correo 867 Montevideo
U.S.A.	1920	1108	National Academy of Sciences Office of the Foreign Secretary 2101 Constitution Ave N.W. Washington, D.C. 20148
U.S.S.R.	1935	390	Academy of Sciences of the U.S.S.R. Leninskij Prospekt 14 Moscow 71
Vatican City State	1932	6	Pontificia Academia delle Scienze Citta del Vaticano
Venezuela	1953	5	Asociacion Venezolana de Astronomia Facultad de Ingenieria Departamento de Geodesia Universidad del Zulia Maracaibo
Yugoslavia	1935	22	Savez društva matematičara, fizičara i astronoma Jugoslavije Attn. Secretary General ul. Dr. Ilije Duričića 4 Novi Sad

APPENDIX

MEMBERSHIP OF COMMISSIONS

ALPHABETICAL LIST OF MEMBERS

COMPOSITION DES COMMISSIONS
MEMBERSHIP OF COMMISSIONS

4.   COMMISSION DES EPHÉMÉRIDES (EPHEMERIDES)

PRÉSIDENT: V. K. Abalakin

VICE-PRÉSIDENT: A. M. Sinzi

COMITÉ D'ORGANISATION: R. L. Duncombe, T. Lederle, J. H. Lieske,
   B. Morando, A. Orte, P. K. Seidelmann, G. A. Wilkins

MEMBRES: Aoki S., Bec A., de Greiff G. A., Deprit A., Dunham D. W.,
   Fricke W., Fursenko M. A., Gondolatsch F., Haupt R. F., Janiczek P. M.,
   Johnston K. J., Klepczinski W. J., Kovalevsky J., Lahiri N. C.,
   Morrison L. V., Oesterwinter C., O'Handley D. A., Rodriguez M.,
   Sadler D. H., Shapiro I. I., Taylor G. E., Ting Y. T., Van Flandern T. C.,
   Wackernagel H. B., Walter H. G., Williams J. G., Winkler G. M. R.,
   Yallop B. D.

5.   COMMISSION DE DOCUMENTATION (DOCUMENTATION)

PRÉSIDENT: J.-C. Pecker

VICE-PRÉSIDENT: W. D. Heintz

COMITÉ D'ORGANISATION D. A. Kemp, J. Kleczek, P. Lantos, J. R. Shakeshaft,
   T. S. Shcherbina-Samojlova, G. A. Wilkins

MEMBRES: Aaoki S., Beer A., Bidelman W., Bouska J., Dewhirst D. W.,
   Dixon R. S., Fleckenstein J. O., Fricke W., Griffin R. F., Hirose H.,
   Martynov D. Ya., Maxwell A., McNally D., Meadows A. J., Mein P.,
   Mitton S. A., Ogorodnikov K. F., Radlova L. N., Remy-Battiau L.,
   Schmadel L., Schmidt K. H., Sykes J. B., Velghe A. G., Weidemann V.,
   Wempe J., Worley C. E.

6.   COMMISSION DES TÉLÉGRAMMES ASTRONOMIQUES (ASTRONOMICAL TELEGRAMS)

PRÉSIDENT: E. Roemer

VICE-PRÉSIDENT: J. Hers

Sans Comité d'Organisation

MEMBRES: Biraud F., Candy M. P., Cesco C. U., Cunningham L. E.,
   Everhart E., Grindlay J. E., Kozai Y., Marsden B. G., (Secrétaire de la
   Comm.),Martynov D. Ya., Mrkos A., Pounds K. A., Rosino L., Simon P.

7. COMMISSION DE LA MÉCANIQUE CÉLESTE (CELESTIAL MECHANICS)

PRÉSIDENT: V. Szebehely

VICE-PRÉSIDENT: Y. Kozai

COMITÉ D'ORGANISATION: E. P. Aksenov, V. A. Brumberg, M. S. Davis, A. Déprit,
    G. N. Duboshin, G. E. O. Giacaglia, E. A. Grebenikov, M. Hénon, G. Hori,
    P. Kustaanheimo, P. J. Message, B. Popovic, J. Schubart, P. K. Seidelmann.

MEMBRES: Abalakin V. K., Aksnes K., Antonocopoulos G., Aoki S., Balmino G.,
    Batrakov Y., Bec A., Bettis D.G., Bohme S., Bozis G., Brouke R., Calame O.,
    Candy M. P., Cesco R. P., Chapront J., Contopoulos G., Cook A.H.,
    Counselman C. C., Cunningham L. E., Danby J. M. A., Demin V. G.,
    Dormand J. R., Duncombe R. L., Everhart E., Eichhorn H. K., FabreH.,
    Ferraz-Mello S., Fiala A., Foreschlé C., Galibina I. V., Garfinkel B.,
    Gaska S., Goldreich P., Goudas C. L., Grouchinsky N. P., Hadjidemetriou J.D.
    Hagihara Y., Hamid S. E., Heggie D. C., Henrard J., Herget P., Jefferys
    W. H., Jeffreys Sir Harold, Jupp A. H., Katsis D. N., Kaula W. M.,
    Kholshevnikov V. K., King-Hele D.G., Kovalevsky J., Lala P., Lazovic J. P.,
    Lundquist C. A., Magnaradze N. C., Marsden B.G., Matas V., Meffroy J.,
    Merman G. A., Møller O., Morando B., Mulholland J. D., Musen P., Nacozy P.E.
    Nahon F., O'Handley D., Omarov T. B., Orlov A. A., Peale S. J.,
    Petrovskaya M. S., Robinson W. J., Roy A. E., Ryabov J. A., Sagnier J. L.,
    Scholl H., Sconzo P., Sehnal L., Shapiro I. I., Sharaf S. G., Shteins K. A.,
    Siegel C. L., Sinclair A. T., Siry J. W., Soultanov G. F., Standish E. M.,
    Stiefel E., Tapley B. D., Tatevyan S. K., Thiry Y., Vinti J. P.,
    Williams C. A., Yarov-Yarovoy M. S., Yoshida J.

8. COMMISSION DE L'ASTRONOMIE DE POSITION (POSITIONAL ASTRONOMY)

PRÉSIDENT: R. H. Tucker

VICE-PRÉSIDENT: E. Høg

COMITÉ D'ORGANISATION: C. A. Anguita, S. Débarbat, W. Fricke, B. L. Klock,
    E. Marcus, A. A. Nemiro, J. L. Schombert, K. N. Tavastsherna, G. Teleki,
    G. van Herk, H. Yasuda.

MEMBRES: Adams A. N., Argyracos J. G., Atkinson R. d'E., Bacchus P.,
    Bagildinski B. K., Billaud G., Bohrmann A., Brouw W. N., Bykov M. F.,
    Chernega N. A., Chollet F., Counselman III Ch., Dambara T., Dejaiffe R.J.,
    De Vegt C., Dieckvoss W., Dravskikh A. F., Duncombe R. L., Eichhorn
    von Wurmb H. K., Fogh Olsen H. J., Gauss F. S., Gay J., Gliese W.,
    Gordon Y. E., Grudler P., Gubanov V. S., Guinot B., Gulyaev A. P.,
    Haas J., Hansson N., Heintz W. D., Hoffleit E. D., Hughes J. A.,
    Isobe S., Jackson P., Johnston K., Kharin A. S., Konin V. V., Korol'A.K.,
    Kosin G. S., Lacroute P., Larink J., Laustsen S., Lederle T., Lévy J. R.,
    López J.A., Manrique W. T., Melchior P., Mitić L. A., Murray C. A.,
    Nefed'eva A. I., Nikoloff I., Noel F., Okuda T., Osório J.J.S.P.,
    Petrov G. M., Pilowski K., Podobed V. V., Polozhentsev D. D.,

Proverbio E., Quijano L., Raimond E., Reiz A., Requième Y., Rhynsburger
R. W., Robertson W. H., Rousseau J.-M., Rybka P., Sandig H.-U.,
Scheepmaker A. C., Schmeidler F., Schuler W., Sevarlić B. M.,
Slaucitajs S., Stoy R. H., Thomas D. V., Tsubokawa I., Tuzi K., Von der
Heide J.E.B., Walter H. G., Wood H. W., Woolsey E. G., Yamazaki A.,
Zverev M. S.

9. COMMISSION DES INSTRUMENTS ASTRONOMIQUES (ASTRONOMICAL INSTRUMENTS)

PRESIDENT: J. Ring

VICE-PRESIDENT: E. H. Richardson

COMITE D'ORGANISATION: P. Connes, I. M. Kopylov, A. Labeyrie,
W.C. Livingston, D. McMullan, N. N. Mikhel'son, L. B. Robinson,
G. Walker, M. F. Walker, G. Wlérick

MEMBRES: Atkinson R.d'E., Babcock H. W., Baum W. A., Beck N. G.,
Bhattacharyya J. C., Blitzstein W., Brahde R., Brück H., Burton W.,
Carruthers G., Charvin P., Clarke D., Crawford D. L., Delbouille L.,
Dobronravin P. P., Duchesne M., Dunham T., Jr., Dunkelman L.,
Fellgett P. B., Ford W. K., Jr., Gay J., Giovanelli R.G., Hallam K. L.,
Hammerschlag R., Hilliard R., Hoekstra R., Honneycutt R. K., Jayarajan A.P.,
Jelley J. V., Karpinskij V. N., Khokhlova V. L., Kovatechev B. J.,
Lasker B. E., Luud L. S., McGee J. D., Moroz V. I., Nikonov V. B.,
Odgers G. I., Petford A. D., Pope J. D., Purgathofer A. Th., Rakos K. D.,
Reay N. K., Roddier F., Rösch J., Schöneich W., Schroeder D. J., Shakhbazyan
Y. L., Smyth M. J., Texereau J., Treanor P. J., S.J., Tull R. C.,
Valníček B., Veismann U. K., Wayman P. A., West R. M., Wynne Ch. G.,

10. COMMISSION DE L'ACTIVITE SOLAIRE (SOLAR ACTIVITY)

PRESIDENT: G. Newkirk, Jr.

VICE-PRESIDENT: V. Bumba

COMITE D'ORGANISATION: R. J. Bray, T. Hirayama, V. A. Krat, J.-L. Leroy,
S. F. Smerd, N. V. Steshenko, M. Stix, P. Sturrock, J. Wilcox,
C. Zwaan

MEMBRES: Altschuler M. D. , Altrock R. C., Aly M. K. M., Ambroz P.,
Anderson K. A., Athay R. G., Avignon Y., Balli E., Banin V., Barrow C. H.,
Beckers J. M., Bell B., Bhatnager A., Billings D. E., Bonov A. D.,
Brown J. C., Bruckner G., Bruzek A., Cannon C. J., Carlquist P., Cimino M.
Coutrez R., Covington A., Culhane J. L., de Groot T., de Jager C.,
Deubner F. L., Dezsö L., Dizer M., Dodson-Prince H., Dollfus A.,
Dryer M., Dubov E. E., Dulk G.A., Dunn R.B., Elste G. H., Erickson W.,

Falciani R., Fokker A. D., Fortini T., Friedman H., Fritzová-Švestkova L., Gabriel A. H., Gaizauskas V., Giovanelli R. G., Gleissberg W., Gnevyshev M. N., Gnevysheva R. S., Godoli G., Gokhale M. H., Gopasjuk S. I., Gurtovenko E. A., Hagan J. P., Hanasz J., Hansen R. T., Hedeman E. R., Howard R. F., Hyder C., Ioshpa B., Jager F. W., Jakimiec J., Jensen E., Jockers K., Kai K., Kanno M., Kjeldseth M. O., Kleczek J., Koeckelenberg A., Kopecky M., Koutmy S., Krause F., Krivsky L., Kruger A., Kuklin G.V., Kundu M. R., Kunzel H., Kuperus M., Letfus V., Lincoln J. V., Livshits M. A., Loughhead R. E., MacQueen R., Macris C. J., McCabe M., McKenna L. S., McLean D., Makita M., Malitson H. H., Maltby P., Malville J. M., Mendel'shtam S. L., Martres M. J., Mattig W., Maxwell A., Mein P., Menzel D. H., Mergentaler J., Michard R., Mogilevskij E. I., Mohler O. C., Moiseev I. G., Moreton G. E., Mustel E. R., Nagasawa S., Nakagawa Y., Namba O., Neupert W. M., Nishi K., Noyes R. W., Obridko V. N. Ohki K., Öhman Y., Orrall F. Q., Parkinson J. P., Parkinson W.H., Pick M., Piddington J. H., Pneuma G. W., Popovici C., Priest E., Rayrole J., Razmadze T. S., Reeves E. M., Rigutti M., Roberts B., Roemer M., Romanchuk P. R., Rompolt B., Rösch J., Roselet J.-P., Saito K., Sakurai K. Sarabhai V. A., Sawyer C., Schatten K. H., Schluter A., Schmidt H. U., Schroeter E. H., Semel M., Servajean R., Severny A. B., Shapley A. H., Sheeley Jr. N. R., Sitnik G. F., Slonim E. M., Smith E.v.P., Smith Henry J., Smith S., Stellmacher G., Stenflo J. O., Stepanov J. O., Suemoto Z., Švestka Z., Sykora J., Takakura T., Tamenaga T., Tanaka H., Tandberg-Hanssen E., Teske R. G., Tifrea E., Tlamicha A., Topolová-Euzicková B., Trellis M., Tritakis B., Tuominen J., Uchida Y., Underwood J. H., Valnicek B., Van Allen J. A., Van't Veer F., Vojtech L, Vorpahl J., Vitinskij Ju. I., Waldmeier M., Wentzel D. G., Wiehr E., Wild J. P., Wilson P., Wöhl H., Xanthakis J., Yakovkin N.A., Yoshimura H., Zelenka A., Zirin H.

12. COMMISSION DE LA RADIATION ET DE LA STRUCTURE DE L'ATMOSPHÈRE SOLAIRE
(RADIATION AND STRUCTURE OF THE SOLAR ATMOSPHERE)

PRESIDENT: M.K.V. Bappu

VICE-PRESIDENT: Y. Uchida

COMITE D'ORGANISATION: J. M. Beckers, G. E. Brueckner, A. N. Cox, R. G. Giovanelli, C. Jordan, W. Mattig, S. I. Gopasyuk, V. M. Sobolev, P. Souffrin.

MEMBRES: Acton L., Adam M. G., Allen C. W., Altrock R. C., Altschuler M., Athay R. G., Beckman J. E., Bel N., Billings D. E., Bhatnagar A., Blackwell D. E., Blamont J. E., Bohm K. H., Bohm-Vitense E., Bonnet R., Brault J. W., Bray R. J., Bruzek A., Bumba V., Cannon C., Delache P., Delbouille L., de Jager C., Deubner F. L., Dezso L., Dubov E. E., Dumont S., Dunn R. B., Edmonds F. N., Elliott J., Elste G., Evans J. W., Falciani R., Fossat E., Frazier E. M., Friedman H., Gabriel A., Gingerich O. J., Godoli G., Gokdogan N., Goldberg L., Gordon C., Gnevyshev M. N., Grevasse N., Harvey J. W., Haupt R. F., Hiei E., Holweger H., Hotinli M., House L., Houtgast J., Jager F. W.,

Jefferies J. T., Jordan S., Kanno M., Kato S., Kawaguchi I., Kneer F.,
Kononovich E. V., Kopecky M., Koutchmy S., Krat V. A., Kubicela A.,
Kuklin G. V., Kuperus M., Laborde G., Labs D., Leighton R. B.,
Leroy J. L., Linsky J. L., Livingston W. C., Locke J. L., Lopez
Arroyo M., Loughhead R. E., Lust R., McKenna-Lawler S., Marik M.,
Matsushima S., Mein P., Mergentaler J., Mewe R., Michard R.,
Migeotte M. V., Mihalas D., Moore-Sitterly Ch., Mouradian Z.,
Müller E.A., Namba O., Nevel L., Newkirk G.A., Nicolet M., Nikolsky G.M.
Nishi K., Noyes R. W., Orral F. O., Oster L., Pande C., Pajdusakova L.,
Parkinson W. H., Pasachoff J. M., Pecker J. C., Peyturaux R.,
Pierce A. K., Reeves E. M., Righini G., Rigutti M., Roddier F.,
Roland C., Schroter E. H., Seaton M. J., Semel M., Severny A. B.,
Sivaraman K. R., Sitnik G. F., Skumanich A., Stellmacher G., Stepanov V.E.,
Stenflo J. O., Suemoto Z., Svestka Z., Swensson J. W., Tanaka K.,
Tandberg-Hanssen E., Teplitskaya R. B., Thomas R. N., Thomas J. H.,
Tousey R., Unno W., Unsold A., Vasil'eva G.Ya., Vukicevic-Karabin M.,
Waldmeier M., Warwick J. W., Wilson P. R., Wlerick G., Wohl H.,
Wyller A., Youssef N. H., Zelenka A., Zirin H., Zirker J. B., Zwaan C.

14. COMMISSION DES DONNEES SPECTROSCOPIQUES FONDAMENTALES (FUNDAMENTAL
SPECTROSCOPIC DATA)

PRESIDENT: E. Trefftz

VICE-PRESIDENT: J. G. Phillips

COMITE D'ORGANISATION: A. H. Gabriel, R. H. Garstang, W. Lochte-Holtgreven,
S. L. Mandel'shtam, R. W. Nicholls, S. Sahal, W. Wiese

MEMBRES: Allen C. W., Andrew K. L., Baird K. M., Barrow R., Bates D. R.,
Bely O , Blaha M., Branscomb L. M., Brault J., Burgess A., Carroll P. K.,
Carver J. H., Cook A. H., Corliss C. H., Czyzak S. J., Dalgarno A.,
Delsemme A. H., Desesquelles J., Dobronravin P. P., Douglas A. E.,
Dressler K., Dufay M., Edlén B., Engelhardt E., Essen L., Feautrier N.,
Felenbok P., Fink U., Flower D., Garton W.R.S., Glagolevskij Ju. V.,
Grant I. P., Green L. C., Heddle D.W.O., Hefferlin R. A., Herman R.,
Herzberg G., Hesser J. E., House L. L., Huber M.C.E., Huebner W.,
Humphreys C. J., Jacquinot P., Johnson D. R., Johnson F. M., Joly F.,
Jordan C., Jordan H. L., Junkes J., Kessler K. G., Khokhlova V. L.,
King R. B., Kipper T., Kohl J. L., Lagerqvist A., Lang J., Lawrence G.,
Layzer D., Martin W. C., Mel'nikov O. A., Mewe R., Migeotte M. V.,
Milford S. N., Mohler O. C., Monfils A., Moore-Sitterly C. E.,
Naqvi A. M., Nevin T. E., Newsom G., Nussbaumer H., Obi S., Oetken L.,
Omont A., Parkinson W. H., Peach G., Petrini D., Pfenning H., Prokof'ev
V. K., Querci F., Richter J., Schadee A., Schrijver J., Seaton M. J.,
Shore B. W., Smit J. A., Smith G., Smith W. H., Somerville W. B.,
Sørensen G., Summers H. P., Swings J.-P., Swings P., Takayanagi K.,
Tatum J. B., Tech J. L., Terrien J., Tousey R., Traving G.,
van Bueren H. G., van Regemorter H., van Rensbergen W., Varsavsky C. M.,
Vujnović V., Wares G. W., Weniger S., Wilson R., Zirin H.

15. <u>COMMISSION POUR L'ETUDE PHYSIQUE DES COMÈTES, DES PETITES PLANÈTES ET DES</u>
    <u>MÉTÉORITES</u> (PHYSICAL STUDY OF COMETS, MINOR PLANETS AND METEORITES)

PRÉSIDENT: N. B. Richter

VICE-PRÉSIDENT: B. D. Donn

SECRÉTAIRE: J. Rahe

COMITÉ D'ORGANISATION: E. Anders, C. R. Chapman, A. H. Delsemme,
    O. V. Dobrovolsky, T. Gehrels, B. J. Levin, D. D. Morrison, J. Rahe,
    E. Roemer, A. A. Yavnel'.

MEMBRES: Arpigny C., Beyer M., Bertaud Ch., Biermann L., Blamont J. E.,
    Bobrovnikoff N. T., Bouska J., Brandt J. C., Brecher A., Brown P.
    Lancaster, Burns J. A., Candy M. P., Ceplecha Z., Chapman C. R.,
    Cherednichenko V. I., Cristescu C., Cruikshank D. P., Demenko A. A.,
    Dossin F., Douglas A. E., Dryer M., Everhart E., Eviatar A., Gerard E.,
    Greenberg J. M., Grossman L., Grudzinska S., Harwitt M. O., Haser L.,
    Haupt H., Herzberg G., Huebner W. F., Hunaerts J. J., Johnson T. V.,
    Keller H. U., Konopleva V. P., Kresak L., Krinov E. L., Larson H. P.,
    Liller W., Lüst R., Lyttleton R. A., McCord T. B., McCroskey R. E.,
    Malaise D. J., Maran S. P., Marsden B. G., Martel-Chossat M. T.,
    Matson D. L., Mendis D. A., Milet B., Miller F. D., Moore E., Mrkos A.,
    Neff J. S., Ness N. F., O'Dell C. R., O'Keefe J. A., Öpik E. J.,
    Proisy P., Remy-Battiau L., Riives W. G., Schmidt H. U., Sekanina Z.,
    Shul'man L. M., Simonenko A. M., Sivaraman K. R., Snyder L. E.,
    Swings P., Tomita U., Vanysek V., Veverka J. F., Vsekhsvjatskij S. K.,
    Wallis M., Wasson J., Waterfield R. L., Wehinger P., Wetherill G. W.,
    Whipple F. L., Wood J. A., Woszcyk A., Wyckoff S., Zellner B. H.

PRÉSIDENT: T. C. Owen

VICE-PRÉSIDENT: B. A. Smith

COMITÉ D'ORGANISATION: M. J. S. Belton, D. Gautier, J. E. Guest, C. H.Mayer,
    S. Miyamoto, D. Morrison, C. Sagan

MEMBRES: Akabane T., Arthur D. W. G., Ashbrook J., Barreto L. M.,
    Barrow C. H., Baum W. A., Berge G. L., Bobrov M. S., Boyce P. B.,
    Brahic A., Brecher A., Broadfoot A. L., Brunk W. E., Bullen K. E.,
    Camichel H., Campbell D. B., Chamberlain J. W., Chapman C., Collinson E. H.,
    Colombo G., Connes J., Cook A. F., Counselman C. C., Davies M. E.,
    de Marcus W. C., de Mottoni G., Dickel J.R., Dollfus A., Drake F. D.,
    Elliot J. L., Eshleman V. R., Fink U., Fox W. E., Gehrels T., Giclas
    H. L., Goldstein R. M., Goody R. M., Greyber H. D., Guerin P., Hagfors T.,
    Hall J. S., Halliday I., Herzberg G., Hide R., Hovenier J. W., Hubbard
    W. B., Hunt G. E., Hunten D. M., Ingrao H. C., Irvine W. M.,
    Iwasaki K., Jeffreys H., Jelly J.,

Johnson T. V., Kaplan L., Karandikar R.V., Kellogg W. W., Kondo Y.,
Kopal Z., Koval'I. K., Kumar S. S., Kuzumin A. D., Larson H. P.,
Leblanc Y., Levin B. J., Liddell U., Link F., Lipskij Y. N.,
McCord T. B., McElroy M. B., Masursky H., Matson D., Menzel D. H.,
Middlehurst B., Millman P. M., Moroz V. E., Narayana J. V., O'Leary B. T.,
Öpik E. J., Ottelet I., Parkinson T., Petropoulos B., Pettengill G. H.,
Rogers A. E. E., Rösch J., Runcorn S. K., Safronov V., Saissac J.,
Salisbury J. W., Shapiro I.I., Shimizu T., Shoemaker E., Sinton W.,
Smith J. J., Smoluchowski R., Strong J. D., Tombaugh C., Trafton L.,
Tyler G. L. Jr., van Allen J. A., Veverka J., Wallace L. V., Whitaker E.A.,
Wildey R. L., Williams I. P., Young A. T.

17. COMMISSION DE LA LUNE (THE MOON)

PRÉSIDENT: E. Anders

VICE-PRÉSIDENT: K. P. Florensky

COMITÉ D'ORGANISATION: R. B. Baldwin, A. Dollfus, V. P. Dzhapiashvili,
W. K. Hartmann, K. Koziel, B. Yu. Levin, J. D. Mulholland, J. A. O'Keefe,
S. K. Runcorn, C. P. Sonett

MEMBRES: Arthur D. W. G., Bastin J., Beckmann J. E., Bender P., Brecher A.,
Calame O., Cameron W.S., Chapman C. R., Counselman C. C., Davies M.E.,
El Baz F., Elston W. E., Evans D. J., Ezersky V. I., Fielder G., Gavrilov I.Y.,
Geake J.E., Geiss J., Gold T., Goldreich P., Gorenstein P., Goudas C.L.,
Green J., Grossman L., Guest J., Gurstein A.A., Habibullin S.T., Hagfors T.,
Jeffreys H., Johnson T. V., Kaula W. M., Kopal Z., Kuzmin A.D., Leikin G.A.,
Link F., Lipsky V.N., Martynov Ja., Matson D. L., McCord T.B., Menzel D.H.,
Meyer C., Middlehurst B. M., Mikhailov A.A., Millman P., Mills A.A.,
Miyamoto S., Moore P.O.B.E., Morrison L.V., Moutsoulas M., Nefed'ev A.A.,
Pettengill G., Potter K.I., Rösch J., Sagan C., Shoemaker E., Strom R.,
Troitsky V.S., Urey H.C., Van Flandern T.C., Walker R.M., Wasson J.T.,
Weimer T., Wetherill G.W., Whitaker E.A., Williams J.G., Wilson L.,
Wood J. A.

19. COMMISSION DE LA ROTATION DE LA TERRE (ROTATION OF THE EARTH)

PRÉSIDENT: R. O. Vicente

VICE-PRÉSIDENT: P. E. G. Pâquet

COMITÉ D'ORGANISATION: H. J. M. Abraham, P. L. Bender, B. Elsmore,
H. Enslin, K. Lambeck, C. Sugawa, G. Teleki, G. Winkler,
Ya. S. Yatskiv, S. Yumi (ex officio)

MEMBRES: Anderle R. J., Atkinson R.d'E., Barlier F., Billaud G.,
Bonanomi J., Brosche P., Buchar E., Cecchini G., Cohen M.H., Davies J.G.

Débarbat S. V., Dejaiffe R., Djurovic D., Drâmbǎ C., Fedorov E. P.,
Fichera E., Fleckenstein J. O., Gaposhkin E. M., Groten E., Guinot B.,
Hall R. G., Hers J., Hosoyama K., Hurukawa K., Iijima S., Jeffreys H.,
Kakuta C., Kalmykov A. M., Knowles S. B.., Kolaczek B., Kostina L. D.,
Kozai Y., Lederle T., Lefebvre M., Levallois J. J., Locke J. L.,
Markowitz Wm., McCarthy D. D., Meinig M., Melchior P. J., Mikhailov A. A.
Milovanović V., Miyadi M., Morgan P., Mueller I. I., Nicolini T.,
O'Hara N. P. G., Okamoto I., Okuda T., Ooe M., Opalski W., Orte A.,
Oterma L., Parijskij N. N., Pavlov N. N., Popelar J., Popov N. A.,
Proverbio E., Randić L., Rochester M. G., Runcorn S. K., Sakharov V. I.,
Sasao T., Sato K., Sekiguchi N., Shapiro I. I., Sheptunov G. S.,
Smith F. G., Smith H. M., Smylie D. E., Stoyko A., Stoyko N., Takagi S.,
Thomas D. V., Torao M., Tsao M., Tsubokawa I., Wako Y., Williams J. G.,
Yokoyama K.

20. COMMISSION DES POSITIONS ET DES MOUVEMENTS DES PETITES PLANETES,
DES COMETES ET DES SATELLITES (POSITIONS AND MOTIONS OF MINOR PLANETS,
COMETS AND SATELLITES)

PRESIDENT: B. G. Marsden

VICE-PRÉSIDENT: G. Sitarski

COMITÉ D'ORGANISATION: F. K. Edmondson, S. Ferraz-Mello, P. Herget,
    E. I. Kazimirchak-Polonskaya, L. Kresák, W. H. Robertson, E. Roemer J.
    Schubart, G. E. Taylor

MEMBRES: Abalakin V. K., Aksnes K., Arend S. J. V., Babadzhanov P. B.,
    Batrakov Yu. V., Bielicki M., Borsenberger-Bec A., Bruwer J. A.,
    Burns J. A., Calame O., Candy M. P., Cristescu C., Cunningham L. E.,
    Debehogne H., Delsemme A. H., De Pascual M., Dirikis M. A.,
    Dunham D. W., Dvorak R., Evdokimov Yu. V., Everhart E., Franklin F. A.,
    Froeschlé C., Galibina I. V., Garfinkel B., Gehrels T., Giclas H. L.,
    Gilmore A. C., Harrington R. S., Hirose H., Itzigsohn M., Kiang T.,
    Klemola A. R., Klepczynski W. J., Kohoutek L., Kovalevsky J., Kozai Y.,
    Lieske J. H., Lindblad B. A., McCrosky R. E., Michkovitch V. V.,
    Milet B., Mintz-Blanco B., Morando B., Mulholland J. D., Nacozy P. E.,
    Orlov A. A., Oterma L., Pascu D., Pittich E., Popović B., Porter J. G.,
    Protitch M. B., Rasmusen H. Q., Sagnier J. L., Schmitt A., Scholl H.,
    Schrutka G., Sekanina Z., Shelus P. J., Shtejns K. A., Sinclair A. T.,
    Strobel W., Sultanov G. F., Tomita Ko., Torroja J. M., Van Houten C.J.,
    Van Houten-Groeneveld I., Vsekhsvyatskij S. K., West R. M., Whipple F. L.
    Wild P., Wilkins G. A., Williams J. G., Yakhontova-Samojlova N.,
    Yeomans D. K., Zadunaisky P. E.

21. COMMISSION DE LA LUMIERE DU CIEL NOCTURNE (THE LIGHT OF THE NIGHT SKY)

PRÉSIDENT: R. Dumont

VICE-PRÉSIDENT: H. Tanabe

COMITÉ D'ORGANISATION: A. S. Asaad, R. H. Giese, C. Leinert, Yu.L. Trutse, G. Weill, J. L. Weinberg, R. D. Wolstencroft

MEMBRES: Anderson K., Angione R., Banos C. J., Bates D. R., Blackwell D. E., Blamont J. E., Chamberlain J. W., Cook A. F., Dachs J., Divari N. B., Dufay M., Dunkelman L., Fracassini M., Galperin G. I., Greenberg J. M., Harwit M., Haug U., Henry R. C., Huruhata M., Hunten D. M., James J. F., Jarrett A. H., Kaplan J., Karandikar R. V., Karyagina Z. V., Kastler A., Knaflich H. B., Kordylewski K., Koutchmy S., Kulkarni P. V., Leinert C., Levasseur A. C., Link F., Mattila K., Megrelishvilli T. G., Mukai T., Neuzil L., Ney E. P., Nicolet M., Pearse R. W. B., Pfleiderer J., Pitz E., Roach F. E., Robley R., Roosen R. G., Rozhkovskij D. A., Sanchez-Martinez F., Saxena P. P., Schmidt Th., Sharov A. S., Soberman R. K., Sparrow J. G., Steiger W. R., van Allen J. A., van de Hulst H. C., Wallace L. V., Weniger S., Witt A. N.

22. COMMISSION DES MÉTÉORES ET DE LA POUSSIÈRE INTERPLANETAIRE (METEORS AND INTERPLANETARY DUST)

PRÉSIDENT: I. Halliday

VICE-PRÉSIDENT: W. G. Elford

COMITÉ D'ORGANISATION: H. Fechtig, C. L. Hemenway, K. N. Kramer, B. A. Lindblad, B. A. McIntosh, Z. Sekanina, A. N. Simonenko, J. Štohl

MEMBRES: Abbott W. N., Anders E., Babadzhanov P. B., Belkovich O. I., Brownlee D. E., Ceplecha Z., Cook A. F., Davies J. G., Davis J., Elford W. G., Ellyett C. D., Fedynsky V. V., Fireman E. L., Giese R. H., Guth V., Hajduk A., Halliday I., Harvey G. A., Hawkins G. S., Hey J. S., Hirose H., Hodge P. W., Hoppe J. A., Hughes D. W., Jacchia L., Javnel' A. A., Jennison R. C., Kaiser T. R., Kashscheev B. L., Katasev L.A. Keay C. S. L., Kizilirmak A., Kostylev K. V., Kresák L., Kresáková M., Krinov E. L., Kvíz Z., Lebedinets V. N., Levin B. J., Lovell B., McCrosky R. E., McDonnell J. A. M., Millman P. M., Nazarova T. N., O'Keefe J. A., Öpik E. J., Padevět V., Plavcová Z., Porubčan V., Rajchl J., Roosen R. G., Russell J. A., Saho C. Y., Šimek M., Soberman R. K., Southworth R. B., Tomita K., Verniani F. F., Wetherill G. W., Whipple F. L., Wood J. A.

24. COMMISSION D'ASTROMÉTRIE PHOTOGRAPHIQUE (PHOTOGRAPHIC ASTROMETRY)

PRÉSIDENT: C. A. Murray

VICE-PRÉSIDENT: H. K. Eichhorn

COMITÉ D'ORGANISATION: Ch. de Vegt, L. W. Frederick, W. Gliese,
    R. S. Harrington, P. Lacroute, H. I. Potter, W. van Altena

MEMBRES: Abhyankar K. D., Blaauw A., Bouigue R., Brosche P., Churms J.,
    Clube S. V. M., Crézé M., Delhaye J., Deutsch A. N., Dieckvoss W.,
    Dommanget J., Elsmore B., Fatchikhin N. V., Firneis M. G.,
    Fracastoro M. G., Franz O. G., Gallouet L., Gatewood G., Giclas H. L.,
    Goyal A. N., Haas J., Heintz W. D., Herget P., Hershey J. L., Hill G.,
    Hoffleit E. D., Ianna P. A., Jones B. F., Jones D. H. P., Klemola A. R.,
    Klock B. L., Lapushka K. K., Latypov A. A., Lavdovskij V. V.,
    Lippincott S. L., Lozinskij A. M., Luyten W. J., Ménessier M. O.,
    Meurers J., Morgan W. W., Nicholson W., O'Connell J. K., Oja T.,
    Onegina A. B., Podobed V. V., Prochazka F. V., Quijano L., Robertson W.H.,
    Roemer E., Sanders W. L., Schilt J., Stoy R. H., Strand K. Aa.,
    Thomas D. V., Upgren A., Valbousquet A., van de Kamp P., Vasilevskis S.,
    Wagman N. E., Wesselink A. J., Wood H. W.

25. COMMISSION DE PHOTOMETRIE ET DE POLARIMETRIE STELLAIRES
    (STELLAR PHOTMETRY AND POLARIMETRY)

PRÉSIDENT: M. F. McCarthy .

VICE-PRÉSIDENT: J. A. Graham

COMITÉ D'ORGANISATION: A. Behr, G. V. Coyne, A. Feinstein, P. W. Hill,
    A. R. Hyland, T. Markkanen, K. Osawa, F. G. Rufener, W. Schoneich,
    V. Straizys, J. Tinbergen.

MEMBRES: Argue A. N., Bahng J. D. R., Becker W., Bigay J. H., Bok B. J.,
    Borgman J., Breger M., Bruck H. A., Chis G., Chugajnov P. F., Cousins
    A. W. J., Crawford D. L., Dachs J., Eelsalu H. T., Fernie J. D., Fitch W.S.,
    Gallouet L., Gehrels T., Golay M., Goy G., Gutierrez-Moreno A.,
    Haffner H., Hardie R. H., Hauck B., Hiltner A., Holmberg E.B., Irwin J. B.
    Jerzykiewicz M., Johnson H. L., Joshi S. C., Kron G. E., Landolt A. U.,
    Lasker B., Lockwood G. W., Masani A., Mayer P., Mendoza V. E. E.,
    Mianes P., Mitchell R. I., Moreno H., Morris S. C., Muller A. B.,
    Mumford G. S., Nikonov V. B., Notni P., Perry C. L., Roslund C., Rybka E.,
    Sarma M. B. K., Schmidt H., Shawl S. J., Smyth M. J., Steinlin U. W.,
    Stock J., Stroheimer W., Todoran I., Tolbert C. R., Ulrich B., Ureche V.,
    Velghe A. G., Wallenquist A. A. E., Walraven T., Wesselink A. J.,
    Willstrop R. V., Young A. T.

26. COMMISSION DES ETOILES DOUBLES (DOUBLE STARS)

PRÉSIDENT: P. Muller

VICE-PRESIDENT: O. G. Franz

COMITÉ D'ORGANISATION: A. Batten, M. Fracastoro, R. Harrington, S. Lippincott, C. Worley

MEMBRES: Abt. H., Arend S., Baize P., Cester B., Couteau P., Da Silva A. V., Deutsch A. N., Djurkovic P. M., Dommanget J., Dunham D. W., Finsen W. S., Freitas- Mourao R. R., Geyer E., Hadjidemetriou J. D., Heintz W. D., Hershey J. L., Hidajat B., Holden F., Johnson M. C., Kiselev A. A., Kulikovsky P. G., Luyten W. J., Meyer C., Poveda A., Rakos K. D., Scarfe C. D., Strand K. Aa., The P. S., van de Kamp P., Walker R. L., Wieth-Knudsen N. P.

27. COMMISSION DES ETOILES VARIABLES (VARIABLE STARS)

PRÉSIDENT: J. Smak

VICE-PRESIDENT: J. D. Fernie

COMITÉ D'ORGANISATION: N. Baker, M. Breger, A. N. Cox, M. F. Feast, H. Gursky, W. E. Kunkel, W. Wenzel

MEMBRES: Aizenman M. L., Alania I. F., Andrews A. D., Arkhipova V. P., Ashbrook J., Bakos G. A., Barnes T. G., Bath G. T., Balazs-Detre J., Bateson F. M., Bertaud Ch., Bessell M. S., Bhatnagar P. L., Bond H. E., Boulon J., Boyarchuk A. A., Butler C. J., Chavira E., Cherepashchuk A. M., Chis G., Christy R. F., Chugainov P. F., de Kock R. P., Demers S., Dickens R. J., Efremov Yu. N., Edwards P. J., Eskioglu A. N., Evans D. S., Fitch W. S., Friedjung M., Frolov M. S., Gahm G., Gaposchkin S., Gascoigne S. C. B., Gershberg R. E., Geyer E. H., Godoli G., Gorbatsky V.G Graham J., Hall D. S., Hansen C. J., Haro G., Harwood M., Heidmann N., Heiser A., Herbig G. H., Herr R. B., Hesser J. E., Heyden F. J., Hoffleit E. D., Houk N., Huffer C. M., Huruhata M., Hutchings J. B., Iben I., Jarzebowski T., Jones A. F., Kippenhahn R., Kholopov P. N., Kopylov I. M., Kordylewski K., Kraft R. P., Krzemiński W., Kuhi L. V., Kukarkin B. V., Kumsishvili J. I., Kwee K. K., Landolt A. U., Ledoux P., Lesh J. R., Leung K. C., Lockwood G. W., Lortet-Zuckermann M. C., Luud L. S., McNamara D. H., Maffei P., Mannino G., Masuni A., Mavridis L.N Mayall M. W., Miller W. J., Milone L. A., Mirzoyan L. V., Moffett T. J., Morguleff N., Mumford G. S., Murdin P. G., Neff J. S., Nicolov N. S., O'Connell D. J. K., Odgers G. J., Oosterhoff P. T., Opolski A., Osawa K., Oskanyan V., Payne-Gaposchkin C. H., Peltier L. C., Percy J. R., Piotrowski S. L., Plaut L., Popova M., Rakos K. D., Renson P., Richter G.,

Rodgers A. W., Romano G., Rosino L., Sarma M. B. K., Sawyer Hogg H. B.,
Schoembs R., Shobbrook R. R., Sinvhal S. D., Smeyers P., Smith Harlan J.,
Starrfield S., Stepień K., Strohmeier W., Swope H. H., Szeidle B.,
Takeuti M., Tammann G. A., Tsesevich V. P., Tempesti P., Terzan A.,
Tijn A., Dije H. R. E., Tremko J., Usher P. D., Vandekerkhove E.,
van Agt S., van Genderen A. M., van Hoof A., Vogt N., Wachmann A. A.,
Walker M. F., Wallerstein G., Walraven T., Warner B., Wehlau A.,
Wesselink A. J., Wilson L. A., Wright F. W.

28. COMMISSION DES GALAXIES (GALAXIES)

PRESIDENT: B. E. Markarian

VICE-PRESIDENT: B. E. Westerlund

COMITÉ D'ORGANISATION: H. C. Arp, G. R. Burbidge, R. D. Davies, E. A. Dibaj,
K. C. Freeman, J. Heidmann, E. B. Holmberg, M. Peimbert, V. C. Rubin,
G. A. Tammann

MEMBRES: Abell G. O., Ables H. D., Alfvén H., Allen R. J. Ambartsumian V.A.,
Andrillat Y., Arakelian M.A., Ardeberg A. L., Bahcall J. N., Barbieri C.,
Barbon R., Baum W. A., Bertola F., Bigay J. H., Bok B. J., Bolton J. G.,
Bondi H., Bonnor W. B., Bottinelli L., Braccesi A., Brecher K.,
Brosche P., Capaccioli M., Carranza G. J., Chincarini G. L., Contopoulos G.
Courtès G., de Vaucouleurs G., Einasto J. E., Ekers R. D., Elvius A. M.,
Esipov V. F., Evans D. A., Fehrenbach C., Field G. B., Ford W. K.,
Fraser C. W., Gascoigne S. C. B., Goss W. M., Gougenheim L., Graham J. A.,
Haro G., Heckmann O., Heeschen D.S., Hodge P. W., Hoyle F., Johnson H. M.,
Kalinkov M., Karachentsev I. D., Kellerman K. I., Kiang T., King I. R.,
Kopylov I. M., Kundu M. R., Kustaanheimo P., Larson R. B., Layzer D.,
Lequeux J., Lin C. C., Lindblad P. O., Low F. J., Lynden-Bell D.,
Lynds B. T., Lynds C. R., McClure R. H., McCrea W. H., Mackay C. D.,
McVittie G. C., Mark J. W.-K., Mayall N. U., Mills B. Y., Morgan W. W.,
Narlikar J. V., Neugebauer G., Neyman J., Noonan T. W., Oke J. B., Oleak H.
Omer G. C., Oort J. H., Oesterbrock D. E., Pachner J., Pacholcyk A. G.,
Page T. L., Peimbert M., Pismish de Recillas P., Poveda A.,
Prendergast K. H., Pronik V. I., Reaves G., Reddish V. C., Richter N.,
Rindler W., Roberts M. S., Roberts W. W., Robinson I., Rood H. J.,
Rosino L., Rudnicki K., Sancisi R., Sargent W. L. W., Saslaw W. C.,
Schmidt M., Schücking E. L., Sciama D. W., Scott E. L., Searle L.,
Sersic J. L., Setti G., Shakeshaft J. R., Shane C. D., Shen B. S. P.,
Smith M. G., Spinrad H., Stein W. A., Stibbs D. W. N., Takarada K.,
Takase B., Tifft W. G., Tinsley B., Toomre A., Tovmassian H. M.,
Treder H. J., Ulrich M. H., Vandekerkhove E., van den Bergh S., van der
Kruit P. C., van der Laan H., van Woerden H., Visvanathan N., Vorontsov-
Velyaminov B.A., Whitford A. E., Wild P., Williams R. E., Wills B. J.,
Wilson A. G., Zel'dovich Ya. B.

29. COMMISSION DES SPECTRES STELLAIRES (STELLAR SPECTRA)

PRESIDENT: M. Hack

VICE-PRESIDENT: W. K. Bonsack

COMITE D'ORGANISATION: Y. Andrillat, A. A. Boyarchuk, C. O. R. Jaschek,
    J. Jugaku, R. J. Kovachev, D. C. Morton, J. P. Swings, K. O. Wright

MEMBRES: Abhyankar K. D., Abt H. A., Aller L. H., Andrillat H., Asaad A. S.,
    Aslanov I. A., Bappu M. K. V., Berger J., Bertaud Ch., Bessel M.A.,
    Bidelman W. P., Bloch M., Boesgaard A. M., Boggess A.III, Bouigue R.,
    Boulon J., Buscombe W., Cayrel R., Cayrel-de-Strobel G., Chalonge D.,
    Climenhaga J. L., Code A. D., Conti P.S., Cowley A. P., Cowley C. R.,
    Divan L., Doazan V., Dobronravin P. P., Dolidze M. V., Dunham Jr T.,
    Edmonds Jr. F. N., Evans D. S., Faraggiana R., Feast M. W., Fernandes M. R.,
    Friedjung M., Fringant A. M., Fujita Y., Garrison R. F., Gershberg R. E.,
    Glagolevskij Ju. V., Goldberg L., Gollnow H., Gorbatskij V. G.,
    Gratton L., Gray D. F., Greenstein J. L., Griffin R. E. M., Griffin R. F.,
    Groth H. G., Guthrie B. N. G., Haro G., Heintze J. R. W., Henize K. G.,
    Herbig G. H., Herman R., Houziaux L., Huang Su-Shu, Hyland A. R.,
    Jaschek M., Johnson H. R., Keenan P. C., Khokhlova V. L., King R. B.,
    Kodaira K., Kogure T., Kopylov I. N., Kraft R. P., Labs D., Lambert D. L.,
    Larsson-Leander G., Locanthi D. D., Luud L. S., McNamara D. H.,
    Maitzen H. M., Mannino G., Mel'nikov O.A., Milligan J. E., Morguleff N.,
    Mustel E. R., Myerscough V. P., Nicholls R. W., Nikitin A. A.,
    Nishimura S., Oetken L., Oke J. B., Osawa K., Pagel B. E. J., Parson S. B.,
    Pasinetti L. E., Payne-Gaposchkin C. H., Peat D. W., Pedoussaut A.,
    Peery B. F., Plavec M., Preston G. W. III, Przybylski A., Querci F.,
    Reimers D., Ringuelet A. E., Rodgers A. W., Sahade J., Sanwal N. B.,
    Schild R. E., Seggewiss W., Sinnerstad U., Slettebak A., Smak J.,
    Smolinski J., Snow Th., Stawikowski A., Stecher T. P., Svolopoulos S. N.,
    Swensson J. W., Swings P., Taffara S., Tech T., Thackeray A. D.,
    Thompson G. I., Tsuji T., Underhill A. B., Utsumi K., Vandekerkhove E.,
    Van't Veer C., Viotti R., Voigt H. H., Vreux J. M., Wallerstein G. N.,
    Waterworth M. D., Wehlau W. H., Wellmann P., Wempe J., Wehinger P.,
    Weniger S., Weiss W. W., Williams P. M., Wilson O. C., Wilson R., Wing R.F.,
    Wolff S. C., Wood H. J., Wyckoff S., Wyller A.A., Yamashita Y.

30. COMMISSION DES VITESSES RADIALES (RADIAL VELOCITIES)

PRESIDENT: A. H. Batten

VICE-PRESIDENT: M. Duflot

COMITE D'ORGANISATION: R. F. Griffin, A. G. D. Philip

MEMBRES: Abt H.A., Barbier-Brossat M., Beardsley W. R., Bertiau F. C.,
    Bolton C. T., Bouigue R., Boulon J., Crampton D., De Jonge J. K.,
    De Vaucouleurs G., Edmondson F. K., Eelsalu H. T., Fehrenbach C.,
    Fletcher J. M., Georgelin Y., Gollnow H., Harding G. A., Heintze J. R. W.,

Hill G., Hube D. P., Imbert M., Kraft R. P., Martin M., Maurice E.,
Mel'nikov O. A., Oetken L., Pédoussat A., Perry C. K., Preston G. W. III,
Prévot L., Rebeirot E., Rubin V. C., Sahade J., Serkowski K.,
Thackeray A. D., Willstrop R. V.

31. COMMISSION DE L'HEURE (TIME)

PRESIDENT: A. Orte

VICE-PRESIDENT: S. Iijima

COMITÉ D'ORGANISATION: D. J. Belocerkovskij, H. Enslin, B. Guinot,
   G. Hemmleb, P. Melchior, J. D. Mulholland, C. J. A. Penny, A. M. Sinzi,
   G. M. R. Winkler

MEMBRES: Abele M. K., Abraham H. J., Benavente J., Bender P. L., Billaud G.,
   Blinov N., Bonanomi J., Brkić Z. M., Caprioli G., Dingle H., Drâmbă G.,
   Fichera E., Fliegel H. F., Gökmen T., Hall R. G., Hers J., Klepczinski W.J
   Lacombe C. G., Lederle T., Lieske J. H., Markowitz Wm., Matsunami N.,
   McCarthy D. D., Mikhailov A. A., Miyadi M., Morrison L. V., Mueller I. I.,
   Noel F., O'Handley D. A., Opalski W., Pâquet P., Parcelier P., Pavlov N.N.
   Pensado J., Postoiev A., Proverbio E., Randić L.,Sadler D. H.,
   Schuler W., Stoyko A., Shcheglov V. P., Shiryaev A. V., Smith H. M.,
   Takagi S., Torao M., Tsao M., Tsuchiya A., Vicente R. O., Wackernagel H.B.
   Webrová L,, Wieth-Knudsen N. P., Wilkins G. A., Yumi S.

33. COMMISSION DE LA STRUCTURE ET DE LA DYNAMIQUE DU SYSTÈME GALACTIQUE. (THE
   STRUCTURE AND DYNAMICS OF THE GALACTIC SYSTEM)

PRÉSIDENT: F.J. Kerr

VICE-PRÉSIDENT: G.G. Kuzmin

COMITÉ D'ORGANISATION: D.L. Crawford, J. Einasto, K.C. Freeman, M. Fujimoto,
   P.O. Lindblad, K.F. Ogorodnikov, L. Perek, R. Wielen

MEMBRES: Aarseth S., Agekjan T.A., Altenhoff W.J., Ambartsumian V.A., Andrle
   P., Aoki S., Ardeberg A.L., Athanassoula E., Baldwin J.E., Barbanis B.,
   Becker W., Beer A., Bigay J.H., Blaauw A., Blanco V.M., Bok B.J., Boulon J.,
   Burke B.F., Burton W.B., Churchwell E.B., Clube S.V.M., Contopoulos G.,
   Courtès G., Crampton D., Crézé M., Cuperman S., Delhaye J., Dickel H.R.,
   Dickel J.R., Dieter N.H., Drilling J.S., Dzigvashvili R.M., Edmondson F.K.,
   Elsässer H., Elvius T., Fehrenbach Ch., Fenkart R.P., FitzGerald M.P., Fricke
   W., Genkin I.L., Georgelin Y.M., Georgelin Y.P., Gliese W., Goldreich P.,
   Gomez A.E., Gordon M.A., Gyldenkerne K., Haffner H., Haug U., Hayli A.,
   Heckmann O., Hénon M., Hohl F., Hori G., Hunter C., Innanen K.A., Irwin J.B.,
   Isobe S., Iwaniszewska-Lubienska C., Iwanowska W., Jaschek C.O.R., Johnson H.

M., Jones D.H.P., Kaburaki M., Kalnajs A.J., Kato S., Kharadze E.K., Kholopov
P.N., King I.R., Kinman T.D., Klare G., Kukarkin B.V., Kulsrud R., Kurth R.,
Larson R.B., Lecar M., Lin C.C., Lodén K., Lodén L.O., Lunel M., Luyten W.J.,
Lynden-Bell D., Lyngå G., McCarthy M.F., McCuskey S.W., MacConnell D.J.,
MacRae D.A., Maksumov M.N., Malmquist K.G., Mark J.W.K., Marochnik L.,
Martinet L., Mavridis L.N., Mayor M., Ménessier M.O., Mezger P.G., Miller R.
H., Mirzoyan L.V., Miyamoto M., Mohr J.M., Monnett G., Morgan W.W., Münch G.,
Murray C.A., Nahon F., Neckel Th., Niimi H., Oja T., Ollongren A., Oort J.H.,
Ostriker J.P., Ovenden M.W., Palmer P.E., Pavlovskaya E.D., Perry C.L., Pesch
P., Philip A.G. Davis, Pilowski K., Plaut L., Priester W., Ramberg J.M.,
Riegel K.W., Roberts W.W., Roman N.G., Rubin V.C., Rybicki G.B., Sandqvist
Aa., Sanduleak N., Schilt J., Schmidt H., Schmidt K.H., Schmidt M., Schmidt-
Kaler Th., Seggewiss W., Shane W.W., Sharov A.S., Sher D., Shimizu T., Shu F.
Simonson S.C., Sletteback A., Spiegel E.A., Steinlin U.W., Stephenson C.B.,
Stibbs D.W.N., Sturch C., Svolopoulos S.N., Szebehely V., Tammann G.A., Thé
Pik Sin, Toomre A., Tosa M., Tsioumis A., Upgren A.R., Vanderlinden H.L.,
Vandervoort P.O., Van Hoof A., Van Woerden H., Varsavsky C.M., Velghe A.G.,
Verschuur G.L., Vetesnik M., Wayman P.A., Weaver H.F., Westerhout G.,
Westerlund B.E., White R.E., Whiteoak J.B., Wilson Th.L., Woltjer L.,
Woolley Sir Richard, Yuan C.

34. COMMISSION DE LA MATIÈRE INTERSTELLAIRE ET DES NEBULEUSES PLANETAIRES (INTERSTELLAR MATTER AND PLANETARY NEBULAE)

PRESIDENT: G. B. Field

VICE-PRESIDENT: V. Radhakrishnan

ORGANIZING COMMITTEE: L. Higgs, G. S. Khromov, J. Lequeux, B. T. Lynds, M. Morimotc
D. C. Morton, M. Peimbert, H. van Woerden, B. Zuckerman

MEMBERS: Acker A., Aitken D. K., Akabane K., Aller L. H., Altenhoff W., Andrew
B. H., Andriesse C. D., Andrillat H., Andrillat Y., Arhipova V. P., Arny T. T.,
Axford W. I., Baars J. W. M., Baker P., Baldwin J. E., Barrett A. H., Bash
F. N., Baudry A., Beckman J., Behr A., Bel N., Bergeron J., Berkhuijsen E. M.,
Biermann L., Bless R. C., Boggess A., Bohlin R. C., Bok B. J., Borgman J.,
Braes L. L. E., Brand P. W. J. L., Brück M. T., Burgess A., Burke B. F., Burton
W. B., Buscombe W., Bystrova N. V., Capriotti E. R., Carruthers R. G., Caswell
J. L., Celnikier L., Cesarsky C., Cesarsky D., Chopinet M., Churchwell W. B.,
Chvojkova E., Clark B. G., Code A. D., Collin-Souffrin S., Courtès G., Cox
D. P., Coyne G. V., Crézé M., Cruvellier P., Cudaback D. D., Cugnon P., Czyzak
S. J., Dalgarno A., Davies R. D., de Jong T., Dibaj E. A., Dickel H. R., Dieter
N. H., Disney M. J., Divan L., Donn B. D., Dopita, M. A. Dorschner J., Downes
D., Drapatz S., Dunham T., Dupree A., Dyson J. E., Elvius A. M., Encrenaz P.,
Esipov V. F., Falle S. A. E. G., Faulkner D. J., Fejes I., Felten J. E., Flower
D., Ford H., Friedemann C., Gardner F. F., Gaustad J. E., Gay J., Gehrels T.,
Georgelin Y., Gerard E., Gilra D. P., Goldreich P., Goldsmith D. W., Goldstein
S. J., Goldsworthy F. A., Goldwire H. C., Gordon C. P., Gordon M. A., Gosachinsky
I. V., Goss W. M., Greenberg J. M., Grewing M., Grubissich C., Grzedzielski S.,
Guélin M., Gull T. R., Gurzadjan G. A., Habing H. J., Hall J. S., Hardebeck
E. G., Harten R. H., Heiles C. E., Helfer H. L., Henize K. G., Herbig G. H.,
Herzberg G., Hidajat B., Hiltner W. A., Hippelein H. H., Hjellming R. M.,
Hobbs L. M., Höglund B., Hua E. T., Hughes V. A., Hulsbosch A. N. M., Hummer

D., Hutchings J. B., Irvine W. M., Isobe S., Jenkins E. B., Jennings R. E.,
Johnson F. M., Johnson H. M., Johnston K. J., Joly F., Jura M., Kaftan-Kassin
M. A., Kahn F. D., Kaler J. B., Kamijo F., Kaplan S. A., Kazès I., Kegel W. H.,
Kerr F. J., Kharadze E. K., Knapp G. R., Ko H. C., Kogure T., Kohoutek L.,
Komesaroff M. M., Kostyakova E. B., Krishna Swamy K. S., Kundu M. R., Lambrecht
H., Lasker B. M., Lee T. J., Leung C. M., Liller W., Lin C. C., Lodén L. O.,
Lortet-Zuckermann M. C., Louise R., Low F. J., Lutz J., McCray R. A., McCrea
W. H., McGee R. X., McKee C., McLeod J., McNally D., Maihara T., Manchester
R. N., Martel-Chossat M. T., Mathews W. G., Mathewson D. S., Mathis J. S.,
Mattila K., Meaburn J., Mebold U., Menon T. K., Meszaros P., Mezger P. G.,
Michel K. W., Miller J. S., Mills B. Y., Morgan D. H., Mouschovias T. P.,
Münch G., Myers P., Nandy K., Neugebauer G., O'Dell C. R., Okuda H., Oort J. H.,
Osaki T., Osterbrock D. E., Ozernoj L. M., Palmer P. E., Parker E. N., Pecker
J. C., Penston M. V., Penzias A., Perinotto M., Petrosian V., Pöppel W., Pottasch
S. R., Price R. M., Pronik I. I., Pskovskij Yu.P., Radhakrishnan V., Raimond E.,
Reynolds R. J., Riegel K. W., Robbins R. R., Roberts M. S., Roberts W. W.,
Robinson B. J., Rogers A. E. E., Rohlfs K., Rose W. K., Rosino L., Rozhkovskij
D. A., Rubin V. C., Rublev S. V., Salpeter E. E., Sancisi R., Sandquist A.,
Sato F., Sato Y., Savage B., Savedoff M. P., Schalén C., Schatzman E., Scherb
F., Scheuer P. A. G., Schmidt K. H., Schmidt Th., Schmidt-Kaler Th., Schwarz
U. J., Seaton M. J., Seddon H., Serkowski K., Shah G. A., Shakeshaft J. R.,
Shane W. W., Shao C. Y., Shapiro M. M., Sharpless S. L., Shaver P. A., Shawl
S. J., Shcheglov P. V., Shu F., Shuter W. L. H., Shklovskij I. S., Silk J.,
Skilling J., Smith M. G., Snow T. P., Sobolev V. V., Sofia S., Solomon P. M.,
Somerville W. B., Sorochenko R. L., Spitzer L., Stecher T. P., Steigman G.,
Strom R. G. (Dwingeloo), Strömgren B., Sullivan W. T., Swings J. P., Takakubo
K., Tamura S., Terzian Y., Thackeray A. D., Thaddeus P., Thompson A. R.,
Tolbert C. R., Torres-Peimbert S., Townes C. H., Treanor P. J., Turner B. E.,
Turner K. C., Ulrich M. H., Unno W., van de Hulst H. C., van den Bout P.,
van der Laan H., van Horn H. M., Vanysek V., Verschuur G. L., Viswanathan N.,
von Hoerner S., Vorontsov-Vel'yaminov B. A., Walker G. A. H., Weaver H. F.,
Webster B. L., Weiler K. W., Wendker H. J., Wentzel D. G., Wesselius P. R.,
Westerhout G., Weymann R. J., Whiteoak J. B., Wickramasinghe N. C., Williams
D., Williams D. A., Williams I. P., Williams R. E., Wilson R. W., Winnberg
A., Witt A. N., Wolstencroft R. D., Woltjer L., Woodward P., Woolf N. J.,
Wright A. E., Wrixon G. T., Wynn Williams G., Yabushita S., York G., Zimmermann
H.

35. COMMISSION DE LA CONSTITUTION DES ETOILES (STELLAR CONSTITUTION)

PRESIDENT: B. Paczyński

VICE-PRESIDENT: R. J. Tayler

COMITE D'ORGANISATION: D. J. Faulkner, P. Giannone, C. Hayashi,
    I. Iben, R. Kippenhahn, A. G. Massevich, G. Ruben, J.-P. Zahn

MEMBRES: Aizenmann M., Anand S. P. S., Arnett W. D., Appenzeller I.,
    Audouze J., Baglin A., Bhatnagar P. L., Biermann L., Bodenheimer P.,
    Böhm K. H., Bondi H., Boury A., Brownlee R. R., Burbidge G. R.,
    Cameron A. G. W., Carson T. R., Castellani V., Castor J. I., Caughlan G.R.
    Chandrasekhar S., Chevalier C., Chin H. Y., Christy R. F., Cohen J. M.,
    Cowling T. G., Cox A. N., Cox J. P., Davis C. G., Deinzer W.,
    Demarque P. R., Dingens P. S. A., Djie T. A., Dluzhnevskaya O. B.,

Dziembowski W. A., Eggleton P. P., Eminzade T. A., Epstein I.,
Ergma E. V., Ezer D., Faulkner J., Forbes J. E., Fowler W. A.,
Frantzman Yu. L., Fricke K. J., Gabriel M., Gough D. O., Graham E.,
Hamada I., Hitotuyanagi Z., Hoshi R., Hoyle F., Imschennik V. S.,
James R. A., Kähler H., King D. S., Kolesnik I. G., Kothari D. S.,
Kovetz A., Kozłowski M., Köster D., Krook M., Kumar S. S., Kushwaha R.,
Larson R. B., Latour J., Lauterborn D., Lebovitz N. R., Ledoux P.,
McCrea W. H., Maeder A., Marx G., Masani A., Mestel L., Meyer-Hofmeister E.,
Michaud G., Monaghan J. J., Moore D. R., Morris S. C., Moss D. L.,
Möllenhoff C., Nadyozhin D. K., Nakano T., Nakazawa K., Narita S.,
Nishida M., Nomoto K., Ohyama N., Okamoto I., Osaki Y., Ostriker J.P.,
Pines D., Plavec M., Popova M. D., Poveda A., Profi jev V. V.,
Rädler K.-H., Reeves H., Reiz A., Renzini A., Rood R., Rosseland S.,
Rouse C. A., Roxburgh I., Sakashita S., Sakurai T., Salpeter E. E.,
Sato K., Savedoff M. P., Schatten K. H., Schatzman E., Schramm D. N.,
Schwartzschild M., Sears R. S., Sengbusch K. V., Shaviv G., Shibata Y.,
Smeyers P., Smith R. C., Souffrin P., Sparks H., Spiegel E. A.,
Starrfield S. S., Stibbs D. W. N., Strittmatter P. A., Strömgren B.,
Suda K., Sugimoto D., Sweet P. A., Sweigart A. V., Temesvary S., Thomas H.C.,
Tinsley B. M., Toomre J., Trimble-Weber V. L., Truran J. W., Tuominen I.V.,
Tuominen J. V., Tutukov A. V., Ichida J., Ulrich R., Uus U., Van der
Borght R., Van Horn H. M., Vardya M. S., Vauclair G., Vilas S. C.,
Vilhu O., Weigert A., Weiss N. O., Wright G. A. E., Zhevakin S. A.,
Ziółkowski J.

36. COMMISSION DE LA THÉORIE DES ATMOSPHÈRES STELLAIRES (THE THEORY OF STELLAR ATMOSPHERES)

PRÉSIDENT: D. Mihalas

VICE-PRÉSIDENT: G. Traving

COMITÉ D'ORGANISATION: A. G. Hearn, J. T. Jefferies, K. Kodaira, L. V. Kuhi,
J. M. Marlborough, B. E. J. Pagel, F. Praderie, A. A. Sapar, M. Vardya

MEMBRES: Abhyankar K. D., Aller L. H., Altrock R. C., Arpigny C., Athay R. G.
Auer L. H., Auman J. R., Avrett E. H., Baschek B., Bell R. A.,
Biermann L., Blanco C., Bless R. C., Boesgaard A., Böhm K. H.,
Böhm-Vitense E., Busbridge I. W., Carbon D., Carson T. R., Castor J.,
Cayrel R., Cayrel de Strobel G., Conti P. S., Cowley C. R., Cuny Y.,
Davis C. G., De Jager C., Delache Ph., Dumont S., Edmonds F. N.,
Elste G. H. E., Evangelidis E., Faraggiana R., Finn G. D., Fischel D.,
Foy R., Frisch U., Gail H., Gebbie K. B., Gingerich O., Gökdogan N.,
Gordon Ch., Grant I. P., Gray D. F., Greenstein J. L., Grevesse N.,
Groth H. G., Gussmann E. A., Gustafsson B., Hack M. H., Hardorp J.,
Heidmann N., Hekela J., Hitotuyanagi Z., Holweger H., Hotinli M.,
House L. L., Houtgast J., Houziaux L., Huang S. S., Hummer D.G.,
Hunger K., Hutchings J. R., Ivanov V. V., Johnson H. R., Kalkofen W.,
Kandel R. S., Khokhlova V. L., Kolesov A. K., Kopylov I. M.,
Krishna Swamy K. S., Kumar S. S., Kushwaha R. S., Lambert D. L.,
Linsky J. L., Matsumoto M., Matsushima S., Menzel D., Miyamoto S.,

Mukai S., Müller E. A., Münch G., Mustel E. R., Mutschlecner J. P., Myerscough V. P., Nariai K., Neff J. S., Neven L., Orrall F. Q., Pasinetti L., Pecker J. C., Phillips J. G., Pottasch S. R., Reimers D., Rucinski S., Rybicki G. B., Saito S., Schatzman E. L., Schmalberger D.C., Seaton M. J., Sedlmayr E., Skumanich A., Sitnik G. F., Sobolev V. V., Sobouti Y., Spiegel E.A., Spite M., Stibbs D. W. N., Strom S. E., Strömgren B., Swihart T. L., Thomas R. N., Thompson R. I., Tsuji T., Ueno S., Uesugi A., Underhill A. B., Unno W., Unsöld A., van Regemorter H., van't Veer C., van't Veer F., Wehrse R., Weidemann V., Wellmann P., Wickramsinghe D. T., Wilson P. R., Wright K. O., Wyller A. A., Zirker J.B., Zwann C.

37. COMMISSION DES AMAS STELLAIRES ET DES ASSOCIATIONS (STAR CLUSTERS AND ASSOCIATIONS

PRESIDENT: S. van den Bergh

VICE-PRESIDENT: G. Lynga

COMITE D'ORGANISATION: T. A. Agekjan, B. Balàzs, K. C. Freeman
I. R. King, C. A. Murray

MEMBRES: Aarseth S., Alksnis A. K., Artiukhina N. M., Barkhatova K. A.,
Becker W., Bijaoui A., Blaauw A., Bouvier P. B., Breger M., Burkhead M.S.
Burnichon M. L., Cannon R. D., Clariá J. J., Cuffey J., Dickens R. J.,
Dluzhnevskaya O. B., Efremov Yu. N., Eggen O. J., Einsato J., Feast M. W.
Feinstein A., FitzGerald M. P., Gascoigne S. C. B., Golay M., Haffner H.,
Hassan S. M., Hawarden T. G., Heggie D. C., Hénon M., Hesser J. E.,
Hills J. G., Iben I., Johnson H. L., Jones D. H. P., Kadla Z.,
Kholopov P. N., Kron G. E., Larsson-Leander G., Laval A., Lavodovskij V.V.
Lodén L. O., Lohmann W., Lynden-Bell D., Maeder A., Menon T. K.,
Menzies J. W., Meurers J., Moffat A. F. J., Morgan W. W., Osborn W.,
Philips A. G. D., Pismis de Recillas P., Popova M. D., Poveda A.,
Reddish V. C., Rosino L., Rountree-Lesh J., Ruprecht J., Salukvadze G.N.,
Sanders W. L., Sawyer Hogg H. B., Serkowski K., Sharov A., Simoda M.,
Swope H. H., Terzan A., Thé P. S., Upgren A., van Altena W. F.,
Vanderlinden H. L., Vogt N., von Hoerner S., Walker G. A. H., Walker M.F.
Wallenquist A. A. E., Weaver H. F., White R. E., Wielen R., Woolley R.

38. COMMISSION POUR l'ECHANGE DES ASTRONOMES (EXCHANGE OF ASTRONOMERS)

PRÉSIDENT: D. A. MacRae

VICE-PRÉSIDENT: J. Delhaye

COMITÉ D'ORGANISATION: M. K. V. Bappu, G. S. Khromov, A. Reiz,
    P. M. Routly, J. Sahade, F. G. Smith, F. B. Wood

MEMBRES: Abetti G., Alsabti A. W., Bok B. J., Haupt H. F., Keller G.,
    Kourganoff V., Marik M., Mohr J. M., Myamoto S., Okoye S. E.,
    Opolski A., Page T. L., Righini G., Rosseland S., Ruben G., Stoy R. H.,
    Swings P., Teleki G., Tolbert C. R., Wild J. P., Wyatt Jr S. P.

40. COMMISSION DE LA RADIOASTRONOMIE (RADIO ASTRONOMY)

PRÉSIDENT: H. van der Laan

VICE-PRÉSIDENT: G. Swarup

COMITÉ D'ORGANISATION: E. J. Blum, R. Fanti, D. J. McLean, P. G. Mezger,
    A. T. Moffet, H. P. Palmer, Yu. N. Parijskij, B. J. Robinson, B. M.
    Zuckermann

MEMBERS: Ables J. G., Akabane K., Alexander J. K., Allen R. J., Aller H.,
    Altenhoff W. J., Andrew B. H., Argyle P. E., Avignon Y., Baars J. W. M.,
    Baart E. E., Baldwin J. E., Barret A. H., Barrow C., Berge G. L.,
    Berkhuijsen E., Biraud F., Blandford R. D., Bohme A., Boischot A., Bolton
    J. G., Booth R. S., Bracewell R. N., Braude S. Ya., Broten N. W., Brouw
    W. N., Brown R. H., Browne I. W. A., Burbidge G. R., Burke B. F.,
    Caroubalos C., Carr T. D., Caswell J. L., Ceccarelli M., Christiansen
    W. N., Clark D. H., Clark G. B., Cohen M. H., Cole T.. Condon J. J..
    Conklin E. K., Conway R. G., Cooper B. F. C., Costain C. H., Courtez R. A. J.
    Covington A. E., Croom D. L., Cudaback D. D., Daintree E. G., Davies J. G.,
    Davies R. D., Davis M. M., de Groot T., de Jager G., Delannoy J., Denisse
    J. F., Dennison P. A., Dent W. A., Dickel J. R., Dickel H. R., Dieter N. H.
    Dixon R., Douglas J. N., Downes D., Drake F. D., Dravskikh A. F., Dulk
    G. A., Dyson F. J., Ekers R. D., Elgaroy O., Ellis G. R. A., Elsmore B.,
    Elwert G., Epstein E. E., Erickson W. C., Eriksen A. D., Evans K. D.,
    Ewing M., Fomalont E. B., Foster P. R., Friedman H., Galt J.. Gardner F. F.
    Gent H., Ginzburg V. L., Godwire H. C., Gold T., Goldstein S. J., Gonze R.,
    Gordon M. A., Gorgolewski S., Gosachinsky I. V., Goss W. M., Gower J. F. R.
    Grewing M., Guelin M., Gulkis S., Hachenberg O., Haddock F. T., Hagen J. P.
    Hamilton P. A., Hanasz J., Harten R. H., Hartz T. R., Haslam C. G. T.,
    Hazard C., Heeschen D. S., Heidmann J., Heiles C., Hewish A., Hey J. S.,
    Higgs L. A., Hill E. R., Hills R. E., Hirabayashi H., Hjellming R. M.,
    Hoang Binh D., Hobbs R. W., Högbom J. A., Höglund B., Hogg D. E., Hooghoudt
    B. G., Howard W. E., Huchtmeier W. K.-H., Hughes V. A., Ihsanova V. N.,
    Jauncey D. L., Jennison R. C., Johnson D. R., Johnston K., Joly F.,
    Joshi M. N., Kaftan-Kassim M., Kahn F. D., Kai K., Kaifu N., Kakinuma T.,
    Kapahi V. K., Kardashev N. S., Kaufmann P., Kawabata Kin-Aki, Kazes I.,

Kellerman K. I., Kenderdine S., Kerr F. J., Kislyakov A. G., Ko Hsien-Ching, Komesaroff M., Korolkov D. V., Kotelnicov V. A., Kraus J. D., Krishnan T., Kundu M. R., Kuypers J. M. E., Kuzmin A. D., Laffineur M., Lang K. R., Lantos P., Large M. I., Leblanc Y., Lequeux J., Le Squeren A. M., Lilley A. E., Little A. G., Little L. Th., Locke J. L., Longair M. S., Lovell Sir Bernard, Lyne A. G., McAdam W. B., McCulloch P. M., McGee R. X., McKenna-Lawlor S., MacDonald G. H., MacLeod D., MacRae D. A., Manchester R. N., Maran S. P., Matveyenko L. I., Maxwell A., Mayer C. H., Mebold U., Meeks M. L., Menon T. K., Miley G. K., Mills B. Y., Milne D. K., Moriyame F., Morimoto M., Morrison I., Muller C. A., Myers P., Nicolson G. D., Okoye S., Oort J. H., Osterbrock D. E., Pacholczyk A. G., Palmer P., Parker E. A., Parrish A., Pasachoff J. M., Pauliny-Toth I. I. K., Pedlar A., Penzias A. A., Pettengill G. H., Pick-Gutmann M., Ponsonby J. E. B., Pooley G. G., Preuss E., Price R. M., Priester W., Rabben H. H., Radhakrishnan V., Raimond E., Ramaty R., Gopala Rao U. V., Razin V. A., Readhead A. C. S., Reber G., Ribes R. S., Riegel K. W., Righini G., Roberts J. A., Roberts M. S., Roeder R. C., Roger R. S., Rogers A. E. E., Rogstad D. H., Rohlfs K., Rowson B., Rydbeck O. E. H., Ryle Sir Martin., Ryzhkov N. F., Salomonovitch A. E., Salpeter E. E., Sanamjan V. A., Sato F., Scalise E. Jr., Scheuer P. A. G., Schilizzi R. T., Schmidt M., Schwarz U. J., Scott P. F., Seaquist E. R., Seielstad G. E., Shaffer D., Shakeshaft J. R., Sheridan K. V., Shimmins A. J., Shklovsky J. S., Sholomitsky G. B., Shuter W. L. A., Simon P., Simonson S. C., Slee O. B., Slysh V. I., Smerd S. F., Smith A. G., Smith F. G., Smith H. J., Soboleva N. S., Sofue Y., Sorochenko R. L., Spencer R. E., Sramek R., Stanley G. J., Steinberg J. L. Stewart P., Stone R. G., Strom R. G., Sullivan W. T. III, Swenson G. E. Jr Tabara H., Takagi K., Takakura T., Tanaka H., Tanaka R., Terzian Y., Thompson A. R., Tlamicha A., Tolbert C. R., Tovmassian G. M., Townes C. H., Tritton K. P., Troitsky V. S., Tsuchiya A., Tsuruta S., Turlo Z., Turner B. E., Turner K. C., Turtle A. J., Ulrich B. T., Demoulin Ulrich M. H., van de Hulst H. C., van den Bout P., van der Kruit P. C., van Nieuwkoop J., van Woerden H., Varsavsky C. M., Véron P., Verschuur G. L., Wade C. M. Wall J. V., Walmsley M. C., Walsh D., Wardle J. F. C., Warwick J. W., Weaver H. F., Weliachew L., Wendker H. J., Westerhout G., Westfold K. C., Whiteoak J. B., Wickramasinghe N. C., Wielebinski R., Wild J. P., Williams D., Willis A. G., Wills B., Wills D., Wilson R. W., Wilson T. L., Wilson W. J., Winnberg A., Winnewisser G., Witzel A., Woltjer L., Zelenzniakov V. V.

41. COMMISSION DE L'HISTOIRE DE L'ASTRONOMIE
     (HISTORY OF ASTRONOMY)

PRESIDENT: J. Dobrzycki

VICE-PRESIDENT: M. A. Hoskin

COMITÉ D'ORGANISATION: S. M. R. Ansari, O. Gingerich, W. Hartner,
     P. G. Kulikovskij, O. Pedersen

MEMBRES: Abetti G., Argyracos J. G., Ashbrook J., Beer A., Berendzen R.,
     Brunet J. P., Chenakal V. L., Cimino M., Dadić Z., Deeming T. J.,
     Dewhirst D. W., Douglas A. V., Eddy J. A., Eelsalu H., Erpylev N. P.,
     Evans D., Ferrari d'Occhieppo D.K., Filliozat J., Fleckenstein J. O.,

Forbes E., Freiesleben H.C., de Freitas-Mourao R. R., Hawkins G. S.,
Hayli A., Heggie D., Hirose H., Horský Z., Howse H. D., Jackisch G.,
Kennedy J. E., Khromov G. S., Kiang T., King H., Kotsakis D., Lang K.,
Levy J. R., Link F., McKenna-Lawlor S. M. P., Meadows A. H.,
Merleau-Ponty J., Michel H., Michkovitch V. V., Nakayama S., Norlind W.,
North J., Omer G. C., Pelseneer J., Petri W., Pogo A., Ronan C. A.,
Rosen E., Rybka P., Shcheglov V. P., Shukla E. S., Sticker B.,
Sullivan W., Taton R., Thoren V., Verdet J.-P., Wattenberg D., Whitaker E.A.,
Whitrow G. J., Wright H., Yabuuti K., Zagar F.

42. COMMISSION DES ÉTOILES DOUBLES SERRÉES (CLOSE BINARY STARS)

PRÉSIDENT: G. Larsson-Leander

VICE-PRÉSIDENT: B. Warner

COMITÉ D'ORGANISATION: A.H. Batten, A.M. Cherepashchuk, M.G. Fracastoro,
   K. Gyldenkerne, T.J. Herczeg, M. Kitamura, Y. Kondo, M. Plavec,
   S.D. Sinvhal, J. Smak, E.P.J. van den Heuvel, J.A.J. Whelan

MEMBRES: Abhyankar K.D., Andersen J., Bappu M.K., Bath G.T., Beer A.,
   Binnendijk L., Blitzstein W., Breinhorst R., Broglia P., Brownlee R.R.,
   Budding E., Catalano S., Cester B., Chambliss R.C., Chen K.-Y., Chis G.,
   Cillié G.G., Cowley A.P., Cristaldi S., Dadaev A.N., de Kort J., Doughty
   N.A., Ebbighausen E.G., Faulkner J., Ferrari d'Occhieppo K., Fracassini M.,
   Fredrick L.W., Frantsman Yu.L., Fresa A., Gaposchkin S., Geyer E.H.,
   Giannone P., Grygar J., Gursky H., Hall D.S., Harmanec P., Hazlehurst J.,
   Heard J.F., Hill G., Hilditch R.W., Hjellming R.M., Horák T., Huang S.-S.,
   Huffer C.M., Hutchings J., Irwin J.B., Jurkevich I., Jaschek C., Kawabata
   S., Kizilirmak A., Koch R.H., Kopal Z., Kraft R.P., Krat V.A., Kristenson
   H., Kříž S., Kruszewski A., Krzeminski W., Kwee K.K., Landolt A.U., Lavrov
   M.I., Leung K.-Ch., Linnell A.P., Lucy L.B., Magalashvili N.L., Mammano A.,
   Martynov D.Ya., Mauder H., Mayer P., McCluskey G.E., Merrill J.E., Mumford
   G.S., Nelson B., O'Connell D.J.K., Ovenden M.V., Paczynski B., Piotrowski
   S., Popovici C., Popper D.M., Purgathofer A., Rahe J., Rakos K.D., Refsdal
   S., Reuning E.G., Rovithis P., Rucinski S., Sahade J., Scarfe C.D., Schmidt
   H., Schöffel E.F., Seggewiss W., Semeniuk I., Serkowski K., Shapley M.B.,
   Shulberg A.R., Sobieski S., Strohmeier W., Szafraniec R., Todoran I.,
   Tremko J., Trimble V., Tsessevich V.P., Ureche V., Van Hoof A., Van't Veer
   F., Vetesnik M., Walter K., Wehlau W.H., Weigert A., Wellmann P., Wesselink
   A.J., Widorn T., Wilson R.E., Wood D.B., Wood F.B., Wright K.O., Ziółkowski J.

44. COMMISSION DES OBSERVATIONS ASTRONOMIQUES AU-DEHORS DE L'ATMOSPHÈRE TERRESTRE (ASTRONOMICAL OBSERVATIONS FROM OUTSIDE THE TERRESTRIAL ATMOSPHERE

PRÉSIDENT: R. M. Bonnet

VICE-PRÉSIDENT: R. J. Van Duinen

COMITÉ D'ORGANISATION: L. W. Acton, R. C. Bless, A. D. Code, R. Giacconi, C. de Jager, C. Jordan, Y. Kondo, P. Léna, V. K. Prokof'ev, P. Valnicek, R. Wilson

MEMBRES: Bohlin R. C., Brinkman A.C.H., Catura R. C., Dolan J. F., Dupres A., Lamers H. J., Leckrone D. S., Mc Cluskey G. E., Mewe R., Michel F. C., Modisette S. P., Morgan T. H., Novick R., Peytremann E., Rosendhal J. P., Sholomitskij G. B., Snow T. P., Van Beek H. F., Wesselius P., Wu C. C.

45. COMMISSION DES CLASSIFICATIONS SPECTRALES ET INDICES DE COULEUR A PLUSIEURS BANDES (SPECTRAL CLASSIFICATION AND MULTI-BAND COLOUR INDICES)

PRÉSIDENT: B. Hauck

VICE-PRÉSIDENT: A. Slettebak

COMITÉ D'ORGANISATION: A. L. Ardeberg, R. A. Bartaya, A. P. Cowley, C. Jaschek, P. C. Keenan, E. Mendoza, V. L. Straizys

MEMBRES: Albers H., Bahng J. D. R., Bappu M. K. V., Barbier-Brossat M., Beer A., Bell R. A., Bernacca G., Bidelman W. P., Blanco V. M., Boyce P. B., Buscombe W., Cester B., Cherepashchuk A. M., Crampton D.., Crawford D. L., Dessy J. L., Divan L., Duflot M., Elvius T., Evans D. S., Feast M. W., Fehrenbach Ch., Fracassini M.., Garrison R. F., Geyer E. H., Glagolevskij Ju. V., Golay M., Gyldenkerne K., Hack M. H., Hallam K. L., Henize K. G., Herman R., Hill P. W., Hiltner W. A., Hoag A. A., Houk N., Humphreys R., Iwanowska W., Jaschek M., Johnson H. L., Kron G. E., Lodén K., Low F. J., Lyngå G., Lutz J., McCarthy M. F., McClure R. H., MacConnell D. J., McCuskey S. W., McNamara D. H., Maeder A., Maehara H., Mavridis L. M., Morgan W. W., Morguleff N., Nandy K., Nikonov V. B., Notni P., Oja T., Osawa K., Osborn W., Pasinetti L. E., Perry C. L., Philip A. G. D., Preston G. W. III, Reddish V. C., Roman N. G., Rountree-Lesh J., Rudkjøbing M., Sanduleak N. Sanwal N. B., Schild R. E., Schmidt-Kaler Th., Seitter W. C., Sharpless S.L., Sinnerstad U., Sinvhal S. D., Steinlin U. W., Stephenson C. B., Strömgren B., Treanor P. J., Upgren A. R., van den Bergh S., Walborn N.R. Walker G.A.H, Warner B., West R. M., Westerlund B., Wesselius P. R., Williams J.A., Wilson O. C., Wing R., Yoss K. M.,

46. COMMISSION POUR L'ENSEIGNEMENT DE L'ASTRONOMIE. (THE TEACHING OF ASTRONOMY)

PRESIDENT: E. V. Kononvich

VICE-PRESIDENTS: D. G. Wentzel, L. Houziaux

COMITE D'ORGANISATION: W. Buscombe, J. Kleczek, D. McNally, L. N. Mavridis,
B. F. Peery, M. Rigutti, A. E. Riguelet-Kaswalder

MEMBRES: Abell G. O., Abhyankar K. D., Andrillat H. L., Asaad A. S.,
Atanasijevic I., Barocas V., Bottinelli L., Catalá-Poch Mª.A.,
Clarke D., Climenhaga J. L., Dinulescu N., Dominko F., Doughty N.A.,
Elgaröy O., Fawell D. R., Ferraz-Mello S., Gerbaldi M., Gouguenheim L.,
Hauck B., Haupt H. F., Hazer S., Hidajat B., Ivanov V. V., Iwaniszewska C.,
Jarrett A. H., Jørgensen H. E., Kennedy J. E., Kourganoff V., Lambrecht H.,
Lumme K., Marik M., Marsh J. C. D., Meadows A. J., Miles H. G., Moreno H.,
Müller E. A., Nicolov N. S., Okoye S. E., Osborn W., Osorio J.,
Ovenden M. W., Owaki N., Pasachoff J. M., Pishmish P., Porfir'ev V. V.,
Proverbio E., Robbins R. R., Rodgers A. W., Sanqvist Aa., Scheffler H.,
Schmidt T., Ševarlić B. M., Steinitz R., Swihart T. L., Torres-Peimbert S.,
Trehan S. K., Zimmermann H., Zwaan C.

47. COMMISSION DE LA COSMOLOGIE (COSMOLOGY)

PRESIDENT: I. D. Novikov

VICE-PRESIDENT: G. O. Abell

COMITE D'ORGANISATION: Dautcourt G., de Vaucouleurs G., Hayakawa S.
Kellermann K. I., Longair M. S., Rowan-Robinson M., Thorne K. S.

MEMBRES: Aizu K., Alfven H., Ambartsumian V. A., Andrillat H., Baldwin J. E.
Bardeen J. M., Barnothy J. M., Beckman J. E., Bel N., Bertola F.,
Bondi H., Bonnor W. B., Brecher K., Burbidge G. R., Cavaliere A.,
Cohen J. M., Condon J. J., Davidson W., Davis M. M., Dicke R. H.,
Dionysiou D., Doroshkevich A. G., Ehlers J., Ellis G. F. R.,
Faber S. M., Felten J. E., Field G. B., Florides P. S., Fujimoto M.,
Godart O., Gold T., Goldsmith D. W., Gratton L., Greyber H. D.,
Grishchuk L. P., Gunn J. E., Harrison E. R., Hawking S. W., Hayashi C.,
Heckmann O., Heidmann J., Heller M., Hoyle F., Icke V., Jauncey D. L.,
Kasper U., Kato S., Kawabata K., Kovetz A., Kozlovksy B. Z.,
Kustaanheimo P. E., Lasota J. P., Layser D., Liebscher D. E.,
McCrea W. H., McVittie G. C., Misner C. W., Nariai H., Narlikar J. V.,
Ne'eman Y., Neyman J., Nishida M., Noerdlinger P.D., O'Connell R.F.,
Okoye S. E., Omer G. C., Omnes R., Oort J. H., Ozernoi L. M.,
Pachner J., Palmer H. P., Peebles P.J.E., Perisdes S., Petrosian V.,

Rees M. J., Reinhardt M., Rindler W., Robinson I., Roeder R. C.,
Rubin V. C., Ruddy V., Ryle M., Sapar A.A., Sargent W. L. W., Sato H.,
Schatzman E., Scheuer P.A.G., Schmidt M., Schramm D. N., Schücking E.,
Sciama D.W., Scott E. L., Seielstad G.A., Sersic J. L., Setti G.,
Shaviv G., Simon R. L. E., Spyrou N., Steigman G., Stewart J. M.,
Sunyaev R. A., Tammann G. A., Tauber G. E., Tayler R. J., Tifft W. G.,
Treder H. J., Tully R. B., Vaidya P. C., van der Laan H., Webster A. S.,
Wheeler J. A., Whitrow G. J., Wilson A. G., Zeldovich Ya. B.,
Zelmanov A. L., Zieba A., Zieba S.

48. COMMISSION DE L'ASTROPHYSIQUE DE GRANDE ENERGIE (HIGH ENERGY ASTROPHYSICS)

PRÉSIDENT: I. S. Shklovski

VICE-PRÉSIDENT: F. Pacini

COMITÉ D'ORGANISATION: J. Audouze, J. L. Culhane, K. I. Kellermann,
L. M. Ozernoi, E. N. Parker, M. J. Rees, J. Shaham

MEMBRES: Adams D. J., Alfven H., Alvarez L., Arons J., Asseo E., Audouze J.,
Avni Y., Axford W. I., Barnothy J., Bergeron J., Biswas S., Blandford R.D.
Bonometto S., Boyd R.L.F., Brecher K., Burbidge G. R., Lameron A. G. W.,
Catura R. C., Caughlan G. K., Cavaliere A., Chandrasekhar S., Chitre S. N.
Chubb T. A., Clark G., Cohen J. M., Collin-Souffrin S.,Condon J. J.,
Cruise M., Culhane L., Dautcourt G., Davidson W., Davis L. Jr.,
Davis M. M., Davison P., De Felke F., De Graaf T., Dewitt B. S.,
De Young D. S., Dicke R. H., Disney M., Dolan J. F., Drake F. D.,
Duthie J. G., Edwards P. J., Fabian A.C., Fazio G., Fichtel C., Felten J.E.
Fenton K. B., Ferrari A., Field G. B., Fisher P. C., Fowler W. A.,
Friedman H., Garmire G. P., Gialloni R., Ginzberg V. L., Gold T.,
Goldsmith D. W., Gratton L., Greisen K. I., Greyber H., Grewing M.,
Grindlay J., Gunn J. E., Gursky H., Harwit M.O., Hawking S.W.,
Hayakawa S., Henriksen R. N., Henry R.C., Hoang Binh D.Y., Hoyle F.,
Ipser J., Israel W., Ito K., Jackson J. C., Jelley J. V., Jokipii, J. R.,
Kafka P., Kahn F. D., Kellermann K. I., Kellogg E., Kondo M.,
Kozlovsky B. Z., Kreisel E., Kristiansson K. G. B., Kulsrud R. M.,
Kurt V. G., Lamb F., Longair M. S., Lüst R., Lynden-Bell D., McCray R.A.,
Matsuoka M., Mestel L., Meszáros P., Meyer F., Michel F. C.,
Miyamoto S., Morrison P., Ne'eman Y., Novick R., O'Connell R. F.,Oda M.,
Okoye S. E., Ostriker J. P., Ozernoi L. N., Pacholczyk A. G., Pacini F.,
Parker E. N., Parkison J. H., Pauliny-Toth I., Perola C., Peterson B.A.,
Peterson L. E., Petrosian V., Piddington J. M., Pinkau K., Porter N.A.,
Pounds K. A., Preuss E., Quintana H., Rees M.J., Reeves H., Reinhart M.,
Rossi B., Salpeter E. E., Sartori L., Saslaw W. C., Savedoff M. P.,
Scargle J. D., Schatten R. H., Schatzman E., Scheurer P.A. G., Schilizzi R.
Schramm D. N., Sciama D.W., Seielstad G., Setti G., Shaham J., Shapiro M.M.
Shaviv G., Shklovskij I.S., Skilling J., Smith F. G., Sofia S.,
Spitzer L., Steigman G., Stepanian A. A., Sturrock P.A., Sunyaev R.A.,
Syrovatskij S.I., Tayler R.J., Thorne K. S., Trimble V. L., Truan J..
Tsuruta S., Vidal N., Webster A. S., Wentzel D. G., Westford K. C.,
Wheeler J. A., Wolstercroft R., Woltjer L., Zeldovich Ya.B.

49. COMMISSION DU PLASMA INTERPLANÉTAIRE ET DE L'HELIOSPHÈRE (THE INTERPLANETARY
PLASMA AND THE HELIOSPHERE)

PRÉSIDENT: A. Hewish

VICE-PRÉSIDENT: H. J. Fahr

COMITÉ D'ORGANISATION: J. L. Bertaux, S. Cuperman, A. Z. Dolginov,
S. Grdzielski, H. U. Keller, F. Paresce, C. S. Weller.

MEMBRES: Anderson K. A., Ansari S. M. R., Antonucci E., Barrow C. H.,
Barth C. A., Blackwell D. E., Blum P. W., Bonnet R. M., Bowyer C. S.,
Brandt J. C., Chamberlain J. W., Crifo F., De Jager C., Delache P.,
De Young D. S., Dryer M., Durney B., Dyson J. E., Eshleman V. R.,
Eviatar A., Feynman J., Field G. B., Gosling J. T., Harvey C. C.,
Heyvaerts J., Jokipii J. R., Kakinuma T., Kaplan S. A., Lafon J. P.,
Levy E., Lotova N. A., Lüst R., Macqueen R., Mangeney A., Mendis D. A.,
Mestel L., Michel F. C., Nakagawa Y., Newkirk G. A., Parker E. N.,
Perkins Jr. F., Pflug K., Pneuman G., Readhead A. C. S., Reay N. K.,
Riddle A. C., Roach F. E., Roxburgh I. W., Sawyer C., Schatzman E.,
Scherb F., Schmidt H. U., Severny A. B., Smerd S. F., Smith D. F.,
Sonnett C. P., Stone R. G., Sturrock P. A., Suess S. T., Syrovatskij S. J.,
Tanaka H., Van Allen J. A., Wallis M. K., Wilcox J. M., Wild J. P.

50. COMMISSION DE LA PROTECTION DES SITES D'OBSERVATOIRE EXISTANT ET POTENTIELS
(Protection of Existing and Potential Observatory Sites)

PRESIDENT : R. Cayrel

VICE-PRESIDENT : F.G. Smith

COMITE D'ORGANISATION : C.A. Anquita, F. Bertola, D.L. Crawford, H. Elsasser,
C.B. Gascoigne, J.T. Jefferies, W. Mattig, A.B. Muller, S. van den Bergh,
M.F. Walker.

MEMBRES : Ardeberg A., Barreto L.M., Bhattacharyya J.C., Blaauw A.,
Boyarchuk A.A., Dommanget J., Gyldenkerne K., Haupt H., Kubicela A.,
Leibowitz E., Sanchez F.M., Lewis B.M., Markkauen T.K.J., Marx S.,
Mavridis L.N., Mendoza E., Osawa K., Osorio J., Treanor P.S.J., Tremko J.,
Wayman P.S., Woolley Sir Richard, Woszkzyk A.

CONSULTANTS: Mr. B. McInnes

GROUPE DE TRAVAIL POUR LA NOMENCLATURE DU SYSTÈME PLANÉTAIRE
(WORKING GROUP FOR PLANETARY SYSTEM NOMENCLATURE)

PRÉSIDENT: P. M. Millman

MEMBRES: B. Ju. Levin, D. Morrison, Tobias C. Owen, G. H. Pettengill,
        B. A. Smith, E. A. Whitaker.

TASK GROUPS:

1) *Lunar Nomenclature*

P. M. Millman (Acting Chairman), A. Dollfus, F. El-Baz, K.P. Florenskij,
Yu. N. Lipskij, H. Masursky, S. K. Runcorn, V. V. Shevchenko

2) *Mercury Nomenclature*

D. Morrison (Chairman), M. E. Davies, A. Dollfus, O. J. Gingerich, J. E. Guest,
Yu. N. Lipskij, B. A. Smith

3) *Venus Nomenclature*

G. H. Pettengill (Chairman), R. M. Goldstein, M. Ya. Marov, H. Masursky

4) *Mars Nomenclature*

B. A. Smith (Chairman), A. Dollfus, M. Ya. Marov, H. Masursky, S. Miyamoto,
A. V. Morozhenko, C. Sagan

5) *Outer Solar System Nomenclature*

T. C. Owen (Chairman), K. Aksnes, M.S. Bobrov, M. E. Davies, D. Gautier,
B. A. Smith, V. G. Tejfel'

LISTE ALPHABETIQUE DES MEMBRES
ALPHABETICAL LIST OF MEMBERS

Les noms figurent à la première lettre du premier mot; par exemple, le nom Van Houten se trouve à la lettre V.

Les noms des nouveaux membres élus par la XVIe Assemblée Générale sont annotés d'un astérisque.

Les noms des pays sont donnés sous leur forme anglaise.

Les noms des astronomes utilisant l'alphabet cyrillique sont donnés selon les règles de translitération publiée dans les TRANSACTIONS XIVA, 13.

Les Membres de l'Union sont priés de bien vouloir informer le Secrétaire Exécutif de l'UAI, Université de Lausanne, Bâtiment des Sciences Physiques, 1015 Lausanne-Dorigny, Suisse.

Names are given under the first letter of the first word: for example, the name Van Houten will be found under V.

The names of new members, elected at the XVIth General Assembly, are prefixed by an asterisk.

The names of the countries are given in the English form.

Names of astronomers using the Cyrillic alphabet are transliterated according to the rules published in TRANSACTIONS XIVA, 13.

Please inform the Executive Secretary of the IAU, Université de Lausanne, Bâtiment des Sciences Physiques, 1015 Lausanne-Dorigny, Switzerland, of any errors you may find in this list.

AARSETH (Dr S.) Institute of Theoretical Astronomy, Madingley Road, Cambridge, U.K.

ABALAKIN (Dr V.K.) Institute of Theoretical Astronomy, 10 Kutuzov Qay, Leningrad, 192187, U.S.S.R.

ABBOTT (Prof. Dr W.N.) 42 Leophoros A. Michalacopoulou, Ilisia, Athens 612, Greece

ABDALA (Dr J.) Avenida Solano, Residencias Sans Souci, Edificio Araguaney, Planta Baja No.1, Chacaito, Caracas 106, Venezuela

ABELE (Dr M.K.) Latvian State University, Riga, bulv. Rainisa, U.S.S.R.

ABELL (Dr G.O.) Department of Astronomy, University of California, Los Angeles 24, California 90024, U.S.A.

ABETTI (Prof. G.) Istituto Naz. Ottica-Arcetri, Largo Enrico Fermi 7, 50125 Firenze, Italy

ABHYANKAR (Dr K.D.) Astronomy Department, Osmania University, Hyderabad, 7 A.P. India

ABLES (Dr H.D.) U.S. Naval Observatory, Flagstaff Station, Flagstaff, Arizona 86001, U.S.A.

*ABLES (Dr J.G.) CSIRO, Division of Radiophysics, P.O. Box 76, Epping, NSW 2121, Australia

ABRAHAM (H.J.) Mount Stromlo Observatory, Canberra, A.C.T., Australia

ABRAHAM (Dr Z.) Instituto de Fisica, UFRGS, Depto. Astronomia, R. Luiz Englert S/N, 9000 Porto-Alegre, RGS, Brazil

ABRAMI (Prof.A.) Osservatorio Astronomico, via G.B. Tiepolo N.11, Trieste (409), Italy

ABREU (Dr J.C. de Brito e) Observatorio Astronomico de Lisboa, Tapada, Lisbon, Portugal

ABT (Dr H.A.) Kitt Peak National Observatory, Box 4130, Tucson, Arizona 85717, U.S.A.

*ACKER (Dr A.) Observatoire de Strasbourg, 11, rue de l'Université, 67000 Strasbourg, France

ACTON (Dr L.W.) Orgn. 52-14, Bldg. 202, Lockheed Palo Alto Research Laboratory, 3251 Hanover Street, Palo Alto, California 94304, U.S.A.

ADAM (Dr M.G.) University Observatory, South Parks Road, Oxford, U.K.

ADAMOPOULOS (Dr G.) Rue Lamias No 8, Athens (6), Greece

ADAMS (A.N.) U.S. Naval Observatory, Washington 25, D.C., U.S.A.

ADAMS (Dr D.J.) Astronomy Department, University of Leicester, University Road, Leicester, U.K.

ADAMS (Dr Th.) Yerkes Observatory, Williams Bay, Wisconsin 53191, U.S.A.

ADEL (Prof. A.) Atmospheric Research Observatory, Arizona State College, Flagstaff, Arizona, U.S.A.

*ADELMAN (Dr S.J.) Department of Astronomy, Boston University, 725 Commonwealth Avenue, Boston, MA 02215, U.S.A.

ADOLFSSON (Dr T.) Lund Observatory, S-222 24 Lund, Sweden

AFANASEVA (Dr P.M.) Main Astronomical Observatory, USSR Academy of Sciences, Pulkovo, USSR

AGEKJAN (Prof. Dr T.A.) University of Leningrad, Leningrad, USSR

AGRINIER (Dr B.) C.E.N. Saclay - S.E.P., B.P. No 2, 91 Gif-sur-Yvette, France

AHMED (Imam Ibrahim) Helwan Observatory, Helwan, near Cairo, Arab Rep. of Egypt

AHNERT (Dr P.) Zentralinstitut für Astrophysik, Sternwarte Sonneberg, DDR 64 Sonneberg, G.D.R.

*AIKAWA (Dr Toshiki) Astronomical Institute, Tohoku University, Sendai, Japan

*AITKEN (Dr D.K.) Dept. of Physics and Astronomy, University College, Gower Street, London WC1E 6BT, U.K.

AIZENMAN (Dr M.L.) JILA, University of Colorado, Boulder, Colorado 80302, U.S.A.

AIZU (Prof. Dr K.) Physics Department, Rikkyo University, Nishi-Ikebukuro 3-chome, Toshima-ku, Tokyo 171, Japan

AKABANE (Dr K.) Tokyo Astronomical Observatory, Mitaka, Tokyo, Japan

*AKABANE (Dr Tokuhide) Hida Observatory, University of Kyoto, Kamitakara,
    Yoshiki-Gun, Gifu, 506-13, Japan

*AKÇAYLI (Dr M.) Kandilli Observatory, Çengelköy - Istanbul, Turkey

AKSENOV (Dr E.P.) Sternberg Astronomical Institute, Moscow, USSR·

AKSNES (Dr K.) Smithsonian Astrophysical Obser., 60 Garden Street,
    Cambridge, Mass. 02138, U.S.A.

AKYOL (Dr M.Ü.) Department of Astronomy, P.K. 21, Bornova, Izmir, Turkey

ALANIA (Dr I.F.) Abastumani Astrophysical Observatory, Abastumani,
    Georgian SSR, USSR·

ALBANO (Dr J.) Observatorio Astronómico, La Plata, Argentina

*ALBERS (Dr H.) Dept. of Physics and Astronomy, Vassar College,
    Poughkeepsie, NY 12601, U.S.A.

ALBRECHT (Dr R.) Univ.-Observatory, Türkenschanzstr. 17,
    A-1180, Vienna, Austria

*ALDROVANDI (Dr R.) Instituto de Fisica Teórica, Rua Pamplona, 145
    01405 São Paulo-SP, Brazil

*ALDROVANDI (Dr S.M.V.) Instituto Astronômico e Geofísico, C.P. 30627,
    São Paulo, Brazil

ALEXANDER (J.B.) Royal Greenwich Observatory, Herstmonceux Castle,
    Hailsham, Sussex, U.K.

ALEXANDER (J.K.) Code 693, Laboratory for Extraterrestrial Physics, NASA-Goddard
    Space Flight Center, Greenbelt, Maryland 20771, U.S.A.

ALFVEN (Prof. H.) Royal Institute of Technology, S-100 44 Stockholm 70, Sweden

ALKSNIS (Dr A.K.), Radioastrophysical Observatory, Latvian Academy of Sciences
    226026 Riga, U.S.S.R.

ALLADIN (S.M.) Department of Astronomy, Osmania University, Hyderabad, India

*ALLEN (Ch.) Instituto de Astronomía, UNAM, Apartado Postal 70-264,
    México 20, D.F. Mexico

ALLEN (Prof. C.W.) University of London Observatory, Mill Hill Park,
    London, N W 7 2 QS, U.K.

*ALLEN (Dr D.A.) Anglo-Australian Observatory, P.O. Box 296,
    Epping NSW 2121, Australia

ALLEN (Dr R.J.) Kapteyn Astronomical Laboratory, Postbus 800,
    Groningen, The Netherlands

ALLER (Dr H.D.) Department of Astronomy, University of Michigan, Ann Arbor,
    Michigan 48104, U.S.A.

ALLER (Prof. L.H.) Astronomy Department, University of California,
    Los Angeles, California, 90024, U.S.A.

ALLER (Dr M.F.) Department of Astronomy, The University of Michigan,
    Ann Arbor, Michigan 48104, U.S.A.

*ALLOIN (Dr D.) Observatoire de Nice, Le Mont-Gros, 06000 Nice, France

ALMAR (Dr I.) Institute of Geodesy (Földmérési Intézet), Budapest 1051, Guszev u 19, Hungary

*AL-SABTI (Dr A.A.) Physics Department, Science College, Baghdad University, Baghdad, Iraq

ALTENHOFF (Dr W.J.) Max-Planck-Institut für Radioastronomie, Auf dem Hügel 69, 53 Bonn 1, F.R.G.

ALTROCK (Dr R.) Sacramento Peak Observatory, Air Force Cambridge Research Labs, Sunspot, New Mexico 88349, U.S.A.

ALTSCHULER (Dr M.) High Altitude Observatory, Boulder, Colorado 80302, U.S.A.

*ALVAREZ (Dr H.) Centro de Investigación de Astronomía "Francisco J. Duarte", Apartado 264, Mérida, Venezuela

ALVAREZ (Dr L.) Lawrence Radiation Laboratory, University of California, Berkeley, California 94720, U.S.A.

ALY (Prof. M.K.) Director, Rabitatul Aalam Al-Islami, Islamic Observatory, Mecca Almukarrama, Saudi Arabia

AMBARTSUMIAN (Prof. Dr V.A.) President of the Academy of Sciences of Armenian S.S.R., Erevan, Armenia, U.S.S.R.

*AMBROZ (Dr P.) Astronomical Institute of the Czechoslovak Academy of Sciences Observatory, 251 65 Ondřejov, Czechoslovakia

ANAND (Prof. S. P. S.) Department of Astronomy, University of Toronto, Toronto 5, Ontario, Canada

ANDERS (Prof. E.) Enrico Fermi Institute, University of Chicago, 5640 S. Ellis Ave Chicago, Illinois 60637, U.S.A.

*ANDERSEN (Dr J.) Copenhagen University Observatory, Brorfelde, DK-4340 Tølløse, Denmark

ANDERSON (Dr B.) Nuffield Radio Astronomy Laboratories, Jodrell Bank, Macclesfield, Cheshire, U. K.

ANDERSON (Dr C. M.) University of Wisconsin-Madison, Madison, Wisconsin 53706, U.S.A.

ANDERSON (Dr K.A.) Space Science Laboratory, University of California, Berkeley, California 94728, U.S.A.

ANDREW (Dr B.H.) Herzberg Institute of Astrophysics, National Research Council of Canada, Ottawa, Ontario K1A OR6, Canada

ANDREW (Prof. K.L.) Purdue University, Department of Physics, Lafayette, Indiana 47907, U.S.A.

ANDREWS (A. D.) Armagh Observatory, Armagh, N. Ireland, U. K.

ANDREWS D.H., Dominion Astrophysical Observatory, Herzberg Institute of Astrophysics, National Research Council of Canada, 5071 W. Saanich Road, Victoria, B. C. V8X 3X3, Canada

ANDREWS (Dr P. J.) Royal Greenwich Observatory, Herstmonceux Castle, Hailsham, Sussex BN27 1RP, U. K.

ANDRIENKO (Dr D.A.), Kiev State University, Astronomical Observatory and Chair of Astronomy, 252053 Kiev, U.S.S.R.

ANDRIESSE (Dr Ir C.D.) Kapteyn Institute, P.O. Box 800,
Groningen, The Netherlands

ANDRILLAT (Prof. Dr H.) Laboratoire d'Astronomie de la Faculté des Sciences de
Montpellier, Hérault, France

ANDRILLAT (Dr Y.) Laboratoire d'Astronomie de la Faculté des Sciences de
Montpellier, Hérault, France

ANDRLE (Dr P.) Astronomical Institute, Czechoslovak Academy of Sciences,
Budečská 6, Prague 2, Czechoslovakia

ANGEL (Dr R.) Columbia University, New York, N.Y., U.S.A.

ANGIONE (Dr R.) California State University, San Diego, California 92115, U.S.A.

ANGUITA (Dr C.A.) Universidad de Chile, Observatório Astronómico Nacional,
Casilla 36 - D, Santiago, Chile

*ANILE (A.) Viale M. Rapisardi 170, Catania, Italy

ANSARI (Dr S.M.R.) Department of Physics, Aligarh Muslim University,
Aligarh, U.P., India

ANTONACOPOULOS (Dr G.) Department of Astronomy, University of Patras,
Patras, Greece

*ANTONOV (Dr V.A.) Astronomical Observatory of Leningrad University,
10 Linia 33, Leningrad 199178, U.S.S.R.

*ANTONUCCI (E.) Instituto di Fisica Generale, Dell' Università di Torino,
Corso Massimo d'Azeglio 46, 10125 Torino, Italy

ANZER (Dr U.) Max-Planck-Institut für Physik und Astrophysik, 8000 München,
Föhringer Ring 6, F.R.G.

AOKI (Prof. Shinko) Tokyo Astronomical Observatory, Mitaka, Tokyo 181, Japan

AOUDOUZE (Dr J.) Institut d'Astrophysique, 98 bis Boulevard Arago,
75014 Paris, France

APPENZELLER (Dr I.) Landessternwarte, 69 Heidelberg-Königstuhl, F.R.G.

ARAKELIAN (Dr M.A.) Byurakan Astrophysical Observatory, Byurakan,
Armenia, U.S.S.R.

*ARDAVAN (Dr H.) Institute of Astronomy, Madingley Road, Cambridge, U.K.

ARDEBERG (Dr A.L.) Institute of Astronomy, S-222 24 Lund, Sweden

AREND (Prof. Dr S.J.V.) Astronome Honoraire, Observatoire Royal de Belgique,
Avenue Circulaire, 3, B-1180 Bruxelles, Belgium

ARGUE (A.N.) The Observatories, Madingley Road, Cambridge, England, U.K.

ARGYLE (Dr P.E.) Dominion Radio Astrophysical Observatory, Herzberg Institute
of Astrophysics, National Research Council of Canada, P.O. Box 248,
Penticton, B.C. V2A 6K3, Canada

ARGYRACOS (Prof. Dr J.G.) 193 Patission Street, Athens (816), Greece

ARIAS-De GREIFF (Prof. J.) Observatorio Astronómico Nacional,
Apartado Nacional 2584, Bogota 1, D.E., Colombia

ARKHIPOVA (Dr V.P.) Sternberg Astronomical Institute, Moscow, U.S.S.R.

ARNETT (Dr W.D.) Department of Space Science, Rice University,
Houston, Texas 77001, U.S.A.

ARNOULD (M.L.L.) Chaussée de Tubize, 146, B-1430 Wauthier-Braine, Belgium

ARNQUIST (Dr W.N.) 8127 Delgany Avenue, Playa Del Rey, California 90291, U.S.A.

ARNY (Dr T.T.) Department of Physics and Astronomy, University of Massachusetts, Amherst, Massachusetts 01002, U.S.A.

ARONS (Dr J.) Berkeley Astronomy Department, University of California, Berkeley, California 94720, U.S.A.

ARP (Dr H.C.) Mount Wilson and Palomar Observatories, 813 Santa Barbara Street, Pasadena, Cal., 91106, U.S.A.

ARPIGNY (Prof. Dr C.) Institut d'Astrophysique, Université de Liège, Avenue de Cointe, 5, B-4200 Cointe-Ougrée, Belgium

ARROYO (Dr M.L.) Observatorio Astronómico, Alfonso XII, 3, Madrid 7, Spain

ARTHUR (D.W.G.) Lunar and Planetary Laboratory, University of Arizona, Tucson, Arizona, U.S.A.

ARTIUKHINA (Dr N.M.) Sternberg Astronomical Institute, Moscow, U.S.S.R.

ASAAD (Dr A.S.) Helwan Observatory, Helwan, near Cairo, Arab Republic of Egypt

ASHBROOK (Dr J.) 16 Summer Street, Weston 93, Mass., U.S.A.

*ASLAN (Dr Z.) Department of Astronomy, University of Ankara, Ankara, Turkey

ASLANOV (Dr I.A.), Shemakha Astrophysical Observatory, Shemakha, 373243 Azerbaidzan, U.S.S.R.

ASSEO (Dr E.) Observatoire de Paris, Section d'Astrophysique de Meudon, 92190 Meudon, France

*ASTERIADIS (Dr G.) Department of Geodetic Astronomy, University of Thessaloniki, Greece

ATANASIJEVIC (Dr I.) Sterrenkundig Instituut, Toernooiveld, 6805 Nijmegen, The Netherlands

*ATHANASSOULA (Dr E.O.) Research Associate Observatoire de Besançon, 2500 Besançon, France

ATHAY (Dr R.G.) High Altitude Observatory of the University of Colorado, Boulder P.O. Box 1470, Colorado 80302, U.S.A.

ATKINSON (R.d'E.) Astronomy Department, Indiana University, Swain Hall West, Bloomington, Indiana 47401, U.S.A.

*AUBIER (Dr M.) DASOP, Observatoire de Meudon, 92190 Meudon, France

AUER (Dr L.H.) Department of Astronomy, Watson Astronomy Center, Box 2023 Yale Station, Yale University, New Haven, Connecticut 06520, U.S.A.

AULUCK (Prof. F.C.) Department of Physics & Astrophysics, University of Delhi, Delhi-110007, India

AUMAN (Dr J.R.) Department of Geophysics and Astronomy, University of British Columbia, Vancouver, B.C. V6T 1W5, Canada

*AUNER (Dr G.) Sternwarte und Astronomisches Institut, Universitäts-Strasse 4, A-6020 Innsbruck, Austria

*AURIEMMA (G.) Laboratorio di Astrofisicà, Casella Postale 67, 00044 Frascati (Roma), Italy

AVERY (Dr L.W.) Herzberg Institute of Astrophysics, National Research Council of Canada, Ottawa, Ontario K1A OR6, Canada

AVIGNON (Dr Y.) Observatoire de Paris, Section d'Astrophysique, 92190 Meudon, France

*AVNI (Dr Y.) Weizmann Institute of Science, Rehovot, Israel

AVRETT (Dr E.H.) Smithsonian Astrophysical Observatory, 60 Garden Street, Cambridge, Massachusetts 02138, U.S.A.

AXFORD (Dr W.I.) Director, Max-Planck-Institute for Aeronomy, 3411 Lindau/Harz, Germany, F.R.

*AYAD (Dr A.Z.) Faculty of Science, Cairo University, Geza Orman, Arab Rep. of Egypt

AYDIN (Dr C.) Faculty of Science, Department of Astronomy, Ankara, Turkey

BAARS (Dr J.W.M.) Max-Planck-Institut für Radioastronomie, Auf dem Hügel 69, 53 Bonn 1, F.R.G.

BAART (Dr E.E.) Department of Physics, Rhodes University, P.O. Box 94, Grahamstown 6140, South Africa

BABADZHANOV (Dr P.B.), Director, Astrophysical Institute of Tadjikian Academy of Sciences, 734670 Dushanbe, U.S.S.R.

BABCOCK (Dr H.W.) Mount Wilson and Palomar Observatories, 813 Santa Barbara Street, Pasadena, California 91106, U.S.A.

BACCHUS (Prof.P.) 1, Impasse de l'Observatoire, 59 Lille (Nord), France

BAEHR (Dr U.) Astronomisches Rechen-Institut, Quickestr. 44a, 69 Heidelberg, F.R.G.

BAERENTZEN (Dr J.) Ole Roemer Observatory, DK-8000 Aarhus C, Denmark

BAGILDINSKIJ (Dr B.K.) Main Astronomical Observatory, USSR Academy of Sciences Pulkovo, 199178 Leningrad, U.S.S.R.

BAGLIN (Dr A.) Institut d'Astrophysique, 98 bis Boulevard Arago, 75014 Paris, France

BAHCALL (Dr N.) Institute for Advanced Study, Princeton, N.J. 08540, U.S.A.

BAHNER (Dr K.) Landessternwarte, Königstuhl-Heidelberg, F.R.G.

BAHNG (Prof. J.D.R.) Dearborn Observatory, Northwestern University, Evanston, Illinois 60201, U.S.A.

*BAIRD (Dr G.A.) Physics Department, University College, Belfield, Dublin 4, Ireland

BAIRD (Dr K.M.) Division of Physics, National Research Council of Canada, Sussex Drive, Ottawa, Ontario K1A OR6, Canada

BAIZE (Dr P.) 6, rue Daubigny, 75017 Paris, France

BAJAJA (Dr E.) Instituto Argentino de Radioastronomía, Casilla No 5, Villa Elisa, Prov. Bs.As., Argentina

BAKER (Dr J.G.) 60 Garden Street, Cambridge, Mass., 02138, U.S.A.

BAKER (Dr N.) Astronomy Department, Columbia University, New York, N. Y. 10027, U.S.A.

BAKOS (Prof. G.A.) Department of Physics, University of Waterloo, Waterloo, Ontario N2L 3G1, Canada

BALÁZS (Dr B.) Konkoly Observatory, P. O.114, Box 67, Budapest XII, Hungary

*BALÁZS (Dr L.G.) P.O.114/67, H-1525 Budapest, Hungary

BALDWIN (Dr J. E.) Cavendish Laboratory, Free School Lane, Cambridge, U.K.

BALDWIN (Dr R.) Oliver Machinery Co., 445 Sixth Street, NW Grand Rapids, Michigan 49502, U.S.A.

*BALICK (Dr B.) Department of Astronomy, University of Washington, Seattle, WA 98195, U.S.A.

BALKLAVS (Dr A.E.) Radioastrophysical Observatory, Latvian Academy of Sciences, 226026 Riga, U.S.S.R.

BALL (Dr J.A.) Harvard College Observatory, Cambridge, Mass. 02138, U.S.A.

BALLARIO (Prof. M.C.) Osservatorio Astrofisico di Arcetri, Via S. Leonardo 75, Firenze, Italy

BALLI (Prof. Dr E.) University Observatory, Beyazit, Istanbul, Turkey

*BALMINO (Dr G.) GRGS/CNES, 18 Ave Edouard Belin, 31055 Toulouse, CEDEX,France

BANIN (Dr V.G.) SIBIZMIR, USSR Academy of Sciences, 664697 Irkutsk, U.S.S.R.

BANOS (Dr C.) Astronomical Institute, National Observatory of Athens, Athens (306), Greece

BANOS (Prof. Dr G.) University of Ioannina, Laboratory of Astronomy, Ioannina, Greece

BAPPU (Dr M.K.V.) Indian Institute of Astrophysics, Bangalore-560034, India

BARANNE (A.), Observatoire de Marseille, 2 place Le Verrier, 13004 Marseille, France

BARBANIS (Dr B.) University of Patras, Department of Astronomy, Patras, Greece

BARBARO (Dr G.) Osservatorio Astronomico, Vicolo dell'Osservatorio, Padova, Italy

BARBIER-BROSSAT (M.) Chargée de Recherche, Observatoire de Marseille, 2 place Le Verrier, 13004 Marseille, France

BARBIERI (Dr C.) Osservatorio Astronomico di Padova, Padova, Italy

BARBON (Dr R.) Osservatorio Astrofisico Asiago (Vicenza) Italy

*BARCZA (Dr S.) Konkoly Observatory, P.O. Box 67, H-1525 Budapest XII, Hungary

BARDEEN (Dr J.M.) Physics Department, Yale University, New Haven, Conn.06520, U.S.A.

BARKAT (Dr Z.) Department of Physics, The Hebrew University, Jerusalem, Israel

BARKER (Dr E.) McDonald Observatory, Fort Davis, Texas 79734, U.S.A.

BARKHATOVA (Dr K.A.) State University, Sverdlovsk, U.S.S.R.

*BARLAI (K.) Box 67, H-1525 Budapest, Hungary

BARLETTI (Ing. R.) Osservatorio Astrofisico di Arcetri, Largo Enrico Fermi 5, Florence, Italy

BARLIER (F.) Observatoire de Paris, Section d'Astrophysique, 92190 Meudon, France

*BARNARD (Dr A.J.) Department of Physics, University of British Columbia, 2075 Wesbrook Place, Vancouver, B.C., Canada V6T 1W5

*BARNES III (Dr T.G.) Department of Astronomy, The University of Texas at Austin, Austin, TX 78712, U.S.A.

BARNEY (Dr I.) Yale University Observatory, Box 2023 Yale Station, New Haven, Connecticut, U.S.A.

BARNOTHY (Prof. Dr J.M.) 833 Lincoln Street, Evanston, Illinois 60201, U.S.A.

*BAROCAS (Dr V.) Jeremiah Horrocks and Wifred Hall Observatories, Moor Park, Preston, U.K.

BARRETO (Prof.L.M.) Observatorio Nacional, Rua General Bruce 586, Sao Cristóvao Rio de Janeiro, Brazil

BARRETT (Prof. A.H.) Research Laboratory of Electronics, Massachusetts Institute of Technology, Cambridge, Mass. 02139, U.S.A.

*BARROW (Dr C.H.) Department of Physics, University of the West Indies, Kingston 7, Jamaica, W.I.

BARROW (Dr R.) Physical Chemistry Laboratory, Oxford University, South Parks Road, Oxford OX1 3OZ, U.K.

BARRY (Dr D.C.) Dept. of Astronomy University of Southern California, Los Angeles, California 90007, U.S.A.

BARTAYA (Dr R.A.) Astrophysical Observatory, Abastumani, Georgia, U.S.S.R.

BARTH (Dr Ch.) Scientific Director LASP, University of Colorado, Boulder, Colorado 80302, U.S.A.

*BARTHOLDI (Dr P.) Observatoire de Genève, CH-1290 Sauverny, Switzerland

BARTOLINI (Dr C.) Osservatorio Astronomico Universitario, via Zamboni 33 Bologna, Italy

BASART (Dr J.P.) Electrical Engineering Dept. Iowa State University, Ames, Iowa 50010, U.S.A.

BASCHEK (Prof. Dr B.) Lehrstuhl für Theoretische Astrophysik der Universität 69 Heidelberg, Im Neuenheimer Feld 294, F.R.G.

BASH (Dr F.) University of Texas at Austin, Austin, Texas 78712, U.S.A.

BASTIN (Dr J.A.) Physics Department, Queen Mary College (London University) Mile End Road, London E. 1., U.K.

*BASU (Dr B.) Reader in Applied Mathematics, Calcutta University, Calcutta, India

BASU (Dr D.) Instituto Astronômico e Geofísico, Universidade de Sao Paulo, C.P. 30 627, Sao Paulo, Brazil

*BATES(Dr B.)Department of Pure and Applied Physics, The Queen's University of Belfast, Belfast, BT7 1NN, N. Ireland, U.K.

BATES (Prof. D.R.) Department of Applied Mathematics, Queen's University, Belfast, U.K.

BATESON (F.M.) 18 Pooles Road, Greerton, Tauranga, New Zealand

BATH (Dr G.T.) Department of Astrophysics, South Parks Road, Oxford OX1 3RQ,U.K.

BATRAKOV (Dr J.W.) Institute of Theoretical Astronomy, 10, Kutuzov Quay, Leningrad 192187, U.S.S.R.

*BATTANER (Dr E.) Pintor López Mezquita 2, 7°F, Granada, Spain

BATTEN (Dr A.H.) Dominion Astrophysical Observatory, Herzberg Institute of Astrophysics, National Research Council of Canada, 5071 W. Saanich Road, Victoria, B.C. V8X 3X3, Canada

BATTISTINI (Dr P.) Osservatorio Astronomico, Universitario, Via Zamboni 33, Bologna, Italy

*BAUDRY (Dr A.) Observatoire de Bordeaux, Avenue Pierre Sémirot, 33270 Floirac, France

BAUER(C.A.) Pennsylvania State University, Department of Physics, State College, Pennsylvania, U.S.A.

BAUM (Dr W.A.) Planetary Research Center, Lowell Observatory, Flagstaff, Arizona 86002, U.S.A.

BAUSTIAN (W.W.) Kitt Peak National Observatory, 950 North Cherry Avenue, Tucson, Arizona, U.S.A.,

BAUTZ (Dr L.P.) 1325 18th Street, N.W., Apt. 506, Washington, D.C. 20036, U.S.A.

*BEALE (Dr J.S.) Herstmonceux Castle, Hailsham, East Sussex, BN27 1RP, U.K.

BEALS (Dr C.S.) Manotic, Ontario K0A 2N0, Canada

BEARDSLEY (W.R.) Allegheny Observatory, Riverview Park, Pittsburg 14, Pa., U.S.A.

BEAUDET (Dr G.) Département de Physique, Université de Montréal, Montéral, P.Q. H3C 3J7, Canada

BEAVERS (Dr W.) Iowa State University, Ames, Iowa, U.S.A.

BEC (A.) Bureau des Longitudes, 3, Rue Mazarine, Paris 8e, France

BECK (H.G.) Carl Zeiss Observatory, Jena, G.D.R.

BECKER (Prof. Dr F.) Weitlestr. 66/2069, 8 München 45, F.R.G.

BECKER (Dr R.A.) P.O.Box 4609, Carmel, California 93921, U.S.A.

BECKER (Prof. W.) Astronomisches Institut der Universität Basel, Venusstr.7 CH-4102 Binningen, Switzerland

BECKERS (Dr J.M.) Sacramento Peak Observatory, Sunspot, New Mexico 83349, U.S.A.

BECKMAN (Dr J.E.) Astronomy Division, E S T E C, Noordwijk, The Netherlands

BEER (Dr A.) The Observatories, Madingley Road, Cambridge, U.K.

*BEER (Dr R.) Jet Propulsion Laboratory, California Institute of Technology, 4800 Oak Grove Drive, Pasadena, CA 91103, U.S.A.

BEGGS (D.W.) The Observatories, Madingley Road, Cambridge, U.K.

BEHR (Prof. Dr A.) Director, Hamburg Observatory, Gojenbergsweg 112 205 Hamburg 80, F.R.G.

*BEINTEMA (Dr D.A.) Sterrekundig Lab. Kapteyn, Hoogbouw WSN, Postbus 800, Groningen, The Netherlands

BEL (Dr N.) Institut d'Astrophysique 98bis, Boulevard Arago, Paris 75014, France

BELKOVICH (Dr O.I.) Engelgardt Astronomical Observatory, Observatory station Kazan 422526, U.S.S.R.

BELL (Dr B.) Harvard College Observatory, Cambridge, Mass.02138, U.S.A.

BELL (M.B.) Herzberg Institute of Astrophysics, National Research Council of Canada, Ottawa, Ontario K1A OR6, Canada

BELL (R.A.) Astronomy Program, University of Maryland, College Park, Maryland 20742, U.S.A.

BELOTSERKOVSKIJ(Dr D.J.) Institute of Physico-Technical Measurements, Moscow, U.S.S.R.

BELTON (Dr M.S.) Kitt Peak National Observatory, 950 North Cherry Avenue, Tucson, Arizona 85717, U.S.A.

*BELVEDERE (G.) Istituto di Astronomia, Città Universitaria, V. le A. Doria, 95125 Catania, Italy

BELY (O.) Observatoire de Nice, Le Mont-Gros, 06 Nice, France

BELY-DUBAU (Dr F.) Observatoire de Nice, Le Mont-Gros, 06300 Nice, France

*BEM (Dr J.) Astronomical Observatory of the Wroclaw, University, Wroclaw, Poland

*BENAVENTE (J.) Instituto y Observatorio de Marina, San Fernando (Cádiz),Spain

BENDER (Dr P.) Joint Inst. for Laboratory Physics, University of Colorado, Boulder, Colorado 80302, U.S.A.

BENEVIDES-SOARES (Dr P.) Instituto Astronónico e Geofisíco, Caixa Postal 30 627 Sao Paulo 00-SP, Brazil

*BENVENUTI (Dr P.) Osservatorio Astrofiscio, 36012 Asiago, Italy

*BENZ (Dr A.) Radio Astronomy Group ETH, Hochstrasse 58 8044 Zürich, Switzerland

BERENDZEN (Dr R.) Dean, College of Arts and Sciences, The American University Washington, D.C. 20016, U.S.A.

*BERG (Dr R.A.) 419 Space Science Center, Department of Physics and Astronomy University of Rochester, Rochester, NY 14627, U.S.A.

BERGE (Dr G.L.) Owens Valley Radio Observatory, California Institute of Technology, Pasadena, Cal. 91109, U.S.A.

*BERGEAT (Dr J.) Observatoire de Lyon, 69230 Saint-Genis-Laval, France

*BERGER (Dr C.) CERGA, Plateau de Roquevignon, Avenue Copernic, 06130 Grasse, France

BERGER (Dr J.) Institut d'Astrophysique, 98bis, Bld Arago, 75014 Paris, France

*BERGER (Dr X) CERGA, Plateau de Roquevignon, Avenue Copernic, 06130 Grasse, France

BERGERON (Dr J.) Observatoire de Paris, Section d'Astrophysique de Meudon 92190 Meudon, France

BERKHUIJSEN (Dr E.M.) Max-Plank-Institut für Radioastronomie, Auf dem Hügel 69, 53 Bonn 1, F.R.G.

BERNACCA (Dr P.) Osservatorio Astrofiscio, Asiago (Vicenza) Italy

BERTAUD (Dr Ch.) Observatoire de Paris, Section d'Astrophysique, 92190 Meudon, France

*BERTAUX (Dr J.L.) Laboratoire d'Aéronomie CNRS, Réduit de Verrières, B.P. No.3, 91370 Verrières-le-Buisson, France

BERTHOMIEU (Dr G.) Observatoire de Nice, Le Mont-Gros, 06300 Nice, France

BERTIAU (Prof. Dr F.C. S.J.) Waverse Baan 220, 3030 Heverlee, Belgium

BERTOLA (Dr F.) Osservatorio Astronomico, 35100 Padova, Italy

BESSELL (Dr M.S.) Mount Stromlo and Siding Spring Observatories, Private Bag,
    Woden P.O., A.C.T. 2606, Australia

*BETTIS (Dr D.G.) Department of Aerospace Engineering and Engineering Mechanics,
    The University of Texas at Austin, Austin, TX 78712, U.S.A.

*BEUERMANN (Dr K.P.) Max-Planck-Institut für Extraterrestrische Physik,
    8046 Garching bei München, F.R.G.

BEYER (Dr M.) Justus-Brinckmann-Strasse 101, Hamburg-Bergedorf, F.R.G.

BHATIA (Dr P.K.) Department of Mathematics, Faculty of Engineering,
    University of Jodhpur, Jodhpur (Rajasthan), India

BHATIA (Dr V.B.) Department of Physics & Astrophysics, University of Delhi,
    Delhi-110007, India

BHATNAGAR (Dr A) Vedhshala (Astronomical Observatory), Naranpura,
    Ahmedabad-13, India

BHATNAGAR (Prof. P.L.) The Mehta Research Institute, 26 Dilkusha, New Katra,
    Allahabad-211002 India

BHATTACHARYYA (J.C.) Astrophysical Observatory, Kodaikanal 3, India

BHONSLE (R.V.) Physical Research Laboratory, Ahmedabad

BIDELMAN (Dr W.P.) Warner and Swasey Observatory, 1975 Taylor Road,
    East Cleveland, Ohio 44112, U.S.A.

BIELICKI (Dr M.) Astronomical Observatory of Warsaw University,
    Al. Ujazdowskie 4, Warsaw, Poland

BIERMANN (Prof. Dr L.) Direktor des Instituts für Astrophysik, Max-Planck
    Institut für Physik und Astrophysik, Institut für Astrophysik,
    Föhringer Ring 6, München 23, F.R.G.

*BIERMANN (Dr P.L.) Institut für Astrophysik der Universität Bonn,
    Auf dem Hügel 71, D-53 Bonn, F.R.G.

BIGAY (Dr J.H.) Directeur de l'Observatoire de Lyon, 69 Saint-Genis-Laval, France

BIJAOUI (Dr A.) Observatoire de Nice, Le Mont-Gros, 06300 Nice, France

BILLAUD (G.) CERGA., 8, Boulevard Emile Zola, 06130 Grasse, France

BILLINGS (Dr D.E.) High Altitude Observatory of the University of Colorado,
    Boulder, Colorado, U.S.A.

BINGHAM (Dr R.G.) Royal Greenwich Observatory, Herstmonceux Castle,
    Hailsham, Sussex, U.K.

BINNENDIJK (Prof. Dr L.) Flower and Cook Observatory, University of
    Pennsylvania, Philadelphia 4, Pennsylvania, U.S.A.

BIRAUD (Dr F.) Observatoire de Paris, Section d'Astrophysique de Meudon,
    92190 Meudon, France

BISNOVATYI-KOGAN (Dr G.S.) Department of Theoritical Astrophysics, Space
    Research Institute, USSR Academy of Sciences, Profsoyuznaja 88,
    Moscow 117485, U.S.S.R.

BISWAS (Prof. S.) Tata Institute of Fundamental Research, Homi Bhabha Road,
    Bombay 5, India

BLAAUW (Prof. Dr A.) Sterrewacht-Huygens Laboratorium, Wassenaarseweg 78,
    Leiden 2405, Netherlands

BLACKETT (Prof. P.M.S.) Imperial College of Science and Technology,
    South Kensington, London, S.W.7, U.K.

BLACKWELL (Prof. D.E.) Dept. of Astrophysics, University Observatory,
    South Parks Road, Oxford, U.K.

BLAHA (Dr M.) Goddard Space Flight Center, Code 680. Greenbelt,
    Maryland 20771, U.S.A.

BLAMONT (Prof. J.E.) Service d'Aéronomie, Réduit de Verrières,
    91 Verrières-le-Buisson, France

BLANCO (Dr C.) Osservatorio Astrofisico, Città Universitaria, Viale A. Doria,
    I-95125 Catania, Italy

BLANCO (Dr V.M.) Observatorio Inter-Americano de Coerro Tololo,
    Casilla 63-D, La Serena, Chile

*BLANDFORD (Dr R.D.) Institute of Astronomy, Madingley Road, Cambridge, U.K.

BLEEKER (Dr Ir J.A.M.) Cosmic Ray Working Group, Kamerlingh Onnes Laboratorium,
    Nieuwsteeg 18, Leiden, The Netherlands

BLESS (Dr R.C.) Washburn Observatory, University Wisconsin,
    475 North Charter Street, Madison, Wisconsin 53706, U.S.A.

BLINOV (Dr N.S.) Sternberg Astronomical Institute, Moscow, U.S.S.R.

BLITZSTEIN (Dr W.) Flower and Cook Observatory, University of Pennsylvania,
    Philadelphia 4, Pennsylvania, U.S.A.

BLOCH (Dr M.) Observatoire de Lyon, 69230 Saint-Genis-Laval (Rhône), France

*BLUDMAN (Dr S.A.) Department of Physics, University of Pennsylvania,
    Philadelphia, PA 19174, U.S.A.

BLUM (Dr E.J.) Observatoire de Paris, Section d'Astrophysique,
    92190 Meudon, France

*BLUM (Dr P.W.) Institut für Astrophysik der Universität Bonn,
    Auf dem Hügel 71, D-53 Bonn, F.R.G.

BOBROV (Dr M.S.), Astronomical Council, USSR Academy of Sciences,
    Pyatnitskaya ul. 48, 109017 Moscow, U.S.S.R.

BOBROVNIKOFF (Dr N.T.) 1623 Visalia Avenue, Berkeley, California 94707, U.S.A.

BODENHEIMER (Dr P.) Lick Observatory, University of California, Santa Cruz,
    California 95060, U.S.A.

BOESGAARD (Dr A.M.) Institute of Astronomy, University of Hawaii,
    2525 Correa Road, Honolulu, Hawai 96822, U.S.A.

BOGGESS (Dr A. III) Goddard Space Flight Center, Washington 25, D.C., U.S.A.

*BOGGESS (Dr N.W.) NASA Headquarters, 400 Maryland Avenue, SW,
    Washington, DC 20546, U.S.A.

BOGORODSKIJ (Dr A.F.) Astronomical Observatory of the Kiev University,
    Kiev, U.S.S.R.

*BOHLIN (Dr J.D.) Code 7141B, E.O. Hulburt Center for Space Research,
    Naval Research Laboratory, Washington, DC 20375, U.S.A.

BOHLIN (Dr R.C.) Goddard Space Flight Center, Greenbelt, Maryland 20771, U.S.A.

BÖHM (Prof. Dr K.H.) Astronomy Department, University of Washington, Seattle,
    Washington 98105, U.S.A.

BÖHM-VITENSE (Prof. Dr E.) Astronomy Department, University of Washington,
    Seattle, Washington 98105, U.S.A.

BÖHME (Dr A.) Zentralinstitut für Solar-Terrestrische Physik, Bereich C 1,
    Telegrafenberg, DDR 15, Potsdam, G.D.R.

BÖHME (Dr S.) Astronomisches Rechen-Institut, Mönchhofstrasse 12-14,
    69 Heidelberg, F.R.G.

BOHRMANN (Prof. Dr A.) Angelhofweg 31, 6901 Wilhelmsfeld, F.R.G.

BOISCHOT (Dr A.) Observatoire de Paris, Section d'Astrophysique,
    92190 Meudon, France

BOK (Prof. B.J.) Steward Observatory, University of Arizona,
    Tucson, Arizona 85721, U.S.A.

BOKSENBERG (Dr A.) University College London, Gower Street, London WC1, U.K.

*BOLCAL (Dr Ç.) University Observatory, Üniversite, Istanbul, Turkey

BOLDT (Dr E.A.) NASA, Goddard Space Flight Center, Greenbelt, Maryland 20771, U.S.A.

*BOLEY (Dr F.I.) Department of Physics and Astronomy, Wilder Laboratory,
    Dartmouth College, Hanover, NH 03755, U.S.A.

*BOLTON (Dr Ch.T.) David Dunlap Observatory, University of Toronto,
    Richmond Hill, Ontario, Canada L4C 4Y6

BOLTON (J.G.) C.S.I.R.O., Division of Radiophysics, Box 76,
    Epping, N.S.W., 2121, Australia

BONANOMI (Dr J.) Directeur de l'Observatoire de Neuchâtel, Neuchâtel, Switzerland

BOND (Dr H.E.) Louisiana State University, Baton Rouge, Louisiana 70803, U.S.A.

BONDARENKO (Dr L.N.) Sternberg Astronomical Institute, Universitetskij,
    Prospekt 13, Moscow 117234, U.S.S.R.

BONDI (Dr H.) King's College, Strand, London, W.C. 2, U.K.

BONEFF (Prof. Dr N.) Bulgarian Academy of Sciences, Section of Astronomy,
    "7 Novembre" 1, Sofia, Bulgaria

BONEV (B.) Observatoire Populaire de Stara-Zagora, Bulgaria

BONNEAU (Prof. M.) Faculté des Sciences, Place Victor Hugo, 13-Marseille, France

BONNET (Dr R.) L.P.S.P., 91-Verrières-le-Buisson, France

BONNOR (Prof. W.B.) Queen Elisabeth College, Campden Hill Road, London W.8., U.K.

BONOMETTO (Dr S.) Osservatorio Astronomico, Vicolo dell'Osservatorio,
    Padova, Italy

BONOV (Dr A.D.) Université de Sofia, Sofia, Bulgaria

BONSACK (Prof. W.K.) Institute for Astronomy, 2840 Koluwalu St.
    University of Hawaii, Honolulu, Hawaii 96822, U.S.A.

*BOOKMYER (Dr B.B.) Department of Physics and Astronomy, Clemson University,
   Clemson, SC 29631, U.S.A.

BOOTH (Dr R.S.) Nuffield Radio Astronomy Laboratory, Jodrell Bank,
   Macclesfield, Cheshire, U.K.

BORGMAN (Dr J.) "Kapteyn" Sterrewacht, Mensingenweg 20,
   Roden (Drente), The Netherlands

BORNER (Dr G.) Max-Planck-Institut für Physik und Astrophysik, 8000 München,
   Föhringer Ring 6, F.R.G.

*BORRA (Dr E.F.) Département de Physique, Université de Laval, Cité Universitaire,
   Québec, Québec, Canada G1K 7P4

*BOSMA (Dr P.B.) Free University, De Boelelaan 1081, Amsterdam, The Netherlands

BOSMAN-CRESPIN (D.) Place d'Italie, 4, B-4000 Liège, Belgium

BOTELHEIRO (Dr A.) Observatorio Astronomico, Lisboa, Portugal

BOTEZ (Dr E.) Pedagogical Institute, Suceava, Roumania

BOTTINELLI (Dr L.) Observatoire de Paris, Section d'Astrophysique de Meudon,
   92190 Meudon, France

BOUIGUE (Prof. R.) Université Paul Sabatier
   31077 Toulouse, Cedex, France

BOULON (Dr J.) Observatoire de Paris, 61, avenue de l'Observatoire,
   75 Paris 14e, France

BOURY (Dr A.J.J.L.) Chef de Travaux à l'Université de Liège,
   Rue de la Chaussée, 60, 4870 Theux, Belgium

BOUŠKA (Dr J.) Department of Astronomy and Astrophysics, Charles University,
   Švédská 8, Prague 5 - Smíchov, Czechoslovakia

BOUVIER (Prof. P.B.) Observatoire de Genève, CH-1290 Sauverny, Switzerland

BOWEN (Dr E.G.) 172 Edinburgh Road, Castlecrag, N.S.W., 2068, Australia

BOWYER (Dr C.S.) Astronomy Department, University of California, Berkeley,
   California 94720, U.S.A.

BOYARCHUK (Dr A.A.) Crimean Astrophysical Observatory, USSR Academy of Sciences,
   P/O Nauchny, Crimea 334413, U.S.S.R.

BOYCE (Dr P.B.) P.O. Box 1269, Flagstaff, Arizona 86001, U.S.A.

BOYD (Prof. R.L.F.) Department of Physics, University College London,
   Mullard Space Science Laboratory, Holmbury St Mary, Dorking, Surrey, U.K.

*BOYNTON (Dr P.E.) Astronomy Department, University of Washington,
   Seattle, WA 98195, U.S.A.

BOZIS (Dr G.) Department of Theoretical Mechanics, University of Thessaloniki,
   Thessaloniki, Greece

BOZKURT (Dr S.) Ege University Observatory, Bornova, Izmir, Turkey

BRACCESI (A.) Laboratorio di Radioastronomia, Istituto di Fisica, Bologna, Italy

BRACEWELL (Prof. R.N.) Radio Astronomy Institute, Stanford University,
   Stanford, California 94305, U.S.A.

BRAES (Dr L.L.E.) Sterrewacht, Leiden, The Netherlands

BRAHDE (R.) Institute of Theoretical Astrophysics, University of Oslo, Blindern, Oslo 3, Norway

*BRAHIC (Dr A.) Observatoire de Paris, Département d'Astrophysique Fondamentale, 92190 Meudon, France

*BRANCH (Dr D.R.) Department of Physics and Astronomy, The University of Oklahoma, 440 West Brooks, Norman, OK 73069, U.S.A.

*BRAND (Dr P.W.J.L.) University Department of Astronomy, Royal Observatory, Edinburgh EH9 3HJ, U.K.

*BRANDIE (Dr G.W.) Astronomy Group, Department of Physics, Queen's University, Kingston, Ontario, Canada K7L 3N6

BRANDT (Dr J.C.) Code 680, Laboratory for Solar Physics, Goddard Space Flight Center, Greenbelt, Maryland 20771, U.S.A.

BRANSCOMB (Dr L.M.) National Bureau of Standards, Washington 25, D.C., U.S.A.

BRANSON (Dr N.J.B.A.) The Old Schools, Cambridge University, Cambridge, U.K.

BRAUDE (Prof. Dr S.Ya.) Institute of Radio Physics and Electronics, Academy of Sciences of the Ukrainian SSR, Khar'kov-85, U.S.S.R.

BRAULT (Dr J.W.) Kitt Peak National Observatory, P.O. Box 26732, 950 North Cherry Ave., Tucson, Arizona 85726, U.S.A.

*BRAUNINGER (Dr H.) Max-Planck-Institut für Physik und Astrophysik, Institut für Extraterrestrische Physik, 8046 Garching, F.R.G.

BRAY (Dr R.J.) CSIRO, Division of Physics, University Grounds, Chippendale, NSW 2008, Australia

BREALEY (G.A.) Dominion Astrophysical Observatory, Herzberg Institute of Astrophysics, National Research Council of Canada, 5071 W, Saanich Road, Victoria, B.C. V8X 3X3, Canada

*BRECHER (Dr A.) Massachusetts Institute of Technology, Room 54-1114, Department of Earth and Planetary Sciences, Cambridge, MA 02139, U.S.A.

BRECHER (Prof. K.) Dept. of Physics, Room 6-201, Massachusetts Institute of Technology, 77 Massachusetts Avenue, Cambridge, Mass. 02139, U.S.A.

*BRECKINRIDGE (Dr J.B.) Jet Propulsion Laboratory, California Institute of Technology, 4800 Oak Grove Drive, Pasadena, California 91103, U.S.A.

BREGER (Dr M.) Dept. of Astronomy, University of Texas, Austin, Texas 78712, U.S.A.

*BREINHORST (Dr R.A.) Sternwarte der Universität Bonn, 53 Bonn 1, Auf dem Hügel 71, Germany, F.R.

BREJDO (Dr I.I.) Main Astronomical Observatory, USSR Academy of Sciences, Pulkovo, 196140 Leningrad, U.S.S.R.

BRIDLE (Dr A.H.) Astronomy Group, Department of Physics, Queen's University, Kingston, Ontario, K7L 3N6, Canada

*BRIEVA (Prof. E.) Director, National Observatory, Bogotá, Colombia

BRINI (Prof. D.) Laboratorio T.E.S.R.E., via Castagnoli 1, Bologna, Italy

BRINKMAN (Dr A.C.) Laboratorium voor Ruimteonderzoek, Beneluxlaan 21, Utrecht, The Netherlands

BRKIĆ (Z.M.) Astronomska Observatorija, Veliki Vracar, Beograd, Yugoslavia

BROADFOOT (Dr L.) Kitt Peak National Observatory, 950 North Cherry Ave., Tucson, Arizona 85721, U.S.A.

*BRODERICK (Dr J.) Department of Physics, Virginia Polytechnic Institute, Blackburg, VA 24061, U.S.A.

BROGLIA (Dr P.) Osservatorio Astronomico, Merate (Como), Italy

BROSCHE (Dr P.) Sternwarte der Universität, D-53 Bonn, Auf dem Hügel 71, F.R.G.

BROSTERHUS (Dr E.B.F.) Dominion Astrophysical Observatory, Herzberg Institute of Astrophysics, National Research Council of Canada, 5071 W. Saanich Road, Victoria, B.C. V8X 3X3, Canada

BROTEN (N.W.) Herzberg Institute of Astrophysics, National Research Council of Canada, Ottawa, Ontario K1A OR6, Canada

BROUCKE (Dr R.) 5603 Stardust Road, La Canada, California 91011, U.S.A.

BROUW (Dr W.N.) Radiosterrenwacht, Dwingeloo, 7514 The Netherlands

BROWN (Dr H.) Division of Geological Sciences, California Institute of Technology, Pasadena 4, California, U.S.A.

*BROWN (Dr J.C.) Department of Astronomy, The University, Glasgow, G12 - 8QW, U.K.

BROWN (P.L.) Karlsvik, 30 Eghams Wood Road, Beaconsfield, Bucks., U.K.

BROWN (Prof. R.H.) Chatterton Astronomy Department, School of Physics, University of Sydney, Sydney, N.S.W. 2006, Australia

BROWN (Dr R.L.) National Radio Astronomy Observatory, Charlottesville, Virginia 22901, U.S.A.

*BROWNE (Dr I.W.A.) Nuffield Radio Astronomy Labs., Jodrell Bank, Macclesfield, Cheshire, U.K.

*BROWNLEE (Dr D.) Department of Astronomy, University of Washington, Seattle, Wash. 98195, U.S.A.

BROWNLEE (Dr R.R.) P.O. Box 1663, Los Alamos, New Mexico, U.S.A.

BRÜCK (Prof. H.A.) Craigower, Penicuik, Midlothian, Scotland, U.K.

BRÜCK (Dr M.T.) Royal Observatory, Blackford Hill, Edinburgh 9, U.K.

BRUCKNER (Dr G.) U.S. Naval Research Laboratory, Plasma Physics Section, Code 7700, Washington D.C. 20390, U.S.A.

BRUMBERG (Dr V.A.) Institute of Theoretical Astronomy, 10, Kutuzov Qay, Leningrad, 192187, U.S.S.R.

BRUNET (Dr J.P.) Observatoire de Marseille, 2, Place le Verrier, 13004 Marseille, France

BRUNK (W.) Lunar and Planetary Programs, NASA Headquarters, Washington, D.C. 20546, U.S.A.

BRUWER (J.A.) South African Astronomical Observatory, P.O. Box 3718, Johannesburg 2000, South Africa

BRUZEK (Dr A.) Fraunhofer Institut, Schöneckstrasse 6, Freiburg 1. Br. Germany F.R.

BUCHAR (Prof. Dr E.) Astronomical Institute, Technical University, Karlovo nám. 13, Prague 2, Czechoslovakia

*BUCHLER (Dr J.R.) Department of Physics and Astronomy, University of Florida, Gainesville, FL 32611, U.S.A.

*BUDDING (Dr E.) Department of Astronomy, University of Manchester, Manchester M13 9PL, U.K.

*BUES (Dr I.D.) I. Mathematisches Institut der Freien Universität, 1 Berlin 33, Hüttenweg 9, Germany F.R.

BUHL (Dr D.) NASA, Goddard Space Flight Center, Code 691, Greenbelt, Maryland 20771, U.S.A.

BULLARD (Sir Edward C.) University of Cambridge, Dept. of Geodesy and Geophysics, Madingley Rise, Madingley Road, Cambridge, U.K.

BULLEN (Prof. K.E.) Department of Applied Mathematics, University of Sydney, Sydney N.S.W., Australia

BUMBA (Dr V.) Astronomical Institute, Czechoslovak Academy of Sciences, Observatory Ondřejov 25165, Czechoslovakia

BURBIDGE (Dr E.M.) Department of Physics, University of California, San Diego, La Jolla, California 92037, U.S.A.

BURBIDGE (Dr G.R.) University of California, P.O. Box 109, La Jolla, California 92037, U.S.A.

BURCH (Dr C.R.) H.H. Royal Fort Wills Physics Laboratory, Bristol 8, U.K.

BURGER (Dr Ir J.J.) Werkgroep Kosmische Straling, Kamerlingh Onnes Laboratorium, Nieuwsteeg 18, Leiden, The Netherlands

BURGESS (Dr A.) University of Cambridge, Department of Applied Mathematics and Theoretical Physics, Silver Street, Cambridge, U.K.

BURGESS (Dr D.D.) Physics Department, Imperial College of Science and Technology, London S.W. 7, U.K.

BURKE (Dr B.F.) Room 26-459, Massachusetts Institute of Technology, Cambridge, Massachusetts 02139, U.S.A.

BURKE (Prof. J.A.) Department of Physics, University of Victoria, Victoria, B.C., Canada

BURKHEAD (Dr M.S.) Dept. of Astronomy, Swain Hall West 319, Indiana University, Bloomington, Indiana 47401, U.S.A.

BURNICHON (Dr M.L.) Institut d'Astrophysique, 98 bis, Boulevard Arago, F - 75014 Paris, France

BURNS (Dr J.A.) Thurston Hall, Cornell University, Ithaca, New York 14850, U.S.A.

BURTON (Dr W.B.) National Radio Astronomy Observatory, Edgemont Road, Charlottesville, Virginia 22901, U.S.A.

BURTON (B.Sc.W.M.) SRC Appleton Laboratory, Astrophysics Research Division, Culham Laboratory, Abingdon Oxon, OX14 3DB, U.K.

BUSBRIDGE (Dr I.W.) "Haremere" Westerham Road, Keston, Kent, BR 2 6HH, U.K.

BUSCOMBE (Dr W.) Astronomy Dept. Northwestern University, Evanston, Illinois 60201, U.S.A.

*BUTI (Dr B.) Physical Research Laboratory, Navrangpura, Ahmedabad 380009, India

BUTLER (Dr C.J.) Armagh Observatory, Armagh, Northern Ireland

BUTLER (Dr H.E.) Royal Observatory, Blackford Hill, Edinburgh 9, U.K.

*BYARD (Dr P.L.) Department of Astronomy, The Ohio State University,
174 West 18th Avenue, Colombus, OH 43210, U.S.A.

BYKOV (Dr M.F.) Astronomical Institute of the Usbek SSR, Academy of Sciences,
Tashkent, U.S.S.R.

BYSTROV (Dr N.F.) Main Astronomical Observatory, USSR Academy of Sciences,
Pulkovo, 196140 Leningrad, U.S.S.R.

BYSTROVA (Dr N.V.) Special Astrophysical Observatory. USSR Academy of Sciences,
Leningrad Branch of the SAO, Pulkovo 196140, U.S.S.R.

CABRITA (E.M.L.) Tapada da Ajuda, Lisboa 3, Portugal

CACCIANI (Dr A.) Osservatorio Astronomico di Roma, Rome, Italy

*CACCIN (B.) Osservatorio Astronomico di Capodimonte, Via Moiariello 16,
80131 Napoli, Italy

CAHN (Dr J.) University of Illinois, Urbana, Illinois, Urbana, Illinois 61801,
U.S.A.

CALAMAI (Prof. G.) Osservatorio astrofisico di Arcetri, Firenze, Italy

*CALAME (Dr O.) CERGA, Plateau de Roquevignon, Avenue Copernic, 06130 Grasse, France

CALOI (Dr V.) Laboratorio di Astrofisica, Casella Postale 67, Frascati (Roma), Italy

*CALVO (Dr M.) Violante de Hungria 5, 42C.- Zaragoza, Spain

*CAMARENA (Dr V.) Departamento de Física de la Tierra y del Cosmos,
Facultad de Ciencias, Zaragoza, Spain

CAMERON (Dr A.G.W.) Belfer Graduate School of Science, Yeshiva University,
Amsterdam Ave & 186th St., New York, N.Y. 10033, U.S.A.

CAMERON (W.) NASA, Goddard Space Flight Center, Greenbelt, Maryland 20771, U.S.A.

CAMICHEL (Dr H.) Observatoire de Toulouse, 31-Toulouse (Hte-Garonne), France

CAMM (Dr G.L.) Department of Mathematics, The University, Manchester 13, U.K.

*CAMPBELL (Dr D.B.) Arecibo Observatory, P.O. Box 995, Arecibo, P.R. 00612, U.S.A.

CAMPBELL (B.Sc. J.W.) Science Research Council, Royal Observatory Edinburgh,
Edinburgh, EH9 3HJ, U.K.

*CANAL (Dr R.) Departamento de Física de la Tierra y del Cosmos,
Universidad de Barcelona, Barcelona, Spain

CANAVAGGIA (De R.) Observatoire de Paris, 61, Avenue de 1'Observatoire,
75014 Paris, France

CANDY (M.P.) Perth Observatory, Bickley, Western Australia 6076

CANFIELD (Dr R.C.) Department of Physics, Code C-011, University of California,
San Diego, La Jolla, California 92093, U.S.A.

*CANIZARES (Dr C.R.) Massachusetts Institute of Technology, Center for Space
Research, Building 37-501, Cambridge, MA 02139, U.S.A.

*CANNON (Dr C.J.) Department of Applied Mathematics, University of Sydney,
Sydney, NSW 2006, Australia

CANNON (Dr R.D.) UK 48-inch Schmidt Telescope Project, Royal Observatory,
Edinburgh EH9 3HJ, U.K.

CANTU' (Dr A.M.) Osservatorio Astrofisico di Arcetri, Largo Enrico Fermi, 5,
    Firenze, Italy

CAPACCIOLI (Dr M.) Osservatorio Astrofisico, Asiago (Vicenza), Italy

*CAPEN (Dr C.F. Jr) Lowell Observatory, Box 1629, Flagstaff, AZ 86001, U.S.A.

*CAPLAN (Dr J.) Observatoire de Marseille, 2, Place Le Verrier,
    13004 Marseille, France

CAPRIOLI (Dr G.) Osservatorio astronomico di Monte Mario, Via Trionfale 204,
    Rome, Italy

CAPRIOTTI (Dr E.R.) Astronomy Department, The Ohio State University,
    174 West 18th Avenue, Columbus, Ohio 43210, U.S.A.

*CAPUTO (F.) Consiglio Nazionale Ricerche, Laboratorio Astrofisica Spaziale,
    Casella Postale 67, 00044 Frascati, Italy

CARDUS (J.O.) Astronome de l'Observatorio del Ebro, Roquetas, Tarragona, Spain

CARLETON (Dr N.) Smithsonian Institution, Astrophysical Observatory,
    Cambridge, Mass. 02138, U.S.A.

CARLQVIST (P.A.) Royal Institute of Technology, Division of Plasma Physics,
    Stockholm 70, Sweden

*CAROFF (Dr L.J.) Space Science Division, Mail Stop 245-3, NASA-Ames
    Research Center, Moffett Field, CA 94035, U.S.A.

CAROUBALOS (Dr C.) Observatoire de Meudon, 92190 Meudon, France

CARPENTER (C.Egn. G.J.) Royal Observatory, Edinburg EH9 3HJ, U.K.

CARPENTER (Dr Lloyd H. NASA, Goddard Space Flight Center, Greenbelt,
    Maryland 20771, U.S.A.

CARR (Dr T.D.) Department of Physics, University of Florida, Gainesville,
    Florida, U.S.A.

CARRANZA (Dr G.J.) Laprida 880, Córdoba, Argentina

*CARRASCO (Dr L.) Instituto de Astronomía, UNAM, Apartado Postal 70-264
    México 20, D.F. México

CARRASCO (Dr R.) Observatorio Astronómico Nacional de Madrid, Alfonso XII,3,
    Madrid 7, Spain

CARROLL (Prof. P.K.) Department of Physics, University College, Belfield,
    Dublin 4, Ireland

*CARRUTHERS (Dr G.) Code 7123 Department of the Navy, Naval Research
    Laboratory, Washington D.C. 20375, U.S.A.

CARSON(Dr T.R.)University Observatory, Buchanan Gardens, St. Andrews,
    Fife, Scotland, U.K.

CARVER (Prof. J.H.) Department of Physics, University of Adelaide,
    Adelaide, Australia

CASANOVAS CORDERROURE (J.) Instituto Universitario de Astrofisica,
    Universidad de La Laguna, Tenerife, Spain

* CASINI(Dr C) Istituto di Astronomia dell'Università di Milano,
    Via Brera 28, Milano, Italy

CASSINELLI (Dr J.) University of Wisconsin, Washburn Observatory, Madison, Wisconsin 53706, U.S.A.

CASTELLANI (Dr V.) Laboratorio di Astrofisica, Casella Postale 67, Frascati (Roma), Italy

CASTELLI (J.P.) AFCRL (LIR/Stop 30) L.G. Hanscom Field, Bedford, MA 01730, U.S.A.

CASTOR (Dr J.I.) Joint Institute for Laboratory Astrophysics, University of Colorado, Boulder, Colorado 80302, U.S.A.

*CASWELL (Dr J.L.) CSIRO, Division of Radiophysics, P.O. Box 76, EPPING NSW 2121, Australia

*CATALA (Dr M.A.) Departamento de Física de la Tierra y del Cosmos, Universidad de Barcelona, Barcelona, Spain

CATALANO (Dr F.A.) Osservatorio Astrofisico, Città Universitaria, Viale A. Doria, I-95125 Catania, Italy

CATALANO (Dr S.) Osservatorio Astrofisico, Città Universitaria, Viale A. Doria 95125 Catania, Sicilia, Italy

*CATO (Dr B.T.) Onsala Space Observatory, S-430 34 Onsala, Sweden

*CATURA (Dr R.C.) Space Sciences Laboratory, Dept. 52-12, Bldg 202, Lockheed, 3251 Hanover Street, Palo Alto, CA 94304, U.S.A.

CAUGHLAN (Dr G.) Department of Physics, Montana State University, Bozeman, Montana 59715, U.S.A.

CAVALIERE (Dr A.) Laboratorio del C.N.E.N., Casella Postale 65, Frascati (Roma), Italy

CAYREL (Dr R.) Observatoire de Paris, Section d'Astrophysique, Meudon 92190, France

CAYREL-de-Strobel (Dr G.) Institut d'Astrophysique, 98bis Boulevard Arago, 75014 Paris, France

*CAZENAVE (Dr A.) GRGS/CNES, 18, Avenue Edouard Belin, 31055 Toulouse CEDEX, France

CAZZOLA (Dr P.) Osservatorio Astronomico, Vicolo dell'Osservatorio, Padova, Italy

CECCARELLI (Prof. M.) Istituto di Fisica generale, Università di Bologna, Italy

CECCHINI (Prof. G.) Via Roma, 56011 Calci (Pisa), Italy

*CELNIKIER (Dr L.) DAPHE, Observatoire de Meudon, 92190 Meudon, France

CEPLECHA (Dr Z.) Astronomical Institute, Czechoslovak Academy of Sciences, Observatory Ondřejov, Czechoslovakia

CESCO (Dr C.U.) Observatorio Félix Aguilar, Facultad de Ingenería, Universidad Nacional de Cuyo, San Juan, Argentina

CESTER (Prof. Dr B.) Osservatorio Astronomico, Via G.B. Tiepolo 15, Trieste (409), Italy

*CEZARSKY (Dr C.) CEA Saclay, B.P. No.2, 91190 GIF-S/-YVETTE, France

*CEZARSKY (Dr D.) DERAD Observatoire de Meudon, 92190 Meudon, France

CHAFFEE (Dr F.) Smithsonian Astrophysical Observatory, Cambridge, Massachusetts 02138, U.S.A.

*CHAISSON (Dr E.J.) Center for Astrophysics, 60 Garden Street, Cambridge,
    Ma. 02138, U.S.A.

CHALONGE (Dr D.) Institut d'Astrophysique, 98bis Boulevard Arago,
    75014 Paris, France

CHAMBERLAIN (Prof. J.W.) Department of Space Physics and Astronomy,
    Rice University, Houston, Texas 77001, U.S.A.

CHAMBLISS (Dr C.) Kutztown State College, Kutztown, Pennsylvania 19530, U.S.A.

CHANDRA (S.) General Electric Co., Room M 1325, P.O. 8555, Philadelphia 19101,
    Pennsylvania, U.S.A.

CHANDRASEKHAR (Prof. S.) Laboratory for Astrophysics & Space Research,
    933 East 56th Street, Chicago, Ill, 60637, U.S.A.

CHANG (Dr Shao-Chang) Geophysics Institute of National Central University,
    Chung-li, Taiwan

*CHANMUGAM (Dr G.) Louisiana State University, Department of Physics and
    Astronomy, Baton Rouge, LA 70803, U.S.A.

*CHAPMAN (Dr C.R.) Planetary Science Institute, 2030 E. Speedway, Suite 201,
    Tucson AZ 85719, U.S.A.

CHAPMAN (Dr R.D.) Solar Physics Group, Goddard Space Flight Center,
    Greenbelt, Maryland 20771, U.S.A.

CHAPRONT (Dr J.) Bureau des Longitudes, 3, Rue Mazarine, Paris 8e, France

CHARVIN (P.) Observatoire de Meudon, 5 place Janssen, 92 Meudon, France

CHAU (Prof. W.Y.) Astronomy Group, Department of Physics, Queen's University,
    Kingston, Ontario, K7L 3N6, Canada

CHAVIRA (E.) Instituto Nacional de Astrofísica Optica y Electrónica,
    Tonantzintla, Pue. Apartados Postales Nos 216 y 51, Puebla,
    Pue. Mexico

CHEKIRDA (Dr A.T.) Kharkov Observatory, Kharkov, U.S.S.R.

CHEN (Dr K.Y.) Astronomy Department, Rosemary Hill Observatory,
    University of Florida, Gainesville, Fla 32601, U.S.A.

CHENAKAL (Dr V.L.) Director of Lomonosov Museum, Universitetskaja nab.3
    Leningrad 199164, U.S.S.R.

CHEREDNICHENKO (Dr V.I.) Kiev Polytechnical Institute, Kiev 56, U.S.S.R.

CHEREPASHCHUK (Dr A.M.) Sternberg Astronomical Institute, Moscow, U.S.S.R.

CHERNEGA (Dr N.A.) Astronomical Observatory, Kiev University, Kiev, U.S.S.R.

CHERTOPRUD (Dr V.E.), Hydrometeorological Center of the USSR,
    123376 Moscow, U.S.S.R.

CHEVALIER (Dr C.) Observatoire de Paris, Section d'Astrophysique de Meudon,
    92190 Meudon, France

*CHEVALIER (Dr R.) Kitt Peak National Observatory, P.O.Box 26732,
    Tucson, AZ 85726, U.S.A.

CHINCARINI (G.L.) Department of Astronomy, P.M.A. Bldg, Room 16.336,
    University of Texas, Austin, Texas 78712, U.S.A.

CHIOSI (Dr C.) Osservatorio Astronomico, Vicolo dell'Osservatorio, Padova, Italy

CHIS (Prof. G.) Observatoire Astronomique, Rue de la République 109,
Cluj, Roumania

*CHISTYAKOV (Dr V.F.) Solar Station, Gornotaezhnoe, Primorsky Kray, 692533
U.S.S.R.

CHITRE (Dr S.M.) Tata Institute of Fundamental Research, Homi Bhabha Road,
Colaba, Bombay 5, India

CHIU (Dr H.Y.) Institute for Space Studies, 475 Riverside Drive,
New York, N.Y. 10027, U.S.A.

CHIUDERI (Dr C.) Osservatorio Astrofisico di Arcetri, Largo Enrico Fermi 5,
Firenze, Italy

CHIUDERI (Dr F.D.) Osservatorio Astrofisico di Arcetri, Largo E. Fermi 5,
Firence, Italy

*CHMIELEWSKI (Dr Y.) Observatoire de Genève, 1290 Sauverny, Switzerland

CHOI (Bon Chol) Pyongyang Astronomical Observatory, Academy of Sciences,
Pyongyang, Democratic People's Republic of Korea

*CHOLLET (F.). DANOF, Observatoire de Paris, 61, Avenue de l'Observatoire,
75014, Paris, France

CHOPINET (M.) Observatoire de l'Université de Bordeaux,
33 Floirac, France

CHOU (Dr K.C.) Department of Astronomy and Meteorology, Yonsei University,
Seoul, Republic of Korea

CHRISTIANSEN (Prof. W.N.) The University of Sydney, School of Electrical
Engineering, Sydney, N.S.W., Australia 2006

CHRISTY (Prof. R.F.) Physics Department, California Institute of Technology,
Pasadena, Cal., U.S.A.

CHUBB (Dr T.A.) 5023 N 38th Street, Arlington 7, Virginia, U.S.A.

CHUGAJNOV (Dr P.F.), Crimean Astrophysical Observatory, USSR Academy of Sciences,
P/O NAUCHNIY, 334413 Crimea, U.S.S.R.

CHURCHWELL (Dr E.B.) Max-Planck-Institut für Radioastronomie, Auf dem Hügel 69,
53 Bonn 1, Germany F.R.

CHURMS (J.) South African Astronomical Observatory, P.O.Box 9, Cape 7935,
South Africa

CHUVAEV (Dr K.K.), Crimean Astrophysical Observatory, USSR Academy of
Sciences, p/o Nauchny, Crimea, U.S.S.R.

CHVOJKOVA (Woyk) (Dr E.) Astronomical Institute, Czechoslovak Academy of
Sciences, Budečská 6, Prague 2, Czechoslovakia

CIATTI (Dr F.) Osservatorio Astrofisico, Asiago (Vicenza) Italy

CID PALACIOS (Dr R.) Facultad de Ciencias, Ciudad Universitaria,
Zaragoza, Spain

CILLIE (Prof. G.G.) 4, Minserie Street, Stellenbosch 7600, South Africa

CIMINO (Prof.Dr M.) Direttore, Osservatorio Astronomico su Monte Mario,
Rome, Italy

*CLARIA (Dr J.) Asociación Venezolana de Astronomía, Apartado de Correos No 264, Mérida, Venezuela

CLARK (Dr A.) Dept. of Mechanical and Aerospace Sciences, University of Rochester, Rochester, New York 14627, U.S.A.

CLARK (Dr B.G.) National Radio Astronomy Observatory, P.O. Box 2, Green Bank, West Virginia 24944, U.S.A.

*CLARK (Dr D.H.) Mullard Space Science Laboratory, Holmbury St. Mary, Dorking, Surrey, U.K.

CLARK (Dr G.) Physics Department, Massachusetts Institute of Technology, Cambridge, Massachusetts 02139, U.S.A.

CLARK (Dr Th.A.) Laboratory for Extraterrestrial Physics, University of Maryland, College Park, Maryland 20742, U.S.A.

CLARKE (Dr D.) Department of Astronomy, The University, Glasgow G12 8QW, U.K.

CLARKE (Dr T.R. Jr.) Mclaughlin Planetarium, Royal Ontario Museum, Toronto, Ontario, Canada

*CLAUSEN (Dr J.V.) Copenhagen University Observatory, Brorfelde, DK-4340 Tølløse, Denmark

CLAYTON (Dr D.D.) Space Science Department, Rice University, Houston, Texas, U.S.A.

CLEGG (Dr P.E.) Physics Department, Queen Mary College, Mile End Road, London E1 4NS, U.K.

CLEMENT (COUTTS) (Dr C.M.) Department of Astronomy, University of Toronto, Toronto, Ontario M5T 2A6, Canada

CLEMENT (Prof. Dr M.J.) David Dunlap Observatory, University of Toronto, Richmond Hill, Ontario L4C 4Y6, Canada

CLIMENHAGA (Dr J.L.) Department of Physics, University of Victoria, Victoria B.C., Canada

CLUBE (Dr S.V.M.) Royal Observatory Edinburgh, U.K.

COCKE (Dr J.) Steward Observatory, University of Arizona, Tucson, Arizona 85721, U.S.A.

CODE (A.D.) Washburn Observatory, University of Wisconsin, 475 North Charter Street, Madison, Wisconsin 53706, U.S.A.

*CODINA LANDABERRY (Prof. Dr S.J.) Instituto Astrônomico e Geofísico, U.S.P. / Caixa Postal 30627, 01000 Sao Paulo-SP, Brazil

COELHO BALSA (Dr M.C.) Observatorio Astronómico de Universidade de Coimbra, Coimbra, Portugal

COFFEN (Dr D.) Institute for Space Studies, 2880 Broadway, New York, N.Y. 10025, U.S.A.

*COHEN (Dr J.G.) Kitt Peak National Observatory, P.O. Box 26732, Tucson, AZ 85726, U.S.A.

COHEN (Dr J.M.) Institute for Advanced Study, School of Natural Sciences, Princeton, New Jersey 08540, U.S.A.

*COHEN (Dr L.) Department of Physics and Astronomy, The University of North Carolina, Chapel Hill, NC 27514, U.S.A.

COHEN (Prof. M.H.) Department of Astronomy, California Institute of Technology, Pasadena, California 91109, U.S.A.

COLBURN (Dr D.S.) Ames Research Center, Moffet Field, California 94035, U.S.A.

*COLE (Dr T.W.) CSIRO, Division of Radiophysics, P.O. Box 76, Epping, NSW 2121, Australia

COLGATE (Dr S.) New Mexico Institute of Mining and Technology, Socorro, New Mexico, U.S.A.

COLLINS II (Dr G.W.) Department of Astronomy, Ohio State University, Columbus, Ohio 43210, U.S.A.

COLLINSON (E.H.) Fox's Corner, Snape, Saxmundham, Suffolk IP17 1SD, U.K.

COLOMB (Dr F.R.) Instituto Argentino de Radioastronomía, Casilla No 5, Villa Elisa, Prov. Bs.As., Argentina

COLOMBO (Prof Dr G.) Instituto Meccanica Applicata Universita di Padova, via F. Marzolo 9, Padova, Italy

*CONDON (Dr J.J.) Department of Physics, Robeson Hall, Virginia Polytechnic Institute, Blacksburg, VA 24061, U.S.A.

*CONKLIN (Dr E.K.) Arecibo Observatory, P.O. Box 995, Arecibo, PR 00612, U.S.A.

CONNES (J.) Service de Calcul Numérique, Observatoire de Meudon, 5, place Janssen, 92190 Meudon

CONNES (P.) Laboratoire du C.N.R.S., Bellevue, France

CONTI (Dr P.S.) JILA, University of Colorado, Boulder, Colorado 80302, U.S.A.

CONTOPOULOS (Prof. Dr G.) Department of Astronomy, Panepistimiopolis, Athens 621, Greece

CONWAY (Dr R.G.) Nuffield Radio Astronomy Laboratories, Jodrell Bank, Macclesfield, Cheshire, U.K.

COOK (Dr A.F.) Smithsonian Astrophysical Observatory, Cambridge, Mass., 02138, U.S.A.

COOK (Prof. A.H.) The Cavendish Laboratory, Madingley Road, Cambridge CB3 OHE, U.K.

COOKE (Dr B.A.) X-Ray Astronomy Group, Physics Department, Leicester University, Leicester, U.K.

COOPER (B.F.C.) CSIRO, Division of Radiophysics, P.O. Box 76, Epping 2121, Australia

CORLISS (Dr C.H.) National Bureau of Standards, Washington D.C. 20234, U.S.A.

CORNEJO (A.) Instituto Nacional de Astrofísica, Optica y Electrónica, Tonantzintla, Puebla, Mexico

*COSMOVICI (Prof. Dr C.B.) Laboratorio di Fisica Cosmica, Università, 73100 Lecce, Italy

COSTAIN (Dr C.C.), Division of Physics, National Research Council of Canada, Ottawa, Canada K1A OR6

COSTAIN (Dr C.H.) Dominion Radio Astrophysical Observatory, Herzberg Institute of Astrophysics, National Research Council of Canada, P.O. Box 248, Penticton, B.C V2A 6K3, Canada

*COSTERO (R.) Instituto de Astronomía, UNAM, Apartado Postal 70-264,
   México 20, D.F. Mexico

COUDER (Dr A.) Observatoire de Paris, 61 Avenue de l'Observatoire,
   F-75014 Paris, France

COUDERC (Dr P.) Observatoire de Paris, 61 Avenue de l'Observatoire,
   F-75014 Paris, France

COUNSELMAN III (Dr Ch.) Dept. of Earth & Planetary Sciences, Massachusetts
   Institute of Technology, Cambridge, Mass. 02139, U.S.A.

*COUPINOT (Dr G.) Observatoire du Pic du Midi, 65200 Bagnères-de-Bigorre, France

COURTES (Dr G.) Observatoire de Marseille, 2 Place Le Verrier,
   13004 Marseille, France

COUSINS (Dr A.W.J.) South African Astronomical Observatory, P.O. Box 9,
   Observatory, Cape 7935, South Africa

COUTEAU (Dr P.C.) Observatoire de Nice, Le Mont-Gros, 06300 Nice, France

COUTREZ (Prof. Dr R.A.J.)   , Institut d'Astronomie et d'Astrophysique,
   Université Libre de Bruxelles, Avenue F.D. Roosevelt, 50,
   B-1050 Bruxelles, Belgium .

COVINGTON (A.E.) National Research Council of Canada, Ottawa,
   Ontario K1A OR6, Canada

COWLEY (Dr A.P.) Astronomy Dept., University of Michigan,
   Ann Arbor, Mich. 48104, U.S.A.

COWLEY (Dr Ch.) Astronomy Department, University of Michigan,
   Ann Arbor, Michigan 48104, U.S.A.

COWLING (Prof. T.G.) Department of Mathematics, The University, Leeds 2, U.K.

COX (Dr A.N.) Box 1663, Los Alamos, New Mexico, U.S.A.

COX (Dr D.P.) Dept. of Physics, University of Wisconsin, Madison,
   Wisconsin 53706, U.S.A.

COX (Dr J.P.) Joint Institute for Laboratory Astrophysics, 1511 University
   Avenue, Boulder, Colorado 80302, U.S.A.

COYNE (Dr G.V., S.J.) Specola Vaticana, Castel Gandolfo, 00040 Rome, Italy

CRAMPTON (Dr D.) Dominion Astrophysical Observatory, Herzberg Institute of
   Astrophysics, National Research Council of Canada, 5071 W. Saanich Road,
   Victoria, B.C. V8X 3X3, Canada

CRAWFORD (Dr D.L.) Kitt Peak National Observatory. 950 North Cherry Avenue,
   Tucson, Arizona 85717, U.S.A.

CRÈZE (Dr M.), Observatoire de Besançon, 41 bis, Avenue de l'Observatoire
   F 25000 Besançon, France

*CRIFO (Dr F.) DASOP, Observatoire de Meudon, 92190 Meudon, France

CRISTALDO (Prof. S.) Viale A.Doria, Città Universitaria, 95123 Catania, Italy

CRISTESCU (Dr C.) Observatoire Astronomique, str. Cutitul de Argint 5,
   Bucarest 28, Roumanie

CROOM (Dr D.L.) Radio and Space Research Station, Ditton Park, Slough, U.K.

CRUIKSHANK (Dr D.) University of Hawaii, Institute of Astronomy,
   Honolulu, Hawaii 96872, U.S.A.

*CRUISE (Dr A.M.) Mullard Space Science Laboratory, Holmbury St. Mary,
    Nr. Dorking, Surrey. U.K.

*CRUTCHER (Dr R.M.) 121 Observatory, University of Illinois, Urbana, IL 61801, U.S.A.

CRUVELLIER (Dr P.) Observatoire de Marseille, Place le Verrier,
    13004 Marseille, France

CRUZ-GONZALEZ (C.) Instituto de Astronomia, UNAM, Apartado Postal 70 - 264,
    México, D.F., México

CSADA (Dr I.K.) Konkoly Observatory, Konkoly Thege U. 13-17, Budapest, XII, Hungary

CUDABACK (Dr D.D.) Radio Astronomy Lab., 633 Campbell Hall, University of
    California, Berkeley, California 94720, U.S.A.

CUFFEY (J.) New Mexico State University, Dept. of Earth Sciences and Astronomy,
    University Park, New Mexico 88001, U.S.A.

CUGNON (Dr P.) Observatoire Royal de Belgique, av. Circulaire 3,
    1180 Bruxelles, Belgium

CULHANE (Dr J.L.) Mullard Space Science Laboratory, Physics Department,
    University College London, Gower Street, London, W.C.1., U.K.

CUNNINGHAM (Dr L.E.) Department of Astronomy, University of California,
    Berkeley 4, California 94720, U.S.A.

CUNY (Dr Y.) Observatoire de Meudon, 92190 Meudon, France

CUPERMAN (Dr S.) Department of Physics and Astronomy, Tel-Aviv University,
    Ramat Aviv, Israel

CUREA (Prof. Dr I.) Recteur de l'Université de Timisoara, Timisoara, Roumania

CZYZAK (Prof. S.J.) Astronomy Department, The Ohio State University,
    Colombus, Ohio 43210, U.S.A.

DACHS (Dr J.) Astronomisches Institut der Ruhr-Universität, Postfach 2148,
    4630 Bochum-Querenburg, Germany F.R.

DADAEV (Dr A.N.) Pulkovo Observatory, Leningrad, U.S.S.R.

DADIC (Dr Z.) Institut zopovijest nauka Jazu, Demetrova 18, Zagreb 3, Yugoslavia

*DAIGNE (Dr G.) DASOP, Observatoire de Meudon, 92190 Meudon, France

DAINTREE (Dr E.J.) Nuffield Radio Astronomy Laboratories, Jodrell Bank,
    Macclesfield, Cheshire, U.K.

DALGARNO (Prof. A.) Smithsonian Astrophysical Observatory, 60 Garden Street,
    Cambridge, Mass., 02138, U.S.A.

DALLAPORTA (Prof. N.) Istituto di Fisica teorica, Universita di Padova, Italy

DALTABUIT (Dr E.) Instituto de Astronomía, UNAM, Apartado Postal 70 - 264,
    México D.F., México

DAMBARA (Dr Takeshi) Geographical Survey Institute, 7-1000, Kamimeguro,
    Meguro-ku, Tokyo, Japan

DANBY (Dr J.M.A.) Yale University Observatory, Box 2023, Yale Station,
    New Haven, Connecticut, U.S.A.

*DANKS (Dr A.C.) Kapteyn Sterrewacht, Mensingheweg 20, Roden, The Netherlands

DANZIGER (Dr I.J.) Harvard College Observatory, 60 Garden Street, Cambridge,
    Mass., 02138, U.S.A.

DA ROCHA VIEIRA (Dr E.) Instituto de Fisica, Universidade Federal do Rio Grande do Sul, 90000 Porto-Alegre, RS, Brazil

DA SILVA (Dr A.V.C.S.) Observatorio Astronomico da Universidade, Santa Clara, Coimbra , Portugal

DA SILVA (Prof. Dr. I.F.) Faculdade de Ciencias, Lisbon, Portugal

*DA SILVA (Dr L.) Observatório Nacional, Rua General Bruce, 586, 20000 Rio de Janeiro, Brazil

DAUBE-KURZEMNIECE (Dr I.A.) Astrophysical Laboratory of the Academy of Sciences of the Latvian SSR, Riga, U.S.S.R.

DAUTCOURT (Dr G.) Zentralinstitut für Astrophysik, Telegrafenberg, 15 Potsdam, Germany F.R.

DAVIDSON (Dr K.) School of Physics and Astronomy, University of Minnesota, Minneapolis, MN 55455, U.S.A.

DAVIDSON (Prof. W.) Mathematics Department, University of Otago, P.O. Box 56, Dunedin, New Zealand

DAVIES (Dr J.G.) Nuffield Radio Astronomy Laboratories, Jodrell Bank, Macclesfield, Cheshire, U.K.

*DAVIES (Dr M.E.) The Rand Corporation, 1700 Main Street, Santa Monica, CA 90406, U.S.A.

DAVIS (Dr C.G.) University of California, Los Alamos Scientific Laboratory, Los Alamos, Box 1663 MS 420, New Mexico 87544, U.S.A.

DAVIS (Prof. L. Jr.) Physics Department, California Institute of Technology, Pasadena, California 91109, U.S.A.

DAVIES (Dr R.D.) Nuffield Radio Astronomy Laboratories, Jodrell Bank, Macclesfield, Cheshire, U.K.

*DAVIS (Dr M.) Department of Astronomy, Harvard University, 60 Garden Street, Cambridge, MA 02138, U.S.A.

DAVIS (Dr J.) Department of Physics, University of Sydney, Sydney, 2006, N.S.W., Australia

DAVIS (Dr M.M.) Arecibo Observatory, P.O. Box 995, Arecibo, Puerto Rico 00612, U.S.A.

DAVIS (Prof. M.S.) The University of North Carolina, Department of Physics, Chapel Hill, N.C. 27514, U.S.A.

DAVIS (Dr R.J.) Smithsonian Astrophysical Observatory, 60 Garden Street, Cambridge, Mass., 02138, U.S.A.

*DAVISON (Dr P.J.N.) Mullard Space Science Laboratory, Holmbury St. Mary, Dorking, Surrey, U.K.

DAWE (Dr J.A.) Royal Observatory, Edinburgh EH 9 3HJ, U.K.

DEBARBAT (S.) Observatoire de Paris, 61, av. de l'Observatoire, 75014 Paris, France

DEBEHOGNE (Dr H.) Observatoire Royal de Belgique, avenue Circulaire, 3, B-1180 Bruxelles, Belgium

DE BIASE (Dr G.A.) Osservatorio Astronomico, Viale del Parco Mellini, 84, Roma, Italy

*De BOER (Dr K.S.) Kapteyn Astronomical Institute, Postbus 800, Groningen, The Netherlands

*DE CASTRO (Dr A.) Observatorio Astronómico Nacional, Madrid, Spain

DECAUX (B.) Bureau International de l'Heure, Observatoire de Paris,
61, Avenue de l'Observatcire, 75014 Paris, France

DEEMING (Dr T.J.) Astronomy Department, 404 Physics Building, University of
Texas, Austin, Texas 78712, U.S.A.

*DEERENBERG (Dr A.J.M.) Cosmic Ray Working Group, Huygens Laboratory,
Wassenaarseweg 78, Leiden, The Netherlands

DE FELICE (Dr F.) Osservatorio Astronomico, Vicolo dell'Osservatorio,
Padova, Italy

DE FREITAS PACHECO (Dr J.A.) Observatorio de Sao Paulo - USP,
Caixa Postal 30627, Sao Paulo, Brazil

DE GRAAF (Dr T.) Kapteyn Astronomical Laboratory, Postbus 800,
Groningen, The Netherlands

DE GRAAFF (Dr W.) Laboratorium voor Ruimte-Onderzoek,
Beneluxlaan 21, Utrecht, The Netherlands

*De GRAAUW (Dr T.) ESTEC, Space Science Dept., Noordwijk,
Domeinweg, The Netherlands

DE GROOT (M.J.H.) Armagh Observatory, Armagh BT61 9DG, N. Ireland

De GROOT (Dr T.) Sterrewacht "Sonnenborgh", Servaas Bolwerk 13,
Utrecht, The Netherlands

DEHARVENG (Dr M.) L.A.S. - Traverse du Siphon, Les Trois Lucs,
13012 Marseille, France

DEINZER (Dr W.) Universitäts-Sternwarte, 34 Göttingen,
Geismarlandstrasse 11, Germany F.R.

DE JAGER (Prof. Dr C.) Director, The Astronomical Institute, Space Research
Laboratory, Beneluxlaan 21, Utrecht, The Netherlands

*De JAGER (Dr G.) Department of Physics, Rhodes University, P.O. Box 94,
Grahamstown 6140, South Africa

DEJAIFFE (Dr R.J.) Observatoire Royal de Belgique, Av. Circulaire, 3,
B-1180 Bruxelles, Belgium

DE JONG (Dr T.) Sterrewacht, Leiden, The Netherlands

DE JONGE (Dr J.K.) Department of Astronomy, University of Pittsburgh,
Riverview Park, Pittsburgh, Pa. 15214, U.S.A.

*DEKKER (Dr E.) Sterrewacht Leiden, Huygens Laboratorium, Wassenaarseweg 78,
Leiden, The Netherlands

De KORT (S.J. Rev. Dr J.) Department of Astronomy, Faculty of Science,
Catholic University, Toernooiveld, Nijmegen, 6805, The Netherlands

*De KORTE (Dr P.A.J.) Cosmic Ray Working Group, Huygens Laboratory,
Wassenaarseweg 78, Leiden, The Netherlands

DELACHE (Dr Ph.) Observatoire de Nice, Le Mont-Gros, 06-Nice, France

*DE LA HERRAN (Ing. J.R.) Instituto de Astronomía, UNAM, Apartado Postal 70-264,
Mexico 20, D.F. Mexico

DELANNOY (J.) Observatoire de Paris, Section d'Astrophysique,
92140 Meudon, France

*DE LA NOE (Dr J.) DASOP, Observatoire de Meudon, 92190 Meudon, France

*DE LA REZA (Dr R.) Observatoire de Genève, 1290 Sauverny, Switzerland

DELBOUILLE (Prof. Dr L.) Institut d'Astrophysique, Université de Liège,
    Avenue de Cointe, 5, B-4200 Cointe-Ougrée, Belgium

DELHAYE (Prof. J.) Observatoire de Paris, 61, Avenue de l'Observatoire,
    75014 Paris, France

DE LOORE (Dr C.W.H.) Astrofysisch Instituut, Vrije Universiteit Brussel,
    F.D. Rooseveltlaan, 50, B-1050 Brussel, Belgium

DELPLACE (Dr A.M.) Observatoire de Paris, Section d'Astrophysique de Meudon,
    92190 Meudon, France

DELSEMME (Dr A.H.) Department of Physics and Astronomy, University of Toledo,
    Toledo, Ohio 43606, U.S.A.

DE MARCUS (Prof. W.C.) Department of Physics, University of Kentucky,
    Lexington, Kentucky, U.S.A.

*DEMARCQ (J.) Laboratoire d'Optique, Observatoire de Nice, B.P. 252,
    06007 Nice-Cedex, France

DEMARQUE (Prof. P.R.) Yale University Observatory, Box 2023, Yale Station,
    New Haven, Connecticut 06520, U.S.A.

DEMENKO (Dr A.A.) Kiev State University, Astronomical Observatory and Chair of
    Astronomy, 252053 Kiev, U.S.S.R.

DEMERS (Dr S.) Institute of Astronomy, Laurentian University, Sudbury,
    Ontario P3E 2C6, Canada

DEMIN (Prof. Dr V.G.) Moscow University, Leninskie Gory, Moscow, V-234, U.S.S.R.

DE MOTTONI Y PALACIOS (Dr G.) Via Fratelli Rosselli 15-23, 16145 Genova, Italy

DENNISON (Dr E.W.) Mount Wilson and Palomar Observatories, 813 Santa Barbara
    Street, Pasadena, California, U.S.A.

DENNISON (Dr P.A.) Department of Physics, The University of Adelaide,
    Adelaide, S.A. 5001, Australia

DENISSE (Dr J.F.) Observatoire de Paris, 61, Avenue de l'Observatoire,
    75014 Paris, France

DENT (Dr W.A.) Physics and Astronomy Dept., University of Massachusetts,
    Amherst, Massachusetts 01002, U.S.A.

DEPRIT (Prof. A.) Department of Mathematical Sciences, University of
    Cincinnati, Cincinnati, Ohio 45221, U.S.A.

DE SABBATA (Prof. Dr V.) Instituto di Fisica Universita di Bologna,
    via Irnerio 46, Bologna, Italy

DESESQUELLES (Dr J.) Laboratoire de Spectrométrie Ionique et Moléculaire de
    l'Université de Lyon, 43, Boulevard du 11 Novembre 1918,
    69 Villeurbanne, France

DEUBNER (Dr F.L.) Fraunhofer Institut, Schöneckstr. 6, 7800 Freiburg,
    Germany F.R.

DEUTSCH (Prof. Dr A.N.) Pulkovo Observatory, Leningrad-M140, U.S.S.R.

DEUTSCHMANN (Dr W.A.) Department of Physics and Astronomy, Dickinson College,
    Carlisle, Pa 17013, U.S.A.

DE VAUCOULEURS (Prof. G.) Department of Astronomy, University of Texas, Austin 12, Texas 78412, U.S.A.

DE VEGT (Dr Ch.) Sternwarte Bergedorf, Gojenbergsweg 112, 205 Hamburg 80, Germany F.R.

DEVINNEY (E.J., Jr.) University of Southern Florida, Tampa, Florida 33612, U.S.A.

DEWHIRST (Dr D.W.) The Observatories, Madingley Road, Cambridge, U.K.

DeWITT (Dr B.S.) Dept. of Physics, University of Texas at Austin, Austin, Texas 78712, U.S.A.

*DeWITT (Dr C.) Department of Astronomy, The University of Texas at Austin, 15.212 R.L. Moore Hall, Austin, TX 78712, U.S.A.

DeWITT (Dr J.H.) Arthur J. Dyer Observatory, Vanderbilt University, Nashville 5, Tennessee, U.S.A.

DeYOUNG (Dr D.S.) National Radio Astronomy Obser., Edgemont Road, Charlottesville, Virginia 22901, U.S.A.

DEZSO (Prof. L.) Heliophysical Observatory of the Hungarian Academy of Sciences, P.O. B. 30, H-4010 Debrecen, Hungary

DIBAY (Dr E.A.) Sternberg Astronomical Institute, Moscow, U.S.S.R.

DICKE (Dr R.H.) P.O. Box 708, Jadwin Hall, Princeton University, Princeton, New Jersey 08540, U.S.A.

DICKEL (Dr H.R.) Department of Astronomy, University of Illinois Observatory, Urbana, Illinois 61801, U.S.A.

DICKEL (Dr J.R.) University of Illinois Observatory, Urbana, Illinois 61801, U.S.A.

DICKENS (R.J.) Royal Greenwich Observatory, Herstmonceux Castle, Hailsham, Sussex, U.K.

*DICKINSON (Dr D.F.) Center for Astrophysics, 60 Garden Street, Cambridge, MA 02138, U.S.A.

DIECKVOSS (Prof. Dr W.) Hamburger Sternwarte, Hamburg-Bergedorf, Germany F.R.

DIETER (Dr N.H.) Radio Astronomy Laboratory, University of California, Berkeley, California 94720

DIMITROFF (Prof. G.Z.) R.F.D. No 1, Hartland, Vermont 05048, U.S.A.

DINESCU (Dr A.) Observatorul Astronomic, Strada Cutitul de Argint 5, Bucuresti 28, Roumania

DINGENS (Prof. Dr P.) Sterrenkundig Observatorium, Rijksuniversiteit Gent Krijgslaan 271, B 9000 Gent, Belgium

DINGLE (Prof. H.) 104, Downs Court Road, Purley, Surrey, U.K.

DINULESCU (Prof. N.) Soseana Kiseleff No 13, Bucarest, secteur 1, Roumania

*DIONYSIOU (Dr D.) Astronomy Department, University of Athens, Panepistimiopolis, Athens - 621, Greece

DIRIKIS (Dr M.A.) Astronomical Observatory of the Latvian University, Latvian SSR, Riga, U.S.S.R.

DISNEY (Dr M.) Mount Stromlo and Siding Spring Observatories, Private Bag, Woden P.O., A.C.T. 2606, Australia

DIVAN (Dr L.) Institut d'Astrophysique, 98bis Boulevard Arago, 75014 Paris, France

DIVARI (Dr N.B.) Odessa Politechnical Institute, Odessa, U.S.S.R.

DIXON (Dr M.E.) Department of Astronomy, University of Edinburgh, Edinburgh, U.K.

DIXON (Dr R.S.) The Ohio State University Radio Observatory, 2015 Neil Avenue, Colombus, Ohio 43210, U.S.A.

DIZER (Prof. Dr M.) Astronomical Observatory, Kandilli, Çengelköy, Istanbul, Turkey

*DJUROVIĆ (Dr D.) Faculté des Sciences naturelles et mathématiques de Belgrade, Volgina 7, 11050, Yugoslavia

DJURKOVIĆ (P.M.) Astronomical Observatory, Volgina 7, Beograd, Yugoslavia

DLUZHNEVSKAYA (Dr O.B.) Astronomical Council, USSR Academy of Sciences, Pyatnitskaya ul. 48, 109017 Moscow. U.S.S.R.

DOAN (Dr N.H.) Observatoire de Lyon, 69 St. Genis Laval, France

DOAZAN (V.) Observatoire de Paris, 61, Avenue de l'Observatoire, 75014 Paris, France

DOBRITSCHEV (V.M.) Department of Astronomy, "7th November" Street 1, Sofia, Bulgaria

DOBRONRAVIN (Dr P.P.) Crimean Astrophysical Observatory, USSR Academy of Sciences, p/o Nauchny, Crimea, U.S.S.R.

DOBROVOLSKIJ (Dr O.V.) Institute of Astrophysics, Dushanbe, U.S.S.R.

DOBRZYCKI (Dr Jerzy) Institute of the History of Science and Technics, Nowy Swiat 72, Warszawa, Poland

*DODD (Dr R.J.) Royal Observatory Edinburgh, Blackford Hill, Edinburgh EH9 3HJ, U.K.

D'ODORICO (Dr S.) Osservatorio Astrofisico, Asiago (Vicenza), Italy

DODSON-PRINCE (Dr H.W.) McMath Hulbert Observatory, University of Michigan, Pontiac 4, Michigan, U.S.A.

DOGAN (Dr N.) Faculty of Science, Department of Astronomy, Ankara, Turkey

DOHERTY (Dr L.H.) Herzberg Institute of Astrophysics, National Research Council of Canada, Ottawa, Ontario K1A OR6, Canada

DOHERTY (Dr L.R.) Astronomy Department, Sterling Hall, University of Wisconsin, Madison, Wisconsin 53706, U.S.A.

*DOKUCHAEVA (Dr O.D.) Sternberg Astronomical Institute, Moscow 117234, U.S.S.R.

DOLAN (Dr J.F.) Code 682, Goddard Space Flight Center, Greenbelt, MD 20771, U.S.A.

DOLGINOV (Prof. Dr A.Z.) Physical-Technical Institute, USSR Academy of Sciences, Leningrad, U.S.S.R.

DOLIDZE (Dr M.V.) Abastumani Observatory, Abastumani, Georgia, U.S.S.R.

DOLLFUS (Dr A.) Observatcire de Paris, Section d'Astrophysique, 92190 Meudon, France

DOMINKO (Prof. Dr F.) Saranovićeva 11, Ljubljana, Yugoslavia

DOMMANGET (Dr J.) Chef du Département d'Astrométrie et de Mécanique Céleste, Observatoire Royal de Belgique, Avenue Circulaire 3, B-1180 Bruxelles, Belgium

DONITCH (N.) address unknown

DONN (Dr B.D.) Code 613, Goddard Space Flight Center, Greenbelt, Maryland, U.S.A.

*DOPITA (Dr M.A.) Mt. Stromlo and Siding Spring Observatories, Private Bag,
Woden P.O. A.C.T. 2606, Australia

*DORMAND (Dr J.R.) Maths Dept., Teesside Polytechnic, Middlesbrough,
Cleveland TS1 3BA, U.K.

DOROSHKEVICH (Dr A.G.) Institute of Applied Mathematics, USSR Academy of
Sciences, Moscow, U.S.S.R.

DORSCHNER (Dr J.) University Observatory, Schillergässchen 2, 69 Jena,
Germany D.R.

*DOSCHEK (Dr G.A.) Naval Research Laboratory, E.O. Hulburt Center for Space
Research, Code 7125.8, Washington DC 20375, U.S.A.

DOS REIS (Prof. M.) Director of the Observatory, Coimbra, Portugal

DOSSIN (Dr F.) European Southern Observatory, Am Bahnhof 21, 205 Hamburg 80,
Germany F.R.

DOUGHTY (Dr N.A.) Department of Physics, University of Canterbury,
Private Bag, Christchurch, New Zealand

DOUGLAS (Dr A.E.) Herzberg Institute of Astrophysics, National Research
Council of Canada, Ottawa, Ontario K1A OR6, Canada

DOUGLAS (Dr A.V.) 127 King Street W., Kingston, Ontario, Canada

DOUGLAS (Prof. N.) Department of Astronomy, The University of Texas,
Austin, Texas, U.S.A.

*DOWNES (Dr D.) Max-Planck-Institut für Radioastronomie, Auf dem Hügel 69,
53 Bonn, Germany F.R.

DRAKE (Prof. F.D.) Department of Astronomy, Cornell University,
Ithaca, New York 14850, U.S.A.

DRÂMBĂ (Prof. Dr C.) Observatoire de Bucarest, 5 Rue Cutitul de Argint,
Bucarest 28, Roumania

*DRAPATZ (Dr S.) Max-Planck-Institut für Physik und Astrophysik, Institut für
Extraterrestrische Physik, 8046 Garching bei München, Germany F.R.

*DRAVINS (Dr D.) Lund Observatory, Svanegatan 9, S-222 24 Lund, Sweden

DRAVSKIKH (Dr A.F.) Leningrad Branch of Special Astrophysical Observatory,
Academy of Sciences, Pulkovo, Leningrad 196140, U.S.S.R.

DRESSLER (Dr K.) Labor. Phys. Chem. E.T.H., 8000 Zürich, Switzerland

DRILLING (Dr J.S.) Louisiana State University, Baton Rouge, Louisiana 70803, U.S.A.

DROFA (Dr V.K.) Astronomical Observatory, Kiev University, Kiev, U.S.S.R.

DRYER (Dr M.) Space Environment Laboratory, National Oceanic & Atmospheric Adm.,
Boulder, Colorado 80302, U.S.A.

DUBINSKIJ (Dr B.A.) Scientific Council for "Radio Astronomy", USSR Academy of
Sciences, Moscow, U.S.S.R.

DUBOSHIN (Prof. Dr G.N.) Sternberg Astronomical Institute, Moscow B-234, U.S.S.R.

DUBOV (Dr E.E.) Crimean Astrophysical Observatory, USSR Academy of Sciences,
p/o Nauchny, Crimea, U.S.S.R.

DUCHESNE (Dr M.) Observatoire de Paris, 61, Avenue de l'Observatoire,
75014 Paris, France

DUFAY (M.) Institut de Physique Générale de la Faculté des Sciences de Lyon
18, Quai Claude Bernard, Lyon, France

DUFLOT (Dr M.) Observatoire de Marseille, 2, Place Le Verrier,
Marseille 4e, France

DUFLOT (R.) Chargée de Recherches CNRS, Observatoire de Marseille,
2, place Le Verrier 13 - Marseille, 4e, France

DULEY (Prof. W.W.) Department of Physics, York University, Toronto,
Ontario M3J 1F3, Canada

DULK (Dr G.A.) Department of Astro-Geophysics, University of Colorado,
Boulder, Colorado 80302, U.S.A.

DUMONT (Dr R.) Observatoire de Bordeaux, 33270 Floirac, France

DUMONT (Dr S.) Institut d'Astrophysique, 98 bis Bd Arago, 75014 Paris, France

DUNCOMBE (Dr R.L.) Dept. of Aerospace Engineering, University of Texas at Austin
Austin, Texas 78712, U.S.A.

*DUNHAM (Dr D.W.) Computer Science Corporation, 10th Floor, 8728 Colesville Road,
Silver Spring, Maryland 20910, U.S.A.

DUNHAM (Dr T. Jr.) P.O.Box 135, Chocorua, New Hampshire 03817, U.S.A.

DUNKELMAN (Dr L.) Astrophysics Branch, Code 613, Goddard Space Flight Center,
Greenbelt, Maryland 20771, U.S.A.

DUNN (Dr R.B.) Sacramento Peak Observatory, Sunspot, New Mexico, U.S.A.

DUPREE (Dr A.K.) Harvard College Observatory, 60 Garden Street,
Cambridge, Mass. 02138, U.S.A.

*DuPUY (Dr D.L.) Department of Astronomy, Saint Mary's University, Halifax,
Nova Scotia, Canada B3H 3C3

DURNEY (Dr B.) High Altitude Observatory, Boulder, Colorado 80302, U.S.A.

DURRANT (Dr Ch. J.) Fraunhofer Institut, Schöneckstrasse 6, D78 Freiburg-im-Breisgau
Germany, F.R.

*DÜRST (Dr J.) Swiss Federal Observatory, Schmelzbergstr.25, CH-8006 Zürich,
Switzerland

DUTHIE (Dr J.G.) The University of Rochester, Department of Physics and
Astronomy, Rochester, New York 14627, U.S.A.

*DVORAK (Dr R.) Universitäts-Sternwarte, Universitätsplatz 5, A-8010 Graz,
Austria

DVORYASHIN (Dr A.S.) Crimean Astrophysical Observatory, USSR Academy of
Sciences, Crimea- U.S.S.R.

*DWORETSKY (Dr M.M.) Dept. of Physics and Astronomy, University College London,
Gower Street, London WC1E 6BT, U.K.

DYCK (Dr M.) Kitt Peak National Observatory, Tucson, Arizona 85717, U.S.A.

DYER (Dr E.R.) National Academy of Sciences, 2101 Constitution Avenue, N.W.,
Washington D.C. 20418, U.S.A.

DYSON (Dr F.J.) Institute for Advanced Studies, Princeton, New Jersey 08540,
U.S.A.

DYSON (Dr J.E.) Department of Astronomy, The University, Manchester M 13 9 PL. U.K.

DZHAPIASHVILI (Dr V.P.) Vice-Director Abastumani, Astrophysical Observatory, Abastumani, Georgia, U.S.S.R.

DZIEMBOWSKI (Dr W.) Polish Academy of Sciences, Institute of Astronomy, Al. Ujazdowskie 4, Warszawa, Poland

DZIGVASHVILI (Dr R.M.) Abastumani Astrophysical Observatory, Georgian Academy of Sciences, Abastumani, U.S.S.R.

EBBIGHAUSEN (Prof. E.G.) Department of Physics, University of Oregon, Eugene, Oregon, U.S.A.

EDDY (Dr J.) High Altitude Observatory, Boulder, Colorado 80302, U.S.A.

EDLÉN (Prof. B.) Institute of Physics, Sölvegatan 14, S-223 63 Lund, Sweden

EDMONDS (Dr F.N., Jr.) Department of Astronomy, The University of Texas, Austin, Texas 78712, U.S.A.

EDMONDSON (Dr F.K.) Swain Hall West 319A, Indiana University, Bloomington, Indiana 47401, U.S.A.

EDWARDS (Prof. P.J.) Dept. of Physics, University of Otago, P.O.Box 56, Dunedin, New Zealand

EELSALU (Dr H.) W. Struve Astrophysical Observatory, Tartu, Tõravere, Estonia, U.S.S.R.

EFREMOV (Dr Yu. I.), Institute for Applied Mathematics, USSR Academy of Sciences 125047, Moscow, USSR.

EFREMOV (Dr Yu. N.)Sternberg Astronomical Institute, Moscow V-234, 117234, U.S.S.R.

EGGLETON (Dr P.P.) Institute of Theoretical Astronomy, Madingley Road, Cambridge, U.K.

*EHLERS (Dr J.) Max-Planck-Institut für Physik und Astrophysik, Föhringer Ring 6, 8000 München 40, Germany, F.R.

*EHRGMA (Dr E.V.) Astronomical Council, Pyatnitskaya 48, Moscow 109017, U.S.S.R.

EICHHORN-von WURMB. (Prof.H.K.) Department of Astronomy, University of South Florida, Tampa, Florida 33549, U.S.A.

EINSATO (Dr J.) W. Struve Astrophysical Observatory, Tartu, Tõravere, Estonia, U.S.S.R.

*EINICKE (Dr O.H.) Copenhagen University Observatory, Østervoldgade 3, DK-1350 Copenhagen, Denmark

EKERS (Dr R.D.) Kapteyn Laboratorium, Postbus 800, Groningen, The Netherlands

EL-BAZ (Dr F.) Center for Earth and Planetary Studies, National Air and Space Museum, Smithsonian Institution ,Washington DC, 20560, U.S.A.

ELFORD (Dr W.G.) Department of Physics, University of Adelaide, Adelaide, Australia

ELGARØY (Dr. Ö.) Institute of Theoretical Astrophysics, University of Oslo, Blindern, Oslo 3, Norway

*ELLDÉR (Dr A.J.) Onsala Space Observatory, S-430 34 Onsala, Sweden

*ELLIOT (Dr H.) Blackett Laboratory (Physics Dept.) Imperial College,
London SW7 2AZ, U.K.

ELLIOTT (Dr I.) Dunsink Observatory, Castleknock, Co. Dublin, Ireland

*ELLIOTT (Dr J.L.) Cornell University, Center for Radiophysics and Space
Research, Space Sciences Building, Ithaca, NY 14853, U.S.A.

ELLIS (Dr G.F.R.) Department of Applied Mathematics, University of Cape Town,
Private Bag, Rondebosch, 7700 Cape, S. Africa

ELLIS (Prof. G.R.A.) University of Tasmania, P.O. Box 2520, Hobart
Tasmania, Australia

ELLYETT (Prof. C.D.) Department of Physics, University of Newcastle,
Newcastle, Australia

*
EL-RAEY (Dr M.E.) Faculty of Science, Alexandria University, Alexandria,
Arab Republic of Egypt

ELSÄSSER (Prof. Dr H.) Max-Planck-Institut für Astronomie und Landessstern-
warte, 6900 Heidelberg-Königstuhl, Germany F.R.

*EL-SHAARAWY (Dr M.B.) Helwan Observatory, Helwan, Arab Republic of Egypt

ELSMORE (B.) Mullard Radio Observatory, Cavendish Laboratory, Free School
Lane, Cambridge, U.K.

*ELST (Dr E.W.) Royal Belgian Observatory, Ringlaan 3, 1180 Brussels, Belgium

ELSTE (Dr G.) Department of Astronomy, University of Michigan, Ann Arbor,
Michigan 48104, U.S.A.

ELSTON (Prof. W.E.) University of New Mexico, Department of Geology,
Albuquerque, New Mexico 87106, U.S.A.

ELVEY (Dr C.T.) 5359 E. Hawthorne Street, Tucson, Arizona 85711, U.S.A.

ELVIUS (Dr A.M.) Stockholm Observatory, Saltsjöbaden, S-133 00 Sweden

ELVIUS (Prof.Dr T.) Director, Astronomical Observatory, S-222 24 Lund, Sweden

ELWERT (Prof. Dr G.) Astronomisches Institut der Universität, Gmelinstr. 6,
Tübingen, Germany, F.R.

EL'YASBERG (Prof. Dr P.E.) Institute of Cosmical Research, USSR Academy of
Sciences, Moscow, USSR

EMINZADE (Dr T.A.) Shemakha Astrophysical Observatory, Azerbajdzh Academy
of Sciences, Baku, U.S.S.R.

ENCRENAZ (Dr P.) Observatoire de Meudon, 92190 Meudon, France

ENGELHARD (Prof. Dr E.J.G.) Sackring 34, 33 Braunschweig, Germany F.R.

ENGIN (Dr S.) Faculty of Science, Department of Astronomy, Ankara, Turkey

ENGVOLD (O.) Institute of Theoretical Astrophysics, University of Oslo
P.O.Box 1o29, Blindern, Oslo 3, Norway

ENOME (Dr S.) Toyokawa Observatory, Toyokawa, 442, Japan

ENSLIN (Dr H.) Deutsches Hydrographisches Institut, 2002 Hamburg 4,
Germany F.R.

*EPPS (Dr H.W.) Department of Astronomy, University of California,
Los Angeles, CA 90024, U.S.A.

EPSTEIN (Dr E.E.) Aerospace Corporation, P.O.Box 95085, Los Angeles, California 90045, U.S.A.

EPSTEIN (Dr I.) Rutherford Observatory, Columbia University, New York 27, N.Y. U.S.A.

*EPSTEIN (Dr R.) Nordita, Blegdamsvej 17, DK- 2100 Copenhagen, Denmark

*ERDI (Dr B.) Astronomical Department of Lorand Eötvös University, 1083 Budapest, Kún Béla tér 2. , Hungary

ERICKSON (Dr W.C.) University of Maryland, Department of Physics and Astronomy, College Park, Maryland, U.S.A.

ERIKSEN (G.) Institute of Theoretical Astrophysics, University of Oslo, Blindern, Oslo 3, Norway

ERPYLEV (Dr N.P.) Astronomical Council, USSR Academy of Sciences, Pyatnitskaya ul. 48, 109017, Moscow, U.S.S.R.

ERUSHEV (Dr N.N.) Crimean Astrophysical Observatory, USSR Academy of Sciences, P/O Nauchniy, 334413 Crimean, U.S.S.R.

ESHLEMAN (Prof. V.R.) Radioscience Laboratory, Stanford University, Stanford, California, U.S.A.

ESIPOV (Dr V.F.) Sternberg Astronomical Institute, Moscow- U.S.S.R.

ESKIOGLU (Dr A.Nihat) P.K.14, Maltepe , Ankara, Turkey

*ESPOSITO (Dr F.P.) Department of Physics, University of Cincinnati, Cincinnati, OH 45221, U.S.A.

ESSEN (Dr L.) National Physical Laboratory, Teddington, Middlesex, U.K.

*EVANGELIDIS (Dr E.) Astronomical Department, University of Patras, Patras, Greece

EVANS (Dr D.S.) Department of Astronomy, University of Texas, Austin, Texas 78712, U.S.A.

EVANS (Dr J.V.) MIT Lincoln Laboratory, Millstone/Haystack Observatory, Lexington, Massachusetts 02173, U.S.A.

EVANS (Dr J.W.) Director, Sacramento Peak Observatory, Sunspot, New Mexico, U.S.A.

*EVANS (Dr K.D.) X-Ray Astronomy Group, Physics Dept., The University, Leicester, LE1 7RN,U.K.

*EVANS (Dr N.J.) Department of Astronomy, The University of Texas at Austin, Austin, Texas 78712, U.S.A.

*EVANS (Dr R.G.) SRC Appleton Laboratory, Astrophysics Research Division, Culham Laboratory, Abingdon,Oxon, U.K.

EVDOKIMOV (Dr Yu.V.) Astronomical Observatory, Kazan University, Kazan, U.S.S.R.

EVERHART (Dr E.) Department of Physics, University of Denver, 2115 S. University Bd. Denver, Colorado 80210, U.S.A.

EVIATAR (Prof. A.) Tel Aviv University, Ramat Aviv, Israel

EWEN (Dr H.I.) 60 Beaver Road, Weston, Massachusetts, U.S.A.

*EWING (Dr M.S.) California Institute of Technology, Owens Valley Radio
   Observatory, Pasadena, CA 91125, U.S.A.

EZER (Prof. Dr. D.)Department of Physics, Middle East Technical University
   Ankara, Turkey

EZERSKIJ (Dr V.I.) Astronomical Observatory, Kharkov University,
   Kharkov, U.S.S.R.

*FABER (Dr S.M.) Lick Observatory, University of California,
   Santa Cruz - CA 95064, U.S.A.

*FABIAN (Dr A.C.) Institute of Astronomy, Madingley Road, Cambridge, CB3 OHA, U.K.

FABRE (Dr H.) Observatoire de Nice, Le Mont-Gros, 06 Nice, France

*FAHLMAN (Dr G.G.) Department of Geophysics and Astronomy, University of British
   Columbia, 2075 Wesbrock Place, Vancouver, B.C., Canada V6T 1W5

*FAIR (Dr H.J.) Institut für Astrophysik der Universität Bonn, Auf dem Hügel 71,
   D-53 Bonn, Germany F.R.

*FAHY (Dr E.F.) Physics Department, University College, Cork, Ireland

FALCIANI (Dr R.) Osservatorio Astronomico di Capodimonte,
   Via Moiariello, 16, 80131 Naples, Italy

*FALLE (Dr S.A.E.G.) Max-Planck-Institut für Physik und Astrophysik
   Föhringer Ring 6, 8 München 40, Germany, F.R.

FALLER (Dr J.) Joint Institute for Laboratory Astrophysics,
   Boulder, Colorado 80302, U.S.A.

FALTHAMMAR (Prof. C.G.) Department of Plasma Physics, Royal Institute of
   Technology, S-100 44 Stockholm, Sweden

*FANTI (C.G.) Laboratorio di Radioastronomia, Via Irnerio 46,
   40126 Bologna, Italy

FANTI (Dr R.) Physical Institute of the University of Bologna,
   via Irnerio 46, Bologna, Italy

FARAGGIANA (Dr R.) Astronomical Observatory Trieste, via G.B. Tiepolo 11,
   Trieste, Italy

FATCHIKHIN (Dr N.V.) Pulkovo Observatory, Leningrad, U.S.S.R.

FAULKNER (Dr D.J.) Mount Stromlo and Siding Spring Observatories, Private Bag,
   Woden P.O., A.C.T. 2606, Australia

FAULKNER (Dr J.) Lick Observatory, University of California,
   Santa Cruz, California 95060, U.S.A.

*FAWELL (Dr D.R.) University of London Observatory, Mill Hill Park,
   London, N.W.7, U.K.

FAY (Dr T.) Indiana University, Department of Astronomy, Bloomington,
   Indiana 47401, U.S.A.

FAZIO (Dr G.G.) Smithsonian Astrophysical Observatory, 60 Garden Street,
   Cambridge, Mass., 02138

FEAST (Dr M.W.) South African Astronomical Observatory, P.O. Box 9,
   Observatory, Cape 7935, South Africa

FEAUTRIER (Dr N.) Observatoire de Paris, Section d'Astrophysique de Meudon,
   92190 Meudon, France

*FECHTIG (Dr H.) Max-Planck-Institut für Kernphysik, Postfach 103980, D-69 Heidelberg, Germany F.R.

FEDOROV (Prof. E.P.) The Main Astronomical Observatory of the Ukrainian Academy of Sciences, Kiev-127, Goloseevo, U.S.S.R.

FEDYNSKIJ (Prof. Dr V.V.), Astronomical Council, USSR Academy of Sciences, Pyatnitskaya ul. 48, 109017 Moscow, U.S.S.R.

FEHRENBACH (Prof. Ch.) Directeur de l'Observatoire de Haute Provence, Saint Michel l'Observatoire, 04300 Forcalquier, France

FEINSTEIN (Dr A.) Observatorio Astronómico, La Plata, Argentina

*FEISSEL (M.) DANOF, Observatoire de Paris, 61, Avenue de l'Observatoire, 75014 Paris, France

FEIX (Dr G.) Astronomisches Institut der Ruhr-Universität, Postfach 2148, 4630 Bochum-Querénburg, Germany F.R.

*FEJES (Dr I.) FÖMI Satellite Geodetic Observatory, H-1373 Budapest Pf.:546, Hungary

FELDMAN (Dr P.A.) Herzberg Institute of Astrophysics, National Research Council of Canada, Ottawa, Ontario K1A OR6, Canada

FELDMAN (Dr U.) Department of Physics and Astronomy, Tel-Aviv University, Ramat-Aviv, Israel

FELENBOK (P.) Observatoire de Paris, Section d'Astrophysique, 92190 Meudon, France

FELLGETT (Prof. P.B.) Department of Applied Physical Sciences, University of Reading, Whiteknights Park, Reading, Berks., U.K.

FELLI (Dr M.) Osservatorio Astrofisico di Arcetri, Largo Enrico Fermi, 5, Firenze, Italy

FELTEN (Dr J.E.) Code 602, Goddard Space Flight Center, Greenbelt, MD 20771, U.S.A.

FENKART (Dr R.P.) Astronomisches Institut der Universität Basel, Venusstrasse 7, CH-4102 Binningen, Switzerland

FENTON (Dr K.B.) Department of Physics, University of Tasmania, Hobart, Tasmania 7001, Australia

FERNIE (Dr J.D.) David Dunlap Observatory, University of Toronto, P.O.Box 360 Richmond Hill, Ontario L4C 4Y6, Canada

*FERRARI (A.) Istituto di Fisica Generale dell' Università di Torino, Corso Massimo d'Azeglio 46, 10125 Torino, Italy

FERRARI D'OCCHIEPPO (Prof. Dr K.) Universitäts-Sternwarte, Türkenschanzstrasse 17, A-1180 Wien, Austria

FERRAZ-MELLO (Prof. Dr S.) Universidade de Sao Paulo, Departamento de Astronomia, Caixa Postal 30627, 01000 - Sao Paulo, SP, Brazil

*FEYNMAN (Dr J.) NCAR, High Altitude Observatory, P.O. Box 3000, Boulder, CO 80303, U.S.A.

*FIALA (Dr A.) Naval Observatory, NAO, Washington, D.C. 20390, U.S.A.

*FICARRA (A.) Lab. Nazionale di Radioastronomia, via Irnerio 46, Bologna, Italy

FICHERA (Prof. E.) Instituto Universitario Navale, Astronomia Generale e Sferica, Via Amm. Acton, I-80100 Napoli, Italy

FICHTEL (Dr C.E.) NASA, Goddard Space Flight Center, Greenbelt, Maryland 20771, U.S.A.

FIELD (Dr G.B.) Harvard College Observatory, Center for Astrophysics, 60 Garden Str., Cambridge, Mass. 02138, U.S.A.

FIELDER (Dr G.) Lunar and Planetary Unit, Department of Environmental Sciences, University of Lancaster, Bailrigg, Lancaster, U.K.

FILLIOZAT (Dr J.) Professeur au Collège de France, 11, Place Marcelin Berthelot, 75005 Paris, France

FINDLAY (Dr J.W.) National Radio Astronomy Observatory, Edgemont Road, Charlottesville, Virginia, U.S.A.

FINK (Dr U.) Lunar and Planetary Laboratory, University of Arizona, Tucson, Arizona 85721, U.S.A.

FINN (Dr G.D.) Institute for Astronomy, 2525 Correa Road, Honolulu, Hawaii 96822, U.S.A.

FINSEN (Dr W.S.) P.O. Box 4204, Johannesburg 2000, South Africa

FIREMAN (Dr E.L.) Smithsonian Astrophysical Observatory, 60 Garden Street, Cambridge, Mass., 02138, U.S.A.

FIRMANI (Dr C.) Instituto de Astronomia, UNAM, Apartado Postal 70 - 264, Mexico D.F., Mexico

FIRNEIS (Dr M.G.) Institut für Theoretische Astronomie, Türkenschanzstr. 17, 1180 Wien, Austria

FIROR (Dr J.W.) High Altitude Observatory of the University of Colorado, Boulder, Colorado, U.S.A.

FISCHEL (Dr D.) Code 671, Goddard Space Flight Center, Greenbelt, Maryland 20771, U.S.A.

FISCHER (Prof. Dr P.L.) Universitäts-Sternwarte Wien, A-1180 Wien, Türkenschantzstrasse 17, Austria

*FISHER (Dr J.R.) National Radio Astronomy Observatory, P.O. Box 2, Green Bank, WV 24944, U.S.A.

FISHER (P.C.) Lockheed Palo Alto Research Laboratory, 3251 Hanover Street, Palo Alto, California 94304, U.S.A.

FISHER (Dr R.R.) Institute for Astronomy, University of Hawaii, 2525 Correa Road, Honolulu, Hawaii 96822, U.S.A.

FISHKOVA (Dr L.M.) Abastumani Astrophysical Observatory, Abastumani, Mt.Canobily, Georgia, U.S.S.R.

FITCH (Dr W.S.) Steward Observatory, University of Arizona, Tucson, Arizona 85721, U.S.A.

FITZGERALD (Dr M.P.) Department of Physics, University of Waterloo, Waterloo, Ontario N2L 3G1, Canada

FIX (Dr J.D.) The University of Iowa, Iowa City, Iowa 52240, U.S.A.

FLECKENSTEIN (Prof. Dr J.O.) Rebgasse 32, 4000 Basel, Switzerland

FLEISCHER (Dr R.) 1733 Church Street, N.W. Washington, D.C. 20036, U.S.A.

FLETCHER (Dr J.M.) Dominion Astrophysical Observatory, Herzberg Institute of Astrophysics, National Research Council of Canada, 5071 W. Saanich Road, Victoria, B.C. V8X 3X3, Canada

*FLIEGEL (Dr H.F.) Jet Propulsion Laboratory, 4800 Oak Grove Drive,
     Pasadena, CA 91103, U.S.A.

 FLORENSKIJ (Dr K.P.) Institute for Space Research, Academy of Sciences USSR,
     Profsojuznaja st., 88, Moscow, W-485, U.S.S.R.

*FLORENTIN-NIELSEN (Dr R.)   Copenhagen University Observatory, Brorfelde,
     DK-4340 Tølløse, Denmark

 FLORIDES (Dr P.S.) School of Mathematics, 39 Trinity College, Dublin 2, Ireland

 FLORSCH (Dr A.) Observatoire de Strasbourg, 11, rue de l'Université,
     67000 Strasbourg, France

 FLOWER (Dr D.) Observatoire de Paris, Section d'Astrophysique de Meudon,
     92190 Meudon, France

*FOGARTY (Dr W.G.) Helsinki University of Technology, Radio Laboratory,
     02150 Otaniemi, Finland

 FOGH OLSEN (H.J.) Copenhagen University Observatory, Brorfelde,
     DK - 4340 Tølløse, Denmark

 FOKKER (Dr A.D.) Sterrewacht Sonnenborgh, Servaas Bolwerk 13,
     Utrecht, the Netherlands

 FOMALONT (Dr E.B.) National Radio Astronomy Observatory, Green Bank,
     W.Va 24944, Box 2, U.S.A.

 FORBES (Dr E.G.) History Department, University of Edinburgh, William Robertson
     Building, 50, George Square, Edinburgh EH8 9JY, U.K.

 FORBES (Dr J.E.) P.O. Box 88120, Indianapolis, Ind. 46208, U.S.A.

*FORD (Dr H.C.) Department of Astronomy, University of California,
     Los Angeles, CA 90024, U.S.A.

 FORD (Jr Dr W.K.) Dept. of Terrestrial Magnetism, Carnegie Institution of
     Washington, Washington, D.C. 30015, U.S.A.

*FORT (Dr B.) LAM, Observatoire de Meudon, 92190 Meudon, France

 FORTI (Dr G.) Osservatorio di Arcetri, Largo Enrico Fermi, 5, Firenze, Italy

 FORTINI (Prof. T.) Osservatorio Astronomico, Rome, Italy

*FOSBURY (Dr R.A.E.) Anglo-Australian Observatory, P.O. Box 296,
     Epping, NSW 2121, Australia

*FOSSAT (Dr E.) Observatoire de Nice, Le Mont-Gros, 06000 Nice, France

 FOUKAL (Dr P.) Harvard College Observatory, 60 Garden Street,
     Cambridge, Mass., 02138, U.S.A.

 FOWLER (Prof. W.A.) California Institute of Technology, 1201 E. California Blvd.
     Pasadena, California, U.S.A.

 FOX (W.E.) 40, Windsor Road, Newark, Nottinghamshire, U.K.

*FOY (Dr R.) DAPHE, Observatoire de Meudon, 92190 Meudon, France

 FRACASSINI (Prof. Dr M.) Osservatorio Astronomico di Milano,
     via Brera 28, Milano, Italy

 FRACASTORO (Prof. M.G.) Direttore, Osservatorio astronomico di Torino,
     Pino Torinese, Torino 10025, Italy

*FRANDSEN (Dr S.) University of Aarhus, Langelandsgade, DK-8000 Aarhus C, Denmark

FRANKLIN (Dr F.A.) Smithsonian Institution Astrophysical Observatory,
    60 Garden Street, Cambridge, Mass., 02138, U.S.A.

FRANTSMAN (Dr Yu. L.) Radio Astrophysical Observatory, Latvian Academy of
    Sciences, Riga, Latvian SSR, U.S.S.R.

FRANZ (Dr O.G.) Lowell Observatory, Flagstaff, Arizona 86002, U.S.A.

FRASER (Dr C.W.) University Observatory, Buchanan Gardens,
    St Andrews Fife KY16 9LZ, U.K.

*FRAZIER (Dr E.N.) Space Physics Laboratory, The Aerospace Corporation,
    P.O. Box 92957, Los Angeles, CA 90009, U.S.A.

FREDGA (Dr K.) Institute of Plasma Physics, Royal Institute of Technology,
    S-100 44 Stockholm 70, Sweden

FREDRICK (Prof. L.W.) Leander McCormick Observatory, University of Virginia,
    Charlottesville, Virginia, 22903, U.S.A.

FREEMAN (Dr K.C.) Mount Stromlo and Siding Spring Observatories, Research
    School of Physical Sciences, The Australian National University
    Private Bag, Canberra 2600, Australia

FREIESLEBEN (Dr H.C.) Hydrografisches Institut der Universität,
    Alt-Oberweg 24, 2000 Hamburg 4, Germany F.R.

FREITAS-MOURAO (Dr R.R.) Observatorio Nacional, Rua General Bruce 586,
    Sao Cristóvão, Rio de Janeiro, Brazil

FRESA (Prof. A.) Via Porto 23, Salerno, Italy

FRICKE (Dr K.J.) Universitätssternwarte, Geismarlandstr. 11,
    3400 Göttingen, Germany F.R.

FRICKE (Prof. Dr W.) Direktor des Astronomischen Rechen-Instituts,
    Mönchhofstrasse 12-14, 69 Heidelberg 1, Germany F.R.

FRIEDEMANN (Dr Ch.) Universitätssternwarte und Astrophysikalisches Institut
    der Friedrich-Schiller-Universität, Jena, Germany D.R.

FRIEDJUNG (Dr M.) Institut d'Astrophysique, 98 bis, Boulevard Arago,
    75014 Paris, France

FRIEDLANDER (Dr M.W.) Washington University, Department of Physics,
    St. Louis, Missouri 63130, U.S.A.

FRIEDMAN (Dr H.) Code 7100, U.S. Naval Research Laboratory,
    Washington D.C. 20390, U.S.A.

FRINGANT (Dr A.M.) Institut d'Astrophysique, 98 bis Boulevard Arago,
    75014 Paris, France

FRISCH (Dr H.) Observatoire de Nice, Le Mont-Gros, 06300 Nice, France

FRISCH (Dr U.) Institut d'Astrophysique, 98 bis Boulevard Arago,
    75014 Paris, France

FRITZOVÁ-Svestková (Dr L.) 368 Singletary Lane, Framingham, Mass., 01701, U.S.A.

FROESCHLE (Dr C.) Observatoire de Nice, Le Mont-Gros, 06300 Nice, France

*FROESCHLE (Dr Ch.) Observatoire de Nice, Le Mont-Gros, 06300 Nice, France

*FROGEL (Dr J.A.) Cerro Tololo Interamerican Observatory, Casilla 63-D,
    La Serena, Chile

FROLOV (Dr M.S.), Astronomical Council, USSR Academy of Sciences, Pyatnitskaya ul. 48, 109017 Moscow, U.S.S.R.

FRYE (Dr G.M.) Case Institute of Technology, Cleveland, Ohio, U.S.A.

FUCHS (Prof. Dr.J.) Goldschlag-Strasse 185/9, A-1140 Wien, Austria

FUJIMOTO (Dr M.) Department of Physics, Faculty of Science, Nagoya University, Furocho, Chikusa-ku, Nagoya, Japan

FUJITA (Dr Y.) Department of Astronomy, Faculty of Science, University of Tokyo, Bunkyo-ku, Tokyo, Japan 113

*FULCHIGNONI (M.) Laboratorio di Astrofisica Spaziale del CNR, Casella Postale 67, Frascati, Italy

FURENLID (Dr K.I.) Kitt Peak National Observatory, P.O. Box 4130, Tucson, Arizona 85717, U.S.A.

FURSENKO (Dr M.A.) Institute of Theoretical Astronomy, 10, Kutuzov Qay, Leningrad, 192187, U.S.S.R.

GABRIEL (Dr A.H.) Astrophysics Research Unit, Culham Laboratory, Abingdon, Berkshire, U.K.

GABRIEL (Dr M.R.L.) Institut d'Astrophysique, Université de Liège, Avenue de Cointe, 5, B-4200 Cointe-Ougrée, Belgium

GAHM (Dr G.F.) Stockholm Observatory, S-133 00 Saltsjöbaden, Sweden

GAIDE (A.) 14, rue de Lausanne, CH-1028 Préverenges, Suisse

*GAIL (Dr H.P.) Lehrstuhl für Theoretische Astrophysik, Im Neuenheimer Feld 294, 69 Heidelberg 1, Germany F.R.

GAIZAUSKAS (Dr V.) Herzberg Institute of Astrophysics, National Research Council of Canada, Ottawa, Ontario K1A OR6, Canada

*GALAL (Dr A.A.) Helwan Observatory, Helwan, Arab Rep. of Egypt

*GALEOTTI (P.) Consiglio Nazionale delle Ricerche, Laboratorio di Cosmo-Geofisica, Corso Massimo d'Azeglio 46, 10125 Torino, Italy

GALIBINA (Dr I.V.) Institute of Theoretical Astronomy, 10, Kutuzov Quay, Leningrad, 192187, U.S.S.R.

GALKIN (Dr L.S.) Crimean Astrophysical Observatory, USSR Academy of Sciences, P/O Nauchniy,334413 Crimea, U.S.S.R.

*GALLAGHER, III (Dr J.S.) University of Minnesota, School of Physics and Astronomy, 148 Physics Building, Minneapolis, MN 55455, U.S.A.

GALLET (R.M.) 964 - 7th Street, Boulder, Colorado 80302, U.S.A.

GALLOUET (Dr L.) Observatoire de Paris, 61, Av. de l'Observatoire, 75014 Paris, France

GALPERIN (Dr Yu.I.), Space Research Institute, USSR Academy of Sciences, 117810 Moscow, U.S.S.R.

GALT (Dr J.A.) Dominion Radio Astrophysical Observatory, Herzberg Institute of Astrophysics, National Research Council of Canada, P.O. Box 248, Penticton, B.C. V2A 6K3, Canada

*GAMMELGAARD (Dr P.) University of Aarhus, Langelandsgade, DK-8000 Aarhus C, Denmark

GAPOSCHKIN (Dr E.) Smithsonian Institution, Astrophysical Observatory,
    Cambridge, Mass. 02138, U.S.A.

GAPOSCHKIN (Dr S.) Harvard College Observatory, Cambridge, Mass. 02138, U.S.A.

GARDNER (Dr F.F.) P.O.Box 76, Epping, N.S.W., 2121, Australia

GARFINKEL (Dr B.) Department of Astronomy, Yale University, New Haven,
    Connecticut 06520, U.S.A.

GARLICK (Dr G.F.J.) Department of Physics, University of Hull,
    Hull HU6 7RX, U.K.

GARMIRE (Dr G.P.) 328 Downs, California Institute of Technology,
    Pasadena, Cal. 91109, U.S.A.

GARNIER (Ing.R.) European Southern Observatory, Casilla 16317 Correo 9,
    Santiago de Chile, Chile

GARRISON (Prof. R.F.) David Dunlap Observatory, University of Toronto,
    Richmond Hill, Ontario L4C 4Y6, Canada

GARSTANG (Dr R.H.) Joint Institute for Laboratory Astrophysics,
    University of Colorado, Boulder, Colorado 80302, U.S.A.

GARTON (Prof. W.R.S.) Department of Physics, Imperial College,
    London S.W.7, U.K.

GARZOLI (Dr S.L.) Marcelo T. de Alvear 2412, 4⁰ H, Buenos Aires, Argentina

GASCOIGNE (Dr S.C.B.) Mount Stromlo Observatory, Canberra, A.C.T. Australia

GASKA (Dr S.) Astronomical Institute, N. Copernicus University,
    Sienkiewicza 30, Toruń, Poland

*GATEWOOD (Dr G.D.) Allegheny Observatory, University of Pittsburgh,
    Riverview Park, Pittsburgh, PA 15214, U.S.A.

*GAUSS (F.S.) Transit Circle Division, U.S. Naval Observatory, Washington D.C. 20390
    U.S.A.
GAUSTAD (Dr J.E.) Astronomy Department, University of California,
    Berkeley, Cal. 94720, U.S.A.

GAUTIER (Dr D.) Observatoire de Meudon, 92- Meudon, France

GAVIOLA (Dr E.) Instituto Fisico Balseiro, Bariloche, Rio Negro, Argentina

GAVRILOV (Dr I.V.) Main Astronomical Observatory, Ukrainian Academy of
    Sciences, Kiev, U.S.S.R.

GAY (Dr J.) Observatoire de Paris, Section d'Astrophysique de Meudon,
    92190 Meudon, France

GEAKE (Dr J.E.) Faculty of Technology, University of Manchester,
    Manchester 1, U.K.

GEBBIE (Dr K.) Joint Institute for Laboratory Astrophysics, University
    of Colorado, Boulder, Colorado 80302, U.S.A.

GEHRELS (Dr T.) Space Science Building, University of Arizona,
    Tucson, Arizona 85721, U.S.A.

GEISS (Prof. J.) Physikalisches Institut der Universität, Sidlerstrasse 5,
    3012 Bern, Switzerland

GEL'FREJKH (Dr G.B.) Main Astronomical Observatory, USSR Academy of Sciences,
    Pulkovo, U.S.S.R.

GENKIN (Dr I.L.) Astrophysical Institute of Academy of Sciences of
    Kazakh SSR, Alma-Ata 480068, U.S.S.R.

GENT (H.) Royal Radar Establishment, St Andrews Road, Great Malvern,
    Worcestershire, U.K.

GENTILI de GIUSEPPE (M.) Observatoire du Pic-du-Midi,
    65- Bagnères-de-Bigorre (Hautes-Pyrénées), France

GEORGELIN (Dr Y.) Observatoire de Marseille,2 Place le Verrier,
    Marseille 13004, France
*GEORGELIN (Dr Yvonne) Observ.de Marseille, 2, Pl.Le Verrier, 13004 Marseille,France

*GERARD (Dr E.). DERAD, Observatoire de Meudon, 92190 Meudon, France

*GERBALDI (M.) IAP, 98bis, Boulevard Arago, 75014 Paris, France

GERLEI (O.) Heliophysical Observatory, Debrecen, Hungary

GEROLA (Prof. H.) Instituto de Astronomía y Fisica del Espacio,
    Casilla de Correo 67, Suc. 28, Buenos Aires, Argentina

GERSHBERG (Dr R.E.) Crimean Astrophysical Observatory, USSR Academy of Sciences
    P/O Nauchniy,334413 Crimea, U.S.S.R.

*GETMANTSEV (Dr G.G.) Radiophysical Research Institute,
    Lyadov Street 25/14 603024 Gorky, U.S.S.R.

GEYER (Dr E.H.) Observatorium Hoher List (Universität Bonn)
    5568 Daun/Eifel, Germany F.R.

GHABRUS (Dr Roushdy Azer),Helwan Observatory, Helwan near Cairo,
    Arab Republic of Egypt

GIACAGLIA (Prof. G.E.O.) 227 Taylor Hall, The University of Texas at Austin,
    Austin, Texas 78712, U.S.A.

GIACCONI (Dr R.) American Science and Engineering Inc. 955 Massachusetts Ave
    Cambridge, Mass. 02139, U.S.A.

*GIACHETTI (R.) Osservatorio Astrofisico di Arcetri, Largo E. Fermi 5,
    Firenze, Italy

GIANNONE (Dr P.) Osservatorio Astronomico, Viale Parco Mellini 84,
    00185 Roma, Italy

GIANNUZZI (Prof. M.A.) Corso Matteotti 190, Albano, Roma, Italy

GICLAS (H.L.) Lowell Observatory, Flagstaff, Arizona 86001, U.S.A.

GIERASCH (Dr P.) Cornell University, Space Sciences Building,
    Ithaca, New York 14850, U.S.A.

GIESE (Dr R.H.) Ruhr-Universität Bochum, Bereich Extraterrestrische Physik
    NA Gebäude, 463 Bochum, Germany F.R.

*GIETZEN (J.W.) Herstmonceux Castle, Hailsham, East Sussex, BN 27 1RP, U.K.

*GILMAN (Dr P.A.) NCAR, P.O. Box 3000, Boulder, CO 80303, U.S.A.

*GILMORE (A.C.) 101 Happy Valley Road, Wellington 2, New Zealand

*GILRA (Dr D.P.) Kapteyn Astronomical Institute, University of Groningen,
    P.O.Box 800, Groningen, The Netherlands

GINGERICH (Prof. O.J.) Smithsonian Astrophysical Observatory, Center for Astro-
    physics, 60 Garden Street, Cambridge, Mass. 02138, U.S.A.

GINZBURG (Acad. V.L.), Physical Institute, USSR Academy of Sciences,
117924 Moscow, U.S.S.R.

GIOVANELLI (Dr R.G.), CSIRO, Division of Physics, University Grounds,
Chippendale, NSW 2008, Australia

*GIOVANNINI FONTI (C.) Osservatorio Astronomico, Via Zamboni 33,
Bologna, Italy

GLAGOLEVSKIJ (Dr Ju.V.) Special Astrophysical Observatory, USSR Academy
of Sciences, St. Zelenchukskaya, Stavropolsky Kraj, U.S.S.R.

GLASER (Dr H.) Solar Physics Program, Physics and Astronomy Programs NASA,
Office of Space Science and Applications, Washington D.C. 20546, U.S.A.

*GLASPEY (Dr J.W.) Département de physique, Université de Montéral,
C.P. 6128, Succursale A. Montréal, P.Q., H3C 3J7 Canada

*GLASS (Dr I.S.) South African Astronomical Observatory, P.O. Box 9,
Observatory, Cape 7935, Republic of South Africa

*GLASSGOLD (Dr A.E.) Department of Physics, New York University,
4 Washington Place, New York, NY 10003, U.S.A.

*GŁEBOCKI (Dr R.) Institute of Physics, Gdánsk University, ul.Wita Stwosza
80-216 Gdánsk, Poland

GLEDHILL (Dr J.A.) Department of Physics, Rhodes University, P.O.Box 94,
Grahamstown 6140, South Africa

GLEISSBERG (Prof. Dr W.) Buchenweg 12
6370 Oberursel 4, Germany, F.R.

*GLENCROSS (Dr W.M.) Department of Physics and Astronomy
University College, Gower Street, London WC1E 6BT, U.K.

GLIESE (Dr W.) Astronomisches Rechen-Institut, Mönchhofstr. 12-14,
69 Heidelberg, Germany F.R.

GLUSHNEVA (Dr I.N.) Sternberg Astronomical Institute, Universitetskij prospekt 13
Moscow 117234, U.S.S.R.

GNEVYSHEV (Dr M.N.) Astronomical Observatory, Leningrad- M140,U.S.S.R.

GNEVYSHEVA (Dr R.S.) Main Astronomical Observatory, USSR Academy of Sciences,
Pulkovo, U.S.S.R.

GÖBEL (Dr E.) Institut für Theoretische Astronomie, Türkenschanzstr.17,
A-1180 Wien, Austria

GODART (Prof. Dr.O.) Chemin du Cyclotron 2, B-1348 Louvain-la-Neuve, Belgium
GODOLI (Prof. G.) Osservatorio Astrofisico di Arcetri, Largo Enrico Fermi no. 5,
50125 Firenze, Italy
GÖKDOĞAN (Prof. Dr N.) Director, University Observatory, Beyazit, Istanbul;
Turkey

*GOKHALE (Dr M.H.) Indian Institute of Astrophysics, Bangalore 560034, India

GÖKMEN (Dr T.) Kandilli Observatory, Çengelköy, Istanbul, Turkey

GOLAY (Prof. Dr M.) Directeur de l'Observatoire de Genève,
1290 Sauverny (GE) Switzerland

GOLD (Prof. T.) Cornell University, Ithaca, N.Y. 14850, U.S.A.
(Space Science Building)

GOLDBERG (Prof. L.) Director, Kitt Peak National Observatory,
P.O.Box 26732, Tucson, Arizona 85726, U.S.A.

GOLDREICH (Dr P.) California Institute of Technology, Pasadena, Cal. 91109, U.S.A.

GOLDSMITH (Dr D.W.) 1655 Twelfth Avenue, San Francisco, Cal. 94122, U.S.A.

GOLDSMITH (Dr S.) Dept. of Physics & Astronomy, Tel Aviv University, Israel

GOLDSTEIN (Dr R.M.) Jet Propulsion Laboratory, California Institute of Technology
4800 Oak Grove Drive, Pasadena, Cal. 91103, U.S.A.

GOLDSTEIN (Dr S.J.) Leander McCormick Observatory, University of Virginia,
Charlottesville, Virginia 22903, U.S.A.

GOLDSWORTHY (Prof. F.A.) Dept. of Mathematics, University of Leeds, Leeds 2,
Yorkshire, U.K.

GOLDWIRE (Dr H.C. Jr.) Space Science Dept. Rice University, Houston,
Texas 77001, U.S.A.

GOLLNOW (Dr H.) Mount Stromlo Observatory, Canberra, A.C.T., Australia

*GOLUB (Dr L.) American Science and Engineering Inc., 955 Massachusetts Ave,
Cambridge, MA 02139, U.S.A.

GOMES (Dr A.M.), University of Brazil, Rua Ipiranga, 25-40 Andar,
Rio de Janeiro, ZC-01, Brazil

*GOMEZ (Dr A.E.) Observatoire de Meudon, 92190 Meudon, France

*GOMEZ (Dr J.) p⁰. Imperial N⁰ 29 6⁰ H. Madrid 5, Spain

GONDOLATSCH (Prof. Dr. F.) Astronomisches Recheninstitut, Mönchhofstr. 12-14,
69 Heidelberg, Germany F.R.

GONZÁLEZ (G.) Instituto Nacional de Astrofísica, Optica y Electrónica,
Tonantzintla, Pue. Apartados Postales nos 216 y 51
Puebla, Pue. Mexico

GONZE (Ir. R.F.J.) Observatoire Royal de Belgique, avenue Circulaire 3,
B-1180 Bruxelles, Belgium

GOODY (Dr E.M.) Director, Blue Hill Observatory, Harvard University,
Cambridge, Mass. 02138, U.S.A.

GOPALA RAO (U.V.) Astrophysical Observatory, Kodaikanal-3, India

GOPASYUK (Dr S.I.) Crimean Astrophysical Observatory, USSR Academy of Sciences,
Crimea, U.S.S.R.

GORBATSKIJ (Dr V.G.) Leningrad University Observatory, Leningrad, U.S.S.R.

GORDON (Dr C.) Department of Astronomy, Hampshire College, Amherst, Mass.01002,
U.S.A.

GORDON (Dr Ch.) PECKER-WIMEL, 60 E 8th Street, New York, N.Y. 10003, U.S.A.

GORDON (Dr I.M.) Institute of Radiophysics and Electronics,
Academy of Sciences of the Ukrainian SSR, Khar'kov 85, U.S.S.R.

GORDON (Dr J.E.) Director of the Nikolaiev Observatory, Nikolaiev, U.S.S.R.

GORDON (Dr K.) Hampshire College, Amherst, Mass. 01002, U.S.A.

GORDON (Dr M.A.) National Radio Astronomy Observatory, Suite 100,
2010 North Forbes Boulevard, Tucson, Arizona 85705, U.S.A.

GORENSTEIN (Dr P.) American Science & Engineering, 955 Massachusetts Avenue, Cambridge, Mass. 02139, U.S.A.

GORGOLEWSKI (Dr S.) Astronomical Observatory, N. Copernicus University, Sienkiewicza 30, Toruń, Poland

GOSACHINSKIJ (Dr I.V.) Special Astrophysical Observatory, USSR Academy of Sciences, Leningrad Branch of the SAO, Pulkovo 196140, U.S.S.R.

GOSLING (Dr J.T.) P-4, MS-436, Los Alamos Scientific Laboratory, Los Alamos, NM 87545, U.S.A.

GOSS (Dr W.M.) CSIRO, Radiophysics Division, P.O. Box 76, Epping, N.S.W. 2121, Australia

GOSSNER (Dr S.D.) 415 E 50th Street, New York, N.Y. U.S.A

*GOTT, III (Dr J.R.) Dept. of Astrophysical Sciences, Princeton University, Princeton, NH 08540, U.S.A.

*GOTTESMAN (Dr S.T.) Department of Physics and Astronomy, University of Florida, Gainesville, FL 32611, U.S.A.

GOTTLIEB (Dr C.A.) Harvard College Observatory, Cambridge, Mass.02138, U.S.A.

GOTTLIEB (K.) Mount Stromlo Observatory, Canberra, A.C.T., Australia

GOUDAS (Dr C.L.) University of Patras, Patras, Greece

GOUGH (Dr D.O.) Institute of Theoretical Astronomy, Madingley Road, Cambridge, U.K.

GOUGUENHEIM (Dr L.) Observatoire de Paris, Section d'Astrophysique de Meudon, 92190 Meudon, France

GOULD (Dr R.) Physics Department, University of California-San Diego, La Jolla, California, U.S.A.

GOWER (A) 4010 Saanich Road, Victoria, B.C., Canada

GOWER (Dr J.F.R.) 4010 Saanich Road, Victoria, B.C., Canada

GOY (G.) Observatoire Cantonal de Genève, Genève, 1290 Sauverny, Switzerland

GOYAL (A.N.) Department of Mathematics, University of Rājasthan, Jaipur, India

GRABOSKE (Dr H.) Lawrence Livermore Laboratory, University of Calif., Livermore, California, U.S.A.

GRABOWSKI (Dr B.) Opole, ul. Oleska 45 m 16, Poland

GRADSZTAJN (Dr E.) Institut de Physique Nucléaire d'Orsay, 91-Gif s/Yvette, France

*GRAHAM (Dr E.) National Center for Atmospheric Research, High Altitude Observatory, P.O. Box 3000, Boulder, Colorado 80303, U.S.A.

GRAHAM (Dr J.A.) Cerro Tololo Inter-American Observatory, Casilla 63-D, La Serena, Chile

GRAHL (Dr B.H.) Max-Planck-Institut für Radioastronomie, D-53 Bonn 1, Auf dem Hügel 69, Germany F.R.

GRAINGER (Dr J.) University of Manchester, Institute of Science & Technology, Sackville Street, Manchester 1, U.K.

GRANT (Dr I.P.) Pembroke College, Oxford, U.K.

GRATTON (Prof. L.) Laboratorio di Astrofisica, Casella Postale, 67, 00044 Frascati (Roma) Italy

GRAY (Dr D.F.) Department of Astronomy, University of Western Ontario, London, Ontario N6A 3K7, Canada

GREBENIKOV (Prof. Dr E.A.) Sternberg Astronomical Institute, Universitetskij prospekt 13, Moscow 117234, U.S.S.R.

GREEN (Dr J.) 941 Via Nogales, Palos Verdes Estate, California 90274, U.S.A.

GREEN (Prof. L.C.) Strawbridge Observatory, Haverford College, Haverford, Pennsylvania 19041, U.S.A.

GREEN (Dr R.M.) Department of Astronomy, The University, Glasgow, W.2, Scotland

GREENBERG (Dr J.M.) Huygens Laboratorium, Rijksuniversiteit te Leiden, Wassenaarseweg 78, Leiden 2405, The Netherlands

*GREENBERG (Dr R.) Planetary Science Institute, 2030 East Speedway, Suite 201, Tucson, Arizona 85719, U.S.A.

GREENSTEIN (Dr G.) Amherst College, Astronomy Dept. Amherst, Mass. 01002, U.S.A.

GREENSTEIN (Prof. J.L.) Hale Observatories, California Institute of Technology, 1201 E. California Street, Pasadena, California 91109, U.S.A.

GREGORY (Dr P.C.) Department of Physics, University of British Columbia, Vancouver, B.C. V6T 1W5, Canada

GREGUL (Dr A.J.) Astronomical Observatory of the University, Kiev, Ukrainian SSR, U.S.S.R.

GREISEN (Dr K.I.) Cornell University, Physics Department, Clark Hall, Ithaca, New York 14850, U.S.A.

GREVESSE (Dr N.) Institut d'Astrophysique, Université de Liège, Avenue de Cointe 5, B-4200 Cointe-Ougrée, Belgium

GREWING (Dr M.) Institut für Astrophysik und extraterrestrische Forschung, Auf dem Hügel 71, 53 Bonn, Germany F.R.

GREYBER (Dr H.D.) 10123 Falls Road, Potomac, Md. 20854, U.S.A.

GRIFFIN (Dr R.E.M.) The Observatories, Madingley Road, Cambridge, CB3 0HA, U.K.

GRIFFIN (Dr R.F.) The Observatories, Madingley Road, Cambridge CB3 0HA, U.K.

GRIFFITH (Prof. J.S.) Department of Mathematical Sciences, Lakehead University, Thunder Bay, Ontario P7B 5R1, Canada

GRIGOREVSKIJ (Dr V.M.) Odessa State University, Astron.Observ.270014 Odessa,USSR.

*GRINDLAY (Dr J.E.) Harvard College Observatory, 60 Garden Street, Cambridge, Mass. 02138, U.S.A.

*GRININ (Dr V.P.) Crimean Astrophysical Observatory of the USSR Academy of Sciences, P/O Nauchny, Crimea 334413, U.S.S.R.

GRISHCHUK (Dr L.P.) Sternberg Astronomical Institute, Universitetskij prospekt 13, Moscow 117234, U.S.S.R.

*GROSS (Dr P.G.) Warner and Swasey Observatory, 1975 Taylor Road, East Cleveland, OH 44112, U.S.A.

GROSSMAN (Dr A.S.) Erwin W. Fick Observatory, Iowa State University, Ames, Iowa 50010, U.S.A.

*GROSSMAN (Dr L.) The University of Chicago, Dept, of the Geophysical Sciences, 5734 South Ellis Avenue, Chicago, IL 60637, U.S.A.

GROSSMANN-DOERTH (Dr U.) Fraunhofer Institut, Schöneckstrasse 6,
7800 Freiburg im Breisgau, Germany F.R.

*GROTEN (Prof. Dr E.) Technische Hochschule, Petersenstrasse 13,
Darmstadt, Germany F.R.

*GROTH, III (Dr E.J.) Physics Department, Jadwin Hall, Princeton University,
Princeton, NJ 08540, U.S.A.

GROTH (Dr H.G.) Universitäts-Sternwarte München, Sternwartstrasse 23,
8 München 27, Germany F.R.

GRUBISSICH (Prof. Dr C.) via Aosta 34/5, 35100 Padova, Italy

*GRUDLER (P.) CERGA, Plateau de Roquevignon, Avenue Copernic, 06130 Grasse, France

*GRUDZINSKA (Dr S.) Institute of Astronomy, N. Copernicus, University,
Chopina 12/18, 87-100 Toruń, Poland

GRUEFF (Dr G.) Laboratorio di Radioastronomia, Via Irnerio 46, Bologna, Italy

GRUSHINSKIJ (Prof. Dr N.P.) Sternberg Astronomical Institute, Moscow, U.S.S.R.

GRYGAR (Dr J.) Astronomical Institute, Czechoslovak Academy of Sciences,
Observatory Ondřejov, Czechoslovakia

GRZEDZIELSKI (Dr S.) Astronomical Observatory of Warsaw University,
Al Ujazdowskie 4, Warsaw, Poland

*GUARNIERI (A.) Osservatorio Astronomico, Via Zamboni 33, Bologna, Italy

GUBANOV (Dr V.S.) Pulkovo Observatory, Leningrad 196140, U.S.S.R.

GUDUR (Dr N.) Ege University Observatory, P.K. 21 Bornova, Izmir, Turkey

GUELIN (Dr M.) Observatoire de Paris, Section d'Astrophysique de Meudon,
92190 Meudon, France

GUERIN (Dr P.) Institut d'Astrophysique, 98 bis Boulevard Arago,
75014 Paris, France

GUEST (Dr J.E.) University of London Observatory, Mill Hill Park,
London N.W.7, U.K.

*GUIBERT (Dr J.) DERAD, Observatoire de Meudon, 92190 Meudon, France

*GUIDICE (Dr D.A.) Ionospheric Research Laboratory, AF Geophysical Labs,
Hanscom AFB, MA 01731, U.S.A.

GUINOT (Dr B.) Observatoire de Paris, 61, Avenue de l'Observatoire,
75014 Paris, France

GULKIS (Dr S.) Jet Propulsion Laboratory, 1836-365, 4800 Oak Grove Drive,
Pasadena, California 91103, U.S.A.

*GULL (Dr T.R.) C23B, Lockheed Electronics Corporation, 16811 El Camino Real,
Houston, TX 77058, U.S.A.

GULMEN (Dr O.) Ege University Observatory, Bornova, Izmir, Turkey

GULYAEV (Dr A.P.) Sternberg Astronomical Institute, Universitetskij prospekt 13,
Moscow 117234, U.S.S.R.

GULYAEV (Dr R.A.) Institute of Terrestrial Magnetism, Ionosphere and Radio Wave
Propagation, Akademicheskij gorodok, Moscow region 142092, U.S.S.R.

GUNN (Dr J.E.) Hale Observatory, California Institute of Technology, Pasadena,
California 91109, U.S.A.

GÜNTZEL-LINGNER (Dr U.) Astronomisches Rechen-Institut, Mönchhofstrasse 12-14, 69 Heidelberg, Germany F.R.

GURM (Dr H.S.) Department of Physics, Panjabi University, Patiala (Panjab), India

GURSHTEIN (Dr A.A.) Institute for Space Research, Academy of Sciences of the U.S.S.R., Profsojuznaya st, 88, Moscow 117810, GSP-312, U.S.S.R.

GURSKY (Dr H.) American Science and Engineering, Inc., 11 Carleton Street, Cambridge, Mass. 02142, U.S.A.

GURTOVENKO (Dr E.A.) Main Astronomical Observatory, Ukrainian Academy of Sciences, Kiev, U.S.S.R.

GURZADIAN (Prof. Dr G.A.), Byurakan Astrophysical Observatory, Byurakan 378433 Armenia, U.S.S.R.

GUSEINOV (Dr O.Kh.), Shemakha Astrophysical Observatory, Shemakha, 373243 Azerbaidzan, U.S.S.R.

GUSEJNOV (Dr R.Eh.), Shemakha Astrophysical Observatory, Shemakha, 37243 Azerbaidzan, U.S.S.R.

GUSSMANN (Dr E.A.) Zentralinstitut für Astrophysik, DDR 15 Potsdam, Telegrafenberg, Germany D.R.

GUSTAFSSON (Dr B.) Astronomical Observatory, Box 515, S-751 20 Uppsala 1, Sweden

GUTH (Dr V.) Astronomical Institute, Czechoslovak Academy of Sciences, Observatory Ondřejov, Czechoslovakia

GUTHRIE (B.N.G.) The Royal Observatory, Edinburgh, 9, U.K.

GUTIERREZ-MORENO (Dr A.) Observatorio Astronómico de la Universidad de Chile, Casilla 36-D, Santiago de Chile, Chile

GYLDENKERNE (K.) Universitets Observatoriet, Brorfelde, DK-4340 Tølløse, Denmark

HAAS (Prof. Dr.J.) Lipschitz-Str. 1, 53 Bonn, Germany F.R.

HABIBULLIN (Prof. Dr Sh. T.) Director, Astronomical Observatory of the State University, Lenina 18, Kazan, U.S.S.R.

HABING (Dr H.J.) Sterrewacht, Leiden, The Netherlands

HACHENBERG (Prof. Dr O.) Max-Planck-Institut für Radioastronomie, Auf dem Hügel 69, 53 Bonn 1, Germany, F.R.

HACK (Prof. M.) Director, Astronomical Observatory, Via Tiepolo 11, 34131 Trieste I, Italy

HACYAN (Dr S.) Instituto de Astronomia, UNAM, Apartado Postal 70-264, Mexico D.F., Mexico

HADDOCK (Dr F.) The Observatory, University of Michigan, Ann Arbor, Michigan, U.S.A.

HADJIDEMETRIOU (Dr J.D.) Department of Theoretical Mechanics, University of Thessaloniki, Thessaloniki, Greece

HAERENDEL (Dr G.) Institut für extraterrestrische Physik des Max-Planck-Instituts für Physik und Astrophysik, 8046 Garching b. München, Germany, F.R.

HAFFNER (Prof. H.) Institut für Astronomie und Astrophysik, Am Hubland, 8700 Würzburg, Germany F.R.

HAGEN (Dr J.P.) Professor of Radio Astronomy, The Pennsylvania State University University Park, Pennsylvania, U.S.A.

HAGEN-THORN (Dr V.A.) Leningrad University, Leningrad, 199178 10 Linija 33 U.S.S.R.

HAGFORS (Dr T.) MIT Lincoln Laboratory, Millstone, Haystack, Lexington, Massachusetts 02173, U.S.A.

HAGIHARA (Prof. Y.) Department of Astronomy, Faculty of Science, University of Tokyo, Bunkyo-ku, Tokyo, Japan

HAJDUK (Dr A.) Astronomical Institute, Slovak Academy of Sciences, Dúbravská Cesta, Bratislava, Czechoslovakia

HALL (Dr D.) Kitt Peak National Observatory, 950 North Cherry Avenue Tucson, Arizona 85726, U.S.A.

HALL (Dr D.S.) Associate Professor, Dyer Observatory, Vanderbilt University, Nashville, Tennessee 37235, U.S.A.

HALL (Dr J.S.) Director, Lowell Observatory, Flagstaff, Arizona, U.S.A.

HALL (Dr R.G.) U.S. Naval Observatory, Washington D.C. 20390, U.S.A.

HALLAM (Dr K.L.) Code 941, Goddard Space Flight Center, Greenbelt, Maryland 20771, U.S.A.

HALLIDAY (Dr I.) Planetary Sciences Section, Herzberg Institute of Astrophysics National Research Council of Canada, Ottawa, Ontario K1A OR6, Canada

HAMADA (Prof. T.) Department of Physics, Ibaraki University, Mito 310, Japan

*HAMDY (Dr M.A.M.) Helwan Observatory, Helwan, Arab Republic of Egypt

HAMEEN-ANTTILA (Prof. Dr A.) Oulun Yliopiston TAhtitieteen Laitos, (The Astronomical Institute of the Oulu University, Oulu, Finland

HAMID (Dr S.El-Din) Department of Astronomy, Faculty of Science, University Fouad, Giza, Cairo, Arab Rep. of Egypt

*HAMILTON (Dr P.A.) University of Tasmania, Hobart, Tasmania

HAMMERSCHLAG (Dr Ir R.H.) Sterrenwacht, Servaas Bolwerk 13, Utrecht, The Netherlands

*HAMZAOGLU (Dr E.) Kandilli Observatory, Çengelköy, Istanbul, Turkey

HANASZ (Dr J.) Astronomical Observatory, Torun- Piwnice, Poland

HANSEN (Dr C.J.) Dept. of Astrophysics and Physics, University of Colorado Boulder, Colorado, U.S.A.

*HANSEN (Dr L.) Copenhagen University Observatory, Østervoldgade 3, DK-1350 Copenhagen K, Denmark

HANSEN (Dr R.) High Altitude Observatory, Boulder, Colorado 80302, U.S.A.

HANSSON (N.) Lund Observatory, S-222 24 Lund, Sweden

HARDEBECK (Dr E.G.) 110 Terrace Drive, Big Pine, California 93513, U.S.A.

HARDIE (Dr R.H.) Dyer Observatory, Vanderbilt University, Nashville 5, Tennessee, U.S.A.

HARDING (G.A.) Royal Greenwich Observatory, Herstmonceux Castle, Hailsham, Sussex, U.K.

HARDORP (Prof. Dr J.) Department of Earth and Space Sciences, State University of New York, Stony Brook, New York 11790, U.S.A.

*HARMANEC (Dr P.) Astronomical Institute of the Czechoslovak Academy of Sciences
     Observatory, 251 65 Ondřejov, Czechoslovakia

HARO (Dr G.) Instituto Nacional de Astrofísica Optica y Electrónica,
     Tonantzintla, Pue. Apartados Postales Nos. 216 y 51, Puebla, Pue, Mexico

HARRINGTON (Dr J.P.) Dept. of Physics and Astronomy, University of Maryland,
     College Park, Maryland 20742, U.S.A.

HARRINGTON (Dr R.S.) U.S. Naval Observatory, Washington, D.C. 20390, U.S.A.

*HARRIS (Dr D.E.) Netherlands Foundation for Radio Astronomy,
     Radiosterrenwacht, Oude Hoogeveensedijk 4,Dwingeloo, The Netherlands

HARRISON (Prof. E.R.) Dept. of Physics and Astronomy, University of Massachusetts,
     Amherst, Mass. 01002, U.S.A.

HARROWER (Prof. G.A.) Astronomy Group, Dept. of Physics, Queen's University
     Kingston, Ontario K7L 3N6, Canada

HARTEN (Dr R.H.) Sterrewacht, Kaiserstraat 57, Leiden, The Netherlands

*HARTL (Dr H.J.) Sternwarte und Astronomisches Institut, Universitäts-Str.4,
     A-6020 Innsbruck, Austria

*HARTLEY (Dr K.F.) Herstmonceux Castle, Hailsham, East Sussex, BN27 1RP, U.K.

HARTNER (Dr W.) Institut für Geschichte der Naturwissenschaften
     Johann Wolfgang Goethe Universität, Frankfurt am Main, Germany, F.R.

HARTWICK (Dr F.D.A.) Department of Physics, University of Victoria,
     Victoria, B.C., Canada

HARTZ (Dr T.R.) Communications Research Center, Dept. of Communications,
     P.O. Box 490, Terminal A, Ottawa, Ontario, Canada

HARVEY (Dr Ch.) Observatoire de Meudon, 92-Meudon, France

*HARVEY (Dr G.) Langley Research Center, Langley Station, Hampton, Va. 23365, U.S.A

HARVEY (Dr J.W.) Kitt Peak National Observatory, 950 North Cherry Avenue,
     Tucson, Arizona 85717, U.S.A.

HARWIT (Dr M.) Cornell University, Center for Radiophysics and Space Research,
     Clark Hall, Ithaca, New York 14850, U.S.A.

HARWOOD (M.) Harvard College Observatory, Cambridge, Mass. 02138, U.S.A.

HASER (Dr L.) Max-Planck-Institut für Physik und Astrophysik,
     Institut für Extraterrestrische Physik, 8046 Garching b/ München
     Germany,F.R.

*HASLAM (Dr G.Ch.T.) Max-Planck-Institut für Radioastronomie, Auf dem Hügel 69,
     5300 Bonn 1, Germany, F.R.

HASSAN (Dr S.M.) Helwan Institute of Astronomy and Geophysics, Helwan, Cairo,A.R.E.

HATTORI (Dr A.) Kwasan Observatory, University of Kyoto, Yamashina, Kyoto, Japan

HAUCK (Prof.Dr B.) Directeur de l'Institut d'Astronomie de l'Université de
     Lausanne, 1290 Chavannes-des-Bois, Switzerland

HAUG (Dr U.) Hamburger Sternwarte, 205 Hamburg-Bergedorf, Germany, F.R.

HAUGE (Dr O.) Institute of Theoretical Astrophysics, University of Oslo,
     P.O.Box 1029, Blindern, Oslo 3, Norway

HAUPT (Prof. Dr H.) Director, Universitäts-Sternwarte, Universitätsplatz 5,
    A- 8010 Graz, Austria

HAUPT (R.) Assistant Director of the American Ephemeris and Nautical Almanac
    U.S. Naval Observatory, Washington, D.C.,U.S.A.

HAUPT (R.F.) U.S. Naval Observatory, Washington, D.C., 20390, U.S.A.

HAUPT (Dr W.) Astronomisches Institut der Ruhr-Universität Bochum,Postfach 2148,
    4630 Bochum-Querenburg, Germany F.R.

HAVLEN (Dr R.J.) European Southern Observatory, Bergedorferstr. 131
    205 Hamburg 80, Germany, F.R.

HAVNES (Dr O.) Institut of Theoretical Astrophysics, University of Oslo,
    P.O. Box 1029, Blindern, Oslo 3, Norway

*HAWARDEN (Dr T.G.) UK 48-inch Schmidt Telescope Unit, Private Bag,
    Coonabarabran, NSW 2857, Australia

HAWKINS (Dr G.S.) 47 Cypress Road, Wellesley Hills, Boston, Mass. 02181, U.S.A.

HAWKING (Dr S.W.) Institute of Theoretical Astronomy, Madingley Road,
    Cambridge, U.K.

HAYAKAWA (Prof. Dr S.) Dept. of Physics, Faculty of Science, Nagoya University
    Furocho, Chikusa-ku, Nagoya, Japan

HAYASHI (Prof. C.) Dept. of Nuclear Science, Faculty of Science, Kyoto University,
    Kyoto, Japan

*HAYES (Dr D.S.) 1807 E. Waverly Street, Tucson, AZ 85719, U.S.A.

HAYLI (Dr A.F.) Observatoire de Besançon, 43, Avenue de l'Observatoire,
    25000 Besançon, France

HAYMES (Dr R.C.) Rice University, Houston, Texas 77001, U.S.A.

HAZARD (Dr C.) Institute of Theoretical Astronomy, Madingley Road,
    Cambridge, U.K.

HAZER (Dr S.) Faculty of Science, Department of Astronomy, P.K. 21
    Bornova, Izmir, Turkey

HAZLEHURST (Dr J.) The Astronomy Centre, University of Sussex, Falmer,Nr.Brighton,
    Sussex, U.K.
*HEAP (Dr S.R.) Code 671, NASA Goddard Space Flight Center
    Greenbelt, MD 20771, U.S.A.

HEARN (Dr A.G.) Sterrewacht Sonnenborgh, Servaas Bolwerk 13, Utrecht, Holland

*HEARNSHAW (Dr J.B.) Dept. of Physics, University of Canterbury, Private Bag
    Christchurch, New Zealand

HECKMANN (Prof. Dr O.) Wohnstift Göttingen, App.B 1403
    Charlottenburgerstr.29, 3400 Göttingen-Geismar, Germany, F.R.

HEDDLE (Prof. D.W.O.) Head, Dept. of Physics, Royal Holloway College,
    Egham, Surrey TW20 0ex, U.K.

HEDEMAN (E.R.) McMath-Hulbert Observatory, Lake Angelus Road,
    Pontiac, Michigan, U.S.A.

HEESCHEN (Dr D.S.) National Radio Astronomy Observatory, Edgemont Road,
    Charlottesville, Virginia 22901, U.S.A.

*HEFFERLIN (Dr R.A.) Drawer H. Collegedale, TN 37315, U.S.A.

*HEGGIE (Dr D.C.) Applied Mathematics, King's Buildings, Mayfield Road, Edinburgh EH9 3JP, U.K.

*HEGYI (Dr D.J.) The Harrison M.Randall Laboratory of Physics, The University of Michigan, Ann Arbor, MI 48104, U.S.A.

HEIDMANN (Dr J.) Observatoire de Paris, Section d'Astrophysique, 92190 Meudon, France

HEIDMANN (Dr N.) Institut d'Astrophysique, 98 bis Boulevard Arago, 75014 Paris, France

HEILES (Dr C.E.) Astronomy Department, University of California, Berkeley, California 94720, U.S.A.

HEINTZ (Prof. W.D.) Sproul Observatory, Swarthmore, Pa. 19081, U.S.A.

HEINTZE (Dr J.R.W.) Sterrewacht "Sonnenborgh", Servaas Bolwerk 13, Utrecht, The Netherlands

HEISER (Dr A.) A.J. Dyer Observatory, Box 1803, Vanderbilt University, Nashville, Tenn. 37203, U.S.A.

*HEKELA (Dr J.) Astronomical Institute of the Czechoslovak Academy of Sciences, Observatory, 251 65 Ondřejov, Czechoslovakia

HELFER (Dr H.L.) Department of Physics and Astronomy, University of Rochester, Rochester, N.Y. 14628, U.S.A.

HELLER (Dr M.) ul. Powstancόw Warszawy 13/94, 33-110 Tarnόw, Poland

HELMKEN (Dr H.) Smithsonian Astrophysical Observatory, 60 Garden Street, Cambridge, Mass. 02138, U.S.A.

*HELT (Dr B.E.) Copenhagen University Observatory, Østervoldgade 3, DK-1350 Copenhagen K, Denmark

HEMENWAY (Dr C.L.) 100 Fuller Road, Albany, New York 12205, U.S.A.

HEMMLEB (Dipl. Ing. G.) Zentralinstitut Physik der Erde, DDR 15 Potsdam, Telegrafenberg, Germany D.R.

HENIZE (Dr K.G.) Astronaut, Code CB Johnson Space Center Houston, Texas 77058, U.S.A.

HENON (M.) Observatoire de Nice, Le Mont-Gros, 06300 Nice, France

HENOUX (Dr J.C.) Observatoire de Paris, Section d'Astrophysique de Meudon, 92190 Meudon, France

HENRARD (Dr J.) Facultés Universitaires de Namur, rue de Bruxelles 61, 5000 Namur, Belgium

HENRIKSEN (Dr R.N.) Astronomy Group, Department of Physics, Queen's University, Kingston, Ontario K7L 3N6, Canada

HENRY (Dr R.C.) E.O. Hulbert Center for Space Research, Department of Physics, The Johns Hopkins University, Baltimore, Maryland 21218, U.S.A.

HERBIG (Dr G.H.) Lick Observatory, University of California, Santa Cruz, California 95060, U.S.A.

HERCZEG (Prof. Dr T.) New York University, Physics Department 4 Washington Place, New York, N.Y. 10003, U.S.A.

HERGET (Dr P.) Cincinnati Observatory, Observatory Place,
    45208 Cincinnati, Ohio, U.S.A.

HERMAN (Dr R.) Observatoire de Paris, Section d'Astrophysique, 92190 Meudon,
    France

HERR (Dr R.B.) Physics Department, University of Delaware, Newark,
    Delaware 19711, U.S.A.

HERS (J.) 48, Central Road, Linden Extension, Randburg 2194, South Africa

*HERSHEY (Dr J.L.) Sproul Observatory, Swarthmore College, Swarthmore,
    PA 19081, U.S.A.

HERZBERG (Dr G.) Herzberg Institute of Astrophysics, National Research Council
    of Canada, Ottawa, Ontario K1A OR6, Canada

HESSER (Dr J.E.) Cerro Tololo Inter-American Observatory, Casilla 63-D,
    La Serena, Chile

HEWISH (Dr A.) Cavendish Laboratory, Free School Lane, Cambridge, U.K.

HEWITT (Dr A.V.) U.S. Naval Observatory, Flagstaff Station, Flagstaff,
    Arizona, U.S.A.

HEY (Dr J.S.) 4 Shortlands Close, Willingdon, Sussex, U.K.

HEYDEN (Dr F.J., S.J.) Director of Solar Division, Manila Observatory,
    P.O. Box 1231, Manila, Philippines D-404

HEYVAERTS (Dr J.F.) Observatoire de Paris, Section d'Astrophysique de Meudon,
    92190 Meudon, France

HIBBS (A.R.) Jet Propulsion Laboratory, California Institute of Technology,
    Pasadena, California 91109, U.S.A.

HIDAJAT (Dr B.) Acting Director, Bosscha Observatory, Lembang, Java, Indonesia

*HIDALGO (Dr M.A.) Facultad de Ciencias Físicas, Universidad de Zaragoza,
    Zaragoza, Spain

*HIDE (Prof. R., F.R.S.) Geophysical Fluid Dynamics Laboratory,
    Meteorological Office (21), Bracknell, Berkshire, U.K.

HIEI (Dr Eijiro) Tokyo Astronomical Observatory, University of Tokyo,
    Mitaka, Tokyo, Japan

HIGGS (Dr L.A.) Herzberg Institute of Astrophysics, National Research Council
    of Canada, Ottawa, Ontario K1A OR6, Canada

*HILDITCH (Dr R.W.) University Observatory, Buchanan Gardens, St. Andrews,
    Fife, KY16 9LZ, U.K.

*HILDNER III, (Dr E.G.) High Altitude Observatory, NCAR, P.O. Box 3000,
    Boulder, CO 80303, U.S.A.

HILL (Dr G.) Dominion Astrophysical Observatory, Herzberg Institute of Astro-
    physics, National Research Council of Canada, 5071 W. Saanich Road,
    Victoria, B.C. V8X 3X3, Canada

HILL (Dr P.W.) University Observatory, St Andrews, Fife, Scotland

HILLIARD (Dr R.) Steward Observatory, University of Arizona, Tucson,
    Arizona 85721, U.S.A.

HILLS (Dr J.G.) The University of Michigan, Ann Arbor, Michigan 48104, U.S.A.

*HILLS (Dr R.E.) Cavendish Laboratory, Cambridge University, Madingley Road,
    Cambridge, U.K.
HILTNER (Prof. W.A.) Department of Astronomy, University of Michigan
    Ann Arbor, Michigan 48109, U.S.A.
HINTEREGGER (Dr H.E.) Hq. AFCRL (CRUU) Stop 30, L.G. Hanscom Field, Bedford,
    Mass. 01731, U.S.A.

*HIPPELEIN (Dr H.H.) Max-Planck-Institut für Astronomie, 6900 Heidelberg 1,
    Königstuhl, Germany F.R.

*HIRABAYASHI (Dr Hisashi) Nobeyama Solar Radio Station, Tokyo, Astronomical
    Observatory, University of Tokyo, Nobeyama, Nagano Prefecture, Japan

*HIRAI (Dr Masanori) Department of Earth Sciences and Astronomy, Fukuoka University
    of Education, Munakata, Fukuoka, Japan

HIRAYAMA (Dr Tadashi) Tokyo Astronomical Observatory, University of Tokyo,
    Mitaka, Tokyo, Japan

HIROSE (Dr H.) Tokyo Astronomical Observatory, Mitaka, Tokyo, Japan

HIRST (P.W.) 1, Clifford Crescent, Pekalmy Township, Bergvliet,
    Cape Province, South Africa

HITOTUYANAGI (Prof. Dr Z.) Department of Astronomy, Faculty of Science,
    Tohoku University, Sendai, Japan

*HJALMARSON (Dr Å.G.) Onsala Space Observatory, S-430 34 Onsala, Sweden

HJELMING (Dr R.M.) National Radio Astronomy Observatory, Edgemont- Road,
    Charlottesville, Va. 22901, U.S.A.

HOAG (Dr A.A.) Lowell Observatory, Flagstaff,
    Arizona 86002, U.S.A.

HOANG BINH (Dr D.) Observatoire de Paris, Section d'Astrophysique de Meudon,
    92190 Meudon, France

HOBBS (Dr L.M.) Yerkes Observatory, Williams Bay, Wisconsin 53191, U.S.A.

HOBBS (Dr R.W.) NASA/Goddard Space Flight Center, Code 614, Solar Physics Branch,
    Laboratory for Space Sciences, Greenbelt, Maryland 20771, U.S.A.

HODGE (Prof. P.W.) Astronomy Department, University of Washington, Seattle,
    Washington 98105, U.S.A.

HOEKSTRA (Dr R.) Laboratorium voor Ruimteonderzoek, Beneluxlaan 21,
    Utrecht, The Netherlands

*HOEY (Dr M.J.) Physics Department, University College, Belfield, Dublin 4, Ireland

HOFFLEIT (Dr E.D.) Yale University Observatory, Box 2023, Yale Station,
    New Haven, Conn., U.S.A.

HØG (E.) Brorfelde-Observatoriet, DK 4340 Tølløse, Denmark

HÖGBOM (Dr J.A.) Stockholm Observatory, S-133 00 Saltsjöbaden, Sweden

HOGG (Dr D.E.) National Radio Astronomy Observatory, Edgemont Road,
    Charlottesville, Virginia 22901, U.S.A.

HÖGLUND (Dr B.) Onsala Space Observatory, 43034 Onsala, Sweden

HOHL (Dr F.) NASA Hampton, NASA Langley Research Center, Virginia, U.S.A.

HOLDEN (F.) 302 S. Market Street, San Jose, California 95113, U.S.A.

HOLMBERG (Prof. E.B.) Eneliden 2, S 43300 Partille, Sweden

HOLWEGER (Dr H.) Institut für Theoretische Physik, Oldhausenstr.,
    2300 Kiel, Germany F.R.

HONEYCUTT (Dr R.K.) Indiana University, Bloomington, Indiana 47401, U.S.A.

HOOGHOUDT (B.G.) University Observatory, Sterrewacht, Leiden, The Netherlands

HOPPE (Prof. Dr J.A.) Sonnenberg Strasse 12, DDR - 69 Jena, Germany D.R.

HORÁK (Dr T.B.) Astronomical Institute, Czechoslovak Academy of Sciences,
    Budečská 6, Prague 2, Czechoslovakia

HORÁK (Dr Z.) Technical University, Technická 4, Prague 6, Czechoslovakia

HORI (Dr G.) Yale University Observatory, Box 2023, Yale Station, New Haven,
    Connecticut, U.S.A.

HORSKÝ (Dr Z.) Astronomical Institute of the Czechoslovak Academy of Sciences,
    Budečská 6, 120 23 Praha 2, Czechoslovakia

HOSHI (Dr R.) Department of Physics, Kyoto University, Kyoto, Japan

HOSKIN (Dr M.A.) Churchill College, Cambridge, U.K.

HOSKING (Dr R.J.) Dept. of Applied Mathematics, University of Waikato,
    Hamilton, New Zealand

HOSOKAWA (Dr Yoshimasa) Yamagata University, Koshirakawa-cho, Yamagata-shi,
    Yamagata, Japan

HOSOYAMA (Dr K.) International Latitude Observatory, Mizusawa, Iwate, Japan

HOTINLI (Dr M.) University Observatory, Beyazit, Istanbul, Turkey

HOUCK (Dr J.) Cornell University, Space Sciences Building, Ithaca,
    New York 14850, U.S.A.

HOUK (Dr N.) Dept. of Astronomy, University of Michigan, Ann Arbor,
    Michigan 48104, U.S.A.

*HOUSE (Dr F.C.) Astronomisches Institut der Ruhr-Universität,
    4630 Bochum, Germany F.R.

HOUSE (Dr L.L.) High Altitude Observatory, Boulder, Colorado 80302, U.S.A.

HOUTGAST (Dr J.) Astronomical Observatory "Sonnenborgh", Zonnenburg 2,
    Utrecht, The Netherlands

HOUZIAUX (Prof. Dr L.N.) Faculté des Sciences, Département d'Astrophysique,
    Plaine de Nimy, B-7000 Mons, Belgium

HOVENIER (J.W.) Wiskundig Seminarium, Vrije Universiteit, De Boelelaan 1081,
    Amsterdam 11, The Netherlands

HOWARD (Dr R.F.) Mount Wilson and Palomar Observatories, 813 Santa Barbara Street,
    Pasadena, California 91106, U.S.A.

HOWARD (Dr W.E. III) National Radio Astronomy Observatory, Edgemont Road,
    Charlottesville, Virginia 22901, U.S.A.

*HOWARD (Dr W.M.) Department of Physics, Northwestern University, Evanston, IL 60 201, U.S.A.

*HOWSE (Commander H.D.) National Maritime Museum, Greenwich, London SE10, U.K.

HOYLE (Prof. F.) St. John's College, Cambridge, U.K.

*HOYNG (Dr P.) Lab. voor Ruimte Onderzoek, Beneluxlaan 21, Utrecht, The Netherlands

*HUA (Dr Chon Trung) LAS, Traverse du Siphon, Les Trois Lucs, 13012 Marseille, France

HUANG (Dr Su-Shu) Dearborn Observatory, Northwestern University, Evanston, Illinois 60201, U.S.A.

HUBBARD (Dr W.B.) Department of Planetary Sciences, University of Arizona, Tucson, Arizona 85721, U.S.A.

HUBE (Dr D.P.) Department of Physics, University of Alberta, Edmonton, Alberta T6G 2E1, Canada

HUBENET (Dr H.) Sterrewacht "Sonnenborgh" Servaas Bolwerk 13, Utrecht, The Netherlands

HUBER (Dr M.C.E.) Atomic Physics and Astrophysics Group, Federal Institute of Technology (ETH-Z), Gloriastrasse 35, CH-8006 Zürich, Switzerland

*HUCHTMEIER (Dr W.K.H.) Hamburger Sternwarte, 205 Hamburg 80, Gojenbergsweg 112, / Germany F.R.

*HUDSON (Dr H.S.) Department of Physics, Revelle College, P.O. Box 109, La Jolla, CA 92037, U.S.A.

HUEBNER (Dr W.F.) University of California, Los Alamos Scientific Laboratory, Los Alamos, New Mexico 87544, U.S.A.

HUFFER (Dr C.M.) 2942 Olive View Road, Alpine, California 92001, U.S.A.

* HUGHES (Dr D.W.) Department of Physics, The University, Sheffield, S10 2TN,U.K.

HUGHES (Dr J.A.) U.S. Naval Observatory, Washington, D.C. 20390, U.S.A.

HUGHES (Dr V.A.) Astronomy Group, Department of Physics, Queen's University, Kingston, Ontario K7L 3N6, Canada

HUGUENIN (Dr G.R.) Dept. of Physics and Astronomy, University of Massachusetts, Amherst, Mass. 01002

HULSBOSCH (Dr A.N.M.) Sterrenkundig Instituut, Katholieke Universiteit, Nijmegen, The Netherlands

HUMMER (Dr D.) Joint Institute for Laboratory Astrophysics, University of Colorado, Boulder, Colorado, U.S.A.

HUMPHREYS (Dr C.J.) Purdue University, Physics Department, Lafayette, Indiana 47907, U.S.A.

HUMPHREYS (Dr R.M.) The University of Minnesota, School of Physics and Astronomy, 148 Physics Building, Minneapolis, Minnesota 55455, U.S.A.

HUMPHRIES (Dr C.M.) Royal Observatory, Blackford Hill, Edinburgh EH9 3HJ, U.K.

HUNAERTS (Dr J.J.) Observatoire Royal de Belgique, Avenue Circulaire, 3, B-1180 Bruxelles, Belgium

HUNDHAUSEN (Dr A.) High Altitude Observatory, Boulder, Colorado, U.S.A.

HUNGER (Prof. Dr K.) Institut für Theoretische Physik und Sternwarte,
    Neue Universität, Olshausenstrasse, 2300 Kiel 1, Germany, F.R.

HUNT (Dr G.E.) Meteorological Office, London Road, Bracknell, Bershire, U.K.

*HUNT (Dr R.) Dept. of Mathematics, University of Strathclyde,
    Livingstone Tower, 26 Richmond Street, Glasgow G.1., U.K.

HUNTEN (Dr D.M.) Kitt Peak National Observatory, Box 26732, Tucson,
    Arizona 85726, U.S.A.

HUNTER (Dr A.) Royal Greenwich Observatory, Herstmonceux Castle,
    Hailsham, Sussex, U.K.

HUNTER (Dr Ch.) Department of Mathematcis, The Florida State University,
    Tallahassee, Fla 32306, U.S.A.

HUNTER (Dr J.H.) Yale University Observatory, Box 2023, Yale Station,
    New Haven, Connecticut 06520, U.S.A.

*HURNIK (Prof. Dr H.) Astronomical Observatory of A. Mickiewicz
    University, Słoneczna 36, 60-286 Poznan, Poland

HURUHATA (Prof. M.) Tokyo Astronomical Observatory, Mitaka, Tokyo, Japan

HURUKAWA (Dr K.) Tokyo Astronomical Observatory, Mitaka, Tokyo, Japan

HUTCHINGS (Dr J.B.) Dominion Astrophysical Observatory, Herzberg
    Institute of Astrophysics, National Research Council, 5071 W. Saanich Road,
    Victoria, B.C. V8X 3X3, Canada

HYDER (Dr C.L.) NASA Solar Group, University of New Mexico, 800 Yale Blvd,N.E.,
    Albuquerque, New Mexico 87106, U.S.A.

* HYLAND (Dr A.R.)Dept. of Astronomy, Australian National University,
    P.O. Box 4, G.P.O., Canberra, A.C.T., Australia

HYNEK (Dr J.A.) Dearborn Observatory, Northwestern University,
    Evanston, Ill. 60201, U.S.A.

* HYSOM (E.J.) 65 Dalkeith Road, Harpenden, Hertfordshire, U.K.

HYUN (Prof. J.J.) Dept. of Astronomy and Meteorology, College of Arts and
    Sciences, Seoul National University, Seoul, Republic of Korea

IANINI (Dr G.) Observatorio Astronomicó, Cordoba, Argentina

*IANNA (Dr P.A.) Dept. of Astronomy, The University of Virginia,
    P.O. Box 3818 University Station, Charlottesville, VA 22903, U.S.A.

IBANOGLU (Dr C.) Ege University Observatory, Bornova, Izmir, Turkey

IBEN (Prof. I.,Jr.) University of Illinois, Urbana-Champaign, Illinois, U.S.A.

ICKE (Dr V.) Institute of Astronomy, University of Cambridge,
    Madingley Road, Cambridge CB3 OEZ, U.K.

IDLIS (Dr G.M.) Institute for History of Science and Technology,USSR Academy
    of Science, 103012, Moscow, U.S.S.R.

IIJIMA (Dr S.) Tokyo Astronomical Observatory, Mitaka, Tokyo, Japan

IKEDA (Dr T.) International Latitude Observatory, Mizusawashi, Iwate-Ken,Japan

IKHSANOV (Prof. Dr R.N.) Main Astronomical Observatory, USSR Academy of Sciences, Pulkovo, 196140 Leningrad, U.S.S.R.

IKSHANOVA (Dr V.N.) Main Astronomical Observatory, USSR Academy of Sciences, Pulkovo, 196140 Leningrad, U.S.S.R.

ILL (Dr M.) Astronomical Observatory, Baja, Hungary

*ILLES-ALMAR (E.) Konkoly Observatory, Budapest XII, P.O. 114. Box 67, Hungary

*IMBERT (Dr M.) Observatoire de Marseille, 2, Place Le Verrier, 13004 Marseille, France

IMSHENNIK (Dr V.S.) Institute of Applied Mathematics, USSR Academy of Sciences, Moscow, U.S.S.R.

INGRAO (H.C.) 58 Hundreds Road, Wellesley Hills, Mass. 02181, U.S.A.

INNANEN (Dr K.A.) CRESS, Dept. of Physics, York University, 4700 Keele Street Downsview, Ontario M3J 1P3, Canada

INOUE (Dr T.) Kyoto-Sangyo-University, Kamigamo, Kyoto, 603 Japan

IONNISIANI (Dr B.K.) Main Astronomical Observatory, Academy of Sciences, Pulkovo, 196140 Leningrad, U.S.S.R.

IOSHPA (Dr B.A.), Institute for Terrestial Magnetism and Ionosphere, Akademgorodok, 142092 Podol'sk, U.S.S.R.

IPSER (Dr J.R.) Laboratory for Astrophysics and Space Research, University of Chicago, Chicago, Ill. 60637, U.S.A.

IRELAND (Dr J.G.) c/o 13, Gordon Road, Belvedere, Kent, U.K.

IRIARTE (B.) Instituto Nacional de Astrofísica Optica y Electrónica Tonantzintla, Pue. Apartados Postales Nos. 216 y 51, Puebla, Pue. Mexico

IRVINE (Dr W.M.) Dept. of Physics and Astronomy, University of Massachusetts, Amherst, Mass. 01002, U.S.A.

IRWIN (Dr J.B.) Dept. of Earth and Planetary Sciences, Newark State College, Union, N.J. 07083, U.S.A.

ISHIDA (Dr K.) Tokyo Astronomical Observatory, The University of Tokyo, 2-21-1 Osawa, Mitaka, Tokyo, Japan

ISHIZAWA (Dr T.) Dept. of Astronomy, University of Kyoto, Kyoto, Japan

ISOBE (Dr S.) Tokyo Astronomical Observatory, Mitaka, Tokyo, Japan

ISRAEL (Prof. W.) Dept. of Physics, University of Alberta, Edmonton, Alberta T6G 2E1, Canada

*ISSERSTEDT (Dr J.) Astronomisches Institut, Am Hubland, 8700 Würzburg, Germany, F.R.

ITO (Dr K.) Dept. of Physics, Rikkyo University, Nishi-Ikebukuro, Tokyo, Japan

ITZIGSOHN (M.) Observatorio Astronómico, La Plata, Argentina

IVANCHUK (Dr V.I.) Kiev University, Observatornaya 3, Kiev 53, Ukrainian SSR, U.S.S.R.

*IVANOV (Dr G.) Astronomical Observatory, Varna, Bulgaria

IVANOV (Dr V.V.) Astronomical Observatory, Leningrad University,
Leningrad, U.S.S.R.

IVANOV-KHOLODNY (Dr G.S.), Institute of Applied Geophysics, USSR Academy of
Sciences, 107150 Moscow, U.S.S.R.

IWANISZEWSKA-LUBIENSKA (Dr C.) Astronomical Observatory, N. Copernicus University,
Sienkiewicza 30, Toruń, Poland

IWANOWSKA (Prof. Dr W.) Institute of Astronomy, Nicolaus Copernicus University,
ul. Chopina 12/18, Pl - 87-100 Toruń, Poland

*IWASAKI (Dr K.) Kwasan Observatory, University of Kyoto, Yamashina,
Kyoto, Japan

IZVEKOV (Dr V.A.) Institute of Theoretical Astronomy, 10, Kutuzov Qay,
Leningrad, 192187, U.S.S.R.

JAAKKOLA (T.S.) University of Helsinki, Observatory and Astrophysics
Laboratory, Tähtitornimäki, SF- 00130 Helsinki 13, Finland

JACCHIA (Dr L.) Smithsonian Astrophysical Observatory, 60 Garden Street,
Cambridge, Mass. 02138, U.S.A.

*JACKISCH (Dr G.) Zentralinstitut für Astrophysik, Sternwarte Sonneberg,
DDR- 64 Sonneberg, Germany D.R.

JACKSON (C.) Observatorio Austral Yale-Columbia, C. de C. 263,
San Juan, Argentina

JACKSON (Dr J.C.)Dept. of Mathematics, University College London, Gower Street,
London WCIE 6BT, U.K.

JACKSON (Dr P.) Universitäts-Sternwarte Wien, Türkenschanzstr. 17,
A 1180 Vienna, Austria

JACKSON (Prof. W.M.) 525 College St. N.W., Dept. of Chemistry, Howard University,
Washington, D.C. 20059, U.S.A.

*JACOBS (Dr K.C.) Max-Planck-Institut für Astrophysik, Föhringer Ring 6
8 München 40, Germany, F.R.

JACOBSEN (Prof. T.S.) University of Washington, Seattle 5, Washington 98115, U.S.A.

JACQUINOT (P.) Laboratoires de Bellevue, 1, Place Aristide-Briand,
Bellevue, 92-Meudon, France

JÄGER (Prof. Dr F.W.) Heinrich-Hertz Institut für Solar-Terrestrische Physik,
DDR 15 Potsdam, Telegrafenberg, Germany D.R.

JAKIMIEC (Dr J.) Wrocław University Astronomical Institute, Kopernika 11,
Wrocław 9, Poland

JAMES (J.F.) The Schuster Physical Laboratory, The University, Manchester M13 9PL,
U.K.

JAMES (Dr R.A.) Department of Astronomy, The University, Manchester 13, U.K.

JAMESON (Dr R.F.) Dept. of Astronomy, The University, Leicester, U.K.

*JANES (Dr K.) Boston University, 725 Commonwealth Avenue, Boston MA 02215, U.S.A.

*JANICZEK (Dr P.M.) Naval Observatory, NAO, Washington, DC 20390, U.S.A.

*JANKOVICS (Dr I.) Box 67, H-1525 Budapest, Hungary

JÄRNEFELT (Prof. G.J.) Ohjaajantie 3 B 18, Helsinki 40, Finland

JARRETT (Prof. A.H.) Boyden Observatory, P.O. Box 334, Bloemfontein 9300
Republic of South Africa

JARZEBOWSKI (Dr T.) Astronomical Observatory, Kopernika 11, Worcław, Poland

JASCHEK (Dr C.) Observatoire, 11, rue de l'Université, 67000 Strasbourg, France

JASCHEK (Dr M.) Observatoire, 11, rue de l'Université, 67000 Strasbourg, France

JASTROW (Dr R.) NASA, Institute for Space Studies, 2880 Broadway,
New York, 10025, N.Y., U.S.A.

JAUNCEY (Dr D.L.) CSIRO Division of Radiophysics, P.O. Box 76,
Epping, N.S.W. 2121, Australia

JAVET (Prof. Dr P.) Chantemerle 19, 1010 Lausanne, Switzerland

JAYARAJAN (A.P.) Astrophysical Observatory, Kodaikanal-3, India

JEFFERIES (Dr J.T.) Institute for Astronomy, University of Hawaii,
2840 Kolowalu Street, Honolulu, Hawaii 96822, U.S.A.

JEFFERS (Dr S.) CRESS, Dept. of Physics, York University, 4700 Keele Street,
Downsview, Ontario M3J 1P3, Canada

JEFFREYS(Lady) Girton College, Cambridge CB3 OJG, U.K.

JEFFREYS (Sir Harold) 160 Huntingdon Road, Cambride, U.K.

JEFFREYS (Dr W.) University of Texas, Dept. of Astronomy, Austin, Texas 78712,
U.S.A.

JELLEY (Dr J.V.) Nuclear Physics Division, Building 8, Atomic Energy
Research Establishment, Harwell, Didcot, Berkshire, U.K.

JENKINS (Dr E.B.) Princeton University Observatory, Princeton, N.J. 08540, U.S.A.

JENKINS (L.F.) Yale University Observatory, Box 2023 Yale Station,
New Haven, Conn., U.S.A.

*JENNER (Dr D.C.) Dept. of Astronomy, University of California,
Los Angeles, CA 90024, U.S.A.

*JENNINGS (Dr R.E.) Dept. of Physics and Astronomy, University College,
Gower St., London WC1E 6BT, U.K.

JENNISON (Prof. R.C., the Electrcnics Laboratory, University of Kent,
Canterbury, Kent, U.K.

JENSCH (A.). Carl Zeiss Observatory, Jena, Germany

JENSEN (Prof. E.) Director, Institute of Theoretical Astrophysics,
University of Oslo, Blindern, Oslo 3, Norway

JERZYKIEWICZ (Dr M.) Astronomical Institute, University of Wrocław,
Ul Kopernika 11, Wrocław 9, Poland

*JOCKERS (Dr K.) Max-Planck-Institut für Physik und Astrophysik
Föhringer Ring 6, Germany, F.R.

JOHANSEN (Dr K.T.) University Observatory, Brorfelde, Tølløse, Denmark

*JOHNSON (Dr D.R.) National Bureau of Standards, Molecular
Spectroscopy Section, Physics B-268, Washington, DC 20234, U.S.A.

JOHNSON (Dr F.M.) Dept. of Physics, California State University, Fullerton
    CA 92634, U.S.A.

JOHNSON (Dr H.L.) 8431 E. Appomattox Street, Tucson, Arizona 85710, U.S.A.

JOHNSON (Dr H.M.) Orgn. 52 12 Bldg. 202, 3251 Hanover St.
    Palo Alto, CA 94304, U.S.A.

JOHNSON (Prof. H.R.) University of Indiana, Dept. of Astronomy,
    Swain Hall West 319, Bloomington, Indiana 47401, U.S.A.

JOHNSON (Dr M.C.) Dept. of Physics, University, Edgbaston, Birmingham 15, U.K.

*JOHNSON (Dr T.V. Jet Propulsion Laboratory
    183-501 4800 Oak Grove Drive, Pasadena, CA 91103, U.S.A.

JOHNSTON (Dr K.) Naval Research Laboratory
    Washington D.C. 20390, U.S.A.

JOKIPII (Dr J.R.) 4981 Calle Luisa
    Tucson, Arizona 85718, U.S.A.

JOLY (Dr F.) Laboratoire d'Astrophysique de l'Université de Bordeaux,
    40, rue Lamartine, 33400 Talence, France

JONES (A.F.) 14 Main Road, Tahunanui, Nelson, New Zealand

JONES (Dr B.F.) Lick Observatory, University of California,
    Santa Cruz, California 95060, U.S.A.

JONES (Dr D.H.P.) Royal Greenwich Observatory, Herstmonceux Castle,
    Hailsham, Sussex, U.K.

*JONES (Dr E.M.) Los Alamos Scientific Laboratory, P.O. Box 1663 , Mail Stop 664
    Los Alamos, NM 87545, U.S.A.

JORDAN (Dr C.) Dept. of Theoretical Physics, University of Oxford,
    12 Parks Road, Oxford, U.K.

JORDAN (Dr H.L.) Director am Institut für Plasmaphysik der Kernforschungs-
    anlage Jülich G.m.b.H., Postfach 365, 517, Jülich 1, Germany, F.R.

JORDAN (Dr S.D.) Goddard Space Flight Center, Greenbelt, Maryland 20771,
    U.S.A.

JØRGENSEN (Dr H.E.) University Observatory, ØsterVoldgade 3 , DK-1350 Copenhagen K
    Denmark

JOSEPH (Dr J. H.) Dept. of Geophysics & Planetary Sciences, Tel-Aviv University,
    Ramat Aviv, Israel

JOSHI (Dr M.N. Radio Astronomy Centre, Post Box 8, Ootacamund 643001, India

JOSHI (S.C.) Uttar Pradesh State Observatory, Manora Peak, Naini Tal, India

JUGAKU (Dr J.) Tokyo Astronomical Observatory, Mitaka- Tokyo, Japan

JUNG (Dr J.) Observatoire de Strasbourg, 11, rue de l'Université
    67000 Strasbourg, France

JUNKES (Dr J.;S.J.) Specola Vaticana, Castel Gandolfo (Rome) Vatican City State

JUPP (Dr A.H.) Dept. of Applied Mathematics, University of Liverpool,
    P.O.Box 147, Liverpool L69 3BX, U.K.

*JURA (Dr M.) Dept. of Astronomy, University of California, Los Angeles,CA 900024
    U.S.A.
JURKEVICH (Dr I.) Naval Research Lab.,Code 7850,Washington, DC 20390, U.S.A.

KABURAKI (Prof. M.) Department of Astronomy, Faculty of Science, University of Tokyo, Yayoicho, Bunkyo-ku, Tokyo, Japan

KADLA (Dr Z.) Pulkovo Observatory, Leningrad M 140, U.S.S.R.

KAFKA (P.) Max-Planck-Institut für Physik und Astrophysik, 8000 München, Föhringer Ring 6, Germany F.R.

KAFTAN-KASSIM (Dr M.A.) State University of New York, 1400 Washington Avenue, Albany, New York 12203, U.S.A.

*KÄHLER (Dr H.) Hamburger Sternwarte, 205 Hamburg 80, Gojenbergsweg 112, Germany F.R.

* KAHLER (Dr S.) American Science and Engineering, Inc., 955 Massachusetts Avenue, Cambridge, MA 02139, U.S.A.

KAHN (Dr F.D.) Department of Astronomy, The University, M13 9PL Manchester, U.K.

KAI (Dr Keizo) Tokyo Astronomical Observatory, University of Tokyo, Mitaka, Tokyo, Japan

*KAIFU (Dr Norio) Department of Astronomy, University of Tokyo, Bunkyo-Ku, Tokyo, Japan

KAIJDANOVSKIJ (Dr N.L.) Special Astrophysical Observatory, USSR Academy of Sciences, Leningrad Branch of the SAO, Pulkovo 196140, U.S.S.R.

KAISER (Prof. T.R.) Department of Physics, The University, Sheffield 10, U.K.

KAKINUMA (Dr T.) Research Institute of Atmospherics, Nagoya University, Toyokawa, Aichi, Japan

KAKUTA (Dr Ch.) International Latitude Observatory of Mizusawa, Mizusawa, Japan

KALANDADZE (Dr N.B.) Astrophysical Observatory, Abastumani, Georgia, U.S.S.R.

KALER (Dr J.B.) University of Illinois Observatory, Urbana, Illinois 61801, U.S.A.

KALINKOV (M.) Section d'Astronomie de l'Académie Bulgare des Sciences, Rue 7 Novembre No 1, Sofia, Bulgaria

KALKOFEN (Dr W.) Smithsonian Astrophysical Observatory, 60 Garden Street, Cambridge, Mass. 02138, U.S.A.

KALLOGLIAN (Dr A.T.) Byurakan Astrophysical Observatory, Byurakan, Armenia, U.S.S.R.

*KÁLMÁN (Dr B.) Heliophysical Observatory of the Hungarian, Academy of Sciences, H-4010 Debrecen, P.O. Box 30, Hungary

KALMYKOV (Dr A.M.) Director, Kitab Latitude Station, Uzbek Academy of Sciences, 731740 Kitab, U.S.S.R.

KALNAJS (Dr A.J.) Mt. Stromlo and Siding Spring Observatories, School of Physical Sciences, The Australian National University, Box 4, P.O., Canberra, A.C.T., 2600, Australia

*KAMEL (Dr O.S.M.) Faculty of Science, Cairo University, Geza Orman, Arab Rep. of Egypt

KAMIJO (Dr Fumio) Department of Astronomy, University of Tokyo, Yayoi, Bunkyo-ku, Tokyo, Japan

KAMINISHI (Prof. Dr K.) Department of Physics, Faculty of Science, Kumamoto University, Kumamoto, Japan

*KAMPER (Dr K.W.) David Dunlap Observatory, University of Toronto, Richmond Hill, Ontario, Canada L4C 4Y6

*KANAEV (Dr I.I.) Main Astronomical Observatory, Academy of Sciences, Pulkovo, U.S.S.R.

KANDEL (Dr R.S.) Observatoire de Meudon 92190 Meudon, France

KANE (Dr S.R.) Space Sciences Laboratory, University of California, Berkeley, California 94720, U.S.A.

*KANEKO (Dr Noboru) Department of Physics, Faculty of Science, Hokkaido University, Japan

KANNO (Dr Mitsuo) Hida Observatory, University of Kyoto, Kamitakara, Gifu-ken, Japan

KANYO (S.) Konkoly Observatory, 1121 Budapest-Szabadsághegy, Hungary

*KAPAHI (Dr V.K.) Radio Astronomy Centre, Post Box No.8, Ootacamund 643001, India

KAPLAN (Dr J.) Department of Physics, University of California at Los Angeles, Los Angeles, California, U.S.A.

KAPLAN (L.) Jet Propulsion Laboratory, 4800 Oak Grove Drive, Pasadena, California, U.S.A.

KAPLAN (Prof. Dr S.A.) Radiophysical Research Institute, Gorkii, U.S.S.R.

KARACHENTSEV (Dr I.D.) Special Astrophysical Observatory, Academy of Sciences USSR, ul. Berezhnogo 167, st. Zelenchukskaja,Stavropolsky kraj, 357140 U.S.S.R.

KARANDIKAR (Dr R.V.) Director, Nizamiah Observatory, Begumpet, Hyderabad-16, India

KARDASHEV (Dr N.S.) Sternberg Astronomical Institute, Moscow, U.S.S.R.

KARLSSON (Dr B.A.R.) Institute of Astronomy, S-222 24 Lund, Sweden

KARPINSKIJ (Dr V.N.) Main Astronomical Observatory, USSR Academy of Sciences, Pulkovo,   196140 Leningrad, U.S.S.R.

KARYAGINA (Dr Z.V.) Astrophysical Institute, Alma-Ata, U.S.S.R.

KASHSCHEEV (Prof. Dr B.L.) Institute of Radio Physics and Electronics, Ukrainian SSR Academy of Sciences, 310085 Khar'kov, U.S.S.R.

KASPER (Dr U.) Zentralinstitut für Astrophysik, Telegrafenberg, DDR 15, Potsdam, Germany D.R.

KASTLER (Prof. A.) Ecole Normale Supérieure, 45, rue d'Ulm, 75014 Paris, France

KATASEV (Dr L.A.), Astronomical Council, USSR Academy of Sciences, Pyatnitskaya ul. 48, 109017 Moscow, U.S.S.R.

KATGERT-MERKELIJN (Dr J.K.) Sterrewacht te Leiden, Huygens Laboratorium, Wassenaarseweg 78, Leiden, The Netherlands

KATO (Dr Shoji) Institute of Astrophysics, College of Science, University of Kyoto, Kyoto, Japan

KATSIS (Dr D.N.) 12, rue Varnis, Nea Smyrne, Athens, Greece

*KAUFMAN (Dr M.) Department of Physics, University of Notre Dame, Notre Dame, IN 46556, U.S.A.

*KAUFMANN (Dr J.P.) Institut für Astrophysik, Technische Universität,
1 Berlin 10, Ernst Reuter Platz 7

KAUFMANN (Prof. P.) Centro de Rádio-Astronomia e Astrofísica, Universidade
Mackenzie, C.P. 8792, São Paulo, Brazil

KAULA (Dr W.M.) Institute of Geophysics and Planetary Physics, University of
California, Los Angeles, California 90024, U.S.A.

KAWABATA (Dr Kin-aki) Tokyo Astronomical Observatory, University of Tokyo,
Mitaka, Tokyo, Japan

KAWABATA (Dr S.) Kyoto Gakuen University, Sogabe-cho, Kameoka, Kyoto, Japan

KAWAGUCHI (Dr I.) Kwasan Observatory, University of Kyoto, Yamashina,
Kyoto, Japan

KAZES (Dr I.) Observatoire de Paris, Section d'Astrophysique, 92190 Meudon, France

KAZIMIRCHAK-POLONSKAYA (Dr E.I.) Institute of Theoretical Astronomy,
10, Kutuzov Qay, Leningrad, 192187, U.S.S.R.

KEAY (Dr C.S.L.) University of Newcastle, New South ·Wales, Australia

KEENAN (Dr P.C.) Perkins Observatory, Delaware, Ohio 43015, U.S.A.

KEGEL (Dr W.H.) Institut für Theoretische Astrophysik der Universität,
69 Heidelberg, Im Neuenheimer Feld 294, Germany F.R.

*KELLER (Dr C.F.) Los Alamos Scientific Laboratory, P.O. Box 1663,
Los Alamos, NM 87544, U.S.A.

KELLER (Dr G.) Department of Astronomy, Ohio State University, 174 West
18th Avenue, Columbus, Ohio 43210, U.S.A.

KELLER (Dr H.U.) Director Observatory and Planetarium, P.O.B. 161, Neckarstr. 47,
D 7000 Stuttgart 1, Germany F.R.

*KELLER (Dr H.Uwe) MPI for Aeronomie
3411 Katlenburg, Lindau 3, Germany, F.R.

KELLERMAN (Dr K.I.) National Radio Astronomical Observatory, Box 2,
Green Bank, West Virginia 24944, U.S.A.

KELLOG (Dr W.W.) National Center for Atmospheric Research, Boulder,
Colorado 80302, U.S.A.

*KELLOGG (Dr E.M.) Smithsonian Astrophysical Observatory, 60 Garden Street,
Cambridge, MA 02138, U.S.A.

KENDERDINE (Dr S.) Mullard Radio Astronomy Observatory, Cavendish Laboratory,
Free School Lane, Cambridge, U.K.

KENNEDY (Prof. J.E.) Department of Physics, University of Saskatchewan,
Saskatoon, Saskatchewan S7N QWO, Canada
KERES (Prof. Dr Kh.P.), Institute for Astrophysics and Physics of the Atmosphere,
Estonian Academy of Sciences, Toravere, 202444 Tartu, U.S.S.R.
KERR (Dr F.J.) Astronomy Program, University of Maryland, College Park,
Maryland 20742, U.S.A.

KERR (Prof. R.P.) Dept. of Mathematics, University of Canterbury, Private Bag,
Christchurch, New Zealand

KESSLER (Dr K.G.) Atomic Physics Division, National Bureau of Standards 13.00,
Washington , D.C., 20234, U.S.A.

KESTEVEN (Prof. M.L.) Astronomy Group, Dept. of Physics,
Queen's University, Kingston, Ontario K7L 3N6, Canada

KHACHIKIAN (Dr E.Y.) Byurakan Astrophysical Observatory, Byurakan,
Armenia, U.S.S.R.

*KHARE (Dr B.N.) Cornell University, Center for Radiophysics and Space
Research, Space Sciences Building, Ithaca, NY 14850, U.S.A.

KHARADZE (Prof. Dr E.K.) Director, Astrophysical Observatory,
Abastumani, Georgia, U.S.S.R.

KHARIN (Dr A.S.) The Main Astronomical Observatory of the Ukrainian
Academy of Sciences, Kiev 252127, U.S.S.R.

KHARITONOV (Dr A.V.) Astrophysical Institute of Kazak, Akad. Sci.
Alma-Ata 68, Kamenskoe plato, U.S.S.R.

KHETSURIANI (Dr T.S.) Abastumani Astrophysical Observatory,
Abastumani, Georgia - USSR

KHOKHLOVA (Mrs Dr V.L.), Astronomical Council, USSR Academy of Sciences,
Pyatnitskaya ul. 48, 109017 Moscow, U.S.S.R.

KHOLPOV (Dr P.N.) Sternberg State Astronomical Institute,
117234 Moscow, U.S.S.R.

KHOLSHEVNIKOV (Dr K.V.) Leningrad University, Faculty of Mathematics,
10 linia 33, Leningrad 199178, U.S.S.R.

KHROMOV (Dr G.S.) Astronomical Council of the USSR Academy of Sciences,
Pyatnitskaya 48, Moscow 109017, U.S.S.R.

KIANG (Dr T.) Dunsink Observatory, Castleknock, Co. Dublin, Ireland

*KIASATPOOR (Prof.A.) Professor of Astronomy, University of Isfahan,
Isfahan, Iran

*KIBBLEWHITE (Dr E.J.) Institute of Astronomy, Madingley Road,
Cambridge CB3 OHA, U.K.

KIKUCHI (Dr S.) Astronomical Institute, Tohoku University, Sendai, Japan

KILADZE (Dr R.I.) Abastumani Astrophysical Observatory,
Georgian Academy of Sciences, Abastumani, U.S.S.R.

KILAR (Dr B.) Faculty of Architecture, Civil Engineering and Geodesy,
Ljubljana University, Jamova 2, 51000 Ljubljana, Yugoslavia

*KILKENNY (Dr D.) University Observatory, St. Andrews, Fife, KY16 9LZ, U.K.

KIM (Y.H.) Pyongyang Astronomical Observatory, Academy of Sciences,
Pyongyang, Korea (D.P.R.)

KING (Dr D.S.) Dept. of Astronomy, University of New Mexico, Albuquerque,
New Mexico 87106, U.S.A.

KING (Dr H.C.) LcLaughlin Planetarium, Royal Ontario Museum, Toronto, Ontario,
Canada

KING (Dr I.R.) Berkeley Astronomy Department, University of California,
Berkeley, California 94720, U.S.A.

KING (Dr R.B.) P.O. Box 725, Mendocino, California 95460, U.S.A.

KING-HELE (D.G., F.R.S.) Royal Aircraft Establishment, Farnborough,
Hants, U.K.

KINMAN (Dr T.D.) Kitt Peak National Observatory, P.O. Box 26732,
950 North Cherry Avenue, Tucson, Arizona 85726, U.S.A.

KINOSHITA (Dr H.) Tokyo Astronomical Observatory, Mitaka, Tokyo 181,
Japan

KIPPENHAHN (Prof. R.) Max-Planck-Institut für Physik und Astrophysik,
Föhringer Ring 6, 8000 München 40, Germany, F.R.

KIPPER (Prof.A.) W. Struve Astrophysical Observatory, Tartu, Tõravere,
Estonia, U.S.S.R.

KIPPER (Dr T.A.) W. Struve Tartu Astrophysical Observatory, Tõravere,
Tartu 202444 Estonia, U.S.S.R.

KIRAL (Dr A.) University Observatory, Beyazit, Istanbul, Turkey

*KIRBIYIK (Dr H.) Physics Department, Middle East Technical University,
Ankara, Turkey

*KIRKPATRICK (Dr R.C.) MS 220, Los Alamos Scientific Laboratory,
Los Alamos, NM 87544, U.S.A.

KISELEV (Dr A.A.) Main Astronomical Observatory, USSR Academy of Sciences,
Pulkovo, U.S.S.R.

KISLYAKOV (Dr A.G.) Radiophysical Research Institute, Gorkij, U.S.S.R.

*KISLYUK (Dr V.S.) Main Astronomical Observatory of the Ukrainian
Academy of Sciences, Kiev 252127, U.S.S.R.

KISSELL (Dr K.E.) Director, AF Avionics Lab./RSO, Wright-Patterson Air Force Base,
Ohio 45433, U.S.A.

KITAMURA (Dr M.) Tokyo Astronomical Observatory, Mitaka, Tokyo, Japan

KITAMURA (Dr S.) Science Education Institute of Osaka, Sumiyoshi-ku,
Osaka, Japan

KIZILIRMAK (Prof. Dr A.) Ege University Observatory, P.K. 21, Bornova
Izmir, Turkey

KJAERGAARD (P.) Copenhagen University Observatory, Øster Voldgade 3,
DK- 1350, Denmark

KJELDSETH MOE (Dr O.) Institute of Theoretical Astrophysics, University of Oslo,
P.O. Box 1029, Blindern, Oslo 3, Norway

KLARE (Dr G.) Bad Landessternwarte, 6900 Heidelberg-Königstuhl, Germany F.R.

KLARMAN (Dr J.) Washington University, Department of Physics,
Saint Louis, Missouri 63130, U.S.A.

KLECZEK (Dr J.) Astronomical Institute, Czechoslovak Academy of Sciences,
Observatory Ondřejov, Czechoslovakia

KLEIN (Dr M.J.) Jet Propulsion Laboratory, Pasadena, Cal. 91103, U.S.A.

KLEINMANN (Dr D.) Smithsonian Astrophysical Obser. 60 Garden Street,
Cambridge, Mass. 02138, U.S.A.

KLEMOLA (Dr A.R.) Lick Observatory, University of California, Santa Cruz,
California 95060, U.S.A.

KLEMPERER (Dr W.K.) Storage Technology Corporation, Room 159, Bldg 3, Dept.816, Broomfield, Colorado 80027, U.S.A.

KLEPCZYNSKI (Dr W.J.) U.S. Naval Observatory, Washington D.C. 20390, U.S.A.

KLIMISHIN (Dr I.A.), Pedagogic Institute, 284000 Ivano-Frankovsk, U.S.S.R.

KLINGLESMITH (Dr D.A.) Goddard Space Flight Center, Laboratory for Optical Astronomy, Greenbelt, Maryland 20771, U.S.A.

KLOCK (Dr B.L.) U.S. Naval Observatory, Dept. of the Navy, Washington DC 20390, U.S.A.

KNAFLICH (H.B.) The Boeing Company, Aerospace Group, M.S. 88-21, P.O. Box 3999, Seattle, Washington 98124, U.S.A.

*KNAPP (Dr G.R.) Owens Valley Radio Observatory, California Institute of Technology, Pasadena, CA 91125, U.S.A.

*KNEER (Dr F.) Fraunhofer-Institut, Schöneckstr.6, 78 Freiburg i.Br., Germany, F.R.

KNOWLES (S.B.) Naval Research Laboratory, Washington D.C.20390, U.S.A.

KO (Dr Hsien-Ching) Dept. of Electrical Engineering, Ohio State University, Columbus, Ohio 43210, U.S.A.

KOBRIN (Prof. Dr M.M.) Radiophysical Research Institute, Gorkii, U.S.S.R.

*KOÇER (Dr D.) University Observatory, Üniversité, Istanbul, Turkey

KOCH (Dr L.) Centre d'Etudes Nucléaires de Saclay, Service Electronique Physique, B.P. no.2, 91 Gif-sur-Yvette, France

KOCH (Dr R.H.) Dept. of Astronomy, University of Pennsylvania, David Rittenhouse Laboratory, 33rd and Walnut Streets, Philadelphia, Pa.19174 U.S.A.

*KOCHAROV (Dr G.E.) 194021 Physico-Technical Institute, Academy of Sciences USSR, Leningrad, U.S.S.R.

KODAIRA (Dr K.) Dept. of Astronomy, Faculty of Sciences, University of Tokyo, Bunkyo-ku, Tokyo, Japan

KOECKELENBERGH (Dr A.) Observatoire Royal de Belgique, ave Circulaire 3, B-1180 Bruxelles, Belgium

KOEHLER (Dr J.A.R.) Department of Physics, University of Saskatchewan, Sakatoon, Saskatchewan S7N OWO, Canada

KOELBLOED (Dr D.) Astronomical Institute, Roetersstraat 1a, Amsterdam, The Netherlands

*KOESTER (Dr D.) Institut für Theoretische Physik und Sternwarte der Universität Kiel, Olshausenstrasse, 23 Kiel 1, Germany, F.R.

KOGURE (Dr T.) Dept. of Physics, Faculty of Science, Ibaraki University, Mito (310) Japan

*KOHL (Dr J.L.) Harvard College Observatory, 60 Garden Street, Cambridge, MA 02138, U.S.A.

KOHLER (Prof. Dr H.) Sauerbruchstrasse 6, Heidenheim a.d.Brenz Germany, F.R.

KOHOUTEK (Dr L.) Hamburger Sternwarte, 205 Hamburg, Bergedorf, Germany F.R.

KOLACZEK (Dr B.) Al. Armii Ludowej 17/55, 00-632 Warszawa, Poland

KOLCHINSKIJ (Dr I.G.) Main Astronomical Observatory, Ukrainian Academy of Sciences, Kiev, U.S.S.R.

KOLESNIK (Dr L.N.) Main Astronomical Observatory, Academy of Sciences Ukrainian SSR, Kiev -252127, U.S.S.R.

KOLESOV (Dr A.K.) Astronomical Observatory, Leningrad, U.S.S.R.

KOLLBERG (Dr E.L.) Chalmers Univ. of Technology, Research Laboratory of Electronics, Fack, S-402 20 Göteborg 5, Sweden

KOMAROV (Dr N.S.) Odessa Astronomical Observatory, Shevchenko Park, Odessa 270014, U.S.S.R.

KOMESAROFF (M.M.) C S I R O Division of Radiophysics, P.O. Box 76, Epping, N.S.W. 2121, Australia

KONDO (Dr M.) Institute of Earth Science and Astronomy, College of General Education, University of Tokyo, Komaba 3-8-1, Meguroku, Tokyo, 153, Japan

KONDO (Dr Y.) Chief, Astrophysics Section (TN23) NASA-Manned Spacecraft Center, Houston, Texas 77058, U.S.A.

KONIN (Dr V.V.) Nikolaev Observatory, Nikolaev, U.S.S.R.

KONOPLEVA (Dr V.P.) Main Astronomical Observatory, Academy of Sciences, Ukrainian SSR, Kiev 127, U.S.S.R.

KONOVOVICH (Dr E.V.) Sternberg Astronomical Institute, Moscow, V-234, U.S.S.R.

KOPAL (Prof. Z.) Dept. of Astronomy, The University, Manchester 13, U.K.

KOPECKÝ (Dr M.) Astronomical Institute, Czechoslovak Academy of Sciences, Observatory Ondřejov, Czechoslovakia

*KOPP (Dr R.A.) High Altitude Observatory, NCAR, P.O.Box 3000, Boulder, CO 80303, U.S.A.

KOPYLOV (Dr I.M.) Director, Special Astrophysical Observatory, U S S R Academy of Sciences, St. Zelenchukskaya, Stavropolsky Kraj, U.S.S.R.

KORCHAK (Dr A.A.) Institute of Terrestrial Magnetism, Ionosphere and Propagation of Radio Waves, USSR Academy of Sciences, Moscow, U.S.S.R.

KORDYLEWSKI (Dr K.) Astronomical Observatory, Kopernika 27, Cracow, Poland

KOROL (Dr A.K.) Main Astronomical Observatory, Ukrainian Academy of Sciences, Kiev, U.S.S.R.

KOROLKOV (Dr D.V.) Special Astrophysical Observatory, USSR Academy of Sciences, Leningrad Branch of the SAO, Pulkovo 196140, U.S.S.R.

KOSIN (Dr G.S.) Main Astronomical Observatory, USSR Academy of Sciences, Pulkovo, 196140 Leningrad, U.S.S.R.

KOSTINA (Dr L.D.) Main Astronomical Observatory, USSR Academy of Sciences, Pulkovo, U.S.S.R.

KOSTYAKOVA (Dr E.B.) Sternberg Astronomical Institute, Moscow, U.S.S.R.

KOSTYLEV (Dr K.V.) Engelhardt Observatory, Kazan, U.S.S.R.

KOTELNIKOV (V.A. Acad.) Commission for Radioastronomy, USSR Academy of Sciences, Moscow, U.S.S.R.

KOTHARI (Dr D.S.) Dept. of Physics, University of Delhi, New Delhi, India

*KOTOV (Dr V.A.) Crimean Astrophysical Observatory of the USSR Academy of
   Sciences, P/O Nauchny, Crimea 334413, U.S.S.R.

KOTSAKIS (Prof. Dr D.) 189 Hippocrates Str., Athens 708, Greece

KOURGANOFF (Prof. V.) Faculté des Sciences, Service d'Astronomie,
   91- Orsay, France

KOUTCHMY (Dr S.) Institut d'Astrophysique, 98 bis, Boulevard Arago,
   75014 Paris, France

KOVACHEV (Dr B.J.) Section d'Astronomie de l'Académie Bulgare des Sciences,
   Rue 7 Novembre No 1, Sofia, Bulgaria

*KOVÁCS (A.) Heliophysical Observatory of the Hungarian Academy of Sciences,
   P.O. Box 30, H-4010 Debrecen, Hungary

KOVAL (Dr I.K.) Main Astronomical Observatory, Ukrainian Academy of Sciences,
   Kiev, U.S.S.R.

KOVALEVSKY (Dr J.) Directeur du Centre d'Etudes et de Recherches Géodynamiques
   et astronomiques (CERGA) 8, Blvd. Emile Zola, 06130 Grasse, France

KOVAR (Dr N.S.) Associate Professor,Physics Dept. University of Houston,
   Texas 77004, U.S.A.

KOVAR (Dr R.P.) 9666 E. Orchard Drive, Englewood, Colorado 80110, U.S.A.

KOVETZ (Dr A.) Dept. of Physics and Astronomy, Tel-Aviv University,
   Ramat-Aviv, Israel

KOYAMA (Dr S.) Kagawa University, Takamatsu, Kagawa, Japan

KOZAI (Dr Y.) Tokyo Astronomical Observatory, Mitaka, Tokyo, Japan

KOZHEVNIKOV (Dr N.I.) Sternberg Astronomical Institute, Moscow, U.S.S.R.

KOZIEL (Prof. Dr K.) Jagellonian University, Observatory,
   Kopernika 27/3, Cracow, Poland

KOZLOVSKY (Dr B.Z.) Department of Physics and Astronomy, Tel-Aviv University
   Ramat Aviv, Israel

*KOZŁOWSKI (Dr M.) Astronomical Observatory of the Warsaw University,
   00478 Warsaw, Aleje Ujazdowskie 4, Poland

KOZYREV (Dr N.A.) Pulkovo Observatory, Leningrad, U.S.S.R.

KRAFT (Dr R.P.) Mount Wilson and Palomar Observatories, 813 Santa Barbara Street,
   Pasadena, Cal., U.S.A.

*KRAICHEVA (Dr Z.) Dept. of Astronomy, Bulgarian Academy of Sciences,
   "7th November" Street 1, 1000 Sofia, Bulgaria

KRAMER (Dr Kh.N.) Astronomical Observatory, Odessa University, Odessa, U.S.S.R.

KRANJC (Dr A.) Osservatorio Astronomico di Bologna, Italy

KRAT (Prof. V.A.) Pulkovo Observatory, Leningrad M-140, U.S.S.R.

KRAUS (Dr J.D.) Radio Observatory, Ohio State University,
   Columbus, Ohio, U.S.A.

KRAUSE (Dr F.) Zentralinstitut für Astrophysik, DDR 15 Potsdam,
   Telegrafenberg, Germany D.R.

KRAUSHAAR (Prof. W.L.) Dept. of Physics, University of Wisconsin, Sterling Hall, Madison, Wisc. 53706, U.S.A.

KREISEL (Dr E.) Zentralinstitut für Astrophysik, Telegrafenberg, DDR 15, Potsdam, Germany, D.R.

KRESÁK, (Dr L.) Astronomical Institute, Slovak Academy of Sciences, 899 30 Bratislava, Czechoslovakia

KRESÁKOVÁ (Dr M.) Astronomical Institute, Slovak Academy of Sciences, Dúbravska cesta A/11, Bratislava, Czechoslovakia

*KRIEGER (Dr A.S.) American Science and Engineering Inc. 955 Massachusetts Ave Cambridge, MA 02139, U.S.A.

KRINOV (Dr E.L.) Committee on Meteorites of the Academy of Sciences of the USSR, ul. M. Ul'ianovoy 3, korpus 1, Moscow W-313, U.S.S.R.

KRISHNAN (Prof. T.) Director, Madras Institute of Technology, Chromepet, Madras 600044, India

KRISHNA SWAMY (Dr K.S.) Theoretical Astrophysics Group, Tata Institute of Fundamental Research, Homi Bhabha Road, Bombay 400 005, India

KRISTENSON (Dr H.) Arkitektvägen 6, S- 245 00 Staffanstorp, Sweden

KRISTIAN (Dr J.) Mount Wilson and Palomar Observatories, Carnegie Institution of Washington, California Institute of Technology, 813 Santa Barbara Street, Pasadena, Cal. 91106, U.S.A.

KRISTIANSSON (Prof. K.G.B.) Dept. of Physics, University of Lund, Sweden

KŘIVSKÝ (Dr L.) Astronomical Institute, Czechoslovak Academy of Sciences, Observatory Ondřejov, Czechoslovakia

KŘIŽ (Dr S.) Astronomical Institute, Czechoslovak Academy of Sciences, Observatory Ondřejov, Czechoslovakia

KROGDAHL (Prof. W.S.) Dept. of Astronomy, University of Kentucky, Lexington, Kentucky, U.S.A.

KRON (Dr G.E.) 416 N. Bertrand Street, Flagstaff, Arizona 86001, U.S.A.

KRON (K.G.) 416 N. Bertrand Street, Flagstaff, Arizona 86001, U.S.A.

KRONBERG (Dr P.P.) David Dunlap Observatory, University of Toronto, Richmond Hill, Ontario L4C 4Y6, Canada

KROOK (Dr M.) Harvard College Observatory, Cambridge, Mass. 02138, U.S.A.

KRUEGER (Dr A.) Heinrich-Hertz-Institut für Solar-Terrestrische Physik, Rudower Chaussee, Berlin-Adlershof, Germany D.R.

KRUEGER (Prof. E.) Via Mauro Macchi 65, Milano, Italy

KRUSZEWSKI (Dr A.) Astronomical Observatory of Warsaw University, Warsaw, Al. Uajzdowskie 4, Poland

KRZEMINSKI (Dr Wojciech) Institute of Astronomy, Al. Uajzdowskie 4, 00-478 Warszawa, Poland

KSANFOMALITI (Dr L.V.) Space Research Institute of Academy of Sciences USSR, Profsojusnaja 88, Moscow, 117485, U.S.S.R.

KUBIAK (Dr M.) Astronomical Observatory of the Warsaw University Al Ujazdowskie 4, 00-478 Warszawa, Poland

KUBICELA (A.) Astronoska opservatorija, Volgina 7, 11050 Beograd, Yugoslavia

KUBOTA (Dr J.) Kwasan Observatory, Yamashina, Kyoto, Japan

KUHI (Dr L.V.) Berkeley Astronomy Department, University of California, Berkeley, California 94720, U.S.A.

KÜHNE (Dr Chr.) Firma Carl Zeiss, 7082 Oberkochen, Postfach 35/36, Germany F.R.

*KUIJPERS (Dr J.M.E.) Rijksuniversiteit Utrecht, Sterrewacht, Servaas Bolwerk 13, Utrecht, The Netherlands

*KUIPER (Dr T.B.H.) 220 San Vicente Blvd. 310, Santa Monica, CA 90402, U.S.A.

KUKARKIN (Prof. Dr B.V.) Sternberg Astronomical Institute, Universitetskij Prospekt 13, Moscow 117234, U.S.S.R.

KUKLIN (Dr G.V.) Post Office Meget, Sibizmiran, Irkutsk, U.S.S.R.

KULIKOV (Prof. Dr K.A.) Moscow State University, 117234 Moscow, U.S.S.R.

KULIKOVSKIJ (Dr P.G.) Sternberg State Astronomical Institute, Moscow V-234, U.S.S.R.

KULKARNI (Dr P.V.) Physical Research Laboratory, Navrangpura, Ahmedabad-9, India

KULSRUD (Dr R.M.) Princeton University, Princeton, New Jersey 08540, U.S.A.

KUMAR (Prof. S.S.) Leander McCormick Observatory, University of Virginia, Charlottesville, Virginia, U.S.A.

KUMSISHVILI (Dr J.I.) Abastumani Astrophysical Observatory, Georgian Academy of Sciences, Abastumani, U.S.S.R.

KUNDU (Dr M.R.) Astronomy Program, University of Maryland, College Park, Maryland 20742, U.S.A.

KUNKEL (Dr W.E.) Observatorio Nacional, Rua General Bruce 586 ZC 08 Rio de Janeiro, GB, Brazil

KÜNZEL (H.) Heinrich-Hertz-Institut für Solar Terrestrische Physik, DDR 15 Potsdam, Telegrafenberg, Germany D.R.

KUPERUS (Dr M.) Astronomical Observatory "Sonnenborgh", Servaas Bolwerk 13, Utrecht, The Netherlands

*KUPO (Dr I.) Dept. of Physics and Astronomy, Tel-Aviv University, Ramat- Aviv, Israel

KUPO (Dr I.D.) Astrophysical Institute of the Kazakh SSR Academy of Sciences, Alma-Ata - 480058, U.S.S.R.

KUPPERIAN (Dr J.E., Jr.) Astrophysics Branch, Goddard Space Flight Center, NASA, Greenbelt, Maryland, U.S.A.

KURIL'CHIK (Dr V.N.) Sternberg Astronomical Institute, Moscow, U.S.S.R.

KUROCHKA (Dr L.N.), Kiev State University, Astronomical Observatory and Chair of Astronomy, 252053 Kiev, U.S.S.R.

KURT (Dr V.G.) Space Research Institute, USSR Academy of Sciences, 117810 Moscow, U.S.S.R.

KURTH (Dr R.) Professor of Mathematics, Georgia Institute of Technology, Atlanta, Ga., U.S.A.

KUSHWAHA (Prof. R.S.) Department of Mathematics, University of Jodhpur, Jodhpur (Raj), India

KUSTAANHEIMO (Prof.P.E.) Dia E. 451, Danmarks Tekniske Hojskole
DK 2800 Lyngby, Denmark

KUTTER (Dr G.S.) Evergreen State College, Olympia, Washington 98505, U.S.A

KUTUZOV (Dr S.A.), Leningrad State University, ul.Smol'nogo, 193124 Leningrad,USSR.

KUZMENKO (Dr K.N.) Khar'kov State University, 310077 Khar'kov, U.S.S.R.

KUZ'MIN (Dr A.D.) Physical Institute, USSR Academy of Sciences, Moscow, U.S.S.R.

KUZMIN (Prof. Dr G.G.) W. Struve Astrophysical Observatory of Tartu,
202 444 Tôravere, Tartu, Estonia, U.S.S.R.

KUZNETSOV (Dr D.A.) Special Astrophysical Observatory Akad. Sci. USSR,
ul Berezhnogo 167, St. Zelenchukskaya Stavropolsky Kraj, 357140, U.S.S.R.

*KVIZ (Dr Z.) University of New South Wales, Broken Hill Division, P.O.Box 334,
Broken Hill, NSW 2880, Australia

KWEE (Dr K.K.) Sterrewacht, Leiden, The Netherlands

LABEYRIE (Dr A.) Observatoire de Paris, Section d'Astrophysique de Meudon,
92190 Meudon, France

LABEYRIE (Dr L.) G.E.A. B.P. No 2, 91 Gif-s/Yvette, France

LABORDE (Dr G.) Observatoire de Paris, Section d'Astrophysique, 92190 Meudon,
France

*LABRUM (N.R.) CSIRO, Division of Radiophysics, Post Office Box 76,
Epping, N.S.W., 2121, Australia

LABS (Dr D.) Landessternwarte, Königstuhl, 69 Heidelberg, Germany F.R.

LACCHINI (Prof. G.B.) Borgo Durbecco, Via A. Cicognani No 10, Faenza (Ravenna)
Italy

LACROUTE (P.) Prof. honoraire, 2, rue d'Alise, 21000 Dijon, France

LACOMBE (Dr C.G.) Observatorio Nacional, Rua General Bruce, 586, São Cristovão
(ZC-08) Rio de Janeiro, Gb., Brazil

LAFFINEUR (Dr M.) Institut d'Astrophysique, 98 bis Boulevard Arago,
75014 Paris, France

*LAFON (Dr J.P.) DESPA, Observatoire de Meudon, 92190 Meudon, France

LAGERQVIST (Prof. A.) Institute of Physics, Stockholms University, Vanadisvägen 9,
S-113 46 Stockholm VA. Sweden

LAHIRI (N.C.) Officer-in-charge, Nautical Almanac Division, Meteorological
Department, Alipore, Calcutta, India

*LAHULLA (Dr J.F.) Observatorio Astronómico Nacional, Madrid, Spain

*LALA (Dr P.) Astronomical Institute of the Czechoslovak Academy of Sciences,
Observatory, 251 65 Ondřejov, Czechoslovakia

LALLEMAND (Dr A.) Observatoire de Paris, 61, Avenue de l'Observatoire,
75014 Paris, France

*LAMB (Dr F.K.) Department of Physics, University of Illinois at Urbana,
Urbana, IL 61801, U.S.A.

LAMBECK (Dr K.) Groupe de Recherches de Géodésie Spatiale, Observatoire de Paris,
Section d'Astrophysique de Meudon, 92190 Meudon, France

LAMBERT (Dr D.) The University of Texas at Austin, College of Natural Sciences, Department of Astronomy, 15.212 R.L. Moore Hall, Austin, Texas 78712, U.S.A.

LAMBRECHT (Prof. Dr H.) Direktor der Universitäts-Sternwarte, Schillergässchen 2, 69 Jena, Germany D.R.

*LAMERS (Dr H.J.G.L.M.) Lab. voor Ruimte-Onderzoek, Beneluxlaan 21, Utrecht, The Netherlands

LAMLA (Dr E.) Astronomische Institute, Universität Bonn, Auf dem Hügel 71, D 53 Bonn, Germany F.R.

*LANDE (Dr K.) University of Pennsylvania, Department of Physics, Philadelphia, PA 19174, U.S.A.

*LANDECKER (Dr T.L.) Dominion Radio Astrophysical, Observatory, Herzberg Institute of Astrophysics, National Research Council of Canada, P.O. Box 248, Penticton, B.C., Canada V2A 6K3

*LANDI DEGL'INNOCENTI (E.) Osservatorio Astrofisico di Arcetri, Largo E. Fermi 5, I-50125 Firenze, Italy

LANDI-DESSY (Dr J.) Observatorio Nacional Argentino, Cordoba, Argentina

LANDINI (M.) Osservatorio Astrofisico di Arcetri, Firenze, Italy

LANDOLT (Dr A.) Department of Astronomy and Physics, Louisiana State University, Baton Rouge, Louisiana 70803, U.S.A.

LANDSTREET (Prof. J.D.) Department of Astronomy, University of Western Ontario, London, Ontario N6A 3K7, Canada

*LANG (Dr J.) Astrophysics Research Division, Appleton Laboratory, Culham, Nr Abingdon, Oxfordshire, OX 14 3DB, U.K.

LANG (Dr K.R.) Physics Department, Tufts University, Medford, Mass. 02155, U.S.A.

LANTOS (Dr P.) Observatoire de Paris, Section d'Astrophysique de Meudon, 92190 Meudon, France

LaPAZ (Dr L.) Director Emeritus, Institute of Meteoritics, Campus P.O. Box 23, University of New Mexico, Albuquerque, New Mexico 87131, U.S.A.

LAPOINTE (Dr S.M.) Département de Physique, Université de Montréal, Montréal, Québec H3C 3J7, Canada

LAPUSHKA (Dr K.K.) Astronomical Observatory, Latvian University, Riga, Latvian SSR, U.S.S.R.

*LAQUES (Dr P.) Observatoire du Pic du Midi, 65200 Bagnères-de-Bigorre, France

LARGE (Dr M.I.) Chatterton Astronomy Department, University of Sydney, Sydney, Australia

LARI (Dr C.) Laboratorio Nazionale di Radioastronomia, Via Irnerio 46, Bologna, Italy

LARINK (Prof. Dr J.) Hamburger Sternwarte, 205 Hamburg-Bergedorf, Germany F.R.

*LARSON (Dr H.P.) The University of Arizona, Lunar and Planetary Laboratory, Tucson AZ 85721, U.S.A.

LARSON (Dr R.B.) Yale University Observatory, New Haven, Connecticut 06520, U.S.A.

LARSSON-LEANDER (Dr G.) Lund Observatory, S-222 24 Lund, Sweden

LASKARIDES (Dr P.G.) Laboratory of Astronomy, University of Athens,
Panepistimiopolis, Athens (621), Greece

LASKER (Dr B.M.) Cerro Tololo Inter-American Observatory, Casilla 63-D,
La Serena, Chile

*LASOTA (Dr J.P.) Institute of Astronomy, Polish Academy of Sciences,
Al. Ujazdowskie 4, 00-478 Warszawa, Poland

LATHAM (Dr D.W.) Smithsonian Institution Observ.,Cambridge, Mass. 02138, U.S.A.

LATOUR (Dr J.J.) Observatoire de Nice, Le Mont-Gros, 06000 Nice, France

LATYPOV (Dr A.A.) Astronomical Institute of the Usbek SSR, Academy of Sciences,
Tashkent, U.S.S.R.

*LAUBERTS (Dr A.) Astronomical Observatory, Box 515, S-751 20 Uppsala, Sweden

LAURENT (Dr B.E.) Department of Physics, University of Stockholm, Vanadisvägen 9,
S-113 46 Stockholm, Sweden

LAUSTSEN (Dr S.) European Southern Observatory, ETP Division, CERN,
CH-1211 Geneva 23, Switzerland

LAUTERBORN (Dr D.) Hamburger Sternwarte, 2050 Hamburg 80, Gojenbergsweg 112,
Germany F.R.

LAUTMAN (Dr D.A.) Smithsonian Astrophysical Observatory, 60 Garden Street,
Cambridge, Mass. 02138, U.S.A.

LAVAL (Dr A.) Observatoire de Marseille, 2, Place le Verrier, 13004 Marseille,
France

LAVDOVSKIJ (Dr V.V.) Pulkovo Observatory, Leningrad, U.S.S.R.

LAVROV (Dr M.I.) Engelhardt Astronomical Observatory, Kazan University,
Kazan, U.S.S.R.

LAVRUKHINA (Prof. Dr A.K.) Institute of Geochemistry and Analytic Chemistry,
USSR Academy of Sciences, Moscow, U.S.S.R.

LAWRENCE (Dr G.M.) LASP, Space Science Building, University of Colorado,
Boulder, Colorado 80302, U.S.A.

LAYZER (Prof. D.) Harvard College Observatory, 60 Garden Street, Cambridge,
Mass. 02138, U.S.A.

LAZOVIC (Dr J.) Dept. of Astronomy, Faculty of Sciences, Belgrade University,
Studentski trg 16, P.O. Box 550, 11001 Beograd, Yugoslavia

LEBEDINETS (Dr V.N.) Commission on Comets and Meteors, Astronomical Council,
USSR Academy of Sciences, Moscow, U.S.S.R.

LEBLANC (Dr Y.) Observatoire de Paris, Section d'Astrophysique de Meudon,
92190 Meudon, France

LEBOVITZ (Dr N.) Department of Mathematics, Eckhart Hall, Room 309,
University of Chicago, 1118-32 East 58th Street, Chicago, Illinois 60637,
U.S.A.

LECAR (Dr M.) Smithsonian Astrophysical Observatory, 60 Garden Street,
Cambridge, Mass. 02138, U.S.A.

*LECKRONE (Dr D.S.) Code 671, NASA Goddard Space Flight Center, Greenbelt,
MD 20771, U.S.A.

LE CONTEL (J.M.) Université de Nice, Observatoire, Le Mont-Gros, 06300 Nice, France

LEDERLE (Dr T.) Astronomisches Rechen-Institut, Mönchhofstrasse 12-14, 69 Heidelberg, Germany F.R.

LEDOUX (Prof. Dr P.) Institut d'Astrophysique, Université de Liège, Avenue de Cointe, 5, B-4200 Cointe-Ougrée, Belgium

LEE (Dr P.D.) Louisiana State University, Baton Rouge, Louisiana 70803, U.S.A.

*LEE (Dr T.J.) Royal Observatory, Blackford Hill, Edinburgh EH9 3HJ, U.K.

*LEER (Prof. Dr E.) Nordlysobservatoriet, University of Tramsø, Norway

LEFEBVRE (M.) C.N.E.S./G.R.G.S., 18, avenue Edouard Belin, F-31055 Toulouse Cedex, France

LEFEVRE (Dr J.) Observatoire de Nice, Le Mont Gros, 06300 Nice, France

LEGG (Dr T.H.) Herzberg Institute of Astrophysics, National Research Council of Canada, Ottawa, Ontario K1A OR6, Canada

LEHNERT (Dr B.P.) Royal Institute of Technology, Department of Electronics, Stockholm 70, Sweden-10044

LEIBOWITZ (Dr E.M.) Tel-Aviv University, Ramat-Aviv, Tel-Aviv, Israel

LEIGHTON (Prof. R.B.) California Institute of Technology, Pasadena, California 91109, U.S.A.

LEIKIN (Dr G.A.) Astronomical Council of the USSR Academy of Sciences, Pyatnitskaya ul. 48, 109017 Moscow, U.S.S.R.

LEINERT (Dr Ch.) Max-Planck-Institut für Astronomie, 6900 Heidelberg 1, Königstuhl, Germany F.R.

LEMAIRE (Dr P.) C.N.R.S., Laboratoire de Physique Stellaire et Planétaire, B.P. No 10, 91 Verrières-le-Buisson, France

*LEMAITRE (Dr G.) Observatoire de Marseille, 2, Place Le Verrier, 13004 Marseille, France

LEMKE (Dr D.) Max-Planck-Institut für Astronomie, 6900 Heidelberg 1, Königstuhl, Germany F.R.

LENA (Dr P.) Observatoire de Meudon, 92190 Meudon, France

LENGAUER (Dr G.G.), Main Astronomical Observatory, USSR Academy of Sciences, Pulkovo, 196140 Leningrad, U.S.S.R.

*LEORAT (Dr J.) DAPHE, Observatoire de Meudon, 92190 Meudon, France

LEQUEUX (J.) Observatoire de Paris, Section d'Astrophysique, 92190 Meudon, France

LEROY (Dr J.L.) Observatoire du Pic-du-Midi, 65 Bagnères-de-Bigorre, Hautes-Pyrénées, France

LE SQUEREN (A.M.) Observatoire de Paris, Section d'Astrophysique, 92190 Meudon, France

LETFUS (Dr V.) Astronomical Institute, Czechoslovak Academy of Sciences, Observatory Ondrějov, Czechoslovakia

LEUNG (Dr Kam-Ching) Behlen Laboratory of Physics, University of Nebraska. Lincoln, Nebraska 68508, U.S.A.

LEVALLOIS (J.J.) Institut Geographique National, 2 Av. Pasteur, 94 Saint-Mandé, France

*LEVASSEUR-REGOURD (Dr A.C.) Service d'Aéronomie du C.N.R.S., B.P. No 3, 91370 Verrières-le-Buisson, France

LEVIN (Dr B.J.) The Astronomical Council, USSR Academy of Sciences, Pyatnitskaya 48, Moscow 109017, U.S.S.R.

*LEVINE (Dr R.H.) Center of Astrophysics, 60 Garden Street, Cambridge, MA 02138, U.S.A.

*LEVY (Dr E.H.) Lunar and Planetary Laboratory, The University of Arizona Tucson, AZ 85721, U.S.A.

LEVY (Dr J.R.) Observatoire de Paris, 61, Avenue de l'Observatoire, 75014 Paris, France

*LEWIN (Dr W.H.G.) Massachusetts Institute of Technology, Center for Space Research, Cambridge, MA 02139, U.S.A.

*LEWIS (Dr B.M.) Carter Observatory, P.O. Box 2909, Wellington, New Zealand

LI (J.Y.) Pyongyang Astronomical Observatory, Academy of Sciences, Pyongyang, Korea (P.D.R.)

LI (Dr Ned C.) California University, 6531 Withworth Rd, Los Angeles, Calif. 90035, U.S.A.

LIDDELL (U.) National Aeronautics and. Space Administration, Lunar and Planetary Program, Office of Space Science and Applications, Washington, D.C. 20546, U.S.A.

*LIEBSCHER (Dr D.E.) Zentralinstitut für Astrophysik, Sternwarte Babelsberg, DDR - 1502 Potsdam-Babelsberg, Rosa-Luxemburg-Str. 17a, Germany D.R.

LIESKE (Dr J.H.) California Institute of Technology, Jet Propulsion Laboratory, 4800 Oak Grove Drive, Pasadena, Calif. 91103, U.S.A.

LILLER (Dr M.) Harvard College Observatory, 60 Garden Street, Cambridge, Mass. 02138, U.S.A.

LILLER (Prof. W.) Harvard College Observatory, 60 Garden Street, Cambridge, Mass. 02138, U.S.A.

LILLEY (Prof. A.E.) Harvard College Observatory, Cambridge, Mass. 02138, U.S.A.

LIMBER (Dr D.N.) Yerkes Observatory, Williams Bay, Wisconsin, U.S.A.

LIN (Dr C.C.) Department of Mathematics, Massachusetts Institute of Technology, Cambridge, Massachusetts, U.S.A.

LINCOLN (J.V.) Solar-Terrestrial Data Ser. , National Geophysical & Solar Terrestrial Data Center, NOAA, Boulder, Colorado 80302, U.S.A.

LINDBLAD (Dr B.A.), Lund Observatory, S-222 24 Lund, Sweden

LINDBLAD (Dr P.O.) Stockholm Observatory, S-133 00 Saltsjöbaden, Sweden

LINDOFF (Dr U.I.G.S.) Institute of Astronomy, S-222 24 Lund, Sweden

LINFOOT (Dr E.H.) The Observatories, Madingley Road, Cambridge, U.K.

LING (Dr Chih-bing) Institute of Mathematics, Academia Sinica, P.O. Box No 143, Taipei, Taiwan

*LINGENFELTER (Dr R.E.) University of California, Department of Astronomy, Los Angeles, CA 90024, U.S.A.

LINK (Dr F.) Institut d'Astrophysique, 98 bis Boulevard Arago, 75014 Paris, France

LINNELL (Dr A.P.) Department of Astronomy, Michigan State University, East Lansing, Michigan 48823, U.S.A.

LINNIK (Prof. Dr V.P.) Pulkovo Observatory, Leningrad, U.S.S.R.

LINSKY (Dr J.) Joint Institute for Laboratory Astrophysics, Boulder, Colorado 80302, U.S.A.

LIPPINCOTT (S.L.) Sproul Observatory, Swarthmore College, Swarthmore, Pennsylvania 19081, U.S.A.

LIPSKIJ (Dr Y.N.) Sternberg Astronomical Institute, Moscow, U.S.S.R.

LISZKA (Dr L.) Geophysical Observatory, S-981 00 Kiruna, Sweden

LITTLE (A.G.) Chatterton Astronomy Department, School of Physics, University of Sydney, Sydney, N.S.W., Australia

LITTLE (C. Gordon) Chief, Division 87, National Bureau of Standards, Room 3001, Radio Bldg. Boulder, Colorado, U.S.A.

LITTLE (Dr L.T.) Electronics Laboratory, University of Kent at Canterbury, Kent, U.K.

LITVAK (Dr M.) Smithsonian Astrophysical Observatory, Cambridge, Mass.02138, U.S.A.

LIU (Dr Sou-Yang) Astronomy Program, University of Maryland, College Park, Maryland 20742, U.S.A.

LIVINGSTON (Dr W.C.) Kitt Peak National Observatory, P.O. Box 4130, Tucson, Arizona 85717, U.S.A.

LIVSHITS (Dr M.A.) Institute of Terrestrial, Magnetism and Ionosphere, USSR Academy of Sciences, Akademgorodok, 142092 Podol'sk, U.S.S.R.

LLOYD-EVANS (Dr T.H.H.) South African Astronomical Observatory, P.O. Box 9, Observatory, Cape 7935, South Africa

*LOCANTHI (Dr D.D.) Jet Propulsion Laboratory, 4800 Oak Grove Drive, Pasadena, CA 91103, U.S.A.

LOCHTE-HOLTGREVEN (Prof. Dr W.) Institut für Experimentalphysik der Universität Kiel, Germany F.R.

LOCKE (Dr J.L.) Herzberg Institute of Astrophysics, National Research Council of Canada, Ottawa, Ontario K1A OR6, Canada

LOCKWOOD (Dr G.W.) Lowell Observatory, Flagstaff, Arizona 86001, U.S.A.

LODÉN (Dr Kerstin) Stockholm Observatory, S-133 00 Saltsjöbaden, Sweden

LODÉN (Dr L.) Astronomiska Observatoriet, Box 515, 751 20 Uppsala, Sweden

LOHMANN (Prof. Dr W.) Astronomisches Rechen-Institut, Mönchhofstrasse 12-14, 69 Heidelberg, Germany F.R.

LONGAIR (Dr M.S.) Mullard Radio Astronomy Observatory, Cavendish Laboratory, Madingley Road, Cambridge CB3 OHE, U.K.

LOPEZ (Dr J.A.) Pedro Echagüe 74 (oeste), San Juan, Argentina

LOPEZ GARCIA (Dr Z.) Observatorio Astronómico, La Plata, Argentina

LORON (Dr M.M.) Director del Observatorio Astronomico, Alfonso XII, No 3, Madrid, Spain

LORTET-ZUCKERMANN (M.C.) Institut d'Astrophysique, 98 bis Boulevard Arago, 75014 Paris, France

*LOSCO (Dr L.) Université de Besançon, 30, Avenue de l'Observatoire, 25000 Besançon, France

LOTOVA (Dr N.A.) Physical Institute, USSR Academy of Sciences, Moscow, U.S.S.R.

LOUGHHEAD (Dr R.E.) CSIRO, Division of Physics, University Grounds, Chippendale, NSW 2008, Australia

LOUISE (Dr R.) L.A.S.- Traverse du Siphon, Les 3 Lucs, 13 Marseille 12, France

LOVAS (M.) Konkoly Observatory of the Ungarian Academy of Sciences, Konkoly Thege Miklós u 13/17, Budapest XII, Hungary

LOVELL (Prof. A.C.B.) (Sir Bernard) Nuffield Radio Astronomy Laboratories, Jodrell Bank, Macclesfield, Cheshire SK11 9DN, U.K.

LOW (Dr F.J.) Lunar and Planetary Laboratory, The University of Arizona, Tucson, Arizona 85721, U.S.A.

*LOZINSKAYA (Dr T.A.) Sternberg Astronomical Institute, Moscow 117234, U.S.S.R.

LOZINSKIJ (Dr A.M.), Astronomical Council, USSR Academy of Sciences, Pyatnitskaya ul. 48, 109017 Moscow, U.S.S.R.

LUBECK (K.) Hamburger Sternwarte, 205 Hamburg-Bergedorf, Germany F.R.

*LUCCHIN (F.) Istituto di Fisica "G. Galilei", Via Marzolo 8, 35100 Padova, Italy

*LUCKE (Dr P.B.) Observatoire de Genève, 1290 Sauverny, Switzerland

LUCY (Dr L.B.) Department of Physics, Faculty of Arts and Sciences, University of Pittsburgh, Pittsburgh, Pennsylvania 15213, U.S.A.

*LUKAČEVIĆ (Dr I.) Institut za mehaniku PMF, Studentski trg 16, P.O. Box 550, 11001 Beograd, Yugoslavia

LUKATSKAJA (Dr F.I.) Main Astronomical Observatory, Ukrainian Academy of Sciences, Kiev, U.S.S.R.

LUMME (Dr K.A.) Observatory and Astrophysics Laboratory, University of Helsinki, Tähtitorninmäki, SF - 00130 Helsinki 13, Finland

LUNDQUIST (Dr C.A.) Director, Space Sciences Lab., Marshall Space Flight Center, Huntsville, AL 35812, U.S.A.

LUNEL (M.) Observatoire de Lyon, 69 Saint-Genis-Laval, France

*LUNGU (Dr N.) Institut Pédagogique d'Oradea, 5, rue Armatei Rosii, Oradea 3.700, Roumania

LÜST (Prof. Dr R.) Max-Planck-Institut für extraterrestrische Physik, 8046 Garching b. München, Germany F.R.

LÜST (Dr Rhea) Max-Planck-Institut für Physik und Astrophysik, Föhringer Ring 6, 8000 München 23, Germany F.R.

LUTZ (Dr B.L.) Dept. of Earth & Space Sciences, State University of New York, Stony Brook, New York 11790, U.S.A.

*LUTZ (Dr J.H.) Dept. of Pure and Applied Mathematics, Washington State University, Pullman, WA 99163, U.S.A.

LUTZ (Dr T.E.) Department of Pure and Applied Mathematics, Washington State University, Pullman, Washington 99163, U.S.A.

LUUD (Dr L.S.), Institute for astrophysics and physics atmosphere,
    Toravere, 202444 Tartu, U.S.S.R.

LUYTEN (Prof. W.J.) University of Minnesota, 211 Space Science Center,
    Minneapolis 55455, Minnesota, U.S.A.

LYNDEN-BELL (Prof. D.) Institute of Astronomy, The Observatories,
    Madingley Road, Cambridge CB3 OHA, U.K.

LYNDS (Dr B.T.) Kitt Peak National Observatory, P.O.Box 26732, Tucson,
    Arizona 85726, U.S.A.

LYNDS (Dr C.R.) Kitt Peak National Observatory, 950 North Cherry Avenue
    P.O. Box 26732, Tucson, Arizona 85726, U.S.A.

LYNE (Dr A.G.) Nuffield Radio Astronomy Laboratories, Jodrell Bank,
    Macclesfield, Cheshire, U.K.

LYNGA (Dr G.) Institutionen för astronomi, 222 24 Lund, Sweden

LYTTKENS (Dr E.) Skolgatan 33 B, S-752 21 Uppsala, Sweden

LYTTLETON (Dr R.A.) St. John's College, Cambridge, U.K.

*McADAM (Dr W.B.) Department of Astrophysics, Sydney University,
    Sydney, N.S.W. 2006, Australia

McCABE (M.) Associate Astronomer, Institute for Astronomy, 2840 Kolowalu Street
    Honolulu, Hawaii 96822, U.S.A.

*McCARTHY (Dr D.D.) Naval Observatory, 34th and Massachusetts Ave., NW,
    Washington, DC 20390, U.S.A.

McCARTHY (Dr M.F., S.J.) Specola Vaticana, Castel Gandolfo, Vatican City State
McCLAIN (E.F.) 225 Maple Road, Morningside, Maryland 20023, U.S.A.

*McCLINTOCK (Dr J.E.) Massachusetts Institute of Technology,
    37-521 Center for Space Research, Cambridge, MA 02139, U.S.A.

McCLURE (Dr R.H.) Yale University, Department of Astronomy, New Haven,
    Connecticut 07520, U.S.A.

McCLUSKEY (Dr G.E.) LeHigh University, Bethlehem, Pa. 18015, U.S.A.

McCORD (Dr T.B.) Institute for Astronomy, 2680 Woodlawn Drive
    Honolulu, Hawaii 96822, U.S.A.

McCRACKEN (Prof. K.G.) C.S.I.R.O., Mineral Physics Section, P.O. Box 136,
    North Ryde, N.S.W. 2113, Australia

McCRAY (Dr R.A.) Joint Institute for Laboratory Astrophysics, Boulder,
    Colorado, U.S.A.

McCREA (Prof.W.H.) Astronomy Centre, Sussex University,
    Brighton BN1 9QH U.K.

McCROSKY (Dr R.E.) Smithsonian Astrophysical Observatory, 60 Garden Street,
    Cambridge, Massachusetts 02138, U.S.A.

*McCULLOCH (Dr P.M.) Department of Physics, University of Tasmania, Hobart,
    Tasmania, Australia

McCUSKEY (Dr S.W.) Warner and Swasey Observatory, Case Institute of
    Technology, East Cleveland, Ohio 44112, U.S.A.

McCUTCHEON (Dr W.H.) Department of Physics, University of British Columbia
Vancouver, B.C. V6T 1W5, Canada

McDONALD (Dr F.B.) NASA, Goddard Space Flight Center, Greenbelt, Maryland 20771, U.S.A.

McDONALD (J.K. Petrie) Dominion Astrophysical Observatory, Herzberg Institute of Astrophysics, National Research Council of Canada, 5071 W. Saanich Road, Victoria, B.C., V8X 3X3, Canada

McDONNELL (Dr J.A.M.) Electronics Laboratories, University of Kent at Canterbury, Canterbury, Kent, U.K.

*McDONOUGH (Dr T.R.) Jet Propulsion Laboratory, 183-301, 4800 Oak Grove Drive, Pasadena, Cal. 91103, U.S.A.

McELROY (Dr M.B.) Kitt Peak National Observatory, 950 North Cherry Avenue, Tucson, Arizona, U.S.A.

McGEE (Prof. J.D.) Physics Department, Imperial College of Science and Technology, Prince Consort Road, London, S.W. 7, U.K.

McGEE (Dr R.X.) CSIRO, Division of Radiophysics, Box 76, Epping, N.S.W. 2121, Australia

McINTOSH (Dr B.A.) Herzberg Institute of Astrophysics, National Research Council of Canada, Ottawa, Ontario K1A OR6, Canada

McKEE (Dr Chr.) University of California, Dept. of Physics, Berkeley, Cal.94720 U.S.A.

McKENNA-LAWLOR (S.M.P.) Vishnu House, Blanchardstown, Dublin, Ireland

McKINLEY (Dr D.W.R.) 1889 Fairmeadow Crescent, Ottawa, Ontario, Canada

McLEAN (Dr D.J.) CSIRO, Division of Radiophysics, P.O. Box 76, Epping, N.S.W.2121 Australia

McMULLAN (Dr D.) Royal Greenwich Observatory, Herstmonceux Castle, Hailsham, Sussex, U.K.

McNALLY (Dr D.) University of London Observatory, Mill Hill Park, London, NW7 2QS, U.K.

McNAMARA (Dr D.H.) Department of Physics, Brigham Young University, Provo, Utah, U.S.A.

McNARRY (L.R.) Herzberg Institute of Astrophysics, National Research Council of Canada, Ottawa, Ontario K1A OR6, Canada

McVITTIE (Prof. G.C.)74 Old Dover Road, Canterbury, Kent, U.K.

McWHIRTER (Dr R.W.P.) Science Research Council, Astrophysics Research Unit, Culham Laboratory, Abingdon, Berkshire, U.K.

MacCONNELL (Dr D.J.) Fundacion Centro de Investigacion de Astronomia, Apartado 264, Merida, Edo. Merida, Venezuela

MacGARROLL (Dr R.) Observatoire de Meudon, 92190 Meudon, France

MacLEOD (Dr J.M.) Herzberg Institute of Astrophysics, National Research Council of Canada, Ottawa, Ontario K1A OR6, Canada

MacQUEEN (Dr R.) High Altitude Observatory, Boulder, Colorado, U.S.A.

MacRAE (Dr D.A.) David Dunlap Observatory, University of Toronto, Richmond Hill, Ontario L4C 4Y6, Canada

MACCHETTO (Dr F.) Space Science Department, ESTEC, Noordwijk, The Netherlands

MACDONALD (Dr G.H.) The Electronics Laboratory, The University,
    Canterbury, Kent, U.K.

MACHADO (J.M.A. Braz) Servicos Radioelectricos dos C.T.T., Lisbon, Portugal

MACKAY (Dr C.D.) The Institute of Astronomy, Madingley Road,
    Cambridge, U.K.

MACRIS (Dr C.J.) Director Research Center for Astronomy and Applied Mathematics,
    Academy of Athens, 14 Anagnostopoulou Street, Athens (136) Greece

MADEIRA (Dr J.A.) Observatorio Astronomico, Tapada, Lisbon, Portugal

MAEDER (Dr A.) Observatoire de Genève, CH-1290 Sauverny, Switzerland

*MAEHARA (Dr Hideo) Kiso Station, Tokyo Observatory, Mitaka-mura, Kiso-gun,
    Nagano-ken, 397-01, Japan

MAFFEI (Dr P.) Laboratorio di Astrofisica, Casella Postale 67,
    00044 Frascati (Roma), Italy

MAGALASHVILI (Dr N.L.) Abastumani Observatory, Abastumani, Georgia, U.S.S.R.

MAGNAN (Dr Chr.) Institut d'Astrophysique, 98 bis, Boulevard Arago,
    75014 Paris, France

MAGNARADZE (Dr N.G.) State University, Tbilisi, Georgia, U.S.S.R.

*MAIHARA (Dr T.) Department of Physics, Kyoto University, Kyoto 606, Japan

*MAILLARD (Dr J.-P.) DOPTO, Observatoire de Meudon, 92190 Meudon, France

*MAITZEN (Dr H.M.) Universitätssternwarte, Türkenschanzstr.17, A-1180 Wien,
    Austria

MAKAROVA (Dr E.A.) Sternberg Astronomical Institute, Moscow, U.S.S.R.

MAKITA (Dr M.) Tokyo Astronomical Observatory, University of Tokyo,
    Mitaka, Tokyo, Japan

MALACARA (Dr D.) Instituto Nacional de Astrofísica Optica y Electrónica
    Tonantzintla, Pue. Apartados Postales Nos. 216 y 51, Puebla, Pue.
    Mexico

MALAISE (Dr D.J.) Institut d'Astrophysique, Université de Liège,
    avenue de Cointe 5, B-4200 Cointe-Ougrée, Belgium

*MALINOVSKY (Dr M.) LPSP, Route des Gatines, 91370 Verrières-le-Buisson, France

MALITSON (H.H.) Code 693, Laboratory for Extraterrestrial Physics,
    NASA-Goddard Space Flight Center, Greenbelt, Maryland 20771, U.S.A.

MALLIA (Dr E.A.) Department of Astrophysics, South Parks Road,
    Oxford OXI 3RQ, U.K.

MALMQUIST (Prof. K.G.) S:t Olofsgatan 10 A, S-753 21 Uppsala, Sweden

MALTBY (Dr P.) Institute of Theoretical Astrophysics, University of Oslo,
    Blindern, Oslo 3, Norway

MALVILLE (Dr J.M.) High Altitude Observatory, Boulder, Colorado, U.S.A.

MAMEDBEJLI (Dr G.D.), Shemakha Astrophysical Observatory,
    Shemakha, 373243 Azerbaidzan, U.S.S.R.

MAMMANO (A.) Osservatorio Astrofisico, Asiago (Vicenza) Italy

MANARA (Dr A.) Osservatorio Astronomico di Milano, via Brera 28,
Milano, Italy

MANCHESTER (Dr R.N.) CSIRO, Division of Radiophysics, P.O.Box 76,
Epping, N.S.W., Australia 2121

*MANCUSO (S.) Osservatorio Astronomico di Capodimonte, Via Moiariello 16,
80131 Napoli, Italy

MANDEL'SHTAM (Prof. Dr S.L.), Lebedev Institute of Spectroscopy ,
Academy of Sciences, 142092 Akademgorodok, U.S.S.R.

MANGENEY (A.) Observatoire de Meudon, 5, place Janssen, 92 Meudon, France

MANNINO (Prof.G.) Osservatorio Astronomico Universitario, Via Zamboni 33,
Bolgona, Italy

*MANRIQUE (Ing. W.T.) Observatorio "Félix Aguilar", Av. Benavídez 8175 Oeste,
Marquesado, San Juan, Argentina

*MANSFIELD (Dr V.N.) Department of Physics & Astronomy, Colgate University,
Hamilton, New York 13346, U.S.A.

MARAN (Dr S.P.) Advanced Systems and Ground Observations Branch, Code 683,
Goddard Space Flight Center, Greenbelt, Maryland 20771, U.S.A.

*MARASCHI (L.) Istituto di Scienze Fisiche, Via Celoria 16, Milano, Italy

MARCUS (Prof. E.) Observatoire de Bucarest, 5 rue Cutitul de Argint,
Bucarest, Roumania

*MARGON (Dr B.) Dept. of Astronomy, University of California, Los Angeles,
California 90024, U.S.A.

MARGONI (R.) Osservatorio Astrofisico, Asiago (Vicenza) Italy

MARIK (Dr M.) Astronomical Institute of the Eötvös Lorénd University,
Kun Béla tér 2, Budapest, Hungary

*MARINO (Eng. B.F.) Auckland Observatory, P.O. Box 72009, Auckland 9, New Zealand

MARK (Dr J.W.K.) Massachusetts Institute of Technology, Dept. of Mathematics
Cambridge, Mass. 02139, U.S.A.

MARKARIAN (Dr B.E.) Byurakan Astrophysical Observatory
Byurakan, Armenia, U.S.S.R.

*MARKKANEN (Dr K.T.J.) Observatory and Astrophysics Laboratory,
University of Helsinki, Tähtitorninmäki, SF-00130 Helsinki 13, Finland

MARKOWITZ (Prof. Dr W.) Nova University, 8000 N. Ocean Drive,
Dania, Fla. 33004, U.S.A.

MARLBOROUGH (Prof. J.M.) Dept. of Astronomy, University of Western Ontario,
London, Ontario N6A 3K7, Canada

MAROCHNIK (Prof. Dr L.S.) Astrophysical Dept.,Rostov University,
Prospekt Stachki 192, Rostov-Don 344061, U.S.S.R.

MARQUES (M.N.) Tapada da Ajuda, Lisboa 3, Portugal

*MARQUES DOS SANTOS (P.) Craam, Universidade Mackenzie, C.P. 8792,
01000 Sao Paulo, SP, Brazil

MARSDEN (Dr B.G.) Smithsonian Astrophysical Observatory,
60 Garden Street, Cambridge, Massachusetts 02138, U.S.A.

*MARSDEN (Prof. P.L.) Department of Physics, University of Leeds, Leeds LS2 9JT, U.K

MARSH (M.Sc. J.C.D.) The Hatfield Polytechnic Observatory, Bayfordbury,
    Near Hertford, Herts, U.K.

*MARŞOGLU (Dr S.) University Observatory, Universite - Istanbul, Turkey

MARTEL-CHOSSAT (Dr M.T.) Observatoire de Lyon, 69 Saint-Genis-Laval, France

*MARTIN (Dr D.H.) Physical Department, Queen Mary College, Mile End Road,
    London E1 4NS, U.K.

MARTIN (Nicole) Astronome adjoint, Observatoire de Marseille, 2, Place Le Verrier,
    13 Marseille 4, France

*MARTIN (Dr P.G.) David Dunlap Observatory, University of Toronto, Richmond Hill,
    Ontario, Canada L4C 4Y6

MARTIN (Jr W.C.) Chief, Spectroscopy Section, National Bureau of Standards,
    Physics Building A-165, Washington D.C. 20234, U.S.A.

MARTINET (L.) Observatoire Cantonal de Genève, Genève, Sauverny 1290, Switzerland

MARTINEZ (Dr M.) Instituto de Astronomía, UNAM, Apartado Postal 70 - 264,
    México D.F., Mexico

MARTINI (Dr A.) Laboratorio di Astrofisica, Casella Postale 67, Frascati (Roma),
    Italy

MARTRES (M.J.) Observatoire de Meudon, Place Jules Janssen, 92190 Meudon, France

MARTYNOV (Prof. Dr D. Ya) Director of the Sternberg Astronomical Institute,
    Moscow University, Leninskije Gory, Moscow - V234, U.S.S.R.

MARVIN (Dr U.B.) Smithsonian Astrophysical Observatory, Cambridge,
    Mass. 02138, U.S.A.

MARX (Prof. G.) Roland Eötvös University, Puskin u. 5-7, Budapest VIII, Hungary

*MARX (Dr S.) Zentralinstitut für Astrophysik, Karl-Schwarzschild-Observatorium,
    DDR - 6901 Tautenburg, Germany D.R.

MASANI (Prof. A.) Osservatorio Astronomico di Brera, Via Brera 28, Milano, Italy

MASLOWSKI (Dr J.) Astronomical Observatory of the Jagellonian University,
    Cracow, Kopernika 27, Poland

MASNOU (Dr F.) Observatoire de Paris, Section d'Astrophysique de Meudon,
    92190 Meudon, France

MASSEVICH (Dr A.G.) Astronomical Council of the USSR Academy of Sciences,
    Pyatnitskaya ul., D.48, 109017, Moscow, Zh-17, U.S.S.R.

*MASURSKY (Dr H.) U.S. Geological Survey, Branch of Astrogeologic Studies,
    601 East Cedar Avenue, Flagstaff, AZ 86001, U.S.A.

*MATAS (Dr V.) Astronomical Institute of the Czechoslovak Academy of Sciences,
    Budečská 6, 120 23 Praha 2, Czechoslovakia

MATHEWS (Dr W.G.) Physics Department, University of California-San Diego,
    La Jolla, California 92038, U.S.A.

MATHEWSON (Dr D.S.) CSIRO, Division of Radiophysics, University Grounds,
    Chippendale, N.S.W., Australia

MATHIS (Dr J.S.) Astronomy Department, University of Wisconsin, Madison,
    Wisconsin 53706, U.S.A.

*MATSON (Dr D.L.) Jet Propulsion Laboratory, 4800 Oak Grove Drive,
    Pasadena, CA 91103, U.S.A.

*MATSUDA (Prof. Takuya) Department of Aeronautical Engineering, Kyoto University,
    Kyoto, 606, Japan

*MATSUMOTO (Dr Masamichi) Faculty of Engineering, Gifu University,
    Kagamigahara City, Gifu Prefecture 504, Japan

*MATSUMOTO (Dr Toshio) Department of Physics, Nagoya University, Chikusa-ku,
    Nagoya, Japan 464

MATSUNAMI (Dr N.) Tokyo Astronomical Observatory, Mitaka, Tokyo, Japan

MATSUOKA (Dr M.) Institute of Space and Aeronautical Science,
    University of Tokyo, Komaba, Meguro-Ku, Tokyo, Japan

MATSUSHIMA (Prof. Dr S.) Department of Astronomy, The Pennsylvania State
    University, University Park, Pennsylvania 16802, U.S.A.

MATTHEWS (Dr T.A.) Department of Physics and Astronomy, University of Maryland,
    College Park, Md. 20740, U.S.A.

MATTIG (Dr W.) Fraunhofer Institut, Schöneckstrasse 6, 78 Freiburg i. Breisgau,
    Germany F.R.

MATTILA (Dr K.) University of Helsinki, Observatory and Astrophysics Laboratory,
    Tähtitornimäki, SF - 00130 Helsinki 13, Finland

MATVEYENKO (Dr L.I.) Institute for Space Research, Profsojuznaja 88,
    Moscow V 485, U.S.S.R.

MATZNER (Dr R.) The University of Texas (Austin) Austin, Texas, U.S.A.

MAUDER (Dr H.) Remeis-Sternwarte, Sternwartstr. 7, 8600 Bamberg, Germany F.R.

MAURICE (Ing. E.) Observatoire de Marseille, 2, Place Le Verrier,
    13004 Marseille, France

MAVRIDIS (Prof. Dr L.N.) Department of Geodetic Astronomy Polytechnic School,
    University of Thessaloniki, Thessaloniki, Greece

*MAX (Dr C.E.) Lawrence Livermore Laboratory, P.O. Box 808, Livermore,
    CA 94550, U.S.A.

MAXWELL (Dr A.) Harvard Radio Astronomy Station, Fort Davis, Texas, U.S.A.

MAY (Ing. J.) Observatorio Radioastronómico de Maipú, Universidad de Chile,
    Casilla 68, Maipú, Chile

MAYALL (Margaret W.) 187 Concord Avenue, Cambridge, Mass. 02138, U.S.A.

MAYALL (Dr N.U.) Kitt Peak National Observatory, 950 North Cherry Avenue,
    Tucson, Arizona, U.S.A.

MAYER (C.H.) Code 7130, Naval Research Laboratory, Washington, D.C. 20375, U.S.A.

MAYER (Dr P.) Department of Astronomy and Astrophysics, Charles University,
    Švédská 8, Prague 5 - Smíchov, Czechoslovakia

MAYER (Dr U.) Astron. Institut der Universität, Waldhäuser Str. 64,
    74 Tübingen, Germany F.R.

MAYFIELD (Dr E.B.) Solar Physics Department, The Aerospace Corporation,
    P.O. Box 92957, Los Angeles, California 90009, U.S.A.

MAYOR (Dr M.) Observatoire de Genève, CH-1290 Sauverny, Switzerland

MAZZUCCONI (Dr F.) Osservatorio Astrofisico di Arcetri, Largo Enrico Fermi, 5, Firenze, Italy

MEABURN (Dr J.) Department of Astronomy, The University,Manchester 13, U.K.

MEADOWS (Dr A.J.) Astronomy Department, University of Leicester, University Road, Leicester, U.K.

*MEBOLT (Dr U.) Max-Planck-Institut für Radioastronomie, 5300 Bonn 1, Auf dem Hügel 69, Germany F.R.

MEEKS (Dr M.L.) MIT, Neroc Haystack Observatory, Westford, Mass. 01886, U.S.A.

MEFFROY (Dr J.) Chef de Travaux, Faculté des Sciences, 34 Montpellier, France

*MEGESSIER (Dr C.) DEPEG, Observatoire de Paris, 61, Avenue de l'Observatoire, 75014 Paris, France

MEGRELISHVILI (Dr T.G.) Abastumani Astrophysical Observatory, Georgian Academy of Sciences, Abastumani, U.S.S.R.

MEHLTRETTER (Dr J.P.) Fraunhofer Institut, Schöneckstrasse 6, 78 Freiburg i. Breisgau, Germany F.R.

MEIN (P.) Observatoire de Meudon, 5, Place Janssen, 92190 Meudon, France

MEINEL (Dr A.B.) Optical Sciences Center, University of Arizona, Tucson, AZ 85721, U.S.A.

MEINIG (Dr M.) Zentralinstitut Physik der Erde, DDR 15 Potsdam, Telegrafenberg, Germany D.R.

MEISEL (Dr D.D.) Dept. of Physics and Astronomy, State University College of Arts and Sciences, Geneseo, N.Y. 14454, U.S.A.

*MEKLER (Dr Y.) Department of Environmental Sciences, Tel-Aviv University, Ramat Aviv, Israel

MELCHIOR (Prof. Dr P.) Chef du Département d'Astronomie de Position et de Géodynamique, Observatoire Royal de Belgique, Avenue Circulaire 3, B-1180 Bruxelles, Belgium

MELNIKOV (Prof. Dr O.A.) Pulkovo Observatory, Leningrad, U.S.S.R.

MEN' (Dr A.V.) Institute of Radiophysics and Electronics, Academy of Sciences, Ukrainian SSR, Kharkov, U.S.S.R.

MENDEZ (Dr M.) Instituto de Astronomía, UNAM, Apartado Postal 70-264, México, D.F., Mexico

*MENDIS (Dr D.A.) Dept. of Applied Physics and Information Science, University of California, San Diego, La Jolla, CA 92037, U.S.A.

MENDOZA V. (Dr E.E.) Instituto de Astronomía, UNAM, Apartado Postal 70-264, México, D.F., Mexico

*MENEGUZZI (Dr M.) Centre d'Etudes Nucléaire de Saclay, Service d'Electronique Physique, B.P. No 2, 91190 Gif-sur-Yvette, France

MENNESSIER (Dr M.O.) Observatoire de Paris, 61, Avenue de l'Observatoire, 75014 Paris, France

MENON (Dr T.K.) Tata Institute of Fundamental Research, Homi Bhaba Road, Colaba, Bombay 5, India

MENZEL (Dr D.H.) Harvard College Observatory, 60 Garden Street, Cambridge, Mass. 02138, U.S.A.

*MENZIES (Dr J.W.) Jeremiah Horrocks Observatory, Moor Park, Preston PR1 6AD, Lancs, U.K.

MERGENTALER (Prof. Dr J.) Astronomical Observatory, Kopernika 11, Wroclaw, Poland

*MERLEAU-PONTY (Prof. J.) Université de Nanterre, 200, Avenue de la République, 92000 Nanterre, France

MERMAN (Dr G.A.) Institute of Theoretical Astronomy, 10, Kutuzov Qay, Leningrad, 192187, U.S.S.R.

MERMAN (Dr N.V.) Main Astronomical Observatory, USSR Academy of Sciences, Pulkovo, U.S.S.R.

MERRILL (Dr J.E.) 7836 Hosbrook Road Cincinnati, Ohio 45243, U.S.A.

MERTON (Dr G.) 17 Holywell, Oxford, U.K.

MERTZ (Dr L.) Harvard College Observatory, 60 Garden Street, Cambridge, Mass. 02138, U.S.A.

*MERZANIDIS (Dr C.) Astronomical Department, University of Thessaloniki, Thessaloniki, Greece

MESSAGE (Dr P.J.) Mathematics Department, University of Liverpool, P.O. Box 147, Liverpool L 69 3BX, U.K.

MESTEL (Prof. L.) Director of Astronomy Centre, Physics Building, University of Sussex, Falmer, Brighton BN1 9QH, U.K.

*METZ (Dr K.) Institut für Astronomie und Astrophysik, 8 München 80, Scheinerstrasse 1, Germany F.R.

MEURERS (Prof. Dr J.) Direktor der Universitätssternwarte, Türkenschanzstrasse 17, Vienna A-1180, Austria

MEWE (Dr R.) Laboratorium voor Ruimteonderzoek, Beneluxlaan 21, Utrecht, The Netherlands

*MEYER (Dr C.) CERGA, Plateau de Roquevignon, Avenue Copernic, 06130 Grasse, France

MEYER (Dr F.) Max-Planck-Institut für Physik und Astrophysik, Föhringer Ring 6, 8000 München 23, Germany F.R.

MEYER-HOFMEISTER (Dr E.) Max-Planck-Institut für Physik und Astrophysik, Föhringer Ring 6, 8000 München 23, Germany F.R.

MEZGER (Dr P.G.) Max-Planck-Institut für Radioastronomie, Auf dem Hügel 69, 53 Bonn 1, Germany F.R.

MIANES (P.) Observatoire de Lyon, 69 Saint-Genis-Laval, France

MICHARD (Prof. R.) Président de l'Observatoire de Paris, 61, Avenue de l'Observatoire, 75014 Paris, France

MICHAUD (Dr G.) Département de Physique, Université de Montréal, Montréal, Québec H3C 3J7, Canada

MICHEL (Prof. F.C.) Department of Space Science, Rice University, Houston, Texas 77001, U.S.A.

MICHEL (H. Ir.) rue Ten Bosch, 54, B-1050 Bruxelles, Belgium

*MICHEL (Prof. K.W.) Institut für Physikalische Chemie, Pfaffenwaldring 55, 7 Stuttgart 80, Germany F.R.

MICHKOVITCH (Prof. V.V.) Serbian Academy of Sciences and Arts, Knez Mihaïlova, 35, Belgrade, Yugoslavia

MICZAIKA (Dr G.R.) TRW Systems, 1 Space Park, Redendo Beach, California, U.S.A.

MIDDLEHURST (B.M.) Portofino Ville, Apt 2, Nassau Bay, Houston, Texas 77058, U.S.A.

MIETELSKI (Dr J.) Astronomical Observatory of the Jagellonian University, Cracow, Kopernika 27, Poland

MIGEOTTE (Prof. Dr M.) Institut d'Astrophysique, Université de Liège, Avenue de Cointe, 5, B-4200 Cointe-Ougrée, Belgium

MIHAILA (Dr I.) Observatorul Astronomic, Strada Cutitul de Argint 5, Bucuresti 28, Roumania

MIHALAS (Dr D.) High Altitude Observatory, P.O. Box 3000, Boulder, Colorado 80303, U.S.A.

MIKESELL (Dr A. H.) US Naval Observatory, Washington, D.C. 20390, U.S.A.

MIKHAIL (Dr J.S.) Helwan Observatory, Near Cairo, Arab Rep. of Egypt

MIKHAILOV (Prof. Dr A.A.) Director of the Pulkovo Observatory, Leningrad M 140, U.S.S.R.

MIKHEL'SON (Dr N.N.) Main Astronomical Observatory, USSR Academy of Sciences, Pulkovo, 196140 Leningrand, U.S.S.R.

*MILANO (L.) Osservatorio Astronomico di Capodimonte, Via Moiariello 16, 80131 Napoli, Italy

*MILES (H.G.) Mathematics Department, Lanchester Polytechnic, Coventry, U.K.

MILET (B.) Aide Astronome, Observatoire de Nice, Le Mont-Gros, 06-Nice, France

MILEY (Dr G.K.) Sterrewacht, Leiden, The Netherlands

MILFORD (Dr S.N.) Department of Physics, University of Queensland, St. Lucia, Brisbane, Queensland 4067, Australia

*MILKEY (Dr R.W.) Kitt Peak National Observatory, P.O. Box 26732, Tucson, AZ 85726, U.S.A.

MILLER (Dr F.D.) Department of Astronomy, The University of Michigan, Physics-Astronomy Building, Ann Arbor, Michigan 48104, U.S.A.

*MILLER (Dr H.R.) Department of Physics, Georgia State University, University Plaza, Atlanta, GA 30303, U.S.A.

MILLER (Dr J.S.) Lick Observatory, University of California, Santa Cruz, California 95060, U.S.A.

MILLER (Dr R.H.) Institute for Computer Research, University of Chicago, 5640 Ellis Ave., Chicago, Ill. 60637, U.S.A.

MILLIGAN (J.E.) Astrophysics Branch, Goddard Space Flight Center, Greenbelt, Maryland, U.S.A.

MILLIS (Dr L.R.) Lowell Observatory, Flagstaff, Arizona 86001, U.S.A.

MILLMAN (Dr P.M.) Herzberg Institute of Astrophysics, National Research Council of Canada, Ottawa, Ontario K1A OR6, Canada

MILLS (Dr A.A.) Department of Astronomy, The University, Leicester LE1 7RH, U.K.

MILLS (Dr B.Y.) School of Physics, University of Sydney, Sydney, N.S.W. 2000, Australia

*MILNE (Dr D.K.) CSIRO Division of Radiophysics, P.O. Box 76, Epping,
  N.S.W. 2121, Australia

*MILONE (Dr E.F.) Department of Physics, University of Calgary, Calgary,
  Alberta, Canada

*MILONE (Dr L.A.) Observatorio Astronómico, Laprida 854, Córdoba, Argentina

MILOVANOVIĆ (Dr V.) 11090 Beograd, Vukasovićeva 10, Yugoslavia

MININ (Dr I.N.) Leningrad University Observatory, Leningrad, U.S.S.R.

*MINN (Dr Young Key) National Astronomical Observatory, Yoksam-Dong, Kangnam-Ku,
  Seoul, 134-03, Republic of Korea

MINNET (H.C.) CSIRO, Division of Radiophysics, P.O. Box 76, Epping, N.S.W. 2121,
  Australia

MINTZ BLANCO MA (B.) Cerro Tololo Interamerican Observatory, Casilla 63-D,
  La Serena, Chile

MIRZOYAN (Dr Prof. L.V.) Byurakan Astrophysical Observatory, Byurakan,
  Armenia, U.S.S.R.

MISNER (Dr Ch.W.) Department of Astronomy, University of Maryland, College Park,
  Maryland 20742, U.S.A.

MISSANA (Prof. N.) Via Puccini 3, 10025 Pino Torinese, Italy

MITALAS (Dr R.) Department of Astronomy, University of Western Ontario,
  London, Ontario N6A 3K7, Canada

MITCHELL (Prof. G.P.) Department of Physics, St-Mary's University,
  Halifax N.S. B3H 3C3, Canada

MITCHELL (R.) c/o C.I.C.E.S.E. Av. Gastelum No 898, Ensenada, Baja California,
  Mexico

MITCHELL (Dr W.E. Jr) Perkins Observatory, Delaware, Ohio, U.S.A.

MITIĆ (L.A.) Astronomical Observatory, Volgina 7, Belgrade, Yugoslavia

MITRA (Dr A.P.) Assistant Director, National Physical Laboratory, New Delhi,
  India

MONSIGNORI FOSSI (Dr B.C.) Osservatorio Astrofisico di Arcetri,
  Largo Enrico Fermi, 5, Firenze, Italy

MITROFANOVA (Dr L.A.), Main Astronomical Observatory, USSR Academy of Sciences,
  Pulkovo, 196140 Leningrad, U.S.S.R.

*MITTON (Dr S.A.) Institute of Astronomy, Madingley Road, Cambridge CB3 OHA, U.K.

MIYADI (Dr M.) 3-6 Senkawa, Chôfu-shi, Tokyo, Japan

MIYAMOTO (Dr M.) Tokyo Astronomical Observatory, Mitaka, Tokyo, Japan

MIYAMOTO (Prof. Dr S.) Director, Kwasan Observatory, Faculty of Science,
  University of Kyoto, Yamashina, Kyoto, Japan

MIYAMOTO (Dr S.) Institute of Space and Aeronautical Science, University of
  Tokyo, Komaba, Meguro-ku, Tokyo, Japan

*MODALI (Dr S.B.) Washington Analytical Services Center, Inc., Wolf R & D.,
  6801 Kenilworth Avenue, Riverdale, MD 20840, U.S.A.

*MODISETTE (Dr J.L.) Houston Baptist University, 7502 Fondren Road, Houston,
  TX 77036, U.S.A.

MOERDIJK (Dr W.G.) Sterrenkundig Instituut, Rijksuniversiteit Gent,
J. Plateaustraat, 22, B-9000 Gent, Belgium

*MOFFAT (Dr A.F.J.) Astronomisches Institut der Ruhr-Universität, 463 Bochum,
Postfach 2148, Germany F.R.

MOFFET (Dr A.T.) Radio Astronomy Laboratory, California Institute of Technology,
Pasadena, California 91109, U.S.A.

*MOFFETT (Dr T.J.) Department of Physics, Purdue University, West Lafayette,
Indiana 47907, U.S.A.

MOGILEVSKIJ (Dr Eh.I.), Institute of Terrestrial Magnetism and Ionosphere,
USSR Academy of Sciences, Akademgorodok, 142092 Podol'sk, U.S.S.R.

MOHLER (Prof. O.C.) 405 Awixa Road,
Ann Arbor, Michigan 48104, U.S.A.

MOHR (Prof. J.M.) Department of Astronomy and Astrophysics, Charles University,
Švedská 8, Prague 5, Czechoslovakia

MOISEEV (Dr I.G.) Crimean Astrophysical Observatory, USSR Academy of Sciences,
p/o Nauchny, Crimea, U.S.S.R.

MOLCHANOV (Prof. A.P.) Radio Astronomical Laboratory, Institute of Physics,
Leningrad State University, V-164, Leningrad, U.S.S.R.

*MÖLLENHOFF (Dr C.) Landessternwarte, D-69 Heidelberg-Königstuhl, Germany F.R.

MØLLER (Dr O.) Ole Rømer Observatory, Arhus C, Denmark

*MOLNAR (Dr M.R.) The Ritter Astrophysical Research Center, The University of
Toledo, Toledo, OH 43606, U.S.A.

MONAGHAN (Dr J.J.) Department of Mathematics, Monash University, Clayton 3168,
Australia

MONFILS (Prof. Dr.A.) Institut d'Astrophysique, Université de Liège,
Avenue de Cointe, 5, B-4200 Cointe-Ougrée, Belgium

MONIN (Dr G.A.) Crimean Astrophysical Observatory, USSR Academy of Sciences,
P/O Nauchniy, 334413 Crimean, U.S.S.R.

MONNETT (Dr G.) Observatoire de Marseille, Pl. Le Verrier, 13004 Marseille,
France

MOOK (Dr D.E.) Dartmouth College, Hanover, New Hampshire, U.S.A.

*MOORE (Dr D.R.) Dept. of Applied Mathematics and Theoretical Physics,
University of Cambridge, Cambridge, U.K.

*MOORE (Dr E.) Joint Observatory for Cometary Research, Department of Physics,
New Mexico Institute of Mining and Technology, Socorro, NM 87801, U.S.A.

MOORE (Patrick O.B.E.) Farthings, 39 West Street, Selsey, Sussex, U.K.

*MOORE (Dr R.L.) Solar Astronomy, California Institute of Technology,
Pasadena, CA 91125, U.S.A.

MOORE-SITTERLY (Dr Ch.) National Bureau of Standards 151.00, Washington,
D.C. 20234, U.S.A.

MORAN (Dr J.) Smithsonian Astrophysical Observatory, Cambridge, Mass. 02138, U.S.A.

MORANDO (Dr B.L.) Bureau des Longitudes, 77 Ave Denfert Rochereau,F75014, Paris

*MORBEY (C.L.) Dominion Astrophysical Observatory, Herzberg Institute of
Astrophysics, National Research Council of Canada, 5071 W. Saanich Road,
Victoria, B.C., Canada V8X 3X3

MOREAU (Dr F.) Astronome Honoraire, Observatoire Royal de Belgique, Avenue Armand Huysmans, 219, B-1050 Bruxelles, Belgium

MORENO (Prof. H.) Observatório Astronómico Nacional, Casilla 36 - D, Santiago, Chile

MORETON (G.E.) 15-5 The Esplanade, Balmoral Beach, N.S.W., 2088, Australia

*MORGAN (Dr D.H.) Royal Observatory, Blackford Hill, Edinburgh, U.K.

*MORGAN (Dr T.H.) Houston Baptist University, 7502 Fondren Road, Houston, TX 77036, U.S.A.

MORGAN (Dr W.W.) Yerkes Observatory, Williams Bay, Wisconsin 53191, U.S.A.

MORGULEFF (Ing. N.) Institut d'Astrophysique, 98 bis, Boulevard Arago, 75014 Paris, France

MORIMOTO (Dr Masaki) Tokyo Astronomical Observatory, University of Tokyo, Mitaka, Tokyo, Japan

*MORISON (Dr I.) University of Manchester, Jodrell Bank, Nr.Macclesfield, Cheshire, U.K.

MORIYAMA (Dr F.) Tokyo Astronomical Observatory, Mitaka, Tokyo, Japan
MOROZ (Prof. Dr V.I.) Space Research Institute, USSR Academy of Sciences, 117810 Moscow, U.S.S.R.
MOROZHENKO (Dr A.V.) The Main Astronomical Observatory of the Ukrainian Academy of Sciences, Kiev- 127, U.S.S.R.

MOROZHENKO (Dr N.N.) Main Astronomical Observatory of Ukrainian Acad. of Sciences, Observatory, Kiev 252127, U.S.S.R.

MORRIS (Dr S.C.) Dominion Astrophysical Observatory, Herzberg Institute of Astrophysics, National Research Council of Canada, 5071 W. Saanich Road, Victoria, B.C. V8X 3X3, Canada

MORRISON(DrD.) Institute for Astronomy, University of Hawaii, Honolulu, Hawaii 96822, U.S.A.

MORRISON (L.V.) Royal Greenwich Observatory, Herstmonceux Castle, Hailsham, Sussex, U.K.

MORRISON (Dr P.) Department of Physics, Massachusetts Institute of Technology, Cambridge, Mass. 02139, U.S.A.

MORTON (Dr D.C.) Princeton University Observatory, Peyton Hall, Princeton, New Jersey 08540, U.S.A.

MORTON (Dr G.A.) 1122 Skycrest Dr., Apt. 6, Walnut Creek, Cal. 94595, U.S.A.

MOSS (Dr D.L.) Department of Mathematics, Manchester University, Manchester M13 9PL, U.K.

MOTZ (Dr L.F.) Columbia University, Department of Astronomy, Pupin Box 57, New York, N.Y. 10027, U.S.A.

MOURADIAN (Z.) Observatoire de Meudon, 5, place Janssen, 92- Meudon, France

*MOUSCHOVIAS (Dr T.Ch.) High Altitude Observatory, P.O.Box 3000 Boulder, Colorado 80303, U.S.A.

MOUTSOULAS (Prof. Dr M.) Department of Astronomy, University of Athens, Athens 621, Greece

MOVAHED-A (Dr R.) Tabriz University, Tabriz, Iran

MRKOS (A.) Dept. of Astronomy and Astrophysics, Charles University,
    Švédská 8, Prague 5- Smíchov, Czechoslovakia

*MUELLER (Dr I.I.) Dept. of Geodetic Science, The Ohio State University,
    1958 Neil Avenue, Columbus, Ohio 43210, U.S.A.

*MUKAI (Prof. Sonoyo) Kanazawa Institute of Technology, P.O. Box Kanazawa-South,
    Ishikawa, 921 Japan

*MUKAI (Dr Tadashi) Kanazawa Institute of Technology, P.O. Kanazawa-South,
    Ishikawa, 921 Japan

MULDERS (Dr G.F.W.) 4519 Everett Street, Kensington, Md. 20795, U.S.A.

MULHOLLAND (Dr J.D.) Dept. of Astronomy, University of Texas, Austin,
    Texas 78712, U.S.A.

MULLALY (Dr R.F.) School of Electrical Engineering, University of Sydney,
    Sydney, Australia
*MULLAN (Dr D.J.) Bartol Research Foundation, The Franklin Institute,
    Swarthmore, PA 19081, U.S.A.

MULLER (Dr A.B.) Kapteyn Astronomical Laboratory, Broerstraat 7, Groningen,
    The Netherlands

MULLER (Prof. Ir.C.A.) Odinksveld 8, Delden, The Netherlands

MÜLLER (Prof. E.A.) Observatoire de Genève, 1290 Sauverny (GE), Switzerland

MÜLLER (Prof. Dr H.) Herzogenmühlestrasse 4, CH-8051 Zürich, Switzerland

MULLER (Dr P.) CERGA, 8 bd. Emile Zola, 06130 Grasse, France

MÜLLER (Prof. Dr R.) Neubeuerer Str.1, 8201 Nussdorf/Inn, Germany F.R.

MUMFORD (G.S.) Department of Mathematics, Tufts University, Medford,
    Massachusetts 02155, U.S.A.

MÜNCH (Prof. G.) California Institute of Technology, 1201 E. California Street,
    Pasadena, California, U.S.A.

MUNRO (Dr R.H.) High Altitude Observatory, NCAR, P.O. Box 3000,
    Boulder, Colorado 80303, U.S.A.

MURDIN (Dr P.G.) Royal Greenwich Observatory, Herstmonceux Castle,
    Hailsham, Sussex, U.K.

*MURDOCH (Dr H.S.) Dept. of Astrophysics, University of Sydney,
    Sydney, N.S.W. 2006, Australia

MURPHY (Dr R.E.) Maryland Academy of Sciences, 119 South Howard Street,
    Baltimore, Maryland 21201, U.S.A.

MURRAY (C.A.) Royal Greenwich Observatory, Herstmonceux Castle,
    Hailsham, Sussex, U.K.

MUSEN (Dr P.) National Aeronautics and Space Administration, Goddard
    Space Flight Center, Greenbelt, Maryland, U.S.A.

MUSMAN (Dr S.) Sacramento Peak Observatory, Sunspot, New Mexico 88349, U.S.A.

MUSTEL (Prof. Dr E.R.) Astronomical Council, Pyatniskaya ul., D.48, 109017 Moscow, Zh-17, U.S.S.R.

MUTSCHLECNER (Dr P.) Dept. of Astronomy, Indiana University, Bloomington, Indiana 47401, U.S.A.

*MUZZIO (Dr J.C.) Observatorio Astronómico, Paseo del Bosque, 1900 La Plata, Argentina

MYACHIN (Dr V.F.) Institute of Theoretical Astronomy, 10, Kutuzov Qay, Leningrad 192187, U.S.S.R.

*MYERS (Dr P.) Massachusetts Institute of Technology, Room 26-349, Dept. of Physics, Cambridge, MA 02139, U.S.A.

MYERSCOUGH (Dr V.P.) Dept. of Mathematcis, Queen Mary College, Mile End Road, London E. 1, U.K.

NACOZY (Dr P.E.) Dept. of Aerospace Engineering, The University of Texas, Austin, Texas 78712, U.S.A.

NADOLSCHI (Prof. Dr Doc. V.) Com. Ardeoani, of. Tescani, jud. Bacǎu, Roumania

NADYOZHIN (Dr D.K.) Institute of Applied Mathematics, Miusskaja Ploszhad 4 Moscow 125047, U.S.S.R.

NAGASAWA (Dr S.) Tokyo Astronomical Observatory, Mitaka, Tokyo, Japan

NAGIRNER (Dr D.I.) Astronomical Observatory of Leningrad University 10 linia 33, Leningrad 199178, U.S.S.R.

*NAGNIBEDA (Dr V.G.) Leningrad University Astronomical Observatory, Leningrad 199178, U.S.S.R.

NAHON (Prof. F.) Observatoire de Paris, 61, Avenue de l'Observatoire 75- Paris 14e, France

NAKAGAWA (Dr Y.)High Altitude Observatory, Boulder, Colorado 80302, U.S.A.

NAKANO (Saburo) 47, Myogadani, Bunkyo-ku, Tokyo, Japan

NAKANO (Dr T.) Dept. of Physics, Kyoto University, Kyoto, Japan

NAKAYAMA (Dr S.) College of General Education, University of Tokyo, Komaba 3-8-1, Meguroku, Tokyo, Japan

*NAKAZAWA (Dr Kiyoshi) Dept. of Physics, Kyoto University, Kyoto, 606 Japan

NAMBA (Dr O.) Astronomical Observatory "Sonnenborgh", Zonnenburg 2, Utrecht, The Netherlands

NANDY (Dr K.) Royal Observatory, Edinburgh 9, U.K.

NAPIER (Dr W.M.) Royal Observatory, Blackford Hill, Edinburgh EH9 3HJ, U.K.

NAQVI (Dr A.M.) 6274 Del Rosa Ave., Apt.11, San Bernardino, Cal. 92404, U.S.A.

NARANAN (Dr S.) Tata Institute of Fundamental Research, Bombay, India

NARAYANA (J.V.) Assistant Meteorologist, Regional Meteorological Centre, 4, College Road, Madras- 600006, India

NARIAI (Dr H.) Research Institute for Theoretical Physics, Hiroshima University, Takehara, Hiroshima-Ken, Japan

NARIAI (Dr K.) Tokyo Astronomical Observatory, Mitaka, Tokyo, Japan

*NARITA (Dr Shinji) Department of Electronics, Doshisha University,
    Kamigyoku, Kyoto, 602 Japan

NARLIKAR (Prof. J.V.) Tata Institute of Fundamental Research, Homi Bhabha Road
    Bombay 5, India

NATHER (Dr R.E.) Dept. of Physics and Astronomy, University of Tel Aviv
    Ramat Aviv, Tel Aviv, Israel

*NATTA (A.) Consiglio Nazionale Ricerche, Laboratorio Astrofisica Spaziale
    Casella Postale 67, 00044 Frascati, Italy

NAUMOV (Dr V.A.) Pulkovo Observatory, Leningrad 196140, U.S.S.R.

NECKEL (Dr H.) Hamburg-Bergedorf, Germany, F.R. (Hamburger Sternwarte)

NECKEL (Dr Th.) Max-Planck-Institut für Astronomie, 6900 Heidelberg-Königstuhl
    Germany, F.R.

NE'EMAN (Dr Y.) Department of Physics and Astronomy, Tel-Aviv University
    Ramat Aviv, Israel

NEFED'EVA (Dr A.I.) Engelhardt Observatory, Kazan, U.S.S.R.

NEFF (Dr J.S.) Dept. of Physics and Astronomy, State University of Iowa,
    Iowa City, Iowa 52240, U.S.A.

NELSON (Prof. B.) Dept. of Astronomy, San Diego State College, San Diego 15,
    California, U.S.A.

NEMIRO (Dr A.A.) Pulkovo Observatory, Leningrad, U.S.S.R.

NESS (Dr N.F.) NASA Goddard Space Flight Center, Greenbelt, Maryland 20771, U.S.A.

NEUGEBAUER (Dr G.) California Institute of Technology, Department of Astrophysics,
    Pasadena, California 91109, U.S.A.

NEUPERT (Dr W.M.) Solar Physics Branch, Goddard Space Flight Center, Greenbelt,
    Maryland, U.S.A.

NEUŽIL (Dr L.) Astronomical Institute, Czechoslovak Academy of Sciences,
    Observatory Ondřejov, Czechoslovakia

NEVEN (Dr L.) Chef du Département d'Astrophysique, Observatoire Royal de
    Belgique, Avenue Circulaire, 3, B - 1180 Bruxelles, Belgium

NEVIN (Prof. T.E.) University College, Dublin, Ireland

NEWKIRK (Jr, Dr G.A.) High Altitude Observatory of the University of Colorado,
    Boulder, Colorado 80302, U.S.A.

NEWSON (Dr G.H.) The Ohio University, Dept. of Astronomy,
    174 W 18th Ave, Columbus, Ohio 43210, U.S.A.

NEWTON (Dr R.R.) Applied Physics Laboratory, Johns Hopkins University,
    8621 Georgia Avenue, Silver Spring, Maryland, U.S.A.

NEY (Dr E.) University of Minnesota, School of Physics & Astronomy, Minneapolis,
    Minn. 55455, U.S.A.

NEYMAN (Prof. J.) Statistical Laboratory, University of California, Berkeley 4,
    California, U.S.A.

NICHOLLS (Prof. R.W.) CRESS, Department of Physics, York University,
    4700 Keele Street, Toronto, Ontario M3J 1P3, Canada

NICHOLSON (W.) Royal Greenwich Observatory, Herstmonceux Castle, Hailsham,
    Sussex, U.K.

NICOLET (Prof. Dr M.) 30 Avenue Den Doorn, 1180 Bruxelles, Belgium

NICOLINI (Prof. T.) Via E. Nicolardi, Palazzo C.U.S.A.N., Colli Aminei,
80131 Napoli, Italy

NICOLOV (Dr N.S.) Université de Sofia, Faculté Physique, Sofia, Bulgaria

NICOLSON (Dr G.D.) c/o National Institute for Telecommunications Research,
P.O. Box 3718, Johannesburg 2000, South Africa

*NIELL (Dr A.E.) Jet Propulsion Laboratory, CPB-300, 4800 Oak Grove Drive,
Pasadena CA 91103, U.S.A.

*NIEMELA (Dr V.S.) I.A.F.E., Casilla Correo 67, Suc. 28, Buenos Aires, Argentina

NIEUWENHUIJZEN (Dr H.) Sterrenwacht, Servaas Bolwerk 13, Utrecht, The Netherlands

NIIMI (Dr H.) Department of Aeronautical Engineering, Kyoto University, Kyoto,
Japan

NIKITIN (Dr A.A.) Observatory, State University, Leningrad, U.S.S.R.

NIKOLOFF (Dr I.) Perth Observatory, Bickley, W.A. 6076, Australia

*NIKOLOV (Dr A.) Department of Astronomy, Sofia University, Anton Ivanov Street 5,
1126 Sofia, Bulgaria

NIKOLSKY (Prof. Dr G. M.), Institute of Terrestrial Magnetism and Ionosphere,
Akademgorodok, 142092 Podol'sk, U.S.S.R.

NIKONOV (Dr V.B.) Crimean Astrophysical Observatory of the Academy of Sciences
of the U.S.S.R., P/O Nauchny, Crimea 334413, U.S.S.R.

*NILSON (Dr P.) Box 1170, Fallen, S-810 65 Skärplinge, Sweden

NILSSON (Dr C.) Smithsonian Astrophysics Observatory, 60 Garden Street,
Cambridge, Mass. 02138, U.S.A.

NININGER (Dr H.H.) Sedona's Meteorite Museum, P.O. Box 146, Sedona, Arizona, U.S.A.

NISHI (Dr K.) Tokyo Astronomical Observatory, University of Tokyo, Mitaka,
Tokyo, Japan

NISHIDA (Prof. M.) Department of Physics, Kyoto University, Kitashirakawa,
Kyoto, Japan

NISHIMURA (Dr S.) Tokyo Astronomical Observatory, Mitaka, Tokyo, Japan

NISSEN (Prof. J.J.) Observatorio Astronomico, San Juan, Argentina

NISSEN (Dr P.E.) Ole Roemer Observatory, DK-8000 Aarhus C, Denmark

NOBILI (Dr L.) Osservatorio Astronomico, Vicolo dell'Osservatorio, Padova, Italy

NOCI (Dr G.) Osservatorio astrofisico di Arcetri, Firenze, Italy

NOËL (F.) Departamento de Astronomía, Universidad de Chile, Casilla 36-D,
Santiago de Chile, Chile

NOERDLINGER (Prof. P.D.) Department of Astronomy, Michigan State University,
East Lansing, Michigan 48823, U.S.A.

*NOMOTO (Dr K.) Department of Astronomy, Faculty of Science, University of Tokyo,
Bunkyo, Tokyo, 113 Japan

NOONAN (Dr Th.W.) State University College at Brockport, Brockport,
New York 14420, U.S.A.

*NORDH (Dr H.L.) Stockholm Observatory, S-133 00 Saltsjöbaden, Sweden

*NORDSTRÖM (Dr B.) Copenhagen University Observatory, Brorfelde, DK-4340 Tølløse, Denmark

NORLIND (Dr W.) Lund Observatory, S-222 24 Lund, Sweden

NØRLUND (Prof. N.E.) Director, Geodetic Institute, Malmögade 6, Copenhagen, Denmark

NORRIS (Dr J.) Yale University, New Haven, Conn. U.S.A.

*NORTH (Dr J.D.) 28 Chalfont Road, Oxford, U.K.

NOTNI (Dr P.) Zentralinstitut für Astrophysik, Sternwarte Babelsberg, DDR 1502 Potsdam-Babelsberg, Rosa-Luxemburg-Strasse 17 a, Germany D.R.

NOTUKI (Dr N.) 1-23 Higashida, Suginami-ku, Tokyo, Japan

NOVICK (Dr R.) Columbia University, Physics Department, 538 West 120th Street, New York, N.Y. 10027, U.S.A.

NOVIKOV (Dr I.D.) Department fo Theoretical Astrophysics, Space Research Institute, USSR Academy of Sciences, Profsoyuznaja 88, Moscow 117485, U.S.S.R.

NOVOSELOV (Prof. Dr V.S.) Astronomical Observatory, Leningrad University, Leningrad, U.S.S.R.

NOYES (Dr R.W.) Smithsonian Institution, 60 Garden Street, Cambridge, Mass. 02138, U.S.A.

NUSSBAUMER (Dr H.) Atomic and Astrophysics Group ETH, Huttenstrasse 34, 8006 Zürich, Switzerland

OBASHEV (S.O.) Astrophysical Institute, Kazakh SSR Academy of Sciences, Alma-Ata 20, U.S.S.R.

OBI (Dr S.) Department of Earth Science, College of General Education, University of Tokyo, Komaba, Shibuya, Tokyo, Japan

OBREGON DIAZ (Dr O.) Departamento de Fisica y Quimica, Unidad Iztapalapa, P.O. Box 55-534, Mexico 13, D.F., Mexico

OBRIDKO (Dr V.N.), Institute for Terrestrial Magnetism and Ionosphere, USSR Academy of Sciences, Akademgorodok, 142092 Podol'sk, U.S.S.R.

OCCHIONERO (Dr F.) Laboratorio de Astrofisica, Casella Postale 67, 00044 Frascati (Roma), Italy

O'CONNEL (Dr D.J.K. S.J.) Borgo S. Spirito 5, 00193 Roma, Italy

O'CONNELL (Dr R.F.) Department of Physics and Astronomy, Louisiana State University, Baton Rouge, Louisiana 70803, U.S.A.

*O'CONNELL (Dr R.W.) Leander McCormick Observatory, The University of Virginia, P.O. Box 3818 University Station, Charlottesville, VA 22903, U.S.A.

*O'CONNOR (Dr S.) Physics Department, University College, Belfield, Dublin 4, Ireland

ODA (Dr M.) Institute of Space and Aeronautical Science, University of Tokyo, Komaba, Meguro-Ku, Tokyo, Japan

O'DELL (Dr C.R.) PD-LST, Marshall Space Flight Center, Huntsville, Alabama 35812, U.S.A.

*O'DELL (Dr S.L.) University of California, San Diego, Code C-011, La Jolla, CA 92093, U.S.A.

ODGERS (Dr G.J.) Dominion Astrophysical Observatory, Herzberg Institute of
  Astrophysics, National Research Council of Canada, 5071 W. Saanich Road,
  Victoria, B.C. V8X 3X3, Canada

OESTERWINTER (Dr C.) Naval Weapons Laboratory, Dahlgren, Virginia, U.S.A.

OETKEN (Dr L.) Zentralinstitut für Astrophysik, DDR 15 Potsdam, Telegrafenberg,
  Germany D.R.

OGORODNIKOV (Prof. Dr K.F.) University, Dept. of Astronomy, Leningrad, U.S.S.R.

O'HANDLEY (Dr D.A.) Caltech/Jet Propulsion Laboratory, 4800 Oak Grove Drive,
  Pasadena, California 91103, U.S.A.

*OHKI (Dr K.) Code 682, NASA, Goddard Space Flight Center, Greenbelt, MD. 20771,
  U.S.A.

ÖHMAN (Prof. Y.) Stockholm Observatory, S-133 00 Saltsjöbaden, Sweden

O'HORA (N.P.J.) Royal Greenwich Observatory, Herstmonceux Castle,
  Hailsham, Sussex, U.K.

*OHRING (Dr G.) Department of Geophysics and Planetary Sciences, Tel-Aviv University,
  Ramat-Aviv, Israel

OHYAMA (Prof. N.) 1-22-26 Hirisawa, Hamamatsu, Japan

OJA (Dr T.) Kvistaberg Observatory, S-19051 Bro, Sweden

*OKA (Dr T.) Herzberg Institute of Astrophysics, National Research Council of Canada,
  Ottawa, Ontario, Canada K1A OR6

OKAMOTO (Dr I.) International Latitude Observatory, Mizusawa, Iwate, Japan

OKE (Dr J.B.) Mount Wilson and Palomar Observatories, 1201 E. California Street,
  Pasadena, California 91104, U.S.A.

O'KEEFE (Dr J.A.) Code 602, Goddard Space Flight Center, Greenbelt,
  Maryland 20771, U.S.A.

OKI (Prof. Dr T.) Department of Earth Science, Faculty of Education,
  Fukushima University, Hamada-cho 12-23, Fukushima, Japan

OKOYE (Dr S.E.) Department of Physics, University of Nigeria, Nsukka, Nigeria

OKUDA (Prof. Dr H.) Department of Physics, Kyoto University, Kitashirakawa,
  Sakyo-ku, Japan

OKUDA (Dr T.) Director of International Latitude Observatory of Mizusawa,
  Mizusawa-shi, Iwate-ken, Japan

OLEAK (Dr H.) Zentralinstitut für Astrophysik, Sternwarte Babelsberg,
  DDR 1502 Potsdam-Babelsberg, Rosa-Luxemburg-Strasse 17 a, Germany D.R.

O'LEARY (Dr B.) Hampshire College, Amherst, Mass. 01002, U.S.A.

OLLONGREN (Dr A.) Centraal Rekeninstituut, Wassenaarseweg 80, Leiden,
  The Netherlands

*OLOFSSON (Dr S.G.) Stockholm Observatory, S-133 00 Saltsjöbaden, Sweden

*OLSEN (Dr E.H.) Copenhagen University Observatory, Brorfelde,
  DK-4340 Tølløse, Denmark

OLSEN (Dr K.H.) Los Alamos Scientific Laboratory, Box 1663 Group J-15,
  Los Alamos, New Mexico 87544, U.S.A.

OLSON (Dr E.C.) University of Illinois Observatory, Urbana, Ill.61801, U.S.A.

OMAROV (Dr T.B.) Astrophysical Institute of the Kazakh Academy of Sciences,
    Alma-Ata, U.S.S.R.

OMER (Dr G.C. Jr.) 1080 SW 11th Terrace, Gainesville, Florida 32601, U.S.A.

*OMNES (Prof.R.) Université d'Orsay, 15, rue Georges Clémenceau,
    91400 Orsay, France

OMONT (Prof.A.) Université de Paris, 7, 2 place Jussieu, 75005 Paris, France

ONDERLIČKA (Dr B.) Department of Astronomy, Purkyně University, Brno,
    Czechoslovakia

ONEGINA (Dr A.B.) Main Astronomical Observatory, Ukrainian Academy of Sciences,
    Kiev, U.S.S.R.

ONO (Dr Yoro) Department of Physics, Hokkaido University, Sapporo,
    Hokkaido, Japan

*OOE (Dr Masatsugu) International Latitude Observatory of Mizusawa,
    Mizusawa-shi, Iwate-ken, 023 Japan

OORT (Prof. Dr J.H.) President Kennedylaan 169, Oegstgeest, The Netherlands

OOSTERHOFF (Prof. Dr P.Th.)Sterrewacht, Ridderhoflaan 107, Koudekerk aan
    de Rijn, The Netherlands

OPALSKI (Dr W.) Institut Astronomique de l'Ecole Polytechnique
    Koszykowa 75, Warsaw, Poland

*OPHER (Dr R.) Faculty of Physics, Israël Institute of Technology, Haïfa, Israël

ÖPIK (Dr E.J.) Armagh Observatory, Armagh, U.K.

OPOLSKI (Dr A.) Astronomical Observatory, Kopernika 11, Wroclaw, Poland

*OPROIU (Dr T.) Str.Bucium 25, Bloc R-5, sc.I, et II, apart.10
    3400 Cluj-Napoca, Roumania

ORLIN (Dr H.)Nactional Academy of Sciences, 2101 Constitution Ave N W
    Washington D.C. 20418, U.S.A.

ORLOV (Dr A.A.) Sternberg Astronomical Institute, Moscow, U.S.S.R.

ORLOVA (Dr N.S.) The Pulkovo Observatory, Leningrad M-140, U.S.S.R.

ORRALL (Prof. Dr F.Q.) Institute for Astronomy, 2840 Koluwalu St.
    University of Hawaii, Honolulu, Hawaii 96822, U.S.A.

ORTE (D. Alberto) Observatorio de Marina, San Fernando, Cádiz, Spain

ORUS (J.J. de Pr.Dr) Barcelona University, Barcelona, Spain

OSAKI (Dr Toru) Ryukoku University, Shimo-gyo-ku, Kyoto, Japan

OSAKI (Dr Y.) Dept. of Astronomy, Faculty of Sciences, University of Tokyo,
    Bunkyo-Ku, Tokyo, Japan

OSAWA (Dr K.) Tokyo Astronomical Observatory, Mitaka, Tokyo, Japan 181

OSBORN (Dr W.) Dept. of Physics, Central Michigan University, Mt.Pleasant,
    Michigan 48859, U.S.A.

OSKANYAN (V.) Astrophysical Observatory, Byurakan, Armenia SSR, U.S.S.R.

OSMER (Dr P.) Cerro Telolo Inter-American Observatory, Casilla 63-D,
    La Serena, Chile

OSORIO (Dr J.J.S.P.) Observatório Astrónomico, Universidade do Porto, Monte da Virgem, Vila Nova de Gaia, Portugal

OSTER (Prof.L.) Joint Institute for Laboratory Astrophysics, University of Colorado, Boulder, Colorado, U.S.A.

OSTERBROCK (Prof. D.E.) Lick Observatory, University of California, Santa Cruz, Santa Cruz, California 95064, U.S.A.

OSTRIKER (Dr J.P.) Princeton University Observatory, Peyton Hall, Princeton, New Jersey 08540, U.S.A.

OTERMA (Dr L.) Sirkkalankatu 31, 20700 Turku 6, Finland

OTTELET (Dr I.J.G.J.) Institut d'Astrophysique, Université de Liège, avenue de Cointe, 5, B-4200 Cointe-Ougrée, Belgium

OVENDEN (Dr M.W.) Dept. of Geophysics and Astronomy, University of British Columbia, Vancouver, B.C. V6T 1W5, Canada

OWAKI (Dr Naoaki) Dept. of Astronomy and Earth Science, Tokyo Gakugei University, 4-1-1 Nukui-kita, Koganei-shi, Tokyo, 184 Japan

OWN (Dr T.) Dept. of Earth and Space Sciences, State University of New York, Stony Brook New York 11790, U.S.A.

OWREN (Dr L.) Dept. of Physics, Div. B., University of Bergen, Allegaten 53-55 5000 Bergen, Norway

OXENIUS (Dr J.) Université Libre de Bruxelles, Faculté des Sciences, Service de Chimie Physique 2, Avenue F.D. Roosevelt 50, 1050 Bruxelles, Belgium

ÖZEMRE (Dr K.) University Observatory, Beyazit, Istanbul, Turkey

OZERNOJ (Dr L.M.) Physical Institute, USSR Academy of Sciences, 117924 Moscow,USSR

OZSVATH (Prof. I.) Earth & Planetary Sciences Laboratory, Graduate Research Center, P.O. Box 30365, Dallas, Texas 75230, U.S.A.

*PAAL G.,H- 1525 Budapest Box 67, Hungary

PACHNER (Prof. J.) Dept. of Physics and Astronomy, University of Regina, Regina, Sakatchewan S4S 0A2, Canada

PACHOLCZYK (Prof. Dr A.G.) Steward Observatory, University of Arizona, Tucson, Arizona, U.S.A.

PACINI (Dr F.) Laboratorio di Astrofisica, Casella Postale 67, Frascati (Roma) Italy

PACZYSNKI (Dr Bogdan) Astronomical Observatory, Al.Uajzdowskie 4, Warszawa, Poland

*PADEVĚT (Dr V.) Astronomical Institute of the Czechoslovak Academy of Sciences, Observatory, 251 65 Ondřejov, Czechslovakia

*PADRIELLI (L.) Laboratorio di Radioastronomia, Via Irnerio 46, 40126 Bologna, Italy

PAGE (Prof. T.L.) Director Van Vleck Observatory, Wesleyan University, Middletown, Connecticut 06457, U.S.A.

PAGEL (Dr B.E.J.) Royal Greenwich Observatory, Herstmonceux Castle, Hailsham, Sussex, U.K.

PAJDUŠAKOVÁ (Dr L.) Director, Skalnaté Pleso Observatory, Slovak Academy of Sciences, Tatranská Lomnica, Czechoslovakia

PAL (Dr Arpad) Université, Cluj, Roumania

*PALLAVICINI (Dr R.) Osservatorio Astrofisico di Arcetri, Largo E. Fermi 5,
    50125 Firenze, Italy

*PALMEIRA (Dr R.A.R.) INPE, C.P.515, 12.200 Sao José dos Campos, Sao Paulo,
    Brazil

PALMER (D.R.) Royal Greenwich Observatory, Herstmonceux Castle,
    Hailsham, Sussex, U.K.

PALMER (Dr H.P.) Nuffield Radio Astronomy Laboratories, Jodrell Bank,
    Macclesfield, Cheshire, U.K.

PALMER (Dr P.E.) Dept. of Astronomy and Astrophysics, Ryerson Laboratory 162
    1100-14 East 58th Street, Chicago, Illinois 60637, U.S.A.

PANAGIA (Dr N.) Laboratorio di Astrofisica, Casella Postale 67,
    Frascati (Roma), Italy

PANAJOTOV (Dr L.A.) Pulkovo Observatory, Leningrad, U.S.S.R.

PANDE (M.C.) U.P. State Observatory, Nainital India

*PANOV (Dr K.) Dept. of Astronomy, Bulgarian Academy of Sciences,
    "7th November" Street 1, 1000 Sofia, Bulgaria

PAPAGIANNIS (Dr M.) Dept. of Astronomy, Boston University, Boston,Mass. 02215,
    U.S.A.

PAPALIOLIOS (Dr C.) Smithsonian Astrophysical Observatory, Cambridge, Mass.02138,
    U.S.A.

PAQUET (Dr P.E.G.) Observatoire Royal de Belgique, avenue Circulaire 3,
    B- 1180 Bruxelles, Belgium

PARCELIER (Dr P.) Observatoire de Paris, 61, Avenue de l'Observatoire,
    75014 Paris, France

*PARESCE (Dr F.) Space Sciences Laboratory, University of California
    Berkeley, California 94720, U.S.A.

PARIJSKIJ (Dr Yu.N.) Special Astrophysical Observatory, USSR Academy of Sciences
    Leningrad Branch of the SAO, Pulkovo 196140, U.S.S.R.

PARIJSKIJ (Dr N.N.) O. Schmidt Institute of Physics of the Earth
    USSR Academy of Sciences, B. Gruzinskaya 10, Moscow, U.S.S.R.

PARKER (Dr E.A.) Electronics Laboratories, University of Kent at Canterbury,
    Canterbury, Kent, U.K.

PARKER (Dr E.N.) Institute for Nuclear Studies, University of Chicago,
    Chicago, Ill. 60637, U.S.A.

PARKER (Dr R.A.R.) Code CB, Johnson Space Center, Houston, Texas 77058, U.S.A.

PARKINSON (Dr J.H.) Mullard Space Science Laboratory, Holmbury St. Mary
    Dorking, Surrey, U.K.

PARKINSON (Dr T.) Kitt Peak National Observatory, 950 North Cherry Avenue,
    Tucson, Arizona 85726, U.S.A.

PARKINSON (Dr W.H.) Harvard College Observatory, 60 Garden Street,
    Cambridge 38, Mass. 02138, U.S.A.

*PARRISH (Dr A.) Dept. of Physics, Massachusetts Institute of Technology
    Cambridge, Mass. 02139, U.S.A.

*PARSAMYAN (Dr E.S.) Byurakan Astrophysical Observatory, 378433 Byurakan,
Armenia, U.S.S.R.

PARSONS (Dr S.B.) Dept. of Astronomy, University of Texas, Austin,
Texas 78712, U.S.A.

*PARTRIDGE (Dr R.B.) Haverford College, Dept. of Astronomy,
Haverford, PA 19041, U.S.A.

PASACHOFF (Dr J.M.) Williams College, Hopkins Observatory, Williamstown,
Mass. 01267, U.S.A.

PASCOAL (A.) Observatório da Universidade do Porto, Monte da Virgem,
Villa Nova de Gaia, Portugal

*PASCU (Dr D.) U.S. Naval Observatory, Washington, DC 20390, U.S.A.

PASCUAL MARTINEZ de (M.) Observatorio Astronómico Nacional
Alfonso XII, 3,Madrid 7, Spain

PASINETTI (Prof. Dr L.E.) Osservatorio Astronomico di Milano-Merate
Via E. Bianchi 46, 22055 Merate, Italy

PASTORIZA (Dr M.G.) Ayacucho 2142, Córdoba, Argentina

PATERNO (Dr L.) Osservatorio Astrofisico, Città Universitaria,
Viale A. Doria, I 95125 Catania, Italy

PATHRIA (Dr R.K.) Dept. of Physics, University of Waterloo,
Waterloo, Ontario N2L 3G1, Canada

PAULINY-TOTH (Dr I.I.K.) Max-Planck-Institut für Radioastronomie
Auf dem Hügel 69, 53 Bonn 1, Germany F.R.

PAVLOV (Prof. Dr N.N.) Pulkovo Observatory, Leningrad, U.S.S.R.

PAVLOVSKAYA (Dr E.D.) Sternberg Astronomical Institute, Moscow, U.S.S.R.

PAYNE-GAPOSCHKIN (Dr C.H.) Harvard College Observatory,
Cambridge, Mass. 02138, U.S.A.

PEACH (Dr G.) Dept. of Physics, University College, Gower Street
London W.C. 1, U.K.

PEACH (Dr J.V.) Dept. of Astrophysics, South Parks Road, Oxford, U.K.

PEALE (Dr S.J.) J.I.L.A. University of Colorado,
Boulder, Colorado, U.S.A.

PEARSE (Prof. R.W.B.) Imperial College of Science and Technology
Prince Consort Road, London, S.W. 7, U.K.

PEAT (Dr D.W.) The Observatories, Madingley Road, Cambridge, U.K.

PECKER (Prof. J.C.) Directeur, Institut d'Astrophysique,
98bis, Boulevard Arago, 75014 Paris, France

PEDERSEN (Prof. O.) History of Science Dept. University of Aarhus,
Ny Munkegade, DK- 8000 C, Denmark

*PEDLAR (Dr A.) Nuffield Radio Astronomy Laboratories, Jodrell Bank
Nr. Macclesfield, Cheshire, U.K.

PEDOUSSAUT (Dr A.) Observatoire de Toulouse, 31 Toulouse (Haute Garonne), France

PEEBLES (Dr P.J.E.) Dept. of Physics, Princeton University, Princeton,
New Jersey 08540, U.S.A.

PEERY (Dr B.F.) Department of Astronomy, Indiana University, Bloomington,
    Indiana 47405, U.S.A.

PEIMBERT (Dr M.) Instituto de Astronomía, UNAM, Apartado Postal 70-264,
    México D.F., Mexico

PEKERIS (Prof. Ch.L.) The Weizmann Institute of Science, Department of Mathematics,
    Rehovoth, Israel

*PEL (Dr J.W.) Max-Planck-Institute für Astronomie, 69 Heidelberg,
    Königstuhl, Germany F.R.

PELLAS (Dr P.) Museum d'Histoire Naturelle- Minéralogie, 61, rue de Buffon,
    75 Paris, France

PELSENEER (Prof Dr J.) Université Libre de Bruxelles, Avenue des Grenadiers, 76,
    B-1050 Bruxelles, Belgium

PELS-KLUYVER (Dr H.A.) Sterrewacht, Leiden, the Netherlands

PELTIER, LESLIE (C.) 327 S. Bredeick Street, Delphos, Ohio, U.S.A.

PENNY (C.J.A.) Royal Greenwich Observatory, Time Department, Herstmonceux Castle,
    Hailsham, Sussex, U.K.

PENSADO (José) Observatoire, Alfonso XII, 3, Madrid, Spain

PENSTON (Dr M.V.) Anglo-Australian Observatory, P.O. Box 296, Epping,
    N.S.W. 2121, Australia

PENZIAS (Dr A.) Bell Telephone Laboratories, Crawford Hill Laboratory, Box 400,
    Holmdel, New Jersey 07733, U.S.A.

PERAIAH (Dr A.) Lehrstuhl für Theoretische Astrophysik der Universität
    Heidelberg, 69 Heidelberg 1, Im Neuheimer Feld 294, Germany F.R.

PERCY (Dr J.R.) David Dunlap Observatory, University of Toronto, Richmond Hill,
    Ontario L4C 4Y6, Canada

*PERDANG (Dr J.) Institut d'Astrophysique, B-4200 Cointe-Ougrée, Belgium

PEREK (Dr L.) Astronomical Institute, Czechoslovak Academy of Sciences,
    Budečská 6, 12023 Prague 2, Czechoslovakia

*PEREZ-PERAZA (Dr J.) Instituto de Astronomía, UNAM, Apartado Postal 70-264,
    México, D.F., Mexico

PERINOTTO (Dr M.) Osservatorio Astrofisico, 50125 Arcetri, Largo E. Fermi 5,
    Firenze, Italy

PERKINS (Dr F.W.) Princeton University, Plasma Physics Laboratory, P.O. Box 451,
    New Jersey 08540, U.S.A.

PEROLA (Dr G.C.) Instituto di Scienze Fisiche, Via Celoria 16, 20133, Milano,
    Italy

*PERRIN (Dr M.N.) IAP, 98 bis, Boulevard Arago, 75014 Paris, France

PERRY (C.L.) Louisiana State University, Baton Rouge, Louisiana 70803, U.S.A.

*PERRY (Dr J.J.) Max-Planck-Institut für Astrophysik, 8 München 40, Föhringer
    Ring 6, Germany F.R.

PERSIDES (Prof. Dr S.C.) Astronomical Department, University of Thessaloniki, Thessaloniki, Greece

PESCH (Prof. P.) Warner & Swasey Observatory, Case Institute of Technology, Taylor and Brunswick Roads, East Cleveland, Ohio 44112, U.S.A.

PETERSEN (Otzen J.) University Observatory, Øster Voldgade 3, Copenhagen K, Denmark

*PETERSON (Dr B.A.) Anglo-Australian Observatory, P.O. Box 296, Epping, N.S.W. 2121, Australia

PETERSON (Dr L.E.) Physics Department, Revelle College, University of California-San Diego, La Jolla, California 92038, U.S.A.

PETFORD (Dr A.D.) University Observatory, South Park Road, Oxford, U.K.

PETRI (Dr W.) Unterleiten 2, Post-Box 106, 8162 Schliersee (OBB), Germany F.R.

PETRINI (Dr D.) Observatoire de Nice, Le Mont-Gros, 06300 Nice, France

*PETROPOULOS (Dr B.) Research Associate, Research Center for Astronomy and Applied Mathematics, Academy of Athens, 14 Anagnostopoulou Street, Athens - 136, Greece

PETROSIAN (Dr V.) Institute of Plasma Physics, Stanford University, Stanford, California 94305, U.S.A.

PETROV (Prof. Dr G.I.) Institute of Cosmical Research, USSR Academy of Sciences, Moscow, U.S.S.R.

PETROV (Dr G.M.) Nikolaev Department of the Main Astronomical Observatory, USSR Academy of Sciences, Nikolaev , Ukrainian SSR, U.S.S.R.

*PETROV (N.) Astronomical Observatory, Varna, Bulgaria

PETROVSKAYA (Dr M.S.) Institute of Theoretical Astronomy, 10, Kutuzov Qay, Leningrad, 192187, U.S.S.R.

PETTENGILL (Prof. G.H.) Department of Earth and Planetary Sciences, Massachusetts Institute of Technology, Cambridge, Mass. 02139, U.S.A

PETTIT (Dr E.) Mount Wilson and Palomar Observatories, 813 Santa Barbara Street, Pasadena 4, California, U.S.A.

*PEYTREMANN (Dr E.) European Space Agency, 114, Avenue Charles de Gaulle, F - 92522 , Neuilly-sur-Seine, France

PEYTURAUX (Dr R.) Institut d'Astrophysique, 98 bis, Boulevard Arago, F - 75014 Paris, France

PFENNING (Dr H.) Max-Planck-Institut für Physik und Astrophysik, 8000 München, Föhringer Ring 6, Germany F.R.

PFLEIDERER (Dr J.) Astronomisches Institut der Universität, Universitätsstr. 4, A-6020 Innsbruck, Austria

PFLUG (Dr K.) Heinrich-Hertz-Institut für Solar-Terrestrische Physik, Berlin-Adlershof, Rudower Chaussee, Germany D.R.

PHILIP (Dr A.G.D.) Dudley Observatory, 100 Fuller Rd., Albany, New York 12208, U.S.A.

PHILLIPS (Prof. J.G.) Dept. of Astronomy, University of California, Berkeley, California, U.S.A.

PICK-GUTMANN (M.) Observatoire de Paris, Section d'Astrophysique, F-92190 Meudon, France

PIDDINGTON (Dr J.H.) CSIRO, Division of Physics, University Grounds, Chippendale, NSW 2008, Australia

PIERCE (Dr A. K.) Kitt Peak National Observatory, 950 North Cherry Avenue, P.O. Box 26732, Tucson, Arizona 85726

*PIIROLA (Dr V. Ee.) Observatory and Astrophysics Laboratory, University of Helsinki, Tähtitorninmäki, SF-00130 Helsinki 13, Finland

*PILKINGTON (Dr J.D.H.) Herstmonceux Castle, Hailsham, East Sussex, BN 27 1 RP, U.K.

PILOWSKI (Prof. Dr K.) Geodätisches Institut der Technischen Hochschule, Nienburger Strasse 1, Hannover,Germany F.R.

*PINES (Dr D.) Department of Physics, University of Illinois at Urbana, Urbana, IL 61801, U.S.A.

PINKAU (Prof. Dr K.) Max-Planck-Institut für Physik und Astrophysik, Institut für Extraterrestrische Physik, 8046 Garching b. München, Germany F.R.

PINTO (Dr G.) Osservatorio Astronomico, Padova, Italy

PIOTROWSKI (Dr S.) Astronomical Observatory, Ujazdowskie 4, Warsaw, Poland

*PIPHER (Dr J.L.) Department of Physics and Astronomy, University of Rochester, Rochester, NY 14627, U.S.A.

PISMIS de Recillas (Dr P.) Instituto de Astronomía, UNAM, Apartado Postal 70-264, México, D.F., Mexico

*PITTICH (Dr E.) Astronomical Institute of the Slovak Academy of Sciences, Dúbravská cesta 2, 899 30 Bratislava, Czechoslovakia

PITZ (Dr E.) Max-Planck-Institut für Astronomie, 6900 Heidelberg 1, Königstuhl, Germany F.R.

PIZZELLA (Prof. Dr G.) Instituto di Fisica "G. Marconi", Università di Roma, Roma, Italy

PLAKIDIS (Prof. Dr S.) 4B Eridanou Street, Athens (612), Greece

PLASKETT (Prof. H.H.) 48 Blenheim Drive, Oxford, U.K.

PLASSARD (Dr J.) Director, Ksara Observatory, Ksara, Lebanon

PLATZECK (Dr R.) Comision Nacional de la Energia Atomica, San Carlos de Bariloche, Argentina

PLAUT (Prof. Dr L.) Kapteyn Astronomical Laboratory, Postbus 800, Groningen, The Netherlands

PLAVCOVA (Dr Z.) Department of Astronomy, University of California, 405 Hilgard Avenue, Los Angeles, California 90024, U.S.A.

PLAVEC (Prof. Dr M.) Department of Astronomy, University of California, 405 Hilgard Avenue, Los Angeles, California 90024, U.S.A.

PNEUMAN (Dr G.) High Altitude Observatory, Boulder, Colorado 80302, U.S.A.

PODOBED (Dr V.V.) Sternberg Astronomical Institute, Moscow V-3234, U.S.S.R.

POGO (Dr A.) Mount Wilson and Palomar Observatories, 813 Santa Barbara Street, Pasadena 4, California, U.S.A.

POHL (Dr E.) Sternwarte Nürnberg, Lützowstr. 10, 8500 Nürnberg, Germany F.R.

*POLAND (Dr A.I.) High Altitude Observatory, NCAR, P.O. Box 3000, Boulder, CO 80303, U.S.A.

POLETTO (Dr G.) Osservatorio Astrofisico Firenze, Largo E. Fermi 5, Firenze, Italy

*POLLACK (Dr J.B.) Space Science Division, NASA-Ames Research Center,
     Moffett Field, CA 94035, U.S.A.

*POLNITZKY (Dr G.) Universitäts-Sternwarte, Türkenschanz-Strasse 17,
     A-1180 Wien, Austria

*POLOSKOV (Prof.Dr S.M.) Institute of Applied Geophysics, USSR Academy of Sciences,
     107150 Moscow, U.S.S.R.

POLOZHENTSEV (Dr D.D.) Pulkovo Observatory, Leningrad M-140, U.S.S.R.

POLUPAN (Dr P.N.) Astronomical Observatory, Kiev University, Kiev, U.S.S.R.

PONSONBY (Dr J.E.B.) Nuffield Radio Astronomy Laboratories, Jodrell Bank,
     Macclesfield, Cheshire, U.K.

POOLEY (Dr G.G.) Mullard Radio Astronomy Observatory, Cavendish Laboratory,
     Cambridge, U.K.

POPE (B. Sc. J.D.) Royal Greenwich Observatory, Herstmonceux Castle, Hailsham,
     Sussex, U.K.

*POPELAR (Dr J.) Gravity and Geodynamics Division, Earth Physics Branch,
     Department of Energy, Mines and Resources, 3 Observatory Crescent,
     Ottawa, Ontario, Canada K1A OE4

POPOV (Dr N.A.) Gravimetrical Observatory, Ukrainian Academy of Sciences,
     Poltava, U.S.S.R.

POPOVA (Dr M.D.) Section d'Astronomie de l'Académie Bulgare des Sciences,
     Rue du 7 Novembre No 1, Sofia, Bulgaria

POPOVIĆ (Prof. Dr B.) Ognjena Price 80, Beograd , Yugoslavia

POPOVICI (Prof. Dr C.) Observatoire de Bucarest, 5, rue Cutitul de Argint,
     Bucarest, Roumania

PÖPPEL (Dr W.) Instituto Argentino de Radioastronomía, Casilla No 5, Villa Elisa,
     Prov. de Bs.As., Argentina

POPPER (Prof. D.M.) Department of Astronomy, University of California,
     Los Angeles 24, California, U.S.A.

PORFIR'EV (Dr V.V.), MOPI, ul. Radio 10, 107846 Moscow, U.S.S.R.

PORTER (Dr J.G.) Whitestones, Hempstead Lane, Hailsham, Sussex. U.K.

PORTER (Prof. N.A.) Department of Physics, University College, Belfield,
     Dublin 4, Ireland

PORUBČAN (Dr V.) Astronomical Institute of the Slovak Academy of Sciences,
     Dúbravská cesta 2, 899 30 Bratislava, Czechoslovakia

POTTASCH (Prof. Dr. S.R.) Kapteyn Astronomical Institute, Postbus 800,
     Groningen, The Netherlands

POTTER (Dr H.I.) Main Astronomical Observatory, Academy of Sciences,
     Pulkovo, U.S.S.R.

POULAKOS (Dr C.) Research Center for Astronomy and Applied Mathematics,
     Academy of Athens, 14 Anagnostopoulou Street, Athens (136), Greece

*POUMEYROL (F.) Observatoire de Bordeaux, Avenue Pierre Sémirot, 33270 Floirac, France

POUNDS (Dr K.A.) University of Leicester, Dept. of Physics, University Road,
     Leicester, LE 17 RH, U.K.

POVEDA (Dr A.) Instituto de Astronomía, UNAM, Apartado Postal 70-264, México, D.F., Mexico

PRADERIE (Dr F.) Institut d'Astrophysique, 98 bis Boulevard Arago, 75014 Paris, France

PRADHAN (Dr) Nizamiah Observatory, Hyderabad, India

*PRATAP (Dr R.) Physical Research Laboratory, Navrangpura, Ahmedabad 380009, India

*PREITE MARTINEZ (A.) Consiglio Nazionale Ricerche, Laboratorio Astrofisica Spaziale, Casella Postale 67, 00044 Frascati, Italy

PRENDERGAST (Dr K.H.) 1402 Pupin, Columbia University, New York, N.Y.

PRESTON (Dr G.W. III) University of California, Santa Cruz, California 95060, U.S.A.

*PREUSS (Dr E.) Max-Planck-Institut für Radioastronomie, Auf dem Hügel 69, 5300 Bonn, Germany F.R.

PREVOT (Dr L.) Observatoire de Marseille, 2, Place le Verrier, 13004 Marseille, France

PRICE (Dr M.J.) Planetary Science Institute, 252 W. Ina Road, Suite D, Tucson, Arizona 85704, U.S.A.

PRICE (Dr R.M.) 26-463 M.I.T., Cambridge, Mass. 02139, U.S.A.

PRIEST (Dr E.R.) Mathematical Institute, North Haugh, St Andrews Fife, U.K.

PRIESTER (Prof. Dr W.) Direktor, Institut für Astrophysik und Extraterrestrische Forschung, Universität Bonn, Auf dem Hügel 71, 53 Bonn, Germany F.R.

PROBSTEIN (Dr R.F.) Department of Mechanical Engineering, Massachusetts Institute of Technology, Cambridge, Mass. 02139, U.S.A.

PROCHAZKA (Dr F.V.) Observatory of the University of Vienna, Türkenschantzstr. 17, A-1180 Vienna, Austria

PRODAN (Dr Y.I.) Sternberg Astronomical Institute, Universitetskij prospects 13, Moscow 117234, U.S.S.R.

PROISY (Dr P.E.) Observatoire de Lyon, 69230 Saint-Genis-Laval, France

PROKAKIS (Dr Th.) Astronomical Institute, National Observatory of Athens, Athens (306) Greece

PROKOF'EV (Prof. V.K.) Crimean Astrophysical Observatory of the Academy of Sciences of the USSR, P/o Nauchny, Crimea 334413, U.S.S.R.

PROKOF'EVA (Dr I.A.) Main Astronomical Observatory, Academy of Sciences, Pulkovo, U.S.S.R.

PRONIK (Dr I.I.) Crimean Astrophysical Observatory of the USSR Acad. of Sci. P/o Nauchny, Crimea 334413, U.S.S.R.

PROTHEROE (Dr W.M.) Dept. of Astronomy, 174 W. Eighteenth Street, Ohio State University, Columbus, Ohio 43210, U.S.A.

PROTITCH (M.B.) Astronomical Observatory, Veliki Vracar, Belgrade, Yugoslavia

PROVERBIO (Prof. E.) Director Institute of Astronomy, Via Ospedale 72, 09100 Cagliari, Italy

PRONIK (Dr V.I.), Crimean Astrophysical Observatory, USSR Academy of Sciences, P/O Nauchniy 334413 Crimean, U.S.S.R.

PRYCE (Dr M.H.L.) Dept. of Physics, University of British Columbia,
    Vancouver, B.C. V6T 1W5, Canada

PRZYBYLSKI (Dr A.) Mount Stromlo Observatory, Canberry, A.C.T., Australia

PSKOVSKIJ (Dr Ju.P.) Sternberg Astronomical Institute, Moscow, U.S.S.R.

*PUCILLO (M.) Observatorio Astronomico, Via G.B. Tiepolo 11,
    I 34131 Trieste, Italy

*PUGET (Dr J.L.) DAPHE Observatoire de Meudon, 92190 Meudon, France

PUNETHA (Dr L.M.) U.P. State Observatory, Manora Peak, Nainital, U.P., India

PURGATHOFER (Dr A. Th.) Universitäts-Sternwarte Wien, Türkenschanzstr.17
    A-1180 Vienna, Austria

PURTON (Prof. C.R.) CRESS, Dept. of Physics, York University, 4700 Keele Street
    Downsview, Ontario M3J 1P3, Canada

PYPER (Dr D.M.) Dept. of Physics and Astronomy, Tel-Aviv University,
    Ramat Aviv, Israël

*QUAST (G.R.) Observatório Nacional, R. General Bruce 586, Sao Cristovao,
    20.000 Rio de Janeiro, RJ, Brazil

QUERCI (Dr F.) Observatoire de Paris, Section d'Astrophysique de Meudon,
    92190 Meudon, France

*QUERCY (Dr M.) DEPEG, Observatoire de Meudon, 92190 Meudon, France

QUIJANO (Luis) Observatoire, San Fernando, Spain

*QUINTANA (Dr H.) Universidad de Chile, Departamento de Astronomia
    Casilla 36-D, Santiago, Chile
*QUINTANA (Dr J.M.) Calvo Sotelo, 28 Granada, Spain

QUIRK (Dr W.) Goddard Institute for Space Studies, 2880 Broadway,
    New York, N.Y. 10025, U.S.A.

QVIST (Dr B.) Abo Akademi, Abo, Finland

RAADU (Dr M.A.) The Royal Institute of Technology, Dept. of Plasma Physics,
    Teknik Ring 31-33, S-100 44 Stockholm 70, Sweden

RABBEN (Dr H.H.) Max-Planck-Institut für Physik und Astrophysik
    Abteilung Extraterrestrische Physik, Garching/ München, Germany, F.R.

RACHKOVSKY (Dr D.N.) Crimean Astrophysical Observatory, USSR Academy of Sciences,
    P/O Nauchniy, 334413 Crimea, U.S.S.R.

RACINE (Dr R.) David Dunlap Observatory, University of Toronto,
    Richmond Hill, Ontario L4C 4Y6, Canada

RACKHAM (Dr T.W.) 39 Meadow Avenue, Goostrey, Cheshire CW4 8LS, U.K.

RADHAKRISHNAN (Prof. V.) Director Raman Research Institute,
    Bangalore- 560 006, India

RÄDLER (Dr K.H.) Zentralinstitut für Astrophysik, Telegrafenberg,
    DDR 15 Potsdam, Germany D.R.
RADLOVA (Dr L.N.) Institute of Science and Technics Information,
    Dept. of Astronomy, USSR Academy of Sciences, Moscow, U.S.S.R.

RAGHAVAN (Dr N.) II/3 I.I.T. Campus Hauz Khas, N. Delhi 110029, India

RAHE (Prof. Dr J.) Astronomical Institute, University Erlangen-Nürnberg
    Sternwartstr. 7, 86 Bamberg, Germany F.R.

RAHIM (Dr M.H.A.) Helwan Observatory, Helwan (near Cairo)Arab Republic of Egypt

*RAIKOVA (Dr D.) Dept. of Astronomy, Bulgarian Academy of Sciences,
    "7th November" Street 1, 1000 Sofia, Bulgaria

RAIMOND (Dr E.) Netherlands Foundation for Radio Astronomy, Radiosterrenwacht,
    Dwingeloo, 7514, The Netherlands

*RAINE (Dr D.J.) Dept. of Astronomy, University of Leicester, University Road,
    Leicester LE1 7RH, U.K.

RAJCHL (Dr J.) Astronomical Institute, Czechoslovak Academy of Sciences,
    Observatory Ondřejov, Czechoslovakia

RAJU (Dr P.K.) Lehrstuhl für Theoretische Astrophysik, Hausserstr.64,
    7400 Tübingen, Germany, F.R.

RAKAVY (Prof. G.) University of Jerusalem, Einstein Institute of Physics,
    Dept. of Theoretical Physics, The Hebrew University of Jerusalem,
    Jerusalem, Israel

RAKOS (Dr. K.D.) Universitätssternwarte, Türkenschanzstr. 17, A-1180 Wien,
    Austria

RAKSHIT (Prof. H.) Bengal Engineering College, Sibpore, Hewrah, India

RAMATY (Dr R.) NASA Goddard Space Flight Center, Greenbelt, Maryland, U.S.A.

RAMBERG (Prof. Dr J.M.) Genvägen 4, S-13300 Saltsjöbaden, Sweden

RANDIĆ (Prof. Dr. L.) Gundulićeva 54, Zagreb, Yugoslavia

*RANK (Dr D.M.) Lick Observatory, University of California, Santa Cruz, CA 95064,
    U.S.A.
*RANKIN (Dr J.M.) Dept. of Astronomy, Cornell University, Space Sciences Building,
    Ithaca, NY 14853, U.S.A.

RAO (Dr U.R.) Physical Research Laboratory, Navrangpura, Ahmedabad-9, India

RASMUSEN (Dr H.Q.) Vaerslevgaarden 4400 Kalundborg, Denmark

RAYROLE (J.) Observatoire de Meudon, 5, place Janssen, 92-Meudon, France

RAZIN (Dr V.A.) Radiophysical Research Institute, Gorkii, U.S.S.R.

*READHEAD (Dr A.C.S.) Radio Astronomy, Cavendish Laboratory, Madingley Road,
    Cambridge, U.K.

REAVES (Dr G.) Astronomy Dept. University of Southern California,
    Los Angeles, Cal. 90007, U.S.A.

REAY (Dr N.K.) Astronomy Group, Physics Dept., Imperial College, London SW 7, U.K.

*REBEIROT (E.) Observatoire de Marseille, 2, Place Le Verrier,
    13004 Marseille, France

REBER (Dr G.) CSIRO Stowell Avenue, Hobart, Tasmania, Australia 7000

REDDISH (Dr V.C.) Royal Observatory, Edinburgh 9, Scotland

REES (Dr M.J.) Institute of Astronomy, Madingley Road, Cambridge CB3 OHA, U.K.

REEVES (Dr E.M.) Harvard College Observatory, 60 Garden Street, Cambridge,
    Massachusetts 02138, U.S.A.

REEVES (Dr H.) SEP- Saclay- BP No.2, Gif sur Yvette (91) France

REFSDAL (Prof. Dr S.) Hamburger Sternwarte, Gojenbergsweg 112,
    D-205 Hamburg 80, Germany, F.R.

REGO FERNANDEZ (Dr M.) Depart. de Astronomia, Universidad Complutense
    Ciudad Universitaria, Madrid 3, Spain

REID (Dr J.H.) 1007 Reseda Drive, Houston, Texas 77058, U.S.A.

REIMERS (Dr D.) Institut für Theoretische Physik und Sternwarte der
    Universität, Olshausenstr. 2300 Kiel, Germany F.R.

*REINHARDT (Dr M.) Astronomisches Institut der Ruhr-Universität,
    Postfach 2148, D-463 Bochum, Germany, F.R.

REINMUTH (Dr K.) Häusserstr.61, Heidelberg, Germany, F.R.

REIZ (Prof. A.) University Observatory, Østervoldgade 3, DK-1350 Copenhagen-K
    Denmark (temporary address: ESO, Telescope Project Division,
    c/o Cern 11, Geneva 23, Switzerland)

REMY-BATTIAU (Dr L.) Conservateur,Institut d'Astrophysique, Université
    de Liège, avenue de Cointe 5, B-4200 Cointe-Ougrée, Belgium

RENSE (W.A.) Physics Dept. University of Colorado, Boulder, Colorado, U.S.A.

RENSON (Dr P.F.M.L.) Institutd'Astrophysique de l'Université de Liège,
    Avenue de Cointe 5, B-4200 Cointe-Ougrée , Belgium

RENZINI (Dr A.) Astronomical Observatory Bologna, via Zamboni 32, Bologna,
    Italy

REQUIÈME (Y) Observatoire de Bordeaux, 33-Floirac, France

REUNING (Dr E.) Dept. of Physics and Astronomy, University of Georgia,
    Athens, Georgia 30601, U.S.A.

REYNOLDS (Dr J.H.) Dept. of Physics, University of California, Berkeley,
    California, U.S.A.

*REYNOLDS (Dr R.J.) University of Wisconsin-Madison, Space Physics Group,
    1150 University Avenue, Madison, WI 53706, U.S.A.

*RHYNSBURGER (Dr R.W.) U.S. Naval Observatory, Washington, D.C. 20390, U.S.A.

*RIBES (Dr E.) DASOP, Observatoire de Meudon, 92190 Meudon, France

RIBES (Dr J.-C.) Observatoire de Paris, Section d'Astrophysique de Meudon,
    92190 Meudon, France

RICE (D.A.) Coast and Geodetic Survey, Rockville, Maryland 20852, U.S.A.

RICE (Prof. J.B.) Dept. of Physics, Brandon University, Brandon, Manitoba,Canada

RICHARDSON (Dr E.H.) Dominion Astrophysical Observatory, Herzberg
    Institute of Astrophysics, National Research Council of Canada
    5071 W. Saanich Road, Victoria, B.C. V8X 3X3, Canada

RICHARDSON (R.S.) Griffith Observatory, P.O. Box 27787, Los Felix Station,
    Los Angeles 27, Cal., U.S.A.

*RICHER (Dr H.B.) Dept. of Geophysics and Astronomy, University of
    British Columbia, 2075 Wesbrook Place, Vancouver, B.C., CanadaV6T 1W5

RICHTER (Dr G.) Zentralinstitut für Astrophysik, Sternwarte Sonneberg
64 Sonneberg, Germany D.R.

RICHTER (Dr J.) Institut für Experimentale Physik, Olshausenstr 40-60,
Neue Universität, Gebäude D6, 23 Kiel 1, Germany F.R.

RICHTER (Prof. Dr N.) Post Ulla bei Welzel, DDR-5301 Weimar, Germany D.R.

*RICKARD (Dr J.J.) European Southern Observatory, Casilla 16317, Correo 9
Santiago, Chile

*RICKER (Dr G.R.) Massachusetts Institute of Technology, 37-527 Center for
Space Research, Cambridge, MA 02139, U.S.A.

*RIDDLE (Dr A.C.) Dept. of Astro-Geophysics, University of Colorado,
Boulder, CO 80302, U.S.A.

RIEGEL (Dr K.W.) Astronomy- UCLA, Los Angeles, Cal. 90024, U.S.A.

RIEU (Dr N.Q.) Observatoire de Meudon, 92- Meudon, France

RIGHINI (Dr A.) Osservatorio Astrofisico di Catania, Città Universitaria,
Catania, Sicilia, Italy

RIGHINI (Prof.G.) Direttore, Osservatorio Astrofisico di Arcetri,
"Giorgio Abetti" Largo Enrico Fermi 5, 50125 Firenze, Italy

RIGUTTI (Prof.M.) Director, Osservatorio Astronomico di Capodimonte,
via Moiariello, 16, 80131 Napoli, Italy

RIIHIMAA (Dr J.J.) Oulun yliopiston tähtitieteen laitos,
Torikatu 7, Oulu, Finland

RIIVES (Dr V.) Tartu State University, Dept. of Theoretical Physics,
Tartu, Estonia, U.S.S.R.

RINDLER (Prof. W.) Graduate Research Center, P.O. Box 30365, Dallas, Texas 75230,
U.S.A.

RINEHART (Dr J.S.) Hyperdynamics, P.O.Box 392, Santa Fe, New Mexico 87501, U.S.A.

RING (Prof. J.) Infrared Astronomy Group, Imperial College of Science and
Technology, 10 Princes Gardens, London SW7, U.K.

RINGNES (Dr T.S.) Institute of Theoretical Astrophysics, University of Oslo,
Blindern, Oslo 3, Norway

RINGUELET (Dr A.E.) 53 416 7A
1900 La Plata, Argentina

ROACH (Dr F.E.) 2987 Kalakaua Avenue (602) Honolulu, Hawaii 96815, U.S.A.

ROARK (Dr T.) Dept. of Astronomy, 174 West 18th Ave, Ohio State University,
Columbus, Ohio 43210, U.S.A.

ROBBINS (Dr R.R.) Associate Professor, Dept. of Astronomy, University of
Texas at Austin, Austin, Texas 78712, U.S.A.

ROBE (Dr H.A. Gh.) Institut d'Astrophysique, Université de Liège, ave de Cointe 5,
4200 Cointe-Ougrée, Belgium

*ROBERTS (Dr B.) Dept. of Applied Mathematics, University of St. Andrews,
St. Andrews, Fife, U.K.

ROBERTS (Dr J.A.) CSIRO, Division of Radiophysics, Post Office Box 76,
Epping, N.S.W., 2121, Australia

ROBERTS (Dr M.S.) National Radio Astronomy Observatory, Edgemont Road, Charlottesville 22901, Virginia, U.S.A.

ROBERTS (Dr W.O.) Director, National Center for Atmospheric Research, Boulder, Colorado, U.S.A.

ROBERTS (Dr W.W.,Jr.) University of Virginia, Dept. of Applied Mathematics and Computer Science, Thornton Hall, Charlottesville, Va 22901, U.S.A.

ROBERTSON (W.H.) Sydney Observatory, Sydney,N.S.W. 2000, Australia

ROBINSON (Dr B.) CSIRO Division of Radiophysics, Post Office Box 76 Epping, N.S.W., 2121, Australia

ROBINSON (Prof.I.) University of Texas at Dallas, Box 688, Mail Station BE 32 Richardson, Texas, 75080, U.S.A.

ROBINSON (Dr L.B.) Lick Observatory, University of California, Santa Cruz,Ca.USA.

*ROBINSON (Dr W.J.) Dept. of Mathematics, The University, Bradford BD7 1DP, U.K.

ROBLEY (Dr R.) Observatoire du Pic-du-Midi, Bagnères-de-Bigorre, Hautes-Pyrénées, France

*ROBSON (Dr E.I.) Dept. of Physics, Queen Mary College, Mile End Road, London E1 4NS, U.K.

ROCHESTER (Prof. M.G.) Dept. of Physics, Memorial University of Newfoundland, St. John's, Newfoundland, Canada

RODDIER (F.) Faculté des Sciences de Nice, Avenue Valrose, 06- Nice, France

RODGERS (Dr A.W.) Mount Stromlo Observatory, Canberra, A.C.T., Australia

*RODMAN (Dr R.B.) 65 Locust Ave, Lexington, MA 02137, U.S.A.

RODONO (Dr M.) Osservatorio Astrofisico, Città Universitaria, viale A. Doria 95123 Catania, Sicilia, Italy

RODRIGUEZ (M.) Director del Instituto y Observatorio de Marina, San Fernando (Cádiz) Spain

ROEDER (Dr R.C.) Scarborough College, University of Toronto, 1265 Military Trail, West Hill, Ontario, Canada

ROEMER (Dr E.) Lunar and Planetary Laboratory, University of Arizona, Tucson, Arizona 85721, U.S.A.

*ROEMER (Dr M.) Institut für Astrophysik und Extraterrestrische Forschung der Universität Bonn, Auf dem Hügel 71,D53 Bonn, Germany, F.R.

ROGER (Dr R.S.) Dominion Radio Astrophysical Observatory, Herzberg Institute of Astrophysics, National Research Council of Canada P.O. Box 248, Penticton, B.C. V2A 6K3, Canada

ROGERS (Dr A.E.E.) Massachusetts Institute of Technology, Lincoln Laboratory Lexington, Mass. 02173, U.S.A.

ROGERSON (Prof. J.B.) Princeton University Observatory, Peyton Hall, Princeton, N.J. 08540, U.S.A.

ROGSTAD (Dr D.H.) Dept. of Radio Astronomy, California Institute of Technology, Pasadena, Cal. 91109, U.S.A.

ROHLFS (Prof.K.) Max-Planck-Institut für Radioastronomie Argelanderstr.3, 5300 Bonn, Germany F.R.

ROLAND (Dr G.) Institut d'Astrophysique, Université de Liège, Ave de Cointe 5
    B- 4200 Cointe-Ougrée , Belgium

ROMAN (Dr N.G.) NASA Headquarters, Washington D.C. 20546, U.S.A.

ROMANA (Dr A.) Directeur, Observatorio de Ebro, Roquetas, Tarragona, Spain

ROMANCHUK (Dr P.R.) Astronomical Observatory, Observatornaya 3,
    Kiev 252053, U.S.S.R.

ROMANO (Dr G.) Viale S. Francesco 7, Treviso, Italy

*ROMANOV (Dr Y.S.) Astronomical Observatory, Park Shevchenko, Odessa, 270014,
    U.S.S.R.
ROMPOLT (Dr B.) Astronomical Institute of Polish Academy of Sciences
    Kopernika 11, Wrocław, Poland

RONAN (C.A.) 39 New Road, Barton, Cambridge CB3 7AY, U.K.

RÖNNÄNG (Dr B.O.) Onsala Space Observatory, S-43034 Onsala, Sweden

ROOD (Dr H.J.) 52 Elizabeth St. So. Bound Brook, N.J. 08880, U.S.A.

ROOD (Dr R.) Astronomy Dept. University of Virginia, Charlottesville,
    Virginia 22903, U.S.A.

ROOSEN (Dr R.G.) NASA, Goddard Space Flight Center, Laboratory for Solar
    Physics, New Mexico Station, 800 Yale Blvd., NE , Albuquerque,
    New Mexico 87131

RÖSCH (Prof.J.) Directeur de l'Observatoire du Pic-du-Midi,
    65 Bagnères-de-Bigorre (Hautes Pyrénées), France

ROSE (Dr W.K.) Physics Dept. Massachusetts Institute of Technology,
    Cambridge, Mass. 02139, U.S.A.

ROSEN (Dr E.) The City College of the City University of New York,
    Dept. of History, New York, N.Y. 10031, U.S.A.

ROSENBERG (Dr J.) Sterrenwacht, Servaas Bolwerk 13, Utrecht, The Netherlands

*ROSENDHAL (Dr J.D.) NASA, Headquarters, Code SA, Washington DC 20546, U.S.A.

ROSINO (Prof. L.) Director of the Astrophysical Observatory,
    Asiago (Vicenza) Italy

ROSLUND (Dr C.) Lund Observatory, S-222 24 Lund, Sweden

ROSS (Dr D.) IOWA State University, Erwin Fick Observatory, Ames, Iowa 50010,USA

ROSSELAND (Prof. S.) Institute of Theoretical Astrophysics, University of Oslo,
    Blindern, Oslo 3, Norway

ROSSI (Dr B.) Dept. of Physics, Massachusetts Institute of Technology,
    Cambridge, Mass. 02139, U.S.A.

ROSTAS (Dr F.) Observatoire de Meudon, 92190 Meudon, France

*ROTS (Dr A.H.) N.R.A.O. P.O.Box 2, Greenbank, W. Va. 24944, U.S.A.

ROTTENBERG (Dr J.A.) 2911- Bayview Ave., Suite 110C, Willowdale,
    Ontario M2K 1E8, Canada

*ROUEFF (Dr E.) DAPHE, Observatoire de Meudon, 92190 Meudon, France

ROUNTREE LESH (Dr J.) Dept. of Physics and Astronomy, Space Sciences Building
    University of Denver, Denver, Colorado 80210, USA

ROUSE (Dr C.A.) General Atomic Co., P.O.Box 81608, San Diego, Cal.92138, U.S.A.

ROUSSEAU (Dr J.) Observatoire de Lyon, 69230 Saint-Genis-Laval, France

ROUSSEAU (J.-M.) Observatoire de l'Université de Bordeaux, 33270 Floirac, France

ROUTLEDGE (Prof. D.) Dept. of Electrical Engineering, University of Alberta, Edmonton, Alberta T6G 2E1, Canada

ROUTLY (Dr P.M.) U.S.Naval Observatory, 34th and Massachusetts Ave., NW., Washington, D.C. 20390, U.S.A.

*ROVITHIS (Dr P.) Astronomical Institute, National Observatory of Athens, Athens 306, Greece

ROWAN-ROBINSON (Dr G.M.) Dept. of Applied Mathematics, Queen Mary College, Mile End Road, London E1, U.K.

*ROWSON (Dr B.) Nuffield Radio Astronomy Laboratories, Jodrell Bank, Macclesfield, Cheshire, U.K.

ROXBURGH (Prof. I.W.) Mathematics Dept. Queen Mary College, Mile End Road, London E.1., U.K.

ROY (Dr A.E.) Dept. of Astronomy, University of Glasgow, Glasgow, W.2, U.K.

*ROY (Dr J.-R.) Herzberg Institute of Astrophysics, National Research Council of Canada, Ottawa, Ontario, Canada K1A OR6

ROZELOT (Dr J.P.) Observatoire du Pic du Midi, 65200 Bagnères-de-Bigorre,France

ROZHKOVSKI (Dr D.A.) Astrophysical Institute, Alma Ata, U.S.S.R.

RUBASHEV (Dr B.M.) Pulkovo Observatory, Leningrad, U.S.S.R.

RUBEN (Dr G.) Zentralinstitut für Astrophysik, DDR 15 Potsdam, Telegrafenberg, Germany D.R.

RUBIN (Dr V.C.) Dept. of Terrestrial Magnetism, Carnegie Institution of Washington, 5241 Broad Branch Road N.W., Washington D.C.

RUCINSKI (Dr S.) Astronomical Observatory of the Warsaw University, Al. Ujazdowskie 4, Warszawa, Poland

*RUDDY (Dr V.P.) Cork Regional Technical College, Rossa Avenue, Cork, Ireland

RUDERMAN (Dr M.) Columbia University, Astronomy Dept., New York, N.Y. 10027, USA.

RUDKJØBING, (Prof.M.) Astronomical Institute, University of Aarhus, DK-8000 Aarhus C, Denmark

RUDNICKI (Dr K.) Jagiellonian University Observatory, ul. Kopernika 26, Kraków, Poland

RUFENER (Dr F.G.) Observatoire de Genève, 1290 Sauverny (GE) Switzerland

*RUGGE (Dr H.R.) Upper Atmospheric Physics Dept., The Aerospace Corporation P.O. Box 92957, Los Angeles, Cal.90009, U.S.A.

RULE (B.R.) Mount Wilson and Palomar Observatories, 813 Santa Barbara Street, Pasadena, Cal. 91106, U.S.A.

RUMSEY (N.J.) Physics and Engineering Laboratory, D.S.I.R., Lower Hutt, New Zealand

RUNCORN (Prof.S.K.) School of Physics, The University, Newcastle upon Tyne, NE 1 7 RU, U.K.

RUPRECHT (Dr J.) Astronomical Institute, Czechoslovak Academy of Sciences,
    Budečská 6, Prague 2, Czechoslovakia

RUSKOL (Dr E.L.) Institute of Physics of the Earth, USSR Academy of Sciences
    Moscow, U.S.S.R.

RUSSELL (Dr J.A.) Astronomy Dept., University of Southern California
    University Park, Los Angeles, Cal. 90007, U.S.A.

RUSSEV (Dr R.) Bulgarian Academy of Sciences, Dept. of Astronomy,
    7th November Street 1, Sofia, Bulgaria

RUST (Dr D.M.) American Science & Engineering Co., 955 Massachusetts Ave.
    Cambridge, Mass. 02139, U.S.A.

RUSU (Dr I.) Astronomical Observatory, Str. Cutitul de Argint 5, Bucarest 28,
    Roumania

RUSU (Dr L.) Observatorul Astronomic, Strada Cutitul de Argint 5,
    Bucarest 28, Roumania

RŮŽIČKOVÁ-TOPOLOVÁ (Dr B.) Astronomical Institute, Czechoslovak Academy
    of Sciences, Observatory Ondřejov, Czechoslovakia

RYABOV (Prof. Dr Yu. A.), MADI, Leningradskij pr-t 64, 125319 Moscow, U.S.S.R.

RYBICKI (Dr G.B.) Smithsonian Astrophysical Observatory, 60 Garden Street
    Cambridge, Mass. 02138, U.S.A.

RYBKA (Prof. Dr E.) Szopena 20/3, Kraków, Poland

RYBKA (Dr P.) Astronomical Institute of the Wroclaw University,
    Dembowskiego 19, Wroclaw 9, Poland

RYDBECK (Dr O.E.H.) Inst. of Electronphysics I, Chalmers Technical University,
    Fack, S-402 20 Göteborg 5, Sweden

RYLE (Prof.Sir M.) F.R.S., Cavendish Laboratory, Free School Lane, Cambridge,U.K.

*RYLOV (Dr V.S.) Special Astrophysical Observatory, St. Zelenchukskaya,
    Stavropolskij Kraj, 357140, U.S.S.R.

RYTER (Dr Ch.) Cen Saclay-Sep, B.P. No.2, 91 Gif sur Yvette, France

RYZHKOV (Dr N.F.) Leningrad Branch of Special Astrophysical Observatory,
    Academy of Sciences, Pulkovo, Leningrad 196140, U.S.S.R.

RZHIGA (Dr O.N.) Institute of Radio and Electronics, USSR Academy of
    Sciences, Moscow, U.S.S.R.

SACK (Dr N.) Einstein Institute of Physics, Dept. of Theoretical Physics,
    The Hebrew University of Jerusalem, Jerusalem, Israël

SACKMANN-CHRISTY (Dr I.J.) California Institute of Technology, Pasadena,
    Cal. 91109, U.S.A.

SADEH (Dr D.) Dept. of Physics and Astronomy, Tel-Aviv University,
    Ramat Aviv, Israel

SADLER (Dr D.H.) 8 Collington Rise, Bexhill-on-Sea, Sussex TN39 3RT, U.K.

SAFRONOV (Dr V.S.) O.J. Schmidt·Institute of the Physics of the Earth
    USSR Academy of Sciences, B. Gruzinskaya 10, Moscow 242, U.S.S.R.

SAGAN (Prof. C.) Center for Radiophysics and Space Research,
    Laboratory for Planetary Studies, Cornell University,
    Ithaca, New York 14850, U.S.A.

SAGGION (Prof. Dr A.) Instituto di Fisica "G. Galilei" Via Marzolo 8
35100 Padova, Italy

SAGITOV (Dr M.U.) Sternberg Astronomical Institute, Moscow, U.S.S.R.

SAGNIER (Dr J.L.) Bureau des Longitudes, 3, Rue Mazarine, 75006 Paris, France

SAHADE (Prof.J.) 53 416 7A, 1900 La Plata, Argentina

SAHAL (Dr S.) Observatoire de Paris, Section d'Astrophysique de Meudon,
92190 Meudon, France

SAÏSSAC (Dr J.) Observatoire du Pic-du-Midi, 65200-Bagnères de Bigorre, France

SAITO (Dr K.) Tokyo Astronomical Observatory, Mitaka, Tokyo, Japan

SAITO (Dr M.) Tokyo Astronomical Observatory, Mitaka, Tokyo, Japan

SAITO (Dr S.) Kwasan Observatory, University of Kyoto, Yamashina, Kyoto, Japan

SAKASHITA (Dr S.) Dept. of Physics, Hokkaido University, Sapporo, Japan

SAKHAROV (Dr V.I.) Main Astronomical Observatory, USSR Academy of Sciences,
Pulkovo, U.S.S.R.

SAKURAI (Prof. K. Kunitomo) Physics Dept. Faculty of Engineering,
Kanagawa University, Kanagawa-ku, Yokohama, 221 Japan

SAKURAI (Prof. Dr T.) Dept. of Aeronautical Engineering, Faculty of
Engineering, Kyoto University, Kyoto, Japan

SALETIĆ (D.) Military Geographical Institute, Mije Kovacevica 5, Beograd 11000.
Yugoslavia

SALIN (Dr A.) Observatoire de Meudon, 92190 Meudon, France

SALISBURY (Dr J.W.) Air Force Cambridge Research Laboratories, Bldg. 1121.
L.G. Hanscom Field, Bedford, Massachusetts 01730, U.S.A.

SALOMONOVICH (Dr A.E.) Physical Institute, USSR Academy of Sciences,
Moscow, U.S.S.R.

SALPETER (Dr E.E.) Laboratory of Nuclear Studies, Cornell University,
Ithaca, N.Y., U.S.A.

SALUKVADZE (Dr G.N.) Abastumani Astrophysical Observatory, Georgian Academy
of Sciences, Abastumani, U.S.S.R.

SAMAHA (Prof. A.H.M.) 5 Wadi El-Nile Street, Maadi, Cairo, Arab Republic of Egypt

*SAMPSON (Dr D.H.) Dept. of Astronomy, The Pennsylvania State University,
525 Davey Laboratory, University Park, PA 16802, U.S.A.

SANAMIAN (Dr V.A.) Byurakan Astrophysical Observatory, Byurakan, Armenia, U.S.S.R.

*SANCHEZ (M.) Instituto y Observatorio de Marina, San Fernando (Cádiz) Spain

SANCHEZ MAGRO (Dr C.) Instituto Universitario de Astrofisica,
Universidad de la Laguna, Tenerife, Spain

SANCHEZ-MARTINEZ (Dr F.) Instituto Universitario de Astrofisica, Universidad
de la Laguna, Tenerife, Spain

SANCISI (R.) Kapteyn Laboratory, Postbus 800, Groningen, The Netherlands

SANDAGE (A.) Mount Wilson and Palomar Observatories, 813 Santa Barbara Street,
Pasadena 4, Cal. 91101, U.S.A.

SANDAKOVA (Dr E.V.) Kiev University Observatory, Kiev, U.S.S.R.

SANDERS (Dr P.) Observatoire Royal de Belgique, avenue Circulaire 3,
    B-1180 Bruxelles, Belgium

SANDERS (Dr R.) European Southern Observatory, Telescope Project Division,
    c/o CERN, 1211 Geneva 23, Switzerland

SANDERS (Dr W.L.) New Mexico State University, Las Cruces, New Mexico, U.S.A.

*SANDFORD II (Dr M.T.) Los Alamos Scientific Laboratory, Mail Stop 664,
    P.O.Box 1663, Los Alamos, NM 87545, U.S.A.

*SANFORD (P.W.) Mullard Space Science Lab. Holmbury St. Mary
    Dorking, Surrey, U.K.

SANDIG (Prof. H.-U.) Observatory of the Technical University, Lohrmann Institut,
    Dresden 8027, Germany D.R.

SANDQVIST (Dr Aa.) Stockholm Observatory, S-133 00 Saltsjöbaden, Sweden

SANDULEAK (Dr N.) Warner and Swasey Observatory, Taylor and Brunswick Roads,
    East Cleveland, Ohio 44112, U.S.A.

SANWAL (Dr N.B.) Dept. of Astronomy, Osmania University, Hyderabad -7 (A.P.)
    India

SAPAR (Dr A.A.), Institute for astrophysics and physics atmosphere,
    Toravere, 202444 Tartu, U.S.S.R.

SARGENT (Dr W.L.) California Institute of Technology, Pasadena, Cal.91109, U.S.A.

SARMA (M.B.K.) Dept. of Astronomy, Osmania University, Hyderabad, India

SARTORI (Dr B.) University of Nebraska, Lincoln, Nebraska 68508, U.S.A.

*SASAO (Dr Tetsuo) International Latitude Observatory of Mizusawa,
    Mizusawa-shi, Iwate-ken, 023 Japan

SASLAW (Dr W.C.) Leander McCormick Observatory of the University of Virginia,
    P.O.Box 3818, University Station, Charlottesville, Virginia 22903,U.S.A.

SASTRY (Ch.V.) Astrophysical Observatory, Kodaikanal- 3, India

SATO (Dr F.) Chiba Prefecture Education Center, Katsuragi 2-10-1, Chiba, Japan

SATO (Dr H.) Dept. of Physics, Kyoto University, Kyoto, Japan

*SATO (Dr Katsuhiko) Research Institute for Fundamental Physics,
    Kyoto University, Kyoto, 606 Japan

*SATO (Dr Koichi) International Latitude Observatory of Mizusawa,
    Mizusawa-shi, Iwate-ken, 023 Japan

SATO (Dr Y.) Tokyo Astronomical Observatory, Mitaka- Tokyo, Japan

SAUVAL (Dr A.J.P.) Observatoire Royal de Belgique, ave Circulaire 3
    B-1180 Bruxelles, Belgium

SAVAGE (Dr B.) Washburn Observatory, University of Wisconsin,
    475 North Charter Street, Madison, Wisconsin 53706, U.S.A.

SAVEDOFF (Prof. M.P.) Dept. of Physics and Astronomy, University of Rochester,
    River Campus Station, Rochester N.Y. 14627, U.S.A.

SAWYER (Dr C.S.) NOAA, AOML, ORSL, 15 Rickenbacker Causeway, Miami, Fla.33149,
    U.S.A.

SAWYER-HOGG (H.B.) David Dunlap Observatory, University of Toronto,
    Richmond Hill, Ontario L4C 4Y6, Canada

SAXENA (Dr P.P.) Uttar Pradesh State Observatory, Manora Peak, Naini Tal, India

SCALISE (E.Jr.) Centro de rádio astronomía e astrofísica, Universidade Mackenzie, CP. 8792, Sao Paulo, SP, Brazil

SCARFE (Dr C.D.) Dept. of Physics, University of Victoria, Victoria, B.C.V6T 1W5 Canada

SCARGLE (Dr J.D.) Lick Observatory, University of California, Santa Cruz, Cal. 94060, U.S.A.

SCHADEE (Dr A.) Astronomical Observatory, "Sonnenborgh", Servaas Bolwerk 13, Utrecht, the Netherlands

SCHAIFERS (Dr K.) Landessternwarte, Heidelberg-Königstuhl, Germany F.R.

SCHALEN (Prof.C.) Lund Observatory, S-222 24 Lund, Sweden

SCHANDA (Dr E.) Institut für angewandte Physik, Sidlerstrasse 5, CH-3012 Bern, Switzerland

*SCHATTEN (Dr K.H.) Physics Dept. Victoria University, Private Bag, Wellington, New Zealand

SCHATZMAN (Prof. E.L.) Observatoire de Paris-Meudon, Département d'Astrophysique Fondamentale, 5, Place Jansen, 92190 Meudon, France

*SCHEEPMAKER (Dr A.) Cosmic Ray Working Group, Huygens Laboratory, Wassenaarseweg 78, Leiden, The Netherlands

SCHEEPMAKER (Dr A.C.) Boerhaavestraat 59, Voorhout, Netherlands

SCHEFFLER (Dr H.) Landessternwarte, 69 Königstuhl-Heidelberg, Germany F.R.

*SCHEIDECKER (Dr J.-P.) Observatoire de Nice, Le Mont-Gros, 06000 Nice, France

*SCHERB (Dr F.) Physics Dept. University of Wisconsin, Madison, Wisc. 53706, U.S.A.

SCHEUER (Dr P.A.G.) Mullard Radio Astronomy Observatory, Cavendish Laboratory, Free School Lane, Cambridge, U.K.

SCHILD (Dr R.E.) Smithsonian Astrophysical Observatory, 60 Garden Street, Cambridge, Mass. 02138, U.S.A.

*SCHILIZZI (Dr R.) Radiosterrenwacht, Oude Hoogeveensdijk 4, Dwingeloo NL 7514, The Netherlands

SCHILLER (Prof. Dr K.) Pirschweg 6, 6079 Buchschlag über Sprendlingen, Germany, F.R.

SCHILT (Prof. J.) 481 Greenville Avenue, Centerdale 11, Rhode Island, U.S.A.

SCHKODROV (Dr V.G.) Dept. of Astronomy, "7th November" Street 1, Sofia,Bulgaria

SCHLESINGER (Dr B.) Indiana University, Bloomington, Indiana 47401, U.S.A.

SCHLOSSER (Dr W.) Astronomisches Institut der Ruhr-Universität Bochum, Postfach 2148, 4630 Bochum-Querenburg, Germany F.R.

SCHLÜTER (Prof. Dr A.) Max-Planck-Institut für Physik und Astrophysik, Institut für Astrophysik, Aumeisterstr.6, München 23, Germany, F.R.

SCHLÜTER (Prof. Dr D.) Institut für Theoretische Physik und Sternwarte der Universität, Olshausenstr. Neue Universität, Geb. 13. 23 Kiel, Germany F.R.

*SCHMADEL (Dr L.) Astronomisches Rechen-Institut, Mönchhofstr. 12-14, 6900 Heidelberg 1, Germany, F.R.

SCHMAHL (Dr G.) Universitätssternwarte, Geismarlandstr.11,
    3400 Göttingen, Germany F.R.

SCHMALBERGER (Dr D.C.) State University of New York at Albany, ES 314,
    Albany, New York 12203, U.S.A.

SCHMEIDLER (Prof. Dr F.) Universitäts-Sternwarte, Sternwartstr.23,
    München 27, Germany F.R.

SCHMID-BURGK (Dr J.) Institut für Theoretische Astrophysik der Universität,
    Im Neuenheimer Feld 294, 69 Heidelberg, Germany, F.R.

*SCHMIDT (Dr E.G.) Behlen Laboratory of Physics, The University of Nebraska,
    Lincoln, NB 68508, U.S.A.

SCHMIDT (Prof. Dr H.) Direktor der Universitäts-Sternwarte, Auf dem Hügel 71,
    53 Bonn, Germany F.R.

SCHMIDT (Dr. H.U.) Max-Planck-Institut für Physik und Astrophysik,
    Föhringer Ring 6, 8 München 23, Germany, F.R.

SCHMIDT (Dr K.-H.) Zentralinstitut für Astrophysik, Sternwarte Babelsberg,
    DDR 1502 Potsdam Babelsberg, Rosa-Luxemburg-Strasse 17a
    Germany D.R.

SCHMIDT (Prof. M.) Mount Wilson and Palomar Observatories,
    1201 East California Blvd, Pasadena, Cal. 91109, U.S.A.

SCHMIDT (Dr Th.) Landessternwarte Heidelberg-Königstuhl, 69 Heidelberg
    Germany F.R.

SCHMIDT-KALER (Prof. Dr Th.) 581 Witten, Steinhügel 105, Germany F.R.

SCHMITTER Y MARTIN DEL CAMPO (Dr E.F.) Institute of Astronomy, UNAM,
    Apartado Postal 70-264, México D.F., Mexico

SCHNEIDER (Dr M.) Observatoire de Nice, Le Mont-Gros, 06-Nice, France

SCHNELL (Dr A.) Universitäts-Sternwarte, Türkenschantzstr.17,
    A-1180 Vienna, Austria

*SCHNOPPER (Dr H.W.) Center for Astrophysics, 60 Garden Street,
    Cambridge, MA 02138, U.S.A.

*SCHNUR (Dr G.F.O.) ESO Alonso de Córdoba 3107, Vitacura
    Casilla 11P- Correo 11, Santiago de Chile

*SCHOBER (Dr H.J.) Universitäts-Sternwarte, Universitätsplatz 5,
    A-8010 Graz, Austria

*SCHOEMBS (Dr R.) Institut für Astronomie und Astrophysik, Scheinerstr.1,
    8 München 80, Germany, F.R.

*SCHÖFFEL (Dr E.F.) Remeis-Sternwarte Bamberg, Astronomisches Institut der
    Universität Erlangen-Nürnberg, Postfach 4044, 86 Bamberg, Germany, F.R.

*SCHOLL (Dr H.) Astronomisches Rechen-Institut, Mönchhofstr.12-14,
    69 Heidelberg, Germany, F.R.

SCHOLZ (Dr M.) Institut für Theoretische Astrophysik der Universität,
    Im Neuenheimer Feld 294, D-6900 Heidelberg, Germany, F.R.

SCHOMBERT (Dr J.L.) U.S. Naval Observatory, Washington D.C. 20390, U.S.A.

SCHÖNEICH (Dr W.) Forschungsgruppenleiter am Zentralinstitut für
    Astrophysik (Bereich Sternphysik) Potsdam, Germany D.R.

*SCHÖNFELDER (Dr V.) Max-Planck-Institut für Physik und Astrophysik
Institut für Extraterrestrische Physik 8046 Garching bei München
Germany, F.R.

SCHRAMM (Dr D.N.) University of Chicago, Fermi Institute (LASR)
933 E. 56th Street, Chicago, Illinois 60637, U.S.A.

*SCHRIJVER (Dr J.) Lab. voor Ruimte-Onderzoek, Beneluxlaan 21, Utrecht,
The Netherlands

*SCHRÖDER (Dr R.) Hamburger Sternwarte, Gojenbergsweg 112, 205 Hamburg 80,
Germany, F.R.

SCHROEDER (Dr D.J.) Beloit College, Beloit, Wisconsin 53511, U.S.A.

*SCHROLL (Dr A.) Sonnenobservatorium Kanzelhöhe, A-9520 Sattendorf, Austria

SCHRÖTER (Dr E.H.) Universitäts-Sternwarte, Geismarlandstr.11,
34 Göttingen, Germany F.R.

SCHRUTKA-RECHTENSTAMM (Prof. Dr G.) Universitäts-Sternwarte,
Türkenschanzstr. 17, A-1180 Vienna, Austria

SCHUBART (Dr J.) Astronomisches Rechen-Institut, Mönchhofstr. 12-14
69 Heidelberg, Germany, F.R.

SCHÜCKING (Dr E.L.) Dept. of Physics, New York University, New York,NY 10012,
U.S.A.

SCHULER (Dr W.) Sternwarte der Kantonsschule, 4500 Solothurn, Switzerland

SCHULTE (Dr D.H.) Itek Corporation, 10 Maguire Road, Lexington, Mass. U.S.A.

SCHÜRER (Prof. Dr M.) Astronomisches Institut der Universität Bern
Sidlerstr.5, Bern, Switzerland

*SCHUTZ (Dr B.F.) Dept. of Applied Mathematics and Astronomy,
University College, Cardiff, U.K.

*SCHWARTZ (Dr D.A.) Center for Astrophysics, 60 Garden Street, Cambridge,MA 02138,
U.S.A.

*SCHWARTZ (Dr R.D.) Dept. of Physics, University of Missouri-St.Louis
St. Louis, MO 95064, U.S.A.

SCHWARZ (Dr U.J.) Kapteyn Astronomical Laboratory, Postbus 800,
Groningen, The Netherlands

SCHWARZSCHILD (Prof.M.) Princeton University Observatory, Princeton, N.J.08540,
U.S.A.

SCIAMA (Prof. D.W.) Dept. of Astrophysics, South Parks Road,
Oxford, OX1 3 RQ, U.K.

SCONZO (Dr P.) 29 Old Mystic Street, Arlington, Mass. 02174, U.S.A.

SCOTT (Prof. E.L.) Statistical Laboratory, University of California
Berkeley, Cal. 94720, U.S.A.

SCOTT (Dr P.F.) Mullard Radio Astronomy Observatory, Cavendish Laboratory,
Free School Lane, Cambridge, U.K.

*SCOVILLE (Dr N.Z.) Dept. of Physics and Astronomy, The Commonwealth of
Massachusetts, Amherst, MA OI002, U.S.A.

SEAQUIST (Dr E.R.) David Dunlap Observatory, University of Toronto,
Richmond Hill, Ontario L4C 4Y6, Canada

SEARLE (Dr L.) Hale Observatories, 813, Santa Barbara Street,
     Pasadena, Cal. 91106, U.S.A.

SEARS (Dr L.) Astronomy Department, University of Michigan, Ann Arbor,
     Michigan 48104, U.S.A.

SEATON (Prof. M.J.) Department of Physics, University College, London,
     Gower Street, London, W.C.1, U.K.

SEDDON (H.) The Royal Observatory, Edinburgh 9, U.K.

*SEDLMAYER (Dr E.) Lehrstuhl für Theoretische Astrophysik, 69 Heidelberg,
     Im Neuenheimer Feld 294, Germany F.R.

SEDMARK (Dr G.) Osservatorio Astronomico, Via Tiepolo, 11, Trieste, Italy

SEEGER (Prof. Ch.) Department of Astronomy, New Mexico State University,
     Box 4500, Las Cruces, N.M. 88001, U.S.A.

SEEGER (Dr P.A.) Los Alamos Scientific Laboratory, University of California,
     Los Alamos, New Mexico 87544, U.S.A.

SEGGEWISS (Dr W.) Observatorium Hoher List, Universitäts-Sternwarte Bonn,
     D-5568 Daun, Germany F.R.

SEHNAL (Dr L.) Astronomical Institute, Czechoslovak Academy of Sciences,
     Observatory Ondřejov, Czechoslovakia

SEIDELMANN (Dr P.K.) U.S. Naval Observatory, Washington, D.C. 20390, U.S.A.

SEIELSTAD (Dr G.A.) California Institute of Technology, Pasadena, California 91109,
     U.S.A.

SEITTER (Dr W.C.) Universitäts-Sternwarte Bonn, Observatorium Hoher List,
     5568 Daun/Eifel, Germany F.R.

SEKANINA (Dr Z.) Smithsonian Astrophysical Observatory, 60 Garden Street,
     Cambridge, Mass. 02138, U.S.A.

SEKIGUCHI (Dr N.) Tokyo Astronomical Observatory, University of Tokyo, Mitaka,
     Tokyo, Japan

SEMEL (Dr M.) Observatoire de Meudon, 92190 Meudon, France

*SEMENIUK (Dr I.) Astronomical Observatory of the Warsaw University,
     Al. Ujazdowskie 4, 00-478 Warszawa, Poland

SEN (Dr H.K.) USAF, Cambridge Research Laboratories, Hanscom AFB, Mass. 01731, U.S.A.

SEN (Dr S.N.) Indian Association for the Cultivation of Science, Jadavpur, India

SENGBUSCH (V. Dr K.) Max-Planck-Institut für Physik und Astrophysik,
     Föhringer Ring 6, 8000 München 23, Germany F.R.

SERKOWSKI (Dr K.) Lunar and Planetary Laboratory, University of Arizona, Tucson,
     Arizona 85721, U.S.A.

SĔRSIC (Dr J.L.) Observatorio Astronómico, Córdoba, Argentina

SERVAJEAN (Dr R.) Observatoire de Paris, Section d'Astrophysique,
     92 Meudon, France

SETTI (G.) Laboratorio di Radioastronomia, Istituto di Fisica, Bologna, Italy

ŠEVARLIĆ (Prof. Dr B.M.) Directeur, Institut d'Astronomie, Faculté des Sc. nat.
     et Math., Université de Beograd, Volgina 7, 11050 Beograd, Yugoslavia

SEVERNYJ (Prof. Dr A.B.) Director, Crimean Astrophysical Observatory of the
     USSR Academy of Sciences, P/O Nauchny, Crimea, U.S.S.R.

*SHAFFER (Dr D.B.) National Radio Astronomy Observatory, P.O. Box 2, Green Bank, WV 24944, U.S.A.

SHAH (Dr G.A.) Indian Institute of Astrophysics, C/o Raman Research Institute, Hebbal, Bangalore-560006, India

*SHAHAM (Dr J.) Department of Theoretical Physics, Hebrew University, Jerusalem, Israel

SHAKESHAFT (Dr J.R.) Mullard Radio Astronomy Observatory, Cavendish Laboratory, Free School Lane, Cambridge, U.K.

SHAKHBAZIAN (Dr Y.L.) Pulkovo Observatory Leningrad 196140, U.S.S.R.

SHAKHOVSKOJ (Dr N.M.) Crimean Astrophysical Observatory, USSR Academy of Sciences, P/O Nauchniy, 334413 Crimea, U.S.S.R.

*SHAKURA (Dr N.I.) Sternberg Astronomical Institute, Moscow, 117234, U.S.S.R.

SHANE (Dr C.D.) P.O. Box 582, Santa Cruz, California 95060, U.S.A.

SHANE (W.W.) Sterrewacht, Leiden 2401, The Netherlands

SHAO (Cheng-yuan) Harvard College Observatory, 60 Garden Street, Cambridge, Mass. 02138, U.S.A.

SHAPIRO (Dr I.I.) Building 54-622, Mass. Institute of Technology, Cambridge, Mass. 02139, U.S.A.

SHAPIRO (Dr M.M.) Laboratory for Cosmic Ray Physics, Code 7020, Naval Research Laboratory, Washington, D.C. 20390, U.S.A.

*SHAPIRO (Dr S.) Department of Astronomy, Cornell University, Space Sciences Building, Ithaca, NY 14853, U.S.A.

SHAPLEY (A.H.) CRPL, National Bureau of Standards, Boulder, Colorado, U.S.A.

SHAPLEY (M.B.) Sharon Cross Road, Peterboro, New Hampshire, U.S.A.

SHARAF (Dr Sh. G.) Institute of Theoretical Astronomy, 10, Kutuzov Qay, Leningrad, 192187, U.S.S.R.

SHAROV (Dr A.S.) Sternberg Astronomical Institute, Moscow, U.S.S.R.

SHARPLESS (Dr S.L.) Department of Physics and Astronomy, The University of Rochester, Rochester, New York 14627, U.S.A.

*SHAVER (Dr P.A.) Kapteyn Astronomical Institute, P.O. Box 800, Groningen, The Netherlands

SHAVIV (Dr G.) Department of Physics and Astronomy, Tel-Aviv University, Ramat Aviv, Israel

SHAW (Prof. J.H.) Department of Physics and Astronomy, Ohio State University, Columbus, Ohio 43210, U.S.A. (174 W. 18th Ave.)

*SHAW (Dr J.S.) The University of Georgia, Dept. of Physics and Astronomy, Athens, GA 30602, U.S.A.

SHAW (Prof. R.W.) Clark Hall of Science, Cornell University, Ithaca, N.Y. U.S.A.

*SHAWL (Dr S.J.) Dept. of Physics and Astronomy, The University of Kansas, Lawrence, KS 66045, U.S.A.

SHCHEGOLEV (Dr D.E.) Pulkovo Observatory, Leningrad 196140, U.S.S.R.

SHCHEGLOV (Dr P.V.) Sternberg Astronomical Institute, Moscow, U.S.S.R.

SHCHEGLOV (Prof. Dr V.P.) Director of the Astronomical Observatory,
    Tashkent, U.S.S.R.

SHCHERBINA-SAMOJLOVA (Dr I.S.) Institute of Science and Technics Information,
    Department of Astronomy, USSR Academy of Sciences, Moscow, U.S.S.R.

SHEELEY (Dr N.R., Jr.) Kitt Peak Observatory, 950 North Cherry Avenue, Tucson,
    Arizona 85717, U.S.A.

*SHELUS (Dr P.J.) Dept. of Astronomy, University of Texas at Austin, Austin,
    TX 78712, U.S.A.

SHEN (Dr B.S.P.) Flower and Cook Observatory, University of Pennsylvania,
    Philadelphia, Pennsylvania 19104, U.S.A.

*SHEN (Prof. Chun-Shan) Tsing-Hwa University, Hsin-Chu, Taiwan

SHEN (Dr Zee) Department of Mathematics, National Taiwan University,
    Taipei, Taiwan

SHEPTUNOV (Dr G.S.), Blagoveshchensk Laboratory, Main Astronom. Observatory,
    USSR Academy of Sciences, 675000 Blagoveshchensk, U.S.S.R.

SHER (Dr D.) 2837 Minto Drive, Apt. 2, Cincinnati, Ohio 45208, U.S.A.

SHERIDAN (K.V.) P.O.Box 76, Epping, N.S.W., 2121, Australia

*SHEVCHENKO (Dr V.V.) Sternberg Astronomical Institute, Moscow 117234, U.S.S.R.

*SHIBATA (Dr Yukio) Research Institute for Scientific Measurements, Tōhoku
    University, Sanjyo-machi 19-1, Sendai, Japan

SHIM (Dr Woon Taik) Department of Astrogeophysics, Sung Kyun Kwan University,
    Seoul, Republic of Korea

SHIMIZU (Prof. Dr M.) The Institute of Space and Aeronautical Science,
    The University of Tokyo, Tokyo, Japan

SHIMIZU (Dr T.) Institute of Astrophysics, College of Science, University of
    Tokyo, Kyoto, Japan

SHIMMINS (A.J.) Australian National Radio Astronomy Observatory, P.O. Box 276,
    Parkes, N.S.W. 2870, Australia

SHIRYAEV (Dr A.V.) Astronomical Observatory, Leningrad University, Leningrad,
    U.S.S.R.

SHKLOVSKY (Dr I.S.), Space Research Institute, USSR Academy of Sciences,
    117810 Moscow, U.S.S.R.

SHOBBROOK (Dr R.R.) The Chatterton Astronomy Department, School of Physiks,
    University of Sydney, Sydney, N.S.W. 2006, Australia

SHOEMAKER (E.) U.S. Geological Survey, Menlo Park, California, U.S.A.

SHOLOMITSKY (Dr G.B.) Institute of. Space Research, Profsoyuznaja ul. 88,
    Moscow, 117485, U.S.S.R.

SHORE (Dr B.W.) T Division, Lawrence Livermore Laboratory, P.O. Box 808,
    Livermore, California 94550, U.S.A.

SHTEINS (Dr K.A.) Latvian University, Riga, U.S.S.R.

SHU (Dr F.) Department of Earth and Space Sciences, State University of
    New York, Stony Brook, New York 11790, U.S.A.

SHUKLA (K.S.) Reader, Department of Mathematics and Astronomy, Lucknow University,
    Lucknow, U.P., India

SHUL'BERG (Dr A.R.) Astronomical Observatory, Odessa University, Odessa, U.S.S.R.

SHUL'MAN (Dr L.M.) Main Astronomical Observatory of the Ukrainian Acad. of Sci. Kiev - 252127, U.S.S.R.

SHULOV (Dr O.S.) Astronomical Observatory of Leningrad University, 10 linia 33, Leningrad 199178, U.S.S.R.

SHUTER (Dr W.L.H.) Department of Physics, University of British Columbia, Vancouver, B.C. V6T 1W5, Canada

*SIBILLE (Dr G.) Observatoire de Lyon, 69000 Saint-Genis-Laval, France

SIDA (Prof. D.W.) Department of Physics, Carleton University, Ottawa, Ontario, Canada

SIEGEL (C.L.) Göttingen University, Bunsenstrasse 3-5, Göttingen, Germany F.R.

SILK (Dr J.) University of California, Astronomy Department, Berkeley, California 94720, U.S.A.

SILVERBERG (Dr E.) University of Texas, McDonald Observatory, Austin, Texas 78712, U.S.A.

ŠIMEK (Dr M.) Astronomical Institute, Czechoslovak Academy of Sciences, Observatory Ondřejov, Czechoslovakia

SIMKIN (Dr S.M.) Department of Astronomy and Astrophysics, Michigan State University, East Lansing, MI 48824, U.S.A.

SIMODA (Dr M.) Department of Astronomy and Earth Science, Tokyo Gakugei University, Koganei-shi, Tokyo, 184 Japan

SIMON (Dr G.W.) Sacramento Peak Observatory, Sunspot, New Mexico 88349, U.S.A.

SIMON (Dr M.) State University of New York, Stony Brook, New York 11790, U.S.A.

SIMON (Dr N.R.) Behlen Observatory, University of Nebraska, Lincoln, Nebraska 68508, U.S.A.

SIMON (Dr P.) Observatoire de Paris, Section d'Astrophysique, 92190 Meudon, France

SIMON (Prof. Dr R.L.E.) Institut d'Astrophysique, Université de Liège, Avenue de Cointe, 5, B-4200 Cointe-Ougrée, Belgium

SIMONENKO (Dr A.N.) Committee on Meteorite,USSR Academy of Sciences, ul. M. Ulianovoj 3, Korp.1., Moscow 117313, U.S.S.R.

SIMONSON (Dr S.C. III) 3 Empire Place, Greenbelt, MD 20770, U.S.A.

SIMOVLJEVIĆ (Dr Jovan) Prirodno matem. fakultet, katedra za astronomiju Beograd, Studentski trg 16, Beograd, Yugoslavia

SINCLAIR (Dr A. Th.) Royal Greenwich Observatory, Herstmonceux Castle, Hailsham, Sussex, U.K.

SINNERSTAD (Dr U.) Stockholm Observatory, S - 133 00 Saltsjöbaden, Sweden

SINTON (Dr W.M.) Institute for Astronomy, 2840 Koluwalu St. University of Hawaii, Honolulu, Hawaii 96822, U.S.A.

SINVHAL (Dr S.D.) Director, Uttar Pradesh State Observatory, Naini Tal, India

SINZI (Dr A.M.) Hydrographic Office, 5 chome, Tsukiji, Cho-ku, Tokyo, Japan

*SIRY (Dr J.W.) NASA Goddard Space Flight Center, Greenbelt, MD 20771, U.S.A.

SISSON (G.M.) Planetrees, Wall, Hexham, Northumberland, U.K.

SISTERO (Dr R.F.) Sucre 1275, 2° E Córdoba, Argentina

SITARSKI (Dr G.) Astronomical Institute of Polish Academy of Sciences, Al. Ujazdowskie 4, Warszawa, Poland

SITNIK (Dr G.F.) Sternberg Astronomical Institute, Moscow, U.S.S.R.

SIVARAMAN (K.R.) Astrophysical Observatory, Kodaikanal-3, India

SKALAFURIS (Dr A.) Department of Physics, Bartol Research Foundation of the Franklin Institute, Swarthmore, Pennsylvania 19081, U.S.A.

*SKILLING (Dr J.) Dept. of Applied Mathematics and Theoretical Physics, Cambridge University, Silver Street, Cambridge CB3 9EW, U.K.

SKUMANICH (Dr A.) High Altitude Observatory, Boulder, Colorado 80302, U.S.A.

SLAUCITAJS (Prof. Dr S.) Observatorio Astronómico, La Plata, Argentina

SLEBARSKI (T.B.) University Observatory, St Andrews, Scotland, U.K.

SLEE (Dr O.B.)   CSIRO , Division of Radiophysics, Post Office Box 76, Epping, N.S.W., 2121, Australia

SLETTEBAK (Prof. Dr A.) Perkins Observatory, Delaware, Ohio, U.S.A.

SLONIM (Dr E.M.) Astronomical Observatory, Tashkent, U.S.S.R.

SLYSH (Dr V.I.) Institute of Cosmical Research, USSR Academy of Sciences, Moscow, U.S.S.R.

SMAK (Dr J.) Astronomical Observatory of Warsaw University, Al. Ujazdowskie 4, Warsaw, Poland

SMERD (Dr S.F.)   CSIRO , Division of Radiophysics, Post Office Box 76, Epping, N.S.W., 2121, Australia

SMEYERS (Abbé Dr P.) Astronomisch Instituut, Katholieke Universiteit Leuven, Naamsestraat, 61, Leuven, B-3000, Belgique

SMILEY (Dr C.H.) Ladd Observatory, Brown University, Providence, R.I., U.S.A.

SMIT (Prof. J.A.) Physical Laboratory of the University, Sorbonnelaan 4, Utrecht (Uithof), The Netherlands

SMITH (Dr A.G.) Department of Physics, University of Florida, Gainesville, Florida, U.S.A.

SMITH (Dr A.M.) Observational Astronomy Branch, Goddard Space Flight Center, Greenbelt, Maryland 20771, U.S.A.

SMITH (Dr B.A.) Department of Planetary Sciences, The University of Arizona, Tucson, Arizona 85721, U.S.A.

SMITH (Dr D.F.) High Altitude Observatory, P.O. Box 3000, Boulder, Colorado, 80303, U.S.A.

SMITH (Dr E.V.P.) Department of Physics and Astronomy, University of Maryland, College Park, Maryland 20742, U.S.A.

SMITH (Prof. F. G.) Director, Royal Greenwich Observatory, Herstmonceux Castle, Sussex BN27 1RP, U.K.

SMITH (Dr G.) Department of Astrophysics, University Observatory, South Parks Road, Oxford, U.K.

SMITH (H.M.) 23 Normandale, Bexhill on Sea, East Sussex, TN39 3 LU, U.K.

SMITH (Prof. Harlan J.) Director, McDonald Observatory, Astronomy Department, Room 407 Physics Bld., The University of Texas, Austin, Texas 78712

*SMITH (Dr Haywood C., Jr) Department of Astronomy, University of South Florida, Tampa, FL 33620, U.S.A.

SMITH (Dr Henry J.) National Aeronautics and Space Administration, Code SG, Washington 25, D.C., U.S.A.

SMITH (Dr L.F.) Max-Planck-Institut für Radioastronomie, Auf dem Hügel 69, 53 Bonn 1

*SMITH (Dr Myron A.) The University of Texas at Austin, Department of Astronomy, Austin, TX 78712, U.S.A.

SMITH (Dr M.G.) Observatorio Interamericano de Cerro Tololo, Casilla 63-D, La Serena, Chile

SMITH (Dr R. Connon) Astronomy Centre, University of Sussex, Falmer, Brighton, Sussex, BN1 9QH, U.K.

SMITH (Dr W.H.) Princeton University, Princeton, New Jersey 08540, U.S.A.

SMOLINSKI (Dr J.) Krasinkskiego 63a m 12, 87-100 Toruń, Poland

*SMOL'KOV (Dr G. Ya.) Siberian Institute of Terrestrial Magnetism, Ionosphere and Radio, Wave Propagation (SibIZMIR), 664033 Irkutsk - 33. P. Box 4, U.S.S.R.

SMOLUCHOWSKI (Prof. R.) Solid State Sciences, Princeton University, Princeton, New Jersey 08540, U.S.A.

SMRIGLIO (Dr F.) Osservatorio Astronomico, Viale del Parco Mellini, 84, Roma, Italy

*SMYLIE (Dr D.E.) Department of Physics, York University, 4700 Keele Street, Downsview, Ontario, Canada M3J 1P3

SMYTH (Dr M.J.) University of Edinburgh, Department of Astronomy, Royal Observatory, Edinburgh EH9 3HJ, U.K.

*SNOW (Dr T.P. Jr) Princeton University Observatory, Peyton Hall, Princeton, NJ 08540, U.S.A.

SNYDER (Dr L.E.) Astronomy Department, University of Virginia, Charlottesville, Virginia 22903, U.S.A.

SOBERMAN (Dr R.K.) General Electric Co., Valley Forge Space Technology Center, Box 8555, Philadelphia, Pa. 19101, U.S.A.

SOBIESKI (Dr S.) Astrophysics Branch, Code 613, NASA, Goddard Space Flight Center, Greenbelt, Maryland 20771, U.S.A.

SOBOLEV (Dr V.M.) Main Astronomical Observatory, USSR Academy of Sciences, Pulkovo, U.S.S.R.

SOBOLEV (Prof. Dr V.V.) The University, Leningrad, U.S.S.R.

SOBOLEVA (Dr N.S.) Special Astrophysical Observatory, USSR Academy of Sciences, Leningrad Branch of the SAO, Pulkovo 196140, U.S.S.R.

*SOBOUTI (Prof. Y.) Department of Physics, Pahlavi University, Shiraz, Iran

SOCHER (Dr H.) Universitäts-Sternwarte Wien, Türkenschanzstrasse 17, A-1180 Vienna, Austria

*SOCHILINA (Dr A.S.) Institute of Theoretical Astronomy, USSR Academy of Sciences, Leningrad 192187, U.S.S.R.

SOFIA (Dr S.) University of South Florida, Department of Astronomy, Tampa, Florida 33620, U.S.A.

*SOFUE (Dr Yoshiaki) Department of Physics, Nagoya University, Chikusa, Nagoya, 464 Japan

*SOIFER (Dr B.T.) Department of Physics, University of California, San Diego, P.O. Box 109, La Jolla, CA 92037, U.S.A.

*SOLF (Dr J.) Max-Planck-Institut für Astronomie, 6900 Heidelberg 1, Königstuhl, Germany F.R.

SOLHEIM (Dr J.E.) Nordlysobservatoriet, Universitetet i Tromsö, 9001 Tromsö, Norway

SOLOMON (Dr P.M.) Astronomy Department, Columbia University, Box 110 Pupin Hall, New York, N.Y. 10027, U.S.A.

SOMERVILLE (Dr W.B.) University College, Gower Street, London, W.C.1, U.K.

SONETT (Dr C.P.) Department of Planetary Sciences and Lunar and Planetary Laboratory, University of Arizona, Tucson, Arizona 85721, U.S.A.

SØRENSEN (N.G.) Department of Physics, University of Aarhus, Aarhus, Denmark

SOROCHENKO (Dr R.L.) Physical Institute, USSR Academy of Sciences, Moscow, U.S.S.R.

*SORU-ESCAUT (I.) DASOP, Observatoire de Meudon, 92190 Meudon, France

*SOTIROVSKY (Dr P.) DASOP, Observatoire de Meudon, 92190 Meudon, France

SOUFFRIN (P.) Observatoire de Nice, Le Mont-Gros, 06 Nice, France

SOUFFRIN (Dr S.) Institut d'Astrophysique, 98 bis Boulevard Arago, 75014 Paris, France

SOULIE (G.) Observatoire de l'Université de Bordeaux, 33270 Floirac, France

SOUTHWORTH (Dr R.B.) Harvard College Observatory, 60 Garden Street, Cambridge, Mass. 02138, U.S.A.

SPADA (G.) Laboratorio di Astrofisica, Frascati, Roma, Casella Postale 67, Italy

SPARKS (Dr W.M.) Laboratory for Optical Astronomy, Goddard Space Flight Center, Greenbelt, Maryland 20771, U.S.A.

*SPARROW (Dr J.G.) Aeronautical Research Laboratories, Box 4331, P.O., Melbourne, Victoria, Australia 3001

SPEER (Dr R.J.) Physics Department, Imperial College, Prince Consort Road, London S.W.7, U.K.

*SPENCER (Dr R.E.) Nuffield Radio Astronomy Laboratories, Jodrell Bank, Macclesfield, Cheshire, U.K.

SPIEGEL (Dr E.) Courant Institute of Mathematical Sciences, 251 Mercer Street, New York, N.Y. 10003, U.S.A.

SPINRAD (Dr H.) Berkeley Astronomical Department, University of California, Berkeley, California, U.S.A.

SPITE (Dr F.) Observatoire, 92190 Meudon, France

SPITE (Dr M.) Observatoire, 92190 Meudon, France

*SPITHAS (Dr E.) Astronomical Department, University of Thessaloniki,
  Thessaloniki, Greece

SPITZER (Dr L. Jr) Director of the Princeton University Observatory,
  Peyton Hall, Princeton, New Jersey 08540, U.S.A.

*SPOELSTRA (Dr T.A.Th.) Sterrewacht, Huygens Laboratorium, Wassenaarseweg 78,
  Leiden - 2405, The Netherlands

*SPYROU (Dr N.) Astronomical Department, University of Thessaloniki,
  Thessaloniki, Greece

*SRAMEK (Dr R.) Arecibo Observatory, P.O. Box 995, Arecibo, PR 00612, U.S.A.

SREEKANTAN (Dr B.V.) Tata Institute of Fundamental Research, Homi Bhabha Road,
  Bombay 400 005, India

SREENIVASAN (Dr S.R.) Department of Physics, University of Calgary, Calgary,
  Alberta T2N 1N4, Canada

STABELL (R.) Institute of Theoretical Astrophysics, University of Oslo,
  P.O. Box 1029, Blindern, Oslo 3, Norway

STAHR-CARPENTER (Dr M.) Leander McCormick Observatory, Box 3818, University
  Station, Charlottesville, Virginia 22903, U.S.A.

*STALIO (R.) Observatorio Astronomico, Via G.B. Tiepolo 11, I 34121 Trieste,
  Italy

STANDISH (Dr E.M. Jr) Jet Propulsion Laboratory, California Institute of
  Technology, 4800 Oak Grove Drive, Mail Stop CPB-104, Pasadena,
  California 91103, U.S.A.

STANILA (Dr G.) Observatoire Astronomique, str. Cutitul de Argint 5,
  Bucarest 28, Roumania

*STANKEVICH (Dr K.S.) Radiophysical Research Institute, Lyadov Str. 25/14,
  Gorky 603024, U.S.S.R.

STANLEY (G.J.) California Institute of Technology, Pasadena 4, California, U.S.A.

STARRFIELD (Dr S.G.) Physics, Arizona State University, Tempe, Arizona 85281, U.S.A.

STARYTSIN (Dr G.V.) Main Astronomical Observatory, USSR Academy of Sciences,
  Pulkovo, U.S.S.R.

*STAUBERT (Dr R.) Astronomisches Institut der Universität Tübingen, 74 Tübingen,
  Waldhäuserstrasse 64, Germany F.R.

STAWIKOWSKI (Dr A.) Astronomical Observatory of Toruń University, Sienkiewicza 30,
  87-100 Toruń, Poland

STEAD (Dr E.A.) University of Wales, Department of Applied Mathematics, University
  College of Swansea, Singleton Park, Swansea, GLAM, SA2 8PP, U.K.

STEARNS (Dr C.L.) Van Vleck Observatory, Middletown, Connecticut, U.S.A.

STECHER (T.P.) Code 672, Goddard Space Flight Center, Greenbelt, Maryland 20771,
  U.S.A.

*STECKER (Dr F.W.) Theoretical Studies Group, NASA-Goddard Space Flight Center,
  Greenbelt, MD 20771, U.S.A.

STEIGER (Prof. W.R.) Hawaii Institute of Geophysics, University of Hawaii,
  2525 Correa Rd., Honolulu, Hawaii 96822, U.S.A.

STEIGMAN (Dr G.) Yale University Observatory, New Haven, Conn. 06520, U.S.A.

STEIN (Dr R.F.) Brandeis University, Waltham, Massachusetts, U.S.A.

STEIN (Dr W.A.) University of California, Physics Department, La Jolla,
California, U.S.A.

STEINBERG (Dr J.L.) Service de Radioastronomie, Observatoire de Paris,
Section d'Astrophysique, 92190 Meudon, France

STEINITZ (Dr R.) Institute of Planetary and Space Physics, Tel-Aviv University,
Ramat-Aviv, Israel

STEINLIN (Dr U.W.) Astronomisches Institute der Universität Basel,
Venusstrasse 7, CH-4102 Binningen, Switzerland

STELLMACHER (Dr G.) Institut d'Astrophysique, 98 bis, Boulevard Arago,
75014 Paris, France

*STELLMACHER (Dr I.) Bureau des Longitudes, 77, avenue Denfert Rochereau,
75014 Paris, France

STENFLO (Dr J.O.) Institute of Astronomy, S-222 24 Lund, Sweden

STEPANIAN (Dr A.A.) Crimean Astrophysical Observatory P/O Nauchny,
Crimea 334413, U.S.S.R.

STEPANIAN (Dr N.N.) Crimean Astrophysical Observatory P/O Nauchny,
Crimea 334413. U.S.S.R.

STEPANOV (Prof. Dr V.E.) Director, SIBIZMIR, USSR Academy of Sciences,
664697 Irkutsk, U.S.S.R.

STEPHENSON (Prof. C.B.) Warner and Swasey Observatory, Taylor and Brunswick Roads,
East Cleveland, Ohio 44112, U.S.A.

*STEPHENSON (Dr F.R.) Institute of Lunar and Planetary Sciences, School of Physics,
University of Newcastle upon Tyne, NE1 7RU, U.K.

STEPIEN (Dr K.) Astronomical Observatory of the Warsaw University,
Al. Ujazdowskie 4, Warszawa 00-478, Poland

ŠTERNBERK (Dr B.) Astronomical Institute, Czechoslovak Academy of Sciences,
Budečská 6, Prague 2, Czechoslovakia

STESHENKO (Dr N.V.) Crimean Astrophysical Observatory, USSR Academy of Sciences,
P/O Nauchny, Crimea, U.S.S.R.

STEVENS (Dr G.A.) Laboratorium voor Ruimteonderzoek, Beneluxlaan 21, Utrecht,
The Netherlands

*STEWART (Dr J.M.) Institute of Astronomy, Madingley Road, Cambridge, U.K.

STEWART (Dr P.) Department of Mathematics, The University, Manchester,
M13 9PL, U.K.

*STEWART (R.T.) CSIRO, Division of Radiophysics, Post Office Box 76, Epping,
N.S.W. 2121, Australia

STEYAERT (Dr H.L.C.) Sterrenkundig Instituut, Rijksuniversiteit Gent,
J. Plateaustraat, 22, B-9000 Gent, Belgium

STIBBS (Prof. D.W.N.) University Observatory, Buchanan Gardens, St. Andrews,
Fife, Scotland

STICKER (Prof.Dr B.) Direktor des Institutes für Geschichte der
Naturwissenschaften, Hartungstr.5, Hamburg 13, Germany F.R.

STICKLAND (Dr D.J.) Royal Greenwich Observatory, Herstmonceux Castle, Hailsham,
East Sussex, BN 27 1RP, U.K.

STIEFEL (E.) Mathematische Abteilung, Eidgenössische Technische Hochschule,
8000 Zürich, Switzerland

STIX (Dr M.) Universitäts-Sternwarte, 3400 Göttingen, Geismarlandstr. 11, Germany F.R.

STOBIE (Dr R.S.) The Royal Observatory, Blackford Hill, Edinburgh EH9 3HJ, U.K.

STOCK (Dr J.) CIDA, Apartado 264, Mérida, Venezuela

STOCKTON (Dr A.) Institute for Astronomy, University of Hawaii, Honolulu,
    Hawaii 96822, U.S.A.

STODÓKIEWICZ (Dr J.S.) Astronomical Observatory of Warsaw University,
    Warszawa, Al. Ujazdowskie 4, Poland

ŠTOHL (Dr J.) Astronomical Institute, Slovak Academy of Sciences, Dúbravská cesta,
    Bratislava, Czechoslovakia

STONE (Dr R.G.) Code 693, Laboratory for Extraterrestrial Physics, NASA-Goddard
    Space Flight Center, Greenbelt, Maryland 20771, U.S.A.

STOY (Dr R.H.) Royal Observatory, Edinburgh 9, Scotland, U.K.

STOYKO (A.) Observatoire de Paris, 61, Avenue de l'Observatoire, 75014 Paris, France

STRAIŽYS (Dr V.L.) Astronomical Observatory, ul. Chiurlionisa 29,
    232031 Vil'nius, U.S.S.R.

STRAND (Dr K.Aa.) U.S. Naval Observatory, Washington, D.C. 20390, U.S.A.

STRASSL (Prof. Dr H.) Direktor des Astronomischen Instituts der Universität,
    Steinfurterstrasse 107, Münster (Westfalen), Germany F.R.

STRAUSS (Dr F.M.) Instituto de Fisica, UFRGS, Depto. Astronomía, R. Luiz Englert
    S/N, 90000 Porto-Alegre, RGS., Brazil

STRITTMATTER (Dr P.A.) Steward Observatory, University of Arizona, Tucson,
    Arizona 85721, U.S.A.

STROBEL (Dr W.) Astronomisches Rechen-Institut, Mönchhofstrasse 12-14,
    69 Heidelberg, Germany F.R.

STROHMEIER (Prof. Dr W.) Direktor der Remeis-Sternwarte, Sternwarstrasse 7,
    Bamberg, Germany F.R.

*STROM (Dr K.M.) Kitt Peak National Observatory, P.O. Box 26732, Tucson,
    AZ 86726, U.S.A.

STROM (Dr R.G.) Lunar and Planetary Laboratory, University of Arizona, Tucson,
    Arizona 75721, U.S.A.

STROM (Dr Richard G.) Radiosterrenwacht, Oudehoogeveensedijk 4, Dwingeloo (Dr.),
    The Netherlands

STROM (Dr S.E.) Kitt Peak National Observatory, P.O. Box 26732, Arizona 85726, U.S.A.

STRÖMGREN (Prof. B.) Observatoriet, Østervoldgade 3, DK-1350 Copenhagen K, Denmark

*STRONG (Dr I.B.) Los Alamos Scientific Laboratory, Mail Stop 436, Los Alamos,
    NM 87544, U.S.A.

STRONG (Prof. J.D.) Astronomy Research Facility, University of Massachusetts,
    Amherst, Mass. 01002, U.S.A.

STUMPFF (Dr P.) Max-Planck-Institut für Radioastronomie, Auf dem Hügel 69,
    53 Bonn 1, Germany F.R.

STURCH (Dr C.R.) 10003 Branch View Court, Silver Spring, MD 20903, U.S.A.

STURROCK (Dr P.A.) Institute for Plasma Research, Stanford University, Stanford,
    California, U.S.A.

SUDA (Dr K.) Astronomical Institute,
    Tohoku University, Sendai, Japan

SUEMOTO (Prof.Dr Z.) Director Tokyo Astronomical Observatory
    Mitaka, Tokyo 181, Japan

*SUESS (Dr S.T.) Space Environment Laboratory, NOAA/ERL,
    Boulder, CO 80302, U.S.A.

SUGAWA (Dr C.) International Latitude Observatory of Mizusawa,
    Mizusawa-shi, Iwate-ken, Japan

SUGIMOTO (Dr D.) Institute of Earth Science and Astronomy, College of General
    Education, University of Tokyo, 3-8-1 Komaba, Meguro-Ku, Tokyo 153, Japan

SUKHAREV (Dr L.E.) Pulkovo Observatory, Leningrad 196140, U.S.S.R.

SULLIVAN (Dr W.T. III)Dept. of Astronomy, University of Washington,
    Seattle, Washington 98195, U.S.A.

SULTANOV (Acad. G.F.), Director, Shemakha Astrophysical Observatory,
    Shemakha, 373243 Azerbaidzan, U.S.S.R.

*SUMMERS (Dr H.P.) Dept. of Applied Mathematics and Theoretical Physics,
    Silverstr., Cambridge CB3 9EW, U.K.

SUSLOV (Dr A.K.) Leningrad Hydrometerological Institute, 195196 Leningrad, U.S.S.R.

SUZUKI (Dr Y.) 23-1 Nakajima Hironocho Uji-si, Kyoto, Japan

SVATOŠ (Dr J.) Dept. of Astronomy and Astrophysics, Charles University,
    Švédká 8,   Prague 5- Smíchov, Czechoslovakia

SVECHNIKOV (Dr M.A.), Astronomical Department of Sverdlovsk State University,
    620083 Sverdlovsk, U.S.S.R.

SVESTKA (Dr Z.) American Science and Engineering, 955 Massachusetts Avenue
    Cambridge, Mass. 02139, U.S.A.

SVOLOPOULOS (Dr S.N.) Professor of Astrophysics, University of Athens
    Panepistimiopolis, Athens (621), Greece

SWANENBURG (Dr B.N.) Cosmic Ray Working Group, Kamerlingh Onnes Laboratorium,
    Nieuwsteeg 18, Leiden, The Netherlands

SWARUP (Prof. G.) Radio Astronomy Centre, Post Box No.8, Ootacamund - 643 001,
    India

SWEET (Prof. P.A.) The University Observatory,
    Glasgow W. 2, Scotland

SWEIGART (Dr A.V.) Yale University,
   New Haven, Conn. 06520, U.S.A.

SWENSON (Dr G.W. Jr.) Vermillion River Observatory, University of Illinois,
   Urbana, Illinois 61801, U.S.A.

SWENSSON (Dr J.W.) Institute of Physics,
   Sölvegatan 14, S- 223 62 Lund, Sweden

SWIHART (Dr T.L.) Steward Observatory, University of Arizona,
   Tucson, Arizona 85721, U.S.A.

SWINGS (Dr J.P.) 23, Avenue Léon Souguenet, B-4050 Esneux, Belgium

SWINGS (Prof. Dr P.) 23, Avenue Léon Souguenet, B- 4050 Esneux, Belgium

SWOPE (H.H.) Mount Wilson and Palomar Observatories, 813, Santa Barbara Street
   Pasadena, California 91106, U.S.A.

SYKES (Dr J.B.) 20 Milton Lane, Steventon,
   Abingdon, Berkshire, OX 13 6SA, U.K.

SYKES-HART (A.B.) University Observatory,
   Oxford, U. K.

*SYKORA (Dr J.) Astronomical Institute of the Slovak Academy of Sciences,
   Observatory Skalnaté Pleso , 059 60 Tatranská Pleso,
   059 60 Tatranská Lomnica, Czechoslovakia

SYROVATSKIJ (Dr S.I.) Physical Institute, USSR Academy of Sciences,
   117924 Moscow, U.S.S.R.

SYUNYAEV (Dr R.A.) Space Research Institute, USSR Academy of Sciences,
   117810 Moscow, U.S.S.R.

SZAFRANIEC (Dr R.) Astronomical Observatory,
   Kopernika 27, Cracow, Poland

SZEBEHELY (Dr V.) Dept. of Aerospace Engineering, The University of Texas,
   2315 Speedway, Austin, Texas 78712, U.S.A.

SZEIDL (Dr B.) Konkoly Observatory, P.O. Box 67,
   Budapest XII, Hungary

TABARA (Dr H.) The Faculty of Education, Utsunomiya University, Mine 350, Utsunomiya, Japan

TADEMARU (Dr E.) University of Massachusetts, Amherst, Mass 01002, U.S.A.

TAFFARA (Prof. S.) Via Calza 5 bis, Padova, Italy

TAGLIAFERRI (Prof. Dr G.) Osservatorio Astrofisico di Arcetri, Largo Enrico Fermi 5, Florence, Italy

TAI (Dr Yuin-Kwei) Director, Institute of Geophysics, Central University, Taipei, Taiwan

*TAKAGI (Dr Kojiro) Department of Physics, Toyama University, Toyama, Japan

TAKAGI (Dr S.) International Latitude Observatory of Mizusawa, Mizusawa-shi, Iwate-ken, Japan

TAKAKUBO (Dr K.) Astronomical Institute, Faculty of Science, Tohoku University, Sendai, Japan

TAKAKURA (Dr T.) Department of Astronomy, Faculty of Science University of Tokyo, Bunkyo-ku, Tokyo, Japan

TAKARADA (Dr K.) Kyoto University of Industrial Arts and Textile Fibers, Matsugasaki, Sa Kyoku, Kyoto, Japan

TAKASE (Dr B.) Tokyo Astronomical Observatory, Mitaka, Tokyo, Japan

TAKAYANAGI (K.) Institute of Space and Aeronautical Science, University of Tokyo, Komaba, Meguro-ku, Tokyo, Japan

*TAKEDA (Dr Hidenori) Department of Aeronautical Engineering, Kyoto University, Kyoto, 606 Japan

TAKENOUCHI (Dr T.) Tokyo Astronomical Observatory, Mitaka, Tokyo, Japan

TAKEUTI (Dr M.) Astronomical Institute, Tohoku University, Sendai, Japan

*TALBOT (Dr R.J. Jr) Dept. of Space Physics and Astronomy, Rice University, Houston, TX 77001, U.S.A.

TALWAR (S.P.) Department of Physics, University of Delhi, Delhi, India

*TAMENAGA (Dr Tatsuo) Faculty of Education, Mie University, Tsu City, Mie 514, Japan

TAMMANN (Dr G.A.) Astronomisches Institut der Universität Basel, Venusstrasse 7, CH-4102 Binningen, Switzerland

TAMURA (Dr S.) Astronomical Institute, Tohoku University, Sendai, Japan

TANABE (Dr Hiroyoshi) Tokyo Astronomical Observatory, University of Tokyo, Mitaka, Tokyo, Japan

TANAKA (Dr Haruo) Research Institute of Atmospherics, Nagoya University, Ichida-machi, Toyokawa-shi, Aichi, Japan

TANAKA (Dr K.) Tokyo Astronomical Observatory, Mitaka, Tokyo, Japan

TANAKA (Dr R.) University of Niigata, Asahimachi, Niigata, Japan

TANAKA (Prof. Y.) Institute of Space and Aeronautical Science, University of Tokyo 4-6-1, Komaba, Meguro-ku, Tokyo, Japan

TANAKA (Dr Y.) Faculty of Education, Ibaraki University, Mito, Japan

TANDBERG-HANSSEN (Dr E.) Space Sciences Laboratory, Marshall Space Flight Center, Alabama 35812, U.S.A.

TANDON (Dr J.N.) Department of Physics & Astrophysics, University of Delhi, Delhi - 110007, India

TANNAHILL (Dr T.R.) University Observatory, Glasgow, W.2, U.K.

TANNER (R.W.) 1046 Fisher Ave. Ottawa, Ontario, Canada

TAPLEY (Dr B.D.) Dept. of Aerospace Engineering, University of Texas at Austin, Texas 78712, U.S.A.

*TARENGHI (M.) Istituto di Fisica, Via Celoria 16, 20133 Milano, Italy

*TARNSTROM (Dr G.) Radio Astronomy Group, Microwave Laboratory ETH, Gloriastrasse 35, CH-8006 Zürich, Switzerland

*TARTER (Dr C.B.) Lawrence Livermore Laboratory L-71, University of California, Livermore, CA 94550, U.S.A.

TASSOUL (Dr J.L.) Département de physique, Université de Montréal, Case postale 6128, Montréal, P.Q. H3C 3J7, Canada

TATEVYAN (Dr S.K.) Astronomical Council, USSR Academy of Sciences, Pyatnitskaya ul. 48, 109017 Moscow, U.S.S.R.

*TATON (Dr R.) Centre d'Histoire des Sciences Exactes, 12, rue Colbert, 75002 Paris, France

TATUM (Prof. J.B.) Department of Physics, University of Victoria, Victoria, B.C., Canada

TAUBER (Prof. G.E.) Department of Physics and Astronomy, Tel-Aviv University, Ramat Aviv, Israel

TAVARES (Dr J.T.L.) Avenida Dias da Silva, 173, R/C Esq., Coimbra, Portugal

TAVASTSHERNA (Dr K.N.) Main Astronomical Observatory, USSR Academy of Sciences, Pulkovo, Leningrad M-140, U.S.S.R.

TAYLER (Dr R.J.) Astronomy Centre, University of Sussex, School of Mathematical and Physical Sciences, Falmer, Sussex, U.K.

TAYLOR (Dr D.J.) Steward Observatory, University of Arizona, Tucson, Arizona 85721, U.S.A.

TAYLOR (G.E.) H.M. Nautical Almanac Office, Royal Greenwich Observatory, Herstmonceux Castle, Hailsham, Sussex, BN27 1 RP, U.K.

TAYLOR (Dr J.H. Jr) University of Massachusetts, Amherst, Mass 01002, U.S.A.

*TAYLOR (Dr K.) Royal Greenwich Observatory, Herstmonceux Castle, Hailsham, East Sussex, BN27 1 RP, U.K.

TECH (Dr J.) U.S. Department of Commerce, National Bureau of Standards, Spectroscopy Section, Washington D.C. 20234, U.S.A.

TÉHÉRANY (D.) 83, avenue Rey, Teheran, Iran

TEJFEL' (Dr V.G.) Astrophysical Institute of the Kazakh Academy of Sciences, Alma-Ata, U.S.S.R.

TELEKI (Dr G.) Astronomical Observatory, Volgina Str. 7, 11050 Belgrade, Yugoslavia

TEMESVARY (Dr S.) Richard Strauss Strasse 127, 8 München 80, Germany F.R.

TEMPESTI (Prof. P.) Osservatorio Astronomico di Collurania, Teramo, Italy

TEPLITSKAYA (Dr R.B.) SIBIZMIR, USSR Academy of Sciences, Irkutsk, U.S.S.R.

TERASHITA (Dr Y.) Technical University of Kanazawa,Nonoichi-cho Minami-Kyoku, Kanazawa, Japan

*TEREBIZH (Dr V. Yu.) Crimean Station of Sternberg State Institute, P/O Nauchny, Crimea 334413, U.S.S.R.

Ter HAAR (Dr D.) Magdalen College, Oxford, U.K.

TERRAZAS (L.R.) Instituto Nacional de Astrofísica, Optica y Electrónica, Tonantzintla, Apartados Posyales Nos. 216 y 51, Puebla, Mexico

TERRIEN (Dr J.) Bureau International des Poids et Mesures, Pavillon de Breteuil, 92 Sèvres, France

TERZAN (Dr A.) Observatoire de Lyon, 69 St-Genis-Laval, France

TERZIAN (Dr Y.) Center for Radiophysics and Space Research, Space Sciences Building, Cornell University, Ithaca, N.Y. 14850, U.S.A.

TESKE (Dr R.G.) Astronomy Department, University of Michigan, Ann Arbor, Michigan, U.S.A.

TEXEREAU (J.) Observatoire de Paris, 61, Avenue de l'Observatoire, 75014 Paris, France

THACKERAY (Prof. A.D.) c/o Department of Astronomy, University of Cape Town, Rondebosch 7700, South Africa

THADDEUS (Dr P.) Institute for Space Studies, 2880 Broadway, New York, N.Y. 10025, U.S.A.

THÉ (Dr Pik Sin) Astronomical Institute, University of Amsterdam, Roeteresstraat 15, Amsterdam - C, The Netherlands

THERNÖE (K.A.) University Observatory, Østervoldgade 3, DK-1350 Copenhagen, Denmark

THIRY (Prof. Y.) 20, rue des Crocheteurs, 75 Antony (Seine), France (Univ. de Paris)

THOMAS (Dr D.V.) Royal Greenwich Observatory, Herstmonceux Castle, Hailsham, Sussex, U.K.

THOMAS (Dr H.C.) Max-Planck-Institut für Physik und Astrophysik, Föhringer Ring 6, 8000 München 23, Germany F.R.

THOMAS (Dr J.H.) Dept. of Mechanical & Aerospace Sciences, The University of Rochester, Rochester, New York 14627, U.S.A.

THOMAS (Dr R.J.) Code 682, NASA-Goddard Space Flight Center, Greenbelt, Maryland 20771, U.S.A.

THOMAS (Dr R.N.) Institut d'Astrophysique, 98bis, Boulevard Arago, 7504 Paris, France

THOMPSON (Dr A.R.) National Radio Astronomy Observatory, Edgemont Road, Charlottesville, Virginia 22901, U.S.A.

THOMPSON (Dr G.I.) The Royal Observatory, Edinburgh 9, Scotland, U.K.

*THOMPSON (Dr R.I.) Steward Observatory, The University of Arizona, Tucson, AZ 85721, U.S.A.

*THOMSEN (Dr B.) University of Aarhus, Langelandsgade, DK-8000 Aarhus C, Denmark

*THOREN (Prof. V.) Department of History and Philosophy of Science, University of Indiana, Bloomington, Indiana 47401, U.S.A.

THORNE (Dr K.) Physics Department, California Institute of Technology, Pasadena, California 91109, U.S.A.

TIFFT (Dr W.G.) Steward Observatory, University of Arizona, Tucson, Arizona 85721, U.S.A.

TIFREA (Dr E.) Observatoire Astronomique, str. Cutitul de Argint 5, Bucarest 28, Roumania

*TIMOTHY (Dr J.G.) Center for Astrophysics, 60 Garden Street, Cambridge, MA 02138, U.S.A.

TINBERGEN (Dr J.) Sterrewacht, Leiden, The Netherlands

TING (Yeou-Tswen) Chief of Astronomical Department, Taiwan Provincial Weather Bureau, 64, Kung Yuen Road, Taipei, Taiwan

TINSLEY (Dr B.) Astronomy Department, Yale University, Box 2023, Yale Station, New Haven, Connecticut 06520, U.S.A.

TIURI (Prof. Dr M.) Technical University, Department of Electrotechnics, Albert Str. 40-42, Helsinki, Finland

TJIN A DJIE (Dr H.R.E.) Sterrenkundig Instituut Universiteit van Amsterdam, Roetersstraat 15, Amsterdam, The Netherlands

TLAMICHA (Dr A.) Astronomical Institute (Czechoslovak Academy of Sciences), Ondřejov, Czechoslovakia

TODORAN (Dr I.) Observatoire, Université, Cluj, Roumania

TOFANI (G. Ing.) Osservatorio Astrofisico di Arcetri, Largo Enrico Fermi, 5, Firenze, Italy

TOLBERT (Dr C.R.) Leander McCormick Observatory, University of Virginia, Charlottesville, Virginia 22903, U.S.A.

TOMASKO (Dr M.G.) Lunar and Planetary Laboratory, University of Arizona, Tucson, Arizona, U.S.A.

TOMBAUGH (C.W.) New Mexico State University, Research Center, University Park, New Mexico, U.S.A.

TOMITA (Dr Kenji) Research Institute for Theoretical Physics, Hiroshima University, Takehara, Hiroshima-ken, Japan

TOMITA (Koichiro) Tokyo Astronomical Observatory, Mitaka, Tokyo, Japan

TOOMRE (Dr A.) Department of Mathematics, Massachusetts Institute of Technology, Cambridge, Mass. 02139, U.S.A.

TOOMRE (Dr J.) University of Colorado, Boulder, Colorado 80302, U.S.A.

*TOPAKTAŞ (Dr L.) University Observatory, Universite - Istanbul, Turkey

TORAO (Prof. Dr M.) Yoyogi 5-7-9-401, Shibuya, Tokyo, Japan

TORELLI (Dr M.) Osservatorio Astronomico, Viale del Parco Mellini, 84, Roma, Italy

TÖRGARD (Dr I.H.M.) Nils Ericssonsgatan 28, 521 00 Falköping, Sweden

TORRES-PEIMBERT (Dr S.) Instituto de Astronomía, UNAM, Apartado Postal 70-264, México D.F., Mexico

TORROJA (Prof. J.M.) University of Madrid, Fac. de Ciencias, Madrid 3, Spain

 * TOSA (Dr Makoto) Dept. of Physics, Nagoya University, Nagoya, Japan

 TOUSEY (Dr R.) U.S. Naval Research Laboratory, Washington 25, D.C., U.S.A.

 TOVMASSIAN (Dr H.M.) Byurakan Astrophysical Observatory, Byurakan,
       Armenia, U.S.S.R.

 TOWNES (Dr C.H.) Physics Dept. University of California, Berkeley, Cal.95720,
       U.S.A.

 *TOZER (Dr D.C.) School of Physics, University of Newcastle upon Tyne,
       Newcastle upn Tyne, U.K.

 TRAFTON (Dr L.M.) Dept. of Astronomy, University of Texas, Austin, Texas 78712,
       U.S.A.

 TRAVING (Prof. Dr G.) Lehrstuhl für Theoretische Astrophysik der
       Universität, Im Neuenheimer Feld 294, 69 Heidelberg, Germany F.R.

 TREANOR (Dr P.J., S.J.) Specola Vaticana, Castel Gandolfo,
       Vatican City State

 TREDER (Prof. Dr H.-J.) Zentralinstitut für Astrophysik, Sternwarte
       Babelsberg, DDR 1502 Potsdam-Babelsberg, Rosa-Luxemburg-Str.17a
       Germany D.R.

 TREFFTZ (Dr E.) Max-Planck-Institut für Physik und Astrophysik
       Föhringer Ring 6, 8 München 40, Germany, F.R.

 TREHAN (Dr S.K.) Dept. of Mathematics, Panjab University, Chandigarh 14,
       India

 TRELLIS (M.) Observatoire de Nice, Le Mont-Gros,

       06000 Nice, France

 TREMKO (Dr J.) Skalnaté Pleso Observatory, Slovak Academy of Sciences,
       Tatranská Lomnica, Czechoslovakia

 TREXLER (J.H.) U.S. Naval Research Laboratory, Washington 25, D.C., U.S.A.

 TRIMBLE (Dr V.L.) Dept.Physics, Univ.California, Irvine, Cal. 92717 USA
       July-Dec.each year: Univ.Maryland, Astron.Progr.College Park,MD 20742,USA.

 *TRITAKIS (Dr B.) Research Center for Astronomy and Applied Mathematics,
       Academy of Athens, 14 Anagnostopoulou Street, Athens -136, Greece

 *TRITTON (Dr K.P.) UK 48-inch Schmidt Telescope Unit, Private Bag,
       Coonabarabran, NSW 2857, Australia

 TROITSKY (Prof. Dr V.S.) Radiophysical Research Institute,
       Gorkii, U.S.S.R.

 *TRULSEN (Dr J.K.) Nordlysobservatoriet, University of Tromsø, Norway

 TRÜMPER (Prof. Dr J.) Astronomisches Institut der Universität Tübingen
       7400 Tübingen, Waldhäuserstr.64, Germany, F.R.

 TRURAN (Dr J.W.) Belfer Graduate School of Science, Yeshiva University,
       Amsterdam Ave and 186th Street, New York 10033, U.S.A.

 TRUTSE (Dr Yu.L.) Institute of Physics of Atmosphere, USSR Academy of
       Sciences, Moscow, U.S.S.R.

 TS'AI, (Chang-hsien) Taipei Observatory, Yuan Shan, Taipei, Taiwan

TSAO (Prof.Mo.) 47, 3rd Section, Hsin -I Road, Taipei, Taiwan

TSAP (Dr T.T.) Crimean Astrophysical Observatory, USSR Academy of Sciences
    Crimea, U.S.S.R.

TSCHARNUTER (Dr W.M.) Institut für Theoretische Astronomie, University of
    Vienna, Türkenschantzstr.17, A-1180 Vienna, Austria

TSESEVICH (Prof. Dr V.P.) Director of the Astronomical Observatory,
    Odessa, U.S.S.R.

*TSIOUMIS (Dr A.) Dept. of Geodetic Astronomy, University of Thessaloniki,
    Thessaloniki, Greece

TSUBAKI (Dr T.) Institute of Earth Science and Astrophysics, University of Shiga
    Ohtsu- Japan

TSUBOKAWA (Prof. Dr I.) Tokyo University, Seismological Observatory Institute,
    No. 1- 1 gou - 1 chôme, Yayoimachi, Bunkyoku, Tokyo, Japan

TSUCHIYA (Atsushi) Tokyo Astronomical Observatory, University of Tokyo,
    Mitaka, Tokyo, Japan

TSUJI (Dr T.) Dept. of Astronomy, Faculty of Science, University of Tokyo,
    Bunkyo-ku, Tokyo 113, Japan

TSURUTA (Dr S.) Max-Planck-Institut, für Physik und Astrophysik,
    Föhringer Ring 6, Postfach 401212, 8 München 40, Germany F.R.

TUCKER (R.H.) Royal Greenwich Observatory, Herstmonceux Castle,
    Hailsham, Sussex, U.K.

*TUCKER (Dr W.H.) Smithsonian Astrophysical Observatory, 60 Garden Street,
    Cambridge, Mass. 02138, U.S.A.

TÜFEKÇIOĞLU (Dr Z.) Astronomy Dept. University of Ankara, Fen Fakültesi,
    Ankara, Turkey

TULL (Dr R.G.) Dept. of Astronomy, University of Texas, Austin, Texas 78712,
    U.S.A.

*TULLY (Dr J.A.) Observatoire de Nice, Le Mont-Gros, 06000 Nice, France

*TULLY (Dr R.B.) University of Hawaii, Institute for Astronomy, Honolulu, HI 96822
    U.S.A.

TUOMINEN (Dr I.) Observatory and Astrophysics Laboratory, University of
    Helsinki, Tähtitorninmäki, SF- 00130 Helsinki 13, Finland

TUOMINEN (Prof. J.V.) University of Helsinki, Observatory and Astrophysics
    Laboratory, Tähtitorninmäki, SF-00130 Helsinki 13, Finland

TURLO (Dr Zygmunt) Astronomical Observatory, Toruń-Piwnice, Poland

TURNER (Dr B.E.) National Radio Astronomy Observatory, Charlottesville,
    Virginia 22901, U.S.A.

TURNER (Dr K.C.) Dept. of Terrestrial Magnetism, Carnegie Institute of
    Washington, 5241 Broad Branch Road N.W., Washington, D.C.20015, U.S.A.

*TURON-LACARRIEU (Dr P.) DESPA, Observatoire de Meudon, 92190 Meudon, France

TURTLE (Dr A.J.) The Chatterton Astrophysics Dept., School of Physics,
    University of Sydney, Sydney 2006, Australia

TUTUKOV (Dr A.V.) Astronomical Council, USSR Academy of Sciences,
    Pyatnitskaya ul. 48, 109017 Moscow, U. S. S. R.

TUVE (Dr M.A.) Carnegie Institution of Washington, Dept. of Terrestrial
    Magnetism, 5241 Broad Branch Road, N.W., Washington, D,C.20015,U.S.A.

TWISS (Dr R.Q.) National Physical Laboratory, Teddington, Middlesex, U.K.

TYLER (Dr G.L. Jr.) Center for Radar Astronomy, Stanford University,
    Stanford, Cal. 94305, U.S.A.

*TZVETKOV (Dr M.) Dept. of Astronomy, Bulgarian Academy of Sciences,
    "7th November" Street 1, 1000 Sofia, Bulgaria

TUZI (Dr K.) 2-2-20 Ôsawa, Mitaka-shi, Tokyo, Japan

UCHIDA (Dr J.) Dept. of Technology, Tohoku-Gakuin University,
    Tagajo-machi, Miyagi-Ken, Japan

UCHIDA (Dr Yutaka) Tokyo Astronomical Observatory, University of Tokyo,
    Mitaka, Tokyo, Japan

UDAL'TSOV (Dr V.A.) Physical Institute, USSR Academy of Sciences,
    Moscow, U.S.S.R.

UENO (Dr S.) Dept. of Astronomy, University of Southern California,
    University Park, Los Angeles, Cal. 90007, U.S.A.

UESUGI (Dr Akira) Dept. of Astronomy, University of Kyoto, Kyoto, Japan

UETA (Dr J.) Dept. of Astronomy, Faculty of Science, Kyoto University,
    Kyoto, Japan

*ULMER (Dr M.P.) Center for Astrophysics, 60 Garden Street, Cambridge, MA 02138,
    U.S.A.

ULMSCHNEIDER (Dr P.) Institut für Astronomie und Astrophysik, Am Hubland,
    8700 Würzburg, Germany F.R.

ULRICH (Dr B.) Astronomy Dept. University of Texas, Austin, Texas 78712,
    U.S.A.

ULRICH (Dr M.H.) Institut d'Astrophysique, 98 bis, Blvd Arago,
    Paris 14e, France

ULRICH (Dr R.K.) Dept. of Astronomy, University of California, LA
    Los Angeles, Cal. 90024, U.S.A.

UNDERHILL (Dr A.B.) Code 670, Goddard Space Flight Center, Greenbelt, MD 20771,
    U.S.A.

UNDERWOOD(J.H.) Institute for Plasma Research, Stanford University,
    Via Crespi, Stanford, CA 94305, U.S.A.

UNNO (Dr W.) Dept. of Astronomy, Faculty of Science, University of Tokyo
    Bunkyo-ku, Tokyo, Japan

UNSÖLD (Prof. Dr A.) Direktor des Institutes für Theoretische Physik
    Olshausenstr., Neue Universität, Bau 13/I, Kiel, Germany, F.R.

UPGREN (Dr A.) Dept. of Astronomy, Wesleyan University, Middletown, Conn.06457,
    U.S.A.

UPTON (Dr E.K.L.) Astronomy Dept. University of California, Los Angeles,
    California 90024, U.S.A.

*URAS (S.) Stazione Astronomica, Internazionale di Latitudine,
    Via Ospedale 72, Cagliari, Italy

*URASIN (Dr L.A.) Astronomical Observatory Engelhardt, Kazan 422526, U.S.S.R.

URBARZ (Dr H.) Astron. Institut der Universität, Waldhäuser Str. 64,
74 Tübingen, Germany F.R.

URECHE (Dr V.) University of Cluj, Cluj, Roumania

UREY (Prof. H.C.) University of California, La Jolla, California, U.S.A.

USHER (Dr P.D.) 507 Davey Lab., Department of Astronomy, Pennsylvania State
University, University Park, Pennsylvania 16802, U.S.A.

UTSUMI (Dr K.) Department of Astronomy, Faculty of General Education,
Hiroshima University, Hiroshima, Japan

UUS (Dr U.Kh.), Institute for Astrophysics and Physics of the Atmosphere,
Estonian Academy of Sciences, 202444 Tartu, U.S.S.R.

VAIANA (Dr G.) American Science & Engineering Inc., 955 Massachusetts Avenue,
Cambridge, Mass. 02139, U.S.A.

*VAIDYA (Dr P.C.) 34, Shardanagar, Ahmedabad-380 007, India
VAINSTEIN (Dr L.A.) Physical Institute, USSR Academy of Sciences, Moscow, U.S.S.R.

*VALBOUSQUET (Dr A.) Observatoire de Strasbourg, 11, rue de l'Université,
67000 Strasbourg, France

VALNIČEK (Dr B.) Astronomical Institute, Czechoslovak Academy of Sciences,
Observatory Ondřejov, Czechoslovakia

*VALTONEN (Dr M.J.) The University of Alabama, Department of Physics and Astronomy,
Box 1921, University, Alabama 35486, U.S.A.

VAN AGT (Dr S.L.Th.J.) Sterrekundig Instituut, Toernooiveld, 6805 Nijmegen,
The Netherlands

VAN ALBADA (Dr T.S.) Kapteyn Astronomical Laboratory, Hoogbouw W.S.N. - Postbus 800,
Groningen 8002, The Netherlands

VAN ALLEN (Dr J.A.) Department of Physics and Astronomy, University of Iowa,
Iowa City, Iowa 52240, U.S.A.

VAN ALTENA (Dr W.F.) Yale University Observatory, Box 2023, Yale Station,
New Haven, CT 06520, U.S.A.

*VAN BEEK (Dr F.) Lab. voor Ruimte-Onderzoek, Beneluxlaan 21, Utrecht,
The Netherlands

VAN BLERKOM (Dr D.J.) University of Massachusetts, Amherst, Mass 01002, U.S.A.

VAN BREDA (Dr I.G.) University Observatory, Buchanan Gardens, St. Andrews, U.K.

VAN BUEREN (Prof. Dr H.G.) Astronomisch Instituut, Zonneburg 2, Utrecht,
The Netherlands

Van de HULST (Prof. Dr H.C.) Sterrewacht, Kaiserstraat, Leiden,
The Netherlands

Van de KAMP (Prof. Dr P.) Director of the Sproul Observatory, Swarthmore College,
Swarthmore, Pennsylvania 19081, U.S.A.

VANDEKERKHOVE (Dr E.) Astronome Honoraire, Observatoire Royal de Belgique,
Avenue de Sumatra, 7, B-1180 Bruxelles, Belgium

VANDEN BOUT (Dr P.) Dept. of Astronomy, College of Natural Sciences,
University of Texas at Austin, Austin, Texas 78712, U.S.A.

VAN DEN HEUVEL (E.P.J.) Astronomical Institute, Roetersstraat 15, Amsterdam, The Netherlands

Van den BERGH (Prof. S.) David Dunlap Observatory, University of Toronto, Richmond Hill, Ontario L4C 4Y6, Canada

Van der BORGHT (Prof. R.) Department of Mathematics, Monash University, Melbourne, Australia

VAN DER KRUIT (Dr P.C.) Sterrenkundig Laboratorium "Kapteyn" Postbus 800, Groningen 8002, The Netherlands

VAN DER LAAN (Prof.Dr H.) Sterrewacht, Huygens Laboratorium, Wassenaarseweg 78, Leiden 2405, The Netherlands

VANDERLINDEN (Prof.ém. Dr H.L.) University Ghent, Hertogsdreef, 15, B-1170 Brussels, Belgium

VANDERVOORT (Prof. P.O.) Ryerson Physical Laboratory, 1100-14 East 58th Street, Chicago, Illinois 60637, U.S.A.

*VAN De STADT (Dr H.) Rijksuniversiteit Utrecht, Servaas Bolwerk 13, Utrecht, The Netherlands

Van DIGGELEN (Dr J.) Sterrewacht "Sonnenborgh" Zonneburg 2, Utrecht, The Netherlands

*VAN DORN BRADT (Dr H.) M.I.T. Rm. 37-581, Center for Space Research, Cambridge, MA 02139, U.S.A.

VAN DUINEN (R.J.) University of Groningen, Astronomy Department, Space Research Division, Groningen, The Netherlands

VAN FLANDERN (Dr T.C.) U.S. Naval Observatory, Washington D.C. 20390, U.S.A.

VAN GENDEREN (Dr A.M.) Leiden Southern Station, P.O. Box 13, Broederstroom, Transvaal 0240, South Africa

Van HERK (Dr G.) Sterrewacht-Huygens Laboratorium, Wassenaarseweg 78, Leiden 2405, The Netherlands

VAN HOOF (Prof. Dr A.) Astronomisch Instituut, Katholieke Universiteit Leuven, Naamsestraat, 61, B-3000 Leuven, Belgium

VAN HORN (Dr H.M.) University of Rochester, River Campus Station, Rochester, New York 14627, U.S.A.

Van HOUTEN (Dr C.J.) Sterrewacht, Kaiserstraat 57, Leiden, The Netherlands

Van HOUTEN-GROENEVELD (Dr I.) Sterrewacht. Kaiserstraat 57, Leiden, The Netherlands

VAN LEER (Dr B.) The Observatory Leiden, Leiden, The Netherlands

VAN NIEUWKOOP (Dr Ir J.) Astronomical Institute, Servaas Bolwerk 13, Utrecht, The Netherlands

*Van PARADIJS (Dr J.A.) University of Amsterdam, Roetersstraat 15, Amsterdam 1004, The Netherlands

Van Regemorter (Dr H.) Observatoire de Paris, Section d'Astrophysique, 92190 Meudon, France

VAN RENSBERGEN (Dr W.) Dorpsstraat, 36, B-2958 Weerde, Belgium

Van t'VEER (Dr C.) Institut d'Astrophysique, 98 bis Boulevard Arago, 75014 Paris, France

Van t'VEER (Dr F.) Institut d'Astrophysique, 98 bis Boulevard Arago, 75014 Paris, France

VAN WOERDEN (Dr H.) Kapteyn Astronomical Laboratory, Postbus 800, Groningen, The Netherlands

VANÝSEK (Prof. V.) Department of Astronomy and Astrophysics, Charles University, Švédská 8, 15000 Prague 5 - Smíchov, Czechoslovakia

*VAPILLON (Dr L.) DASOP, Observatoire de Meudon, 92190 Meudon, France

VARDANIAN (Dr R.A.) Byurakan Astrophysical Observatory, Byurakan, Armenia, U.S.S.R.

VARDYA (Prof. M.S.) Astrophysics Division, Tata Institute of Fundamental Research, Homi Bhabha Road, Bombay 400 005, India

VARSAVSKY (Dr C.M.) Las Heras 1975 (6°A), Buenos Aires, Argentina

*VARSHALOVICH (Dr D.A.) A.F. Ioffe Physico-technical Institute, USSR Academy of Sciences, Politechnicheskaya, 26, Leningrad, 194021, U.S.S.R.

*VASHKOV'YAK (Dr S.N.) Sternberg Astronomical Institute, Moscow 117234, U.S.S.R.

VASIL'EV (Dr V.M.) Main Astronomical Observatory, USSR Academy of Sciences, Pulkovo, 196140 Leningrad, U.S.S.R.

VASIL'EVA (Dr G.J.) Main Astronomical Observatory, USSR Academy of Sciences, Pulkovo, 196140 Leningrad U.S.S.R.

VASILEVSKIS (Dr S.) Lick Observatory, University of California, Santa Cruz, California 95064, U.S.A.

VAUCLAIR (Dr G.) Observatoire de Paris, Section d'Astrophysique de Meudon, 92190 Meudon, France

*VAUCLAIR (Dr S.) DAPHE, Observatoire de Meudon, 92190 Meudon, France

VAUGHAN (Dr A.H. Jr) Mount Wilson and Palomar Observatory, 813 Santa Barbara Street, Pasadena, California 91106, U.S.A.

VEISMANN (Dr U.K.) Tartu Astrophysical Observatory, Tôravere, 202444 Estonia, U.S.S.R.

VELGHE (Prof. Dr A.G.) Ringlaan 3, 1180 Brussels, Belgium

VELTMAN (Dr Yu.K.), Institute for Astrophysics and Physics of the Atmosphere, Toravere, 202444 Tartu, U.S.S.R.

VENUGOPAL (Dr V.R.) Radio Astronomy Centre, T.I.F.R., P. Box 8, Ootacamund, Tamil Nadu, India

*VERDET (Dr J.P.) DANOF, Observatoire de Paris, 61, Avenue de l'Observatoire, 75014 Paris, France

VERGNANO (Prof. A.) Osservatorio Astronomico, Pino Torinese, Italy

VERNIANI (Prof. F.) Instituto di Fisica, Via Irnerio, 46, Bologna 40126, Italy

*VERON (Dr M.P.) DOPTO, Observatoire de Meudon, 92190 Meudon, France

VERON (Dr P.) Observatoire de Meudon, 92190 Meudon, France

VERSCHUUR (Dr G.L.) Fiske Planetarium, Department of Astro-geophysics, University of Colorado, Boulder, Colorado 80302, U.S.A.

VESECKY (Dr J.F.) Center for Radar Astronomy, Durand Building, Stanford University, Stanford, California 94305, U.S.A.

VETEŠNÍK (Dr M.) Astronomical Institute, Purkyně University, Kotlářská 2, Brno, Czechoslovakia

VEVERKA (Dr J.F.) Center for Radiophysics and Space Research, Cornell University, Ithaca, New York 14850, U.S.A.

VICENTE (Prof. Dr R.O.) R. Mestre Aviz, 30, R/c, Lisboa 3, Portugal

VIDAL (Dr J.M.C.) Avenida José Antonio 677, Barcelona, Spain

*VIDAL (Dr N.V.) Dept. of Physics and Astronomy, Tel-Aviv University, Ramat-Aviv, Israel

VIDAL-MADJAR (A.) Laboratoire de Physique Stellaire et Planétaire, B.P. No 10, 91370 Verrières-le-Buisson, France

*VIGIER (Dr J.P.) Institut H. Poincaré, 11, rue Pierre et Marie Curie, 75005 Paris, France

*VIGOTTI (M.) Laboratorio di Radioastronomia, Via Irnerio 46, 40126 Bologna, Italy

VILA (Dr S.C.) Department of Astronomy, University of Pennsylvania, Philadelphia, Pa. 19104, U.S.A.

VILHU (Dr O.) Observatory and Astrophysics Laboratory, University of Helsinki, Tähtitornimäki, SF - 00130 Helsinki 13, Finland

VINTI (Dr P.J.) Rm. 37-340, M.I.T. Measurement Systems Laboratory, 70 Vassar Street, Cambridge, Massachusetts 02139, U.S.A.

VIOTTI (Dr R.) Laboratorio di Astrofisica Spaziale, C.P. 67, 00044 Frascati, Italy

VIRGOPIA (Dr N.) Osservatorio Astronomico, Via Trionfale 204, 00136 Rome, Italy

VISVANATHAN (Dr N.) Hale Observatories, 813 Santa Barbara Str. Pasadena, Calif. 91101, U.S.A.

VITSINKIJ (Dr Yu.I.) Main Astronomical Observatory, USSR Academy of Sciences, Pulkovo, 196140 Leningrad, U.S.S.R.

*VITON (Dr M.) LAS, Traverse du Siphon, Les Trois Lucs, 13004 Marseille, France

VLACHOS (Dr D.G.) Laboratory of Astronomy, Technical University of Athens, 5 K. Zografou Street, Zografou, Athens (624), Greece

*VOGT (Dr N.) European Southern Observatory, Casilla 16317, Santiago 9, Chile

VOIGT (Prof. Dr H.H.) Direktor der Universitäts-Sternwarte, Geismarlandstr. 11, 34 Göttingen, Germany F.R.

*VOLK (Dr H.J.) Max-Planck-Institut für Kernphysik, 69 Heidelberg, P.O. Box 103980, Germany F.R.

VOLLAND (Dr H.) Astronomisches Institut der Universität, Poppelsdorfer Allee 49, 53 Bonn, Germany F.R.

Von der HEIDE (Dr J.E.B.) Hamburger Sternwarte, Hamburg-Bergedorf, Germany F.R.

Von HOERNER (Dr S.) National Radio Astronomy Observatory, P.O. Box 2, Green Bank, West Virginia, U.S.A.

VON WEIZSÄCKER (Prof. C.F.) Max-Planck-Institut, Riemerschmidstrasse 7, 813 Starnberg, Germany F.R.

VORONTSOV-VEL' YAMINOV (Prof. Dr B.A.) Sternberg Astronomical Institute, Moscow V-234, U.S.S.R.

VOROSHILOV (Dr V.I.) Main Astronomical Observatory, Ukrainian Academy of Sciences, Kiev, U.S.S.R.

*VORPAHL (Dr J.A.) The Aerospace Corporation, P.O. Box 92957, Los Angeles, CA 90009, U.S.A.

*VREUX (Dr J.M.G.J.C.) 28, rue Lambert Lepage, 4134 Hermalle sous Huy, Belgium

VSEKHSVYATSKIJ (Prof. Dr S.K.) Astronomical Observatory, Observatornaja 3, 252053 Kiev, U.S.S.R.

VUJNOVIĆ (Dr V.) Institute of Physics of the University, P.Box 304, 41001 Zagreb, Yugoslavia

VUKICEVIĆ-KARABIN (Dr M.) Dept. of Astronomy, Faculty of Sciences, Belgrade University, Studentski trg 16, P.O. Box 550, 11001 Beograd, Yugoslavia

WACHMANN (Prof. Dr A.A.) Augustastr. 2., 205 Hamburg 80, Germany F.R.

WACKERNAGEL (Dr H.B.) 2939 Country Club Drive, Colorado Springs, Colorado 80909, U.S.A.

WADDINGTON (Dr C.J.) University of Minnesota, Minneapolis, Minn.55455, U.S.A.

WADE (Dr C.M.) National Radio Astronomy Observatory, P.O. Box 2, Green Bank, West Virginia, U.S.A.

WAGMAN (Dr N.E.) Allegheny Observatory, Riverview Park, Pittsburgh, Pennsylvania 15214

*WAGNER (Dr R.L.) Louisiana State University, Dept. of Physics and Astronomy, Baton Rouge, LA 70803, U.S.A.

WAGNER (Dr W.J.) Sacramento Peak Observatory, Sunspot, New Mexico 88349, U.S.A.

*WAGONER (Dr R.V.) Department of Physics, Stanford University, Stanford, CA 94305, U.S.A.

WAKO (Dr Y.) International Latitude Observatory, Mizusawa, Iwate, Japan

WALBORN (Dr N.R.) Cerro Tololo Inter-American Observatory, Casilla 63-D, La Serena, Chile

WALDMEIER (Prof. Dr M.) Director Swiss Federal Observatory, Schmelzbergstrasse 25, 8006 Zürich, Switzerland

WALKER (Prof. A.B.C. Jr) Institute for Plasma Research, Stanford University, Via Crespi, Stanford, California 94305, U.S.A.

WALKER (Dr G.A.H.) Dept. of Geophysics and Astronomy, University of British Columbia, Vancouver, B.C. V6T 1W5, Canada

WALKER (Dr M.F.) Lick Observatory, University of California, Santa Cruz, CA 95060 U.S.A.

WALKER (Dr R.L.) U.S. Naval Observatory, Flagstaff Station, Flagstaff, Arizona 86001, U.S.A.

WALKER (Dr R.M.) Washington University, St. Louis, Missouri 63130, U.S.A.

WALKER (W.S.G.) Variable Star Section, RAS of New Zealand, 18 Pooles Road, Greerton, Taurangan, New Zealand

*WALL (Dr J.V.) Radio Astronomy Group, Cavendish Laboratory, Madingley Road, Cambridge CB3 OHE, U.K.

WALLACE (Dr L.V.) Kitt Peak National Observatory, 950 North Cherry Avenue, Tucson, Arizona, U.S.A.

WALLENQUIST (Prof. Dr Å.A.E., Norrlandsgatan 34D, S-752 29 Uppsala, Sweden

WALLERSTEIN (Dr G.N.) Dept. of Astronomy , University of Washington,Seattle,WA98195
U.S.A.

WALLIS (Dr M.K.) Mathematical Institute, 24 St Giles, Oxford, U.K.

*WALMSLEY (Dr M.C.) Max-Planck-Institut für Radioastronomie
Auf dem Hügel 69, 53 Bonn 1, Germany, F.R.

WALRAVEN (Dr Th.) Leiden Southern Station, P.O. Box 13, Broederstroom,
Transvaal 0240, South Africa

WALSH (Prof.D.) Nuffield Radio Astronomy, Laboratories, Jodrell Bank,
Macclesfield, U.K.

WALTER (Dr H.G.) Astronomisches Rechen-Institut, Mönchhofstr. 12-14,
D-6900 Heidelberg 1, Germany, F.R.

WALTER (Prof. Dr K.) Astronomisches Institut der Universität,
Waldhauserstr. 64, 74 Tübingen, Germany, F.R.

WAMPLER (Dr E.J.) Lick Observatory, Mount Hamilton, California 95140, U.S.A.

*WAMSTEKER (Dr W.) European Southern Observatory, Bergedorfer Str. 131,
205 Hamburg 80, Germany, F.R.

*WANAS (Dr M.I.) Faculty of Science, Cairo University, Geza Orman,
Arab Republic of Egypt

WAPSTRA (Prof. A.) Van Woensel Kooylaan 48, Naarden, The Netherlands

*WARDLE (Dr J.F.) Dept. of Physics, Brandeis University, Waltham, MA 02154
U.S.A.

WARES (Dr G.W.) CRF Astrophysics, Space Physics Laboratory, Air Force
Cambridge Research Laboratory, Laurence G. Hanscom Field, Bedford,
Massachusetts 01730, U.S.A.

WARMAN (Dr J.) Instituto de Astronomia, UNAM, Apartado Postal 70- 264,
México D.F., Mexico

WARNER (Prof. B.) Dept. of Astronomy, University of Cape Town,
Rondebosch 7700, South Africa

WARWICK (Dr J.W.) High Altitude Observatory of the University of Colorado
Boulder, Colorado, U.S.A.

WARZEE (Dr J.) Astronome Honoraire, Observatoire Royal de Belgique,
rue Mattot, 115, B-1410 Waterloo, Belgium

*WASHIMI (Dr Haruichi) The Research Institute of Atmospherics, Nagoya University,
Toyokawa, Aichi, 442, Japan

*WASSON (Dr J.T.) University of California, Los Angeles, Institute of
Geophysics and Planetary Physics, Los Angeles, CA 90024, U.S.A.

WATERFIELD (Dr R.L.) The Observatory, Woolston, Yeovil, Somerset, U.K.

WATERWORTH (Dr M.D.) University of Tasmania, Dept. of Physics, Box 252C.,
G.P.O. Hobart, Tasmania 7001, Australia

WATSON (Dr W.) Dept. of Physics and Dept. of Astronomy, University of
Illinois, Urbana, Illinois 61801, U.S.A.

WATTENBERG (Prof. D.) Direktor der Archenhold-Sternwarte, Berlin-Treptow,
Germany, D.R.

WAYMAN (Prof. P.A.) Director, Dunsink Observatory, Castleknock,
    Co. Dublin, Ireland

WEAVER (Dr H.F.) Radio Astronomy Laboratory, University of California
    Berkeley 4, California, U.S.A.

WEBBER (Dr J.) University of Illinois, Urbana, Illinois 61801, U.S.A.

WEBROVA (Dr L.) Astronomical Institute, Czechoslovak Academy of Sciences,
    Budečská 6, Prague 2, Czechoslovakia

*WEBSTER (Dr A.S.) 14 Clare Hall, Herschel Road, Cambridge, U.K.

WEBSTER (Dr B.L.) Radcliffe Observatory, P.O. Box 373, Pretoria, South Africa

WEEDMAN (Dr D.) Dyer Observatory, Vanderbilt University, Box 1803 Station B,
    Nashville, Tenn. 37203, U.S.A.

WEEKES (Dr T.C.) Mount Hopkins Observatory, Box 97, Amado, Arizona 85640, U.S.A.

WEHINGER (Dr P.A.) Royal Greenwich Observatory, Herstmonceux Castle,
    Hailsham, East Sussex BN27 1RP, U.K.

WEHLAU (Dr A.) Dept. of Astronomy, University of Western Ontario
    London, Ontario N6A 3K7, Canada

WEHLAU (Dr W.H.) Dept. of Astronomy, University of Western Ontario,
    London, Ontario N6A 3K7, Canada

*WEHRSE (Dr R.) Lehrstuhl für Theoretische Astrophysik, Im Neuenheimer Feld 294,
    D-69 Heidelberg, Germany, F.R.

WEIDEMANN (Prof. Dr V.) Institut für Theoretische Physik und Sternwarte
    der Universität, Olshausenstrasse, Neue Universität, Haus C 4/1,
    23 Kiel, Germany, F.R.

WEIGERT (Dr A.) Director, Hamburg Observatory, 205 Hamburg 80, Germany, F.R.

WEILER (Dr K.W.) Kapteyn Astronomy Laboratory, University of Groningen,
    P.O. Box 800, Groningen, The Netherlands

WEILL (G.) Institut d'Astrophysique, 98bis, boulevard Arago,
    75014 Paris

WEIMER (Dr Th.) Observatoire de Paris, 61, avenue de l'Observatoire,
    Paris 14ème, France

WEINBERG (Dr J.L.) Space Astronomy Laboratory, Executive Park East,
    Stuyvesant Plaza, Albany, New York 12203, U.S.A.

*WEISHEIT (Dr J.C.) Geraldine Street 781, Livermore, CA 94550, U.S.A.

WEISS (Dr N.O.) Department of Applied Mathematics and Theoretical Physics,
    Silver Street, Cambridge CB3 9EW, U.K.

WEISS (Dr W.W.) Universitätssternwarte Wien, Figl -Observatorium für
    Astrophysik, Türkenschanzstrasse 17, A-1180 Wien, Austria

*WEISTROP (Dr D.) Kitt Peak National Observatory, P.O. Box 26732,
    Tucson, AZ 85726, U.S.A.

WELCH (Dr W.J.) Radio Astronomy Laboratory, University of California,
    Berkeley, California 94720, U.S.A.

WELIACHEW (Dr L.) Observatoire de Meudon, 92190 Meudon, France

* WELLER (Dr C.S., Jr.) Naval Research Laboratory, Space Science Division,
    Washington, DC 20375, U.S.A.

WELLGATE (G.B.) Royal Greenwich Observatory, Herstmonceux Castle, Hailsham,
    Sussex, U.K.

WELLMANN (Prof.P.) Universitäts-Sternwarte, Sternwartstrasse 23,
    München   27, Germany, F.R.

WEMPE (Prof. Dr J.) Zentralinstitut für Astrophysik, DDR 15 Potsdam,
    Telegrafenberg, Germany, D.R.

WENDKER (Dr H.J.) Hamburger Sternwarte, Gojenbergsweg 112,
    205 Hamburg 80, Germany, F.R.

WENIGER (Dr S.) Observatoire de Paris, Section d'Astrophysique de Meudon,
    92190 Meudon, France

WENTZEL (Dr D.G.) Astronomy Program, University of Maryland, College Park,
    Maryland 20740, U.S.A.

WENZEL (Dr W.) Zentralinstitut für Astrophysik, Sternwarte Sonneberg,
    DDR 64 Sonneberg, Germany D.R.

WESSELINK (Dr A.J.) Yale University Observatory, Box 2023 Yale Station,
    New Haven, Connecticut 06520, U.S.A.

WESSELIUS (Dr P.R.) Dept. of Space Research, University of Groningen,
    P.O. Box 800, Groningen, The Netherlands

WEST(Dr R.M.) ESO Sky Atlas Laboratory, ESO TP Division, CERN, Ch-1211 Genève 23
    Switzerland

WESTERHOUT (DrG) Astronomy Program, University of Maryland, College Park,
    Maryland 20740, U.S.A.

WESTERLUND (Prof.B.E.) Astronomical Observatory
    S 751 20 Uppsala, Sweden

WESTFOLD (Prof. K.C.) Monash University, Clayton 3168, Australia

WETHERILL (Dr G.W.) University of California LA, Los Angeles, CA, U.S.A.

WEYMANN (Prof. R.J.) Steward Observatory, University of Arizona,
    Tucson, Arizona, U.S.A.

WHEELER (Dr J.A.) Joseph Henry Laboratory, Princeton University,
    Princeton, New Jersey 08540, U.S.A.

WHEELER (Dr J.C.) Dept. of Astronomy, University of Texas, Austin,Texas 78712,
    U.S.A.

WHELAN (Dr J.A.J.) Institute of Astronomy, University of Cambridge,
    Cambridge, U.K.

WHIPPLE (Dr F.L.) Director, Smithsonian Astrophysical Observatory,
    60 Garden Street, Cambridge, Massachusetts 02138, U.S.A.

WHITAKER (E.A.) Lunar and Planetary Laboratory, University of Arizona,
    Tucson, Arizona, U.S.A.

WHITE (Dr O.R.) High Altitude Observatory, P.O. Box 1558, Boulder,
    Colorado 80302, U.S.A.

WHITE (Dr R.E.) Steward Observatory, the University of Arizona,
Tucson, Arizona 85721, U.S.A.

*WHITE (Dr R.S.) Institute of Geophysics and Planetary Physics, University
of California, Riverside, CA 92502, U.S.A.

WHITEOAK (Dr J.B. CSIRO Division of Radiophysics, Post Office Box 76
Epping, N.S.W., 2121, Australia

WHITFORD (Dr A.E.) Director of the Lick Observatory, University of California,
Santa Cruz, California 95060, U.S.A.

WHITNEY (Prof. B.S.) University of Oklahoma, Observatory, Norman,
Oklahoma, U.S.A.

WHITNEY (Dr C.A.) Smithsonian Astrophysical Observatory, 60 Garden Street,
Cambridge, Mass. 02138, U.S.A.

WHITROW (Dr G.J.) Dept. of Mathematics, Imperial College of Science and
Technology, Prince Consort Road, London, S.W. 7, U.K.

*WICKRAMASINGHE (Dr D.T.) Institute of Astronomy, Madingley Road,
Cambridge CB3 OHA, U.K.

WICKRAMASINGHE (Prof. N.C., Sc.D.) University College, Dept. of Applied
Mathematics and Mathematical Physics, P.O.Box 78, Cardiff CF1 XL, U.K.

WIDING (Dr K.D.) Code 7144, U.S. Naval Research Laboratory,
Washington D.C. 20390, U.S.A.

WIDORN (Dr T.) Universitäts-Sternwarte Wien, Türkenschanzstr. 17,
A- 1180 Vienna, Austria

WIEDLING (Dr T.) Villavägen 15,, S-611 00 Nyköping, Sweden

WIEHR (Dr E.) Universitäts-Sternwarte, Geismarlandstr. 11, 3400 Göttingen,
Germany F.R.

WIELEBINSKI (Dr R.) Max-Planck-Institut für Radioastronomie,
Auf dem Hügel 69, 53 Bonn 1, Germany F.R.

WIELEN (Dr R.) Astronomisches Recheninstitut, Mönchhofstr. 12-14,
6900 Heidelberg, Germany F.R.

WIESE (W.L.) Chief, Plasma Spectroscopy Section, National Bureau
of Standards, Physics Building A-149, Washington, D.C. 20234, U.S.A.

WIETH-KNUDSEN (Dr N.P.) Svend Trostsvej 12 III, Copenhagen-V, Denmark

WILCOX (Dr J.M.) Institute for Plasma Research, Stanford University,
Via Crespi, Stanford, California 94305, U.S.A.

WILCOCK (Dr W.L.) Dept. of Physics, University College of North Wales,
Bangor, Caernarvonshire, U.K.

WILD (Dr J.P.) CSIRO Division of Radiophysics, P.O. Box 76,
Epping, N.S.W., 2121, Australia

WILD (P.) Astronomisches Institut der Universität Bern, Sidlerstr.5,
Bern, Switzerland

WILD (Dr P.A.T.) Dept. of Astronomy, University of Cape Town,
Rondebosch 7700, South Africa

WILDEY (Dr R.L.) Center of Astrogeology, U.S. Geological Survey,
601 E.- Cedar Avenue, Flagstaff, Arizona 86001, U.S.A.

WILLIAMS (Dr J.G.)
    Jet Propulsion Laboratory, 4800 Oak Grove Drive, Pasadena
    California 91103, U.S.A.

WILKINS (Dr G.A.) H. M. Nautical Almanac Office, Royal Greenwich Observatory
    Herstmonceux Castle, Hailsham, Sussex, U.K.

*WILLIAMON (Dr R.M.) Fernbank Science Center, 156 Heaton Park Drive, N.E.
    Atlanta, GA 30307, U.S.A.

WILLIAMS (Dr C.A.) Department of Astronomy, University of South Florida
    Tampa, Florida 33620, U.S.A.

WILLIAMS (Dr D.) Radio-Astronomy Laboratory, University of California
    Berkeley, California, U.S.A.

WILLIAMS (Dr D.A.) Department of Mathematics, U.M.I.S.T., Manchester 1, U.K.

WILLIAMS (Dr I.P.) Department of Mathematics, Queen Mary College, Mile End Road,
    London E 1, U.K.

WILLIAMS (Dr J.A.) Physics Department, Albion College, Albion,
    Michigan 49224, U.S.A.

WILLIAMS (Dr P.M.) Royal Observatory, Edinburgh EH 9 3 HJ, U.K.

WILLIAMS (Dr R.E.) Steward Observatory, University of Arizona, Tucson 85721, U.S.A.

*WILLIS (Dr A.G.) Sterrewacht Leiden, Huygens Laboratorium, Wassenaarseweg 78,
    Leiden, The Netherlands

*WILLMORE (Dr A.P.) University of Birmingham, P.O. Box 363, Edgbaston,
    Birmingham, B15 2TT, U.K.

WILLS (Dr B.J.) Department of Astronomy, University of Texas, Austin,
    Texas 78712, U.S.A.

WILLS (Dr D.) Department of Astronomy, University of Texas, Austin,
    Texas 78712, U.S.A.

*WILLSON (Dr L.A.) Department of Physics, Iowa State University,
    12 Physics Building, Ames, IA 50011, U.S.A.

WILLSTROP (Dr R.V.) University Observatories, Madingley Road, Cambridge, U.K.

WILSON (Dr A.G.) P.O. Box 113, Topanga, California, U.S.A.

*WILSON (Dr A.S.) Astronomy Centre, University of Sussex, Physics Building,
    Falmer, Brighton BN1 9QH, U.K.

WILSON (Prof. B.G.) Department of Physics, Simon Fraser University,
    Burnaby, B.C., Canada

WILSON (Dr L.) Department of Environmental Sciences, University of Lancaster,
    Bailrigg, Lancaster, U.K.

WILSON (Dr O.C.) Mount Wilson and Palomar Observatories,
    813 Santa Barbara Street, Pasadena 4, California, U.S.A.

WILSON (Dr P.R.) Dept. of Applied Mathematics, University of Syndey,
    Sydney 2006, N.S.W., Australia

WILSON (Prof. R.) Dept. of Physics & Astronomy, University College London,
    Gower Street, London WC1E 6BT, U.K.

WILSON (Dr R.E.) Department of Astronomy, University of South Florida,
Tampa, Florida 33620, U.S.A.

WILSON (Dr R.W.) Bell Laboratories, Crawford Hill Laboratory, Holmdel,
New Jersey 07733, U.S.A.

WILSON (Dr T.L.) Max-Planck-Institut für Radioastronomie, Auf dem Hügel 69,
53 Bonn 1, Germany F.R.

WILSON (Dr W.J.) The University of Texas at Austin, Department of Electrical
Engineering, Austin, Texas 78712

WINCKLER (Dr J.R.) School of Physics and Astronomy, University of Minnesota,
Minnesota 55455, U.S.A. (Minneapolis)

WING (Dr R.) The Ohio State University, Columbus, Ohio 43210, U.S.A.

WINKLER (Dr G.M.R.) U.S. Naval Observatory, Department of the Navy,
Washington D.C. 20390, U.S.A.

WINNBERG (Dr A.) Max- Planck-Institut für Radioastronomie, Auf dem Hügel 69,
53 Bonn 1, Germany, F.R.

*WINNEWISSER (Dr G.) Max-Planck-Institut für Radioastronomie, Auf dem Hügel 69,
53 Bonn, Germany, F.R.

WISNIEWSKI (Dr W.) The University of Arizona, Lunar and Planetary Laboratory,
Tucson, Arizona, U.S.A.

WITHBROE (Dr G.) Harvard College Observatory, 60 Garden Street,
Cambridge, Massachusetts 02138, U.S.A.

WITT (Dr A.N.) Department of Physics and Astronomy, University of Toledo,
2801 W. Bancroft St. Toledo, Ohio 43606, U.S.A.

*WITTEN (Dr L.) Department of Physics, University of Cincinnati,
Cincinnati, OH 45221, U.S.A.

*WITZEL (Dr A.) Max-Planck-Institut für Radioastronomie, Auf dem Hügel 69
D-53 Bonn, Germany, F.R.

WLERICK (Dr G.) Observatoire de Paris, Section d'Astrophysique,
92190 Meudon, France

*WÖHL (Dr H.) Universitäts-Sternwarte, Geismarlandstrasse 11, D-34 Göttingen,
Germany, F.R.

*WOLFE (Dr A.M.) University of Pittsburgh, Department of Physics,
Pittsburgh, PA 15260, U.S.A.

*WOLFENDALE (Dr A.W.) Physics Department, University of Durham,
South Rd., Durham, U.K.

WOLFF (Dr S.C.) Institute of Astronomy, University of Hawaii, 2525 Correa Road
Honolulu, Hawaii 96822, U.S.A.

WOLSTENCROFT (Dr R.D.) Institute for Astronomy, University of Hawaii,
2840 Kolowalu Street, Honolulu, Hawaii 96825, U.S.A.

WOLTJER (Prof.L.) ESO Telescope Project Division, c/o CERN, 1211 Geneva 23,
Switzerland

WOOD (Dr D.B.) Goddard Space Flight Center, Greenbelt, Maryland, U.S.A.

WOOD (Prof. F.B.) University of Florida, Dept. of Physics and Astronomy
Gainesville, Florida 32601, U.S.A.

WOOD (Dr H.J. III) Dept. of Astronomy, University of Virginia,
Charlottesville, Va. 22903, U.S.A.

WOOD (Dr H.W.) 178 Kenthurst Road, Kenthurst, N.S.W. 2154, Australia

WOOD (Dr J.A.) Smithsonian Astrophysical Observatory, 60 Garden Street,
Cambridge, Massachusetts 02138, U.S.A.

*WOODWARD (Dr P.R.) Sterrewacht, Wassenaarseweg, Leiden, The Netherlands

WOOLF (Dr N.J.) School of Physics and Astronomy, University of Minnesota,
Minneapolis, Minnesota 55455, U.S.A.

WOOLFSON (Dr M.M.) Physics Department, University of York, Heslington,
York YO1 5DD, U.K.

WOOLLEY (Sir Richard) South African Astronomical Observatory, P.O. Box 9
Observatory, Cape 7935, South Africa (as from June 1977: Magnolia
House, Hankham, Sussex, U.K.

WOOLSEY (E.G.) 695 Brierwood, Ottawa, Ontario, Canada

*WOOSLEY (Dr S.E.) Lick Observatory, University of California,
Santa Cruz, CA 95064, U.S.A.

WORLEY (C.E.) U.S. Naval Observatory, Washington D.C. 20390, U.S.A.

WORRALL (Dr G.) 110 North End Bassingbourn, Royston, Herts, U.K.

WOSZCYK (Dr A.) Astronomical Observatory of Torún University, Sienkiewicza 30,
Torún, Poland

WRAY (Dr J.D.) Astronomy Department, Northwestern University, Evanston,
Illinois 60201, U.S.A.

WRIGHT (Dr F.W.) Harvard College Observatory, Cambridge, Massachusetts 02138,
U.S.A.

*WRIGHT (Dr G.A.E.) Institut für Astrophysik und Extraterrestrische
Forschung, Auf dem Hügel 71, 53 Bonn, Germany, F.R.

WRIGHT (Dr H.) 579 Forest Lakes Drive, Forest Lakes, Andover, N.J. 07821, U.S.A.

WRIGHT (Dr J.P.) National Science Foundation, 1800 G. Street, N.W.,
Washington, D.C. 20550, U.S.A.

WRIGHT (Dr K.O.) Dominion Astrophysical Observatory, Herzberg Institute
of Astrophysics, National Research Council of Canada, 5071 W. Saanich Road,
Victoria, B.C. V8X 3X3, Canada

WRIGHT (Dr M.C.H.) Radio Astronomy Laboratory, University of California,
Berkeley, California 94720, U.S.A.

WRIXON (Dr G.T.) Bell Telephone Labs., Crawford Hill Laboratory, Holmdel,
N.J. 07733, U.S.A.

WU (Dr Chi-Chao) University, Kapteyn Laboratorium, Hoogbouw WSN, Postbus 800,
Groningen, The Netherlands

WYATT (Dr S.P., Jr.) University of Illinois Observatory, Urbana, Illinois, U.S.A.

WYCKOFF (Dr S.) Royal Greenwich Observatory, Herstmonceux Castle,
    Hailsham, East Sussex BN27 1RP, U.K.

WYLLER (Prof. A.A.) Stockholm Observatory, S-13300 Saltsjöbaden, Sweden

*WYNN-WILLIAMS (Dr C.G.)Mullard Radio Astronomy Observatory, Cavendish Laboratory,
    Madingley Road, Cambridge CB3 OHE, U.K.

WYNNE (Prof. Ch.G.) Royal Greenwich Observatory, Herstmonceux Castle,
    Hailsham, Sussex, U.K.

XANTHAKIS (Prof.J.) Research and Computing Center, Academy of Sciences
    of Athens, Athens, Greece

YABUSHITA (Dr S.) Department of Applied Mathematics and Physics,
    Kyoto University, Kyoto, Japan

YABUUTI (Dr K.) Research Institute of Humanistic Science, Kyoto University,
    Kyoto, Japan

*YAHIL (Dr A.) Department of Astronomy, California Institute of Technology,
    Pasadeny, California 91125, U.S.A.

YAKHONTOVA (Dr N. Samojlova) Institute of Theoretical Astronomy,
    10, Kutuzov Qay, Leningrad, 192187, U.S.S.R.

YAKOVKIN (Dr N.A.), Kiev State University, Astronomical Observatory and
    Chair of Astronomy, 252053 Kiev, U.S.S.R.

*YALLOP (Dr B.D.) Royal Greenwich Observatory, Herstmonceux Castle,
    Hailsham, East Sussex, BN27 1 RP, U.K.

YAMASHITA (Dr K.) Department of Physics, Nagoya University, Chikusa-ku,
    Nagoya, Japan

YAMASHITA (Dr Yasumasa) Dept. of Astronomy, University of Tokyo, Yayoi,
    Bunkyo-ku, Tokyo, Japan

YAMAZAKI (Dr A.) Hydrographic Department, 1st Regional Maritime Safety
    Headquarters, Otaru, Japan

YAPLEE (B.S.) Code 7134, U.S. Naval Research Laboratory, Washington D.C. 20390,
    U.S.A.

YAROV-YAROVOJ (Dr M.S.), MVTU, B Baumanskaya ul. 5, 107005 Moscow, U.S.S.R.

YASUDA (Dr H.) Tokyo Astronomical Observatory, Mitaka, Tokyo, Japan

YATSKIV (Dr Y.S.) Main Astronomical Observatory of the Ukrainian Academy
    of Sciences, Kiev- 252127, U.S.S.R.

YAVNEL' (Dr A.A.) Meteorite Committee, USSR Academy of Sciences, Moscow, U.S.S.R.

YEIVIN (Prof. Y.) Tel-Aviv University, Ramat Aviv, Israel

YEN (Prof. J.L.) Department of Electrical Engineering, University of Toronto,
    Toronto, Ontario, Canada

*YEOMANS (Dr D.K.) Jet Propulsion Laboratory, 4800 Oak Grove Drive,
    Pasadena, CA 91103, U.S.A.

YILMAZ (Dr Fatma) Observatoire de l'Université, Istanbul, Turkey

YILMAZ (Dr N.) Faculty of Science, Department of Astronomy, Ankara, Turkey

*YOKOYAMA (Dr Koichi) International Latitude Observatory of Mizusawa,
Mizusawa-shi, Iwate-ken, 023 Japan

*YORK (Dr D.G.) Princeton University Observatory, Peyton Hall, Princeton, NJ 08540,
U.S.A.

*YOSHIDA (Prof. Junzo) Kyoto Sangyo University, Kamigamo Motoyama,
Kita-ku, Kyoto-shi, 603 Japan

*YOSHIMURA (Dr Hirokazu) Department of Astronomy, Faculty of Science,
University of Tokyo, Tokyo, 113 Japan

YOSS (Dr K.M.) University of Illinois Observatory, Urbana, Illinois, U.S.A.

*YOUNG (Dr A.) Department of Astronomy, San Diego State University,
San Diego, CA 92182, U.S.A.

YOUNG (Dr A.T.) Department of Physics, Texas A+M University, College Station,
Texas 77843, U.S.A.

*YOUNG (Dr L.G.) Department of Physics, Texas A+M University, College Station,
Texas 77843, U.S.A.

YOUSEF (Dr M.A.M.) Astronomy Department, Faculty of Science, Cairo University,
Cairo, Arab Republic of Egypt

*YOUSSEF (Dr H.N.) Faculty of Science, Cairo University, Geza Orman, Arab Republic
of Egypt

YÜ (Dr Ching-Sung) Williams Observatory, Hood College, Frederick, Maryland, U.S.A.

YU (Prof. Kyung Loh) Dept. of Earth Sciences, Teachers College,
Seoul National University, Seoul, Republic of Korea

YUAN (Dr Chi) Physics Department, City College of New York,
New York, N.Y. 10031, U.S.A.

YUMI (Dr S.) International Latitude Observatory of Mizusawa, Mizusawa-shi,
Iwate-ken, Japan

YUN (Dr H.S.) Department of Astronomy, College of Natural Science,
Seoul National University, Seoul, Republic of Korea

ZABRISKIE (Prof. Dr F.R.) Department of Astronomy, 102 Whitmore Laboratory
Pennsylvania State University, University Park, Pennsylvania 16802, U.S.A.

ZACHAROV (Dr I.) Astronomical Institute, Czechoslovak Academy of Sciences,
Observatory Ondřejov, Czechoslovakia

ZADUNAISKY (Prof. P.E.) Observatorio Nacional de Física Cósmica
Depto. de Matemática Aplicada, Av. Mitre 3.100- San Miguel, F.C.G.S.M.
Pcia. de Buenos Aires, Argentina

ZAHN (Dr J.-P.) Observatoire de Nice, Le Mont-Gros, 06000 Nice, France

ZANDER (Dr R.J.) Institut d'Astrophysique de l'Université de Liège,
B-4200 Cointe-Ougrée, Belgium

*ZASOV (Dr A.V.) Sternberg Astronomical Institute, Moscow, 117234, U.S.S.R.

ZEL'DOVICH (Acad.Ya.B.) Department of Theoretical Astrophysics, Space Research Institute, USSR Academy of Sciences, Profsoyuznaja 88, Moscow 117485, U.S.S.R.

*ZELENKA (Dr A.) Swiss Federal Observatory, Schmelzbergstr.25, CH-8006 Zürich, Switzerland

*ZELLNER (Dr B.H.) Lunar and Planetary Laboratory, The University of Arizona, Tucson, AZ 85721, U.S.A.

ZEL'MANOV (Dr A.L.) Sternberg Astronomical Institute, Moscow, U.S.S.R.

ZHELEZNIAKOV (Prof. Dr V.V.) Radiophysical Research Institute, Gorkii, U.S.S.R.

ZHEVAKIN (Prof. Dr S.A.) Radiophysical Research Institute, Gorkii, U.S.S.R.

ZHONGOLOVICH (Prof. Dr I.D.) Institute of Theoretical Astronomy, Leningrad, U.S.S.R.

ZIEBA (Prof.A.) Astronomical Observatory, Jagellonian Observatory, Cracow 31-501 Krakow, Kopernika 27, Poland

*ZIEBA (Dr S.) Obserwatorium Astronomiczne, Uniwersytetu Jagiellońskiego 30-244 Kraków, ul. Orla 171, Poland

ZIKIDES (Dr M.C.) Laboratory of Astronomy, University of Athens, Panepistimiopolis, Athens (621), Greece

ZIMMERMANN (Dr H.) University Observatory, Jena, Schillergässchen 2, Germany, D.R.

ZIOŁKOWSKI (Dr J.) Polish Academy of Sciences, Institute of Astronomy, Al. Ujazdowskie 4, Warsaw, Poland

ZIRIN (Dr H.) Hale Observatories, California Institute of Technology, 1201 E. California Blvd., Pasadena, California 91109, U.S.A.

ZIRKER (Dr J.B.) Institute for Astronomy, 2840 Koluwalu Street, University of Hawaii, Honolulu, Hawaii 96822, U.S.A.

ZUCKERMANN (Dr B.M.) Department of Physics and Astronomy, University of Maryland, College Park, Maryland, U.S.A.

ZVEREV (Prof. Dr M.S.) Pulkovo Observatory, Leningrad- M140, U.S.S.R.

ZWAAN (Dr C.) Astronomical Observatory, "Sonnenborgh", Zonnenburg 2, Utrecht, The Netherlands

REPORT OF THE EXECUTIVE COMMITTEE
1973 - 1975

Reproduced from IAU Information Bulletin No. 36 of June 1976 pp. 8.

EXECUTIVE COMMITTEE, 1973 - 1976

PRESIDENT

Professor L. Goldberg, Kitt Peak National Observatory, P.O. Box 26732, Tucson, Arizona 85762, U.S.A.

VICE-PRESIDENTS

Professor B.J. Bok, Steward Observatory, University of Arizona, Tucson, Arizona 85721, U.S.A. (resigned in August 1974).

Mr J.G. Bolton, C.S.I.R.O., Division of Radiophysics, P.O. Box 76, Epping, N.S.W. 2121, Australia.

Professor Ch. Fehrenbach, Observatoire de Haute Provence, Saint Michel l'Observatoire, 04300 Forcalquier, France.

Professor W. Iwanowska, Institute of Astronomy, Nicolaus Copernicus University, ul. Chopina 12/18, P1-87-100 Toruń, Poland.

Dr P.O. Lindblad, Stockholm Observatory, S-133 00 Saltsjöbaden, Sweden (succeeded Professor B.J. Bok in August 1974).

Professor Sir B. Lovell, Nuffield Radio Astronomy Laboratories, Jodrell Bank, Macclesfield, Cheshire SK11 9DN, U.K.

Professor E.R. Mustel, Astronomical Council, USSR Academy of Sciences, Pyatnitskaya ul. D.48, 109017 Moscow, Zh-17. U.S.S.R.

GENERAL SECRETARY

Professor G. Contopoulos, Astronomy Department, Panepistimiopolis, Athens 621, Greece.

ASSISTANT GENERAL SECRETARY

Professor Edith A. Müller, Observatoire de Genève, CH 1290 Sauverny, Switzerland.

INTRODUCTION

This report covers the period from 1 January 1973 to 31 December 1975. The period under report includes the first eight months of the year 1973 (until 21 August) which fell under the responsibility of the former Executive Committee, but it does not include the first eight months of the year 1976, still within the responsibility of the present Executive Committee. An oral report for the period from 1 January to 23 August 1976 will be presented at the XVIth General Assembly.

ADMINISTRATIVE MATTERS

Executive Committee

The Executive Committee, in its former composition, held its 37th meeting in Sidney during the XVth General Assembly, having had sessions on 19, 20 and 28 August 1973. The meeting was chaired by the President of the Union Prof. B. Strömgren. All members and both advisers were present.

The 38th meeting of the Executive Committee, in its present composition, was held in Sidney on 30 August 1973, under the presidency of Prof. L. Goldberg. All members and both advisers attended the meeting. Further, the Executive Committee held the following meetings:

The 39th meeting at Observatoire de Haute-Provence, Saint Michel l'Observatoire, Forcalquier, France, from 27 to 29 August 1974. Dr P.O. Lindblad, Vice-President, and Prof. C. de Jager, adviser, were excused.

The 40th meeting in Hotel Lagonissi, near Athens, Greece, from 9 to 11 September 1975. Prof. B. Strömgren, adviser, was excused.

Officers' meetings

The President of the IAU, the General Secretary accompanied by the Staff, and the Assistant General Secretary (Officers), met as follows: in Thessaloniki, Greece, on 2 and 3 November 1973; in Athens, Greece, on 4 and 5 February 1974; in Paris, France, on 8 May 1974; in Haute Provence, France, on 26 August 1974; in Geneva, Switzerland, on 10 and 11 February 1975; and on Lagonissi, Greece, on 8 September 1975. The President and the General Secretary met in Chicago, U.S.A., on 28 and 29 May 1975.

Adhering Countries

The Republic of Korea has adhered to the IAU since 1973 thus bringing the number of Adhering Countries to 47.

Members of the IAU

The number of IAU Members was 2509 as on 1 February 1973. A further 689 new Members were admitted at the XVth General Assembly, and 13 Members were deleted from the membership list. The General Secretary was informed of the decease of 58 Members during the period under report. Thus, the IAU had 3127 Members as on 1 April 1976.

Commissions of the IAU

The Transactions of the IAU, volume XVIA, subtitled Reports on Astronomy 1976, include the reports prepared by each of the 40 Presidents of IAU Commissions, assisted by their Vice-Presidents, Organizing Committees, those responsible for the various Working Groups within the Commissions, and by the Commission members themselves.

The XVth General Assembly formed Commissions No. 49 "The Interplanetary Plasma and the Heliosphere" and No. 50 "Protection of Existing and Potential Observatory Sites", the later as a Committee of the Executive Committee. IAU Commission No. 43 "Astrophysical Plasmas and Magnetohydrodynamics" was abolished by the General Assembly. This brought the number of IAU Commissions to 40.

The General Assembly renamed IAU Commission No. 15 to "Physics of Comets, Minor Planets and Meteorites", IAU Commission No. 22 to "Meteors and Interplanetary Dust", IAU Commission No. 25 to "Stellar Photometry and Polarimetry", and IAU Commission No. 42 to "Close Binary Stars".

In view of the untimely decease of Professor K.O. Kiepenheuer, Dr G.A. Newkirk was appointed President ad Interim of IAU Commission No. 10 "Solar Activity", and Dr V. Bumba its Vice-President ad Interim.

Between the General Assemblies, Commissions have co-opted a number of new members from among IAU Members. These co-options have been approved by the Executive Committee and published in the Information Bulletin.

The Executive Committee created, in 1973, a Working Group for Planetary System Nomenclature with the following Task Groups: Lunar Nomenclature, Mercury Nomenclature, Venus Nomenclature, Mars Nomenclature, and Outer Solar System Nomenclature. The composition of the Working Group and of its Task Groups has been published in IAU Information Bulletin No. 33.

Between the General Assemblies, various Commissions organized several Symposia and Colloquia and other projects financially supported by the Union.

SYMPOSIA

IAU Symposia (held between 1 January 1973 and 31 December 1975)

IAU Symposium No. 56 "Chromospheric Fine Structure", Surfers' Paradise, Queensland, Australia, 3-7 September 1973

IAU Symposium No. 57 "Coronal Disturbances", Surfers' Paradise, Queensland, Australia, 7-11 September 1973

IAU Symposium No. 58 "The Formation and Dynamics of Galaxies", Canberra, Australia, 12-15 August 1973

IAU Symposium No. 59 "Stellar Instability and Evolution", Mount Stromlo, Canberra, Australia, 16-18 August 1973

IAU Symposium No. 60 "Galactic Radio Astronomy", Maroochydore, Queensland, Australia, 3-7 September 1973

IAU Symposium No. 61 "New Problems in Astrometry", Perth, Western Australia, 13-17 August 1973

IAU Symposium No. 62 "The Stability of the Solar System and of Small Stellar Systems", Warsaw, Poland, 5-8 September 1973

IAU Symposium No. 63 "Confrontation of Cosmological Theories with Observational Data", Cracow, Poland, 10-12 September 1973

IAU Symposium No. 64 "Gravitational Radiation and Gravitational Collapse", Warsaw, Poland, 5-8 September 1973

IAU Symposium No. 65 "Exploration of the Planetary System", Torún, Poland, 5-8 September 1973

IAU Symposium No. 66 "Late Stages of Stellar Evolution", Warsaw, Poland, 10-12 September 1973

IAU Symposium No. 67 "Variable Stars and Stellar Evolution", Moscow, USSR, 29 July-4 August 1974

IAU Symposium No. 68 co-sponsored by COSPAR "Solar Gamma-, X-, and EUV-Radiation", Buenos Aires, Argentina, 11-14 June 1974

IAU Symposium No. 69 "Dynamics of Stellar Systems", Besançon, France, 9-13 September 1974

IAU Symposium No. 70 "Be and Shell Stars", Bass River, Massachusetts, USA, 15-18 September 1975

IAU Symposium No. 71 "Basic Mechanisms of Solar Activity", Prague, Czechoslovakia, 24-30 August 1975

IAU Symposium No. 72 "Abundance Effects in Classification", Lausanne, Switzerland, 8-11 July 1975

IAU Symposium No. 73 "Structure and Evolution of Close Binaries", Cambridge, U.K., 29 July-1 August 1975

Symposia with IAU Participation

IAU/IUHPS Third Joint Symposium "The Astronomy of Copernicus and its Background",

Torún, Poland, 7 and 8 September 1973

IAU/IUGG/IUPAP/URSI/COSPAR/SCOSTEP International Symposium "Solar-Terrestrial Physics", Sao Paulo, Brazil, 17-22 June 1974

COSPAR/IUTAM/UAL Symposium "La Dynamique des Satellites", Sao Paulo, Brésil 19-21 juin 1974

COSPAR/IAU Symposium "Transient X- and Gamma-Ray Sources", Varna, Bulgaria, 29-31 May 1975

IAU/IUHPS Fourth Joint Symposium on the History of Astronomy "The Origins and Achievements of the Royal Greenwich Observatory 1675-1975", Greenwich, U.K., 14-18 July 1975

SCOSTEP/COSPAR/IAU/IUGG Workshop on "Flare Build-up Study", Falmouth, Massachusetts, U.S.A., 8-11 September 1975

COLLOQUIA

IAU Colloquia (held between 1 January 1973 and 31 December 1975)

IAU Colloquium No. 24 co-sponsored by COSPAR, "Lunar Dynamics and Observational Coordinate Systems", Houston, Texas, U.S.A., 15-17 January 1973

IAU Colloquium No. 25 co-sponsored by COSPAR, "Study of Comets", Greenbelt, Maryland, 28 October-1 November 1974

IAU Colloquium No. 26 co-sponsored by COSPAR, "Reference Coordinate Systems for Earth Dynamics", Torún, Poland, 26-31 August 1974

IAU Colloquium No. 27 "UV- and X-Ray Spectroscopy of Astrophysical and Laboratory Plasmas", Cambridge, Massachusetts, U.S.A., 9-11 September 1974

IAU Colloquium No. 28 co-sponsored by COSPAR, "Planetary Satellites", Ithaca, N.Y., U.S.A., 18-21 August 1974

IAU Colloquium No. 29 "Multiple Periodic Variable Stars", Budapest, Hungary, 1-5 September 1975

IAU Colloquium No. 30 "Jupiter", Tucson, Arizona, U.S.A., 18-23 May 1975

IAU Colloquium No. 31 co-sponsored by COSPAR, "Interplanetary Dust and Zodiacal Light", Heidelberg, F.R.G., 10-13 June 1975

IAU Colloquium No. 32 "Physics of Ap Stars", Vienna, Austria, 8-10 September 1975

IAU Colloquium No. 33 "Observational Parameters and Dynamical Evolution of Multiple Stars", Oaxtepec, Mexico, 13-16 October 1975

Specific Projects of Commissions

The list of Specific Projects of Commissions as approved by the XVth General Assembly is given, together with the corresponding expenditure, in the report on the financial situation of the IAU (page 19). These projects are continuations of previous projects and thus no further comments are necessary.

Regional Astronomical Meetings held under the auspices of the IAU

Second European Astronomical Meeting under the Auspices of the IAU, Trieste, Italy, 2-5 September 1974.

Third European Astronomical Meeting under the Auspices of the IAU, "Stars and Galaxies from Observational Points of Views", Tbilisi, Georgia, U.S.S.R., 1-5 July 1975.

Young Astronomers Schools

Commission No. 46 "Teaching of Astronomy" organized two schools for young astronomers in the period under report.

The school in San Miguel, Argentina, held from 11 May to 7 June 1974, was attended by 60 students, of which 39 were local, 9 from Brazil, 7 from Chile, 2 from Peru, and 1 each from Bolivia, Paraguay and Venezuela. Lectures were given by 20 professors. The curriculum, divided in three courses, included solar plasma physics, the Sun and its environment, and solar energetics.

The school, held in Athens, Greece, from 8 September to 4 October 1975, was attended by 74 students from 16 countries (Bulgaria, Czechoslovakia, Egypt, France, G.F.R., Greece, Hungary, Iraq, Italy, Norway, Poland, Portugal, Spain, Turkey, U.K., Yugoslavia). Lectures were given by 9 invited professors and seminars were held by another 7 eminent scientists. The curriculum included theoretical astronomical lectures and practical courses emphasizing research work with medium-size telescopes.

IAU Publications

Free distribution of IAU publications to selected institutions of developing countries had to be discontinued for financial reasons. Instead, 14 such institutions were given the opportunity to purchase IAU Symposium volumes, Transactions and Highlights of Astronomy at 30% of their retail price, the remaining 70% having been paid by the Union.

All IAU publications continue to be produced and distributed for the account and at the risk of the publisher, who pays the Union a 5% royalty.

The production and distribution of the IAU Information Bulletin are financed by the Union.

Sales of IAU Publications

The Transactions of the IAU sold as follows in the period under report:

	Number of copies
Vol. XIIIA	46
Vol. XIIIB	39
Vol. XIVA	60
Vol. XIVB	66
Vol. XVA	739
Vol. XVB	536

The volumes of Highlights of Astronomy were sold as follows in 1973, 1974 and 1975:

	Number of copies
Highlights I	36
Highlights II	122
Highlights III	364

The sales of IAU Symposium volumes over the same period were:

Symposium volume No. (in number of copies)

32-52	42- 70	52-639	62-542
33-29	43- 97	53-643	63-702
34-35	44-148	54-548	64-667
35-43	45-140	55-709	65-492
36-32	46- 91	56-465	66-558
37-45	47-328	57-467	67-383
38-65	48-358	58-556	68-409
39-57	49-602	59-545	69-388
40-63	50-676	60-525	
41-88	51-631	61-563	

Information Bulletin

The Information Bulletin of the IAU was being prepared and edited by the General Secretary and printed, published and distributed by the publishers of the IAU. It appeared twice a year and was sent, free of charge, to IAU Members, Consultants, members of the International Science Writers Association, selected international organizations, and institutions on a special distribution list.

IAU Publication Policy

The Executive Committee has resolved at its meeting on Lagonissi in September 1975 that all IAU publications will, from the summer of 1976 on, be produced by offset (camera-ready copies) which will considerably reduce their prices.

RELATIONS TO OTHER ORGANIZATIONS

IAU Representation

   1. International Council of Scientific Unions (ICSU). The General Secretary attended, together with the former General Secretary Professor C. de Jager, the 2nd meeting of the ICSU General Committee held in Leningrad, U.S.S.R., from 19 to 21 September 1973. He also represented the IAU at the 3rd and 4th meetings of the ICSU General Committee held in Ankara, Turkey, on 20 and 21, and in Istanbul, Turkey, on 27 September 1974, as well as at the 5th meeting of the ICSU General Committee held in Schloss Laxenburg near Vienna, Austria, on 19 and 20 September 1975. The General Secretary represented the Union at the 15 General Assembly of ICSU in Istanbul, Turkey, 23-26 September 1974.

   2. ICSU Organizations. The Union participates in the activity of a number of Special and Scientific Committees, and Inter-Union Commissions, formed and sponsored by ICSU. For more details please consult pages 13 onwards of Transactions volume XIIC.

   The following list only gives the present state of IAU representation in these organizations.
   (a) Committee on Space Research (COSPAR): C. de Jager, President of COSPAR
   (b) ICSU Abstracting Board (IAB): J.-C. Pecker
   (c) Federation of Astronomical and Geophysical Services (FAGS): H. Enslin, G. Contopoulos
   (d) Inter-Union Committee on Frequency Allocation for Radio Astronomy and Space Science (IUCAF): F.G. Smith and R. Wielebinski. F.G. Smith was succeeded by G. Westerhout in October 1975
   (e) Special Committee on Solar-Terrestrial Physics (SCOSTEP): Z. Svestka
   (f) Scientific Committee on Problems of the Environment (SCOPE): M.F. Walker and J.-C. Pecker
   (g) Committee on the Teaching of Science (CTS): D. McNally
   (h) Committee for Data on Science and Technology (CODATA): G.A. Wilkins
   (i) Inter-Union Commission on Spectroscopy (IUCS): B. Edlén, J.C. Philips and M.J. Seaton.

Other Organizations

The Union is represented in the following organizations:
   (a) La Fondation Internationale du Pic-du-Midi: A. Lallemand

(b) Le Comité Consultatif pour la Définition de la Seconde (CCDS) du Bureau
    International des Poids et Mesures: Wm. Markowitz
(c) Le Comité Consultatif pour la Définition du Mètre (CCDM) du Bureau Inter-
    national des Poids et Mesures: A.H. Cook
(d) Le Comité Consultatif International des Radiocommunications (CCIR): F.G.
    Smith and H.M. Smith
(e) European Physical Society (EPS): Edith A. Müller
(f) Various Services of FAGS
    Bureau International de l'Heure (BIH): H. Enslin and H.M. Smith (B. Guinot
    is Director of the Bureau)
(g) Quarterly Bulletin on Solar Activity (QBSA): Organizing Committee of IAU
    Commission No. 10
(h) International Ursigrams and World Day Service (IUWDS): F.W. Jäger
(i) Scientific Ballooning and Radiation Monitoring Organization (SBARMO): L.D.
    de Feiter (until his untimely death) and P. Simon

REPORT ON IAU FINANCES

In accord with the policy adopted in 1961, the Executive Committee presents to the
General Assembly a "Summarized Account of Receipts and Payments" for the period under report. All accounts are in Swiss francs. They are also given in U.S. dollars
converted at the following ICSU conversion rates, treated as exact:

For the year 1973:   U.S.$ = 2.75 Dutch guilders,
For the year 1974:   U.S.$ = 2.60 Dutch guilders,
For the year 1975:   U.S.$ = 2.67 Dutch guilders = 2.66 Swiss francs.

The conversion rates varying over the three years under report account for the dis-
parities between the closing balances in 1972, 1973 and 1974, and the opening balances in 1973, 1974 and 1975 respectively:

Closing balance	Opening balance	Difference
1972 $ 69 050.34	1973 $ 75 774.07	$ 6 723.73
1973 $ 107 341.62	1974 $ 108 205.91	$ 864.29
1974 $ 104 667.45	1975 $ 104 099.42	$ 568.03

Certain bank operations, such as transfers from one IAU account to the other,
and reimbursements, were made at current rates of exchange and thus resulted in exchange differences.
The summarized account has been derived from 10 (4 in 1973, 3 each in 1974 and
1975) accounts of receipts and payments audited and certified as correct by the
Unions's professional accountant. These accounts will be available for examination
by the Finance Committee at the XVIth General Assembly.

SUMMARIZED ACCOUNT OF RECEIPTS AND PAYMENTS

The statement on the following pages is a simple account of receipts and payments
collected for convenience under a few headings. The subsequent explanatory notes
provide more detailed information on each item of the accounts.

SUMMARIZED ACCOUNT OF RECEIPTS AND PAYMENTS (in U.S. Dollars converted into Swiss francs) FOR THE PERIOD 1973-1975

RECEIPTS

	U.S.Dollars	U.S.Dollars	Swiss Francs
1. Contribution from Adhering Countries			
1973	86 098.58		
1974	83 077.66		
1975	118 382.67	287 558.91	764 906.70
2. Revenue from IAU Publications			
1973	13 528.10		
1974	269.89		
1975	17 892.58	31 690.57	84 296.92
3. Interest			
1973	2 465.79		
1974	5 747.91		
1975	2 839.09	11 052.79	29 400.42
4. Grant from UNESCO Subvention to ICSU			
1973	14 000		
1974	14 130.58		
1975	10 850.00	38 980.58	103 688.34
5. Other Receipts			
1973	74 098.60		
1974	29 119.03		
1975	48 194.14	151 411.77	402 755.31
Total Receipts		520 694.62	1 383 047.69
TOTAL		520 694.62	1 383 047.69

PAYMENTS

	U.S.Dollars	U.S.Dollars	Swiss Francs
1. Administrative Office			
1973	39 266.01		
1974	37 715.82		
1975	38 885.62	115 867.45	308 207.42
2. Contribution to ICSU			
1973	1 503.07		
1974	2 144.53		
1975	2 076.94	5 724.54	15 227.28
3. Commission Expenses			
1973	253.29		
1974	562.32		
1975	241.55	1 057.16	2 812.04
4. Specific Projects of Commissions Approved by the General Assembly			
4.1 Exchange of Astronomers			
1973	1 291.80		
1974	7 664.23		

1975	7 145.00	16 101.03	42 828.74
4.2 Other Projects			
1973	2 067.01		
1974	4 725.92		
1975	3 810.30	10 603.23	28 204.59
5. General Assembly			
1973 (XV)	43 302.91		
1974 (XV)	281.81		
1975	-	43 584.72	115 935.36
6. IAU Publications			
1973	1 974.53		
1974	10 987.11		
1975	3 218.99	16 180.63	43 040.47
7. Distribution of IAU Publications (free and at reduced price)			
1973	2 723.02		
1974	9 569.71	20 251.01	53 867.69
1975	7 958.28		
8. Executive Committee Meetings			
1973	-		
1974	7 584.14		
1975	8 349.52	15 933.66	42 383.53
9. Officers Meetings			
1973	3 282.14		
1974	4 639.80		
1975	2 335.84	10 257 78	27 285.69
10. IAU Symposia and Colloquia and Cosponsorships of Other Scientific Meetings			
1973	16 235.63		
1974	9 577.17		
1975	10 001.00	35 813.80	95 264.71
11. Inter-Union Commissions			
1973	2 600.00		
1974	2 330.00		
1975	2 800.00	7 730.00	20 561.80
12. Projects Authorized by the Executive Committee			
1973	-		
1974	-		
1975	-		
13. Representation			
1973	567.00		
1974	1 039.57		
1975	818.98	2 425.55	6 451.96

14. Bank Charges

```
1973                          134.30
1974                          201.81
1975                          163.49         499.60          1 328.94

15.Young Astronomers Schools
1973                        6 866.52
1974                        8 000.00
1975                        8 000.00       22 866.52        60 824.94

16.Regional Meetings
1973                            -
1974                        2 111.87
1975                        2 245.52        4 357.39        11 590.66

17.Other Payments
1973                       36 556.26
1974                       26 747.72
1975                       87 829.11      151 133.09       402 014.02
              Total Payments             480 387.16      1 277 829.84 (5)
              Excess of Receipts of Payments  40 307.46     107 217.84 (3)
                                                                   68 (8)
              TOTAL                       520 694.62      1 385 047.69
```

BANK ACCOUNTS as on 1 January 1973

	U.S.Dollars	Swiss Francs
Savings Accounts	73 236.13	194 808.10
Current Accounts	75 774.04	201 558.95
TOTAL Balance on IAU Accounts as on 1 January 1973	149 010.17	396 367.05

BANK ACCOUNTS as on 31 December 1975

	U.S.Dollars	Swiss Francs
Savings Accounts	77 424.37	205 948.82
Current Accounts	116 377.76	309 564.84
TOTAL Balance on IAU Accounts as on 31 December 1975	193 802.13	515 513.66

Note:

The difference of $ 296.26 between the amount of $ 40 307.46 in item "Excess of Receipts over Payments", and the balance of $ 40 603.72 on the Current Accounts 1973-1975 is due to the changed rate of exchange between U.S.Dollar and Dutch Guilder which was

```
        in 1973:   1 U.S.$ = 2.75 Dfl.
        in 1974:   1 U.S.$ = 2.60 Dfl.
        in 1975:   1 U.S.$ = 2.67 Dfl.
```

Thus the difference in Dfl. Account (AMRO)   1973/1974 = $ 864.29 U.S.
                                               1974/1975 = $ 568.03

which results in $ 296.26 U.S.

4. SPECIFIC PROJECTS OF COMMISSIONS

Comm.No.	Project	Allocation in $ for 1974-75	Paid in 1973 ($)	Paid in 1974 ($)	Paid in 1975 ($)	Total 1973/74 ($)	Total 1974/75 ($)	Balance for 1976 from allocations
4.1-38	Exchange of Astronomers	29 000.00	1 291.80	7 664.23	7 145.00	8 956.03	14 809.23	14 190.77
4.2	Other Projects:							
4	Ephemerides	880.00	200.00	293.33	293.33	493.33	586.66	293.34
6	IAU Telegram Bureau	1 980.00	533.34	660.00	660.00	1 193.34	1 320.00	660.00
16	Documentation Meudon	1 870.00	500.00	623.33	623.33	1 123.33	1 246.66	623.34
17	Special Working Group	1 870.00	333.67	623.33	623.33	957.00	1 246.66	623.34
20	Minor Planet Center	2 200.00	-	733.33	733.34	733.33	1 466.67	733.33
27	Catalogue of Variable Stars	3 300.00	-	1 242.60	126.97	1 242.60	1 369.57	1 930.43
Inf.Bull.	Southern Hemisphere	1 650.00	500.00	550.00	550.00	1 050.00	100.00	550.00
TOTAL of 4.2		13 750.00	2 067.01	4 725.92	3 610.30	6 792.93	8 336.22	5 413.78
TOTAL Projects		42 750.00	3 358.81	12 390.15	10 755.30	15 748.96	23 145.45	19 604.55

Receipts

1. Contributions from Adhering Countries. The XVth General Assembly of the IAU, held in Sydney, resolved that the unit of contribution be increased to 1 125 gold francs until 1976. This unit of contribution equaled $ 438.92 U.S. and 1 329.93 Swiss francs at the time of the General Assembly. On the recommendation of the Finance Committee, the XVth General Assembly also resolved that the currency in which the affairs of the Union should be conducted should be the Swiss franc. The contributions payable by Adhering Countries were therefore collected in Swiss francs from 1974 on.

As on 31 December 1975, the IAU had 47 Adhering Countries paying annualy a total of 215 units of contribution. Seventeen countries adhered in category 1, fourteen in category 2, six in category 3, two in category 4, three in category 5, one in category 6, three in category 7, and one in category 8.

Thirty one countries had paid their dues before 31 December 1975, and further eight countries remitted their contributions until 1 April 1976. One Adhering Country owes its dues to the IAU for 5 years, one country for 3 years, two countries for 2 years, and four countries for 1 year, or part.

2. Revenue from publications. The bulk of the receipts consisted of royalties from the Reidel Publishing Company for volumes sold. Minor payments came in from institutions of developing countries for publications sold to them at reduced price, from Academic Press Ltd. (London) for sales of publications, and from University Microfilms Ltd. in the form of royalties for sales of "Bibliography of Astronomy 1881-1889".

The yearly average of budgeted receipts from publications is $ 5 500, that is $ 11 000 from 1974, 1975. The excess of $ 7 162.47 is due to exceptionally good sales which were not foreseen at the time the budget was set up. For example, the price of $ 80.50 for Transactions XVA was considered prohibitive, and yet 739 copies were sold in the course of 1973, 1975.

3. Interest on accounts. The amount of $ 11 052.97 are interest on the U.S. dollars account at the Amsterdam-Rotterdam-Bank, Utrecht, which was opened in 1973, when the Union's savings in the U.S.A. were transferred to the Netherlands. However, the interest on this account dropped to 3% in 1975, so that $ 40 000 were retransferred from it to two one-year deposits of $ 20 000 each at two U.S. savings banks. The IAU has, in addition, a savings account in Dutch guilders at the Amsterdam-Rotterdam Bank. The total interest earned on these three savings accounts from 1973 to 1975 is $ 6 988.45. This amount is not shown in the summarized account, but appears, itemized, in the synopses of the IAU accounts as on 31 December of the years in question. Thus, the total interest earned over the period under report is $ 18 041.

4. Grant from UNESCO subvention to ICSU. Detailed accounts of the expenditure of this grant are submitted to UNESCO, through ICSU, every year. The distribution of the grant is left to the discretion of the IAU within the permitted UNESCO categories.

The allocation was $ 14 000 in 1973. In 1974, the allocation of $ 12 000 was increased by $ 2 000 to $ 14 000. In 1975 the allocation of $ 12 000 was not increased. The loss of $ 1 020.42 is due to exchange differences.

5. Other receipts. The entries under this heading include internal bank transfers, transient items, reimbursements, etc. The following is an analysis of the amounts of $ 74 098.60, $ 29 119.03 and $ 48 149.14 for the years 1973, 1974 and 1975 respectively:

Other receipts in 1973

	U.S.$
Transfer of balance on the IAU account opened on the occasion of the XVth General Assembly in Sydney, refund of unspent grant towards Symposium 61, and exchange gain	4 119.27
Transfer of balance on closed savings account at the Chemical Bank, New York	22 194.07

	U.S.$
Transfer of balance on closed savings account at the American Savings Bank	22 600.98
Transfer of DFL. 61.033,70, equivalent to the closing balance on the savings account at the Chemical Bank, to a newly opened U.S.dollars account at the Amsterdam-Rotterdam Bank	21 175.45
Contribution of COSPAR to General Secretary's trip to the U.S.S.R. and Australia	664.15
Refund by Wagon-Lits of price for 3 cancelled air tickets to Sydney	2 294.18
Refund by Presidents of Commissions of unused travel grants to the General Assembly in Sydney	784.95
Transfer to Petty-Cash Utrecht	207.20
Refund of unused part of Symposium travel grant	54.85
Rectified erroneous debit	3.50
TOTAL	74 098.60

Other receipts in 1974

	U.S.$
Refunds of unspent grants	3 013.73
Erroneous credits	26 105.30
TOTAL	29 119.03

Other receipts in 1975

	U.S.$
Erroneous credits	46 149.89
Rectification of erroneous debits	1 679.22
Refunds of unused grants	·360.00
Untraceable receipt	5.03
TOTAL	48 194.14

Recapitulation:		U.S.$
Other receipts in	1973	74 098.60
	1974	29 119.03
	1975	48 194.14
	TOTAL	151 411.77

Clearly, the bulk of this sum consists of internal bank transfers and erroneous credits which are balanced by corresponding payments and rectifications, and cannot therefore be regarded as effective receipts.

Payments

The headings have been numbered to correspond to those of the budget approved by the XVth General Assembly.

   1. Administrative Office. The expenses of the administrative office include salaries, social security and pension payments of the Staff of the IAU Secretariat, expenditure for secretarial help, stationery, removal costs, etc. It should be noted that thanks to the generosity of the Astronomical Institute in Utrecht, the Kitt Peak National Observatory and the Universities of Thessaloniki and Athens, the Union paid no rental, had the free use of communication facilities, and only spent a negligible sum on stationery. The Greek Ministry of Culture and Science, and the University of Thessaloniki and Athens, paid a total of more than U.S.$ 9 000 yearly for salaries of two secretaries, postage, telegrams, etc.

   2. Subscriptions to ICSU. The annual dues of the IAU to ICSU amounted to 2.5% of the income from contributions from Adhering Countries for the year preceding

payment.
     3. Commission expenses. An amount of $ 741.50 was paid to Commission No. 50 within the framework of its site testing programme. $ 315.61 are travel grants.

     4. Projects of Commissions
     4.1 Exchange of Astronomers, and 4.2 Other projects. Details of the allocations made by the XVth General Assembly are given on page 50 of IAU Transactions, volume XVB, for 1974-1976, and on page 56 of IAU Transaction, volume XIVB for 1973. Actual payments were listed in the present Report. In addition $ 200 were paid as a contribution towards the publication of Bibliography and Program Notes by Commission No. 42.
     The balances of $ 14 190.77, $ 1 930.43 and $ 550.00 in favour of Commissions Nos. 38, 27, and the Information Bulletin for the Southern Hemisphere respectively may, but need not be spent in 1976.

     5. General Assembly. For details please see the 1973 accounts (account at Chemical Bank, 2 accounts at Amsterdam-Rotterdam Bank) and the 1974 accounts at the Chemical Bank and Amsterdam-Rotterdam Bank (DFl.).
     The sum of $ 43 584.72 is by $ 2 584.72 higher than the budgetary estimate of the XIVth General Assembly, but by $ 5 915.28 lower than approved by the XVth General Assembly for the year 1973. It includes travel and living expenses of the members of the Executive Committee, travel grants to Presidents of Commissions, young astronomers' grants, local expenses in Sydney, etc.

     6. IAU Publications. The IAU only financed the production and distribution of the Information Bulletin. The Transactions and Symposium volumes were financed by the publisher. The payments under this heading were as follows:

in 1973
D. Reidel Publishing Company, Dordrecht, Holland   U.S.$

     postage for Information Bulletin No. 30 in U.S.A.   141.49
     printing and distribution of Information Bulletin No.28   1 833.04

     Total   1 974.53

in 1974
D. Reidel

     printing and distribution of Information Bulletin No.31   2 900.44
     printing and distribution of Information Bulletin Nos.29,
     30, 32   6 602.05
     printing and distribution of Agenda and Report of Executive
     Committee to the XVth General Assembly   1 484.62

     Total   10 987.11

in 1975
D. Reidel

     printing and distribution of Information Bulletin No.33   3 218.99

     Grand total for 1973-1975   16 180.63

     7. Symposium volumes for developing countries. This item actually covers purchases by the IAU of its own publications to be distributed as courtesy copies to members of the Executive Committee and members of the Scientific Organizing Committees of Symposia, and sold at reduced price to selected institutions of some developing countries. Expenditure was as follows over the period under report:

Courtesy copies   Copies for institutions of developing countries
                 $ $
1973   1 280.49   1 442.53
1974   3 839.92   5 729.79
1975   3 333.04   4 625.24

    Total   8 453.45   11 797.56

Recapitulation: $
For copies to institutions of developing
countries   11 797.56
For courtesy copies   8 453.45

              Total 1973-1975   20 251.01

8. Executive Committee meetings. The costs of the 37th and 38th meetings of the Executive Committee are included in the costs of the XVth General Assembly, item 4.

The amount of $ 7 584.14 as costs of the 39th meeting of the Executive Committee gives the following net expenses:

$ 7 584.14 withdrawn from accounts
$ 145.00 reimbursed as unused
$ 7 439.14 net expenses.

The total net expenses for the 39th and 40th meetings of the Executive Committee only exceed by about 2% the amount foreseen in the approved budget.

9. Officers' meetings. The sequence of the figures $ 3 282.14 for 1973, $ 4 639.80 for 1974, and $ 2 335.84 for 1975 shows the special effort in 1975 to keep expenses as low as possible, in spite of the large travel distances involved. This resulted in the total sum of $ 10 257.78 for the 3 years under report.

10. Symposia and Colloquia, sponsored meetings included. The figure of $ 16 235.63 represents expenses incurred with IAU Symposia Nos. 56 through 64, IAU Colloquia Nos. 18 and 23, and the IAU/IUHPS Symposium in Toruń. $ 215.00 were paid for the translation of 8 Russian papers presented at the Kepler Symposium in Leningrad, 1971.

$ 9 577.17 were spent on travel grants to IAU Symposia Nos. 65, 67, 68 and 69, IAU Colloquia Nos. 25 through 28, and the International Symposium on Solar-Terrestrial Physics held in Sao Paulo.

The sum of $ 10 001 covers travel grants to participants in IAU Symposia No. 70 through 73, IAU Colloquia Nos. 29, 30, 32 and 33, and contribution to the COSPAR/IAU Symposium in Varna and the IUHPS/IAU Symposium in Greenwich.

11. Inter-Union Commissions. The amount of $ 7 730.00 consists of annual contributions to IUCAF, SCOSTEP, ICSU-AB and ICSU-CTS. In 1973 $ 100 were paid to IUCM.

12. Projects authorized by the Executive Committee. No such projects were organized in the period under report.

13. Representation. The sum of $ 2 425.55 was spent for travel grants to IAU representatives on IUCM in 1973, on EPS, IUCM and ICSU-AB, and for a framed document in 1974, and at the Third Regional Meeting in Tbilisi, and the FAGS Council in 1975.

14. Bank charges. They are well within the budget.

15. Young Astronomers Schools. The budgets approved for 1973 and 1974-76 allowed for $ 23 100 for Young Astronomers Schools. The actual amount of $ 22 866.52 spent is well within the budget. The Executive Committee resolved that no such school would be held in 1976.

16. Regional meetings. The Regional Astronomical Meetings held in 1974 and 1975 only cost $ 4 357.39 while the budgetary estimate was $ 5 500. No Regional meeting will be held in 1976.

17. Other payments. Reference is made to the comment of heading 6 under Receipts. The following is an analysis of the amounts of $ 36 556.26, $ 26 747.72 and $ 87 829.11 for the years 1973, 1974 and 1975 respectively:

1973		U.S.$
Transfer to U.S.Dollars account at the Amsterdam-Rotterdam Bank		22 394.00
Transfer to Petty Cash Utrecht		207.20
Transfer from Sydney account		3 251.33
Travel and removal expenses		10 008.26
Insurance		136.36
Penalty interest		181.82
Refund of erroneous payment		373.79
Erroneous debit		3.50
	Total	36 556.26
1974		
Travel expenses		642.42
Rectification of erroneous credits		26 105.30
	Total	26 747.72
1975		
Transfer to savings accounts		40 000.00
Rectification of erroneous credits		46 149.89
Erroneous debits		1 679.22
	Total	87 829.11

Recapitulation:	U.S.$	
Other expenses in 1973	36 556.26	
1974	26 747.72	
1975	87 829.11	
Total 1973-'75	151 133.09	

Note: Item 17 "Refund to Reserves" of the approved budget for 1974-1976 does not come into consideration, as it was not necessary to have recourse to the reserves in 1973.

Audit for the Accounts 1973-1976

Messieurs,

Conformément à la mission que vous avez bien voulu nous confier, nous vous présentons le résultat de notre examen des comptes de l'UNION ASTRONOMIQUE INTERNATIONALE pour la période du 1er janvier 1973 au 31 décembre 1975.

L'examen notamment, a porté sur les documents suivants:
- comptes des recettes et dépenses pour l'année 1973, vérifiés par nous-mêmes
- comptes des recettes et dépenses pour l'année 1974, vérifiés par nous-mêmes
- comptes des recettes et dépenses pour l'année 1975, vérifiés par nous-mêmes
- comptes des recettes et dépenses et situation des comptes bancaires pour la période de trois ans: 1973, 1974 et 1975.

Les comptes ont été comparés avec les cahiers de comptes reçus et pièces comptables diverses.
Le taux employé pour convertir les U.S. DOLLARS en FRANCS SUISSES a été le suivant d'après les règles comptables du C.I.U.S.:

1 U.S.DOLLAR = 2.66 FRANCS SUISSES

Nous certifions que ces comptes sont en accord complet avec les informations et ex-
plications que nous avons demandées pour nous permettre d'effectuer notre examen.

Le compte ci-inclus des recettes et dépenses résumant les compte des années 1973,
1974 et 1975 est la représentation exacte des recettes et dépenses de l'UNION ASTRO-
NOMIQUE INTERNATIONALE pour la période du premier janvier 1973 au trente et un décembre 1975.

A Paris, le 4 février 1975

ROGER BACLE

H.E.C.

Diplômé Expert-Comptable

Conclusion

The assets of $ 193 802.13 U.S. as on 31 December 1975 compare very well with those
of $ 149 010.17 U.S. as on 1 January 1973: the balance is $ 44 791.96 U.S.
    The budget approved by the XVth General Assembly for the years 1973-1976 pro-
vided for an expenditure of $ 351 760 balanced by an income of $ 353 800 leaving an
excess of $ 2 040 of receipts over payments. This budget compared with the actual
receipts and payments of the years 1974 and 1975 gives the balance for the year
1976 as follows:

	Receipts in U.S.$	Payments in U.S.$
Budget 1974-1976 as approved by the XVth General Assembly	353 800.00	351 760.00
Actual figures for 1974 and 1975 (adjusted)	-264 490.00	-247 187.00
	89 310.00	104 573.00
Excess of receipts over payments	15 263.00	
Total	104 573.00	

Available for payments in 1976 $ 15 263.00

The budgetary estimates taking into account the present rate of exchange of US$ 1 =
2.66 Swiss francs, are as follows for 1976:

Budget 1976

Receipts in U.S.Dollars		Payments in U.S.Dollars	
1. Contributions from Adhering Countries	107 500	1. Administrative Office	47 700
2. Revenue from IAU Publications	11 600	2. Contribution to ICSU	2 960
3. Interest on accounts	5 400	3. Commissions expenses	400
4. Grant from UNESCO subvention to ICSU	12 000	4. Projects of Commissions	
		4.1 Exchange of Astronomers	14 191
		4.2 Other Projects	5 414
Total	136 500	5. General Assembly	39 000
		6. IAU Publications	11 000
		7. Free copies to institutions of developing countries	7 000
		8. Executive Committee meeting	0
		9. Officers' meetings	3 300
		10. Symposia and Colloquia	12 000
		11. Contributions to Inter-Union Commissions	2 800

12. Projects approved by the Executive Committee		800
13. Representation		1 500
14. Bank charges		300
15. Young Astronomers' Schools		0
16. Regional meeting		0
	Total	148 365
Receipts		136 500
Excess of payments over receipts		11 865
Available from 1975		15 263
Excess of funds available over payments (as reserve)		3 398

Explanatory notes:

Receipts

1. The figure of $ 107 500 has been obtained by multiplying 215 units of contributions by 1 329.93 Swiss francs and dividing the result by 2.66, the rate of exchange of the U.S.Dollar to Swiss francs.
2. The amount of about $ 5 000 paid by the D. Reidel Publishing Company on royalties before 31 December 1975 only passed the accounts in January 1976. Another $ 6 600 are expected to be payable as royalties for sales in 1975, and for publications sold to institutions in developing countries.
3. The two savings accounts in the U.S.A. will earn $ 2 600 of interest, the U.S.Dollars account at the Amsterdam-Rotterdam Bank $ 1 000, and the savings account in Dutch guilders at the latter $ 1 800, giving a total of $ 5 400.
4. The grant from the UNESCO subvention to ICSU has been allocated in the amount of $ 12 000 for 1976.

Payments

1. The figure of $ 47 700 is the difference between the amount budgeted for 1974-1976 and the one actually paid in 1974, 1975. In addition to the normal payments it will have to cover removal expenses of the office and staff.
2. The contribution to ICSU is 2.5% of the total of contributions paid by Adhering Countries in 1975.
3. The $ 400 have been set aside for the site-testing programme of Commission No. 50. Not included are the $ 200 paid towards the publication of Biography and Programme Notes (Commission No. 42).
4. Specific Projects of Commissions: see table 4.
5. General Assembly. The sum of $ 39 000 is to cover, in the main, the travel and living expenses of the members of the Executive Committee, travel grants to Presidents of Commissions, a hundred grants to young astronomers, travel and living expenses of the staff, local secretarial help and other local expenses, which have considerably increased since 1973, when the amount of $ 33 000 had been approved by the General Assembly.
6. IAU Publications. The amount of $ 11 000 represent the costs of Information Bulletin Nos. 34 through 37, and the production and distribution of the Report of the Executive Committee to the XVIth General Assembly.
7. This item includes expenditure for courtesy copies to members of the Executive Committee and members of the Scientific Organizing Committees of Symposia, as well as costs for IAU publications purchased by the IAU for and sold to selected institutions of developing countries at reduced prices.
8. The costs of the 41st and 42nd meetings of the Executive Committee are included in the $ 39 000 foreseen for the XVIth General Assembly.
9. At least 3 Officers' meetings will be held in 1976. $ 1 100 for each may turn out to be an underestimate.

10. Symposia and Colloquia. Three symposia and 7 colloquia will be held in 1976. Moreover, the IAU will cosponsor 2 COSPAR meetings. The amount of $ 12 000 includes a small reserve.

11. Contributions to Inter-Union Commissions. They will be as in 1975.

12. Projects approved by the Executive Committee. $ 800 is the balance available for such projects.

13. Experience has shown that the amounts foreseen for purposes of representation were inadequate. The sum of $ 1 500 is not therefore exaggerated.

14. Bank charges are expected to be higher than the yearly average in view of the numerous payments to be made in connection with the General Assembly.

The Finance Committee, to convene during the XVIth General Assembly, will be asked to endorse the budgetary estimate for 1976, as proposed above, and to approve a comprehensive budget for the years 1977-1979 to be submitted to the General Assembly.

LIST OF DECEASED MEMBERS

The Executive Committee regrets to record the loss by death of the following Members:

ABBOT, C.G. - 17 December 1973
ASTAPOVICH, J.S. - 1 January 1976
BOK, P.F. - 19 November 1975
BOURGEOIS, P.E. - 11 May 1974
CARROLL, Sir John A. - 2 May 1974
CESKO, R.P. - 11 March 1974
CHEBOTAREV, G.A. - 4 August 1975
CHUDOVICHEVA, N.A. - 3 August 1972
CLEMENCE, G.M. - 22 November 1974
COLLINDER, P. - 6 December 1975
CONDON, E.U. - 26 March 1974
DAENE, H. - 7 December 1975
DE FEITER, L.D. - 2 September 1975
DETRE, L. - 15 October 1974
DOS SANTOS, A.J. Baptista - 14 July 1973
FERRARO, V.C.A. - 4 January 1974
GJELLESTAD, G. - 11 January 1972
GORSHKOV, P.M. - 30 July 1975
HARRIS, B.J. - 23 December 1974
HERRICK, S. - 20 March 1974
HINDMARSH, W.R. - 29 December 1973
HOPMANN, J. - 11 October 1975
HOUCK, T.E. - 1 June 1974
KALIKHEVICH, N.S. - 9 March 1975
KIENLE, H. - 15 February 1975
KIEPENHEUER, K.O. - 23 May 1975
KNIPE, G.F.G. - 19 December 1973
KUIPER, G.P. - 23 December 1973
LINDSAY, E.M. - 27 July 1974
LINES, A.W. -
LITTLEFIELD, T.A. - 26 August 1974
MADWAR, M.R. - 10 December 1973
MAITRE, V. - 9 December 1975
MILLER, W.J. - 30 November 1973
MINKOWSKI, R.L. - 4 January 1976
MUGGLESTONE, D. - 20 March 1976
NOVOPASHENNYJ, E.V. - 18 February 1975

OLIVIER, C.P. - 14 August 1975
PATON, J. - 26 August 1973
PIKEL'NER, S.B. - 19 November 1975
RABE, E.K. - 11 July 1974
REDMAN, R.O. - 7 March 1975
ROSEN, B. - 2 January 1974
RUBLEV, S.V. - 10 November 1974
SALPETER, E.W. - 6 January 1976
SCOTT, F.P. - 1 November 1974
SHCHIGOLEV, B.M. - 6 February 1976
SMART, W.M. - 17 September 1975
SYNTINSKAYA, N.N. - 4 July 1974
VAN BIESBROECK, G. - 23 February 1974
VAN DEN BOS, W.H. - 30 March 1974
VERBAANDERT, M.J. - 4 September 1974
VYSSOTSKY, A.N. - 31 December 1973
WILDT, R. - 9 January 1976
WURM, K. - 16 February 1975
YAKOVKIN, N.A. - 18 November 1974
ZAGAR, F. - 17 February 1976
ZONN, W. - 28 February 1975
ZWICKY, F. - 8 February 1974

LIST OF IAU PUBLICATIONS

Transactions of the IAU, volume XVA, pp. VIII+762, price $ 80.50 (Members $ 53.67).
    Editor: C. de Jager. Published by D. Reidel Publishing Company, Dordrecht,
    Holland.
Transactions of the IAU, volume XVB, pp. IX+334, price $ 39.00 (Members $ 26.00).
    Editors: G. Contopoulos and A. Jappel. Published by D. Reidel.
Highlights of Astronomy as presented at the XVth General Assembly and the Extra-
    ordinary General Assembly of the IAU, 1973, pp. VII+574, price $ 65.00 (Mem-
    bers $ 42.50).
    Editor: G. Contopoulos. Published by D. Reidel.

IAU Symposium Volumes

No. 49 "Wolf-Rayet and High Temperature Stars", pp. XIV+263, price $ 30.50 (Members
    $ 20.33), paperback $ 22.50 (Members $ 15.00). Editors: M.K.V. Bappu and J.
    Sahade. Published by D. Reidel Publishing Company, Dordrecht, Holland.
No. 50 "Spectral Classification and Multicolour Photometry", pp. XIV+314, price $
    35.50 (Members $ 23.67), paperback $ 26.00 (Members $ 17.33). Editors: Ch.
    Fehrenbach and B.E. Westerlund. Published by D. Reidel.
No. 51 "Extended Atmospheres and Circumstellar Matter in Spectroscopic Binary Sys-
    tems", pp. XIV+291, price $ 34.00 (Members $ 22.67), paperback $ 24.50 (Mem-
    bers $ 16.33). Editor: A.H. Batten. Published by D. Reidel.
No. 52 "Interstellar Dust and Related Topics", pp. XVII+584, price $ 56.00 (Members
    $ 37.33), paperback $ 40.50 (Members $ 27.00). Editors: J.M. Greenberg and
    H.C. van de Hulst. Published by D. Reidel.
No. 53 "Physics of Dense Matter", pp. X+327, price $ 36.50 (Members $ 24.33). Editor:
    C.J. Hansen. Published by D. Reidel.
No. 54 "Problems of Calibration of Absolute Magnitudes and Temperature of Stars",
    pp. VII+304, price $ 42.50 (Members $ 28.33), paperback $ 31.50 (Members $
    21.00). Editors: B. Hauck and B.E. Westerlund. Published by D. Reidel.

No. 55 "X- and Gamma-Ray Astronomy", pp. X+323, price $ 30.50 (Members $ 20.33),
        paperback $ 22.00 (Members $ 14.67). Editors: H. Bradt and R. Giacconi. Pu-
        blished by D. Reidel.
No. 56 "Chromospheric Fine Structure", pp. XVII+310, price $ 38.00 (Members $ 25.33),
        paperback $ 29.00 (Members $ 19.33). Editor: R.G. Athay. Published by D. Reidel
        Publishing Company.
No. 57 "Coronal Disturbances", pp. XVIII+508, price $ 57.00 (Members $ 38.00), paper-
        back $ 45.00 (Members $ 30.00). Editor: G. Newkirk, Jr. Published by D. Reidel.
No. 58 "The Formation and Dynamics of Galaxies", pp. XV+441, price $ 54.00 (Members
        $ 36.00) paperback $ 39.50 (Members $ 26.33). Editor: J.R. Shakeshaft. Published
        by D. Reidel.
No. 59 "Stellar Instability and Evolution", pp. XI+201, price $ 27.00 (Members $
        18.00), paperback $ 19.50 (Members $ 13.00). Editors: P. Ledoux, A. Noels, and
        A.W. Rodgers. Published by D. Reidel.
No. 60 "Galactic Radio Astronomy", pp. XIII+654, price $ 78.00 (Members $ 52.00),
        paperback $ 60.00 (Members $ 40.00). Editors: F.J. Kerr and S.C. Simonson III.
        Published by D. Reidel.
No. 61 "New Problems in Astrometry", pp. XII+335, price $ 30.00 (Members $ 20.00),
        paperback $ 22.50 (Members $ 15.00). Editors: W. Gliese, C.A. Murray, and R.H.
        Tucker. Published by D. Reidel.
No. 62 "The Stability of the Solar System and of Small Stellar Systems", pp. VIII+
        313, price $ 38.00 (Members $ 25.35), paperback $ 28.00 (Members $ 18.67).
        Editor: Y. Kozai. Published by D. Reidel.
No. 63 "Confrontation between Cosmological Theories and Observational Data", pp.
        XI+381, price $ 39.50 (Members $ 26.33), paperback $ 34.00 (Members $ 22.67).
        Editor: M.S. Longair. Published by D. Reidel.
No. 64 "Gravitational Radiation and Gravitational Collapse", pp. XVI+224, price $
        28.00 (Members $ 18.66), paperback $ 20.00 (Members $ 13.33). Editor: C. De
        Witt-Morette. Published by D. Reidel.
No. 65 "Exploration of the Planetary System", pp. XIII+566, price $ 67.00 (Members
        $ 44.67), paperback $ 39.50 (Members $ 26.33). Editors: A. Woszczyk and C.
        Iwaniszewska. Published by D. Reidel.
No. 66 "Late Stages of Stellar Evolution", pp. IX+269, price $ 32.00 (Members $
        21.33), paperback $ 24.00 (Members $ 16.00). Editor: R.J. Tayler. Published by
        D. Reidel.
No. 67 "Variable Stars and Stellar Evolution", pp. XVII+614, price $ 84.00 (Members
        $ 56.00), paperback $ 56.00 (Members $ 37.35). Editors: V.E. Sherwood and L.
        Plaut. Published by D. Reidel.
No. 68 "Solar Gamma-, X-, and EUV Radiation", pp. XII+439, price $ 58.00 (Members
        $ 38.65), paperback $ 39.50 (Members $ 26.35). Editor: S.R. Kane. Published by
        D. Reidel.
No. 69 "Dynamics of Stellar Systems", pp. X+461, price $ 56.00 (Members $ 37.35),
        paperback $ 39.50 (Members $ 26.35). Editor: A. Hayli. Published by D. Reidel.

IAU Colloquium Volumes

No. 1 "The Problem of the Variation of the Geographical Coordinates in the Southern
        Hemisphere, pp. VII+24, price $ 4.00. Editor: O. Cáceres. La Plata Observatory.
No. 16 "Analytical Procedures for Eclipsing Binary Light Curves". Invited contribu-
        tions appeared in *Astrophysics and Space Science*.
No. 17 "L'âge des Etoiles", price Fr.Frs. 50,00. Editors: G. Cayrel de Scrobel and
        A.M. Delplace. Published by Observatoire de Paris, Meudon.
No. 18 "Orbital and Physical Parameters of Double Stars". *Journal of the Astronomi-
        cal Society of Canada*, volume 67, pp. 49-87, April 1973.
No. 20 "Meridian Astronomy". Detailed report in IAU Symposium volume No. 61 New
        Problems in Astrometry.

No. 21 "Variable Stars in Globular Clusters and in Related Systems", pp. X+234. Editor: J.D. Fernie.
No. 24 "Lunar Dynamics and Observational Coordinate System". Editor: M. Moutsoulas. *The Moon* 8, No. 4, 1973.

Publications by, or on behalf of, IAU Commissions and by Astronomical Services

Commission No. 6. IAU (Telegram Bureau) circulars Nos. 2476-2875. Issued by the IAU Central Bureau for Astronomical Telegrams.
Commission No. 10. Cartes Synoptiques de la Chromosphère Solaire (Volume V, fascicule IV). Prepared by Observatoire de Paris à Meudon.
Commission No. 16. Newsletter No. 11. Published by the President of Commission.
Commission No. 19. Monthly Notes of the IPMS Nos. 11 (1972)-10 (1975). Published by the International Polar Motion Service, Mizusawa-shi, Japan.
Commission No. 24. Astrographic Catalogues. Information regarding the different sections of the Catalogue can be obtained from the Observatories listed in Transactions IAU XIVB, 176, 177, 1971.
Commission No. 20. Minor Planet Circulars Nos. 3407-3908. Mimeographed circulars issued by the Minor Planet Center, Cincinnati.
Commission No. 26. Circulaire d'Information Nos. 59-67. President: Miss S.L. Lippincott, Sproul Observatory, Swarthmore College, Swarthmore, Pa., U.S.A.
Commission No. 27. Information Bulletin on Variable Stars Nos. 753-997. Prepared by Konkoly Observatory, Budapest, Hungary.
Commission No. 31. Annual Report of the BIH. Published by the Bureau International de l'Heure, Paris, France.
Commission No. 41. Information Circulars Nos. 22-25. Published by the President of the Commission.
IAU Information Bulletin Nos. 30-34. Prepared by the General Secretary.

Other Publications

Proceedings of the First European Astronomical Meeting, Athens, 4-9 September 1972, Published by Springer Verlag.
Volume 1 "Solar Activity and Related Interplanetary and Terrestrial Phenomena". Price 94 DM. Edited by J. Xanthakis.
Volume 2 "Stars and the Milky Way System". Price 138 DM. Edited by L.N. Mavridis.
Volume 3 "Galaxies and Relativistic Astrophysics". Price 126 DM. Edited by B. Barbanis and J.D. Hadjidemetriou.
Proceedings of the Second European Astronomical Meeting, Trieste, 2-5 September 1974, in 2 volumes. Price $ 60 U.S. *Memorie della Società Astronomica Italiana* Vol. XLV, 1975. Edited by L. Gratton.